IEEE Recommended Practice for Electric Power Systems in Commercial Buildings

Published by
The Institute of Electrical and Electronics Engineers, Inc.

IEEE

IEEE
Std 241-1990
(Revision of IEEE
Std 241-1983)

IEEE Recommended Practice for Electric Power Systems in Commercial Buildings

Sponsor

**Power Systems Engineering Committee
of the
IEEE Industry Applications Society**

Approved December 6, 1990

IEEE Standards Board

Approved May 17, 1991

American National Standards Institute

Abstract: A guide and general reference on electrical design for commercial buildings is provided. It covers load characteristics; voltage considerations; power sources and distribution systems; power distribution apparatus; controllers; services, vaults, and electrical equipment rooms; wiring systems; systems protection and coordination; lighting; electric space conditioning; transportation; communication systems planning; facility automation; expansion, modernization, and rehabilitation; special requirements by occupany; and electrical energy management. Although directed to the power oriented engineer with limited commercial building experience, it can be an aid to all engineers responsible for the electrical design of commercial buildings. This recommended practice is not intended to be a complete handbook; however, it can direct the engineer to texts, periodicals, and references for commercial buildings and act as a guide through the myriad of codes, standards, and practices published by the IEEE, other professional associations, and governmental bodies.
Keywords: Commercial buildings, electric power systems, load characteristics.

Second Printing
July 1995

The Institute of Electrical and Electronics Engineers, Inc.
345 East 47th Street, New York, NY 10017-2394, USA

Copyright © 1991 by
The Institute of Electrical and Electronics Engineers, Inc.
All rights reserved. Published 1991
Printed in the United States of America

ISBN 1-55937-088-2

Library of Congress Catalog Number 91-073747

September 18, 1991   *SH13912*

Institute of Electrical and Electronics Engineers.
    IEEE recommended practice for electric power systems in commercial buildings / sponsor, Power Systems Engineering Committee of the IEEE Industry Applications Society ; approved December 6, 1990 [by] IEEE Standards Board, approved May 17, 1991 [by] American National Standards Institute.
    p.   cm.
    "IEEE Std 241-1990 (Revision of IEEE Std 241-1983)."
    "Recognized as an American national standard (ANSI)."
    Includes bibliographical references and index.
    ISBM 1-55937-088-2
    1. Commercial buildings--Power supply. 2. Commercial buildings--Electric equipment. I. IEEE Industry Applications Society. Power Systems Engineering Committee. II. IEEE Standards Board. III. American National Standards Institute. IV. Title. V. Title: Institute of Electrical and Electronics Engineers recommended practice for electric power systems in commercial buildings.
TK4035.M37I57   1991
621.319'24 -- dc20
                                                                     91-24427
                                                                     CIP

Foreword

(This Foreword is not a part of IEEE Std 241-1990, IEEE Recommended Practice for Electric Power Systems in Commercial Buildings.)

The purpose of IEEE Std 241-1990, the "Gray Book," is to promote the use of sound engineering principles in the design of commercial buildings. It is hoped that it will alert the electrical engineer or designer to the many problems that can be encountered in designing electrical systems for commercial buildings and to develop a concern for the professional aspects of commercial building engineering. The Gray Book is not intended to be a complete handbook; however, it will direct the engineer to texts, periodicals, and references pertaining to commercial buildings and will also act as a guide through the myriad of codes, standards, and practices published by the IEEE and other professional associations and governmental bodies.

The fourth edition of this recommended practice has been written to update readers on the state of the art and to ensure quality electrical engineering design for commercial buildings. Material contained in previous editions of this book has been reused or updated, where practical.

All of the previous contributors to the Gray Book are hereby thanked by the present working group for their diligence and dedication; there would not have been a fourth edition without their many contrbutions to the technical accuracy and substance of this recommended practice.

At the time it recommended these practices, the working group of the Power Systems Committee had the following members:

Thomas E. Sparling, *Chair*

Tarif Abboushi	Steve Goble	James W. Patterson
Robert Atkinson	Daniel L. Goldberg	James R. Pfafflin
Jerry Baskin	Brad Gustafson	Michael Poliacof
C. E. Becker	Ed Hammer	Simon Rico
Sid Benjamin	John Hennings	Donald R. Ross
Doug Bors	Barry N. Hornberger	Vincent Saporita
D. S. Brereton	Lawrence Hogrebe	Steve Schaffer
Arthur Bushing	R. Gerald Irvine	S. Sengupta
René Castenschiold	Andrew Juhasz	Robert L. Seymour
Kao Chen	Douglas R. Kanitz	Lester Smith, Jr.
James M. Daly	Larry Kelly	Robert L. Smith, Jr.
Tom DeGenaro	E. A. Khan	Gary T. Smullin
M. J. DeLerno	Isaac Kopec	Donald Sparling
Edward A. Donoghue	F. W. Kussy	Wayne Stebbins
Ralph Droste	Richard Lennig	John D. Stolshek
James R. Duncan	Dan Love	Paul Strodes
Joseph Eto	Zach R. McCain	Elmer Sumka
Jerry Frank	David Moore	S. I. Venugopalan
Arthur Freund	Hugh O. Nash	Rudolph R. Verderber
Philip E. Gannon	Patrick Nassif	Leonard D. Whalen
W. G. Genné	Philip Nobile	Donald W. Zipse

When the IEEE Standards Board approved this standard on December 6, 1990, it had the following membership:

Marco W. Migliaro, Chair   **James M. Daly,** *Vice Chair*

Andrew G. Salem, *Secretary*

Dennis Bodson
Paul L. Borrill
Fletcher J. Buckley
Allen L. Clapp
Stephen R. Dillon
Donald C. Fleckenstein
Jay Forster*
Thomas L. Hannan

Kenneth D. Hendrix
John W. Horch
Joseph L. Koepfinger*
Irving Kolodny
Michael A. Lawler
Donald J. Loughry
John E. May, Jr.

Lawrence V. McCall
L. Bruce McClung
Donald T. Michael*
Stig Nilsson
Roy T. Oishi
Gary S. Robinson
Terrance R. Whittemore
Donald W. Zipse

*Member Emeritus

IEEE Recommended Practice for Electric Power Systems in Commercial Buildings

Working Group Members and Contributors

Thomas E. Sparling, *Chair*

Chapter 1 — Introduction: Daniel L. Goldberg, *Chair*; Arthur Freund, R. Gerald Irvine, Isaac Kopec, Philip Nobile, Donald W. Zipse

Chapter 2 — Load Characteristics: Douglas R. Kanitz, *Chair*; Arthur Bushing, R. Gerald Irvine, Lester Smith, Jr.

Chapter 3 — Voltage Considerations: Gary T. Smullin, *Chair*; C. E. Becker, D. S. Brereton, S. I. Venugopalan

Chapter 4 — Power Sources and Distribution Systems: R. Gerald Irvine, *Chair*; Vincent Saporita, Tom DeGenaro

Chapter 5 — Power Distribution Apparatus: Jerry Baskin and Jerry Frank, *Co-Chairs*; C. E. Becker, Daniel L. Goldberg, Lawrence Hogrebe, Simon Rico, S. Sengupta

Chapter 6 — Controllers: René Castenschiold, *Chair*; M. J. DeLerno, W. G. Genné, F. W. Kussy, S. Sengupta

Chapter 7 — Services, Vaults, and Electrical Equipment Rooms: James W. Patterson, *Chair*; Barry N. Hornberger, Thomas E. Sparling

Chapter 8 — Wiring Systems: James M. Daly, *Chair*; R. Gerald Irvine, Larry Kelly

Chapter 9 — System Protection and Coordination: Robert L. Smith, Jr., *Chair*; Jerry Baskin, Tom DeGenaro, Steve Goble, R. Gerald Irvine, E. A. Khan, Dan Love, Steve Schaffer, Thomas E. Sparling

Chapter 10 — Lighting: John D. Stolshek, *Chair*; Ed Hammer, Lawrence Hogrebe, Donald R. Ross, Rudolph R. Verderber

Chapter 11 — Electric Space Conditioning: James R. Pfafflin, *Chair*; Daniel L. Goldberg, John Hennings, Thomas E. Sparling

Chapter 12—Transportation: Edward R. Donaghue and Robert L. Seymour, *Co-Chairs*; Sid Benjamin, René Castenschiold, Ralph Droste, Andrew Juhasz, Zach R. McCain, Elmer Sumka

Chapter 13—Communication and Signal System Planning: James R. Duncan, *Chair*; Doug Bors, R. Gerald Irvine, David Moore

Chapter 14—Facility Automation: Philip E. Gannon, *Chair*; Daniel L. Goldberg, Brad Gustafson, R. Gerald Irvine, Richard Lennig, Patrick Nassif, Paul Strodes

Chapter 15—Expansion, Modernization, and Rehabilitation: Daniel L. Goldberg, *Chair*; Arthur Freund, R. Gerald Irvine, Michael Poliacof, Leonard D. Whalen

Chapter 16—Special Requirements by Occupancy: Thomas E. Sparling, *Chair* (excluding 16.13) and Hugh O. Nash, *Chair* (16.13); Tarif Abboushi, Robert Atkinson, Daniel L. Goldberg, Donald Sparling

Chapter 17—Electrical Energy Management: C. E. Becker, *Chair*; Kao Chen, Joseph Eto, Daniel L. Goldberg, Wayne Stebbins

Contents

CHAPTER   PAGE

1. Introduction 41
  1.1   Scope.. 41
       1.1.1 Voltage Levels...................................... 41
  1.2   Commercial Buildings 43
       1.2.1 System Requirements for Commercial, Residential, and
              Institutional Buildings 43
       1.2.2 Electrical Design Elements 44
  1.3   The Industry Applications Society (IAS) 45
       1.3.1 Grouping of Commercial Buildings 45
  1.4   IEEE Publications 45
  1.5   Professional Registration 46
       1.5.1 Professional Liability................................. 46
  1.6   Codes and Standards 46
       1.6.1 National Electrical Code (NEC)....................... 46
       1.6.2 Other NFPA Standards 47
       1.6.3 Local, State, and Federal Codes and Regulations 47
       1.6.4 Standards and Recommended Practices 48
  1.7   Handbooks... 50
  1.8   Periodicals 51
  1.9   Manufacturers' Data 52
  1.10 Safety 52
       1.10.1 Appliances and Equipment........................... 53
       1.10.2 Operational Considerations 54
  1.11 Maintenance 54
  1.12 Design Considerations...................................... 54
       1.12.1 Coordination of Design 55
       1.12.2 Flexibility... 56
       1.12.3 Specifications 56
       1.12.4 Drawings 57
       1.12.5 Manufacturers' or Shop Drawings......................... 58
  1.13 Estimating 58
  1.14 Contracts 60
  1.15 Building Access and Loading 60
  1.16 Contractor Performance 61
  1.17 Environmental Considerations 61
  1.18 Technical Files 62
  1.19 Electronic Systems... 62
  1.20 Power Supply Disturbances................................. 62
       1.20.1 Harmonics 62
       1.20.2 Electromagnetic Interference (EMI).................. 63
       1.20.3 Programmable Logic Controller (PLC).................. 63
  1.21 Definitions 63
  1.22 References 70

CHAPTER   PAGE

2. Load Characteristics... 73
    2.1   General Discussion.. 73
          2.1.1   Load Estimates 75
          2.1.2   Load Tabulation 76
          2.1.3   Relation to Power Company 78
          2.1.4   Relation to the NEC 81
    2.2   Load Characteristics 81
          2.2.1   Lighting 81
          2.2.2   General-Purpose Receptacles for Appliance Loads........... 82
          2.2.3   Space-Conditioning and Associated Auxiliary Equipment 82
          2.2.4   Plumbing and Sanitation 86
          2.2.5   Fire Protection 87
          2.2.6   Transportation Systems................................. 88
          2.2.7   Data Processing 88
          2.2.8   Food Preparation 89
          2.2.9   Miscellaneous or Special Loads 90
    2.3   Electromagnetic Hazards, Pollution, and Environmental Quality 91
    2.4   Additions to Existing Systems 91
    2.5   Total Load Considerations.................................... 92
          2.5.1   Estimation of Building Load............................. 94
    2.6   Example — Sample Partial Load Calculation for an Office Building .. 96
    2.7   References 99

3. Voltage Considerations 101
    3.1   General Discussion... 101
          3.1.1   Definitions.. 101
          3.1.2   Standard Nominal System Voltages for the United States 102
          3.1.3   Application of Voltage Classes 102
          3.1.4   Voltage Systems Outside of the United States 106
          3.1.5   Voltage Standard for Canada............................. 106
    3.2   Voltage Control in Electric Power Systems 106
          3.2.1   Principles of Power Transmission and Distribution on Utility
                 Systems 106
          3.2.2   System Voltage Tolerance Limits 108
          3.2.3   Development of Voltage Tolerance Limits for ANSI
                 C84.1-1989.. 109
          3.2.4   Voltage Profile Limits for a Regulated Distribution System ... 111
          3.2.5   Nonstandard Nominal System Voltages 112
          3.2.6   System Voltage Nomenclature 114
          3.2.7   Use of Distribution Transformer Taps to Shift the Utilization
                 Voltage Spread Band 114
    3.3   Voltage Selection 116
          3.3.1   Selection of Utilization Voltage of 600 V and Below 116
          3.3.2   Utility Service Supplied from a Primary Distribution
                 System 117

3.3.3   Utility Service Supplied from Transmission Lines 118
3.3.4   Utility Policy for Supplying Tenants in Commercial
        Buildings 118
3.4   Voltage Ratings for Utilization Equipment 119
3.5   Effect of Voltage Variation on Utilization Equipment 120
      3.5.1   Induction Motors 121
      3.5.2   Synchronous Motors 122
      3.5.3   Incandescent Lamps 122
      3.5.4   Fluorescent Lamps 122
      3.5.5   High-Intensity Discharge Lamps (Mercury, Sodium, and
             Metal Halide) 122
      3.5.6   Infrared Heating Processes 123
      3.5.7   Resistance Heating Devices 123
      3.5.8   Electron Tubes 123
      3.5.9   Capacitors 123
      3.5.10 Solenoid Operated Devices 123
      3.5.11 Solid-State Equipment 123
3.6   Calculation of Voltage Drops 124
      3.6.1   General Mathematical Formulas 124
      3.6.2   Cable Voltage-Drop Tables 125
      3.6.3   Busway Voltage-Drop Charts and Tables 127
      3.6.4   Transformer Voltage-Drop Charts 130
3.7   Improvement of Voltage Conditions 134
3.8   Voltage-Drop Considerations in Locating the Low-Voltage Power
      Source 134
3.9   Momentary Voltage Variations — Voltage Dips 135
3.10 Calculation of Voltage Dips 138
      3.10.1 The Effect of Motor Starting on Standby or Emergency
             Generators 138
      3.10.2 Effect of Motor Starting on a Distribution System 138
      3.10.3 Motor Starting Voltage Drop — Transformers 138
      3.10.4 Motor Starting Voltage Drop — Cables and Busways 139
3.11 Phase Voltage Unbalance in Three-Phase Systems 139
      3.11.1 Causes of Phase Voltage Unbalance 139
      3.11.2 Measurement of Phase Voltage Unbalance 141
      3.11.3 Effect of Phase Voltage Unbalance 143
3.12 Harmonic Voltages 143
      3.12.1 Nature of Harmonics 143
      3.12.2 Effect of Harmonics 144
      3.12.3 Harmonic Producing Equipment 145
      3.12.4 Reduction of Harmonic Interference 146
3.13 Transient Overvoltages 146
3.14 References 146
3.15 Bibliography 148

4. Power Sources and Distribution Systems 149
    4.1   General Discussion.. 149
    4.2   Electric Power Supply... 149
            4.2.1   Selecting a Power Source 149
            4.2.2   Planning for Utility Service 152
            4.2.3   Electric Rates 154
    4.3   Interrelated Utility and Project Factors That Influence Design 156
            4.3.1   Grounding on AC Services 157
    4.4   Electric Utility Metering and Billing......................... 157
            4.4.1   Metering by Type of Premises......................... 158
            4.4.2   Metering by Service Voltage Characteristics 158
            4.4.3   Meter Location....................................... 158
            4.4.4   Meter Mounting, Control, and Associated Equipment 158
            4.4.5   Types of Metering 160
            4.4.6   Utility Billing...................................... 163
    4.5   Transformer Connections 165
    4.6   Principal Transformer Secondary Connections 166
    4.7   System Grounding 167
            4.7.1   Grounding of Low-Voltage Systems (600 V and Below)....... 168
            4.7.2   Grounding of Medium-Voltage Systems (Over 600 V) 168
            4.7.3   Ground-Fault Circuit Interrupter (GFCI) 169
    4.8   Distribution Circuit Arrangements 169
            4.8.1   Radial Feeders 170
            4.8.2   Primary-Selective Feeders 173
            4.8.3   Secondary-Selective Feeders 174
            4.8.4   Secondary Network................................... 175
            4.8.5   Looped Primary System............................... 183
    4.9   Emergency and Standby Power Systems 185
            4.9.1   Lighting 186
            4.9.2   Power Loads... 187
            4.9.3   Power Sources 187
            4.9.4   Transfer Methods 189
            4.9.5   Special Precautions.................................. 191
            4.9.6   Transfer of Power 192
    4.10 Uninterruptible Power Supply (UPS) Systems 192
            4.10.1   Application of UPS 197
            4.10.2   Power System Configuration for 60 Hz Distribution......... 197
            4.10.3   Power System Configuration for 400 Hz Distribution........ 206
    4.11 Voltage Regulation and Power Factor Correction 207
            4.11.1   Voltage Regulation 207
            4.11.2   Power Factor Correction 208
    4.12 System Reliability Analysis................................... 208
            4.12.1   Reliability Data for Electrical Equipment 208
            4.12.2   Reliability Analysis and Total Owning Cost.......... 209

4.13 References 209

5. Power Distribution Apparatus 211
  5.1 General Discussion... 211
  5.2 Transformers.. 212
    5.2.1 Transformer Types 212
    5.2.2 Transformer Specifications............................... 215
    5.2.3 Transformer Construction 220
  5.3 Medium- and High-Voltage Fuses................................ 223
    5.3.1 Fuse Ratings 225
    5.3.2 Fuse Applications 227
  5.4 Metal-Enclosed 5–34.5 kV Interrupter Switchgear 228
    5.4.1 Automatic Control Devices 229
    5.4.2 Auxiliary Equipment and Features......................... 229
    5.4.3 Capability Required 230
  5.5 Metal-Clad 5–34.5 kV Circuit Breaker Switchgear............... 230
    5.5.1 Circuit Breakers... 231
    5.5.2 Instrument Transformers and Protective Relaying.......... 232
    5.5.3 Control.. 232
    5.5.4 Main Bus Current Selection 232
    5.5.5 Ground and Test Devices 232
  5.6 Metal-Enclosed, Low-Voltage 600 V Power Switchgear and
       Circuit Breakers 233
    5.6.1 Drawout Switchgear 233
    5.6.2 Low-Voltage Power Air-Type Circuit Breakers 234
    5.6.3 Selection of Circuit Breaker Tripping Characteristics ... 234
  5.7 Metal-Enclosed Distribution Switchboards 235
    5.7.1 Components 235
    5.7.2 Construction Features 235
  5.8 Primary-Unit Substations..................................... 236
  5.9 Secondary-Unit Substations................................... 236
    5.9.1 Basic Circuits 237
    5.9.2 Incoming Line Section 237
    5.9.3 Transformer Section 237
    5.9.4 Low-Voltage Switchgear Section 237
  5.10 Panelboards... 237
    5.10.1 Lighting and Appliance Panelboards 237
    5.10.2 Power Distribution Panelboards 238
    5.10.3 Motor Starter Panelboards 238
    5.10.4 Multiple-Section Panelboards 238
    5.10.5 Panelboard Data 238
  5.11 Molded-Case Circuit Breakers 239
    5.11.1 Types of Molded-Case Circuit Breakers 240
    5.11.2 Use of Molded-Case Circuit Breakers 240

5.12 Low-Voltage Fuses 241
      5.12.1 Fuse Ratings 241
      5.12.2 Current Limitation 242
      5.12.3 NEC Categories of Fuses 242
      5.12.4 UL Listing Requirements 243
      5.12.5 Fuses Carrying Class Letter 243
      5.12.6 Cable Limiters (Protectors) 245
5.13 Service Protectors 246
5.14 Enclosed Switches 246
      5.14.1 NEMA Requirements 246
      5.14.2 Application 247
5.15 Bolted Pressure Switches and High-Pressure Contact Switches 247
      5.15.1 Manual Operations 247
      5.15.2 Electrical Trip 248
5.16 Network Protectors 248
5.17 Lightning and System Transient Protection 250
      5.17.1 Surge Arresters and Capacitors 250
      5.17.2 Apparatus and Electromagnetic Interference (EMI) 251
5.18 Load Transfer Devices 252
      5.18.1 Automatic Transfer Switches 254
      5.18.2 Automatic Transfer Circuit Breakers 255
      5.18.3 Automatic Load Transfer Devices 255
5.19 Interlock Systems 256
5.20 Remote Control Contactors 257
      5.20.1 Remote Control Lighting Contactors 257
      5.20.2 Remote Control Switches for Power Loads 258
5.21 Equipment Ratings 260
5.22 References 261
5.23 Bibliography 263

6. Controllers 265
    6.1 General Discussion 265
    6.2 Starting 266
         6.2.1 Part-Winding Starters 266
         6.2.2 Resistor or Reactor Starters 267
         6.2.3 Autotransformer Starters 269
         6.2.4 Wye-Delta (Y–Δ) Starters 270
         6.2.5 Series-Parallel Starters 271
         6.2.6 Solid-State Starters 271
         6.2.7 Cost Comparison 271
    6.3 Protection 271
    6.4 Special Features 276
    6.5 Control Systems 276
         6.5.1 Panelboard-Type Constructions 276

        6.5.2 Separate Enclosures 277
        6.5.3 Combination Starters 277
        6.5.4 Simplified Control Centers 277
        6.5.5 Complex Panels 277
        6.5.6 Motor Control Centers 277
    6.6 Low-Voltage Starters and Controllers 281
    6.7 Multiple-Speed Controllers....................................... 282
    6.8 Fire Pump Controllers 284
    6.9 Medium-Voltage Starters and Controllers 284
    6.10 Synchronous Motor Starters 285
    6.11 DC Motor Controls.. 286
    6.12 Pilot Devices 286
        6.12.1 Manually Operated Devices................................. 286
        6.12.2 Automatically Operated Devices 288
        6.12.3 Programmable Controllers 289
    6.13 Speed Control of DC Motors 290
    6.14 Speed Control of AC Motors 291
    6.15 Power System Harmonics from Adjustable Speed Motor Controls ... 292
    6.16 References 292

7. Services, Vaults, and Electrical Equipment Rooms 293
    7.1 Incoming Lines and Service Laterals 293
        7.1.1 Overhead Service... 293
        7.1.2 Lines Over and On Buildings 294
        7.1.3 Weather and Environmental Considerations 295
        7.1.4 Underground Service 295
        7.1.5 Service Entrance Conductors Within a Building 296
    7.2 Service Entrance Installations 297
        7.2.1 Number of Services 297
        7.2.2 Physical Arrangement 298
        7.2.3 Low- and Medium-Voltage Circuits.......................... 298
        7.2.4 Load Current and Short-Circuit Capacity................... 301
        7.2.5 Limiting Fault Current..................................... 303
        7.2.6 Ground-Fault Protection 306
    7.3 Vaults and Pads for Service Equipment 306
        7.3.1 Vaults... 306
        7.3.2 Outdoor Pads... 307
        7.3.3 Safety and Environment 307
    7.4 Network Vaults for High-Rise Buildings 308
        7.4.1 Network Principles 308
        7.4.2 Preliminary Vault Design 309
        7.4.3 Detailed Vault Design 309
    7.5 Service Rooms and Electrical Closets 311
        7.5.1 Space Requirements 312

	7.5.2	Illumination	312
	7.5.3	Ventilation...	312
	7.5.4	Foreign Facilities	312
7.6		References	312

8. Wiring Systems	315
	8.1	Introduction	315
	8.2	Cable Systems	315
	8.3	Cable Construction	316
	8.3.1	Conductors	316
	8.3.2	Comparison Between Copper and Aluminum	317
	8.3.3	Insulation	318
	8.3.4	Cable Design	322
	8.4	Cable Outer Finishes	328
	8.4.1	Nonmetallic Finishes................................	328
	8.4.2	Metallic Finishes	328
	8.4.3	Single-Conductor and Multiconductor Constructions	331
	8.4.4	Physical Properties of Materials for Outer Coverings	332
	8.5	Cable Ratings	332
	8.5.1	Voltage Rating	332
	8.5.2	Conductor Selection	332
	8.5.3	Load Current Criteria	333
	8.5.4	Emergency Overload Criteria	335
	8.5.5	Voltage-Drop Criteria	335
	8.5.6	Fault Current Criteria	337
	8.5.7	Frequency Criteria	337
	8.5.8	Elevated Ambient Temperature........................	339
	8.5.9	Hot-Spot Temperature Criteria	339
	8.5.10	Termination Criteria................................	339
	8.6	Installation	339
	8.6.1	Layout	341
	8.6.2	Open Wire	341
	8.6.3	Aerial Cable	341
	8.6.4	Open Runs	342
	8.6.5	Cable Tray..	342
	8.6.6	Cable Bus	343
	8.6.7	Conduit	344
	8.6.8	Direct Burial......................................	344
	8.6.9	Hazardous (Classified) Locations	345
	8.6.10	Installation Procedures	346
	8.7	Connectors..	347
	8.7.1	Types Available	347
	8.7.2	Connectors for Aluminum............................	348
	8.7.3	Connectors for Cables of Various Voltages	352

      8.7.4   Performance Requirements...................................... 352

      8.7.5   Electrical and Mechanical Operating Requirements 352

8.8   Terminations 353

      8.8.1   Purpose 353

      8.8.2   Definitions... 353

      8.8.3   Cable Terminations...................................... 354

      8.8.4   Jacketed and Armored Cable Connectors................... 361

      8.8.5   Separable Insulated Connectors 362

      8.8.6   Performance Requirements................................ 362

8.9   Splicing Devices and Techniques 362

      8.9.1   Taped Splices 363

      8.9.2   Preassembled Splices 365

8.10 Grounding of Cable Systems 365

      8.10.1 Sheath Losses.. 366

8.11 Protection from Transient Overvoltage........................... 367

8.12 Testing 367

      8.12.1 Application and Utility.................................... 367

      8.12.2 Alternating-Current versus Direct-Current.................. 368

      8.12.3 Factory Tests 368

      8.12.4 Field Tests.. 369

      8.12.5 Procedure 372

      8.12.6 DC Corona and Its Suppression 372

      8.12.7 Line Voltage Fluctuations 372

      8.12.8 Resistance Evaluation 373

      8.12.9 Megohmmeter Test 373

8.13 Locating Cable Faults 373

      8.13.1 Influence of Ground-Fault Resistance 373

      8.13.2 Equipment and Methods 374

      8.13.3 Selection 376

8.14 Cable Specification 376

8.15 Busway... 377

8.16 Busway Construction 377

8.17 Feeder Busway 378

8.18 Plug-in Busway 379

8.19 Lighting Busway 380

8.20 Trolley Busway 382

8.21 Standards... 382

8.22 Selection and Application of Busways 383

      8.22.1 Current-Carrying Capacity 383

      8.22.2 Short-Circuit Current Rating 383

      8.22.3 Voltage Drop.. 384

      8.22.4 Thermal Expansion 386

      8.22.5 Building Expansion Joints 386

      8.22.6 Welding Loads 386

8.23 Layout... 387
8.24 Installation 389
    8.24.1 Procedure Prior to Installation 389
    8.24.2 Procedure During Installation 390
    8.24.3 Procedure After Installation.......................... 390
8.25 Field Testing 390
8.26 Busways Over 600 V (Metal-Enclosed Bus) 391
    8.26.1 Standards 391
    8.26.2 Ratings... 391
    8.26.3 Construction.. 391
    8.26.4 Field Testing 392
8.27 References 392
8.28 Bibliography 393

9. System Protection and Coordination 397
9.1 General Discussion.. 397
    9.1.1 Single- and Multiple-Pole Interrupters 398
    9.1.2 Sources of Short-Circuit Currents......................... 399
    9.1.3 Rotating Machine Reactance 399
    9.1.4 Utility Source 401
    9.1.5 Symmetrical and Asymmetrical Currents 401
    9.1.6 Why Are Short-Circuit Currents Asymmetrical? 401
    9.1.7 The DC Component of Asymmetrical Short-Circuit
           Currents 404
    9.1.8 Total Short-Circuit Current 406
9.2 Short-Circuit Calculations 408
    9.2.1 Type of Power System Faults 408
    9.2.2 Three-Phase Bolted Faults 408
    9.2.3 Line-to-Line Bolted Faults.............................. 408
    9.2.4 Line-to-Ground Bolted Faults 408
    9.2.5 Arcing Faults 409
9.3 Selection of Equipment.. 409
    9.3.1 Equipment Rating 410
9.4 Basis of Short-Circuit Current Calculations 410
    9.4.1 Total Current Basis of Rating............................ 411
    9.4.2 Symmetrical Current Basis of Rating 412
    9.4.3 Comparison of Duty Calculation Methods 413
9.5 Details of Short-Circuit Current Calculations 413
    9.5.1 Step-by-Step Procedure 413
    9.5.2 System Conditions for Most Severe Duty.................. 414
    9.5.3 Preparing Impedance Diagrams 415
    9.5.4 Component Impedance Values 415
    9.5.5 Combining Impedances 415
    9.5.6 Use of Per Unit, Percent, or Ohms 417

9.5.7 Per Unit Representation 419
9.5.8 Electric Utility System 422
9.5.9 Transformers 423
9.5.10 Busways, Cables, and Conductors 423
9.5.11 Rotating Machines 424
9.5.12 Motors Rated 600 V or Less 424
9.5.13 Other Circuit Impedances 424
9.5.14 Shunt Connected Impedances 425
9.5.15 System Driving Voltage 425
9.5.16 Determination of Short-Circuit Currents 425
9.5.17 Longhand Solution 425
9.5.18 Determination of Line-to-Ground Fault Currents 427
9.5.19 Effect of Low Available Utility kVA 439
9.5.20 General Discussion of Short-Circuit Current Calculations 439
9.6 Method of Reducing the Available Short-Circuit Current 441
    9.6.1 Effect of Distribution Circuit Lengths on Short-Circuit
           Current 441
    9.6.2 Current-Limiting Fuses 441
    9.6.3 Current-Limiting Reactors 441
    9.6.4 Current-Limiting Busways 442
    9.6.5 Examples of Reducing Available Short-Circuit Current 444
9.7 Selective Coordination 446
    9.7.1 Coordination of Protective Devices 447
    9.7.2 Preliminary Steps in a Coordination Study 447
    9.7.3 Mechanics of Achieving Coordination 447
    9.7.4 Coordination Examples with Explanations 451
9.8 Fuses 460
    9.8.1 Fuse Coordination 460
    9.8.2 Fuse Selectivity Ratio Tables 460
    9.8.3 Fuse Time Current Characteristic Curves 463
    9.8.4 I^2t Values for Coordination 463
    9.8.5 Coordinating Fusible Unit Substation 464
    9.8.6 Summary 465
    9.8.7 Fuse Current-Limiting Characteristics 465
    9.8.8 Application of Fuses 471
    9.8.9 Bus Bracing Requirements 471
    9.8.10 Circuit Breaker Protection 472
    9.8.11 Wire and Cable Protection 472
    9.8.12 Motor Starter Short-Circuit Protecton 472
    9.8.13 Transformer Fuse Protection 473
    9.8.14 Motor Running Overcurrent Protection 474
    9.8.15 Fuse Device Maintenance 475
9.9 Current-Limiting Circuit Breakers 476
9.10 Ground-Fault Protection 477

9.11   References 479
9.12   Bibliography 480

10. Lighting 481
    10.1   General Discussion 481
            10.1.1   Lighting Objectives 481
            10.1.2   Lighting Regulations 482
    10.2   Lighting Terminology 483
    10.3   Illumination Quality 485
            10.3.1   Visual Comfort 485
            10.3.2   Veiling Reflections 485
            10.3.3   Room Finishes 486
            10.3.4   Color 487
            10.3.5   Psychological Factors 488
    10.4   Illumination Quantity 488
    10.5   Light Sources 491
            10.5.1   Incandescent Lamps 491
            10.5.2   Fluorescent Lamps 493
            10.5.3   High-Intensity Discharge (HID) Lamps 495
    10.6   Ballasts 498
            10.6.1   Fluorescent Lamp Ballasts 498
            10.6.2   High-Intensity Discharge (HID) Lamp Ballasts .. 501
    10.7   Luminaires 501
            10.7.1   General Considerations That Affect the Selection of a
                      Lighting System 501
            10.7.2   Special Lighting Distributions 502
            10.7.3   Lighting and Other Building Subsystems 503
            10.7.4   Visual Comfort Probability (VCP) 503
            10.7.5   Luminaires and Air Conditioning 503
    10.8   Lighting Application Techniques 505
            10.8.1   Uniform Lighting 506
            10.8.2   Non-Uniform Lighting 507
            10.8.3   Task-Ambient Lighting 513
            10.8.4   Special Lighting Considerations for Stores 519
            10.8.5   Electric Lighting and Daylighting 525
            10.8.6   Outdoor and Sports Area Lighting 525
    10.9   Control of Lighting 529
            10.9.1   Switching for 480Y/277 V Distribution Systems . 529
            10.9.2   Remote Control Switching Relays and Lighting
                      Contactors 530
            10.9.3   Dimming and Flashing of Lamps 533
            10.9.4   Dimming of HID Lamps 533
    10.10   Lighting Maintenance 534
    10.11   Voltage 535

10.12 Power Factor 537
10.13 Temperature 537
10.14 Ballast Sound 539
10.15 Lighting Economics 539
      10.15.1 First Costs 540
      10.15.2 Type and Quality of Lighting Desired 540
      10.15.3 Energy Costs 541
      10.15.4 Maintenance... 541
      10.15.5 Effect on Personnel 541
10.16 Illuminance Calculations 541
10.17 Lighting and Thermal Considerations 542
10.18 References 544

11. Electric Space Conditioning 547
   11.1 General Discussion 547
      11.1.1 Space-Conditioning Control 548
   11.2 Primary Source of Heat 548
      11.2.1 Electric Space Heating................................ 548
      11.2.2 Heat Pumps 549
      11.2.3 Resistance Heaters 550
      11.2.4 Types of Air-Conditioning Systems 553
   11.3 Energy Conservation 553
      11.3.1 Practical Energy Conservation 554
      11.3.2 Reduction in Demand for Space Conditioning 555
      11.3.3 Standards and Codes 556
   11.4 Definitions 556
   11.5 References 559

12. Transportation 561
   12.1 General Discussion 561
      12.1.1 Use 561
      12.1.2 Design Factors.. 561
      12.1.3 Consultations.. 562
      12.1.4 Efficiency and Economy 562
      12.1.5 Responsibilities 562
   12.2 Types of Transportation 563
      12.2.1 Electric Elevators 563
      12.2.2 Hydraulic Elevators 563
      12.2.3 Dumbwaiters 564
      12.2.4 Escalators and Moving Walks 564
      12.2.5 Material Lifts 566
      12.2.6 Manlifts 568
      12.2.7 Pneumatic Tubes 568
   12.3 Elevator Control, Motors, and Motor Generators 568

12.3.1 Control Systems 569

12.3.2 DC Rheostatic Control................................. 569

12.3.3 AC Resistance Control................................ 569

12.3.4 AC Servo 569

12.3.5 Generator Field Control 570

12.3.6 DC Direct Drive 570

12.3.7 Variable Frequency Control 570

12.3.8 Hoisting Motors 570

12.3.9 Rating of Motor-Generator Sets 571

12.3.10 Starting Motor-Generator Sets 571

12.3.11 Control Design Fundamentals 571

12.3.12 Rectifiers for Power Conversion 571

12.4 Elevator Horsepower and Efficiency........................... 572

12.4.1 Horsepower Calculation 572

12.4.2 Ratings 572

12.4.3 Variations 572

12.4.4 Efficiency 572

12.5 Elevator Energy Consumption and Heat Release 572

12.5.1 Energy Consumption 572

12.5.2 Heat Release.. 573

12.6 Elevator Conductors and Diversity Factor 573

12.6.1 Conductor Size 573

12.6.2 Current Ratios...................................... 576

12.6.3 Diversity of Operation 576

12.6.4 Disconnecting Means 576

12.7 Elevator Operation 577

12.7.1 Manual and Automatic Operations 577

12.7.2 Single Automatic Pushbutton Operation 577

12.7.3 Selective Collective Operation 577

12.7.4 Duplex Collective Operation 577

12.7.5 Group Automatic Operation 578

12.8 Quality of Elevator Performance 578

12.8.1 Passenger Comfort 578

12.8.2 Leveling 578

12.8.3 Handicap Requirements 579

12.9 Elevator Doors and Automatic Door Operation 579

12.9.1 Interlocks 579

12.9.2 Power Operated Doors 579

12.9.3 Passenger Elevator Doors 579

12.9.4 Freight Elevator Doors 579

12.9.5 Fire Rating of Elevator Entrances 579

12.10 Group Supervisory Control 580

12.11 Specification of Elevator Plant.............................. 580

12.11.1 Selection of an Elevator 580

12.11.2   Car Capacity 580
12.11.3   Double-Deck Elevators and Sky Lobbies 581
12.11.4   Selection of Elevator Car Speed 582
12.11.5   Quantity of Service.................................. 582
12.11.6   Building Classes 582
12.11.7   Quality of Service 582
12.11.8   Number of Elevators Required 582
12.11.9   Roundtrip Time...................................... 583
12.11.10   Interval 583
12.11.11   Grouping of Elevators in the Plant 583
12.12   Regenerated Energy 583
12.13   Standby Power Operation of Elevators 584
12.13.1   Standby Power Requirements 584
12.13.2   Elevator Standby Power System 585
12.14   Operation of Elevators During Fire Conditions 586
12.14.1   Sprinklers in Elevator Machine Rooms and Hoistways ... 586
12.15   Emergency Signals and Communications 587
12.16   Car Lighting 587
12.17   References 587

13. Communication and Signal System Planning 589
13.1   General Discussion 589
13.1.1   Cable Plant.. 590
13.1.2   Riser Conduits and Sleeves 590
13.1.3   Riser Shafts and Slots 591
13.1.4   Underfloor Distribution Systems..................... 591
13.1.5   Underfloor Raceways................................ 591
13.1.6   Ceiling Distribution Systems 591
13.1.7   Undercarpet Wiring Systems 592
13.2   Telephone Facilities.. 592
13.2.1   Service Entrance Cables 593
13.2.2   Main Terminal Room 594
13.2.3   Riser Systems..................................... 595
13.2.4   Apparatus Closets 595
13.2.5   Satellite Closets 596
13.2.6   Telephone Equipment Rooms 596
13.2.7   Public Telephones................................. 597
13.3   Data Facilities... 597
13.3.1   Data Terminals 597
13.3.2   Cable Systems and Support Structures 598
13.3.3   Data Transmission and Network Equipment........... 599
13.3.4   Data Center Installations 600
13.4   Fire Alarm Systems... 602
13.4.1   Design Criteria................................... 602

        13.4.2 Power Supply 603
  13.5 Security and Access Control Systems 603
  13.6 Audio Communication Systems 603
        13.6.1 Public Address Systems 604
        13.6.2 Intercom Systems 604
        13.6.3 Dictation Systems 605
        13.6.4 Sound Reinforcement Systems 605
        13.6.5 Audio Production and Recording Equipment 606
  13.7 Image Communication Systems 606
        13.7.1 Video Systems 607
        13.7.2 Television Signal Distribution 607
        13.7.3 Projection Systems 608
  13.8 Nurse Call Systems 608
  13.9 Pocket Paging Systems 608
  13.10 References 609

14. Facility Automation 611
  14.1 Introduction 611
        14.1.1 Adjuncts to the Basic System 611
  14.2 Functions 611
  14.3 Establishment of System Requirements 612
  14.4 System Description and Equipment 613
        14.4.1 Terminology 613
        14.4.2 General Description of a FAS and Its Components 614
  14.5 Central Monitoring and Control Equipment 616
  14.6 HVAC and Energy Management 620
        14.6.1 HVAC Monitoring and Control 621
        14.6.2 Direct Digital Control 622
        14.6.3 Energy Management 625
  14.7 Fire Management 626
        14.7.1 Life Safety Communications 628
  14.8 Security 629
        14.8.1 Security Systems 629
        14.8.2 Access Control 630
  14.9 Transportation and Traffic 631
  14.10 Pollution and Hazardous Waste 631
  14.11 Electric Systems 632
        14.11.1 Lighting Automation Systems 632
        14.11.2 Normal and Emergency Power Sources 632
        14.11.3 Uninterruptible Power Supply (UPS) Systems 633
        14.11.4 Electric Distribution Systems 633
  14.12 Mechanical Utilities 633
        14.12.1 Functions 633
        14.12.2 Integration 633

14.13 Communications 634
    14.13.1 Maintenance 634
    14.13.2 Central Supervisory Station 634
14.14 Miscellaneous Systems 634
    14.14.1 Computerized Maintenance Management System
             (CMMS) 634
14.15 FAS Design and Installation 636
    14.15.1 System Specification 636
    14.15.2 Selection of Data Transmission Media (DTM) 637
    14.15.3 Interfacing to Existing Equipment or Equipment
             Provided by Other Disciplines 637
    14.15.4 Procurement Documents 637
    14.15.5 Installation 638
    14.15.6 Testing.. 638
    14.15.7 Warranty 638
    14.15.8 Software 638
    14.15.9 System Documentation 638
    14.15.10 Programming.. 639
    14.15.11 Physical Installation................................. 639
    14.15.12 Backup Modes 640
14.16 Training 641
14.17 Maintenance and Operation 642
    14.17.1 Maintenance Approaches 642
    14.17.2 Operations 643
    14.17.3 Maintenance Contracts 644
    14.17.4 Special Considerations 645
14.18 References 645
14.19 Bibliography 646

15. Expansion, Modernization, and Rehabilitation....................... 649
  15.1 General Discussion 649
  15.2 Preliminary Study 650
    15.2.1 Examples of Poor Planning 652
    15.2.2 Making Choices 655
  15.3 Design Considerations 656
    15.3.1 Utility Considerations 656
    15.3.2 Ground-Fault Protection 657
    15.3.3 Sizing Feeders 657
    15.3.4 Staging Operations................................... 658
    15.3.5 Scrap Materials 659
    15.3.6 Hazardous Scrap Materials 659
  15.4 Retaining Old Service Equipment 659
  15.5 Completely New Service Equipment 661
  15.6 Additional New Service Point 661

15.7   Voltage Transformation 662
15.8   Distribution of Power to Main Switchboards 663
15.9   Existing Plans... 664
15.10 Scheduling and Service Continuity 665
       15.10.1 Example 1—Building Extension, Medium-Voltage
               Service 666
       15.10.2 Example 2—Building Extension, Low-Voltage Service 667
       15.10.3 Example 3—Retention of Existing System and
               Expansion.. 670
       15.10.4 Example 4—Rehabilitation on a College Campus 670
15.11 Wiring Methods 673
       15.11.1 Smoke and Fire Considerations 673
15.12 References 674

16. Special Requirements by Occupancy 675
16.1   General Discussion 675
16.2   Auditoriums 676
16.3   Automobile Areas 677
       16.3.1 Sales Offices and Showrooms 678
       16.3.2 Service Shops...................................... 678
       16.3.3 Parking Lots and Garages 678
       16.3.4 Car Wash Facilities................................ 679
       16.3.5 Automated Parking Facilities........................ 679
16.4   Banks 679
16.5   Brokerage Offices... 681
16.6   Places of Worship 681
16.7   Athletic and Social Clubs................................. 682
16.8   Colleges and Universities 683
       16.8.1 Central Power Plants 683
       16.8.2 Classrooms 683
       16.8.3 Laboratories...................................... 683
       16.8.4 Dormitories 684
       16.8.5 Miscellaneous Requirements 684
16.9   Computer Centers 685
16.10 Department Stores 686
       16.10.1 Distribution Systems 686
       16.10.2 Lighting Systems 686
       16.10.3 Communication and Signaling Systems................ 688
16.11 Fire Stations 688
16.12 Gymnasiums 689
16.13 Health Care Facilities 689
       16.13.1 Hospitals 689
       16.13.2 Outpatient Facilities.............................. 695
       16.13.3 Nursing Homes and Limited Care Facilities 695

16.13.4 Psychiatric Facilities 695
16.14 Hotels 696
16.15 Libraries.. 697
16.16 Museums 698
16.17 Newspaper Buildings 698
16.18 Office Buildings 699
    16.18.1 Large and High-Rise Office Buildings.................. 701
16.19 Parks and Playgrounds 705
16.20 Piers, Docks, and Boat Marinas 705
16.21 Police Stations 705
16.22 Prisons 706
    16.22.1 Administration Buildings 707
    16.22.2 Cell Blocks 707
    16.22.3 Manufacturing Buildings........................... 707
    16.22.4 Dining Halls 707
    16.22.5 Hospitals 707
    16.22.6 Power Plants 707
    16.22.7 Warehouse Buildings 708
16.23 Radio Studios 708
16.24 Recreation Centers 708
16.25 Residential Occupancies (Commercial) 709
16.26 Restaurants... 710
16.27 Schools (Kindergarten Through 12th Grade) 710
    16.27.1 General Requirements 711
    16.27.2 Mechanical Systems 711
    16.27.3 Laboratories....................................... 711
    16.27.4 Manual Training Departments....................... 711
    16.27.5 Kitchen Facilities 711
16.28 Shopping Centers.. 711
16.29 Supermarkets 711
16.30 Swimming Pools and Fountains 713
16.31 Telephone Buildings 714
16.32 Television Studios 714
16.33 Theaters 715
    16.33.1 General Lighting Systems 715
    16.33.2 Stage Lighting Systems 716
16.34 Transportation Terminals 716
    16.34.1 Requirements for All Terminals 716
    16.34.2 Railroad Terminals................................. 719
    16.34.3 Bus Terminals 719
    16.34.4 Airports... 719
    16.34.5 Rapid Transit Stations............................. 720
16.35 References 720
16.36 Bibliography 722

CHAPTER   PAGE

17. Electrical Energy Management.................................... 723
    17.1   Energy Management Requirements............................ 723
    17.2   Energy Conservation Opportunities 723
           17.2.1   Energy Audit 724
           17.2.2   Project Lists 724
    17.3   The Energy Management Process 724
           17.3.1   Obtain Management Approval and Commitment........ 724
           17.3.2   Embarking on an Energy Conservation Program 725
           17.3.3   Equipment Audit 726
           17.3.4   Tracking Progress 727
           17.3.5   Overall Considerations 727
    17.4   Lighting 728
           17.4.1   Introduction................................... 728
           17.4.2   Energy-Efficient Light Sources 728
           17.4.3   Energy-Efficient Ballasts...................... 731
           17.4.4   New Luminaires for Energy-Efficient Light Sources 732
           17.4.5   Energy-Saving Lighting Techniques 734
           17.4.6   Lighting and Energy Standards.................. 736
    17.5   Calculating Energy Savings 737
           17.5.1   Introduction................................... 737
           17.5.2   Tariffs 737
           17.5.3   Billing Calculations........................... 737
           17.5.4   Declining Block Rate and Example 738
           17.5.5   Demand Rate.................................... 739
           17.5.6   Time-of-Use Rate............................... 739
           17.5.7   Time Value of Money 740
           17.5.8   Evaluating Motor Losses 740
           17.5.9   Transformer Losses 742
           17.5.10 Evaluating Losses in Other Equipment 742
    17.6   Load Management... 743
           17.6.1   Introduction................................... 743
           17.6.2   Controllers.................................... 743
           17.6.3   Equipment Audit and Load Profile 744
    17.7   Efficiencies of Electrical Equipment 745
           17.7.1   Losses 745
           17.7.2   Efficiency 746
           17.7.3   Oversizing Electrical Equipment 746
           17.7.4   Motors 746
           17.7.5   Transformers 748
           17.7.6   Capacitors 748
           17.7.7   Equipment Overview 748
    17.8   Metering.. 748
    17.9   Operations.. 750
           17.9.1   Metering of Tenant Areas and Building Operations....... 750

17.9.2   Interlocking or Key Switches 750
17.9.3   Load Control 751
17.9.4   Load Shedding....................................... 751
17.9.5   Load or Peak Shaving with On-Site Generation 751
17.10   Energy Conservation Equipment 752
17.11   References 753
17.12   Bibliography 754
17.13   Projects 755

FIGURES

Fig 1   Groups of Loads in a Typical 10-Story Commercial Building 74
Fig 2   Load Profiles 95
Fig 3   Principal Transformer Connections to Supply the System
          Voltages of Table 17... 105
Fig 4   Typical Utility Power Generation, Transmission, and Distribution
          System 106
Fig 5   Effect of Regulator Compensation on Primary Distribution
          System Voltage 107
Fig 6   Voltage Profile of the Limits of Range A, ANSI C84.1-1989 112
Fig 7   Vector Diagram of Voltage Relations for Voltage-Drop
          Calculations 124
Fig 8   Voltage-Drop Curves for Typical Interleaved Construction of
          Copper Busway at Rated Load, Assuming 70 °C (158 °F)
          Operating Temperature 131
Fig 9   Voltage-Drop Curves for Typical Plug-in Type Busway at
          Balanced Rated Load, Assuming 70 °C (158 °F) Operating
          Temperature 132
Fig 10   Voltage-Drop Curves for Typical Feeder Busways at Balanced
          Rated Load Mounted Flat Horizontally, Assuming 70 °C (158 °F)
          Operating Temperature 132
Fig 11   Voltage-Drop Curve versus Power Factor for Typical Light-Duty
          Trolley Busway Carrying Rated Load, Assuming 70 °C (158 °F)
          Operating Temperature 133
Fig 12   Voltage-Drop Curves for Three-Phase Transformers,
          225–10 000 kVA, 5–25 kV 133
Fig 13   Effect of Low-Voltage Source Location on Area That Can Be
          Supplied Under Specific Voltage-Drop Limits 136
Fig 14   Flicker of Incandescent Lamps Caused by Recurrent Voltage
          Dips 137
Fig 15   Typical Voltage Behavior of a Generator When an Induction
          Motor Is Started by a Full Voltage Starting Device 139
Fig 16   Minimum Generator Voltage Due to Full Voltage Starting
          of a Motor 140

FIGURES   PAGE

Fig 17   Voltage Drop in a Transformer Due to Full Voltage Starting
          of a Motor 142
Fig 18   De-Rating Factor for Motors Operating with Phase Voltage
          Unbalance 144
Fig 19   Transformer Secondary Connections 167
Fig 20   Radial-Circuit Arrangements in Commercial Buildings 171
Fig 21   Radial-Circuit Arrangement—Common Primary Feeder to
          Secondary-Unit Substations.................................. 172
Fig 22   Radial-Circuit Arrangement—Individual Primary Feeders to
          Secondary-Unit Substations.................................. 172
Fig 23   Primary-Selective Circuit Arrangements 174
Fig 24   Secondary-Selective Circuit Arrangement (Double-Ended
          Substation with Single Tie) 175
Fig 25   Secondary-Selective Circuit Arrangement (Individual Substations
          with Interconnecting Ties) 176
Fig 26   Distributed Secondary Network 177
Fig 27   Basic Spot Network 178
Fig 28   Four-Unit Spot Network...................................... 179
Fig 29   Enclosure Monitor 182
Fig 30   Looped Primary-Circuit Arrangement 184
Fig 31   Loop-Loop Primary-Circuit Arrangement 185
Fig 32   Typical Transfer Switching Methods 189
Fig 33   Service Transfer Using Circuit Breakers....................... 190
Fig 34   Secondary Transfer Using Circuit Breakers..................... 191
Fig 35   Nonredundant UPS System Configuration........................ 194
Fig 36   "Cold" Standby Redundant UPS System......................... 195
Fig 37   Parallel Redundant UPS System............................... 196
Fig 38   Isolated Redundant UPS System 197
Fig 39   Single-Module UPS System 198
Fig 40   Parallel Capacity UPS System 199
Fig 41   Parallel Redundant UPS System............................... 199
Fig 42   Dual Redundant UPS System 200
Fig 43   Isolated Redundant UPS System 201
Fig 44   Parallel Tandem UPS System 202
Fig 45   Hot Tied Bus UPS System 203
Fig 46   Super-Redundant Parallel System—Hot Tied Bus UPS
          System 204
Fig 47   Uninterruptible Power with Dual Utility Sources and Static
          Transfer Switches 205
Fig 48   Current-Limiting Action of Fuses............................. 242
Fig 49   Typical Circuit for Cable Limiter Application 245
Fig 50   Diagram Illustrating Multiple Automatic Double-Throw Transfer
          Switches Providing Varying Degrees of Emergency and Standby
          Power 253

Fig 51   Mechanically Held, Electrically Operated Lighting Contactor
          Controlled by Multiple Momentary Toggle-Type Control Stations ... 257
Fig 52   Remote Control Lighting Contactors Controlled from Local and
          Remote Locations 258
Fig 53   Various Control Means for Remote Control Lighting Contactors.... 259
Fig 54   Principles of the Most Common Reduced Voltage Starters for
          Squirrel-Cage Motors, Part 1 267
Fig 55   Principles of the Most Common Reduced Voltage Starters for
          Squirrel-Cage Motors, Part 2 268
Fig 56   Principles of the Most Common Reduced Voltage Starters for
          Squirrel-Cage Motors, Part 3 269
Fig 57   Simplified Wiring Diagram of a Solid-State Controller 272
Fig 58   Typical Solid-State Controller with Fusible Disconnect 273
Fig 59   NEMA versus IEC Size Comparison 283
Fig 60   Typical Pushbutton Control Circuits 287
Fig 61   Conductor Stranding... 317
Fig 62   Typical Values for Hardness versus Temperature 320
Fig 63   Electrical Field of Shielded Cable............................... 326
Fig 64   Electrical Field of Conductor on Ground Plane in Uniform
          Dielectric 327
Fig 65   Unshielded Cable on Ground Plane.............................. 327
Fig 66   Commonly Used Shielded and Nonshielded Constructions 329
Fig 67   Curves Showing the AC/DC Resistance Ratio That Exists on a
          400 Hz System.. 340
Fig 68   Procedures for Connecting Aluminum Conductors 351
Fig 69   Stress-Relief Cone... 356
Fig 70   Typical Class 1 Porcelain Terminator (for Solid Dielectric Cables) .. 357
Fig 71   Typical Class 1 Molded-Rubber Terminator (for Solid Dielectric
          Cables) 358
Fig 72   Typical Class 3 Molded-Rubber Terminator (for Solid Dielectric
          Cables) 359
Fig 73   Typical Taped Splice in Shielded Cable 364
Fig 74   Illustration of Versatility of Busways, Showing Use of Feeder,
          Plug-in, Lighting, and Trolley Types............................ 379
Fig 75   Feeder Busway 380
Fig 76   Installation View of Small Plug-in Busway Showing Individual
          Circuit Breaker Power Tapoff and Flexible Bus Drop Cable 381
Fig 77   Lighting Busway Supporting and Supplying Power to High-
          Intensity Discharge Fixture 381
Fig 78   Diagram Illustrating Voltage Drop and Indicating Error When
          Approximate Voltage-Drop Formulas Are Used 385
Fig 79   Short Circuit on Load Side of Main Switch 399
Fig 80   Total Short-Circuit Current Equals Sum of Sources 400
Fig 81   Symmetrical AC Wave... 401

Fig 82   Asymmetrical AC Wave 402
Fig 83   Typical Short Circuit 402
Fig 84   Phase Relations of Voltage and Short-Circuit Currents (Medium-
          Voltage Generator Feeding a Distribution Line).................. 403
Fig 85   Symmetrical Current and Voltage in a Zero Power Factor Circuit .. 404
Fig 86   Asymmetrical Current and Voltage in a Zero Power Factor
          Circuit 405
Fig 87   Components of Current Shown in Fig 86......................... 405
Fig 88   Decay of DC Component and Effect of Asymmetry of Current 406
Fig 89   Symmetrical Short-Circuit Currents from Four Sources Combined
          into Total 407
Fig 90   Typical System Single-Line Diagram 414
Fig 91   Equivalent Impedance Diagram for System in Fig 90 415
Fig 92   Impedance Vectors 416
Fig 93   How Series Impedances Are Added 417
Fig 94   Combining Impedances in Parallel 417
Fig 95   Single-Line Diagram — Specific 480Y/277 V Network Served
          Building 428
Fig 96   Impedance Diagram for the System in Fig 95.................... 429
Fig 97   Current-Limiting Action of a Current-Limiting Fuse.............. 442
Fig 98   Typical Current-Limiting Fuse Peak Let-Through Curves 443
Fig 99   Typical Available Short-Circuit Currents on Large Office Building
          Distribution Systems 445
Fig 100   Some Possible Arrangements to Limit and Control Available
          Short-Circuit Current 446
Fig 101   Typical Distribution Single-Line Diagram (Coordination Example) .. 448
Fig 102   Time Current Curves for 125–600 A Molded-Case Circuit
          Breakers 449
Fig 103   Time Current Curves for 600–4000 A Power Circuit Breakers 450
Fig 104   Through-Fault Protection Curves for Liquid Immersed or Dry-
          Type Category II Transformers (501–1667 kVA, Single-Phase;
          501–5000 kVA, Three-Phase) 453
Fig 105   Time Current Curves Showing Primary Fuse and Transformer
          Protection Criteria for a Delta-Wye, Solidly Grounded Transformer
          for·the System Shown in Fig 101 454
Fig 106   Time Current Curves Showing Primary Fuse and Transformer
          Protection Criteria for a Delta-Delta Transformer for the System
          Shown in Fig 101 455
Fig 107   Time Current Curves Showing Primary Fuse and Transformer
          Protection Criteria for a Wye-Wye Transformer Solidly Grounded
          on Both Primary and Secondary for the System Shown in
          Fig 101 456
Fig 108   Time Current Curves Showing Complete Phase Coordination for
          the System Shown in Fig 101 459

Fig 109   Time Current Curves Showing Complete Ground Coordination for
          the System Shown in Fig 101 461
Fig 110   Typical Total Clearing Time versus Current Curves for Type RK5
          Fuses 462
Fig 111   Selectivity of Fuses... 463
Fig 112   Typical Application Example Using the Data in
          Table 71 464
Fig 113   Coordination Study of Primary and Secondary Fuses Showing
          Selective System 465
Fig 114   Fusible Unit Substation 466
Fig 115   Completed Coodination Study of Low-Voltage Fusible Substation
          Shows Complete Selectivity 467
Fig 116   Effects of I, I^2, and I^2t 468
Fig 117   Peak Let-Through Current as a Function of Available Symmetrical
          RMS Fault Current 470
Fig 118   Application Example of Fuse Let-Through Charts 470
Fig 119   Example for Determining Bracing Requirements for 800 A
          Motor Control Center 471
Fig 120   Selection of Fuses to Provide Short-Circuit Protection and
          Backup Protection for Motor Starters 473
Fig 121   Typical Low-Voltage Distribution Transformer Secondary
          Protection 474
Fig 122   Typical Protection for 225 kVA Lighting Transformer 475
Fig 123   Protection for Typical Motor Circuit 476
Fig 124   Detecting Ground-Fault Current 477
Fig 125   Typical Ground-Fault Relaying 478
Fig 126   Ceiling System Providing Illumination, Air Supply, Air Return,
          Space Partitioning, and Acoustical Treatment 504
Fig 127   Near-Horizontal Drafting Boards Positioned So That Draftsmens'
          Views Are Parallel with the Luminaires 508
Fig 128   Two U-Shaped Fluorescent Lamps, with Efficient, Low-Brightness
          Anodized Aluminum Luminaire, Are Employed in Each Unit to
          Supply Lighting from the Ceiling 509
Fig 129   Two Foot by Two Foot Luminaires Containing U-Shaped
          Fluorescent Lamps 511
Fig 130   Secretarial Area with Uniform Lighting That Has Non-Uniform
          Possibilities 512
Fig 131   Open-Plan Office Area with Task-Ambient Lighting 514
Fig 132   Task-Ambient Lighting in a Workstation 515
Fig 133   Task-Ambient Lighting with a Single Luminaire Providing Both
          Components 516
Fig 134   Installation for Luminaires Supplying Ambient
          Lighting 517
Fig 135   Parabolic Wedge Louver 520

FIGURES   PAGE

Fig 136   Building Retrofitted with Energy-Conserving Lamp Products
          Substantially Reduce Operating Costs for the Building, Yet Do Not
          Sacrifice the Environment 521
Fig 137   Special Lighting in a Store 522
Fig 138   Illumination in a Supermarket by Metal-Halide Lamps in Recessed
          Luminaires.. 523
Fig 139   Indirect Lighting in a Store 524
Fig 140   Parking Area Lighted with "High-Mast" 90 Foot Poles 526
Fig 141   Parking Garage Lighted with 150 W High-Pressure Sodium
          Lamps.. 527
Fig 142   High School Football Field Lit with 1500 W Metal-Halide Lamps ... 528
Fig 143   Typical Data Available for the Evaluation of High-Pressure
          Sodium Lamps Used for Floodlighting 529
Fig 144   Typical Components for the Remote Control of Lighting........... 531
Fig 145   Comparison of Distribution Required for 277 V and 120 V
          Lighting Systems 537
Fig 146   Opaque Balustrade Escalator 565
Fig 147   Transparent Balustrade Escalator................................ 566
Fig 148   Basic Electric Elevator Speed Control Systems 569
Fig 149   Typical Standby Power Transfer System 585
Fig 150   Block Diagram of a Medium-Level FAS 617
Fig 151   Example 1 — Building Extension, Medium-Voltage Service 666
Fig 152   Example 2 — Building Extension, Low-Voltage Service, Simplified
          Single-Line Diagrams of Seven Possible Ways of Handling a Major
          Expansion 668
Fig 153   Example 3 — Retention of Existing System and Expansion 671
Fig 154   Example 4 — Rehabilitation on a College Campus 672
Fig 155   Fluorescent PL Lamp with Socket Adapter........................ 730
Fig 156   Peak Demand Reduction Using Multiple On-Site Generators 752

TABLES

Table 1   Voltage Classes 42
Table 2   Load Tabulation of Equipment Utility Requirements 77
Table 3   Licensed Electrical Contractor Work Request 78
Table 4   Prescriptive Unit Lighting Power Allowance (ULPA) (W/ft^2) —
          Gross Lighted Area of Total Building 83
Table 5   Typical Appliance/General-Purpose Receptacle Loads (Excluding
          Plug-in-Type A/C and Heating Equipment)....................... 84
Table 6   Typical Apartment Loads 84
Table 7   Total Connected Electrical Load for Air Conditioning Only 85
Table 8   All-Weather Comfort Standard Recommended Heat Loss Values ... 86
Table 9   Typical Power Requirement (kW) for High-Rise Building Water
          Pressure Boosting Systems 87

TABLES   PAGE

Table 10   Typical Power Requirement (kW) for Electric Hot Water Heating
           System 87
Table 11   Typical Power Requirement (kW) for Fire Pumps in Commercial
           Buildings (Light Hazard) 88
Table 12   Typical Loads in Commercial Kitchens 90
Table 13   Types of Electrical Load Equipment........................ 92
Table 14   Comparison of Maximum Demand 94
Table 15   Connected Load and Maximum Demand by Tenant
           Classification 96
Table 16   Electrical Load Estimation 97
Table 17   Standard Nominal System Voltages and Voltage
           Ranges 103
Table 18   Standard Voltage Profile for a Regulated Power Distribution
           System, 120 V Base 110
Table 19   Nominal System Voltages................................. 113
Table 20   Tolerance Limits from Table 17, Range A in Volts........ 115
Table 21   Tolerance Limits for Standard Three-Phase Induction Motors
           in Volts 116
Table 22   Tolerance Limits for Standard Fluorescent Lamp Ballasts in
           Volts 116
Table 23   Low-Voltage Motors..................................... 118
Table 24   Voltage Ratings of Standard Motors 119
Table 25   General Effect of Voltage Variations on Induction Motor
           Characteristics 121
Table 26   Effect of Voltage Variations on Incandescent Lamps 122
Table 27   Balanced Three-Phase, Line-to-Line Voltage Drop for 600 V
           Single-Conductor Cable per 10 000 A × feet 60 °C (140 °F)
           Conductor Temperature, 60 Hz 128
Table 28   Voltage-Drop Values for Three-Phase Busways with Copper
           Bus Bars, in V per 100 feet, Line-to-Line, at Rated Current with
           Balanced Entire Load at End........................... 129
Table 29   Voltage-Drop Values for Three-Phase Busways with Aluminum
           Bus Bars, in V per 100 feet, Line-to-Line, at Rated Current with
           Balanced Entire Load at End........................... 130
Table 30   Areas That Can Be Supplied for Specific Voltage Drops and
           Voltages 136
Table 31   Comparison of Motor Starting Methods for Squirrel-Cage
           Induction Motors...................................... 141
Table 32   Cost Comparison of Purchased and Generated Power 151
Table 33   Preferred Kilovoltampere Ratings 215
Table 34   Transformer Approximate Impedance Values 217
Table 35   Insulation Temperature Ratings in °C 217
Table 36   Voltage Insulation Classes and Dielectric Tests 218
Table 37   Sound Levels for Dry-Type Transformers in dB.......... 218

TABLES   PAGE

Table 38   Sound Levels for Single-Phase and Three-Phase Oil Cooled
           Transformers in dB 219
Table 39   Maximum Continuous Current and Interrupting Ratings for
           Horn Fiber Lined, Expulsion-Type Fuses 225
Table 40   Maximum Continuous Current and Interrupting Ratings for
           Solid Material, Boric Acid Fuses (Fuse Units) 226
Table 41   Maximum Continuous Current and Interrupting Ratings for
           Solid Material, Boric Acid Fuses (Refill Units) 226
Table 42   Maximum Continuous Current and Interrupting Ratings for
           Current-Limiting Fuses 227
Table 43   Maximum Continuous Current and Interrupting Ratings for
           Current-Limiting Fuses (Motor Starters) 227
Table 44   Maximum Continuous Current and Interrupting Ratings for
           Electronic Fuses 228
Table 45   Fuse Classification 243
Table 46   Cost Comparison of Reduced Voltage Starters 274
Table 47   NEMA Standardized Starters and Contactors 282
Table 48   Properties of Copper and Aluminum 318
Table 49   Commonly Used Insulating Materials 319
Table 50   Rated Conductor Temperature Ratings 321
Table 51   Properties of Jackets and Braids 332
Table 52   Uprating for Short-Time Overloads 336
Table 53   Minimum Conductor Sizes, in AWG or kcmil, for Indicated Fault
           Current and Clearing Times 338
Table 54   Wiring Methods for Hazardous Locations 345
Table 55   ICEA DC Test Voltages (kV) After Installation Pre-1968 Cable 369
Table 56   ICEA DC Test Voltages (kV) After Installation 1968 and Later
           Cable 370
Table 57   Cable DC Test Voltages (kV) for Installation and Maintenance 370
Table 58   Busway Ratings as a Function of Short-Circuit Power Factor 383
Table 59   Voltage-Drop Values for Three-Phase Busways with Copper Bus
           Bars, in Volts per 100 Feet, Line-to-Line, at Rated Current with
           Entire Load at End 387
Table 60   Voltage-Drop Values for Three-Phase Busways with Aluminum
           Bus Bars, in Volts per 100 Feet, Line-to-Line, at Rated Current
           with Entire Load at End 388
Table 61   Typical Busway Parameters, Line-to-Neutral, in Milliohms per
           100 Feet, 25 °C 389
Table 62   Voltage, Insulation, Continuous Current, and Momentary Current
           Ratings of Nonsegregated Phase Metal-Enclosed Bus 391
Table 63   Approximate Minimum Values of Arcing Fault Currents in Per
           Unit of Bolted Values 409
Table 64   Transformers 418
Table 65   Approximate Impedance Data—Insulated Conductors—60 Hz ... 420

TABLES   PAGE

Table 66   Busways 421

Table 67   System Impedances in Ohms and Per Unit for Fig 96 430

Table 68   Short-Circuit Calculations for Bus 1 in Ohms for Fig 96 433

Table 69   Computer-Generated Short-Circuit Calculation Results for All
Buses That Are Similar to the System in Fig 96 434

Table 70   Line-to-Ground Fault Calculations in Ohms for Fig 96 440

Table 71   Typical Selectivity Schedule 458

Table 72   Recommended Surface Reflectances for Offices.................. 487

Table 73   Illuminance Recommended for Use in Selection of Values for
Interior Lighting Design 490

Table 74   Weighting Factors to Be Considered in Selecting a Specific
Illuminance with the Ranges of Values for Each Category in
Table 73 491

Table 75   Appropriate Initial Efficacies for the lm/W Range of Commonly
Used Lamps 492

Table 76   Relative Merit of Uniform, Non-Uniform, and Task-Ambient
Lighting Systems 506

Table 77   Fluorescent Ballast Sound Ratings 540

Table 78   Moving Walk Capacities 567

Table 79   Maximum Treadway Slope and Speed 567

Table 80   Riser Sleeves or Conduits for Telephone Service 590

Table 81   Apparatus Closet Specifications 595

Table 82   Centralized Electronic Key Telephone Equipment Space
Specifications in Linear Feet 596

Table 83   Satellite Closet Specifications 596

Table 84   Examples of Existing Conditions and Replacement Strategies
and the Justification of Remedial Action 653

Table 85   Approximate Illumination Footcandles at 10 Feet Mounting
Height (Elliptical versus Standard Reflector) 728

Table 86   Energy-Saving versus Standard Fluorescent Lamps 729

Table 87   Typical Input Watts for Fluorescent Lamp Ballastss 732

Table 88   Typical Input Watts for HID Lamp Ballasts 733

Table 89   Energy Savings and Cost Comparisons for Four Alternative
Fluorescent Systems 735

INDEX 756

Chapter 1
Introduction

1.1 Scope. IEEE Std 241-1990, IEEE Recommended Practice for Electric Power Systems in Commercial Buildings, commonly known as the "Gray Book," is published by the Institute of Electrical and Electronics Engineers (IEEE) to provide a recommended practice for the electrical design of commercial buildings. It has been prepared on a voluntary basis by engineers and designers functioning as the Gray Book Working Group within the IEEE Power Systems Engineering Committee.

This recommended practice will probably be of greatest value to the power oriented engineer with limited commercial building experience. It can also be an aid to all engineers responsible for the electrical design of commercial buildings. However, it is not intended as a replacement for the many excellent engineering texts and handbooks commonly in use, nor is it detailed enough to be a design manual. It should be considered a guide and general reference on electrical design for commercial buildings.

Tables, charts, and other information that have been extracted from codes, standards, and other technical literature are included in this recommended practice. Their inclusion is for illustrative purposes; where the correctness of the item is important, the latest referenced document should be used to assure that the information is complete, up to date, and correct. It is not possible to reproduce the full text of these items in this recommended practice.

1.1.1 Voltage Levels. It is important to establish, at the outset, the terms describing voltage classifications. Table 1, which is adapted from IEEE Std 100-1988, IEEE Standard Dictionary of Electrical and Electronics Terms, Fourth Edition (ANSI) [5],[1] indicates these voltage levels. ANSI/NFPA 70-1990, National Electrical Code (NEC) [3],[2] described in 1.6.1, uses the term "over 600 volts" generally to refer to what is known as "high voltage." Many IEEE Power Engineering Society (PES) standards use the term "high voltage" to refer to any voltage higher than 1000 V. All

---

[1] The numbers in brackets correspond to those in the references at the end of each chapter. IEEE publications are available from the Institute of Electrical and Electronics Engineers, IEEE Service Center, 445 Hoes Lane, P.O. Box 1331, Piscataway, NJ 08855-1331.

[2] ANSI publications are available from the Sales Department of the American National Standards Institute, 11 West 42nd Street, 13th Floor, New York, NY 10036. NFPA publications are available from Publications Sales, National Fire Protection Association, 1 Batterymarch Park, P.O. Box 9101, Quincy, MA 02269-9101.

Table 1
Voltage Classes

	NOMINAL SYSTEM VOLTAGE			
	TWO WIRE	THREE WIRE	FOUR WIRE	MAXIMUM VOLTAGE
		Single-Phase Systems		127
	(120)	120/240		127/254
		Three-Phase Systems		
			208Y/120	220Y/127
		(240)	240/120	245/127
		480	480Y/277	508Y/293
		(600)		635
		(2400)		2540
		4160	4160Y/2400	4400Y/2540
		(4800)		5080
		(6900)		7260
			(8320Y/4800)	8800Y/5080
			(12000Y/6930)	12700Y/7330
			12470Y/7200	13200Y/7620
			13200Y/7620	13970Y/8070
		13800	(13800Y/7970)	14520Y/8380
			(20780Y/12000)	22000Y/12700
			(22860Y/13200)	24200Y/13970
		(23000)		24340
			24940Y/14400	26400Y/15240
		(34500)	34500/19920	36510Y/21080
		(46 kV)		48.3 kV
		69 kV		72.5 kV
		115 kV		121 kV
		138 kV		145 kV
		(161 kV)		169 kV
		230 kV		242 kV
		345 kV		362 kV
		500 kV		550 kV
		765 kV		800 kV
		1100 kV		1200 kV

Left-side vertical bracket labels:
- IEEE Std for Industrial & Commercial Power Systems
- LOW VOLTAGE SYSTEMS
- MEDIUM VOLTAGE — ANSI C84.1-1989 no voltage class stated
- HIGH VOLTAGE — ANSI C84.1-1989 — HIGHER VOLTAGE SYSTEMS
- EHV EHV EHV — ANSI C92.2-1987

NOTE: See Table 17 in Chapter 3 for a complete listing of system voltages.

nominal voltages are expressed in terms of rms. For a detailed explanation of voltage terms, see Chapter 3. ANSI C84.1-1989, Voltage Ratings for Electric Power Systems and Equipment (60 Hz) [2][3] lists voltage class designations applicable to industrial and commercial buildings where medium voltage extends from 1000 V to 69 kV nominal.

1.2 Commercial Buildings. The term "commercial, residential, and institutional buildings," as used in this chapter, encompasses all buildings other than industrial buildings and private dwellings. It includes office and apartment buildings, hotels, schools, and churches, marine, air, railway, and bus terminals, department stores, retail shops, governmental buildings, hospitals, nursing homes, mental and correctional institutions, theaters, sports arenas, and other buildings serving the public directly. Buildings, or parts of buildings, within industrial complexes, which are used as offices or medical facilities or for similar nonindustrial purposes, fall within the scope of this recommended practice. It is not possible to cover each type of occupancy in this text; however, many are covered in Chapter 16. Medical areas are covered in IEEE Std 602-1986, IEEE Recommended Practice for Electric Systems in Health Care Facilities (ANSI) [10] (the "White Book").

The specific use of the commercial building in question, rather than the nature of the overall development of which it is a part, determines its electrical design category. While industrial plants are primarily machine- and production-oriented, commercial, residential, and institutional buildings are primarily people- and public-oriented. The fundamental objective of commercial building design is to provide a safe, comfortable, energy-efficient, and attractive environment for living, working, and enjoyment. The electrical design must satisfy these criteria if it is to be successful.

Today's commercial buildings, because of their increasing size and complexity, have become more and more dependent upon adequate and reliable electric systems. One can better understand the complex nature of modern commercial buildings by examining the systems, equipment, and facilities listed in 1.2.1.

1.2.1 System Requirements for Commercial, Residential, and Institutional Buildings. The systems, equipment, and facilities that must be provided to satisfy functional requirements will vary with the type of facility, but will generally include some or all of the following:

(1) Building electric service
(2) Power distribution system
(3) Lighting—Interior and exterior, both utilitarian and decorative; task and general lighting.
(4) Communications—Telephone, facsimile, telegraph, satellite link, building-to-building communications (including microwave, computer link, radio, closed-circuit television, code call, public address, paging, fiber-optic and electronic intercommunication, pneumatic tube, doctors' and nurses' call, teleconferencing), and a variety of other signal systems.

---

[3]ANSI publications are available from the Sales Department of the American National Standards Institute, 11 West 42nd Street, 13th Floor, New York, NY 10036.

(5) Fire alarm systems — Fire pumps and sprinklers, smoke and fire detection, alarm systems, and emergency public address systems.

(6) Transportation — Elevators, moving stairways, dumbwaiters, and moving walkways.

(7) Space conditioning — Heating, ventilation, and air conditioning.

(8) Sanitation — Garbage and rubbish storage, recycling, compaction, and removal; incinerators; sewage handling; and document shredders and pulpers.

(9) Plumbing — Hot and cold water systems and water treatment facilities.

(10) Security watchmen, burglar alarms, electronic access systems, and closed-circuit surveillance television

(11) Business machines — Typewriters, computers, calculators, reproduction machines, and word processors.

(12) Refrigeration equipment

(13) Food handling, catering, dining facilities, and food preparation facilities

(14) Maintenance facilities

(15) Lightning protection

(16) Automated building control systems

(17) Entertainment facilities and specialized audiovisual systems

(18) Medical facilities

(19) Recreational facilities

(20) Legally required and optional standby/emergency power and peak-shaving systems

(21) Signing, signaling, and traffic control systems; parking control systems including automated parking systems

1.2.2 Electrical Design Elements. In spite of the wide variety of commercial, residential, and institutional buildings, some electrical design elements are common to all. These elements, listed below, will be discussed generally in this section and in detail in the remaining sections of this recommended practice. The principal design elements considered in the design of the power, lighting, and auxiliary systems include:

(1) Magnitudes, quality, characteristics, demand, and coincidence or diversity of loads and load factors

(2) Service, distribution, and utilization voltages and voltage regulation

(3) Flexibility and provisions for expansion

(4) Reliability and continuity

(5) Safety of personnel and property

(6) Initial and maintained cost ("own and operate" costs)

(7) Operation and maintenance

(8) Fault current and system coordination

(9) Power sources

(10) Distribution systems

(11) Legally required and optional standby/emergency power systems

(12) Energy conservation, demand, and control

(13) Conformance with regulatory requirements

(14) Special requirements of the site related to: seismic requirements (see IEEE Std 693-1984 [12]), altitude, sound levels, security, exposure to physical

elements, fire hazards (see IEEE Std 979-1984 [13]), hazardous locations, and power conditioning and uninterruptible power supply (UPS) systems

1.3 The Industry Applications Society (IAS). The IEEE is divided into 35 groups and societies that specialize in various technical areas of electrical engineering. Each group or society conducts meetings and publishes papers on developments within its specialized area.

The IEEE Industry Applications Society (IAS) currently encompasses 26 technical committees covering aspects of electrical engineering in specific areas (petroleum and chemical industry, cement industry, glass industry, power systems engineering, and others). Papers of interest to electrical engineers and designers involved in the field covered by this recommended practice are, for the most part, contained in the transactions of the IAS.

The Gray Book is published by the IEEE Standards Department on behalf of the Power Systems Engineering Committee. Individuals who desire to participate in the activities of the committees, subcommittees, or working groups in the preparation and revision of texts such as this should write the IEEE Standards Department, 445 Hoes Lane, P.O. Box 1331, Piscataway, NJ 08855-1331.

1.3.1 Grouping of Commercial Buildings. The principal groupings of commercial buildings are:

(1) Multiple-story buildings, office buildings, and apartment buildings
(2) Public buildings and stores, such as retail shops and supermarkets
(3) Institutional buildings, such as hospitals, large schools, colleges, corporate headquarters
(4) Airport, railroad, and other transportation terminals
(5) Large commercial malls and shopping centers
(6) Competitive and speculative buildings of types (1) and (2) above where minimum costs are essential, and where interior finishes are left to future tenants

The Production and Application of Light, Power Systems Engineering, Power Systems Protection, Codes and Standards, Energy Systems, and Safety Committees of the IAS are involved with commercial building activities and some publish material applicable to many types of commercial facilities.

1.4 IEEE Publications. The IEEE publishes several recommended practices that are similar in style to the IEEE Gray Book, prepared by the Industrial and Commercial Power Systems Department of the IEEE Industry Applications Society.

(1) IEEE Std 141-1986, IEEE Recommended Practice for Electric Power Distribution for Industrial Plants (ANSI) (the "Red Book").
(2) IEEE Std 142-1982, IEEE Recommended Practice for Grounding of Industrial and Commercial Power Systems (ANSI) (the "Green Book").
(3) IEEE Std 242-1986, IEEE Recommended Practice for Protection and Coordination of Industrial and Commercial Power Systems (ANSI) (the "Buff Book").
(4) IEEE Std 399-1990, IEEE Recommended Practice for Industrial and Commercial Power Systems Analysis (ANSI) (the "Brown Book").

(5) IEEE Std 446-1987, IEEE Recommended Practice for Emergency and Standby Power Systems for Industrial and Commercial Applications (ANSI) (the "Orange Book").

(6) IEEE Std 493-1990, IEEE Recommended Practice for the Design of Reliable Industrial and Commercial Power Systems (ANSI) (the "Gold Book").

(7) IEEE Std 602-1986, IEEE Recommended Practice for Electric Systems in Health Care Facilities (ANSI) (the "White Book").

(8) IEEE Std 739-1984, IEEE Recommended Practice for Energy Conservation and Cost-Effective Planning in Industrial Facilities (ANSI) (the "Bronze Book").

1.5 Professional Registration. Most regulatory agencies require that the designs for public and other commercial buildings be prepared under the jurisdiction of state-licensed professional architects or engineers. Information on such registration may be obtained from the appropriate state agency or from the local chapter of the National Society of Professional Engineers.

To facilitate obtaining registration in different states under the reciprocity rule, a National Professional Certificate is issued by the National Council of Engineering Examiners, Records Department[4] to engineers who have obtained their home state license by examination. All engineering graduates are encouraged to start on the path to full registration by taking the engineer-in-training examination as soon after graduation as possible. The final written examination in the field of specialization is usually conducted after 4 years of progressive professional experience.

1.5.1 Professional Liability. Recent court and regulatory decisions have held the engineer and designer liable for situations that have been interpreted as malpractice. These decisions have involved safety, environmental concerns, specification and purchasing practice, and related items. Claims for accidents, purportedly resulting from poor design or operating practices (e.g., too low lighting levels) or nonconformance to applicable codes and standards have resulted in awards against engineering firms and design staff. It is a good idea for the practicing engineer to determine policies for handling such claims, and to evaluate the need for separate professional liability insurance.

1.6 Codes and Standards

1.6.1 National Electrical Code (NEC). The electrical wiring requirements of the NEC [3] are vitally important guidelines for commercial building electrical engineers. The NEC is revised every 3 years. It is published by and available from the National Fire Protection Association (NFPA), the American National Standards Institute (ANSI) and from each state's Board of Fire Underwriters (usually located in each state's capital). It does not represent a design specification but does identify minimum requirements for the safe installation and utilization of electricity. It is strongly recommended that the introduction to the NEC, Article 90 [3], which covers purpose and scope, be carefully reviewed.

---

[4] For more information, write to the National Council of Engineering Examiners, Records Department, P.O. Box 1686, Clemson, SC 29633-1686.

NFPA CY-70HB90, *NFPA National Electrical Code Handbook* [16][5] contains the complete NEC text plus explanations. This book is edited to correspond with each edition of the NEC. McGraw-Hill's *Handbook of the National Electrical Code* [20][6] and other handbooks provide explanations and clarification of the NEC requirements.

Each municipality or jurisdiction that elects to use the NEC must enact it into law or regulation. The date of enactment may be several years later than issuance of the code; in which event, the effective code may not be the latest edition. It is important to discuss this with the inspection or enforcing authority. Certain requirements of the latest edition of the code may be interpreted as acceptable by this authority.

1.6.2 Other NFPA Standards. NFPA publishes the following related documents containing requirements on electrical equipment and systems:

(1) NFPA/SFPE GL-HFPE-88, The SFPE Handbook of Fire Protection Engineering — 1988 Edition.
(2) NFPA GL-101ST91, Life Safety Code Handbook — 1991 Edition.
(3) NFPA 20-1990, Installation of Centrifugal Fire Pumps.
(4) NFPA 70B-1990, Electrical Equipment Maintenance.
(5) NFPA 70E-1988, Electrical Safety Requirements for Employee Workplaces.
(6) NFPA 71-1989, Installation, Maintenance, and Use of Signaling Systems for Central Station Service.
(7) NFPA 72-1990, Installation, Maintenance, and Use of Protective Signaling Systems.
(8) NFPA 72E-1990, Automatic Fire Detectors.
(9) NFPA 72G-1989, Installation, Maintenance, and Use of Notification Appliances for Protective Signaling Systems.
(10) NFPA 75-1989, Protection of Electronic Computer/Data Processing Equipment.
(11) NFPA 77-1988, Static Electricity.
(12) NFPA 78-1989, Lightning Protection Code.
(13) NFPA 92A-1988, Smoke Control Systems.
(14) NFPA 99-1990, Health Care Facilities — Chapter 8: Essential Electrical Systems for Health Care Facilities, and Appendix E: The Safe Use of High Frequency Electricity in Health Care Facilities.
(15) NFPA 101-1988, Life Safety Code — 1988 Edition.
(16) NFPA 110-1988, Emergency and Standby Power Systems.
(17) NFPA 110A-1989, Stored Energy Emergency and Standby Power Systems.
(18) NFPA 130-1990, Fixed Guideway Transit Systems.

1.6.3 Local, State, and Federal Codes and Regulations. While most municipalities, counties, and states adopt the NEC [3] without change or with modifications,

---

[5]NFPA publications are available from Publications Sales, National Fire Protection Association, 1 Batterymarch Park, P.O. Box 9101, Quincy, MA 02269-9101.

[6]McGraw-Hill publications are available from McGraw-Hill Book Company, 1221 Avenue of the Americas, New York, NY 10020.

some have their own codes. In most instances, the NEC is adopted by local ordinance as part of the building code. Deviations from the NEC may be listed as addenda. It is important to note that only the code adopted by ordinance as of a certain date is official, and that governmental bodies may delay adopting the latest code. Federal rulings may require use of the latest NEC rulings, regardless of local rulings, so that reference to the enforcing agencies for interpretation on this point may be necessary.

Some city and state codes are almost as extensive as the NEC. It is generally accepted that, in the case of conflict, the more stringent or severe interpretation applies. Generally, the entity responsible for enforcing (enforcing authority) the code has the power to interpret it. Failure to comply with the NEC or local code provisions, where required, can affect the owner's ability to obtain a certificate of occupancy, may have a negative effect on insurability, and may subject the owner to legal penalty.

Legislation by the U.S. federal government has had the effect of giving standards, such as certain ANSI Standards, the impact of law. The Occupational Safety and Health Act, administered by the U.S. Department of Labor, permits federal enforcement of codes and standards. The Occupational Safety and Health Administration (OSHA) adopted the 1971 NEC for new electrical installations and also for major replacements, modifications, or repairs installed after March 5, 1972. A few articles and sections of the NEC have been deemed to apply retroactively by OSHA. The NFPA created the NFPA 70E (Electrical Requirements for Employee Workplaces) Committee to prepare a consensus standard for possible use by OSHA in developing their standards. Major portions of NFPA 70E-1988, Electrical Safety Requirements for Employee Workplaces [15] have been included in OSHA regulations.

OSHA requirements are published in the *Federal Register* [18].[7] OSHA rules for electric systems are covered in 29 CFR Part 1910 of the *Federal Register* [18].

The U.S. National Institute of Occupational Safety and Health (NIOSH) publishes *Electrical Alerts* [17][8] to warn of unsafe practices or hazardous electrical equipment.

The U.S. Department of Energy, by encouraging building energy performance standards, has advanced energy conservation standards. A number of states have enacted energy conservation regulations. These include ASHRAE/IES legislation embodying various energy conservation standards such as ASHRAE/IES 90.1-1989, Energy Efficient Design of New Buildings Except New Low Rise Residential Buildings [4]. These establish energy or power budgets that materially affect architectural, mechanical, and electrical designs.

1.6.4 Standards and Recommended Practices. In addition to NFPA, a number of organizations publish documents that affect electrical design. Adherence to these documents can be written into design specifications.

---

[7] The *Federal Register* is available from the Superintendent of Documents, U.S. Government Printing Office, Washington, DC 20402 (telephone 202-783-3238) on a subscription or individual copy basis.

[8] Copies of this bulletin are available from NIOSH Publications Dissemination, 4676 Columbia Parkway, Cincinnati, OH 45226.

The American National Standards Institute (ANSI) coordinates the review of proposed standards among all interested affiliated societies and organizations to assure a consensus approval. It is in effect a clearinghouse for technical standards. Not all standards are ANSI approved.

Underwriters Laboratories, Inc. (UL)[9] and other independent testing laboratories may be approved by an appropriate jurisdictional authority (e.g., OSHA) to investigate materials and products including appliances and equipment. Tests may be performed to their own or to another agency's standards; and a product may be "listed" or "labeled." The UL publishes an *Electrical Construction Materials Directory, Electrical Appliance and Utilization Equipment Directory, Hazardous Location Equipment Directory,* and other directories. It should be noted that other testing laboratories (where approved) and governmental inspection agencies may maintain additional lists of approved or acceptable equipment; the approval must be for the jurisdiction where the work is to be performed.

The Electrification Council (TEC)[10] representative of the investor-owned utilities publishes several informative handbooks, such as

(1) *Industrial and Commercial Power Distribution*
(2) *Industrial and Commercial Lighting*
(3) An energy analysis computer program, AXCESS, for forecasting electricity consumption and costs in existing and new buildings.

The National Electrical Manufacturers Association (NEMA)[11] represents equipment manufacturers. Its publications serve to standardize certain design features of electrical equipment, and provide testing and operating standards for electrical equipment. Some NEMA Standards contain important application information for electrical equipment, such as motors and circuit breakers.

The IEEE publishes several hundred electrical standards relating to safety, measurements, equipment testing, application, maintenance, and environmental protection.

The following three publications from the IEEE and ANSI are general in nature:

(1) IEEE Std 100-1988, IEEE Standard Dictionary of Electrical and Electronics Terms, Fourth Edition (ANSI).
(2) IEEE Std 315-1975 (Reaff. 1989), IEEE Standard Graphic Symbols for Electrical and Electronics Diagrams (ANSI, CSA Z99-1975), and Supplement IEEE Std 315A-1986 (ANSI).
(3) ANSI Y32.9-1972 (Reaff. 1989), American National Standard Graphic Symbols for Electrical Wiring and Layout Diagrams Used in Architecture and Building Construction. (Important for the preparation of drawings.)

---

[9] UL publications are available from Underwriters Laboratories, 333 Pfingsten Road, Northbrook, IL 60062.

[10] TEC publications are available from the Electrification Council, 1111 19th Street, N.W., Washington, DC 20036.

[11] NEMA publications are available from the National Electrical Manufacturers Association, 2101 L Street, N.W., Washington, DC 20037.

The Electrical Generating Systems Association (EGSA)[12] publishes performance standards for emergency, standby, and co-generation equipment.

The Intelligent Buildings Institute (IBI)[13] publishes standards on the essential elements of "high-tech" buildings.

The Edison Electric Institute (EEI)[14] publishes case studies of electrically space-conditioned buildings as well as other informative pamphlets.

The International Electrotechnical Commission (IEC) is an electrical and electronics standards generating body with a multinational membership. The IEEE is a member of the U.S. National Committee of the IEC.

1.7 Handbooks. The following handbooks have, over the years, established reputations in the electrical field. This list is not intended to be all-inclusive. Other excellent references are available, but are not listed here because of space limitations.

(1) Fink, D. G. and Beaty, H. W. *Standard Handbook for Electrical Engineers*, 12th edition, New York: McGraw-Hill, 1987. Virtually the entire field of electrical engineering is treated, including equipment and systems design.

(2) Croft, T., Carr, C. C., and Watt, J. H. *American Electricians Handbook*, 11th edition, New York: McGraw-Hill, 1987. The practical aspects of equipment, construction, and installation are covered.

(3) *Lighting Handbook*, Illuminating Engineering Society (IES).[15] This handbook in two volumes (Applications, 1987; Reference, 1984) covers all aspects of lighting, including visual tasks, recommended lighting levels, lighting calculations, and design, which are included in extensive detail.

(4) *Electrical Transmission and Distribution Reference Book*,[16] Westinghouse Electric Corporation, 1964. All aspects of transmission and distribution, performance, and protection are included in detail.

(5) *Applied Protective Relaying*, Westinghouse Electric Corporation, 1976. The application of protective relaying to commercial-utility interconnections, protection of high-voltage motors, transformers, and cable are covered in detail.

(6) *ASHRAE Handbook*,[17] American Society of Heating, Refrigerating, and Air-Conditioning Engineers. This series of reference books in four volumes, which are periodically updated, detail the electrical and mechanical aspects of space conditioning and refrigeration.

---

[12] EGSA publications are available from the Electrical Generating Systems Association, 10251 West Sample Road, Suite D, P.O. Box 9257, Coral Springs, FL 33075-9257.

[13] IBI publications are available from the Intelligent Buildings Institute, 2101 L Street, N.W., Washington, DC 20037.

[14] EEI publications are available from the Edison Electric Institute, 1111 19th Street, N.W., Washington, DC 20036.

[15] IES publications are available from the Illuminating Engineering Society, 345 East 47th Street, New York, NY 10017.

[16] Westinghouse publications are available from Westinghouse Electric Corporation, Printing Division, Forbes Road, Trafford, PA 15085.

[17] ASHRAE publications are available from the American Society of Heating, Refrigerating, and Air-Conditioning Engineers, 1791 Tullie Circle, N.E., Atlanta, GA 30329.

(7) *Motor Applications and Maintenance Handbook*, second edition, 1987, Smeaton, R. S., ed., McGraw-Hill, 1987. Contains extensive, detailed coverage of motor load data and motor characteristics for coordination of electric motors with machine mechanical characteristics.

(8) *Industrial Power Systems Handbook*, Beeman, D. L., ed., McGraw-Hill, 1955. A test on electrical design with emphasis on equipment, including that applicable to commercial buildings.

(9) *Electrical Maintenance Hints*, Westinghouse Electric Corporation, 1984. The preventive maintenance procedures for all types of electrical equipment and the rehabilitation of damaged apparatus are discussed and illustrated.

(10) *Lighting Handbook*,[18] Philips Lighting Company, 1984. The application of various light sources, fixtures, and ballasts to interior and exterior commercial, industrial, sports, and roadway lighting projects.

(11) *Underground Systems Reference Book*, Edison Electric Institute, 1957. The principles of underground construction and the detailed design of vault installations, cable systems, and related power systems are fully illustrated, and cable splicing design parameters are also thoroughly covered.

(12) *Switchgear and Control Handbook*, Smeaton, R. S., ed., McGraw-Hill, 1977 (second edition 1987). Concise, reliable guide to important facets of switchgear and control design, safety, application, and maintenance including high- and low-voltage starters, circuit breakers, and fuses.

(13) *Handbook of Practical Electrical Design*, McPartland, J. M., ed., McGraw-Hill, 1984.

A few of the older texts may not be available for purchase, but are available in most professional offices and libraries.

1.8 Periodicals. *IEEE Spectrum*, the monthly magazine of the IEEE, covers all aspects of electrical and electronics engineering with broad-brush articles that bring the engineer up to date. It contains references to IEEE books; technical publication reviews; technical meetings and conferences; IEEE group, society, and committee activities; abstracts of papers and publications of the IEEE and other organizations; and other material essential to the professional advancement of the electrical engineer.

The transactions of the IEEE Industrial Applications Society are directly useful to commercial building electrical engineers. Following are some other well-known periodicals:

(1) *ASHRAE Journal*, American Society of Heating, Refrigerating, and Air-Conditioning Engineers.

(2) *Electrical Construction and Maintenance (EC&M)*, Intertec Publishing Corp.[19]

(3) *Fire Journal*, National Fire Protection Association (NFPA).

---

[18] Phillips Lighting publications are available from Phillips Lighting Company, 200 Franklin Square Drive, P.O. Box 6800, Somerset, NJ 08875-6800.

[19] EC&M publications are available from EC&M, Intertec Publishing Corporation, 1221 Avenue of the Americas, New York, NY 10020.

(4) *IAEI News*, International Association of Electrical Inspectors.[20]
(5) *Lighting Design and Application (LD&A)*, Illuminating Engineering Society
(6) *Electrical Systems Design*, Andrews Communications Inc.[21]
(7) *Engineering Times*, National Society of Professional Engineers (NSPE).[22]
(8) *Consulting-Specifying Engineer*, Cahners Publishing Company.[23]
(9) *Plant Engineering*, Cahners Publishing Company.

1.9 Manufacturers' Data. The electrical industry, through its associations and individual manufacturers of electrical equipment, issues many technical bulletins, data books, and magazines. While some of this information is difficult to obtain, copies should be available to each major design unit. The advertising sections of electrical magazines contain excellent material, usually well illustrated and presented in a clear and readable form, concerning the construction and application of equipment. Such literature may be promotional; it may present the advertiser's equipment or methods in its best light and should be carefully evaluated. Manufacturers' catalogs are a valuable source of equipment information. Some manufacturers' complete catalogs are quite extensive, covering several volumes. However, these companies may issue condensed catalogs for general use. A few manufacturers publish regularly scheduled magazines containing news of new products and actual applications. Data sheets referring to specific items are almost always available from marketing offices.

1.10 Safety. Safety of life and preservation of property are two of the most important factors in the design of the electric system. This is especially true in commercial buildings because of public occupancy, thoroughfare, and high occupancy density. In many commercial buildings, the systems operating staff have very limited technical capabilities and may not have any specific electrical training.

Various codes provide rules and regulations as minimum safeguards of life and property. The electrical design engineer may often provide greater safeguards than outlined in the codes according to his or her best judgment, while also giving consideration to utilization and economics.

Personnel safety may be divided into two categories:
(1) Safety for maintenance and operating personnel
(2) Safety for the general public

Safety for maintenance and operating personnel is achieved through the proper design and selection of equipment with regard to enclosures, key-interlocking, circuit breaker and fuse-interrupting capacity, the use of high-speed fault detection

---

[20] IAEI publications are available from the International Association of Electrical Inspectors, 930 Busse Highway, Park Ridge, IL 60068.

[21] This publication is available from Andrews Communications, Inc., 5123 West Chester Pike, P.O. Box 556, Edgemont, PA 19028.

[22] NSPE publications are available from the National Society of Professional Engineers, 1420 King Street, Alexandria, VA 22314.

[23] This publication is available from Cahners Publishing Company, Cahners Plaza, 1350 East Touhy Avenue, P.O. Box 5080, Des Plaines, IL 60017-8800.

and circuit-opening devices, clearances, grounding methods, and identification of equipment.

Safety for the general public requires that all circuit-making and circuit-breaking equipment, as well as other electrical apparatus, be isolated from casual contact. This is achieved by using dead-front equipment, locked rooms and enclosures, proper grounding, limiting of fault levels, installation of barriers and other isolation (including special ventilating grills), proper clearances, adequate insulation, and similar provisions outlined in this recommended practice.

The U.S. Department of Labor has issued the "Rule on Lockout/Tagout" published in the *Federal Register* (53 FR 1546) [18], January 2, 1990, which is concerned with procedures for assuring the safety of workers who are directly involved in working with or near energized conductors or conductors that, if energized, could be hazardous.

ANSI C2-1990, National Electrical Safety Code (NESC) [1] is available from the IEEE. It covers basic provisions for safeguarding from hazards arising from the installation, operation, or maintenance of (1) conductors in electric supply stations, and (2) overhead and underground electrical supply and communication lines. It also covers work rules for construction, maintenance, and operation of electrical supply and communication equipment. Part 4 of the NESC deals specifically with safe working methods.

Circuit protection is a fundamental safety requirement of all electric systems. Adequate interrupting capacities are required in services, feeders, and branch circuits. Selective, automatic isolation of faulted circuits represents good engineering practice. Fault protection, which is covered in Chapter 9, should be designed and coordinated throughout the system. Physical protection of equipment from damage or tampering, and exposure of unprotected equipment to electrical, chemical, and mechanical damage is necessary.

1.10.1 Appliances and Equipment. Improperly applied or inferior materials can cause electrical failures. The use of appliances and equipment listed by the Underwriters Laboratories (UL), Inc., or other approved laboratories is recommended. The Association of Home Appliance Manufacturers (AHAM)[24] and the Air-Conditioning and Refrigeration Institute (ARI)[25] specify the manufacture, testing, and application of many common appliances and equipment. High-voltage equipment and power cable is manufactured in accordance with IEEE, UL, NEMA, and ANSI Standards, and the engineer should make sure that the equipment he or she specifies and accepts conforms to these standards. Properly prepared specifications can prevent the purchase of inferior or unsuitable equipment. The lowest initial purchase price may not result in the lowest cost after taking into consideration operating, maintenance, and owning costs. Value engineering is an organized approach to identification of unnecessary costs utilizing such methods as life-cycle cost analysis, and related techniques.

---

[24] AHAM publications are available from the Association of Home Appliance Manufacturers, 20 North Wacker Drive, Chicago, IL 60606.

[25] ARI publications are available from the Air-Conditioning and Refrigeration Institute, 1815 North Fort Meyer Drive, Arlington, VA 22209.

1.10.2 Operational Considerations. When the design engineers lay out equipment rooms and locate electrical equipment, they cannot always avoid having some areas accessible to unqualified persons. Dead-front construction should be utilized whenever practical. Where dead-front construction is not available (as in existing installations), all exposed electrical equipment should be placed behind locked doors or gates or otherwise suitably "guarded."

In commercial buildings of modern design, the performance of work on live power systems should be prohibited unless absolutely necessary, and then only if qualified personnel are available to perform such work.

A serious cause of failure, attributable to human error, is unintentional grounding or phase-to-phase short circuiting of equipment that is being worked upon. By careful design, such as proper spacing and barriers, and by enforcement of published work safety rules, the designer can minimize this hazard. Unanticipated backfeeds through control circuitry from capacitors, instrument transformers, or test equipment presents a danger to the worker.

Protective devices, such as ground-fault relays and ground-fault detectors (for high-resistance or ungrounded systems), will minimize damage from electrical failures. Electrical fire and smoke can cause staff to disconnect all electric power, even if there is not direct danger to the occupants. Electrical failures that involve smoke and noise, even though occurring in nonpublic areas, may panic occupants. Nuisance tripping can be minimized by careful design and selection of protective equipment.

1.11 Maintenance. Maintenance is essential to proper operation. The installation should be designed so that maintenance can be performed with normally available maintenance personnel (either in-house or contract). Design details should provide proper space, accessibility, and working conditions so that the systems can be maintained without difficulty and excessive cost.

Generally, the external systems are operated and maintained by the electric utility, though at times they are a part of the commercial building distribution system. Where continuity of service is essential, suitable transfer equipment and alternate source(s) should be provided. Such equipment is needed to maintain minimum lighting requirements for passageways, stairways, and critical areas as well as to supply power to critical loads. These systems usually include automatic or manual equipment for transferring loads on loss of normal supply power or for putting battery- or generator-fed equipment into service.

Annual or other periodic shutdown of electrical equipment may be necessary to perform required electrical maintenance. Protective relaying systems, circuit breakers, switches, transformers, and other equipment should be tested on appropriate schedules. Proper system design can facilitate this work.

1.12 Design Considerations. Electrical equipment usually occupies a relatively small percentage of total building space, and, in design, it may be "easier" to relocate electrical service areas than mechanical areas or structural elements. Allocation of space for electrical areas is often given secondary consideration by architectural and related specialties. In the competing search for space, the electrical engineer is responsible for fulfilling the requirements for a proper electrical installation while at the same time recognizing the flexibility of electric systems in terms of layout and placement.

Architectural considerations and appearances are of paramount importance in determining the marketability of a building. Aesthetic considerations may play an important role in the selection of equipment, especially lighting equipment. Provided that the dictates of good practice, code requirements, and environmental considerations are not violated, the electrical engineer may have to negotiate design criteria to accommodate the desires of other members of the design team.

1.12.1 Coordination of Design. The electrical engineer is concerned with professional associates such as the architect, the mechanical engineer, the structural engineer, and, where underground services are involved, the civil engineer. They must also be concerned with the builder and the building owner or operator who, as clients, may take an active interest in the design. More often, the electrical engineer will work directly with the coordinator of overall design activities, usually the architect, or the project manager; and must cooperate with the safety engineer, fire protection engineer, perhaps the environmental enginner, and a host of other concerned people, such as space planners and interior decorators, all of whom have a say in the ultimate design. The electrical designer must become familiar with local rules and know the authorities having jurisdiction over the design and construction. It can be inconvenient and embarrassing to have an electrical project held up at the last moment because proper permits have not been obtained, for example, a permit for a street closing to allow installation of utilities to the site or an environmental permit for an on-site generator.

Local contractors are usually familiar with local ordinances and union work rules and can be of great help in avoiding pitfalls. In performing electrical design, it is essential, at the outset, to prepare a checklist of all the design stages that have to be considered. Major items include temporary power, access to the site, and review by others. Certain electrical work may appear in nonelectrical sections of the specifications. For example, the furnishing and connecting of electric motors and motor controllers may be covered in the mechanical section of the specifications. For administrative control purposes, the electrical work may be divided into a number of contracts, some of which may be under the control of a general contractor and some of which may be awarded to electrical contractors. Among items with which the designer will be concerned are: preliminary cost estimates, final cost estimates, plans or drawings, technical specifications (which are the written presentation of the work), materials, manuals, factory inspections, laboratory tests, and temporary power. The designer may also be involved in providing information on electrical considerations that affect financial justification of the project in terms of owning and operating costs, amortization, return on investment, and related items.

Many electrical designs follow the concept of competitiveness in the commercial sense. Here, cost is a primary consideration, and such designs tend toward minimum code requirements. There is great pressure on the designer to consider cost above maintainability and long life. However, the experienced designer can usually adopt effective compromises.

In cases where the owner or builder is the ultimate occupant, and in buildings such as libraries, municipal buildings, and hospitals, considerations of safety, long life, use by the public, and even prestige may dictate a type of construction often referred to as "institutional." Such design emphasizes reliability, resistance to wear and use, safety to public, and special aesthetic considerations, such as the "age-lessness" of the structure. Smaller buildings, shops, and stores may provide more latitude to the designers in that they are, within budget limitations, subject to a minimum of control in selecting lighting fixtures, equipment, and accessories.

1.12.2 Flexibility. Flexibility of the electric system means the adaptability to development and expansion as well as to changes to meet varied requirements during the life of the building. Often a designer is faced with providing utilities where the loads may be unknown. For example, many office buildings are constructed with the tenant space designs incomplete ("shell and core" designs). In some cases, the designer will provide only the core utilities available for connection by others to serve the working areas. In other cases, the designer may lay out only the basic systems and, as tenant requirements are developed, fill in the details. Often the tenant provides all of his or her own working space designs.

Because it is usually difficult and costly to increase the capacity of risers and feeders, it is important that provisions for sufficient capacity be provided initially. Extra conductors or raceway space should be included in the design stage if additional loads may be added later. This consideration is particularly important for commercial buildings with the increasing use of electronic equipment and air conditioning. The cost and difficulties in obtaining space for new feeders and larger switchgear, which would be required when modernizing or expanding a building, may well be considered in the initial design. A load growth margin of 50% applied to the installed capacity of the major feeders is often justified where expansion is anticipated. Each project deserves careful consideration of the proper load growth margin to be allowed.

Flexibility in an electric wiring system is enhanced by the use of oversize or spare raceways, cables, busways, and equipment. The cost of making such provisions is usually relatively small in the initial installation.

Empty riser shafts and holes through floors may be provided at relatively low cost for future work. Consideration should be given to the provision of satellite electric closets initially for future expansion. Openings through floors should be filled in with fireproof, easily removed materials to prevent the spread of fire and smoke between floors. For computer rooms and the like, flexibility is frequently provided by raised floors made of removable panels, providing access to a wiring space between the raised floor and the slab below.

1.12.3 Specifications. A contract for installation of electric systems consists of a written document and drawings. These become part of the contract, which contains legal and engineering sections. The legal nontechnical sections contain the general terms of the agreement between contractor and owner, such as payment, working conditions, and time requirements; and it may include clauses on performance bonds, extra work, penalty clauses, and damages for breach of contract.

The engineering section consists of the technical specifications. The specifications give descriptions of the work to be done and the materials to be used. In larger installations, it is common practice to use a standard outline format listing division, section, and subsection titles or subjects of the Construction Specifications Institute (CSI).[26] Where several specialties are involved, CSI Division 16 covers the electrical installation and CSI Division 15 covers the mechanical portion of the work. The building automation system, integrating several building control systems, is covered in CSI Division 13, Special Construction. It is important to note that some electrical work will almost always be included in CSI Divisions 13 and 15. Each division has a detailed breakdown of various items, such as switchgear, motor starters, and lighting equipment as specified by CSI.

In order to assist the engineer in preparing contract specifications, standard technical specifications (covering construction, application, technical, and installation details) are available from technical publishers and manufacturers (which may require revision to avoid proprietary specifications). Large organizations, such as the U.S. General Services Administration (GSA) and the Veterans Administration (VA) develop their own standard specifications. In using any prepared or computer-generated specification, it should be understood that a detailed review of the generated document will be necessary to ensure a meaningful product. Where a high degree of unique considerations are involved, these tools may be useful only as a guide.

MASTERSPEC, issued by the American Institute of Architects (AIA),[27] permits the engineer to issue full-length specifications in a standardized format. SPECTEXT II, which is an abridged computer program with similar capabilities, is issued by CSI. The U.S. Army Corps of Engineers publishes *The Corps of Engineers Guide Specifications*, known as "CEGS;" and the U.S. Navy publishes the *U.S. Naval Facilities Engineering Command Guide Specifications*, known as "NFGS" (Naval Facilities Guide Specifications).

Computer-aided specifications (CAS) are being developed that will automatically develop specifications as an output from the computer-aided engineering/computer-aided design and drafting (CAE/CADD) process.

1.12.4 Drawings. Designers will usually be given preliminary architectural drawings as a first step. These will permit them to arrive at the preliminary scope of the work; roughly estimate the requirements for, and determine in a preliminary way, the location of equipment; and the methods and types of lighting. In this stage of the design, such items as hung ceilings, recessed or surface-mounted fixtures, and general types of distribution will be decided. It is important to discuss the plans with the senior engineer, and with the architect who has the advantage of knowing the type of construction and building finishes. The mechanical engineer will indicate the mechanical loads that will exist.

It is during this early period that the designer should emphasize the need for: room to hang conduits and other raceways, crawl spaces, structural reinforcements for heavy equipment, special floor loadings; clearances around switchgear,

---

[26] CSI publications are available from the Construction Specifications Institute, 601 Madison Street, Industrial Park, Alexandria, VA 22314.

[27] AIA publications are available from the American Institute of Architects, 1735 New York Avenue, N.W., Washington, DC 20006.

transformers, busways, cable trays, panelboards, and switchboards; and other items that may be required. It is much more difficult to obtain such special requirements once the design has been committed.

The single-line diagrams should then be prepared in conformity with the utility's service requirements. Based on these, the utility will develop a service layout. Electrical drawings are based on architectural drawings and, while prepared at the same time as the structural and mechanical drawings, they are usually the last ones completed because of the need to resolve physical interferences.

Checking is an essential part of the design process. The checker looks for design deficiencies in the set of plans. It is usually a shock to the young designer or drafter when he or she receives their first drawing marked up in red to indicate all kinds of corrections that are required. The designer can help the checker by having at hand reference and catalog information detailing the equipment he or she has selected. The degree of checking is a matter of design policy.

CAE and CADD are tools by which the engineer/designer can perform automatic checking of interferences and clearances with other trades. The development of these computer programs has progressed to the level of automatically performing load flow analysis, fault analysis, and motor starting analysis from direct entry of the electrical technical data of the components and equipment.

1.12.5 Manufacturers' or Shop Drawings. After the design has been completed and contracts are awarded, manufacturers and other suppliers will submit manufacturers' or shop drawings for approval or information. It is important to return these shop drawings as quickly as possible, otherwise the contractor may claim that his or her work was delayed by failure to receive approval or other permission to proceed. Unless drawings are unusable, it is a good idea not to reject them but to stamp the drawings approved as noted and mark them to show changes and corrections. The supplier can then make whatever changes are indicated and will not have to wait for a completely approved set of drawings before commencing work.

Unless otherwise directed, communications with contractors and suppliers is always through the construction (often inspection) authority. In returning corrected shop drawings, remember that the contract for supplying the equipment usually rests with the general contractor and that the official chain of communication is through him or her. Sometimes, direct communication with a subcontractor or a manufacturer may be permitted; however, the content of such communication should always be confirmed in writing with the general contractor. Recent lawsuits have resulted in the placing of responsibility for shop drawing correctness (in those cases and possibly future cases) on the design engineer, leaving no doubt that checking is an important job.

1.13 Estimating. A preliminary estimate is usually requested. Sometimes, the nature of a preliminary estimate makes it nothing more than a good guess. Enough information is usually available, however, to perform the estimate on a square foot or similar basis. The preliminary estimate becomes part of the overall feasibility study for the project.

A second estimate is often provided after the project has been clearly defined, but before any drawings have been prepared. The electrical designer can determine the type of lighting fixtures and heavy equipment that is to be used from sketches and architectural layouts. Lighting fixtures as well as most items of heavy equipment can be priced directly from catalogs, using appropriate discounts.

The most accurate estimate is made when drawings have been completed and bids are about to be received or the contract negotiated. In this case, the estimating procedure of the designer is similar to that of the contractor's estimator. It involves first the take-offs, that is, counting the number of receptacles, lighting fixtures, lengths of wire and conduit, determining the number and types of equipment, and then applying unit costs for labor, materials, overhead, and profit. The use of standard estimating sheets is a big help. Various forms are available from the National Electrical Contractors Association (NECA).[28] For preliminary estimates, there are a number of general estimating books that give unit cost (often per square foot) figures and other general costs, such as *Building Construction Cost Data, Mechanical Cost Data, and Electrical Cost Data* by R. S. Means.[29] Several computer programs permit the streamlining and standardizing of engineering estimating.

The estimator/designer must include special costs, such as vehicles, temporary connections, temporary or construction power, rental of special tools, scaffolding, and many other items. Because of interference with local operations, as at a public terminal, work may have to be performed during overtime periods. Electricians generally receive overtime premium pay, usually at a rate of time-and-a-half or double-time.

Electrical base pay may represent about half the total cost when considering employee benefits, overhead, and supervision. The designer will typically estimate 15% to 25% for overhead and 10% for profit, with possibly an additional 5% to 10% markup when the electrical contractor is a subcontractor.

In pricing equipment and materials, manufacturers' catalogs can be used. There is often an appropriate discount to be applied that may be listed in the front of the catalog. The determination of the correctness of this discount and which discount table is to be used must be made by the distributor or manufacturer. Many companies publish a catalog with list prices and simply issue revised discount lists to take care of price changes. Certain items, for example, copper cable, vary in price from day to day, dependent upon the cost of base materials. When the owner purchases equipment or materials for installation by the contractor, costs for the installation, handling, overhead, and profit will be added on by the contractor.

Extra work ("extras") is that work performed by the contractor that has to be added to the contract for unforeseen conditions or changes in the scope of work. The contractor is not usually faced with competition in making these changes, therefore, extra work is expected to be more costly than the same work if included

---

[28] Write to the National Electrical Contractors Association, 7315 Wisconsin Avenue, Bethesda, MD 20814.

[29] To obtain this publication, write to R. S. Means Company, 100 Construction Plaza Avenue, Kingston, MA 02364.

in the original contract. Extra cost on any project can be minimized by greater attention to design details in the original planning stage. On rehabilitation or modification work, extras are more difficult to avoid; however, with careful field investigation, extras can be held to a minimum.

1.14 Contracts. Contracts for construction may be awarded on either a lump-sum or a unit-price basis, or on a cost-plus (time-and-material) basis. Lump sum involves pricing the entire job as one or several major units of work.

The unit-price basis simply specifies so much per unit of work, for example, so many dollars per foot of 3 inch conduit. The lump-sum contract is usually preferable when the design can be worked out in sufficient detail. The unit-price contract is desirable when it is not possible to determine exactly the quantities of work to be performed and where a contractor, in order to provide a lump-sum contract, might have to overestimate the job to cover items that he or she could not accurately determine from the drawings.

If the unit-price basis is used, the estimated quantities should be as accurate as possible, otherwise it may be advantageous for the contractor to quote unit prices of certain items as high as possible and reduce other items to a minimum figure. It could be to the contractor's advantage to list those items on which he or she would receive payment first or which would be most likely to increase in quantity at their highest prices.

The time-and-material basis is valuable for emergency or extra work where it would be impractical to use either of the above two methods. It has the disadvantage of requiring a close audit of manpower and material expenditures of the contractor. Where only part of the work is not clearly defined, a combination of these three methods of pricing might be in order.

1.15 Building Access and Loading. It is imperative that the equipment fit into the area specified, and that the floor load rating is adequate for the weight of the equipment. Sizes of door openings, corridors, and elevators for the moving of equipment (initially and for maintenance and replacement purposes) should be checked. However, it is easy to forget that equipment has to be moved across floors, and that the floor load ratings of the access areas for moving the equipment should be adequate. If floor strengths are not adequate, provision should be made to reinforce the floor, or, if practical, to specify that the load be distributed so that loading will not exceed structural limitations.

It is important to review weights and loadings with the structural engineers. Sometimes, it is necessary to provide removable panels, temporarily remove windows, and even to make minor structural changes in order to move large and heavy pieces of equipment or machinery. Provisions should also be made for removal of equipment for replacement purposes. Clearances should be in accordance with code provisions regarding working space. Clearance should also be provided for installation, maintenance, and such items as cable pulling, transformer replacement, maintenance/testing, and switchgear-drawout space. It is often essential to phase items of work in order to avoid conflict with other electrical work or the work of other trades.

1.16 Contractor Performance. Contractors may be selected on the basis of bid or quoted price or by negotiation. Governmental or corporate requirements may mandate the selection of the lowest qualified bidder. Where the relative amount of electrical work is large, the contract may be awarded to an electrical contractor. In other instances, the electrical work may be awarded as a subcontract by the overall or general contractor, except where prohibited by state law, as in New York, for certain public works.

The performance of the work will usually be monitored and inspected by representatives of the owner or architect/engineer. The work is subject to the inspection of governmental and other assigned approval agencies, such as insurance underwriters. The designer may communicate with the contractor only to the extent permitted by the agency exercising control over the contract: the architect, builder, or general contractor, as may be appropriate. It is essential that the designers, in attempting to expedite the contract, not place themselves in the position of requesting or interpreting into the contract things that are not clearly required by the specifications or drawings without proper authorization.

The contract may require the contractor to deliver, at the end of the work, revised contract drawings, known as "as-built" drawings. These show all changes in the work that may have been authorized or details that were not shown on the original drawings.

1.17 Environmental Considerations. In all branches of engineering, an increasing emphasis is being placed on social and environmental concerns. Today's engineer must consider air, water, noise, lighting, and all other items that have an environmental impact. For example, see IEEE Std 980-1987, IEEE Guide for Containment and Control of Oil Spills in Substations (ANSI) [14] and IEEE Std 640-1985, IEEE Guide for Power Station Noise Control [11]. The limited availability of energy sources and the steadily increasing cost of electric energy require a concern with energy conservation.

These items are becoming more than just a matter of conscience or professional ethics. Laws, codes, rules, and standards issued by legislative bodies, governmental agencies, public service commissions, insurance, and professional organizations (including groups whose primary concern is the protection of the environment and conservation of natural resources) increasingly require an assessment of how the project may affect the environment. Energy conservation is covered in Chapter 17. Environmental studies, which include the effects of noise, vibration, exhaust gases, lighting, and effluents must be considered in their relationship to the immediate and the general environment and the public.

Pad-type transformers (see Chapter 5) can eliminate unsightly fences and walls, where design considerations permit. Landscape architects can provide pleasing designs of trees and shrubbery to completely conceal outdoor substations, and, of course, overhead lines may be replaced by underground systems. Substations situated in residential areas must be carefully located so as not to create a local nuisance. Pre-case sound barriers can reduce transformer and other electrical equipment noise. Floodlighting and parking lot lighting must not spill onto adjacent areas where it may provide undesirable glare or lighting levels (see IES Committee

Report CP-46-85, *Astronomical Light Pollution and Light Trespass* [21]). The engineer should keep up to date on developments in the areas of environmental protection and energy conservation; Environmental Protection Agency (EPA) guidelines and judicial rulings, and local environmental litigation are generally covered in the *Federal Register* [18] and in the periodicals previously listed.

1.18 Technical Files. Drawings and other technical files are often kept in file cabinets as originals or copies. A system of filing and reference is essential where many such items are involved. A computerized database may be a valuable method of referencing and locating the proper document.

When drawings are produced by computer graphic systems, such as CADD, magnetic tape may be used for storage. Plotters can be used with computer systems to produce hard copy. Original drawings (often prepared on "tracing" material) can be stored photographically on film; the drawings can also be made available on viewers or enlarger-printers. Microfiche involves the placing of microfilm on computer-type cards for handling manually or in data-processing-type systems.

1.19 Electronic Systems. Electronic systems are a major item in commercial buildings: for control purposes, motor control, lighting ballasts, communication systems, data processing, computer applications, data management, and building management systems.

1.20 Power Supply Disturbances. The power supply to equipment may contain transients and other short-term undervoltages or overvoltages, which result primarily from switching operations, faults, motor starting, (particularly large air-conditioning chiller motors), and lightning disturbances.

The system may also contain a harmonic content as described in 1.20.2 below. These electrical disturbances may be introduced anywhere on an electric system or in the utility supply, even by other utility customers connected to the same circuits. A term frequently applied to describe the absence or presence of these power supply deficiencies is "power quality." P1100 (Recommended Practice for Power and Grounding Sensitive Electronic Equipment in Industrial and Commercial Power Systems [the "Emerald Book"]) will examine in detail the effects of the power supply on equipment performance. It will also cover methods of diagnosing and correcting performance problems related to the power supply.

1.20.1 Harmonics. Chapter 9 of IEEE Std 141-1986 (ANSI) [6], Chapter 10 of IEEE Std 399-1990 (ANSI) [7], P1100 (the "Emerald Book"), and IEEE Std 519-1981 [9] all contain discussions of harmonics. Harmonics are integral multiples of the fundamental (line) frequency involving nonlinear loads or control devices, including electromagnetic devices (transformers, lighting ballasts), and solid-state devices (rectifiers, thyristors, phase controlled switching devices). In the latter grouping are power rectifiers, adjustable speed electronic controllers, switching-mode power supplies (used in smaller computers), and UPS systems.

Harmonics can cause or increase electromagnetic interference in sensitive electronic systems, abnormal heating of cables, motors, transformers, and other electromagnetic equipment, excessive capacitor currents, and excessive voltages because of system resonances at harmonic frequencies.

Recently, it has been determined that the harmonic content of multiwire systems having a high proportion of switching-mode power supplies is very high. The neutral conductors of these systems should be sized at greater than full rating; transformers derated or designed for high-harmonic content should be used.

A full discussion of harmonics is beyond the scope of this section; reference should be made to the above listed texts.

1.20.2 Electromagnetic Interference (EMI). EMI is the impairment of a wanted electromagnetic signal by an electromagnetic disturbance. EMI can enter equipment either by conduction through power, grounding, control, data, or shielding conductors; or by induction from local electromagnetic or electrostatic fields. The most common causes of EMI problems in sensitive equipment, such as computers, communications equipment, and electronic controllers, are poor inherent design of the equipment or power supply, poor grounding, and unsound design of the equipment interfaces.

It can be reduced by the use of effective grounding (both electronic and equipment grounds); shielding, twisted conductors (pairs), and coaxial cables; effective use of conduit (especially steel conduit) for control and power circuits (where practical) (see IEEE Std 518-1982 [8] and Reference [19]). EMI and other power problems can cause control and equipment malfunctions, slowing of computer operations, lack of reliability, and failure of critical systems. These failures can affect product quality and, in some cases, worker safety.

The use of filters, voltage regulators, surge capacitors, surge arresters, isolation transformers (particularly with electrostatic shielding between coils), power conditioners, UPS systems, or motor-generator sets used for isolation are all methods of reducing EMI. Fiber-optic cables and electro-optical isolation at interfaces are extremely effective methods of providing isolation between systems.

1.20.3 Programmable Logic Controller (PLC). The PLC is a microprocessor designed for control and telemetering systems. It is programmed to accept ladder-type logic, which enables the operator to use relay-type logic; which, in turn, avoids the need to use the conventional software languages. The equipment can be housed in weather and environmental contaminant-resistant housings for field use.

1.21 Definitions. The following definitions should be used in conjunction with this recommended practice:

air, ambient. *See* ambient air.

air, recirculated. *See* recirculated air.

air, return. *See* return air.

air conditioning. The process of treating air so as to simultaneously control temperature, humidity, and distribution to the conditioned space.

air ventilation. The amount of supply air required to maintain the desired quality of air within a designated space.

ambient air. The air surrounding or occupying a space or object.

ballast. An electrical device that is used with one or more discharge lamps to supply the appropriate voltage to a lamp for starting, to control lamp current while it is in operation, and, usually, to provide for power factor correction.

branch-circuit load. The load on that portion of a wiring system extending beyond the final overcurrent device protecting the circuit.

brightness. The subjective attribute of any light sensation, including the entire scale of the qualities "bright," "light," "brilliant," "dim," and "dark."

British thermal unit (Btu). The quantity of heat required to raise one pound of water 1 °C.

cable tray. A unit or assembly of units or sections, and associated fittings, made of metal or other noncombustible material forming a continuous rigid structure used to support cables.

calorie. The quantity of heat required to raise one gram of water 1 °F.

capacity, heat. *See* heat capacity.

chromaticity. The measure of the warmth or coolness of a light source, which is expressed in the Kelvin (K) temperature scale.

coefficient of performance (heat pump). Ratio of heating effect produced to the energy supplied.

coefficient of utilization (CU). For a specific room, the ratio of the average lumens delivered by a luminaire to a horizontal work plane to the lumens generated by the luminaire's lamps alone.

coincident demand. Any demand that occurs simultaneously with any other demand; also the sum of any set of coincident demands.

"cold" standby redundant UPS configuration. Consists of two independent, non-redundant modules with either individual module batteries or a common battery.

commercial, residential, and institutional buildings. All buildings other than industrial buildings and residential dwellings.

conductivity, thermal. *See* thermal conductivity.

connected load. The sum of the continuous ratings of the power consuming apparatus connected to the system or any part thereof in watts, kilowatts, or horsepower.

contrast. Indicates the degree of difference in light reflectance of the details of a task compared with its background.

control. Any device used for regulation of a system or component.

creep. Continued deformation of material under stress.

critical load. That part of the load that requires continuous quality electric power for its successful operation.

degree day. A unit based upon temperature difference and time, which is used for estimating fuel consumption and for specifying nominal heating loads of buildings during the heating season. Degree days = Number of degrees (°F) that the mean temperature is below 65 °F × days.

dehumidification. Condensation of water vapor from the air by cooling below the dew point, or removal of water vapor from air by physical or chemical means.

demand (or demand load). The electrical load at the receiving terminals averaged over a specified interval of time. Demand is expressed in kilowatts, kilovolt-amperes, kilovars, amperes, or other suitable units. The interval of time is generally 15 minutes, 30 minutes, or 60 minutes.

NOTE: If there are two 50 hp motors (which drive 45 hp loads) connected to the electric power system but only one load is operating at any time, the demand load is only 45 hp but the connected load is 100 hp.

demand factor. The ratio of the maximum demand of a system to the total connected load of the system.

NOTES: (1) Since demand load cannot be greater than the connected load, the demand factor cannot be greater than unity.
(2) Those demand factors permitted by the NEC (for example, services and feeders) must be considered in sizing the electric system (with few exceptions, this is 100%); otherwise, the circuit may be sized to support the anticipated load.

diversity factor. The ratio of the sum of the individual maximum demands of the subdivisions of the system to the maximum demand of the complete system.

NOTE: Since maximum demand of a system cannot be greater than the sum of the individual demands, the diversity factor will *always* be equal to or greater than unity.

efficacy. *See* lumens per watt (lm/W).

efficiency. The power (kW) output divided by the power (kW) input at rated output.

electric power cable shielding. The practice of confining the electric field of the cable to the insulation surrounding the conductor by means of conducting or semiconducting layers, or both, which are in intimate contact or bonded to the inner and outer surfaces of the insulation.

electromagnetic interference (EMI). The impairment of a wanted electromagnetic signal by an electromagnetic disturbance.

equivalent sphere illumination (ESI). The measure of the effectiveness with which a practical lighting system renders a task visible compared with the visibility of the same task that is lit inside a sphere of uniform luminance.

extra work (extras). Work performed by the contractor that has to be added to the contract for unforeseen conditions or changes in the scope of work.

fixture. *See* luminaire.

flexibility of the electric system. The adaptability to development and expansion as well as to changes to meet varied requirements during the life of the building.

footcandle (fc). A unit of illuminance (light incident upon a surface) that is equal to 1 lm/ft^2. In the international system, the unit of illuminance is lux (1 fc = 10.76 lux).

footlambert (fL). The unit of illuminance that is defined as 1 lm uniformly emitted by an area of 1 ft^2. In the international system, the unit of luminance is candela per square meter (cd/m^2).

fuse. An overcurrent protective device with a circuit opening, fusible element part that is heated and severed by the passage of overcurrent through it. (To re-energize the circuit, the fuse should be replaced.)

glare. The undesirable sensation produced by luminance within the visual field.

gross demand load. The summation of the demands for each of the several group loads.

heat, specific. The ratio of the quantity of heat required to raise the temperature of a given mass of a substance 1° to the heat required to raise the temperature of an equal amount of water by 1°.

heat capacity. The amount of heat necessary to raise the temperature of a given mass of a substance 1°—the mass multiplied by the specific heat.

heat pump. A refrigerating system employed to transfer heat into a space or substance. The condenser provides the heat, while the evaporator is arranged to pick up heat from the air, water, etc. By shifting the flow of air or other fluid, a heat pump system may also be used to cool a space.

heating system, radiant. A heating system in which the heat radiated from panels is effective in providing heating requirements. The term "radiant heating" includes panel *and* radiant heating.

heating unit, electric. A structure containing one or more heating elements, electrical terminals or leads, electric insulation and a frame or casing, all of which are assembled into one unit.

high-intensity discharge (HID) lamps. A group of lamps filled with various gases that are generically known as mercury, metal halide, high-pressure sodium, and low-pressure sodium.

high voltage. A class of nominal system voltages equal to or greater than 100 000 V and equal to or less than 230 000 V.

humidity. Water vapor within a given space.

humidity, relative. The ratio of the mole fraction of water vapor that is present in the air to the mole fraction of water vapor that is present in saturated air.

infiltration. Leakage of outside air into a building.

illuminance. The unit density of light flux (lm/unit area) that is incident on a surface.

institutional design. Emphasizes reliability, resistance to wear and use, safety to public, and special aesthetic considerations, such as the "agelessness" of the structure.

insulation, thermal. *See* thermal insulation.

interrupter switch. An air switch equipped with an interrupter that makes or breaks specified currents.

isolated redundant UPS configuration. Uses a combination of automatic transfer switches and a reserve system to serve as the bypass source for any of the active systems.

isothermal. A process that occurs at a constant temperature.

kilowatt. A measure of the instantaneous power requirement.

lag. The delay in action of a sensing element of a control element.

lamp. Generic term for a manmade source of light.

load, estimated maximum. The calculated maximum heat transfer that a heating or cooling system will be called upon to provide.

load factor. The ratio of the average load over a designated period of time to the peak load occurring in that period.

load profile. The graphic representation of the demand load, usually on an hourly basis, for a particular day.

low voltage. A class of nominal system voltages 1000 V or less.

lumen (lm). The international unit of luminous flux or the time rate of the flow of light.

lumens per watt (lm/W). The ratio of lumens generated by a lamp to the watts consumed by the lamp. *See also* efficacy.

luminaire. A complete lighting unit that consists of parts designed to position a lamp (or lamps) in order to connect it to the power supply and to distribute its light.

luminaire efficiency. The ratio of lumens emitted by a luminaire and of the lumens generated by the lamp (or lamps) used.

luminance. The light emanating from a light source or the light reflected from a surface (the metric unit of measurement is cd/m^2).

lux. The metric measure of illuminance that is equal to 1 lm uniformly incident on $1\ m^2$ (1 lux = 0.0929 fc).

maximum demand. The greatest of all the demands that have occurred during a specified period of time; determined by measurement over a prescribed time interval.

maximum system voltage. The highest system voltage that occurs under normal operating conditions, and the highest system voltage for which equipment and other components are designed for satisfactory continuous operation without derating of any kind.

medium voltage. A class of nominal system voltages greater than 1000 V and less than 100 000 V.

nominal system voltage. The voltage by which a portion of the system is designated, and to which certain operating characteristics of the system are related. Each nominal system voltage pertains to a portion of the system that is bounded by transformers or utilization equipment.

nominal utilization voltage. The voltage rating of certain utilization equipment used on the system.

nonredundant UPS configuration. Consists of one or more UPS modules operating in parallel with a bypass circuit transfer switch and a battery.

parallel redundant UPS configuration. Consists of two or more UPS modules with static inverter turn-off(s), a system control cabinet, and either individual module batteries or a common battery.

peak load. The maximum load of a specified unit or group of units in a stated period of time.

radiator. A heating unit that provides heat transfer to objects within a visible range by radiation and by conduction to the surrounding air, which is circulated by natural convection.

rated life of a ballast or a lamp. The number of burning hours at which 50% of the units have burned out and 50% have survived.

recirculated air. Return air passed through the air conditioner before being supplied again to the conditioned space.

return air. Air returned from the conditioned space.

reflectance. The ratio of the light reflected by a surface to the light incident.

relative visual performance (RVP). The potential task performance based upon the illuminance and contrast of the lighting system performance.

service voltage. The voltage at the point where the electric system of the supplier and the electric system of the user are connected.

short-circuit current. An overcurrent resulting from a fault of negligible impedance between live conductors having a difference in potential under normal operating conditions. The fault path may include the path from active conductors via earth to neutral.

solar constant. The solar intensity incident on a surface that is oriented normal to the sun's rays and located outside the earth's atmosphere at a distance from the sun that is equal to the mean distance between the earth and the sun.

subtransient reactance. The apparent reactance of the stator winding at the instant the short circuit occurs.

symmetrical. The shape of the ac current waves about the zero axis (when both sides have equal value and configuration).

synchronous reactance. The reactance that determines the current flow when a steady-state condition is reached.

system voltage. The root-mean-square phase-to-phase voltage of a portion of an ac electric system. Each system voltage pertains to a portion of the system that is bounded by transformers or utilization equipment.

task-ambient lighting. A concept involving a component of light directed toward tasks from appropriate locations by luminaires located close to the task for energy efficiency.

temperature, dew point. The temperature at which condensation of water vapor begins in a space.

temperature, dry bulb. The temperature of a gas, or a mixture of gases, that is indicated by an accurate thermometer after correction for radiation.

temperature, effective. An arbitrary index that combines, into a single value, the effects of temperature, humidity, and air movement on the sensation of hot or cold felt by the human body.

temperature, wet bulb. The temperature at which liquid or solid water, by evaporating into the air, can bring the air into saturation adiabatically at the same temperature.

therm. A quantity of heat that is equal to 100 000 Btu.

thermal conductivity. The time rate of heat flow through a unit area of a homogeneous substance under steady conditions when a unit temperature gradient is maintained in the direction that is normal to the area.

thermal diffusivity. Thermal conductivity divided by the product of density and specific heat.

thermal insulation. A material having a high resistance to heat flow and used to retard the flow of heat to the outside.

thermal transmittance (U factor). The time rate of heat flow per unit temperature difference.

thermostat. A device that responds to temperature and, directly or indirectly, controls temperature in a building.

ton of refrigeration. Is equal to 12 000 Btu/hour.

transient overvoltages (or spikes). Momentary excursions of voltage outside of the normal 60 Hz voltage wave.

transient reactance. Determines the current flowing during the period when the subtransient reactance is the controlling value.

transmittance, thermal. *See* thermal transmittance.

uninterruptible power supply (UPS). A device or system that provides quality and continuity of an ac power source.

uninterruptible power supply (UPS) module. The power conversion portion of the uninterruptible power system.

utilization equipment. Electrical equipment that converts electric power into some other form of energy, such as light, heat, or mechanical motion.

utilization voltage. The voltage at the line terminals of utilization equipment.

veiling reflections. Reflected light from a task that reduces visibility because the light is reflected specularly from shiny details of the task, which brightens those details and reduces the contrast with the background.

velocity, room air. The average sustained residual air velocity in the occupied area in the conditioned space.

visual comfort probability (VCP). A rating of a lighting system expressed as a percentage of people who, if seated at the center of the rear of a room, will find the lighting visually acceptable in relation to the perceived glare.

visual task. Work that requires illumination in order for it to be accomplished.

work plane. The plane in which visual tasks are located.

1.22 References. The following references shall be used in conjunction with this chapter:

[1] ANSI C2-1990, National Electrical Safety Code.

[2] ANSI C84.1-1989, Voltage Ratings for Electric Power Systems and Equipment (60 Hz).

[3] ANSI/NFPA 70-1990, National Electrical Code.

[4] ASHRAE/IES 90.1-1989, Energy Efficient Design of New Buildings Except New Low-Rise Residential Buildings.

[5] IEEE Std 100-1988, IEEE Standard Dictionary of Electrical and Electronics Terms, Fourth Edition (ANSI).

[6] IEEE Std 141-1986, IEEE Recommended Practice for Electric Power Distribution for Industrial Plants (ANSI).

[7] IEEE Std 399-1990, IEEE Recommended Practice for Industrial and Commercial Power Systems Analysis (ANSI).

[8] IEEE Std 518-1982, IEEE Guide for the Installation of Electrical Equipment to Minimize Noise Inputs to Controllers from External Sources (ANSI).

[9] IEEE Std 519-1981, IEEE Guide for Harmonic Control and Reactive Compensation of Static Power Converters (ANSI).

NOTE: When the revision of IEEE Std 519-1981 is published, it will supersede IEEE Std 519-1981, and will become a recommended practice.

[10] IEEE Std 602-1986, IEEE Recommended Practice for Electric Systems in Health Care Facilities (ANSI).

[11] IEEE Std 640-1985, IEEE Guide for Power Station Noise Control.

[12] IEEE Std 693-1984, IEEE Recommended Practices for Seismic Design of Substations (ANSI).

[13] IEEE Std 979-1984 (Reaff. 1988), IEEE Guide for Substation Fire Protection (ANSI).

[14] IEEE Std 980-1987, IEEE Guide for Containment of Oil Spills in Substations (ANSI).

[15] NFPA 70E-1988, Electrical Safety Requirements for Employee Workplaces.

[16] NFPA CY-70HB90, NFPA National Electrical Code Handbook, 1990 Edition.

[17] *Electrical Alerts*, U.S. National Institute of Occupational Safety and Health (NIOSH), 4676 Columbia Parkway, Cincinnati, OH 45226.

[18] *Federal Register* (53 FR 1546), U.S. Government Printing Office, Washington, DC 20402 (Telephone: 202-783-3238).

[19] Griffith, D. C. "Uninterruptible Power Supplies," New York: Marcel Decker, 1989.

[20] *Handbook of the National Electrical Code*, New York: McGraw-Hill.

[21] IES Committee Report CP-46-85, *Astronomical Light Pollution and Light Trespass*.

Chapter 2
Load Characteristics

2.1 General Discussion. The electric power distribution system in a building exists solely to serve the loads — the electrical utilization devices. The power distribution system should accomplish that assignment safely and economically, provide sufficient reliability to adequately satisfy the requirements of the building (and its users), and incorporate sufficient flexibility to accommodate changing loads during the life of the building.

This chapter is intended to provide typical load data and a suggested method for determining individual and total connected and total demand load characteristics of a commercial building. The engineer should make provisions for load growth as well as building expansion in order to provide adequate electrical capacity or provision for electrical equipment expansion during the expected life of the building.

The steadily increasing sophistication of some of the load devices (complex communication systems; electronic data processing equipment; fire protection equipment; closed-circuit television security systems; heating, ventilation, and air-conditioning systems; centralized automated building control systems; etc.) increases the difficulty of determining initial load, forecasting future loads, and establishing realistic demand factors.

The electrical engineer should determine a building's electrical load characteristics early in the preliminary design stage of the building in order to select the proper power distribution system and equipment having adequate power capacity with proper voltage levels, and sufficient space and ventilation to maintain proper ambients. Once the power system is determined, it is often difficult to make major changes because of the coordination required with other disciplines. Architects and mechanical and structural engineers will be developing their designs simultaneously and making space and ventilation allocations. It is imperative, therefore, from the start that the electric systems be correctly selected based on realistic load data or *best possible* typical load estimates, or both because all final, finite load data are not available during the preliminary design stage of the project. When using estimated data, it should be remembered that the *typical data* applies only to the condition from which the data was taken and most likely an adjustment to fit the particular application will be required.

While much of the electrical requirements of building equipment, such as ventilating, heating/cooling, lighting, etc., are furnished by other disciplines, the electrical engineer should also furnish to the other disciplines such data as space, accessibility, weight, and heat dissipation requirements for the electric power distribution apparatus. This involves a continuing exchange of information that starts as preliminary data and is upgraded to be increasingly accurate as the design progresses. Documentation and coordination throughout the design process is imperative.

At the beginning of the project, the electrical engineer should review the utility's rate structure and the classes of service available. Information pertaining to demand, energy, and power factor should be developed to aid in evaluating, selecting, and specifying the most advantageous utility connection. As energy resources become more costly and scarce, items such as energy efficiency, power demand minimization, and energy conservation should be closely considered to reduce both energy consumption and utility cost.

System power (that is, energy) losses should be considered as part of the total load in sizing mains and service equipment. ANSI/NFPA 70-1990, National Electrical Code (NEC) [3][30] recommends that the total voltage drop from electrical service entrance to the load terminals of the furthest piece of equipment served should not exceed 5% of the system voltage and, thus, the energy loss, I^2R, will correspondingly be limited.

Listed below are typical load groups and examples of classes of electrical equipment that should be considered when estimating initial and future loads (see also Fig 1).

   (1) Lighting — Interior (general, task, exits, and stairwells), exterior (decorative, parking lot, security), normal, and emergency

Fig 1
Groups of Loads in a
Typical 10-Story Commercial Building

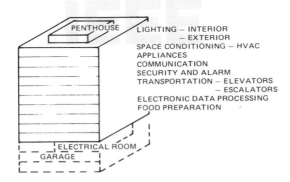

[30]The numbers in brackets correspond to those in the references at the end of each chapter. ANSI publications are available from the Sales Department of the American National Standards Institute, 11 West 42nd Street, 13th Floor, New York, NY 10036. NFPA publications are available from Publications Sales, National Fire Protection Association, 1 Batterymarch Park, P.O. Box 9101, Quincy, MA 02269-9101.

(2) Appliances—Business and copying machines, receptacles for vending machines, and general use

(3) Space Conditioning—Heating, cooling, cleaning, pumping, and air-handling units

(4) Plumbing and Sanitation—Water pumps, hot water heaters, sump and sewage pumps, incinerators, and waste handling

(5) Fire Protection—Fire detection, alarms, and pumps

(6) Transportation—Elevators, dumbwaiters, conveyors, escalators, and moving walkways

(7) Data Processing—Desktop computers, central processing and peripheral equipment, and uninterruptible power supply (UPS) systems, including related cooling

(8) Food Preparation—Cooling, cooking, special exhausts, dishwashing, disposing, etc.

(9) Special Loads—For equipment and facilities in mercantile buildings, restaurants, theaters, recreation and sports complexes, religious buildings, terminals and airports, health care facilities, laboratories, broadcasting stations, etc. (See Chapter 16 for more information.)

(10) Miscellaneous Loads—Security, central control systems, communications, audio-visual, snow melting, recreational or fitness equipment, incinerators, shredding devices, waste compactors, shop or maintenance equipment, etc.

2.1.1 Load Estimates. There are several load estimates that should be made during the course of a project including

(1) A preliminary load estimate, generally based on a projection of available data on existing buildings of the same usage and the square footage or volume. This information is used in preliminary engineering studies for determining feasibility and cost and for very preliminary discussions with the utility.

(2) An early design load estimate, of higher accuracy than (1) above, to determine the types of service required, to present more realistic information to the utility, to begin formal utility negotiations, and to determine the type of distribution system and voltages to be selected. At this point, the areas required for electrical rooms and substations will be determined. Once preliminary architectural decisions have been made, it may be difficult to obtain additional space, access, and floor loading requirements for the electric system. Typical figures that could be used for this type of estimate are included in this chapter.

(3) The NEC [3] specifies minimum service and feeder sizes based on the areas involved and the types of loads. The intent is to prevent the design of an unsafe electric system, which could result from the undersizing of feeders, panelboards, and services (either erroneously or for cost saving purposes). In many modern buildings, the actual maximum demand load will be substantially less than that calculated under NEC methodology; but, where the NEC or equivalent code is in effect, the code calculations must be used in sizing service, feeders, switchboards, and panelboards.

(4) Energy codes, primarily those enacted into law by political subdivisions, may provide budgets for the allocation of electric power. These are part of legislated energy conservation programs and are usually based on ASHRAE/IES Standards in the ASHRAE 90 Series. These codes develop overall energy conservation standards for the building including mechanical and electric systems and building insulation. While specifying the maximum energy usage for different areas of occupancy, they permit the increase of the allowable consumption in certain areas if other areas use less than the allowable limit. Most codes use power allocations as a base (e.g., unit power densities in W/ft^2); others may include energy budgets in which time of usage is one of the variables.

(5) The final load estimates are based on actual take-offs from the final electrical and mechanical drawings. These include "as designed" motor sizes, sizes of permanently connected appliances, lighting loads, estimated loads for receptacles, and heating equipment loads. Even these figures should be reviewed when the requirements of the actual equipment are furnished by contractors and manufacturers.

It is important to distinguish between loads expressed in voltamperes and watts (VA, W, kVA, kW, MVA, or MW). The energy codes are primarily concerned with real energy or watts, while the NEC [3] often requires the use of apparent power in voltamperes.

2.1.2 Load Tabulation. Power systems for different buildings are seldom the same because load requirements differ from building to building. Therefore, the design of an electric distribution system should begin with a load survey to identify the size, location, and nature of the various loads. In assembling this information, Table 2 may be helpful. This is not an easy task; it should not be undertaken lightly.

Table 3 illustrates the manner in which one major utility requires the electrical load data from Table 2 to be consolidated when applying for electrical service.

Most of the data for making the load is usually obtained from those involved in designing the building and its integral systems (for example, lighting, heating, ventilating, and air conditioning, and transportation). Useful information may be obtained from meter readings or measurements for similar buildings, from electric utility companies, from equipment manufacturers and associations, or from some governmental agencies.

The load tabulation provides an opportunity to identify the load of the utilization equipment and the voltages at which it can be served. The lighting load may be 30%–50% of the electrical load in office-type buildings; in contrast, it may be only incidental in restaurants or hotels. Ultimately, the power system must serve all the loads. The load tabulation allows definition of the continuity of operation that is required (for example, for safety or security of occupants, such as for stairway or exit lighting, certain ventilating fans, fire pumps, availability of certain elevators for fire fighting and rescue personnel, etc.).

In addition, the load tabulation may identify those loads that can be considered for load shedding during emergency operation or to minimize energy consumption or peak energy demands. The load tabulation can be used to identify utilization equipment having special requirements (for example, computers or certain lighting

Table 2
Load Tabulation of Equipment Utility Requirements

Room Number	Equipment Load Description	Electrical Requirements						Load Duration Continuous or Cyclical	Mechanical Requirements							
		volts	amperes	kW	hp	phase	hertz		Steam	Hot Water	Cold Water	Waste	Gas	Air	Ex-haust	Others

Table 3
Licensed Electrical Contractor Work Request

ELECTRICAL LOAD INFORMATION
(Enter total connected load in each category)

		3 Phase	1 Phase	Voltage	Largest hp	LRA / FLA	Application
Air Cond.	hp	————	————	————	————	— / —	
Computer	kW	————	————	————			
Computer Air	hp	————	————	————		— / —	
Cooking	kW	————	————	————			
Lighting	kW	————	————	————			
Motors	hp	————	————	————	————	— / —	————
Other	hp	————	————	————	————	— / —	————
Other	kW	————	————	————			————
Refrig.	hp	————	————	————	————	— / —	
Room AC	hp	————	————	————	————	— / —	
Water Heating	kW	————	————	————			
Welding	kW	————	————	————			

Gross sq. ft. —————— Emergency Generator kW ——————

circuits, etc., may impose special requirements, such as extreme reliability or continuity of supply, low noise levels, or ungrounded operation, etc.). These load characteristics or requirements need to be identified as early in the project as possible because they may necessitate special power distribution apparatus.

The location and magnitude of major loads must be carefully noted since such information may have considerable influence on the economic justification for the location and reliability aspects of the power service selected.

A load tabulation should also be made for building expansion projects with care taken to identify existing loads, those to be removed, and those to be added. A review of utility bills is important.

2.1.3 Relation to Power Company. Just as the individual and collective load requirements of one building differ from all other buildings, each electric utility differs to some degree from every other utility in its rate structure, service policies, and requirements, which makes it important for the electrical engineer to contact the utility company early in the design phase. But, before beginning to discuss rate structure and availability of service, the engineer should develop a load survey to estimate initial and future loads and their electrical characteristics, in order to convey to the electric utility the following data:

(1) Initial demand and connected load, and possible expansion
(2) Average usage or load factor
(3) Seasonal and time-of-day variations
(4) Power factor of total load
(5) Ratings of largest loads and associated switching (that is, starting) requirements
(6) Required reliability and expected continuity of service

(7) Identification of interruptible loads, to permit consideration of demand limiting

(8) Identification of loads sensitive to voltage and frequency transients

A detailed discussion on the various aspects of planning for utility service and the many factors affecting electric utility rates is presented in Chapter 4.

The electrical engineer should establish, in consultation with the electric utility, the special service classifications and incentive tariffs that are available to customers employing heat recovery, space-conditioning systems; thermal storage designs; solar energy; off-peak space-conditioning systems; or similar special systems to minimize electric power consumption.

The electrical engineer should analyze the features of rate structures that serve to penalize poor loads. Ratchet clauses cause utility customers to pay a demand charge on the highest demand established during a number of preceding months and is an incentive to control demand. Increased seasonal and time-of-day rates may result in higher electric rates during the high rate periods.

Several techniques are available to the electrical engineer to reduce the cost of electric power. These techniques include the following:

(1) Load Limiters—Load limiters, demand limiters, programmable energy controllers, or load shedding controllers are devices programmed to control building loads in such a sequence or manner that the billing demand remains at an optimized value.

(2) Power Factor Correcting Equipment—Many utilities have the authority to levy *power factor penalties* or surcharges on those users whose power factor is below some specified level, often 85% (but sometimes as high as 95%). Whenever economically feasible, synchronous motors should be selected or capacitors used to compensate for the *lagging power factor*, particularly caused by induction motors and certain lamp ballasts, e.g., "normal power factor" ballasts, to improve the overall power factor of the system.

(3) Power Factor Improvement Techniques—Because the power factor of an induction motor is lowered considerably when the motor is loaded to less than 75%–80% of rated load (even though motor efficiency remains relatively high and constant down to about 25% load), proper sizing of induction motors for the respective application serves to improve the load power factor and minimize the investment in power factor correcting equipment. Power factor correcting capacitors are commonly installed to be switched with the respective motor starter, that is, connected at the motor terminals or at the motor control center. Capacitor correction may not be acceptable for all motor applications. (e.g., motors with electronic speed control or overhauling loads). (Refer to Reference [12] for additional information.) The use of high power factor ballasts in lighting equipment can improve power factor significantly in buildings where lighting is an appreciable part of the total load.

(4) High-Efficiency Motors—Use of high-efficiency motors, which utilize improved materials and modified (from standard motor) construction, may result in a considerable reduction in energy consumption. Due to the lack of uniform testing procedures among the various suppliers, the electrical engi-

neer should exercise caution when evaluating various motor sources solely on the basis of published values of efficiency. Motor efficiency testing should be performed in accordance with IEEE Std 112-1984, IEEE Standard Test Procedure for Polyphase Induction Motors and Generators (ANSI) [6],[31] Method B. Also refer to NEMA MG10-1983, Energy Management Guide for Selection and Use of Polyphase Motors [8][32] and NEMA MG11-1977 (Reaff. 1982), Energy Management Guide for Selection and Use of Single-Phase Motors [9] for additional information. These motors may have higher starting current than standard motors. Starting characteristics and method of motor fault protection should be evaluated carefully, especially where instantaneous or motor short-circuit protectors are used.

(5) Motor-Speed Control—For certain motor applications, such as pumps and blowers, where energy can be saved by reduced speed operation when rated output is not needed, ac induction motors with solid-state, adjustable frequency controllers or multiple-speed motors may be economically justified. Use of adjustable frequency controllers may induce harmonics into the system. Necessary evaluation of the impact and possible solutions should be considered.

(6) Regenerative Systems—Energy can be saved in some motor-driven applications where, under certain operating conditions, the load is capable of driving the motor by utilizing regenerative systems. A loaded descending elevator or an empty ascending elevator, for example, can return energy to the building power system. The designer should analyze system performance during abnormal conditions to prevent equipment malfunction and damage.

(7) Programmed Loads—Certain loads can be programmed to save energy by being switched off during the hours when the space is unoccupied, or the systems are not required.

(8) Switched Loads—The need to provide flexible lighting systems should be satisfied by the choice of luminaire systems and lighting circuitry design. Engineering analyses of lighting systems should consider the following:

(a) Luminaires with the capability of having individual lamps or pairs of lamps switched so that the illumination levels can be set to match the task

(b) Ceiling and luminaire systems that allow the individual luminaires to be removed or installed as the illumination levels vary for the task being performed

(c) Use of photoelectric controls for exterior lighting and sunlit interiors

(d) Use of separate circuits for lighting along the interior perimeter of the building so that as more light is supplied by sunlight during the day, the interior perimeter lighting can be reduced either manually or by automatic controls

---

[31] IEEE publications are available from the Institute of Electrical and Electronics Engineers, IEEE Service Center, 445 Hoes Lane, P.O. Box 1331, Piscataway, NJ 08855-1331.

[32] NEMA publications are available from the National Electrical Manufacturers Association, 2101 L Street, N.W., Washington, DC 20037.

(9) Medium-Voltage Service (2.4–72.5 kV)—It may be possible to reduce billing costs by connecting the building loads through a transformer to the utility's primary service lines.

(10) Medium-Voltage Distribution (2.4–35 kV)—Energy losses within the building may be reduced through designing the distribution system for some voltage above the utilization level of the smaller loads.

(11) Redistribution—Building owners may redistribute electricity through meters to office and apartment tenants, as the utility's regulations allow. Energy consumption is usually less in buildings where the tenants are paying electricity costs directly than in master-metered buildings.

2.1.4 Relation to the NEC. The first section of the NEC [3] calls for the practical safeguarding of persons and property from hazards arising from the use of electricity. It further states that compliance will not necessarily result in a load serving electric system that is *efficient, convenient, adequate,* or *expandable.* It also states that it does not represent a design specification but only identifies minimum requirements. For example, the NEC [3] establishes certain minimum electric system capacity requirements for general lighting, receptacles, etc., based on the type of occupancy and demand factors. It is essential that the electric power system designer, therefore, be very knowledgeable of the contents of the latest edition of the NEC [3], along with any local electrical codes in effect in the area of the project.

Section 2 (200 Series articles) of the NEC [3] covers minimum design requirements for sizing of feeder and service equipment. The NEC, Article 220 [3] deals specifically with branch-circuit, feeder, and service calculations.

2.2 Load Characteristics.

During the process of determining the total capacity of the electric power distribution system for the building, in addition to noting the size and location of each load, much consideration must be given to the various operating or load characteristics, for example, repetitive starting or cycling of a load from lightly loaded to full load, etc. The possibility of noncoincidence of many of the loads often invites consideration of diversity or demand factors. A method for using typical data for load estimation and system sizing for power systems in commercial buildings is introduced in 2.5.

2.2.1 Lighting. As a result of research and development by manufacturers, many highly efficient light sources, luminaires, and auxiliary equipment have been introduced. Research in basic seeing factors has provided greater knowledge of many of the fundamental aspects of the quality and quantity of lighting. Consequently, it is now possible to utilize considerably less lighting energy than in the past. Chapter 10 concentrates on these factors, with considerable attention to the ways to reduce energy consumption of lighting while providing adequately for the seeing requirements and the well-being of the occupants and the objectives of the owners. Additional information about system design to properly serve lighting loads can be found in 4.9.1.

Traditionally, lighting loads have accounted for 20%–50% of the load in air-conditioned commercial buildings. The total lighting load for various buildings has commonly ranged from 3–6 VA/ft^2. Recent energy conservation regulations (where adopted) substantially reduce permitted lighting loads. The individual area lighting loads (either in watts or voltamperes per square foot) vary directly with the required illumination level and inversely with the efficiency of the lighting fixtures and lamps. While stressing that the NEC [3] is not a design manual, the electrical engineer must be aware that the NEC [3] does include, for example, Article 220, "Branch-Circuit and Feeder Calculations" for various types of occupancies in commercial buildings. Minimum power allowance for lighting loads for each square foot of floor area, which help to identify the minimum capacities for the associated feeder-circuit panelboards of the power distribution system, are specified.

The engineer should recognize a consistently increasing trend in exterior lighting for security and decorative effect and then provide service and feeder capacity for the resulting future increases in loads. Not only should added circuit capacity be provided, but consideration should also be given to space in distribution equipment for the added branch circuits.

Criteria for controlling the energy consumption of lighting systems in, and connected with, building facilities have been prepared by the American Society of Heating, Refrigerating, and Air-Conditioning Engineers (ASHRAE) in concert with the Illuminating Engineering Society (IES). They are identified in Section 6 of ASHRAE/IES 90.1-1989, Energy Efficient Design of New Buildings Except New Low-Rise Residential Buildings [4],[33] which establishes an upper limit of power to be allowed for lighting systems plus guidelines for designing and managing those systems. A simplified method based on the above standard for determining the unit lighting power allowance for each building type is shown in Table 4.

2.2.2 General-Purpose Receptacles for Appliance Loads. Power required for appliances depends largely on the type of space usage. Commercial building appliances include such loads as typewriters, desktop computers, copiers, communication equipment, and office automation equipment. Loads for large computers, plug-in-type air conditioners, cooking and laundry equipment, etc., should be considered separately. In contrast to lighting, the overall demand factor for appliances is very low. The NEC, Article 220 (Branch-Circuit and Feeder Calculations) [3] provides information on the allowable (for minimum safety) use of *demand factors for nondwelling receptacle loads*. In general, 1 VA/ft^2 of net demand is adequate for most commercial buildings; however, the wiring (feeders and branch circuits) to serve the connected load is often installed with capacity for 1.5 A per duplex outlet, or 180 VA/100 ft^2 of office area. Typical unit load data for various occupancies are given in Table 5 and for apartments in Table 6.

2.2.3 Space-Conditioning and Associated Auxiliary Equipment. Building design engineers are increasingly using the concept of controlled environment. Space conditioning generally refers to heating, ventilating, cleaning, and cooling systems. The connected and demand power required for space-conditioning systems depends largely on the climatic conditions (that is, the building's geographical location) and the building's envelope design; interior load, such as lighting and

---

[33] ASHRAE publications are available from the American Society of Heating, Refrigerating, and Air-Conditioning Engineers, 1791 Tullie Circle, N.E., Atlanta, GA 30329. IES publications are available from the Illuminating Engineering Society, 345 East 47th Street, New York, NY 10017.

Table 4
Prescriptive Unit Lighting Power Allowance (ULPA) (W/ft²) —
Gross Lighted Area of Total Building

Building Type or Space Activity	0 to 2000 ft²	2001 to 10 000 ft²	10 001 to 25 000 ft²	25 001 to 50 000 ft²	50 001 to 250 000 ft²	>250 000 ft²
Food Service						
Fast Food/Cafeteria	1.50	1.38	1.34	1.32	1.31	1.30
Leisure Dining/Bar	2.20	1.91	1.71	1.56	1.46	1.40
Offices	1.90	1.81	1.72	1.65	1.57	1.50
Retail*	3.30	3.08	2.83	2.50	2.28	2.10
Mall Concourse	1.60	1.58	1.52	1.46	1.43	1.40
Multiple-Store Service						
Service Establishment	2.70	2.37	2.08	1.92	1.80	1.70
Garages	0.30	0.28	0.24	0.22	0.21	0.20
Schools						
Preschool/Elementary	1.80	1.80	1.72	1.65	1.57	1.50
Jr. High/High School	1.90	1.90	1.88	1.83	1.76	1.70
Technical/Vocational	2.40	2.33	2.17	2.01	1.84	1.70
Warehouse/Storage	0.80	0.66	0.56	0.48	0.43	0.40

NOTE: *Includes general, merchandising, and display lighting.

This prescriptive table is intended primarily for core-and-shell (i.e., speculative) buildings or for use during the preliminary design phase (i.e., when the space uses are less than 80% defined). The values in this table are not intended to represent the needs of all buildings within the types listed.

number of occupants; appliances; and special process loads. Data processing centers in commercial buildings require substantial space conditioning using air-, water-, or glycol-cooled systems with supplemental heating. These systems require standby energy sources, such as generators with automatic transfer. All of the above factors can have a major influence on a space-conditioning load. The actual electrical requirements can best be obtained from those responsible for the design of the space-conditioning system. Detailed discussions of definitions, equipment ratings, selection factors, system operation, calculation methods, etc., are included in Chapter 11.

When exact loads are not known or cannot be determined, an approximate preliminary load can be determined as outlined in 2.2.3.1, 2.2.3.2, and 2.2.3.3.

2.2.3.1 Air Conditioning. The air-conditioning load will consist of the motor drives for compressors, chilled water pumps, condensate pumps, evaporative condensers or cooling towers, air distribution fans or blowers, motorized dampers and valves, and associated control circuits. For rough estimation purposes, it may be assumed that 1 ton of refrigeration equipment will require 1 hp of motor drive for refrigeration units only, or approximately 1 kVA of load. The refrigeration unit or compressor will usually constitute about 55%–70% of the total connected air-

Table 5
Typical Appliance/General-Purpose Receptacle Loads
(Excluding Plug-in-Type A/C and Heating Equipment)

Type of Occupancy	Unit Load (VA/ft^2)		
	Low	High	Average
Auditoriums	0.1	0.3	0.2
Cafeterias	0.1	0.3	0.2
Churches	0.1	0.3	0.2
Drafting rooms	0.4	1.0	0.7
Gymnasiums	0.1	0.2	0.15
Hospitals	0.5	1.5	1.0
Hospitals, large	0.4	1.0	0.7
Machine shops	0.5	2.5	1.5
Office buildings	0.5	1.5	1.0
Schools, large	0.2	1.0	0.6
Schools, medium	0.25	1.2	0.7
Schools, small	0.3	1.5	0.9

Other Unit Loads:
  Specific appliances — ampere rating of
    appliance
  Supplying heavy-duty lampholders —
    5 A/outlet

Table 6
Typical Apartment Loads

Type	Load
Lighting and convenience outlets (except appliance)	$3 \ VA/ft^2$
Kitchen, dining appliance circuits	1.5 kVA each
Range	8 to 12 kW
Microwave oven	1.5 kW
Refrigerator	0.3 to 0.6 kW
Freezer	0.3 to 0.6 kW
Dishwasher	1.0 to 2.0 kW
Garbage disposal	0.33 to 0.5 hp
Clothes washer	0.33 to 0.5 hp
Clothes dryer	1.5 to 6.5 kW
Water heater	1.5 to 9.0 kW
Air conditioner (0.5 hp/room)	0.8 to 4.6 kW

conditioning load. The remaining load may consist of pumps, fans, and other auxiliaries. It is customary, therefore, to apply a factor of from 1.6–2.0 to the total tonnage involved, and the result will be a fair estimate of the total connected load to be expected. The above factors would apply in most cases for systems of 100 tons and larger. On systems below this figure, a factor of about 2.3 may be used for preliminary estimates. Where many small-unit air conditioners are used, a factor of 2.8 is suggested.

In air-conditioning systems utilizing refrigeration machines that operate on the absorption principle, the heavy compressor load is eliminated but the auxiliary equipment load is still present. This type of system will usually reduce the electrical load to about 40%–50% of that required for a full electric drive system, or to about 0.7–1.0 kVA/ton.

Table 7 gives the approximate air-conditioning load that might occur in the average commercial building. Loads include compressors and all auxiliary equipment involved in the cooling and ventilating system.

Actual air-conditioning loads are dependent on the internal heat load, which can vary considerably with building design and usage. Unit air conditioners are often used in older commercial buildings or in buildings where the tenant is fully responsible for the air-conditioning load (commonly used in apartment houses). These may be window- or floor-mounted units and these should be treated as fixed-appliance loads.

2.2.3.2 Auxiliary Equipment. The electrical load for boiler room and mechanical auxiliary equipment does not normally constitute a large portion of the building load. Usually, it will not exceed 5% of the total load (not including air conditioning); but it may be as high as 10% in schools. In small commercial buildings, the auxiliary equipment load will consist of small units, many of which may be served by fractional horsepower motors.

While larger buildings will have some fractional horsepower equipment, some of the fans and pumps required may be relatively large, 10–20 hp being the most common and 30–75 hp or more being quite possible. The electrical engineer should consult mechanical designers on the possible use of large motors or electrical heating loads that might affect the preliminary load estimate.

The major pieces of equipment frequently encountered are

(1) Induced draft or forced draft fans
(2) Ventilation or exhaust fans
(3) Pumps for boiler feed, condensate return, sumps, sewage ejectors, and water circulation
(4) Fire and house service tank pumps
(5) Air compressors and service equipment
(6) Electrical heating and auxiliary heating elements

Table 7
Total Connected Electrical Load for Air Conditioning Only

Type of Building	Conditioned Area (VA/ft^2)
Bank	7
Department store	3 to 5
Hotel	6
Office building	6
Telephone equipment building	7 to 8
Small store (shoe, dress, etc.)	4 to 12
Restaurant (not including kitchen)	8

(7) Control devices and circuits

(8) Electronic air cleaners

The induced draft or forced draft fans are normally located in the boiler room and range in size from small fractional horsepower units up to 25 hp or more. Exhaust fans are usually small units scattered throughout the building; although, in some cases, exhausting is handled by a single large fan of 20 hp or more. Where fans are supplied with adjustable sheaves for speed control, the horsepower requirements of most centrifugal fans vary as the cube of the speed.

2.2.3.3 Heating. Electrical heating loads may range in size from many 10 kW or larger units, comprising the building's total heat source and amounting to one-third to one-half of the total electrical load, down to relatively small loads serving specific areas as supplemental heaters rated 10 kW or less. Other units may provide the building's hot water supply, again ranging from large electric boilers to small (1–4 kW) units.

A building surrounded by air colder than its interior air temperature is constantly dissipating heat. The rate of dissipation is controlled by many factors, such as outside temperature, wind velocity, area of exposed surfaces, types of construction materials, amounts of insulation used, fresh air requirements, and the type of usage. The amount of heat required to maintain comfort in a structure may be determined by taking all these factors into consideration. See Chapter 11 for more information.

With a known heat loss, the electrical load in kW can be obtained by dividing the estimated heat loss (Btu/hour) by 3413 since there are 3413 Btu in 1 kWh of electricity. Usually, it is necessary to use a demand factor of 100% for electrical heating loads.

Loads larger than a few hundred watts should be connected to the power panels in order to prevent excessive voltage drop on the lighting circuits. Installed heating loads should not be supplied from lighting panelboards.

Table 8 is based on a building with the insulation necessary to provide proper comfort and operating economy. These values are used for the all-weather comfort standard.

2.2.4 Plumbing and Sanitation. Generally, for a commercial building, the loads of plumbing and sanitation equipment are not large. Typical loads for water pres-

Table 8
All-Weather Comfort Standard Recommended Heat Loss Values

Degree Days	Design Heat Loss per Square Foot of Floor Area (Btu/h)	(watts)
Over 8000	40	11.7
7001 to 8000	38	11.3
6001 to 7000	35	10.3
5001 to 6000	32	9.4
3001 to 5000	30	8.8
Under 3001	28	8.2

sure boosting systems and electric hot water heating are identified in Tables 9 and 10. Sump and sewage pumps are usually small, often applied in pairs with an electrical or mechanical alternator control, so that allowing for several 2 hp duplex units is a satisfactory allowance for the basement (that is, boiler room) of most buildings.

2.2.5 Fire Protection. The largest load for fire protection will usually be a fire pump, which is required to maintain system pressure beyond the capacity of the city water system. (Pertaining to the design of the power system rather than the load magnitude, the fire pump is one of the few loads ever required or permitted to be connected to the power source ahead of the service disconnect device.) Typical power load data for fire pumps are given in Table 11.

Fire detection and alarm systems are highly critical loads; but their magnitudes are generally so small that they can usually be neglected when identifying the total load of the building.

Table 9
Typical Power Requirement (kW) for
High-Rise Building Water Pressure Boosting Systems

Building Type	Unit Quantity	Number of Stories			
		5	10	25	50
Apartments	10 apt./floor	—	15	90	350
Hospitals	30 patients/floor	10	45	250	—
Hotels/Motels	40 rooms/floor	7	35	175	450
Offices	10 000 ft^2/floor	—	15	75	250

Table 10
Typical Power Requirement (kW) for
Electric Hot Water Heating System

Building Type	Unit Quantity	Load
Apartments/Condominiums	20 apt/condo	30
Dormitories	100 residents	75
Elementary schools	100 students	6
High schools	100 students	12
Restaurant (full service)	100 servings/h	30
Restaurant (fast service)	100 servings/h	15
Nursing homes	100 residents	60
Hospitals	100 patient beds	200
Office buildings	10 000 ft^2	5

**Table 11
Typical Power Requirement (kW) for
Fire Pumps in Commercial Buildings (Light Hazard)***

Area/Floor (ft^2)	Number of Stories			
	5	10	25	50
5000	40	65	150	250
10 000	60	100	200	400
25 000	75	150	275	550
50 000	120	200	400	800

*Based on zero pressure at floor 1.

2.2.6 Transportation Systems. Transportation equipment for commercial buildings includes elevators, escalators, conveyors, dumbwaiters, and pneumatic-conveying systems. There is no simple rule-of-thumb method for determining the number and type of elevators or escalators required in a particular size or type (occupancy) of commercial building. Manufacturers of the equipment or specialized consultants are the best source of load information. For additional information on these loads, see Chapter 12. When determining this total load, typical demand factors might be 0.85 for two elevators, 0.75 for four elevators, and a somewhat lower value for additional elevators.

2.2.7 Data Processing. The power requirements of data processing equipment will vary over a wide range. For smaller installations, consisting of appliance-type loads, single-phase power at 120 V may be adequate. For larger installations, including computer and peripheral (or support, or auxiliary) equipment, it may be necessary to supply 208Y/120 V or 480Y/277 V power.

Data processing installations may also be categorized as requiring continuity of high-quality power supply with the flexibility to facilitate changing the loads or location of the equipment, or both. Continuity is of prime importance to avoid loss of information stored in memory units. Power supply distortions, such as voltage dips, spikes, and harmonics, are considered noise to the computer since the input voltage or signal is modified in an undesired manner. These installations will also require specialized lighting (see ANSI/IES RP1-1982, Practice for Office Lighting [2] and IES RP24-1989, IES Recommended Practice for Lighting Offices Containing Computer Visual Display Terminals (VDTs) [7]) and air conditioning, and probably a raised floor to accommodate air-handling plus power, signal, and communications conductors.

Some installations may include a computer central processing unit (CPU), which may utilize high-frequency power distribution. These special requirements include the use of 60/415 Hz motor-generator sets or static inverters. The electrical engineer must be aware of the reduced ampacity of conductors and increased voltage drop at the higher frequency. Refer to Chapter 4 for details.

It should be stressed that, if there is any possibility that electronic computers will be installed, the manufacturer of such equipment must be consulted well in advance to determine specific electrical requirements.

A few illustrative excerpts taken from one computer manufacturer's specifications follow:

(1) Air-conditioned space should be provided for the general machine room, and magnetic tape storage and engineering areas.

(2) Raceways should be 6–10 inches deep and 10–12 inches wide and should be provided with removable covers.

(3) It is recommended that a minimum average illumination of 50 fc be maintained 30 inches above the floor in the general machine room and engineering areas; specific local areas should be illuminated at 70–85 fc.

Typical electronic data processing machine power service requirements are listed below.

(1) Electric service can be any commercially available voltage with an insulated equipment grounding conductor.

(2) Voltage variations to be limited to +6% and –13% (see ANSI C84.1-1989, Electric Power Systems and Equipment—Voltage Ratings (60 Hz) [1])

(3) Line-to-line voltage balance is not usually specified; but 2.5% is a conservative figure.

(4) Frequency variation not greater than ±0.5 Hz.

(5) Maximum total harmonic content of the power system waveforms on the electric power feeders is not in excess of 5% with the equipment not operating.

It may be desirable to provide a separate transformer bank, motor-alternator set, or complete rectifier battery inverter assembly for the electronic data-processing machinery. Analysis of filtering equipment and surge-protective equipment on the incoming utility power line is also required to minimize the likelihood of improper operation due to line transients. If the area is fed by a low-voltage secondary-network power system, the customer should consult the local equipment manufacturer regarding the advisability of a separate transformer bank. Line inductive reactance at the wall box should not exceed 0.0173 ohms/line and can consist of either the overall reactance of the entire power system or the subtransient reactance of the separate alternator. The reactance-to-resistance ratio may be as low as 2 with no upper limit.

Typical loads for a medium to large installation could be as follows:

Central processing unit	75 kVA
Miscellaneous (tape, disks, printers)	175 kVA
400 Hz motor-generator set	64 kVA
Air conditioning	30 tons

Additional comments on electrical power requirements for these loads can be found in Chapter 16. An early check with the utility company supplying electrical service may provide valuable data on supply reliability.

2.2.8 Food Preparation. The magnitude of the electrical load depends more upon the number of meals served at one time than upon the total size of the space. The load also depends upon whether electricity or gas is used to provide the heat for the main equipment (ovens and ranges). However, the additional devices using electric power (that is, fryers, microwave ovens, stock kettles, warming tables, meat slicers and saws, coffee pots, toasters, waffle irons, mixers, potato peelers, etc.) may present a sizable load and should not be overlooked in system design. Besides the equipment directly involved in food preparation, there will be additional service equipment including lighting, dishwashing and garbage disposal equipment, exhaust fans, make-up air heaters, hot water booster heaters, etc. In addition, there may also be refrigeration equipment, varying from walk-in-type refrigerators to freezer units or deep-freeze lockers. (Commercial freezer or cold storage plants present different system design problems; they are considered an industrial type of building, and consequently are excluded from this recommended practice.) When the utility power supply is subject to prolonged outages, freezer or refrigerated loads may require transfer to an alternate or standby power source. (This may also apply to laboratories where senstitive experimental materials are kept under refrigeration.) Table 12 provides some approximate total load data for commercial kitchens that might be located in a commercial building.

The cooling load for the kitchen should not be overlooked since heat gain in the kitchen is often large. This heat can be removed by exhaust fans (for example, range hoods, room exhausts), air conditioning, or a combination of the two. When ventilation alone is used, fan capacity to provide one air change per minute may be necessary. There are so many variables in heat gains for kitchen equipment that a general rule-of-thumb cannot be used for the load required to air condition a commercial kitchen.

Additional comments on electric power requirements for these loads can be found in Chapter 16.

2.2.9 Miscellaneous or Special Loads. There are many loads that do not qualify in the preceding schedule of major load groups that could possibly appear in any new commercial building. These additional loads will generally be small (but could be major in size, e.g., broadcasting equipment) and occur only occasionally in commercial buildings. Therefore, these loads can best be categorized as "miscel-

Table 12
Typical Loads in Commercial Kitchens

	Number Served	Connected Load (kW)
Lunch counter (gas ranges, with 40 seats)		30
Cafeteria	800	150
Restaurant (gas cooking)		90
Restaurant (electric cooking)		180
Hospital (electric cooking)	1200	300
Diet kitchen (gas cooking)		200
Hotel (typical)		75
Hotel (modern, gas ranges, three kitchens)		150
Penitentiary (gas cooking)		175

laneous" or "special loads" since they vary so widely with regard to size and frequency of appearance. However, if such load apparatus will eventually be included in the building, they should be considered (even if only in the form of a spare feeder or branch circuit or as space for a future protective device) when the power system is initially being designed. Multiple-story office buildings may require approximately $1-2$ VA/ft^2 for such general or miscellaneous loads. A partial checklist of such loads is provided in Table 13. Some apparatus, including electric typewriters, desktop computers, or visual-aid equipment, can be served (operated) from the usual 15 A or 20 A receptacles, neither creating any appreciable voltage fluctuations nor demanding critical voltage regulation or an emergency power source. Other kinds of equipment, such as intercommunication, photographic reproduction, or x-ray equipment, may amount to small loads, yet they require a high-quality (for example, stabilized voltage) power source. Some of these special loads, such as welders, may draw heavy currents for short times in repetitive cycles so that voltage variation (potential light flicker) should be investigated.

2.3 Electromagnetic Hazards, Pollution, and Environmental Quality. The increasing use of electronic equipment calls for some consideration of the electromagnetic environment created by this equipment and also the effect of external electromagnetic influences on its performance. Specifically, computers, communications equipment, and other low-level electronic systems require special analysis of the grounding system. Inadequate grounding can be both a shock hazard and a source of noise input to the computer.

Since the cost of providing shielding after construction is quite high, the electrical engineer should analyze shielding requirements for sensitive equipment *before* construction. Some of the following applications may require the degree of control of electromagnetic energy that is only achieved by a shielded enclosure:

(1) Research and development laboratories for low-noise circuitry work
(2) Research and development laboratories using high-energy radio-frequency devices
(3) Special computer facilities
(4) Test and measurement laboratories
(5) Terminal equipment facilities for both line and radio-frequency transmission systems
(6) Hospital and other biomedical research and treatment rooms
(7) Control and monitoring equipment in strong fields of other emitters or strong radio-frequency fields from industrial sources

2.4 Additions to Existing Systems. Whenever it is contemplated that the occupancy of a commercial/industrial/institutional building is to be renovated or if the building is to be expanded or modernized, depending upon the nature and magnitude of the changes in the total and individual loads, an engineering study of the existing electric power distribution system should be included in the initial planning of the building renovation. Additional comments on this subject are offered in Chapter 15 (that is, there is a discussion in Chapter 15 concerning the need for accurate drawings of the details of the existing building and loads).

Table 13
Types of Electrical Load Equipment

(1) Broadcasting equipment (radio, television, microwave communications)	(c) duplicating machines
(2) Control and monitoring systems (centralized, local)	(i) large-scale copiers
(a) security	(ii) small-scale copiers
(b) fire alarm	(d) dictating and transcribing machines
(c) building management systems	(e) data processing center
(3) Fire pumps	(i) mainframe equipment
(4) Health care facilities	(ii) printers and printing equipment
(a) sterilizers	(iii) storage and processing equipment
(b) x-ray machines	(iv) dedicated air-conditioning equipment
(c) laser equipment	(f) typewriters, dedicated word processors, and desktop computers
(d) operating rooms	(11) Public address systems (radio and wired paging systems)
(e) examination offices	(12) Radio and television receivers
(f) miscellaneous	(13) Recreational and fitness equipment
(5) Waste disposal equipment	(a) swimming pool heaters and pumps
(a) incinerators	(b) saunas
(b) central compactors	(c) exercise equipment
(6) Intercommunication systems (facsimile system — FAX)	(d) pool laundry equipment
(7) Kitchen equipment	(14) Shop equipment
(a) cooking	(15) Shredding devices
(b) disposal	(16) Snow melting equipment
(c) refrigeration and freezing	(17) Space conditioning (central or individual)
(d) ventilation	(a) heating
(e) washing	(b) cooling
(f) microwave ovens	(c) ventilating
(g) garbage disposals	(d) air filtering
(8) Laboratory equipment	(18) Television equipment, closed circuit
(a) air compressors and vacuum pumps	(19) Transporting equipment
(b) centrifuges	(a) dumbwaiters
(c) furnaces and refrigerators	(b) conveyors
(d) incubation and cold rooms	(c) elevators
(e) sterilizers	(d) escalators and moving walkways
(f) miscellaneous	(20) Visual-aid equipment
(9) Lighting	(a) motion picture, slide and overhead projectors
(a) general	(b) audiotape and videotape recorders/players
(b) task	(21) Plumbing equipment
(c) exits and stairwells	(a) water pumps
(d) emergency	(b) sewage pumps
(e) exterior	(c) water heaters
(10) Office equipment	(d) water treatment systems
(a) addressing machines	(22) Others
(b) copying cameras, micro file equipment	

2.5 Total Load Considerations. If all the connected loads in the building are arithmetically totaled (that is, all expressed in hp, kW, kVA, or A at some specified voltage) to identify the total building load, the resultant number will in most cases seem to require a larger power system capacity than will be realistically needed to adequately serve the loads. The average load on the power system is usually less than the total connected load; this is termed the "demand load." It may vary depending on the time interval over which the load is averaged. Certain loads may at times be turned off or operated at reduced power levels, reducing the system power requirements (that is, total load). This effect is termed "diversity," and it may be expressed as a diversity factor.

The value of the demand or diversity to be used is highly dependent upon the location of the load in the particular power system being considered. Diversity factors become larger as the loads are totaled nearer to the power source and include more of the diverse building components. The following factors (or definitions) are commonly used when totaling loads to facilitate system planning. See IEEE Std 100-1988, IEEE Standard Dictionary of Electrical and Electronics Terms, Fourth Edition (ANSI) [5]:

branch-circuit load. The load on that portion of a wiring system extending beyond the final overcurrent device protecting the circuit. (See the NEC, Article 220 [3] for complete details and exceptions.)

coincident demand. Any demand that occurs simultaneously with any other demand; also the sum of any set of coincident demands.

connected load. The sum of the continuous ratings of the power-consuming apparatus connected to the system or any part thereof in watts, kilowatts, or horsepower.

demand (or **demand load**). The electrical load at the receiving terminals averaged over a specified interval of time. Demand is expressed in kilowatts, kilovolt-amperes, kilovars, amperes, or other suitable units. The interval of time is generally 15 minutes, 30 minutes, or 60 minutes.

NOTE: If there are two 50 hp motors (which drive 45 hp loads) connected to the electric power system but only one load is operating at any time, the demand load is only 45 hp but the connected load is 100 hp.

demand factor. The ratio of the maximum demand of a system to the total connected load of the system.

NOTES: (1) Since demand load cannot be greater than the connected load, the demand factor cannot be greater than unity.
(2) Those demand factors permitted by the NEC [3] (for example, services and feeders) should be considered when sizing the electric system (with few exceptions, this is 100%); otherwise, the circuit may be sized to support the load.

diversity factor. The ratio of the sum of the individual maximum demands of the subdivisions of the system to the maximum demand of the complete system.

NOTE: Since maximum demand of a system cannot be greater than the sum of the individual demands, the diversity factor will *always* be equal to or greater than unity.

gross demand load. The summation of the demands for each of the several group loads.

load factor. The ratio of the average load over a designated period of time to the peak load occurring in that period.

load profile. The graphic representation of the demand load, usually on an hourly basis, for a particular day. The demand load for typical groups of loads (for example, heating, cooling, lighting, etc.) may be accumulated to determine the demand load of the system; the highest point of the load profile will be the maximum demand load of the system. See Fig 2 for typical load profile representations.

Information on these factors for the various loads and groups of loads is essential in designing the system. For example, the sum of the connected loads on a feeder, multiplied by the demand factor of these loads, will give the maximum demand that the feeder should carry. The sum of the individual maximum demands on the circuits associated with a load center or panelboard divided by the diversity factor of those circuits, will give the maximum demand at the load center and on the circuit supplying it. By the use of the proper factors, as outlined, the maximum demands on the various parts of the system from the load circuits to the power source can be estimated. Tables 14 and 15 provide typical maximum demand and demand factor data for various types of occupancies.

2.5.1 Estimation of Building Load. A suggested procedure for determining the demand load of a building is given in the following steps. Calculations can be summarized in tabulated form as shown in Table 16.

(1) Determine the quantity of load units and the power requirement of each load.

Table 14
Comparison of Maximum Demand

Type of Store	Shopping Center A, New Jersey No Refrigeration*		Shopping Center B, New Jersey Refrigeration		Shopping Center C, New York Refrigeration	
	Gross Area (ft^2)	(W/ft^2)	Gross Area (ft^2)	(W/ft^2)	Gross Area (ft^2)	(W/ft^2)
Bank					4000	9.0
Book	3700	6.0	2500	6.7		
Candy	1600	6.9			2000	10.8
Department	343 500	4.7	222 000	7.3	226 900	8.0
	84 000	3.1	114 000	5.6		
Drug	7000	6.1	6000	7.7		
Men's wear	17 000	5.5	17 000	9.9	2000	10.8
	28 000	4.9	9100	8.8		
Paint					15 600	8.5
Pet					2000	12.1
Restaurant					4000	9.0
Shoe	11 000	6.3	7000	12.5	3300	15.4
	4000	8.0	4400	12.9	2100	9.0
Supermarket	32 000	5.7	25 000	8.6	37 600	11.5
Variety	31 000	4.6	24 000	6.8	37 400	7.1
	30 000	4.4			30 000	7.0
Women's wear	20 400	4.7	19 300	8.9	1360	13.0
	1000	5.8	4500	9.6	1000	11.7

*Loads include all lighting and power, but no power for air-conditioning refrigeration (chilled water), which is supplied from a central plant.

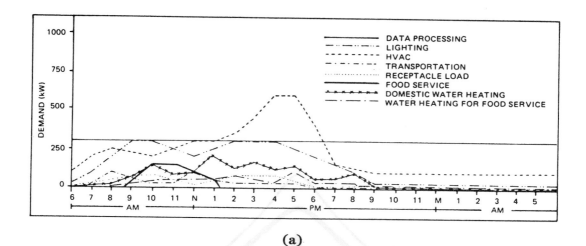

(a)

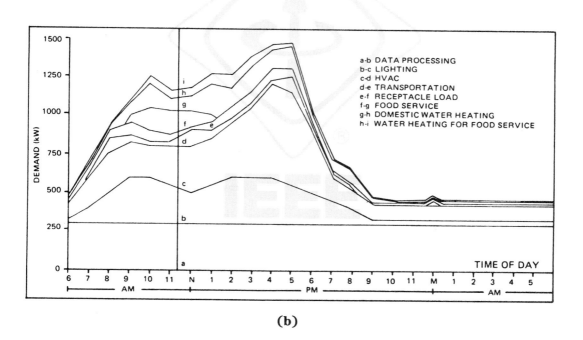

(b)

Fig 2
Load Profiles
(a) Individual Group
(b) Cumulative

Table 15
Connected Load and Maximum Demand by Tenant Classification

Classification	Connected Load (W/ft^2)	Maximum Demand (W/ft^2)	Demand Factor
10 Women's wear	7.7	5.9	0.75
3 Men's wear	7.2	5.6	0.78
6 Shoe store	8.5	6.9	0.79
2 Department store	6.0	4.7	0.74
2 Variety store	10.5	4.5	0.45
2 Drug store	11.7	6.7	0.57
5 Household goods	5.4	3.9	0.76
10 Specialty shop	8.1	6.8	0.79
4 Bakery and candy	17.1	12.1	0.71
3 Food store (supermarkets)	9.9	5.9	0.60
5 Restaurant	15.9	7.1	0.45

NOTE: Connected load includes an allowance for spares.

(2) Determine the demand factor (DF) of the load or group of loads by the definition given in 2.5, or from Table 15, or from the NEC [3] (see 2.1.3).

(3) (a) Determine the demand load (DL) for present and future operating conditions; it is the product of connected load (CL) and demand factor (DF).

    (b) Estimate (for column 8) the power factor (decimal value) of the particular load when operating at its intended rated capacity. The various loads divided by their respective power factor (decimal value) will determine the required *source capacity* in kilovoltamperes.

(4) Compute the gross demand load (GDL) of the building, which is equal to the sum of all the demands of individual and group loads.

(5) Determine the diversity factor (DIF) of the system by estimation or reference from similar projects or reference to the NEC [3].

(6) Estimate the spare capacity to be provided for load growth and identified future loads, such as data processing, food service, air conditioning, etc. Use either a *blanket* percent against the gross demand load or apply the estimated percentage against each load (or group of loads) and, for c, utilize the sum of these increments.

(7) Determine the required capacity from steps (4), (5), and (6). (When the load profile is used, step (5) can be eliminated.)

(8) Select a system with capacity, which will satisfy the required capacity determined in step (7).

CAUTION: The system capacity cannot be less than the minimum permitted by the NEC [3] when adopted by the controlling utility or political subdivision.

2.6 Example—Sample Partial Load Calculation for an Office Building

NOTE: Calculations according to the NEC [3] are required practice in only some jurisdictions. NEC calculations represent the minimum design loadings permitted. Lighting must be included at designed capacity but not less than the minimum shown in the NEC Table. Demand factor is also shown. For

Table 16
Electrical Load Estimation

1	2	3	4	5 = 3·4	6	7 = 5·6	8	9 = 7/8	10
GROUP LOAD	LOAD DESCRIPTION	QUANTITY • NO • SQ FT (SF) • CU FT (CF) • TONS • ETC	UNIT LOAD • W/UNIT • W/SF • W/CF • W/TON • ETC	CONNECTED LOAD (CL) 1000W	EST DEMAND FACTOR (DF)	EST DEMAND LOAD (DL) kW	EST POWER FACTOR (PF)	REQUIRED CAPACITY IN kVA	
								FOR EACH LOAD	FOR GROUP LOAD
	SYSTEM LOSS								
a	TOTAL CONNECTED LOAD (TCL) AND GROSS DEMAND LOAD (GDL)								
b	ESTIMATED DIVERSITY FACTOR (DIF)								
c	SPARE CAPACITY (SC) FOR GROWTH (__ % X GDL IN kVA)								
d	GROSS SYSTEM CAPACITY TO BE PROVIDED [GSC = (a/b) + c]								

example, in office buildings, the lighting minimum is 3.5 VA/ft^2, 100% demand factor. General-purpose receptacles must be included at 180 VA per strap (single, duplex, or triplex) with a minimum of 1 VA/ft^2 with a demand factor of 100% for the first 10 kVA, 50% for all over 10 kVA, and all other loads (with a few exceptions) must be included at nameplate rating. See the examples at the end of the NEC for more detailed examples.

Outside Dimensions:

100 ft × 160 ft (16 000 ft^2/floor), 18 floors of office space

Per Floor:

Corridors (including stairways)	1050 ft^2
Electrical and janitor closets	150 ft^2
Subtotal	1200 ft^2
Elevator and vent shafts	800 ft^2
Total non-office area	2000 ft^2

(1) Code Calculations

NEC Table 220-3(b) [3]
Area lighting	3.5 VA/ft^2
Closets	3.5 VA/ft^2
Stairwells	0.5 VA/ft^2
Receptacles	1.0 VA/ft^2
Diversity on receptacles	50% over first 10 kVA

Lighting:
$(16\,000 - 2000) \times 3.5 = 49\,000$ VA $= 49.0$ kVA Office
$1200 \times 0.5 = 600$ VA $\quad\quad = 0.6$ kVA Other

General-purpose receptacles:
$(16\,000 - 800) \times 1.0 \quad\quad\quad = 15.2$ kVA
$\quad (15.2 - 10) \times 0.5 + 10 \quad\quad = 12.6$ kVA

Total NEC requirements (per floor) $= 62.2$ kVA

(2) Actual Load (All Electric Building)

Usable office area: $13\,000$ ft^2/floor

Lighting:
Lighting fixture 2 ft. $\times$ 4 ft. 3–34 watt energy-saving lamps in parabolic lay-in troffers, one fixture per 80 ft^2. Calculated lighting level 55 fc. Areas other than offices use two lamp fixtures.
Offices: $13\,000 / 80 = 163$ fixtures at 119 VA $= 19.4$ kVA
Other areas: 34 fixtures at 75 VA $\quad\quad = 2.6$ kVA
Lighting total $\quad\quad\quad\quad\quad\quad\quad\quad\quad = 22.0$ kVA

General-purpose receptacles:
One receptacle per 100 ft^2
$\quad 13\,000$ ft^2 / 100 $= 130$ receptacles
$\quad$ Other areas $\quad = 9$ receptacles
$\quad$ Total $\quad\quad\quad = 139$ receptacles

Estimated 180 VA / receptacle $\times$ 139 $= 25\,020$
Using NEC formula,
$\quad$ receptacle load $= (25-10) \times 0.5 + 10 = 17.5$ kVA

Total actual floor load $= 22 + 17.5 = 39.5$ kVA

(3) For the entire building (per NEC)
Service capacity must be calculated load but not less than the required minimum.
Lighting (minimum):	$49.6 \times 18 =$	892.8 kVA
General-purpose receptacles (actual):		
$\quad 25 \times 18 = 450 - 10$ at 100%, 440 at 50%	$=$	230.0 kVA
Other (approximate)	$=$	4200.0 kVA
Total	$=$	5322.8 kVA

2.7 References. The following references shall be used in conjunction with this chapter:

[1] ANSI C84.1-1989, Electric Power Systems and Equipment—Voltage Ratings (60 Hz).

[2] ANSI/IES RP1-1982, Practice for Office Lighting.

[3] ANSI/NFPA 70-1990, National Electrical Code.

[4] ASHRAE/IES 90.1-1989, Energy Efficient Design of New Buildings Except New Low-Rise Residential Buildings.

[5] IEEE Std 100-1988, IEEE Standard Dictionary of Electrical and Electronics Terms, Fourth Edition (ANSI).

[6] IEEE Std 112-1984, IEEE Standard Test Procedure for Polyphase Induction Motors and Generators (ANSI).

[7] IES RP24-1989, IES Recommended Practice for Lighting Offices Containing Computer Visual Display Terminals (VDTs).

[8] NEMA MG10-1983, Energy Management Guide for Selection and Use of Polyphase Motors.

[9] NEMA MG11-1977 (Reaff. 1982), Energy Management Guide for Selection and Use of Single-Phase Motors.

[10] Bauer, G. M. Users' Needs: Space Conditioning, Fire Protection, Data Processing, Life Support and Life Safety Systems, Communication Systems and Signal Circuits, *IEEE Transactions on Industry and Applications*, vol. 1A—10, Mar./Apr. 1974.

[11] McWilliams, D. W. Users' Needs: Lighting, Start-up Power, Transportation, Mechanical Utilities, Heating, Refrigeration and Production, *IEEE Transactions* on *Industry and Applications*, vol. 1A—10, Mar./Apr. 1974.

[12] *Power Factor Correction Capacitors Catalog and Selection Guide*, Commonwealth Sprague Capacitor, Inc., Brown Street, North Adams, MA 01247.

Chapter 3
Voltage Considerations

3.1 General Discussion. An understanding of system voltage nomenclature and the preferred voltage ratings of distribution apparatus and utilization equipment is essential to ensure the proper design and operation of a power distribution system. The dynamic characteristics of the system should be recognized and the proper principles of voltage control applied so that satisfactory voltages will be supplied to all utilization equipment under all normal conditions of operation.

3.1.1 Definitions. The following terms and definitions are quoted from ANSI C84.1-1989, Voltage Ratings for Electric Power Systems and Equipment (60 Hz) [2][34] are used to identify the voltages and voltage classes used in electric power distribution.

3.1.1.1 System Voltage Terms

NOTE: The nominal system voltage is near the voltage level at which the system normally operates. To allow for operating contingencies, systems generally operate at voltage levels about 5%–10% below the maximum system voltage for which system components are designed.

system voltage. The root-mean-square phase-to-phase voltage of a portion of an ac electric system. Each system voltage pertains to a portion of the system that is bounded by transformers or utilization equipment. (All voltages hereafter are root-mean-square phase-to-phase or phase-to-neutral voltages.)

nominal system voltage. The voltage by which a portion of the system is designated, and to which certain operating characteristics of the system are related. Each nominal system voltage pertains to a portion of the system that is bounded by transformers or utilization equipment.

maximum system voltage. The highest system voltage that occurs under normal operating conditions, and the highest system voltage for which equipment and other components are designed for satisfactory continuous operation without derating of any kind. In defining maximum system voltage, voltage transients and temporary overvoltages caused by abnormal system conditions, such as faults, load rejection, and the like, are excluded. However, voltage transients and temporary

---

[34]The numbers in brackets correspond to those in the references at the end of this chapter. ANSI publications are available from the Sales Department of the American National Standards Institute, 11 West 42nd Street, 13th Floor, New York, NY 10036.

overvoltages may affect equipment operating performance and are considered in equipment application.

service voltage. The voltage at the point where the electric system of the supplier and the electric system of the user are connected.

utilization voltage. The voltage at the line terminals of utilization equipment.

nominal utilization voltage. The voltage rating of certain utilization equipment used on the system.

The nominal system voltages contained in Table 17 apply to all parts of the system, both of the supplier and of the user. The ranges are given separately for service voltage and for utilization voltage because they are normally at different locations. It is recognized that the voltage at utilization points is normally somewhat lower than at the service point. In deference to this fact, and the fact that integral horsepower motors, or air-conditioning and refrigeration equipment, or both, may constitute a heavy concentrated load on some circuits, the rated voltages of such equipment and of motors and motor control equipment are usually lower than nominal system voltage. This corresponds to the range of utilization voltages in Table 17. Other utilization equipment is generally rated at nominal system voltage.

3.1.1.2 System Voltage Classes

low voltage. A class of nominal system voltages 1000 V or less.

medium voltage. A class of nominal system voltages greater than 1000 V and less than 100 000 V.

high voltage. A class of nominal system voltages equal to or greater than 100 000 V and equal to or less than 230 000 V.

3.1.2 Standard Nominal System Voltages for the United States. These voltages and their associated tolerance limits are listed in ANSI C84.1-1989 [2] for voltages from 120–230 000 V, and in ANSI C92.2-1987, Power Systems — Alternating-Current Electrical Systems and Equipment Operating at Voltages Above 230 kV Nominal — Preferred Voltage Ratings [3]. The nominal system voltages and their associated tolerance limits and notes in the two standards have been combined in Table 17 to provide a single table, listing all the standard nominal system voltages and their associated tolerance limits for the United States. Preferred nominal system voltages and voltage ranges are shown in boldface type, while other systems in substantial use that are recognized as standard voltages are shown in medium type. Other voltages may be encountered on older systems; but they are not recognized as standard voltages. The transformer connections from which these voltages are derived are shown in Fig 3.

3.1.3 Application of Voltage Classes

(1) Low-voltage class voltages are used to supply utilization equipment.
(2) Medium-voltage class voltages are used as primary distribution voltages to supply distribution transformers that step the medium voltage down to a low voltage to supply utilization equipment. Medium voltages of 13 800 V and below are also used to supply utilization equipment, such as large motors. (See Table 24.)

Table 17
Standard Nominal System Voltages and Voltage Ranges
(Preferred system voltages in boldface type.)

Voltage Class (Note 1)	Nominal System Voltage (Note a) Two-wire	Three-wire	Four-wire	Nominal Utilization Voltage (Note j) Two-wire	Three-wire / Four-wire	Voltage Range A (Note b) Max — Utilization and Service Voltage (Note c)	Range A Min — Service Voltage	Range A Min — Utilization Voltage	Voltage Range B (Note b) Max — Utilization and Service Voltage	Range B Min — Service Voltage	Range B Min — Utilization Voltage
Single-Phase Systems											
Low Voltage (Note 1)	120			115		126	114	110	127	110	106
		120/240			**115/230**	**126/252**	**114/228**	**110/220**	**127/254**	**110/220**	**106/212**
Three-Phase Systems											
			208Y/120 (Note d)	200		**218Y/126**	**197Y/114**	**191Y/110**	**220Y/127**	**191Y/110**	**184Y/106** (Note 2)
			240/120 (Note d)		230/115	**252/126**	**228/114**	**220/110**	**254/127**	**220/110**	**212/106** (Note 2)
		240			230	252	228	220	254	220	212
			480Y/277		460	**504Y/291**	**456Y/263**	**440Y/254**	**508Y/293**	**440Y/254**	**424Y/245**
		480			460	504	456	440	508	440	424
		600 (Note e)			575	630 (Note e)	570	550	635 (Note e)	550	530
Medium Voltage		2 400				2 520	2 340	2 160	2 540	2 280	2 080
			4 160Y/2 400			4 370Y/2 520	4 050Y/2 340	3 740Y/2 160	4 400Y/2 540	3 950Y/2 280	3 600Y/2 080
		4 160				4 370	4 050	3 740	4 400	3 950	3 600
		4 800				5 040	4 680	4 320	5 080	4 560	4 160
		6 900				7 240	6 730	6 210	7 260	6 560	5 940
			8 320Y/4 800			8 730Y/5 040	8 110Y/4 680	(Note f)	8 800Y/5 080	7 900Y/4 560	(Note f)
			12 000Y/6 930			12 600Y/7 270	11 700Y/6 760	(Note f)	12 700Y/7 330	11 400Y/6 580	(Note f)
			12 470Y/7 200			13 090Y/7 560	12 160Y/7 020	(Note f)	13 200Y/7 620	11 850Y/6 840	(Note f)
			13 200Y/7 620			13 860Y/8 000	12 870Y/7 430	(Note f)	13 970Y/8 070	12 504Y/7 240	(Note f)
			13 800Y/7 970			14 490Y/8 370	13 460Y/7 770	(Note f)	14 520Y/8 380	13 110Y/7 570	(Note f)
		13 800				14 490	13 460	12 420	14 520	13 110	11 880
			20 780Y/12 000			21 820Y/12 600	20 260Y/11 700	(Note f)	22 000Y/12 700	19 740Y/11 400	(Note f)
			22 860Y/13 200			24 000Y/13 860	22 290Y/12 870	(Note f)	24 200Y/13 970	21 720Y/12 540	(Note f)
		23 000				24 150	22 430	(Note f)	24 340	21 850	(Note f)
			24 940Y/14 400			26 190Y/15 120	24 320Y/14 040	(Note f)	26 400Y/15 240	23 690Y/13 680	(Note f)
			34 500Y/19 920			36 230Y/20 920	33 640Y/19 420	(Note f)	36 510Y/21 080	32 780Y/18 930	(Note f)
		34 500				36 230	33 640		36 510	32 780	

Voltage Class	Nominal System Voltage (Three-wire)	Maximum Voltage (Note g)
High Voltage	46 000	48 300 } 72 500
	69 000	
	115 000	121 000
	138 000	145 000
	161 000	169 000
	230 000	242 000
Extra-High Voltage	**345 000**	362 000
	500 000	550 000
	765 000	800 000
Ultra-High Voltage	**1 100 000**	1 200 000

NOTES: (1) Minimum utilization voltages for 120–600 volt circuits not supplying lighting loads are as follows:

Nominal System Voltage	Range A	Range B
120	108	104
120/240	108/216	104/208
208Y/120 (Note 2)	187Y/108	180Y/104
240/120	216/108	208/104
240	216	208
480Y/277	432Y/249	416Y/240
480	432	416
600	540	520

(2) Many 220 volt motors were applied on existing 208 volt systems on the assumption that the utilization voltage would not be less than 187 volts. Caution should be exercised in applying the Range B minimum voltages of Table 17 and Note (1) to existing 208 volt systems supplying such motors

Table 17 *(Continued)*

NOTES FOR TABLE 17:

(a) Three-phase, three-wire systems are systems in which only the three-phase conductors are carried out from the source for connection of loads. The source may be derived from any type of three-phase transformer connection, grounded or ungrounded. Three-phase, four-wire systems are systems in which a grounded neutral conductor is also carried out from the source for connection of loads. Four-wire systems in this table are designated by the phase-to-phase voltage, followed by the letter Y (except for the 240/120 V delta system), a slant line, and the phase-to-neutral voltage. Single-phase services and loads may be supplied from either single-phase or three-phase systems. The principal transformer connections that are used to supply single-phase and three-phase systems are illustrated in Fig 3.

(b) The voltage ranges in this table are illustrated in ANSI C84.1-1989, Appendix B [2].

(c) For 120–600 V nominal systems, voltages in this column are maximum service voltages. Maximum utilization voltages would not be expected to exceed 125 V for the nominal system voltage of 120, nor appropriate multiples thereof for other nominal system voltages through 600 V.

(d) A modification of this three-phase, four-wire system is available as a 120/208Y-volt service for single-phase, three-wire, open-wye applications.

(e) Certain kinds of control and protective equipment presently available have a maximum voltage limit of 600 V; the manufacturer or power supplier or both should be consulted to assure proper application.

(f) Utilization equipment does not generally operate directly at these voltages. For equipment supplied through transformers, refer to limits for nominal system voltage of transformer output.

(g) For these systems, Range A and Range B limits are not shown because, where they are used as service voltages, the operating voltage level on the user's system is normally adjusted by means of voltage regulation to suit their requirements.

(h) Standard voltages are reprinted from ANSI C92.2-1987 [3] for convenience only.

(i) Nominal utilization voltages are for low-voltage motors and control. See ANSI C84.1-1989, Appendix C [2] for other equipment nominal utilization voltages (or equipment nameplate voltage ratings).

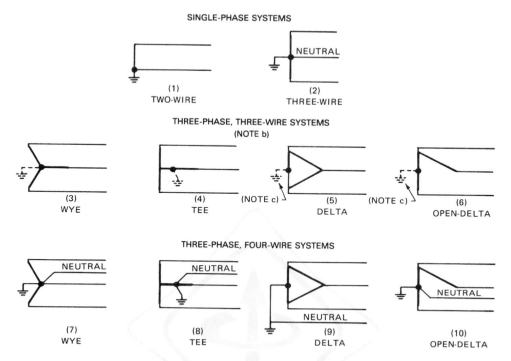

SINGLE-PHASE SYSTEMS

(1) TWO-WIRE

(2) THREE-WIRE — NEUTRAL

THREE-PHASE, THREE-WIRE SYSTEMS
(NOTE b)

(3) WYE

(4) TEE

(NOTE c)

(5) DELTA

(NOTE c)

(6) OPEN-DELTA

THREE-PHASE, FOUR-WIRE SYSTEMS

(7) WYE — NEUTRAL

(8) TEE — NEUTRAL

(9) DELTA — NEUTRAL

(10) OPEN-DELTA — NEUTRAL

NOTES: (a) The above diagrams show connections of transformer secondary windings to supply the nominal system voltages of Table 17. Systems of more than 600 V are normally three-phase and supplied by connections (3), (5) ungrounded, or (7). Systems of 120–600 V may be either single-phase or three-phase, and all the connections shown are used to some extent for some systems in this voltage range.

(b) Three-phase, three-wire systems may be solidly grounded, impedance grounded, or ungrounded, but are not intended to supply loads connected phase-to-neutral (as are the four-wire systems).

(c) In connections (5) and (6), the ground may be connected to the midpoint of one winding as shown (if available), to one phase conductor (*corner* grounded), or omitted entirely (ungrounded).

(d) Single-phase services and loads may be supplied from single-phase or three-phase systems. They are connected phase-to-phase when supplied from three-phase, three-wire systems and either phase-to-phase or phase-to-neutral from three-phase, four-wire systems.

**Fig 3
Principal Transformer Connections
to Supply the System Voltages of Table 17**

(3) High-voltage class voltages are used to transmit large amounts of electric power over transmission lines that interconnect transmission substations. Transmission substations located adjacent to generating stations step the generator voltage up to the transmission voltage. Other transmission substations located in the load area step the transmission voltage down to a primary distribution voltage to supply distribution transformers that step the primary distribution voltage down to a utilization voltage. Transmission lines also interconnect transmission substations to provide alternate paths for power transmission to improve the reliability of the transmission system.

3.1.4 Voltage Systems Outside of the United States. Voltage systems in other countries generally differ from those in the United States; for example, 416Y/240 V is widely used as a utilization voltage even for residential service. Also the frequency in many countries is 50 Hz instead of 60 Hz, which affects the operation of some equipment, such as motors, which will run approximately 17% slower. Plugs and receptacles are generally different, which helps to prevent utilization equipment from the United States from being connected to the wrong voltage.

In general, equipment rated for use in the United States cannot be used outside of the United States, and equipment rated for use outside of the United States cannot be used in the United States. If electrical equipment made for use in the United States must be used outside the United States, information on the voltage, frequency, and type of plug required should be obtained. If the difference is only in the voltage, transformers are generally available to convert the supply voltage to the equipment voltage.

3.1.5 Voltage Standard for Canada. The voltage standard for Canada, CAN3-C235-83, Preferred Voltage Levels for AC Systems, 0 to 50 000 V [6],[35] differs from the United States standard both in the list of standard nominal voltages and in the tolerance limits.

3.2 Voltage Control in Electric Power Systems

3.2.1 Principles of Power Transmission and Distribution on Utility Systems. To understand the principles of voltage control required to provide satisfactory voltage to utilization equipment, a general understanding of the principles of power transmission and distribution in utility systems is necessary since most commercial buildings obtain their electric power requirements from the local electric utility company. Figure 4 is a single-line diagram of a typical utility power generation, transmission, and distribution system.

---

[35] In the U.S., Canadian Standards Association (CSA) Standards are available form the Sales Department, American National Standards Institute (ANSI), 11 West 42nd Street, 13th Floor, New York, NY 10036. In Canada, they are available at the Canadian Standards Association (Standards Sales), 178 Rexdale Boulevard, Rexdale, Ontario, Canada M9W 1R3.

Fig 4
Typical Utility Power Generation,
Transmission, and Distribution System

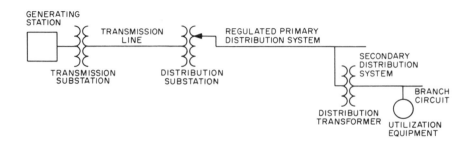

Generating stations are located near convenient sources of fuel and water. The generated power, except for station requirements, is transformed in a transmission substation at the generating station up to a transmission voltage of 69 000 V or higher for transmission to major load areas. Transmission lines are classified as unregulated because the voltage is usually controlled only to keep the lines operating within normal voltage limits and to facilitate power flow. ANSI C84.1-1989 [2] specifies only the nominal and maximum values of voltage for systems over 34 500 V.

Transmission lines supply distribution substations equipped with transformers that step the transmission voltage down to a primary distribution voltage, generally from 4160–34 500 V, although 12 470 V and 13 200 V are in the widest use. However, there is an increasing trend toward the use of 34 500 V for primary distribution as the average load density increases.

Voltage control is applied, when necessary, for the purpose of providing satisfactory voltage to the terminals of utilization equipment. The transformers used to step the transmission voltage down to the primary distribution voltage are generally equipped with tap changing underload equipment, which changes the ratio of the transformer under load in order to maintain the primary distribution voltage within a narrow band regardless of fluctuations in the transmission voltage. Separate step or induction regulators may also be used.

Generally, the regulator controls are equipped with compensators that raise the voltage as the load increases and lower the voltage as the load decreases to compensate for voltage excursions in the primary distribution system. This prevents the voltage from rising to excessive values during light load conditions when the voltage drop along the primary distribution system is low. This is illustrated in Fig 5. Note that buildings close to the distribution substation will receive voltages that average higher than those received by buildings at a distance from the distribution substation. Switched or fixed-shunt capacitors are also used by utility distribution companies to improve the voltage on the primary feeders.

Fig 5
Effect of Regulator Compensation
on Primary Distribution System Voltage

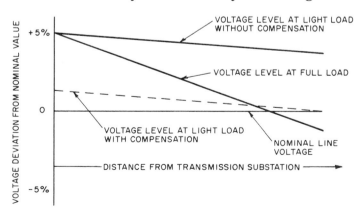

The primary distribution system supplies distribution transformers that step the primary distribution voltage down to utilization voltages, generally in the range of 120–600 V, to supply a secondary distribution system to which the utilization equipment is connected. Small transformers used to step a higher utilization voltage down to a lower utilization voltage, such as 480 V to 280Y/120 V, are considered part of the secondary distribution system.

The supply voltages available to a commercial building depend upon whether the building is supplied by a distribution transformer, the primary distribution system, or the transmission system that, in turn, depends on the electric power requirements of the building.

Small buildings with up to several hundred kilovoltamperes of load and all buildings supplied from secondary networks are supplied from the distribution transformer. The secondary distribution system consists of the connections from the distribution transformer to the building service and the building wiring.

Medium-sized buildings and multiple-building complexes with loads of a few thousand kilovoltamperes are generally connected to the primary distribution system. The building owner provides the section of the primary distribution system within the building, the distribution transformers, and the secondary distribution system.

Large buildings and multiple-building complexes with loads of more than a few thousand kilovoltamperes may be connected to the transmission system. The building owner provides the primary distribution system, the distribution transformers, the secondary distribution system, and may provide the distribution substation.

If power is supplied by local generation at the building, the generators will replace the primary distribution system up to the distribution transformer where generation voltage is over 600 V, and will also replace the distribution transformer where generation is at 600 V or below.

3.2.2 System Voltage Tolerance Limits. Table 17 indicates the range for all standard nominal system voltages of 120–34 500 V for two critical points on the distribution system: the point of delivery by the supplying utility company and the point of connection to utilization equipment. The voltage tolerance limits at the point of delivery by the supplying utility provide the voltage limits within which the supplying utility should maintain the supply voltage in order to provide satisfactory operation of the user's utilization equipment. The voltage tolerance limits at the point of connection of utilization equipment provide the voltage limits within which the utilization equipment manufacturer must design the utilization equipment in order for it to operate satisfactorily.

Table 17 lists two voltage ranges in order to provide a practical application of voltage tolerance limits to distribution systems.

Electric supply systems are to be designed and operated so that most service voltages fall within the Range A limits. User systems are to be designed and operated so that, when the service voltages are within Range A, the utilization voltages are within Range A. Utilization equipment is to be designed and rated to give fully satisfactory performance within the Range A limits for utilization voltages.

Range B is provided to allow limited excursions of voltage outside the Range A limits that necessarily result from practical design and operating conditions. The supplying utility is expected to take action within a reasonable time to restore service voltages to Range A limits. The user is expected to take action within a reasonable time to restore utilization voltages to Range A limits. Insofar as is practical, utilization equipment may be expected to give acceptable performance at voltages outside Range A but within Range B. When voltages occur outside the limits of Range B, prompt corrective action should be taken.

For transmission voltages over 34 500 V, only the maximum voltage is specified because these voltages are normally unregulated, and only a maximum voltage is required to establish the design insulation level for the line and associated apparatus.

The actual voltage measured at any point on the system will vary depending on the location of the point of measurement and the system load at the time the measurement is made. Fixed voltage changes take place in transformers in accordance with the transformer ratio, while voltage variations occur from the operation of voltage control equipment and the changes in voltage drop between the supply source and the point of measurement due to changes in the current flowing in the circuit.

3.2.3 Development of Voltage Tolerance Limits for ANSI C84.1-1989. The voltage tolerance limits in ANSI C84.1-1989 [2] are based on ANSI/NEMA MG1-1978, Motors and Generators (1987 Edition) [4],[36] which establishes the voltage tolerance limits of the standard low-voltage induction motor at $\pm 10\%$ of nameplate voltage ratings of 230 V and 460 V. Since motors represent the major component of utilization equipment, they were given primary consideration in the establishment of this voltage standard.

The best way to show the voltages in a distribution system is by using a 120 V base. This cancels the transformation ratios between systems, so that the actual voltages vary solely on the basis of the voltage drops in the system. Any voltage may be converted to a 120 V base by dividing the actual voltage by the ratio of transformation to the 120 V base. For example, the ratio of transformation for the 480 V system is 480 / 120 or 4, so 460 V in a 480 V system would be 460 / 4 or 115 V.

The tolerance limits of the 460 V motor as they relate to the 120 V base become 115 V + 10% or 126.5 V and 115 V - 10% or 103.5 V. The problem is to decide how this tolerance range of 23 V should be divided between the primary distribution system, the distribution transformer, and the secondary distribution system that make up the regulated distribution system. The solution adopted by American National Standards Committee C84 is shown in Table 18.

The tolerance limits of the standard motor on the 120 V base of 126.5 V maximum and 103.5 V minimum were raised 0.5 - 127 V maximum and 104 V minimum to eliminate the fractional volt. These values became the tolerance limits for Range B in ANSI/NEMA MG1-1978 [4]. An allowance of 13 V was allotted for the voltage drop in the primary distribution system. Deducting this voltage drop from 127 V establishes a minimum of 114 V for utility company services supplied directly from

---

[36] ANSI publications are available from the Sales Department of the American National Standards Institute, 11 West 42nd Street, 13th Floor, New York, NY 10036. NEMA publications are available from the National Electrical Manufacturers Association, 2101 L Street, N.W., Washington, DC 20037.

Table 18
Standard Voltage Profile for a
Regulated Power Distribution System, 120 V Base

	Range A	Range B
Maximum allowable voltage	126(125*)	127
Voltage-drop allowance for the primary distribution feeder	9	13
Minimum primary service voltage	117	114
Voltage-drop allowance for the distribution transformer	3	4
Minimum low-voltage service voltage	114	110
Voltage-drop allowance for the building wiring	6(4†)	6(4†)
Minimum utilization voltage	108(110†)	104(106†)

*For utilization voltages of 120–600 V.
†For building wiring circuits supplying lighting equipment.

the primary distribution system. An allowance of 4 V was provided for the voltage drop in the distribution transformer and the connections to the building's low-voltage wiring. The actual drop will depend on the load, its power factor, and the transformer impedance. Deducting this voltage drop from the minimum distribution voltage of 114 V provides a minimum of 110 V for the utility company supply from 120–600 V. An allowance of 6 V, or 5%, was made for the voltage drop in the building wiring, which is the same as provided in ANSI/NFPA 70-1990, National Electrical Code (NEC), Articles 210-19(a) and 215-2 [5][37] for the maximum voltage drop in the building's low-voltage wiring. This completes the distribution of the 23 V tolerance zone down to the minimum utilization voltage of 104 V on the 120 V base.

The Range A limits for the standard were established by reducing the maximum tolerance limits from 127 V to 126 V and increasing the minimum tolerance limits from 104 V to 108 V. This spread band of 18 V was then allotted: 9 V for the voltage drop in the primary distribution system to provide a minimum primary service voltage of 117 V; 3 V for the voltage drop in the distribution transformer and secondary connections to provide a minimum low-voltage service voltage of 114 V; and 6 V for the voltage drop in the building wiring to provide a minimum utilization voltage of 108 V.

Four additional modifications were made in this basic plan to establish ANSI C84.1-1989 [2]. The maximum utilization voltage in Range A was reduced from 126 V to 125 V for low-voltage systems in the range from 120–600 V because there should be sufficient load on the distribution system to provide at least a 1 V drop on the 120 V base under most operating conditions. This maximum voltage of 125 V is also a practical limit for lighting equipment because the life of the 120 V incandescent lamp is reduced by 42% when operated at 125 V (see Table 26). The voltage-drop allowance of 6 V on the 120 V base for the drop in the building wiring was

---

[37]ANSI publications are available from the Sales Department of the American National Standards Institute, 11 West 42nd Street, 13th Floor, New York, NY 10036. NFPA publications are available from Publications Sales, National Fire Protection Association, 1 Batterymarch Park, P.O. Box 9101, Quincy, MA 02269-9101.

reduced to 4 V for circuits supplying lighting equipment, which raised the minimum voltage limit for utilization equipment to 106 V in Range B and 110 V in Range A because the minimum limits for motors of 104 V in Range B and 108 V in Range A were considered too low for satisfactory operation of lighting equipment. The utilization voltages for the 6900 V and 13 800 V systems in Range B were adjusted to coincide with the tolerance limits of ±10% of the nameplate rating of the 6600 V and 13 200 V motors used on these respective systems.

To convert the 120 V base voltage to equivalent voltages in other systems, the voltage on the 120 V base is multiplied by the ratio of the transformer that would be used to connect the other system to a 120 V system. In general, liquid-filled distribution transformers for systems below 15 000 V have nameplate ratings that are the same as the standard system nominal voltages, and the ratio of the standard nominal voltages may be used to make the conversion (see ANSI C57.12.20-1988, Transformers—Overhead-Type Distribution Transformers, 500 kVA and Smaller: High Voltage, 34 500 V and Below; Low Voltage, 7970/13 800Y V and Below [1]). However, for primary distribution voltages over 15 000 V, the primary nameplate rating of liquid filled distribution transformers is generally not the same as the standard system nominal voltages. Also, distribution transformers may be equipped with taps to change the ratio of transformation. Thus, if the primary distribution voltage is over 15 000 V, or taps have been used to change the transformer ratio, the actual transformer ratio must be used to convert the base voltage to that of another system. Other types of distribution transformers, such as dry-type transformers, have the same voltage ratios as the liquid-filled distribution transformers used by utility companies.

For example, the maximum tolerance limit of 127 V on the 120 V base for the service voltage in Range B is equivalent, on the 4160 V system, to $(4160 / 120) \times 127$ = 4400 V to the nearest 10 V. However, if the 4160 V to 120 V transformer is set on the +2.5% tap, the voltage ratio would be $4160 + (4160 \times 0.025) = 4160 + 104 = 4264$ V to 120 V. The voltage on the primary system equivalent to 127 V on the secondary system would be $(4264 / 120) \times 127 = 35.53 \times 127 = 4510$ V to the nearest 10 V. If the maximum distribution voltage of 4400 V is applied to the 4264 V to 120 V transformer, then the secondary voltage would become $4400 / 4260 \times 120$ = 124 V. So the effect of using a +2.5% tap is to lower the secondary voltage range by 2.5%.

3.2.4 Voltage Profile Limits for a Regulated Distribution System. Figure 6 shows the voltage profile of a regulated power distribution system using the limits of Range A in Table 17. Assuming a standard nominal distribution voltage of 13 200 V, Range A in Table 17, shows that this voltage should be maintained by the supplying utility between a maximum of 126 V and a minimum of 117 V on a 120 V base. Since the base multiplier for converting from the 120 V system to the 13 200 V system is 13 200 / 120 or 110, the actual voltage limits for the 13 200 V system are 110×126 or 13 860 V maximum and 110×117 or 12 870 V minimum.

If a distribution transformer with a ratio of 13 200 V to 480 V is connected to the 13 200 V distribution feeder, Range A of Table 17 requires that the nominal 480 V supply must be maintained by the supplying utility between a maximum of 126 V and a minimum of 114 V on the 120 V base. Since the base multiplier for the 480 V

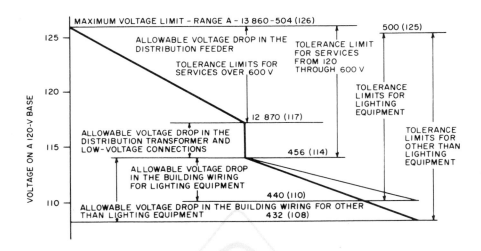

Fig 6
Voltage Profile of the Limits of Range A, ANSI C84.1-1989 [2]

system is 480 / 120 or 4, the actual values are 4 × 126 or 504 V maximum and 4 × 114 or 456 V minimum.

Range A of Table 17 as modified for utilization equipment from 120–480 V provides for a maximum utilization voltage of 125 V and a minimum of 110 V for lighting equipment and 108 V for other than lighting equipment on the 120 V base. Using the base multiplier of 4 for the 480 V system, the maximum utilization voltage would be 4 × 125 or 500 V, and the minimum for other than lighting equipment would be 4 × 108 or 432 V. For lighting equipment connected phase-to-neutral, the maximum voltage would be 500 V divided by $\sqrt{3}$ or 288 V and the minimum voltage would be 4 × 110 or 440 V divided by $\sqrt{3}$ or 254 V.

3.2.5 Nonstandard Nominal System Voltages. Since ANSI C84.1-1989 [2] lists only the standard nominal system voltages in common use in the United States, system voltages will be frequently encountered that differ from the standard list. A few of these may be so widely different as to constitute separate systems that are in too limited use to be considered standard. However, in most cases, the nominal system voltages will differ by only a few percentage points, as shown in Table 19. A closer examination of this table shows that these differences are due mainly to the fact that some voltages are multiples of 110 V, others are multiples of 115 V, and a few are multiples of 120 V.

The reasons for these differences go back to the original development of electric power distribution systems. The first utilization voltage was 100 V. However, the supply voltage had to be raised to 110 V in order to compensate for the voltage drop in the distribution system. This led to overvoltage on equipment connected close to the supply, and the utilization equipment rating was also raised to 110 V. As generator sizes increased and distribution and transmission systems developed, an effort to keep transformer ratios in round numbers led to a series of utilization

Table 19
Nominal System Voltages

Standard Nominal System Voltages	Associated Nonstandard Nominal System Voltages
Low Voltage	
120	110, 115, 125
120/240*	110/220, 115/230, 125/250
208Y/120*	216Y/125
240/120*	
240	230, 250
480Y/277*	416Y/240, 460Y/265
480*	440, 460
600	550, 575
Medium Voltage	
2400	2200, 2300
4160Y/2400	
4160*	4000
4800	4600
6900	6600, 7200
8320Y/4800	
12 000Y/6930	11 000, 11 500
12 470Y/7200*	
13 200Y/7620*	
13 800Y/7970	
13 800*	14 400
20 780Y/12 000	
22 860Y/13 200	
23 000	
24 940Y/14 400*	
34 500Y/19 920*	
34 500	33 000

*Preferred standard nominal system voltages.

voltages of 110 V, 220 V, 440 V, and 550 V, and a series of distribution voltages of 2200 V, 4400 V, 6600 V, and 13 200 V.

As a result of the effort to maintain the suppy voltage slightly above the utilization voltage, the supply voltages were raised again to multiples of 115 V, which resulted in a new series of utilization voltages of 115 V, 230 V, 460 V, and 575 V, and a new series of distribution voltages of 2300 V, 4600 V, 6900 V, and 13 800 V.

As a result of the development of the 208Y/120 V network system, the supply voltages were rasied again to multiples of 120 V. This resulted in a new series of utilization voltages of 120 V, 208Y/120 V, 240 V, 480 V, and 600 V, and a new series of primary distribution voltages of 2400 V, 4160Y/2400 V, 4800 V, 12 000 V, and 12 470Y/7200 V. However, most of the existing primary distribution voltages continued to be used, and no 120 V multiple voltages developed at the transmission level.

In the case of low-voltage systems, the associated nominal system voltages in the right column of Table 19 are obsolete and should not be used. Manufacturers are encouraged to design utilization equipment to provide acceptable performance within the utilization voltage tolerance limits specified in the standard, when possible. Some numbers listed in the right column of Table 19 are used in equipment ratings; but these should not be confused with the numbers designating the nominal system voltage on which the equipment is designed to operate (see 3.4).

In the case of medium voltages, the numbers in the right column of Table 19 may designate an older system in which the voltage tolerance limits are maintained at a different level than the standard nominal system voltage, and special consideration must be given to the distribution transformer ratios, taps, and tap settings (see 3.2.7).

3.2.6 System Voltage Nomenclature. The nominal system voltages in Table 17 are designated in the same way as the designation on the nameplate of the transformer for the winding or windings supplying the system.

(1) Single-Phase Systems

120 V — Indicates a single-phase, two-wire system in which the nominal voltage between the two wires is 120 V.

120/240 V — Indicates a single-phase, three-wire system in which the nominal voltage between the two phase conductors is 240 V, and from each phase conductor to the neutral is 120 V.

(2) Three-Phase Systems

240/120 V — Indicates a three-phase, four-wire system supplied from a delta connected transformer. The midtap of one winding is connected to a neutral. The three phase conductors provide a nominal 240 V three-phase, three-wire system, and the neutral and two adjacent phase conductors provide a nominal 120/240 V single-phase, three-wire system.

Single number — Indicates a three-phase, three-wire system in which the number designates the nominal voltage between phases.

Two numbers separated by Y/ — Indicates a three-phase, four-wire system from a wye-connected transformer in which the first number indicates the nominal phase-to-phase voltage and the second the nominal phase-to-neutral voltage.

NOTES: (1) All single-phase systems and all three-phase, four-wire systems are suitable for the connection of phase-to-neutral load.
(2) See Chapter 4 for methods of system grounding.
(3) See Fig 3 for transformer connections.

3.2.7 Use of Distribution Transformer Taps to Shift the Utilization Voltage Spread Band. Except for small sizes, distribution transformers are normally provided with five taps on the primary winding, generally two at 2.5% above and below rated voltage and one at rated voltage. These taps permit the transformer ratio to be changed to raise or lower the secondary voltage spread band to provide a closer fit to the tolerance limits of the utilization equipment. There are two general situations that require the use of taps.

(1) Where the primary distribution system voltage spread band is above or below the limits required to provide a satisfactory secondary voltage spread band. This occurs under two conditions:

(a) When the primary voltage has a slightly different nominal value than the transformer primary nameplate rating. For example, if a 13 200 V

—480 V transformer is connected to a nominal 13 800 V system, the nominal secondary voltage would be 13 800/13 200 × 480 = 502 V. However, if the 13 800 V system was connected to the +5% tap of the 13 200 V—480 V transformer at 13 860 V, the secondary voltage would be (13 800 / 13 860) × 480 = 478 V, which is practically the same as would be obtained from a transformer having the proper ratio of 13 800 V—480 V.

(b) When the primary voltage spread is in the upper or lower part of the tolerance limits provided in ANSI C84.1-1989 [2]. For example, a 13 200 V—480 V transformer is connected to a 13 200 V primary distribution system close to the distribution substation so that the primary voltage spread band falls in the upper half of the tolerance zone for Range A in the standard or 13 200–13 860 V. This would result in a nominal secondary voltage under no-load conditions of 480–504 V. By setting the transformer on the +2.5% tap at 13 530 V, the secondary voltage would be lowered 2.5% to a range of 468–492 V. This would significantly reduce the overvoltage on the utilization equipment.

(2) By adjusting the utilization voltage spread band to provide a closer fit to the tolerance limits of the utilization equipment. For example, Table 20 shows the shift in the utilization voltage spread band for the +2.5% and +5% taps as compared to the utilization voltage tolerance limits for Range A of ANSI C84.1-1989 [2] for the 480 V system. Table 21 shows the voltage tolerance limits of the old standard 440 V and the new standard 460 V three-phase induction motors. Table 22 shows the tolerance limits for the old standard 265 V and the new standard 277 V fluorescent lamp ballasts. A study of these three tables shows that the normal (100%) tap setting will provide the best fit with the tolerance limits of the 460 V motor and the 277 V ballast; but a setting on the +5% tap will provide the best fit for the 440 V motor and the 265 V ballast. For older buildings that have appreciable numbers of both ratings of motors and ballasts, a setting on the +2.5% tap may provide the best compromise.

Note that these examples assume that the tolerance limits of the supply and utilization voltages are within the tolerance limits specified in ANSI C84.1-1989 [2]. However, this may not always be true. In this case, the actual voltages should be measured with a recording voltmeter for a 7 day period to obtain readings during the night and over weekends when maximum voltages occur. These actual voltages can then be compared with the data in Tables 20–22 to check the proposed transformer ratios and tap settings.

**Table 20
Tolerance Limits from Table 17, Range A in Volts**

Nominal System Voltage	Transformer Tap Voltage	Minimum Utilization Voltage	Minimum Service Voltage	Maximum Utilization and Service Voltage
480Y/277 468Y/270 457Y/264	Normal +2.5% +5%	440Y/254 429Y/248 419Y/242	454Y/262 443Y/256 432Y/250	500Y/289 488Y/282 476Y/275

Table 21
Tolerance Limits for Standard Three-Phase Induction Motors in Volts

Motor Rating	10% Minus	10% Plus
460	414	506
440	396	484

Table 22
Tolerance Limits for Standard Fluorescent Lamp Ballasts in Volts

Ballast Rating	10% Minus	10% Plus
277	249	305
265	238	292

Where a building has not yet been built so that actual voltages can be measured, the supplying utility company should be requested to provide the expected spread band for the supply voltage, preferably supported by a 7 day graphic chart from the nearest available location. When the building owner is to furnish the distribution transformers, recommendations should also be obtained from the supplying utility company on the transformer ratios, taps, and tap settings. With this information, a voltage profile can be prepared to check the expected voltage spread at the utilization equipment.

In general, distribution transformers should have the same primary voltage rating as the nominal voltage of the primary distribution system. Taps should be provided at +2.5% and +5% and –2.5% and –5% to provide adjustment in either direction.

Note that the voltage spread at the secondary terminals of the transformer is equal to the voltage spread at the primary terminals plus the voltage drop in the transformer. Taps only serve to move the secondary voltage up or down in the steps of the taps. They cannot correct for excessive spread in the supply voltage or excessive drop in the building's distribution system. Therefore, if the voltage spread at the utilization equipment exceeds the tolerance limits of the equipment, action must be taken to improve voltage conditions (see 3.7).

3.3 Voltage Selection

3.3.1 Selection of Utilization Voltage of 600 V and Below. Generally, the preferred utilization voltage for large commercial buildings is 480Y/277 V. The three-phase power load is connected directly to the system at 480 V, and fluorescent ceiling lighting is connected phase-to-neutral at 277 V. Small dry-type transformers rated 480 V—208Y/120 V are used to provide 120 V for convenience outlets and 208 V, single-phase and three-phase for office machinery.

Single-phase transformers with secondary ratings of 120/240 V may also be used to supply lighting and small office equipment. However, single-phase transformers should be connected in sequence on the primary phases to maintain balanced load on the primary system.

Where the supplying utility furnishes the distribution transformers, the choice of voltages will be limited to those the utility will provide. Most utilities provide all of the voltages listed in the standard from 120–480 V, although all may not be available at any one location.

The built-up downtown areas of many large cities are supplied from low-voltage networks (see Chapter 4). Originally, the only voltage available was 208Y/120 V. Now, most utilities will provide spot-network installation at 480Y/277 V for large buildings. For tall buildings, space will be required on upper floors for transformer installations and the primary distribution cables supplying the transformers.

Apartment buildings generally have the option of using either 208Y/120 V, three-phase, four-wire systems, or 120/240 V, single-phase systems, since the major load in residential occupancies consists of 120 V lighting fixtures and appliances. The 208Y/120 V systems should be more economical for large apartment buildings, and 120/240 V systems should be satisfactory for small apartment buildings. However, large single-phase appliances, such as electric ranges, which are rated for use on 120/240 V single-phase systems cannot be used efficiently on 208Y/120 V, three-phase, four-wire systems because the line-to-line voltage is appreciably below the rated voltage of the appliance. Where central air conditioning is provided, a large motor is required to drive the refrigeration compressor (see 3.10 on the effects of starting large motors).

3.3.2 Utility Service Supplied from a Primary Distribution System. When a commercial building or commercial complex becomes too large to be supplied at utilization voltage from a single distribution transformer installation, the utility company's primary distribution line should be tapped to supply distribution transformer installations. These transformers may be dry-type, solid-cast and/or resin-encapsulated, or nonflammable liquid-filled types located inside the building. These transformers may also be liquid-filled (mineral oil, silicone, or high flashpoint hydrocarbon) transformers located outside the building or in transformer vaults (see IEEE Std C57.12.00-1987, IEEE Standard General Requirements for Liquid-Immersed Distribution, Power, and Regulating Transformers (ANSI) [7],[38] IEEE C57.12.01-1989, IEEE Standard General Requirements for Dry-Type Distribution and Power Transformers Including Those with Solid Cast and/or Resin-Encapsulated Windings (ANSI) [8], and IEEE C57.94-1982 (Reaff. 1987), IEEE Recommended Practice for Installation, Application, Operation, and Maintenance of Dry-Type General Purpose Distribution and Power Transformers (ANSI) [9]). Refer to Chapter 7 of this book and the NEC, Article 450, Part B [5] for different types and specific provisions.

Whatever primary distribution voltage the supplying utility provides in the area should be accepted. The utility company's primary distribution voltages that are in the widest use are 12 470Y/7200 V and 13 200Y/7620 V. Some utilities may provide a transformation to a lower distribution voltage, such as 4160Y/2400 V, which may

---

[38] IEEE publications are available from the Institute of Electrical and Electronics Engineers, IEEE Service Center, 445 Hoes Lane, P.O. Box 1331, Piscataway, NJ 08855-1331.

be more economical for the building's wiring system. However, the voltage drop in the additional transformation will increase the voltage spread at the utilization equipment. In this case, the voltage spread band should be checked to make sure that it is within satisfactory limits.

Special consideration should always be given when starting larger motors to minimize the voltage dip so as not to affect the operation of other utilization equipment on the system supplying the motor (see 3.9). Larger motors, generally over 150 hp, may be supplied at medium voltage, such as 2400 V or 4160 V from a separate transformer, to eliminate the voltage dip on the low-voltage system. However, these motors and control may be more expensive, and consideration should be given to the fact that the maintenance electricians in commercial buildings may not be qualified to maintain medium-voltage equipment. A contract with a qualified electrical firm may be required for maintenance.

Standard voltages and preferred horsepower limits for polyphase induction motors that are likely to be used for air conditioning are shown in Table 23.

NOTE: The data used in Table 23 were taken from Reference [19], Table 18-5.

In recent years, many utilities have begun leasing transmission voltages in the range of 15 000–35 000 V for distribution circuits. However, equipment costs in the range of 25 000–35 000 V are quite high, and a transformation down to a lower voltage may prove to be the most economical. Note that Table 18 provides for only one transformation between the primary and secondary distribution voltages, so that a voltage profile taking into account both transformations should be prepared to make sure that the voltage at the utilization equipment will fall within acceptable limits.

3.3.3 Utility Service Supplied from Transmission Lines. Normally, commercial buildings and building complexes are not supplied directly from utility transmission lines. When details for these supply voltages are required, see IEEE Std 141-1986, IEEE Recommended Practice for Electric Power Distribution for Industrial Plants (ANSI) [11].

3.3.4 Utility Policy for Supplying Tenants in Commercial Buildings. Some utilities have a policy of providing all or part of the building wiring up to the point

Table 23
Low-Voltage Motors

Motor Nameplate Voltage	Preferred Horsepower Limits
115	No minimum—15 hp maximum
200	No minimum—200 hp maximum
460 and 575	1 hp minimum—1000 hp maximum
	Medium-Voltage Motors
2300	50 hp minimum—6000 hp maximum
4000	100 hp minimum—7500 hp maximum
4500	250 hp minimum—No maximum
6600	400 hp minimum—No maximum

of connection with individual tenants of commercial buildings in return for the right to provide utilization voltage directly to each tenant. When a commercial building is to be rented to more than one tenant, the local utility should be contacted to determine if they wish to supply electric service directly to the tenants. Such an arrangement will not only save the building owner the cost of the building feeders, but also either the cost of the submetering and billing or the problems involved in dividing the electric bill among the tenants and the building owner, or including it in the rent.

The cost of electricity to tenants billed individually is higher than the pro-rata share of a common bill because of the charges in the lower steps of the rate strucure to cover metering, meter reading, and billing. However, individual billing encourages conservation on the part of the tenant.

3.4 Voltage Ratings for Utilization Equipment. Utilization equipment is defined as "electrical equipment that converts electric power into some other form of energy, such as light, heat, or mechanical motion." Every item of utilization equipment should have a nameplate listing, which includes, among other things, the rated voltage for which the equipment is designed. With one major exception, most utilization equipment carries a nameplate rating that is the same as the voltage system on which it is to be used, that is, equipment to be used on 120 V systems is rated 120 V; for 208 V systems, 208 V; for 240 V systems, 240 V; for 480 V systems, 480 V; and for 600 V systems, 600 V; and so on. The major exception is motors and equipment containing motors. See Table 24 for proper selection of the motor nameplate voltage that is compatible with the specific available nominal system voltage. Motors are also about the only utilization equipment used on systems over 600 V.

Prior to the late '60s, low-voltage three-phase motors were rated 220 V for use on both 208 V and 240 V systems, 440 V for use on 480 V systems, and 550 V for use on 600 V systems. The reason for this was that most three-phase motors were used in

Table 24
Voltage Ratings of Standard Motors

Nominal System Voltage	Nameplate Voltage
Single-phase motors	
120	115
240	230
Three-phase motors	
208	200
240	230
480	460
600	575
2400	2300
4160	4000
4800	4600
6900	6600
13 800	13 200

large industrial plants where relatively long circuits resulted in voltages considerably below nominal at the ends of the circuits. Also, utility supply systems had limited capacity, and low voltages were common during heavy load periods. As a result, the average voltage applied to three-phase motors approximated the 220 V, 440 V, and 550 V nameplate ratings.

In recent years, the supplying utilities have made extensive changes to higher distribution voltages. Increased load densities have resulted in shorter primary distribution systems. Distribution transformers have been moved inside buildings to be closer to the load. Lower impedance wiring systems have been used in the secondary distribution system. Capacitors have been used to improve power factors and reduce voltage drop. All of these changes have contributed to reducing the voltage drop in the distribution system and raising the average voltage applied to utilization equipment. By the mid '60s, surveys indicated that the average voltage supplied to motors on 240 V and 480 V systems was 230 V and 460 V, respectively, and there was an increasing number of complaints about overvoltage on motors.

At about the same time, the Motor and Generator Committee of the National Electrical Manufacturers Association decided that improvements in motor design and insulation systems would allow a reduction of two frame sizes in the standard low-voltage three-phase induction motor. As a part of this re-rate program, the nameplate voltage of the low-voltage motors was increased from 220 V, 440 V, and 550 V to 230 V, 460 V, and 575 V, respectively. Subsequently, a motor rated 200 V for use in 208 V systems was added to the program. Table 24 shows the present voltage ratings of standard motors as specified in ANSI/NEMA MG1-1978 [4].

The difference between the nameplate rating of utilization equipment and the system nominal voltage sometimes causes confusion. A recurring request is to make two voltage ratings identical. However, the difference in voltage ratings is necessary because the performance guarantee for utilization equipment is based on the nameplate rating and not on the system nominal voltage. For utilization equipment, such as motors where the performance peaks in the middle of the tolerance range of the equipment, better performance can be obtained over the tolerance range specified in ANSI C84.1-1989 [2] by selecting a nameplate rating closer to the middle of this tolerance range.

3.5 Effect of Voltage Variation on Utilization Equipment. Whenever the voltage at the terminals of utilization equipment varies from its nameplate rating, the performance of the equipment and its life expectancy changes. The effect may be minor or serious, depending on the characteristics of the equipment and the amount of voltage deviation from the nameplate rating. NEMA Standards provide tolerance limits within which performance will normally be acceptable. In precise operations, however, closer voltage control may be required. In general, a change in the applied voltage causes a proportional change in the current. Since the effect on the load equipment is proportional to the product of the voltage and the current and since the current is proportional to the voltage, the total effect is approximately proportional to the square of the voltage.

However, the change is only approximately proportional and not exact because the change in the current affects the operation of the equipment, so the current will continue to change until a new equilibrium position is established. For example, when the load is a resistance heater, the increase in current will increase the temperature of the heater, which will increase its resistance which will in turn reduce the current. This effect will continue until a new equilibrium current and temperature are established. In the case of an induction motor, a reduction in the voltage will cause a reduction in the current flowing to the motor, causing the motor to start to slow down. This reduces the impedance of the motor, causing an increase in the current until a new equilibrium position is established between the current and the motor speed.

3.5.1 Induction Motors. The variations in characteristics as a function of voltage are given in Table 25.

The most significant effects of low voltage are a reduction in starting torque and an increased full-load temperature rise. The most significant effects of high voltage are increased torque, increased starting current, and decreased power factor. The increased starting torque will increase the accelerating forces on couplings and driven equipment. An increased starting current causes greater voltage drop in the supply circuit and increases the voltage dip on lamps and other equipment. In general, voltages slightly above nameplate rating have less detrimental effect on motor performance than voltages slightly below nameplate rating.

Table 25
General Effect of Voltage Variations on Induction Motor Characteristics

Characteristic	Function of Voltage	Voltage Variation 90% Voltage	Voltage Variation 110% Voltage
Starting and maximum running torque	$(\text{Voltage})^2$	Decrease 19%	Increase 21%
Synchronous speed	Constant	No change	No change
Percent slip	$1/(\text{Voltage})^2$	Increase 23%	Decrease 17%
Full-load speed	Synchronous speed-slip	Decrease 1.5%	Increase 1%
Efficiency			
Full load	—	Decrease 2%	Increase 0.5 to 1%
¾ load	—	Practically no change	Practically no change
½ load	—	Increase 1 to 2%	Decrease 1 to 2%
Power factor			
Full load	—	Increase 1%	Decrease 3%
¾ load	—	Increase 2 to 3%	Decrease 4%
½ load	—	Increase 4 to 5%	Decrease 5 to 6%
Full-load current	—	Increase 11%	Decrease 7%
Starting current	Voltage	Decrease 10 to 12%	Increase 10 to 12%
Temperature rise, full load	—	Increase 6 to 7 °C	Decrease 1 to 2 °C
Maximum overload capacity	$(\text{Voltage})^2$	Decrease 19%	Increase 21%
Magnetic noise - no load in particular	—	Decrease slightly	Increase slightly

3.5.2 Synchronous Motors. Synchronous motors are affected in the same way as induction motors, except that the speed remains constant, unless the frequency changes, and the maximum or pull-out torque varies directly with the voltage if the field voltage remains constant (e.g., the field is supplied by a generator on the same shaft with the motor). If the field voltage varies with the line voltage, as often occurs with a static rectifier source, then the pull-out torque varies as the square of the voltage.

3.5.3 Incandescent Lamps. The light output and life of incandescent filament lamps are critically affected by the impressed voltage. The variation of life and light output with voltage is given in Table 26. The variation figures for 125 V and 130 V lamps are also included because these ratings are useful in locations where long life is more important than light output.

3.5.4 Fluorescent Lamps. Fluorescent lamps, unlike incandescent lamps, operate satisfactorily over a range of $\pm 10\%$ of the ballast nameplate voltage rating. Light output varies approximately in direct proportion to the applied voltage. Thus, a 1% increase in applied voltage will increase the light output by 1% and, conversely, a decrease of 1% in the applied voltage will reduce the light output by 1%. The life of fluorescent lamps is affected less by voltage variation than the life of incandescent lamps.

The voltage-sensitive component of the fluorescent fixture is the ballast, which is a small reactor or transformer that supplies the starting and operating voltages to the lamp and limits the lamp current to design values. These ballasts may overheat when subjected to above normal voltage and operating temperature, and ballasts with integral thermal protection may be required. See the NEC, Article 410-73(e) [5].

3.5.5 High-Intensity Discharge Lamps (Mercury, Sodium, and Metal Halide). Mercury lamps that use the conventional unregulated ballast will have a 30% decrease in light output for a 10% decrease in terminal voltage. When a constant wattage ballast is used, the decrease in light output for a 10% decrease in terminal voltage will be about 2%.

Mercury lamps require 4–8 minutes to vaporize the mercury in the lamp and reach full brillance. At about 20% undervoltage, the mercury arc will be extinguished and the lamp cannot be restarted until the mercury condenses, which

Table 26
Effect of Voltage Variations on Incandescent Lamps

Applied Voltage (volts)	120 V Percent Life	120 V Percent Light	Lamp Rating 125 V Percent Life	125 V Percent Light	130 V Percent Life	130 V Percent Light
105	575	64	880	55	—	—
110	310	74	525	65	880	57
115	175	87	295	76	500	66
120	100	100	170	88	280	76
125	58	118	100	100	165	88
130	34	132	59	113	100	100

takes from 4–8 minutes, unless the lamps have special cooling controls. The lamp life is related inversely to the number of starts; so that, if low-voltage conditions require repeated starting, lamp life will be affected adversely. Excessively high voltage raises the arc temperature, which could damage the glass enclosure when the temperature approaches the glass softening point.

Sodium and metal-halide lamps have similar characteristics to mercury lamps; although the starting and operating voltages may be somewhat different. See the manufacturers' catalogs for detailed information.

3.5.6 Infrared Heating Processes. Although the filaments in the lamps used in these installations are resistance-type, the energy output does not vary with the square of the voltage because the resistance varies at the same time. The energy output does vary roughly as some power of the voltage, however, slightly less than the square. Voltage variations can produce unwanted changes in the process heat available unless thermostatic control or other regulating means are used.

3.5.7 Resistance Heating Devices. The energy input and, therefore, the heat output of resistance heaters varies approximately as the square of the impressed voltage. Thus, a 10% drop in voltage will cause a drop of approximately 19% in heat output. This, however, holds true only for an operating range over which the resistance remains approximately constant.

3.5.8 Electron Tubes. The current-carrying ability or emission of all electron tubes is affected seriously by voltage deviation from nameplate rating. The cathode life curve indicates that the life of the cathode is reduced by half for each 5% increase in cathode voltage. This is due to the reduced life of the heater element and to the higher rate of evaporation of the active material from the surface of the cathode. It is extremely important that the cathode voltage be kept near nameplate rating on electron tubes for satisfactory service. In many cases, this will necessitate a regulated power source. This may be located at or within the equipment, and often consists of a regulating transformer having constant output voltage and limited current.

3.5.9 Capacitors. The reactive power input of capacitors varies with the square of the impressed voltage. A drop of 10% in the supply voltage, therefore, reduces the reactive power by 19%. When users make a sizable investment in capacitors for power factor improvement, they lose a large part of the benefit of this investment.

3.5.10 Solenoid Operated Devices. The pull of ac solenoids varies approximately as the square of the voltage. In general, solenoids are designed liberally and operate satisfactorily on 10% overvoltage and 15% undervoltage.

3.5.11 Solid-State Equipment. Silicon-controlled rectifiers, transistors, etc., have no thermionic heaters, and thus are not nearly as sensitive to long-time voltage variation as the electron tube components they are replacing. Sensitive equipment is frequently provided with internal voltage regulators, so that it is independent of supply system regulation. Sensitive equipment and power solid-state equipment are, however, generally limited regarding peak-inverse voltage. They can, therefore, be adversely affected by abnormal voltages of even microsecond duration in the reverse direction. An individual study of the voltage capabilities of the equipment, including surge characteristics, is necessary to determine if abnormal voltage will result in a malfunction.

3.6 Calculation of Voltage Drops. Building wiring designers should have a working knowledge of voltage-drop calculations, not only to meet NEC, Articles 210-19(a) and 215-2 [5] requirements, but also to ensure that the voltage applied to utilization equipment is maintained within proper limits. Due to the vector relationships between voltage, current, resistance, and reactance, voltage-drop calculations require a working knowledge of trigonometry, especially for making exact computations. Fortunately, most voltage-drop calculations are based on assumed limiting conditions, and approximate formulas are adequate.

3.6.1 General Mathematical Formulas. The vector relationships between the voltage at the beginning of a circuit, the voltage drop in the circuit, and the voltage at the end of the circuit are shown in Fig 7. The approximate formula for the voltage drop is

$$V = IR \cos \theta + IX \sin \theta \qquad \text{(Eq 1)}$$

where

V = Voltage drop in circuit, line-to-neutral.
I = Current flowing in conductor.
R = Line resistance for one conductor, Ω.
X = Line reactance for one conductor, Ω.
θ = Angle whose cosine is the load power factor.
$\cos \theta$ = Load power factor, in decimals.
$\sin \theta$ = Load reactive factor, in decimals.

The voltage drop V obtained from this formula is the voltage drop in one conductor, one way, commonly called the "line-to-neutral voltage drop." For balanced three-phase systems, the line-to-line voltage drop is computed by multiplying the line-to-neutral voltage drop by the following constants:

Voltage System	Multiply by
Single-phase	2
Three-phase	1.732

Fig 7
Vector Diagram of Voltage Relations for Voltage-Drop Calculations

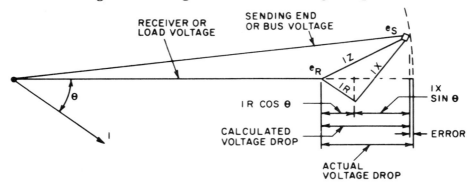

In using the voltage-drop formula, the line current I is generally the maximum or assumed load current, or the current-carrying capacity of the conductor.

The resistance R is the ac resistance of the particular conductor used, considering the particular type of raceway in which it is installed. It depends on the size of the conductor (measured in U.S. wire gauge [AWG] for smaller conductors and in thousands of circular mils [kcmil] for larger conductors), the type of conductor (copper or aluminum), the temperature of the conductor (normally 60 °C [140 °F] for average loading and 75 °C [167 °F] or 90 °C [194 °F], depending on the conductor rating, for maximum loading), and whether the conductor is installed in a magnetic (steel) or nonmagnetic (aluminum or nonmetallic) raceway.

The reactance X also depends on the size and material of the conductor, whether the raceway is magnetic or nonmagnetic, and on the spacing between the conductors of the circuit. The spacing is fixed for multiconductor cable, but may vary with single-conductor cables so that an average value must be used. Reactance occurs because the alternating current flowing in the conductor causes a magnetic field to build up and collapse around each conductor in synchronism with the alternating current. This magnetic field cuts across the conductor itself and the other conductors of the circuit, causing a voltage to be induced into each in the same way that current flowing in the primary of a transformer induces a voltage in the secondary of the transformer. Since the induced voltage is proportional to the rate of change of the magnetic field, which is maximum when the current is passing through zero, the induced voltage will be at maximum when the current is passing through zero or, in vector terminology, the voltage wave is 90° out of phase with the current wave.

θ is the angle between the load voltage and the load current. Cos θ is the power factor of the load expressed as a decimal and may be used directly in the computation of $IR \cos \theta$. Sin θ and cos θ can be obtained from a trigonometric table or calculator.

$IR \cos \theta$ is the resistive component of the voltage drop and is in phase or in the same direction as the current. $IX \sin \theta$ is the reactive component of the voltage drop and is 90° out of phase, or displaced from the current. Sin θ is positive when the current lags the voltage (lagging power factor) and negative when the current leads the voltage (leading power factor).

The approximate calculation of the voltage at the receiving end as shown in Fig 7 is

$$e_R = e_s - V = e_s - (IR \cos \theta + IX \sin \theta) \tag{Eq 2}$$

For exact calculations, the following formula may be used:

$$e_R = \sqrt{e_s^2 - (IX \cos \theta - IR \sin \theta)^2} - (IR \cos \theta + IX \sin \theta)$$

3.6.2 Cable Voltage-Drop Tables. Voltage-drop tables and charts are sufficiently accurate to determine the approximate voltage drop for most problems. Table 27 contains four sections, which give the balanced three-phase, line-to-line voltage drop per 10 000 A × feet for both copper and aluminum conductors in both magnetic and nonmagnetic conduits. The figures are for single-conductor cables operating at 60 °C (140 °F). However, the figures are reasonably accurate up to a conductor temperature of 75 °C (167 °F) and for multiple-conductor cable. In most commercial buildings, the voltage drops in the high-voltage primary distribution system will be insignificant in comparison with the low-voltage system voltage drops. However, the table may be used to obtain approximate values. For border-line cases, the exact values obtained from the manufacturer for the particular cable should be used. The resistance is the same for the same wire size, regardless of the voltage; but the thickness of the insulation is increased at the higher voltages, which increases the conductor spacing, which results in increased reactance, which causes increased errors at the lower power factors. For the same reason, the table cannot be used for open-wire or other installations, such as trays where there is appreciable spacing between the individual phase conductors.

In using the table, the normal procedure is to look up the voltage drop for 10 000 A × feet and multiply this value by the ratio of the actual number of ampere-feet to 10 000. Note that the distance in feet is the distance from the source to the load.

Example 1
500 kcmil copper conductor in steel (magnetic) conduit
Circuit length—200 feet
Load—300 A at 80% power factor
What is the voltage drop?
Using Table 27, Section 1, the intersection between 500 kcmil and the power factor gives a voltage drop of 0.85 V for 10 000 A × feet:

200 feet × 300 A = 60 000 A × feet

$$(60\,000/10\,000) \times 0.85$$

$$= 6 \times 0.85$$

$$= 5.1 \text{ V drop}$$

Voltage drop, phase-to-neutral = 0.577×5.1

$$= 2.9 \text{ V}$$

Example 2
Size No. 12 AWG aluminum conductor in aluminum (nonmagnetic) conduit
Circuit length—200 feet
Load—10 A at 70% power factor
What is the voltage drop?
Using Table 27, Section 4, the intersection between No. 12 AWG aluminum conductor and 0.70 power factor is 37 V for 10 000 A × feet:

200 feet × 10 A = 2000 A × feet

Voltage drop   = $(2000/10\,000) \times 37$

$$= 0.2 \times 37 = 7.4 \text{ V}$$

Example 3

Determine the wire size in Example 2 to limit the voltage drop to 3 V.

Voltage drop in 10 000 A × feet would be (10 000/2000) × 3 = 5 × 3 = 15 V

Using Table 27, Section 4, move along the 0.70 power factor line to find the voltage drop not greater than 15 V. Size No. 8 AWG aluminum conductor has a drop of 15 V for 10 000 A × feet; so it is the smallest aluminum conductor in aluminum conduit that could be used to carry 10 A for 200 feet of circuit length with a voltage drop of not over 3 V, line-to-line.

3.6.3 Busway Voltage-Drop Charts and Tables. Tables 28 and 29 and Figs 8–10 show voltage drops per 100 feet at rated current (end loading) for the entire range of lagging power factors. The actual voltage drop for a balanced three-phase system at a given load power factor equals

$$(\text{rated load voltage drop}) \times \left[\frac{(\text{actual load}) \times (\text{actual length})}{(\text{rated load}) \times 100 \text{ feet}} \right] \qquad (\text{Eq 3})$$

The voltage drop for a single-phase load connected to a three-phase system busway is 15.5% higher than the values shown in the tables. For a two-pole busway serving a single-phase load, the voltage-drop values in Tables 28 and 29 should be multiplied by 1.08.

The previous discussion concerning uniformly distributed loading and concentrated load, of course, applies to a busway. Since plug-in types of busways are particularly adapted to serving the distributed blocks of load, care should be exercised to ensure proper handling of such voltage-drop calculations. Thus, with uniformly distributed loading, the values in the tables should be divided by 2. When several separate blocks of load are tapped off the run at various points, the voltage drop should be determined for the first section using total load. The voltage drop in the next section is then calculated using the total load minus what was tapped off at the first section, etc.

Figure 11 shows the voltage drop curve versus power factor for typical light-duty trolley busway carrying rated load.

Example Using Fig 9: Find the line-to-line voltage drop on a 300 foot run of 800 A plug-in type busway with rated load at 80% power factor.

Solution: Enter the chart at 80% on the horizontal scale. Follow a vertical line to its intersection with the curve for 800 A and extend a line horizontally to its intersection with the vertical scale.

This intersection gives the voltage drop per 100 feet. Multiply this value by 3 to find the voltage drop for 300 feet:

Line-to-line voltage drop = 4.5 × 3

= 13.5 V

Example Using Fig 10: Find the line-to-line voltage drop on a 200 foot run of 1500 A busway carrying 90% rated current at 70% power factor.

Solution: Enter the chart at 70% on the horizontal scale. Follow a vertical line to its intersection with the curve for 1500 A and extend a line horizontally to its

Table 27
Balanced Three-Phase, Line-to-Line Voltage Drop for 600 V Single-Conductor Cable per 10 000 A × feet 60 °C (140 °F) Conductor Temperature, 60 Hz

Load Power Factor Lagging	| Wire Size (AWG or kcmil)																						
	1000	900	800	750	700	600	500	400	350	300	250	4/0	3/0	2/0	1/0	1	2	4	6	8*	10*	12*	14*
Section 1: Copper Conductors in Magnetic Conduit																							
1.00	0.28	0.31	0.34	0.35	0.37	0.42	0.50	0.60	0.68	0.78	0.92	1.1	1.4	1.7	2.1	2.6	3.4	5.3	8.4	13	21	33	53
0.95	0.50	0.52	0.55	0.57	0.59	0.64	0.71	0.81	0.88	1.0	1.1	1.3	1.5	1.9	2.3	2.8	3.5	5.3	8.2	13	20	32	50
0.90	0.57	0.59	0.62	0.64	0.66	0.71	0.78	0.88	0.95	1.1	1.2	1.3	1.6	1.9	2.3	2.8	3.4	5.2	8.0	12	19	30	48
0.80	0.66	0.68	0.71	0.73	0.74	0.80	0.85	0.95	1.0	1.1	1.2	1.4	1.6	1.9	2.3	2.6	3.2	4.8	7.3	11	17	27	43
0.70	0.71	0.73	0.76	0.78	0.80	0.83	0.88	0.97	1.0	1.1	1.2	1.3	1.5	1.8	2.1	2.5	3.0	4.4	6.6	9.9	15	24	38
Section 2: Copper Conductors in Nonmagnetic Conduit																							
1.00	0.23	0.26	0.28	0.29	0.33	0.38	0.45	0.55	0.62	0.73	0.88	1.0	1.3	1.6	2.1	2.6	3.3	5.3	8.4	13	21	33	53
0.95	0.40	0.43	0.45	0.47	0.50	0.54	0.62	0.71	0.80	0.92	1.0	1.1	1.5	1.8	2.2	2.7	3.4	5.3	8.2	13	20	32	50
0.90	0.47	0.48	0.52	0.54	0.55	0.59	0.68	0.76	0.85	0.95	1.1	1.1	1.5	1.8	2.2	2.7	3.3	5.1	7.9	12	19	30	48
0.80	0.54	0.55	0.57	0.59	0.62	0.66	0.73	0.81	0.88	0.97	1.1	1.1	1.4	1.7	2.1	2.5	3.1	4.7	7.2	11	17	27	43
0.70	0.57	0.59	0.62	0.64	0.66	0.69	0.74	0.83	0.88	0.97	1.1	1.1	1.4	1.6	2.0	2.4	2.8	4.3	6.4	9.7	15	24	38
Section 3: Aluminum Conductors in Magnetic Conduit																							
1.00	0.42	0.45	0.49	0.52	0.55	0.63	0.74	0.91	1.0	1.2	1.4	1.7	2.1	2.6	3.3	4.2	5.2	8.4	13	21	33	52	—
0.95	0.62	0.65	0.70	0.73	0.76	0.83	0.94	1.1	1.2	1.4	1.6	1.8	2.3	2.7	3.4	4.2	5.3	8.2	13	20	32	50	—
0.90	0.69	0.72	0.76	0.79	0.82	0.88	0.99	1.2	1.3	1.4	1.6	1.9	2.3	2.7	3.4	4.1	5.1	7.9	12	19	30	48	—
0.80	0.76	0.80	0.83	0.85	0.88	0.95	1.0	1.2	1.3	1.4	1.6	1.8	2.2	2.6	3.2	3.9	4.7	7.3	11	17	27	43	—
0.70	0.80	0.83	0.87	0.89	0.92	0.98	1.1	1.2	1.3	1.4	1.6	1.7	2.1	2.4	2.9	3.6	4.3	6.5	10	15	24	37	—
Section 4: Aluminum Conductors in Nonmagnetic Conduit																							
1.00	0.36	0.39	0.44	0.47	0.51	0.59	0.70	0.88	1.0	1.2	1.4	1.7	2.1	2.6	3.3	4.2	5.2	8.4	13	21	33	52	—
0.95	0.52	0.56	0.60	0.63	0.67	0.74	0.85	1.0	1.1	1.3	1.5	1.8	2.2	2.7	3.4	4.2	5.2	8.2	13	20	32	50	—
0.90	0.57	0.61	0.65	0.68	0.71	0.79	0.89	1.1	1.2	1.3	1.5	1.8	2.2	2.6	3.3	4.1	5.0	7.9	12	19	30	48	—
0.80	0.63	0.66	0.71	0.73	0.76	0.83	0.92	1.1	1.2	1.3	1.5	1.7	2.1	2.5	3.1	3.8	4.6	7.2	11	17	27	42	—
0.70	0.66	0.69	0.73	0.75	0.78	0.83	0.92	1.1	1.1	1.3	1.4	1.6	1.7	2.3	2.8	3.4	4.2	6.4	9.9	15	24	37	—

To convert voltage drop to	Multiply by
Single-phase, three-wire, line-to-line	1.18
Single-phase, three-wire, line-to-neutral	0.577
Three-phase, line-to-neutral	0.577

*Solid conductor. Other conductors are stranded.

intersection with the vertical scale. This intersection gives the voltage drop for a 100 foot run at rated load. For 200 feet at 90% load:

Line-to-line voltage drop $=$ $6.4 \times 2 \times 0.9$

$$= 11.5 \text{ V}$$

Example Using Fig 11: Find the line-to-line voltage drop on a 500 foot run with 50 A load at 80% power factor. The load is concentrated at the end of the run.

Solution: Enter the chart at 80% on the horizontal scale. Follow a vertical line to its intersection with the curve and extend a line horizontally to its intersection with the vertical scale. This intersection gives the voltage drop for 100 feet. For 500 feet:

Table 28
Voltage-Drop Values for Three-Phase Busways with Copper Bus Bars, in V per 100 feet, Line-to-Line, at Rated Current with Balanced Entire Load at End

NOTE: Divide values by 2 for distributed loading.

Rating (amperes)	Power Factor									
	20	30	40	50	60	70	80	90	95	100
Low-voltage-drop ventilated feeder										
800	3.66	3.88	4.04	4.14	4.20	4.20	4.16	3.92	3.60	2.72
1000	1.84	2.06	2.22	2.40	2.54	2.64	2.72	2.70	2.62	2.30
1350	2.24	2.44	2.62	2.74	2.86	2.94	2.96	2.90	2.78	2.30
1600	1.88	2.10	2.30	2.46	2.62	2.74	2.82	2.84	2.76	2.42
2000	2.16	2.34	2.52	2.66	2.78	2.84	2.90	2.80	2.68	2.30
2500	2.04	2.18	2.38	2.48	2.62	2.68	2.72	2.62	2.50	2.14
3000	1.96	2.12	2.28	2.40	2.52	2.58	2.60	2.52	2.40	2.06
4000	2.18	2.36	2.54	2.68	2.80	2.80	2.90	2.80	2.68	2.28
5000	2.00	2.16	2.30	2.40	2.50	2.60	2.68	2.60	2.40	2.10
Low-voltage-drop ventilated plug-in										
800	6.80	6.86	6.92	6.86	6.72	6.52	6.04	5.26	4.64	2.76
1000	2.26	2.56	2.70	2.86	2.96	3.00	3.00	2.92	2.80	2.28
1350	2.98	3.16	3.32	3.38	3.44	3.46	3.40	3.22	3.00	2.32
1600	2.28	2.44	2.62	2.78	2.90	3.00	2.96	2.94	2.88	2.44
2000	2.58	2.78	2.92	3.02	3.10	3.16	3.08	3.00	2.82	2.28
2500	2.32	2.50	2.66	2.76	2.86	2.90	2.86	2.78	2.66	2.18
3000	2.18	2.34	2.48	2.60	2.70	2.74	2.72	2.66	2.58	2.10
4000	2.42	2.56	2.76	2.88	3.00	3.02	3.00	2.96	2.84	2.36
5000	2.22	2.30	2.48	2.60	2.70	2.76	2.74	2.68	2.60	2.16
Plug-in										
225	2.82	2.94	3.04	3.12	3.18	3.18	3.10	2.86	2.70	2.04
400	4.94	5.08	5.16	5.18	5.16	5.02	4.98	4.30	3.94	2.64
600	5.24	5.34	5.40	5.40	5.36	5.00	4.50	2.10	3.62	2.92
800	5.06	5.12	5.16	5.06	5.00	4.74	4.50	3.84	3.32	1.94
1000	5.80	5.88	5.84	5.76	5.56	5.30	4.82	4.12	3.52	1.94
Trolley busway										
100	1.2	1.38	1.58	1.74	1.80	2.06	2.20	2.30	2.30	2.18
Current-limiting ventilated										
1000	12.3	12.5	12.3	12.2	11.8	11.1	10.1	8.65	7.45	3.8
1350	15.5	15.6	15.4	15.3	14.7	13.9	12.6	10.7	9.2	4.7
1600	18.2	18.2	18.0	17.5	16.6	15.6	14.1	11.5	9.5	4.0
2000	20.4	20.3	20.0	19.4	18.4	17.0	13.9	12.1	10.1	3.8
2500	23.8	23.6	23.0	22.2	21.0	19.2	17.2	13.5	10.7	3.8
3000	26.0	26.2	25.8	24.8	23.4	21.5	19.1	15.1	12.0	4.0
4000	29.1	28.8	28.2	27.2	25.6	25.2	21.0	16.6	13.0	4.1

Line-to-line voltage drop = 3.03×5

$$= 15.15 \text{ V}$$

3.6.4 Transformer Voltage-Drop Charts. Figure 12 may be used to determine the approximate voltage drop in single-phase and three-phase 60 Hz liquid filled, self-cooled transformers. The voltage drop through a single-phase transformer is found by entering the chart at a kVA value three times the rating of the single-phase transformer. Figure 12 covers transformers in the following ranges:

Single-phase

250 kVA — 500 kVA, 8.6 – 15 kV insulation classes

833 kVA — 1250 kVA, 2.5 – 25 kV insulation classes

Table 29
Voltage-Drop Values for Three-Phase Busways with Aluminum Bus Bars, in V per 100 feet, Line-to-Line, at Rated Current with Balanced Entire Load at End

NOTE: Divide values by 2 for distributed loading.

Rating (amperes)	20	30	40	50	Power Factor 60	70	80	90	95	100
Low-voltage-drop ventilated feeder										
800	1.68	1.96	2.20	2.46	2.68	2.88	3.04	3.12	3.14	2.90
1000	1.90	2.16	2.38	2.60	2.80	2.96	3.06	3.14	3.12	2.82
1350	1.88	2.20	2.48	2.74	3.02	3.24	3.44	3.56	3.58	2.38
1600	1.66	1.92	2.18	2.42	2.64	2.84	3.02	3.12	3.16	2.94
2000	1.82	2.06	2.30	2.50	2.70	2.88	3.02	3.10	3.04	2.80
2500	1.86	2.10	2.34	2.56	2.74	2.90	3.04	3.10	3.08	2.78
3000	1.76	2.06	2.26	2.52	2.68	2.86	2.98	3.06	3.04	2.78
4000	1.74	1.98	2.24	2.48	2.70	2.88	3.04	3.08	3.12	2.88
5000	1.72	1.98	2.20	2.42	2.62	2.80	2.92	3.02	3.02	2.80
Low-voltage-drop ventilated plug-in										
800	2.12	2.38	2.58	2.80	3.00	3.16	3.26	3.30	3.24	2.90
1000	2.44	2.66	2.86	3.06	3.22	3.36	3.42	3.38	3.28	2.84
1350	2.22	2.48	2.78	3.00	3.24	3.46	3.60	3.68	3.64	3.30
1600	1.82	2.12	2.38	2.62	2.80	2.96	3.08	3.16	3.14	2.88
2000	2.00	2.30	2.50	2.76	2.92	3.06	3.12	3.18	3.12	2.80
2500	2.00	2.28	2.50	2.70	2.92	3.02	3.12	3.16	3.08	1.78
3000	1.98	2.26	2.44	2.66	2.86	3.00	3.10	3.18	3.14	2.82
4000	1.94	2.20	2.48	2.64	2.86	3.00	3.12	3.18	3.16	2.88
5000	1.90	2.16	2.38	2.58	2.76	2.92	3.06	3.10	3.08	2.52
Plug-in										
100	1.58	2.10	2.62	3.14	3.56	4.00	4.46	4.94	5.10	5.20
225	2.30	2.54	2.76	3.68	3.12	3.26	3.32	3.32	3.26	2.86
400	3.38	3.64	3.90	4.12	4.22	4.34	4.38	4.28	4.12	3.42
600	3.46	3.68	3.84	3.96	4.00	4.04	3.96	3.74	3.52	2.48
800	3.88	4.02	4.08	4.20	4.20	4.14	4.00	3.66	3.40	2.40
1000	3.30	3.48	3.62	3.72	3.78	3.80	3.72	3.50	3.30	2.50
Small plug-in										
50	2.2	2.6	3.0	3.5	3.8	4.1	4.5	4.7	4.8	4.6
Current-limiting ventilated										
1000	12.3	12.3	12.1	11.8	11.2	10.9	9.5	8.0	6.6	3.1
1350	16.3	16.3	16.1	15.6	14.7	13.7	12.1	8.1	8.0	3.1
1600	18.0	17.9	17.7	17.0	16.1	14.9	13.4	10.7	8.6	3.3
2000	22.5	22.4	21.8	21.2	19.9	18.2	16.0	12.7	9.9	3.1
2500	25.0	24.6	23.9	23.1	21.7	19.9	17.5	13.7	10.8	3.0
3000	26.2	25.8	25.1	24.1	22.7	20.8	18.2	14.2	10.9	2.9
4000	31.4	31.0	30.2	28.8	27.4	24.8	21.5	16.5	12.7	2.9

Three-phase
>   225 kVA — 750 kVA, 8.6 – 15 kV insulation classes
>   1000 kVA — 10 000 kVA, 2.5 – 25 kV insulation classes

Example Using Fig 12: Find the percent voltage drop in a 2000 kVA three-phase 60 Hz transformer rated 4160 V — 480 V. The load is 1500 kVA at 0.85 power factor.

Solution: Enter the chart at 2000 kVA on the horizontal scale. Follow a vertical line to its intersection with the 0.85 power factor curve. From this point, extend a line horizontally to its intersection with the vertical scale. This intersection gives the percent voltage drop for rated load. Multiply this value by the ratio of actual load to rated load:

Percent voltage drop at rated load   =   3.67

Percent voltage drop at 1500 kVA   = $3.67 \times 1500 / 2000 - 2.75$

Actual voltage drop = $2.75\% \times 480$ = 13.2 V

Fig 8
Voltage-Drop Curves for Typical Interleaved
Construction of Copper Busway at Rated Load,
Assuming 70 °C (158 °F) Operating Temperature

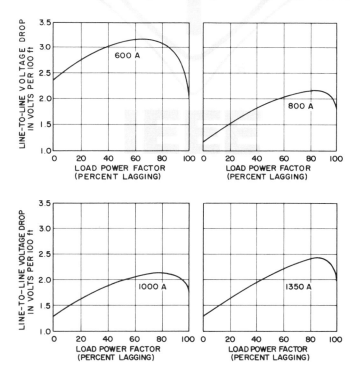

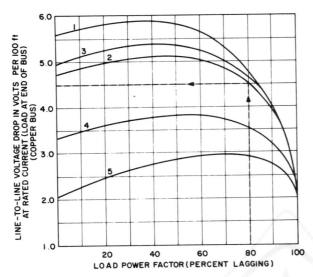

Curve	Bars per Phase	Size of Bar	Rated Current (amperes)
1	1	$\frac{1}{4} \times 4$	1000
2	1	$\frac{3}{16} \times 4$	800
3	1	$\frac{1}{4} \times 2$	600
4	1	$\frac{3}{16} \times 2$	400
5	1	$\frac{3}{16} \times 1$	225

Fig 9
**Voltage-Drop Curves for Typical Plug-in Type Busway at
Balanced Rated Load, Assuming 70 °C (158 °F) Operating Temperature**

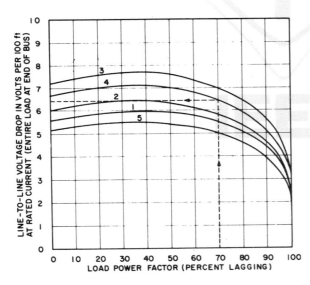

Curve	Bars per Phase	Size of Bar	Rated Current (amperes)
1	1	$\frac{1}{4} \times 2$	600
2	1	$\frac{1}{4} \times 4$	1000
3	2	$\frac{1}{4} \times 2\frac{1}{2}$	1200
4	2	$\frac{1}{4} \times 4$	1500
5	4	$\frac{1}{4} \times 3$	2300

Fig 10
**Voltage-Drop Curves for Typical Feeder Busways at
Balanced Rated Load Mounted Flat Horizontally,
Assuming 70 °C (158 °F) Operating Temperature**

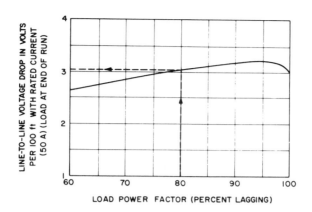

Fig 11
Voltage-Drop Curve versus Power Factor for
Typical Light-Duty Trolley Busway Carrying Rated Load,
Assuming 70 °C (158 °F) Operating Temperature

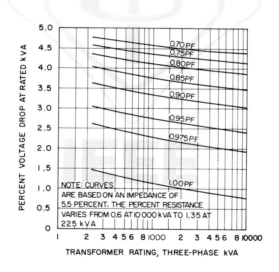

NOTE: Figure 12 applies to 5.5% impedance transformers. For transformers of substantially different impedance, the information for the calculation should be obtained from the manufacturer.

Fig 12
Voltage-Drop Curves for Three-Phase Transformers,
225–10 000 kVA, 5–25 kV

3.7 Improvement of Voltage Conditions. Poor equipment performance, overheating, nuisance tripping of overcurrent protective devices, and excessive burnouts are signs of unsatisfactory voltage. Low voltage occurs at the end of long low-voltage circuits. High voltage occurs at the beginning of low-voltage circuits close to the source of supply.

In cases of low voltage, the first step is to make a load survey to measure the current taken by the affected equipment, the current in the circuit supplying the equipment, and the current being supplied by the distribution transformer under peak load conditions to make sure that the low voltage is not due to overloaded equipment. When the low voltage is due to overload, then corrective action should be taken to relieve the overloaded equipment.

If overload is ruled out or if the utilization voltage is excessively high, a voltage survey should be made, preferably by using recording voltmeters to determine the voltage spread at the utilization equipment under all load conditions and the voltage spread at the utility supply, for comparison with ANSI C84.1-1989 [2], to determine if the unsatisfactory voltage is caused by the plant distribution system or the utility supply. If the utility supply exceeds the tolerance limits specified in ANSI C84.1-1989 [2], the utility should be notified. Most utilities will assist in making this voltage survey by providing the recording voltmeters required to determine the voltage during maximum and minimum load conditions.

When low voltage is caused by excessive voltage drop in the low-voltage wiring, the conductor size may be increased. When the conductor size is 1/0 AWG or larger, the NEC [5] may be referenced for specific provisions concerning the addition of equally sized circuit conductors in parallel. Another solution may be the installation of a separate circuit in order to split the load. When the power factor of the load is low, capacitors may be considered to improve the power factor and reduce the voltage drop. Refer to IEEE Std 141-1986 (ANSI), Chapter 8 [11] for proper application of capacitors. Where low voltage affects a large area, the best solution may be to convert to primary distribution when the building is supplied from a single transformer station, or to install an additional transformer in the center of the affected area when the building has primary distribution. Buildings wired at 208Y/120 V or 240 V may be changed over to 480Y/277 V or 480 V economically when an appreciable section of the wiring system is rated 600 V and motors are dual rated 220 × 440 V or 230 × 460 V.

When the voltage is consistently high or low and the building has primary distribution, the distribution transformer taps may be changed in direction to improve the voltage (see 3.2.7). When the building is supplied from a single distribution transformer furnished by the supplying utility, then a voltage complaint should be made to the utility.

3.8 Voltage-Drop Considerations in Locating the Low-Voltage Power Source. One of the major factors affecting the design of the low-voltage distribution system is the proper location of the low-voltage supply, which should be as close as possible to the center of load. This applies in every case, from a service drop from a distribution transformer on the street to a distribution transformer located outside or inside the building. Frequently, building aesthetics or available space require the low-voltage power supply to be installed at a corner of a building, without regard to what this adds to the cost of the building wiring to keep the voltage drop within satisfactory limits.

Reference [21] shows that, if a power supply is located in the center of a horizontal floor area at point O (see Fig 13), the area that can be supplied from circuits run radially from point O with specified circuit constants, and voltage drop would be the area enclosed by the circle of radius O - X. However, conduit systems are run in rectangular coordinates; thus, with this restriction, the area that can be supplied is reduced to the square X - Y - X′ - Y′ when the conduit system is run parallel to the axes X - X′ - Y - Y′. But the limits of the square are not parallel to the conduit system, and, to fit the conduit system into a square building with walls parallel to the conduit system, the area must be reduced to F - H - B - D.

If the supply point is moved to the center of one side of the building, which is a frequent situation when the transformer is placed outside the building, the area that can be served with the specified voltage drop and specified circuit constants is E - A - B - D. If the supply station is moved to a corner of the building, a frequent location for buildings supplied from the rear or from the street, the area is reduced to O - A - B - C.

Every effort should be made to place the low-voltage supply point as close as possible to the center of the load area. Note that this study is based on a horizontal wiring system, and any vertical components should be deducted to establish the limits of the horizontal area that can be supplied.

Using an average value of 30 feet/V for a fully loaded conductor, the distances in Fig 13 for 5% and 2.5% voltage drops are shown in Table 30. For a distributed load, the distances will be approximately twice the values shown.

3.9 Momentary Voltage Variations — Voltage Dips. The previous discussion covered relatively slow changes in voltage associated with steady-state voltage spreads and tolerance limits. However, sudden voltage changes should be given special consideration. Lighting equipment output is sensitive to applied voltage, and people are sensitive to sudden changes in light. Intermittently operated equipment, such as compressor motors, elevators, x-ray machines, and flashing signs, may produce a flicker when connected to lighting circuits. Care should be taken to design systems that will not irritate building occupants with flickering lights. In extreme cases, sudden voltage changes may even disrupt sensitive electronic equipment.

As little as a 0.5% voltage change produces a noticeable change in the output of an incandescent lamp. The problem is that individuals vary widely in their susceptibility to light flicker. Tests indicate that some individuals are irritated by a flicker that is barely noticeable to others. Studies show that sensitivity depends on how much the illumination changes (magnitude), how often it occurs (frequency), and the type of work activity undertaken. The problem is further compounded by the fact that fluorescent and other lighting systems have different response characteristics to voltage changes (see 3.5). Illumination flicker can be especially objectionable if it occurs often and is cyclical.

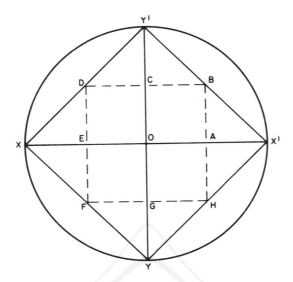

Fig 13
Effect of Low-Voltage Source Location
on Area That Can Be Supplied Under
Specific Voltage-Drop Limits

Table 30
Areas That Can Be Supplied for Specific Voltage Drops and Voltages

| Nominal System Voltage (volts) | Distance (feet) | | | |
| | 5% Voltage Drop | | 2.5% Voltage Drop | |
	OX	OA	OX	OA
120/240	360	180	180	90
208	312	156	156	78
240	360	180	180	90
480	720	360	360	180

Figure 14 shows acceptable voltage-dip limits for incandescent lights. Two curves show how the acceptable voltage flicker magnitude depends on the frequency of occurrence. The lower curve shows a borderline where people begin to detect the flicker. The upper curve is the borderline where some people will find the flicker objectionable. At 10 dips per hour, people begin to detect incandescent lamp flicker for voltage dips larger than 1% and begin to object when the magnitude exceeds 3%.

One source of voltage dips in commercial buildings is the inrush current while starting large motors on a distribution transformer that also supplies incandescent lights. A quick way to estimate flicker problems from motor starting is to multiply the motor locked-rotor starting kVA by the supply transformer impedance. A typical motor may draw 5 kVA/hp and a transformer impedance may be 6%. The equation below estimates flicker while starting a 15 hp motor on a 150 kVA transformer.

$$15 \text{ hp} \times \frac{5 \text{ kVA}}{\text{hp}} \times \frac{6\%}{150 \text{ kVA}} = 3\% \text{ flicker} \qquad \text{(Eq 4)}$$

The estimated 3% dip associated with starting this motor reaches the borderline of irritation at 10 starts/hour. If the voltage dip combined with the starting frequency approaches the objectionable zone, more accurate calculations should be made using the actual locked-rotor current of the motor. Accurate locked-rotor kVA for motors is available from the motor manufacturer and from the starting code letter on the motor nameplate. The values for the code letters are listed in ANSI/NEMA MG1-1978 [4] and in the NEC, Article 430 [5]. Section 3.10 describes more accurate methods for calculating motor-starting voltage dips.

When the amount of the voltage dip in combination with the frequency falls within the objectionable range, then consideration should be given to methods of reducing the dip to acceptable values, such as using two or more smaller motors, providing a separate distribution transformer for motors, or using reduced voltage starting.

Fig 14
Flicker of Incandescent Lamps Caused by Recurrent Voltage Dips

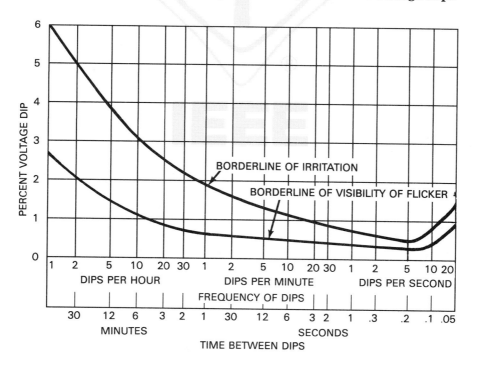

When a commercial building is supplied from a single electric utility company and the building owner or a tenant causes flicker problems for another tenant, the building owner is responsible for correcting the flicker problem. (The supplying electric utility may assist in the investigation.) When one customer of the electric utility causes flicker for another customer, the affected customer should file a complaint with the local electric utility company. Most electric utility companies have guidelines on what is considered an objectionable flicker; but these guidelines vary widely among companies. Flexibility in approach and effective communications between the customer and the utility can be invaluable in resolving potential flicker problems.

3.10 Calculation of Voltage Dips. The following methods are good approximations for the calculation of voltage dips. A more accurate method would be to convert the motor locked-rotor kVA to an equivalent impedance and build a voltage divider network between the motor and the source. This method is more complicated and often employs the assistance of computer programs.

3.10.1 The Effect of Motor Starting on Standby or Emergency Generators. Figure 15 shows the behavior of the voltage of a generator when an induction motor is started. Starting a synchronous motor has a similar effect up to the time of pull-in torque. The case used for this illustration utilizes a full voltage starting device, and the full voltage motor starting kVA is about 100% of the generator rating. For curves A and B, it is assumed that the generator is provided with an automatic voltage regulator.

As shown in Fig 15, the minimum voltage of the generator is an important quantity because it is a determining factor affecting undervoltage devices and contactors connected to the system and the stalling of motors running on the system. The curves of Fig 16 can be used for estimating the minimum voltage occurring at the terminals of a generator supplying power to a motor being started.

3.10.2 Effect of Motor Starting on a Distribution System. It is a characteristic of most ac motors that the current that they draw upon when starting is much higher than their normal running current. Synchronous and squirrel-cage induction motors started on full voltage may draw a current as high as seven or eight times their full-load running current. This sudden increase in the current drawn from the power system may result in an excessive drop in voltage unless it is considered in the design of the system. The motor starting kVA, which is imposed on the power supply system, and the available motor torque are greatly affected by the starting method used. Table 31 gives a comparison of several common methods (see Reference [22]).

3.10.3 Motor Starting Voltage Drop — Transformers. In the case of purchased power, there are frequently transformers or cables, or both, between the starting motor and the generator. Most of the drop in this case is within the distribution equipment. When all of the voltage drop is in this equipment, the voltage falls immediately (because it is not influenced by a regulator as in the case of the generator) and does not recover until the motor approaches full speed. Since the transformer is usually the largest single impedance in the distribution system and,

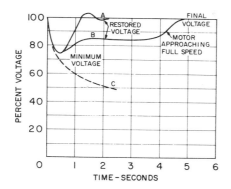

Motor starting kVA = 100% of generator rating.
A = No initial load on generator.
B = 50% initial load on generator.
C = No regulator.

**Fig 15
Typical Voltage Behavior of a Generator When an
Induction Motor Is Started by a Full Voltage Starting Device**

therefore, incurs most of the total drop, Fig 17 has been plotted in terms of motor starting kVA that are drawn if rated transformer secondary voltages are maintained.

3.10.4 Motor Starting Voltage Drop — Cables and Busways. The motor starting voltage drop due to the impedance of cables and busways may be calculated by using Figs 8–11 and Tables 28 and 29, which use the locked-rotor current and power factor as the load. Note that, in computing the circuit length, only the common section of the circuit between the supplying transformer and the motor and the point at which the voltage drop is being calculated should be used. For very large motors of several hundred horsepower, the voltage drop in the system supplying the transformer may have to be considered.

Table 27 cannot normally be used since the locked-rotor power factor in it is below the 70% limit. In this case, the voltage drop should be calculated using the approximate formula. However, since the power factor is quite low, the resistance component is generally negligible, and only the reactance component needs to be computed.

3.11 Phase Voltage Unbalance in Three-Phase Systems

3.11.1 Causes of Phase Voltage Unbalance. Most utilities use four-wire, grounded-wye distribution systems so that single-phase distribution transformers can be connected phase-to-neutral to supply a single-phase load, such as in residences and street lights. Variations in single-phase loading cause the currents in the three-phase conductors to be different, producing different voltage drops and causing the phase voltages to become unbalanced. Normally, the maximum phase voltage unbalance will occur at the end of the primary distribution system; but the

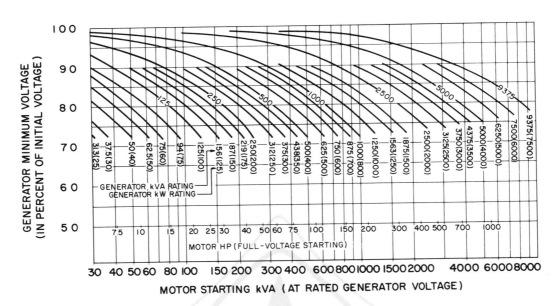

NOTES: (1) The scale of motor horsepower is based on the starting current being equal to approximately 5.5 times normal.
(2) If there is no initial load, the voltage regulator will restore voltage to 100% after dip to values given by curves.
(3) Initial load, if any, is assumed to be of constant-current type.
(4) Generator characteristics are assumed as follows. (a) Generators rated 1000 kVA or less; Performance factor $k = 10$; transient reactance $X_d' = 25\%$; synchronous reactance $X_d = 120\%$.
(b) Generators rated above 1000 kVA: Characteristics for 3600 rev/min turbine generators.

Fig 16
Minimum Generator Voltage Due to Full Voltage Starting of a Motor

actual amount will depend on how well the single-phase loads are balanced between the phases on the system. However, a perfect balance can never be maintained because the loads are continually changing, causing the phase voltage unbalance to also be continually changing. Blown fuses on three-phase capacitor banks will also unbalance the load and cause phase voltage unbalance. Most distribution transformers used to step the distribution voltage down to a utilization voltage have delta connected primaries. Unbalanced primary voltages will introduce a circulating current into the delta winding, which tends to rebalance the secondary voltage. Under these conditions, phase voltage unbalance in the primary distribution system tends to correct itself and should not be a problem.

Commercial buildings make extensive use of four-wire wye utilization voltages to supply lighting loads that are connected phase-to-neutral. Proper balancing of single-phase loads among the three phases on both branch circuits and feeders is necessary to keep the load unbalance and the corresponding phase voltage unbalance within reasonable limits.

Table 31
Comparison of Motor Starting Methods for Squirrel-Cage Induction Motors [22]

Type of Starter (Settings Given Are the Most Common for Each Type	Motor Terminal Voltage (% Line Voltage)	Starting Torque (% Full Voltage Starting Torque)	Line Current (% Full Voltage Starting Current)
Full voltage starter	100	100	100
Reduced voltage: Autotransformer			
80% tap	80	64	68
65% tap	65	42	46
50% tap	50	25	30
Primary resistor starter, single step (adjusted for motor voltage to be 80% of line voltage)	80	64	80
Primary reactor			
50% tap	50	25	50
45% tap	45	20	45
37.5% tap	37.5	14	37.5
Reduced in rush: Part-winding starter (low-speed motors only, × 514 rev/min and below)			
75% winding	100	75	75
50% winding	100	50	50
Wye-delta	100	33	33

NOTE: For a line voltage not equal to the motor-rated voltage, multiply all values in the first and last columns by the ratio (actual voltage) / (motor-rated voltage). Multiply all values in the second column by the ratio [(actual voltage) / (motor-rated voltage)]^2.

3.11.2 Measurement of Phase Voltage Unbalance. The simplest method of expressing the phase voltage unbalance is to measure the voltages in each of the three phases (see Reference [18]). The voltage unbalance is the maximum deviation from the average of the three phase voltages.

percent voltage unbalance =

$$100 \times \frac{\text{maximum deviation from average phase voltage}}{(\text{average phase voltage})} \qquad \text{(Eq 5)}$$

The phase voltage unbalance may also be expressed in symmetrical components as the ratio of the negative-sequence voltage to the positive-sequence voltage.

percent voltage unbalance =

$$100 \times \frac{\text{negative-sequence voltage}}{\text{positive-sequence voltage}}$$

The second formula defines the negative-sequence component of the voltage, which is a more accurate indication of the effect of phase voltage unbalance.

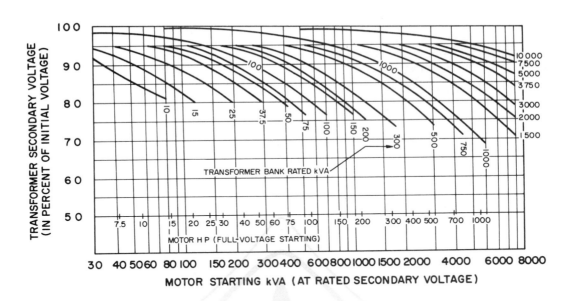

NOTES: (1) The scale of motor horsepower is based on the starting current being equal to approximately 5.5 times normal.

(2) Short-circuit kVA of primary supply is assumed to be as follows:

Transformer Bank kVA	Primary Short-Circuit kVA
0—300	25 000
500—1000	50 000
1500—3000	100 000
3760—10 000	250 000

(3) Transformer impedances are assumed to be as follows:

Transformer Bank kVA	Transformer Bank Impedance (%)
10—50	3
75—150	4
200—500	5
750—2000	5.5
3000—10 000	6

(4) Representative values of primary system voltage drop as a fraction of total drop, for the assumed conditions, are as follows:

Transformer Bank kVA	System Drop / Total Drop
100	0.09
1000	0.25
10 000	0.44

Fig 17
Voltage Drop in a Transformer Due to Full Voltage Starting of a Motor

3.11.3 Effect of Phase Voltage Unbalance. When unbalanced phase voltages are applied to three-phase motors, the phase voltage unbalance causes additional circulating currents to flow in the motor and generate additional heat loss (see References [16] and [20]). Figure 18, which is from ANSI C84.1-1989 [2], shows the nonlinear relationship between percent voltage unbalance and the associated de-rating factor for motors.

Hermetic (sealed) compressor motors that are used in air conditioners seem most susceptible to phase voltage unbalance. Originally, hermetic compressor motors were limited to small sizes; but they are now being built in units up to 1000 hp or more. These motors operate with higher current densities in the windings because of the added cooling effect of the refrigerant. Thus, the same percent increase in the heat loss due to circulating current caused by phase voltage unbalance will have a greater effect on the sealed compressor motor than it will on a standard air-cooled motor.

Since the windings in hermetic compressor motors are inaccessible, they are normally protected by thermally operated switches embedded in the windings and set to open and disconnect the motor when the winding temperature exceeds the set value. The motor cannot be restarted until the winding has cooled down to the point at which the thermal switch will reclose.

When a motor trips out, the first step in determining the cause is to check the running current after it has been restarted to make sure that the motor is not overloaded. The next step is to measure the three phase voltages to determine the amount of phase voltage unbalance. Figure 18 indicates that where the phase voltage unbalance exceeds 2%, the de-rating factor may be 95%, and the motor is likely to become overheated if it is operating close to full load.

Some electronic equipment, such as computers, may also be affected by phase voltage unbalance of more than 2% or 2.5%. The equipment manufacturer can supply the necessary information.

In general, single-phase loads should not be connected to three-phase circuits supplying equipment sensitive to phase voltage unbalance. A separate circuit should be used to supply this equipment.

A large single-phase transformer may be connected in open delta with a small single-phase transformer to supply a large single-phase load and a small three-phase load. Such installations can produce phase voltage unbalance due to the unequal impedance and loads in the two transformers. If objectionable phase voltage unbalance occurs in such an installation, a second small single-phase transformer should be added to complete the delta connection or the three-phase load should be connected to a separate three-phase transformer.

3.12 Harmonic Voltages

3.12.1 Nature of Harmonics. Harmonics are integral multiples of the fundamental frequency. For example, for 60 Hz power systems, the second harmonic would be 2×60 or 120 Hz and the third harmonic would be 3×60 or 180 Hz.

Harmonics are caused by devices that change the shape of the normal sine wave of voltage or current in synchronism with the 60 Hz supply. In general, these

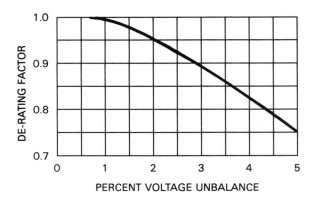

Fig 18
De-Rating Factor for Motors Operating with Phase Voltage Unbalance

include three-phase devices in which the three-phase coils are not exactly symmetrical, and single- and three-phase loads in which the load impedance changes during the voltage wave to produce a distorted current wave. This distortion creates harmonics since all harmonics, being integral multiples of the fundamental frequency, must pass through zero at the same points as the fundamental. Therefore, a distorted wave should be made up of a fundamental frequency and harmonics of various frequencies and magnitudes.

Inductive reactance varies directly as the frequency so that the current in an inductive circuit is reduced in proportion to the frequency for a given harmonic voltage. Conversely, capacive reactance varies inversely as the frequency so that the current in a capacitive circuit is increased in proportion to the frequency for a given harmonic voltage. If the inductive reactance and the capacitive reactance are the same, they will cancel each other out and a given harmonic voltage will cause a large current to flow, limited only by the resistance of the circuit. This condition is called "resonance" and is more apt to occur at the higher harmonic frequencies.

3.12.2 Effect of Harmonics. The harmonic content and magnitude existing in any power system is difficult to predict and effects will vary widely in different parts of the same system because of the different effects of different frequencies. Since the distorted wave is in the supply system, harmonic effects may occur at any point on the system where the distorted wave exists and are not limited to the immediate vicinity of the harmonic producing device.

Harmonics may be transferred from one circuit or system to another by direct connection or by inductive or capacitive coupling. Since 60 Hz harmonics are in the low-frequency audio range, the transfer of these frequencies into communications, signaling, and control circuits employing frequencies in the same range may cause objectionable interference.

3.12.3 Harmonic Producing Equipment

(1) Arc Equipment — Arc furnaces and arc welders supplied from transformers have widely fluctuating loads and produce harmonics. Normally, these do not cause very much trouble unless the supply conductors are in close proximity to communication and control circuits or there are large capacitor banks on the system.

(2) Gaseous Discharge Lamps — Fluorescent, high-pressure sodium, and mercury lamps produce small arcs and, in combination with the ballast, produce harmonics, particularly the third harmonic. Experience shows that the third harmonic current may be 30% of the fundamental in the phase conductors, and accordingly, 90% in the neutral where the third harmonics in each phase add directly since they are displaced one-third of a cycle.

Note that the NEC, Article 210-22(b) [5] requires that the computed load be based on the total ampere ratings of the units and not on the total watts of the lamp for circuits supplying this type of load. Also, Article 310-15 has specific requirements for neutral conductor ampacity.

(3) Variable Speed Drives, Power Supplies, and Rectifiers — Half-wave rectifiers that suppress alternate half-cycles of current generate both even- and odd-numbered harmonics. Full-wave rectifiers tend to eliminate the even-numbered harmonics and usually diminish the magnitude of the odd-numbered harmonics.

The major producer of harmonics is the controlled rectifier whose input current waveform is a variation of a square wave, which is rich in odd-numbered harmonics. Most rectifiers used in commercial buildings are six-pulse types producing harmonic numbers 5, 7, 11, 13, 17, 19, ..., in steadily decreasing magnitudes with increasing frequency (see IEEE Std 100-1988, IEEE Standard Dictionary of Electrical and Electronics Terms, Fourth Edition (ANSI) [10]). Controlled rectifiers are often used in adjustable speed drives, regulated power supplies for electronic equipment, and uninterruptible power supplies.

(4) Rotating Machinery — Normally, the three-phase coils of both motors and generators are sufficiently symmetrical that any harmonic voltages generated are too small to cause any interference.

(5) Induction Heaters — Induction heaters use 60 Hz or higher frequency power to induce circulating currents in metals in order to heat the metal. Harmonics are generated by the interaction of the magnetic fields caused by the current in the induction heating coil and the circulating currents in the metal being heated. Large induction heating furnaces may create objectionable harmonics.

(6) Capacitors — Capacitors do not generate harmonics. However, the reduced reactance of the capacitor to the higher frequencies may cause excessive

harmonic current in the circuit containing the capacitors. In cases of resonance, this current may be very large and may overheat the capacitors. In addition, the high currents may induce interference with communication, signal, and control circuits.

3.12.4 Reduction of Harmonic Interference. Where harmonic interference exists, the conventional reduction measures, such as increasing the separation between the power and communication conductors and the use of shielded communication conductors, should be considered. When capacitor banks are involved, the capacitors may need to be reduced in size or removed. Where resonant conditions exist, the capacitor bank should be changed in size to shift the resonant point to another frequency or small reactors should be connected in series with the capacitors to de-tune the circuit. Where harmonics pass from a power system to a communication, signal, or control circuit through a direct connection, such as a power supply, filters may be used to suppress or short circuit the harmonic frequencies.

Objectionable harmonic currents can be isolated with a series resonant circuit to ground numerically close to the harmonic frequency. Locate the resonant circuit physically near the source of the harmonic (which may be a rectifier unit). This resonant circuit should be sized to carry all the harmonic current the system is generating (see Reference [23]). Third harmonic currents may be isolated by using delta-wye transformers to serve the load.

3.13 Transient Overvoltages. Transient overvoltages (sometimes called "spikes") are momentary excursions of voltage outside of the normal 60 Hz voltage wave. Originally, the major sources of transient overvoltages were lightning strokes on or near overhead supply lines and intermittent ground contacts on ungrounded systems. However, in recent years, the switching of heavily loaded circuits, especially those involving large amounts of capacitance or inductance with devices, such as vacuum switches, controlled rectifier devices, and current-limiting fuses that chop the ac wave, has resulted in a proliferation of transient overvoltages to the extent that they are frequently called "electrical noise" because of their similarity to the noise in communication circuits that obscures the desired signal.

At the same time, solid-state devices, especially in microminiature sizes, which are introduced into computers, control systems, and other electronic equipment, are very susceptible to transient overvoltages, particularly in the reverse direction. When such electronic equipment is used, every effort should be made to minimize possible sources of transient overvoltages and to protect the equipment against the transient overvoltages that may occur with proper surge-protective devices.

3.14 References. The following references shall be used in conjunction with this chapter:

[1] ANSI C57.12.20-1988, Transformers—Overhead-Type Distribution Transformers, 500 kVA and Smaller: High Voltage, 34 500 V and Below; Low Voltage, 7970/13 800Y V and Below.

[2] ANSI C84.1-1989, Voltage Ratings for Electric Power Systems and Equipment (60 Hz).

[3] ANSI C92.2-1987, Power Systems — Alternating-Current Electrical Systems and Equipment Operating at Voltages Above 230 kV Nominal — Preferred Voltage Ratings.

[4] ANSI/NEMA MG1-1978, Motors and Generators (1987 Edition).

[5] ANSI/NFPA 70-1990, National Electrical Code.

[6] CAN3-C235-83, Preferred Voltage Levels for AC Systems, 0 to 50 000 V (Canadian Standards Association).

[7] IEEE C57.12.00-1987, IEEE Standard General Requirements for Liquid-Immersed Distribution, Power, and Regulating Transformers (ANSI).

[8] IEEE C57.12.01-1989, IEEE Standard General Requirements for Dry-Type Distribution and Power Transformers Including Those with Solid Cast and/or Resin-Encapsulated Windings.

[9] IEEE C57.94-1982 (Reaff. 1987), IEEE Recommended Practice for Installation, Application, Operation, and Maintenance of Dry-Type General Purpose Distribution and Power Transformers (ANSI).

[10] IEEE Std 100-1988, IEEE Standard Dictionary of Electrical and Electronics Terms, Fourth Edition (ANSI).

[11] IEEE Std 141-1986, IEEE Recommended Practice for Electric Power Distribution for Industrial Plants (ANSI).

[12] IEEE Std 142-1982, IEEE Recommended Practice for Grounding of Industrial and Commercial Power Systems (ANSI).

[13] IEEE Std 242-1986, IEEE Recommended Practice for Protection and Coordination of Industrial and Commercial Power Systems (ANSI).

[14] IEEE Std 446-1987, IEEE Recommended Practice for Emergency and Standby Power Systems for Industrial and Commercial Applications (ANSI).

[15] IEEE Std 519-1981, IEEE Guide for Harmonic Control and Reactive Compensation of Static Power Converter (ANSI).

[16] Arnold, R. E. NEMA Suggested Standards for Future Design of AC Integral Horsepower Motors, *IEEE Transactions on Industry and General Applications*, vol. IGA-6, Mar./Apr. 1970, pp. 110–114.

[17] *Electrical Data Book*, Minneapolis, MN: Electric Machinery Manufacturing Company.

[18] *Electrical Utility Engineering Reference Book*, vol. 3, Distribution Systems, East Pittsburgh, PA: Westinghouse Electric Corporation, 1965.

[19] Fink, D. G. and Beaty, H. W. *Standard Handbook for Electrical Engineers*, 11th Edition, New York: McGraw-Hill, 1978.

[20] Linders, J. R. Effects of Power Supply Variations on AC Motor Characteristics, *Conference Record of the 1971 IEEE Industry and General Applications Group Annual Meeting*, pp. 1055–1068.

[21] Michael, D. T. Proposed Design Standard for the Voltage Drop in Building Wiring for Low-Voltage Systems, *IEEE Transactions on Industry and General Applications*, vol. IGA-4, Jan./Feb. 1968, pp. 30–32.

[22] *Motor Application and Maintenance Handbook*, 2nd ed., Smeaton, R. W., editor, New York: McGraw-Hill, 1987.

[23] Stratford, R. P. Rectifier Harmonics in Power Systems, *Conference Record of the 1978 IEEE Industry Applications Society Annual Meeting*.

[24] Walker, Michael K. Electric Utility Flicker Limitations, *IEEE Transactions on Industry Applications*, vol. IA-15, no. 6, Nov./Dec. 1979, pp. 644–655.

3.15 Bibliography. The references in this bibliography are listed for informational purposes only.

[B1] Brereton, D. S. and Michael, D. T. Developing a New Voltage Standard for Industrial and Commercial Power Systems, *Proceedings of the American Power Conference*, vol. 30, 1968, pp. 733–751.

[B2] Brereton, D. S. and Michael, D. T. Significance of Proposed Changes in AC System Voltage Nomenclature for Industrial and Commercial Power Systems: I — Low-Voltage Systems, *IEEE Transactions on Industry and General Applications*, vol. IGA-3, Nov./Dec. 1967, pp. 504–513.

[B3] Brereton, D. S. and Michael, D. T. Significance of Proposed Changes in AC System Voltage Nomenclature for Industrial and Commercial Power Systems: II — Medium-Voltage Systems, *IEEE Transactions on Industry and General Applications*, vol. IGA-3, Nov./Dec. 1967, pp. 514–520.

Chapter 4
Power Sources and Distribution Systems

4.1 General Discussion. In other chapters, basic engineering, loads, voltages, apparatus, and circuit protection features for commercial buildings are discussed. This chapter considers electric power supplies, metering and billing, primary and secondary connections of transformers, system grounding, distribution circuit arrangements, emergency systems and equipment, and power factor correction. It is the responsibility of the engineer to develop an efficient and economical means of receiving electric power and distributing it to each area to be served. This function can be carried out in many ways. His or her selection of system arrangements, components, and voltages should be engineered to perform the function reliably and safely, and to deliver the power at correct voltages without hazard to personnel, the building, or equipment.

4.2 Electric Power Supply
4.2.1 Selecting a Power Source. In most cases, the selection of a power source will be determined by the joint action of design engineers and utility engineers.

Economics usually dictate the selection. With the exception of large high-load factor complexes, costs still favor the purchase of electricity for the prime power requirements. On-site total energy and co-generation systems may, in the future, appear more attractive as the energy situation and environmental restrictions impact the utilities. Standby electric generating equipment may be provided, in addition to prime purchased power, to produce the emergency power necessary for critical loads upon failure of the prime source or, where permitted, to reduce the monthly utility bill.

The following criteria are of prime importance in the selection of the power source:

(1) Operating Staff—One of the most important, and often neglected, considerations is the ability of building staff to operate and maintain the proposed system. Commercial buildings often have a very limited number of staff who may not be able to properly and safely maintain medium-voltage systems, complex protective relaying systems, etc.

(2) Availability—Most commercial buildings are located where electric utility service is available or can be made available. For purchased power, the voltage selected and its characteristics for either primary service or secondary service is based on the utility's distribution standards and the particular

services available in the specific area of the facility or facilities being planned. All utility company charges associated with the installation of a new service, or the expansion of an existing service, should be included. Special or nonstandard service requests can be expensive; failure to notify the customer of additional costs would not give the customer a true picture of the project.

(3) Reliability — Generally, the reliability, voltage, and frequency regulation of electric utility service in many areas of the United States is superior to self-generation (see IEEE Std 493-1990, IEEE Recommended Practice for the Design of Reliable Industrial and Commercial Power Systems (ANSI) [15][39]). The reliability of utility service is, of course, dependent not only on the generating facilities, but on the exposure of the feeders from the generating plant [see item (4) below]. The reliability of the electric service is also dependent upon the other loads on the same distribution system. For instance, a site located next to a rock-crushing operation may be of lower quality and less reliable than another site.

(4) Standby Power — Standby power (see IEEE Std 446-1987, IEEE Recommended Practice for Emergency and Standby Power Systems for Industrial and Commercial Applications (ANSI) [14]), as opposed to prime power and emergency power, is made available in case of the failure of the prime source for systems other than emergency systems. Emergency systems provide the minimum required for life safety and should be made available automatically upon failure of the prime power. Standby systems provide service to equipment generally considered essential for facility operation or to prevent loss of critical systems or computer data. Certain codes (see ANSI/NFPA 70-1990, National Electrical Code (NEC) [6][40]) and jurisdictions recognize two types of standby power: legally required and optional. Legally required standby systems are intended to automatically supply an alternate source of power to selected loads (other than emergency) where power failure could create a hazard or hamper rescue or fire fighting operations. Alternate power sources include generator sets, storage batteries, uninterruptible power supplies (UPS system), or a second utility company service. Optional standby systems are intended to automatically or manually supply an alternate on-site generated power source to selected loads (other than emergency and legally required standby) when a power outage could cause discomfort or damage to a process or product. When standby systems are to furnish loads that are less than the prime power load of the facility, all equipment should be selected in advance and an additional system of distribution provided, including circuits, panelboards, feeders, transformers, switchboards,

---

[39]The numbers in brackets correspond to those in the references at the end of this chapter. IEEE publications are available from the Institute of Electrical and Electronics Engineers, IEEE Service Center, 445 Hoes Lane, P.O. Box 1331, Piscataway, NJ 08855-1331.

[40]ANSI publications are available from the Sales Department of the American National Standards Institute, 11 West 42nd Street, 13th Floor, New York, NY 10036. NFPA publications are available from Publications Sales, National Fire Protection Association, 1 Batterymarch Park, P.O. Box 9101, Quincy, MA 02269-9101.

etc. The standby distribution systems generally interface with the prime power source at the service entrance or selected feeders and include some means of transfer.

(5) Purchased versus Generated Power for Prime Power — The installation of electricity generating equipment should be considered only after a thorough analysis of the total owning and operating costs of each system. Many factors will influence this analysis, such as the type of heating system (electric, steam, or hot water) and the comparative cost of the alternatives. An evaluation should be made of the type of electric system to be used in the building, which may vary depending on whether it distributes purchased or generated power.

Generated power will have a much larger new investment cost because of larger and higher priced boilers, generating equipment, space requirements, and pollution and noise control devices. Generating equipment will have to be adequate to handle all loads, including the starting of all motors, and with ample capacity for necessary maintenance and emergencies, with considerations for additional standby capacity in case of contingencies. Generated power may also have a higher cost in taxes, depreciation, and insurance. Table 32 compares some of the investment and operating items that are involved in the two power methods. Federal Department of Energy (DoE), and Environmental Protection Agency (EPA) regulations may impose severe limitations on the building of fossil fuel plants.

(a) Purchased Power — Before discussing power costs and availability with the electric utility, the following load data for initial and future requirements should be estimated as accurately as possible (see Chapter 2 and Reference [30]):

Table 32
Cost Comparison of Purchased and Generated Power

Purchased Power	Generated Power
Investment (if applicable)	
Substation or vault	Generating equipment
Metering and service	Additional boiler capacity
Standby equipment	Additional building space
	Heat-recovery equipment
	Additional pollution control
	Water treatment equipment
	*Additional land for fuel storage/handling
Operating costs	
Electric utility billing	Fuels
*Maintenance, labor, and	Maintenance, labor, and supplies
supplies of medium-voltage	Operating labor
equipment	Insurance
	Taxes
	Depreciation
	Standby utility service
	*Ash disposal charges

*If applicable

(i) Connected load at commissioning and in the future

(ii) Maximum demand at initial year and in the future

(iii) Motor load by categories and by single-phase and three-phase, including horsepower and starting/running requirements of the largest motor in each category

(iv) Required reliability (see IEEE Std 493-1990 (ANSI) [15]), i.e., maximum length of time before an interruption to electrical service is considered critical

(v) Power factor at maximum demand

(vi) Average power factor, i.e., monthly kWh divided by kVAh

(vii) Load factor, i.e., daily and monthly

(viii) Quality of power (harmonic content, transients, etc.)

When the electric utility has to install additional facilities to achieve the service quality needed, an extra charge by the electric utility may be applied. The annual operating cost can be ascertained from the utility tariff. When building expansion is likely, check the ability of the purchased power facilities to supply the increased load.

(b) Generated Power—In addition to the electrical data required when considering purchased power, the hourly, daily, and seasonal steam/hot water requirements should be examined when evaluating generated power. In the case where steam or hot water is to be used for heating and cooling, the exhaust steam or heat reclaimed from gas engines, diesel engines, or gas turbine generating equipment can be utilized at certain times. Only an hour-by-hour study of the coincidence of the electric energy demand and the building requirements for heating and cooling can determine the economics. An extensive study by engineers fully familiar with the comparative economics of purchased electric service versus generated power is essential. A discussion of fuel costs and selection is beyond the scope of this book.

When steam facilities are installed to generate electric energy only, the full investment cost should be charged against the electric system. These costs include: building, boilers, turbine generators and switchgear, fuel handling facilities, water treatment, condensing water, ash handling, and pollution control equipment (air, water, noise, environmental), when required. Under these conditions, generated power cannot generally be justified on an economic basis.

As with purchased power, the engineer should check the feasibility and cost of facilities to meet anticipated future growth. Environmental considerations, including future, tightened standards, are strong deterrents to in-house generation for most commercial buildings. Union work rules, especially with regard to staffing, and rigid safety and operating rules and standards should be considered. Required levels and types of staffing could make an otherwise favorable plan uneconomical. Often large expenditures are required with on-site generation for a comparatively small building or for load expansions.

4.2.2 Planning for Utility Service. Each utility differs to some degree from every other utility in its service policies and requirements. Therefore, communication should be established by the builder with the local supplying utility through the utility's Customer Service Department or Electric Marketing Department as soon as possible so that local requirements can be incorporated in building plans. The utility engineers will need the following information (see Reference [30]):

(1) Plot plan of the area, which shows the buildings (both present and future), roadways, and other structures. Underground utility lines should be documented.

(2) Preferred point of delivery for electric service

(3) Estimated connected load, maximum demand and power factor, and any requirements for future increases

(4) Preferred voltage

(5) Any special equipment, such as large motors, electric furnaces, welders, and x-ray apparatus, which may disturb the supply system

(6) Any requirements for alternate, emergency, or standby service

(7) A single-line diagram of service equipment and, for primary service, the primary distribution system including the sizes and ratings of switches, breakers, and protective devices. Transformer descriptions should include taps and whether the tap changer is a non-load or on-load type, and should show system phase rotation and transformer winding connections.

(8) A load tabulation that indicates the portions of the total load designated for each of the following: central air conditioning, data processing, data processing air conditioning, food service, lighting, motors, refrigeration, room air conditioning, water heating, welding (e.g., maintenance shop), receptacles, heating, elevators, and miscellaneous power.

(9) Rating of Customers' Emergency Generator(s) — The electrical load information required by the utility should be separated by phases and voltages. The largest motor in each category should be shown by horsepower, together with its locked-rotor and full-load amperes.

When requested, the utility will provide the following information:

(1) Rate or rates available

(2) Voltage or voltages available and voltage ranges (see ANSI C84.1-1989, Voltage Ratings for Electric Power Systems and Equipment (60 Hz) [3])

(3) For medium voltages, insulation coordination data
   (a) Basic impulse insulation level (BIL) of equipment
   (b) Ratings of surge arresters
      (i) Duty-cycle voltage rating
      (ii) Pressure relief current rating

(4) Point of delivery of electric service, when preferred point is not acceptable

(5) Line route from the property line to the point of delivery for any portion of the line installed by the utility

(6) Any charges for service, including cost of any underground portion of the line. The utility may provide options for the underground service: direct burial cable, or conduit and manhole system

(7) Requirements for connections at the point of delivery

(8) Requirements for metering

(9) Available short-circuit capacity of the supply system and protective device time current coordination information

(10) Space requirements for a transformer station when required and furnished by the utility

(11) Any utility requirements for service entrance equipment

(12) Any special local exception to the NEC [6], which applies to utility-associated equipment

(13) Any limitations on the starting of motors and the specification for motor controllers

(14) Recommended ratios and taps for transformers provided by the customer

(15) Availability and cost of an alternate or standby electric supply

(16) Available historical data as to the reliability of transmission and distribution feeders from which the new service will be derived. For instance, the number of reclosures on the distribution line that will supply the new facility and the number of reclosures on other distribution lines from the same substation.

(17) Limitations on carrier current signals

(18) Limitations on total harmonic distortion (THD)

(19) Limitations on welders, x-rays, and other inherently single-phase equipment requiring excessive inrush current

(20) Advice on supply system details and essential coordination details concerning installation of power factor correction capacitors

(21) Rules concerning the installation and operation of standby generators

(22) Limitations on load balancing

(23) Phase rotation and service entrance conductor phase identification

4.2.3 Electric Rates. Each electric utility has a series of rate schedules for supplying power to customers under various conditions. To arrive at the most economical condition for obtaining power, the engineer should compare and evaluate the following tariff conditions:

(1) Maximum demand in kW or kVA

(2) Energy consumption in kWh

(3) Adjustment for low power factors

(4) Voltages available

(5) Transformer or substation ownership

(6) Fuel cost adjustment clause

(7) Demand interval

(8) Minimum bill stipulations

(9) Multiple-metering provisions

(10) Auxiliary or standby service charges

(11) Seasonal or time-of-day service rates or charges, or both

(12) Prompt payment savings

(13) Provision for off-peak loads and interruptible loads

(14) Limitations on resale of power to tenants by building owner

(15) Incentives for utility service direct to tenants

(16) Total electric construction

(17) Elimination of multiple-metering service points

(18) Conjunctional billing arrangements

(19) Incentives for energy conservation equipment installation

(20) Incentives for acceptance of excess utility facilities in a particular area, especially depressed areas

A few of the major factors above are the following:

(1) Demand or Fixed Charges — Demand charges cover all generally predictable utility costs, such as depreciation, interest, and insurance. Capital investments for land, buildings, generating equipment and switchgear, transmission lines, and structure transformation and distribution equipment are depreciated over the estimated or specified life of the equipment. Demand charges reflect the investment required by the electric utility to serve the customer's maximum rate of consumption (demand). The demand is usually determined by a demand meter.

(2) Energy or Variable Charges — Energy charges include cost items such as fuel, operating labor, maintenance, raw materials, etc., and should cover all costs involved in operating the electric utility plus a reasonable profit.

(3) Power Factor — Some electric utilities penalize the customer when the power factor of the load drops below a stipulated value, and some utilities provide a credit for high power factor.

(4) Voltage — Most buildings are supplied by electric utilities at utilization voltages, such as 208Y/120 V or 480Y/277 V, which are directly usable by the load equipment. When the building area or load becomes too large to be supplied at the utilization voltage due to excessive cost or excessive voltage drop, or both, the building should be supplied at a distribution voltage that might be typically 4160 V or 13 200 V. In some instances, voltages as high as 34 500 V have been supplied to transformers in building vaults. With a customer-owned distribution system, the associated equipment, including transformers, should be installed by the builder. The only exceptions are when the utility provides all or part of the primary system, including transformers, in return for the right to sell power directly to tenants, or when the utility and the local inspection authority permits more than one service in the building. In the latter case, the utility may treat each service as a separate customer, and, therefore, the bills will be higher than a single bill (see 4.4).

(5) Fuel Cost — Since the major component of the cost of electric energy is the cost of fuel, most utility rates have a fuel adjustment in their energy charges for the average current cost of fuel based on its actual heat content.

The greatest part of the fuel cost is included in the "base" rate, and the supplementary fuel charge represents only the deviation from this base rate; it may be a credit when fuel costs are low.

(6) Other Factors — The remainder of the rate factors listed as having a bearing on the customer's bill are usually well defined on the utility's rate schedule or will be provided on request by the utility's Electric Sales Department.

Some utilities offer preferential rates or other incentives in order to promote a larger and more diversified load base.

Since the oil embargo of 1973, and as a direct consequence of the National Energy Act of 1978, changes in conventional electric utility rate practices have resulted in rate reforms. Essentially, rate reform usually incorporates factors such as

(1) Inverted rates
(2) Flat rates
(3) Lifeline rates
(4) Marginal cost pricing (MCP)
(5) Long-range incremental costing (LRIC)
(6) Construction work in progress (CWIP)
(7) Interruptible rates
(8) Time-of-day/time-of-use pricing
(9) Reduction in the number of declining blocks
(10) Modification of fuel rate adjustments to the fuel rates
(11) Deletion of electric heat discounts

When comparing new reform rates versus old rates for customers, it is usually difficult to determine the individual impact of rate reform and the customer's rate increase. Each customer will be affected differently based upon the type and schedule of operation and the extent of load management equipment that is either existing or planned.

4.3 Interrelated Utility and Project Factors That Influence Design. Factors that influence the design of electric systems are many and are covered in other chapters. However, this brief checklist contains some of the factors that may be helpful in planning system design.

(1) Type, size, shape, and occupancy purposes of the building or buildings
(2) Voltages and voltage ranges of the electric utility system that are available at the building site (see ANSI C84.1-1989 [3])
(3) Electrical rate plans available from the electric utility company
(4) Availability of aerial or underground service and of radial loop or network sources from the electric utility company
(5) Type and rating of building utilization equipment
(6) Economics of utilization voltage distribution
(7) Necessity of including a change to a higher voltage, such as changing from a 480Y/277 V system or engineering a medium-voltage distribution system in a modernization project
(8) Complete or partial replacement of old or obsolete equipment in a modernization project
(9) Application of modern lighting and space-conditioning principles to modernization projects
(10) Reliability of the source or sources of supply. Consistency in maintaining needed reliability throughout the entire electric system is essential to the overall solution. The engineer should carefully evaluate each part of the design for reliability as well as that of the electric utility feeders and their sources. For instance, some incoming feeders may be tapped for other customers; some may be exposed to hazards or have a history of outages, including substation or line circuit breaker/recloser operations, so that they need to be reinforced, or backed up, with an alternate set of feeders. The

engineer should consider all possibilities of planned and inadvertent outages to determine the justification for such reinforcements or alternate feeders. Fortunately, most electric utility feeders have a high degree of reliability. The electric utility's future plans for all feeders involved in the building's electric service should be considered.

(11) Economics of the distribution system as a consequence of available fault levels of utility services and customer-furnished limiters, such as transformers, reactors, system neutral grounding, and current-limiting protective devices.

A mutual and congenial understanding and appreciation of each other's problems is highly desirable at all stages of negotiation among the electric utility, the customer, and the engineer. This is true for both new buildings and expansions. Most commercial building projects, of course, pose no conflicts of interest between the parties. But there are projects where it would be a great advantage to one party if existing plans, designs, procedures, or rules could be modified. The modifications required might include the customer's need for consideration of a more beneficial rate structure, a more economical incoming voltage, the location of incoming service equipment at a more desirable point, or a different step-down transformer winding connection, such as a primary delta to grounded-wye secondary instead of a grounded-wye primary to grounded-wye secondary. The electric utility may want the customer to limit motor-starting currents, or to provide more convenient access to routes for service, or to install more adequate fault protection equipment, which can be of mutual advantage. The electric utility's and the customer's engineer can appreciate the importance of each other's viewpoint by an open and cooperative exchange of information at all stages of the project.

4.3.1 Grounding on AC Services. The utility may require that, in ac service installations, the ground conductor, the service neutral, and the metal housing of the service equipment be connected together at the service equipment. When a continuous metallic underground water piping system is present, the grounding connection should be on the street side of the main valve. Gas service piping should not be used as a ground. The NEC [6] specifies how and where the service neutral (grounded conductor) shall be grounded, and how and where the equipment grounding conductor shall be tied to the ground.

4.4 Electric Utility Metering and Billing. An understanding of utility metering and billing practices is important for evaluating service arrangements. Practices vary depending upon local utility and regulatory body requirements. The design, usage, and load characteristics for a given application should be carefully weighed before selecting service voltage and metering characteristics. When large momentary high demand loads, high seasonal loads, or low power factor loads are involved, billing penalties may also be incurred. On the other hand, high load factor or high power factor loads may merit a billing allowance or credit.

It is considered good practice to consult the electric utility early in the design stages. Late utility negotiations may result in increased costs or delays in service, or both. A complete discussion of service, metering, and billing requirements is always in order, no matter how preliminary. This should provide time for the consideration of various proposals and the selection of the one best suited to a given application.

4.4.1 Metering by Type of Premises. The availability of a particular kind of metering and billing generally depends upon the nature and characteristics of the premises, type of load involved, and local utility and government regulatory requirements. Due to the important influences of the metering scheme on the economics and design of the distribution system, especially in multiple-occupancy buildings, an early decision on the system to be employed is essential.

(1) A single-occupancy building, such as a hospital, a school, or an office building occupied by a single tenant, will be metered by the utility at the service entrance with a watthour-demand meter. With multiple services (where permitted by the electric utility), watthour-demand meter readings may be added together to take advantage of lower rates, and the demands on two or more services may be totalized so that customers may benefit from the diversity of their demand.

(2) Multiple-occupancy buildings, such as apartment houses, shopping centers, condominiums, and large office buildings, are generally equipped with an individual meter for each customer (owner and tenants), except in cases where light and power are included in the tenant's rent, in which case a master metering may be utilized. In some localities, such buildings may be submetered or otherwise distributed with the utility's customer buying power at wholesale rates on electric utility master metering and reselling it to the tenants at legally prescribed rates using private meters.

(3) Where tenants are individually metered, by the electric utility or by the owner, it is important to provide sufficient flexibility in the metering arrangement to facilitate metering changes as tenant changes occur.

4.4.2 Metering by Service Voltage Characteristics. Metering of incoming electric service may be located either on the high-voltage or on the low-voltage side of the transformer, depending on the terms of the contract with the electric utility. When the metering is on the high-voltage side of the transformer, the losses of the transformer will be metered and charged to the customer. In some cases, the customer is given a discount in his or her billing to offset this loss. Requirements vary according to individual utility tariffs.

4.4.3 Meter Location. Subject to agreement with the utility, meters may be installed indoors at the customer's secondary distribution point, in a suitable meter room, or in a separate control building that may also contain control, relay, and associated primary service switchgear. Outdoor installation, including pole mounting, exterior wall attachment, or pad mounting, may also be used, subject to utility approval. It is good practice to review meter locations with the utility early in the design stage. In general, utilities require accessibility for meter reading and maintenance purposes, and suitable meter protection. Where remote reading of meters is performed, the utility may require a telephone line at the meter point.

4.4.4 Meter Mounting, Control, and Associated Equipment. Utilities publish regulations and standards requirements covering meter mounting, control, and associated equipment. Utility revenue metering may be broadly grouped into three categories

(1) Self-contained metering
(2) Instrument transformer metering
(3) Special metering

The descriptions and general requirements for each category follow:

(1) Self-Contained Metering — Meters are connected directly to the system wiring being metered. The customer is generally required to furnish and install an approved meter socket and its associated wiring, conduits, devices, fittings, and bonding. This metering is normally used up to a maximum load of 400 A for low-voltage systems. Guidelines on the metering facilities to be installed are (see Reference [30])

   (a) Single-phase 120Y/240 V service — Five-jaw meter socket
   (b) Three-phase 208Y/120 V service — Seven-jaw meter socket
   (c) Meter bypass facilities should be used for health care facilities and places where continuity of service is important.
   (d) Three-phase 480Y/277 V service — Five- or seven-jaw meter sockets may be used depending upon the three- or four-wire load wiring, respectively.
   (e) Three-phase, four-wire 480Y/277 V service with line-side loads — May require use of a bottom-connected watthour meter and current transformers. Exit lights are an example of load on the line side of the service disconnect. A caution notice indicating the exit lights or other equipment connected to the line side of the service disconnect equipment.
   (f) Meter sockets are usually required to be installed on the line side of the customer's disconnect facilities.
   (g) Meter mountings for bottom-connected, self-contained meters are usually installed on the line side of the customer's disconnect facilities.

(2) Instrument Transformer Metering — The utility will require the use of instrument transformers between system wiring and meter wiring when the service rating exceeds currents or voltages on the order of 200 A or 480 V. The customer is generally required to furnish and install the instrument transformer cabinet or mounting assembly, meter box, conduit, fittings, and bonding. The utility generally furnishes instrument transformers that are installed and the meter-wiring connections made by the utility or the customer, according to utility requirements. Current transformers are usually installed on the line side of the service disconnect and within 10 feet of the associated metering.

(3) Special Metering — Includes totalized metering, impulse metering, telemetering, reactive component or power factor metering, etc. For complete requirements in all such cases, the utility should be consulted.

(4) Motor Control and Protection — Utility rules usually state that the utility assumes no responsibility for failures, equipment, or operations due to service problems. Therefore, utility customers are responsible for equipping motor controllers with suitable undervoltage tripping devices to prevent sustained undervoltage operation. Such devices should be of a time delay

type to avoid unnecessary tripping during momentary disturbances or service interruptions. Customers are also responsible for equipping polyphase motors with protection from single phasing, improper rotation due to phasing, and overheating due to current unbalance.

4.4.5 Types of Metering. Various types of metering, including application, are summarized below. Utility requirements cover the types of metering available for a given application.

(1) Watthour Metering—Measures energy consumption only. Quantities are indicated on a dial or on a cyclometer register. Watthour meters are available with special features, such as

  (a) Differential watthour metering—Kilowatthours from loads up to a certain value are registered on one dial and kWh for loads in excess of the preselected value are registered on another dial or register.

  (b) Time-of-day/time-of-use watthour metering—Consumption for certain periods of time is registered on one dial and other periods on other dials.

(2) Demand Metering—Kilowatt demands are measured by several types of meters. Indicating demand meters of the mechanical type have a pusher arm that remains at the maximum demand of the period, until reset by the meter reader. Dials or cyclometer registers are accessory devices that accumulate the maximum demands.

  (a) Integrating Demand Meters—Energy consumption for a specific time (usually 15, 30, or 60 minutes) is accumulated and shown as a rate of use. A 15 minute demand interval means the meter is accumulating kWh for 15 minutes and multiplying by the ratio of 60:15 minutes or 4 (by means of gears in mechanical meters) and displaying the result as kW.

  (b) Thermal Demand Meters—Uses bimetallic coils as an actuating device. Steady-state loads will be read in approximately the same way as integrating demand meters; however, varying loads may result in different results.

  (c) Graphic Demand Meters

    (i) Watt Meter—A graphic watt meter shows, on chart paper, the loads at each instant. Meter sensitivity determines the percentage of the varying load plotted or recorded.

    (ii) Integrating Meter—A graphic record of the loads for a predetermined interval (say 15 minutes) is shown as well as all loads throughout the metering period (say 30 days).

    (iii) Thermal Meter—A graphic record of the loads as measured by the thermal demand instrument.

(3) Kilovoltampere Demand

    (i) Ammeter and nominal voltage—An ammeter calibrated to read kVA for the nominal supply voltage is sometimes used.

    (ii) Reactive kVA—A kilowatt meter with the voltage element shifted (lagged) 90 electrical degrees can measure the reactive component.

(iii) Reactive kVAh — Average power factor rates-use meters for kWh and reactive kVAh.

(4) Master Metering — Is a single-metered electric service to multiple-occupancy premises. Tenant service costs are either included in the rent as a flat charge or determined by submeters, depending upon local utility regulations.

(5) Multiple Metering — A separate meter is established for each tenant's requirements in a multiple-occupancy building. Each tenant is separately metered and billed by the utility.

(6) Primary Metering — Is medium-voltage metering up to 72 000 V. The customer generally owns and maintains service transformer(s) and meter-mounting equipment. Metering is generally owned and maintained by the utility.

(7) Secondary Metering — Under 600 V, the utility generally owns and maintains service transformers, metering transformers, meter wiring, and meter-mounting equipment.

(8) Totalized Metering — Coincident demand of multiple services is metered by either pulse or integrating demand metering to provide diversified demand registration that is equivalent to that for a single meter. It is generally required by the utility when the service to a single switchboard or panelboard is impractical.

(9) Pulse Metering — Is used to determine coincident demand. Meter registration is affected by the use of electric pulses. Each pulse is a function of load and time. Pulses are received from several sources (that is, metering points) and counted by a totalizing meter. The totalizing meter integrates the received pulses over a given period of time (characteristically, the demand interval) to provide a readout of the total demand. Printed tape demand meters and totalizing demand meters utilize pulse metering. Magnetic tapes are also now being used.

(10) Compensated Metering — Is applicable to primary metered service to a single transformer bank. Rather than primary metering, secondary metering together with a transformer loss compensator is calibrated to compensate for the service transformer losses equivalent to that of a primary meter, saving the cost of high-voltage instrument transformers.

(11) Submetering — Additional metering is installed on a building distribution system for the purpose of determining demand or energy consumption, or both, for certain building load subdivisions, and where the same metering is preceded by a master billing meter. When submetering is required for billing tenants of a commercial building, the metering may be at the medium- or low-voltage distribution point when all loads on the feeder are for one customer. When a feeder supplies more than one customer or when power costs are to be accurately apportioned among various departments, the metering should be installed at each unit substation or low-voltage feed.

(12) Subtractive Metering — Is an application of submetering. Readings of sub-meters are subtracted from associated master meter readings for billing purposes. Subtractive metering permits the determination of the load taken

by an unmetered area when the total energy into the applicable system is known and all other services are metered.

(13) Coincident Demand — (See item (8).) For totalized service without coincident demand, the demand is known as "additive."

(14) Telemetering — Metering pulses are transmitted from one location to another for the purpose of meter reading at a remote location. It may also be used to totalize two or more distant locations.

(15) Power Factor Metering — Either reactive kVA, or kVA and kW are metered to determine the power factor for utility billing purposes. Coincident or cumulative metering is used, depending upon the utility rate schedule. Low power factor loads are often subject to a billing penalty, whereas high power factor loads may merit a billing discount. The exact contract terms can be obtained from the supplying utility.

(16) Electronic Metering — Revenue metering by use of electronic meters is preferred when tariffs contain complex schedules, such as time-of-day/time-of-use, sliding demand windows, and power factor penalties. Sliding demand measurements are performed by continuously analyzing the demand and capturing the maximum demand during *any* demand interval, that is, any 15, 30, or 60 minute interval per schedule. There is no predetermined synchronizing pulse that the demand is measured against since it is a continuous analysis. Electronic meters record all the pertinent information on magnetic tape, computer disks, or retain it within memory. Periodically, the data is obtained by manual replacement of the tape or disk or by a portable recorder that temporarily plugs into the meter for a data capture, see ANSI C12.1-1988, Code for Electricity Metering [2].

4.4.5.1 Metering for Energy Conservation. ASHRAE/IES 90.1-1989, Energy Efficient Design of New Buildings Except New Low-Rise Residential Buildings [11][41] requires a provision be made for check metering in buildings with a connected load of over 250 kVA. Check metering is customer-supplied metering in addition to utility revenue metering. Provision for check metering by category is required for

(1) Lighting and receptacles
(2) Heating, ventilating, and air-conditioning (HVAC) systems
(3) Water heaters, elevators, special occupant equipment, and systems of more than 20 kW, except where 10% or less of the load on a feeder may be from another usage category.

This does not mean that only three meters are required. When several categories of loads are supplied by a common feeder, then where the loads are subdivided, as at a panelboard, check metering provisions would be required.

Multiple-tenant buildings should have a provision for check metering as described above. When the tenant's connected load is 100 kVA or greater, then a provision for check metering should be installed.

---

[41] ASHRAE publications are available from the American Society of Heating, Refrigerating, and Air-Conditioning Engineers, 1791 Tullie Circle, N.E., Atlanta, GA 30329. IES publications are available from the Illuminating Engineering Society, 345 East 47th Street, New York, NY 10017.

Check metering may be permanent or portable. When portable metering is used, then instrument transformer access should be provided. Preferred metering is a local kilowatthour-demand meter or a remote meter supplied by the building automation system.

Types of check metering equipment include

(1) Permanently mounted socket-type or switchboard-type kWh or kilowatt-hour-demand meters for visible indications

(2) Watt-type transducers, which provide analog output signals (e.g., 4–20 mA) to the building automation system

(3) Kilowatthour-type transducers, which provide pulse-type signals to the building's automation system

(4) Portable instrumentation using clamp-on or split-core current transformers (or current test blocks) and voltage test blocks in the enclosure (switchboard, panelboard, etc.)

4.4.6 Utility Billing. It is customary for utilities to meter and bill each customer individually. Utility rates usually consider fixed and variable cost requirements to provide service. Hence, rate schedules generally take the form of a block rate, wherein incremental service costs usually vary as a function of customer usage. Electric service costs generally comprise two components: the demand charge and the energy charge.

Many utility rate schedules do not include a demand charge for smaller customers (for example, less than 50 kW) and instead use a block meter rate that specifies a certain price per kWh, which decreases for succeeding blocks (residential rates frequently employ an inverted rate structure).

The Hopkinson demand rate consists of separate charges for demand and for energy, thus recognizing the load factor. The demand charge or the energy charge, or both may be blocked to give lower prices for higher loads and greater consumption. A rate in this form is called the "block Hopkinson demand rate."

The Wright demand rate, or hours-use rate, consists of a number of energy blocks with decreasing prices for succeeding blocks and in which the sizes of the energy blocks increase with the size of the load. Effectively, this produces demand and energy charges and thus recognizes the load factor.

Many utilities have been granted permission to add provisions for variable cost factors to their rates. Examples of this include purchased fuel differential and real estate tax differential costs. Under these provisions, the utility may pass along to its customers increased costs; however, the utility should also compensate for decreased costs as well.

(1) Master Metering, Rent Inclusion — Offers a saving to the owner in first cost for metering equipment. Savings in operating costs depend upon the type of multiple-occupancy building and applicable utility rates. The owner has to determine tenant electric service costs, usually as a flat sum for purposes of incorporation in the tenant's lease or rental agreements. The flat rate encourages excessive power usage by the tenants.

(2) Multiple Metering and Billing — Generally requires a higher first cost to the owner for multiple-occupancy buildings over the cost of master metering.

On the other hand, the utility is responsible for collecting all tenant electric service costs.

(3) Conjunctional Billing — Large commercial or instructional customers that have several buildings within a territory should explore the availability of conjunctional billing. This consists of adding together the readings of two or more individual billing meters for a single billing. Due to the usual practice of decreasing rates for larger demands, conjunctional billing can result in lower billing than individual billing. Conjunctional billing will generally result in a higher bill than a master or totalizing meter reading because the maximum demand readings on the individual meters rarely occur simultaneously, so that the arithmetic sum will be greater than the simultaneous sum, unless a provision is made for coincident demand measurement.

(4) Power Factor Billing — If the type of load to be installed in the commercial building will result in poor power factor, e.g., less than 90%, then an evaluation should be made to determine when power factor improvement can be justified to avoid penalty payments or other related costs.

(5) Flat Billing — Certain applications involve service to the load of a fixed characteristic. For such loads, the supplying utility may offer no-meter or flat-connected service. Billing is based upon time and load characteristics. Examples include street lighting, traffic signals, and area lighting.

(6) Off-Peak Billing — Is reduced billing for service utilized during utility off-peak periods, such as water heating and ice making loads. The utility monitors may control off-peak usage through control equipment or special metering. Off-peak billing is also based upon on-peak and time-of-day, or time-of-use, metering for all billing loads.

(7) Standby Service Billing — Also known as "breakdown" or "auxiliary service," this service is applicable to utility customers whose electric requirements are not supplied entirely by the utility. In such cases, billing demand is determined either as a fixed percentage of the connected load or by meter, whichever is higher. This applies to loads that are electrically connected to some other source of supply and for which breakdown or auxiliary service is requested.

(8) Backup Service Billing — Is provided through more than one utility circuit, solely for a utility customer's convenience. The utility customer customarily bears the cost of establishing the additional circuit and associated supply facilities. Generally, each backup service is separately metered and billed by the utility.

(9) Demand Billing — Usually represents a significant part of electric service billing and a good understanding of kW demand metering and billing is important. An electric-demand meter measures the average rate of use of electric energy over a given period of time, usually 15 minute, 30 minute, or 60 minute intervals. A demand register records the maximum demand since the last reading. The demand register is reset when recorded for billing purposes.

(10) Minimum Billing Demand — A utility customer may be subject to minimum demand billing, generally consisting of a fixed amount or a fixed percentage

of the maximum demand established over a prior billing period. This type of charge usually applies to customers with high instantaneous demand loads, such as users of welding or x-ray equipment, customers whose operations are seasonal, or those who have contracted for a given service capacity. Equipment requirements and service usage schedules should be carefully reviewed to reduce or avoid minimum billing demand charges.

(11) Load Factor Billing — The ratio of average kW demand to peak kW demand during a given time period is referred to as the "load factor." Utilities may offer a billing allowance or credit for high load factor usage, a qualification usually determined by evaluating how many hours during the billing period the metered demand was used. As an example of such a credit, the utility may provide a reduced rate for the number of kWh that are in excess of the maximum (metered) demand multiplied by a given number of hours (after 360 hours for a 720 hour month or a 50% load factor).

(12) Interruptible or Curtailable Service — Another form of peak-load shaving used by the utilities is interruptible or curtailable service. Primarily available for large facilities with well-defined loads that can be readily disconnected, the utility offers the customer a billing credit for the capability of requesting a demand reduction to a specified contract level during a curtailment period. The monthly credit for each billing month is determined by applying a demand charge credit to the excess of the maximum measured demand used for billing purposes over the contract demand. Should the customer fail to reduce the measured demand during any curtailment period, at least to the contract demand, severe financial penalties may be incurred. An alternative to disconnecting loads is to supply power from in-plant generation.

4.5 Transformer Connections. Commercial building utilization of low-voltage, three-phase systems of recent vintage in the United States fall into either of two nominal voltage levels: 208Y/120 V or 480Y/277 V. Either of these systems can supply three-phase or single-phase loads; both frequently exist in the typical commercial building. The transformer connection used to derive these voltages is almost exclusively delta-wye or a specially constructed (such as five-legged core) wye-wye transformer commonly used in pad-mounted transformers. The delta primary cancels out virtually all third harmonic components and multiples thereof that may be introduced in electrical transformation equipment or in lighting ballasts. The secondary wye connection provides a tap for the neutral and convenient grounding point as described in 4.7.1.

When power loads are fed from a separate transformer, the delta-delta connection is excellent from the harmonic and unbalanced load standpoints; but a convenient balanced grounding point is not provided (and, in some instances, may not be desired). There is little need to consider this connection under normal circumstances in new commercial building electric systems.

When systems are to be expanded, existing conditions may dictate the use of other connections than delta-wye or delta-delta. It is important to understand that certain transformer connections are less desirable than others for given applications; and that some connections, such as three single-phase transformers supplying a three-phase, four-wire unbalanced load from a three-wire supply, can actually be destructive (in terms of a floating neutral).

Occasionally, service requirements of the utility may dictate the use of a system with a four-wire wye primary. The following paragraphs cover a few of the limitations of the connections in the special circumstances when the preferred connections listed above cannot be used.

When it is desired to use a wye primary and a wye secondary, consideration should be given to using a shell-type core construction that will carry zero-sequence flux.

The primary or secondary windings of a three-phase transformer can be connected using either delta or wye. It is recommended that at least one of the windings be connected to provide a path for third harmonic currents to circulate.

The wye four-wire primary with the wye four-wire secondary and the wye four-wire primary with the delta three-wire secondary are not to be recommended for use without proper engineering consideration. In three-legged core construction, if one leg of the primary line is lost, the presence of the neutral will provide three-phase flux conditions in the core. The phase that has lost its primary will then become a very high reactance winding, resulting in fringing flux conditions. The flux will leave the core and enter the surrounding magnetic materials, such as the clamping angles, tie rods, enclosure, etc. This produces an effective induction heater and results in a high secondary voltage across the load of the faulty phase. In a matter of seconds, this induction effect can destroy the transformer. It is also possible that, should the fault occur by the grounding of one of the primary lines, the primary winding at fault could then act as a secondary and feed back to the ground, thereby causing high current to flow in this part of the circuit. These conditions are inherent with this type of connection. Whether the transformer is of the dry or liquid type makes no difference.

4.6 Principal Transformer Secondary Connections. Systems of more than 600 V are normally three-phase wye or delta ungrounded, or wye solid or resistance grounded. Systems of 120 – 600 V may be either single-phase or three-phase. Three-phase, three-wire systems are usually solidly grounded or ungrounded, but may also be impedance grounded. They are not intended to supply loads connected phase-to-ground. Three-phase, four-wire solidly grounded wye systems are used in most modern commercial buildings. Single-phase services and loads may be supplied from single-phase systems, or from three-wire systems and either phase-to-phase loads (e.g., 208 V) or phase-to-neutral loads (e.g., 120 V) from three-phase, four-wire systems (see Fig 19).

Transformers may be operated in parallel and switched as a unit, provided that the overcurrent protection for each transformer meets the requirements of the NEC, Section 450 [6]. To obtain a balanced division of load current, the transformers should have the same characteristics (rated percent IR and rated percent IX) and be operated on the same voltage-ratio tap. Both IR and IX should be equal in order for two transformers to divide the load equally at all power factors of loads.

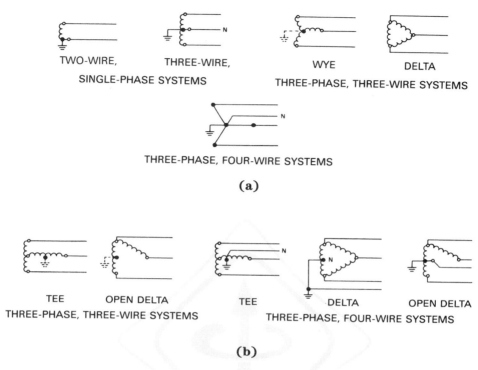

TWO-WIRE,

THREE-WIRE,

SINGLE-PHASE SYSTEMS

WYE

DELTA

THREE-PHASE, THREE-WIRE SYSTEMS

THREE-PHASE, FOUR-WIRE SYSTEMS

(a)

TEE   OPEN DELTA

THREE-PHASE, THREE-WIRE SYSTEMS

TEE   DELTA   OPEN DELTA

THREE-PHASE, FOUR-WIRE SYSTEMS

(b)

Fig 19
Transformer Secondary Connections
(a) Most Commonly Used
(b) Least Commonly Used

4.7 System Grounding. IEEE Std 142-1982, IEEE Recommended Practice for Grounding of Industrial and Commercial Power Systems (ANSI) [13] recommends grounding practices for most systems involving grounding of one conductor of the supply, and the NEC [6] requires grounding of certain systems, as described below. The conductor connected to ground is called the "grounded conductor" and should be distinguished from the grounding conductor (equipment grounding conductor), which is the conductor used to connect noncurrent-carrying conductive parts of electrical equipment to ground. This prevents these parts from acquiring a potential above ground as a result of an insulation failure and causing injury to a person who might come in contact with them. System grounding has the following advantages:

(1) It limits the voltages due to lightning, line surges, or unintentional contact with higher voltage lines and stabilizes the voltage to ground during normal operation.

(2) It limits or prevents the generation of transient overvoltages by changes in the electrostatic potential to ground caused by an intermittent ground on one of the conductors of an ungrounded system.

(3) In combination with equipment grounding, it can be designed to provide a safe method of protecting electric distribution systems by causing the over-current or ground-fault protective equipment to operate to disconnect the circuit in case of a ground fault.

(4) It stabilizes the voltage to ground of line conductors should one of the line conductors develop a fault to ground.

4.7.1 Grounding of Low-Voltage Systems (600 V and Below). The NEC [6] requires that the following low-voltage systems be grounded:

(1) Systems that can be grounded so that the voltage to ground of any un-grounded conductor does not exceed 150 V. This makes grounding manda-tory for the 208Y/120 V three-phase, four-wire system and the 120/240 V single-phase, three-wire system.

(2) Any system in which load is connected between any ungrounded conductor and the grounded conductor. This extends mandatory grounding to the 480Y/277 V three-phase, four-wire system. The 240/120 V, i.e., 240Δ/120 V three-phase, four-wire, open-delta, center-tap ground system, is sometimes supplied for small commercial buildings, where the single-phase load is high and the three-phase load is minimal.

(3) The NEC [6] has special requirements for grounding dc systems and ac systems under 50 V.

The grounded conductor is called the "neutral" on three-phase wye connected systems and single-phase, three-wire systems since it is common to all ungrounded conductors. The NEC [6] requires the grounded conductor to be identified to pre-vent confusion with the ungrounded conductors.

A few utilities provide 240 V and 480 V three-phase, three-wire systems with one phase grounded (corner grounded). This type of grounding is not recommended for commercial buildings and should be accepted only if a suitable alternative system will not be provided.

The NEC [6] requires that separately derived systems be grounded in accord-ance with its rules. An example of a separately derived system is one in which a transformer is used to derive another voltage. The best examples of this are the transformation from a 480 V system to 208Y/120 V or 240/120 V to supply a 120 V load.

An exception to the NEC's grounding requirements is permitted for health care facilities (see the NEC, Article 517 [6]) where the use of a grounded system might subject a patient to electrocution or a spark might ignite an explosive atmosphere in case of an insulation failure (see Chapter 16).

The 240 V, 480 V, and 600 V three-phase, three-wire systems are not required to be grounded; but these systems are not recommended for commercial buildings. When they are used, consideration should be given to providing a derived ground by using a zigzag transformer or delta-wye grounding transformer to obtain the advantages of grounding and limit the damage as described above.

4.7.2 Grounding of Medium-Voltage Systems (Over 600 V). Medium-voltage systems are encountered in commercial buildings when the building becomes too large to be supplied from a single transformer station and the utility primary distribution voltage should be taken through or around the building or buildings to supply the various transformers. Many utility distribution systems are solidly grounded to permit single-phase transformers to be connected phase-to-neutral to supply residences and other small loads, although ungrounded or impedance-grounded systems may occasionally be encountered. The designer should accept whatever grounding system the supplying utility provides. About the only time that the designer has a choice in the grounding of medium-voltage systems is when the supplying utility provides a voltage over 15 000 V and the designer elects to step this voltage down to a lower voltage to distribute through the building, or where large motors (several hundred horsepower) are required, such as in large air-conditioning installations, and it is more economical to use an intermediate voltage, such as 4160 V.

Under these conditions, one method to use is a wye-connected system and then ground the neutral through a resistance that is low enough to stabilize the system voltages but high enough to limit the ground-fault current to a value that will not cause extensive equipment damage before the protective devices can operate. (See IEEE Std 142-1982 (ANSI) [13] for more details.) Since the ground-fault current is limited, ground-fault protection should be installed in addition to phase over-current protection to disconnect the circuit in case of a ground fault.

Whenever a primary substation is customer-owned, the customer has complete say over the grounding methods for his or her own medium-voltage distribution system. Even if the utility owns the transformer, but it is dedicated to the customer, the utility will almost always have specifications for a grounded-wye installation. Although ungrounded or resistance-grounded primary distribution systems may have the advantage of continuity during fault conditions and have low-fault current availability under a single-phase-to-ground fault (safety for electricians and limited machinery damage under faulted conditions), careful consideration should be made before selecting such a system.

4.7.3 Ground-Fault Circuit Interrupter (GFCI). The so-called *people or personnel* protector, GFCI is a very sensitive device that responds to a ground leakage current of 5 mA as a typical design standard. The two circuit wires are both passed through the core of a window current transformer; any difference between these two currents results in a current in the current transformer secondary. Any difference between the two line currents also represents a leakage current to ground; the small current transformer output current, through amplification, trips the integral circuit breaker, which de-energizes the circuit. It does not, however, in itself, provide overcurrent protection. It simply interrupts the very small ground faults as rapidly as it does very large ones. This system, which is designed primarily for the protection of people, should not be confused with the ground-fault protection of equipment whose primary function is to detect ground faults at greater magnitude and which is primarily designed to limit the destructive effects of a ground fault, as contrasted with a shock hazard. The NEC [6] requires GFCI protection for a number of receptacle locations where a significant shock hazard could exist.

4.8 Distribution Circuit Arrangements. Many factors should be considered in the design of the electric power distribution system for a modern commercial building. Some of the most important factors that will influence system design and circuit arrangement are the characteristics of the electric service available at the building site, the characteristics of the load, the quality of service required, the size and configuration of the building, and costs.

Electric service for commercial buildings is available from secondary-network systems in the downtown areas of many large cities in the United States. This service is usually provided from the general distributed street network at a nominal voltage of 208Y/120 V. In cases where the kVA demand of the building load is sufficiently high to justify the establishment of a spot-network system, service may be available at 480Y/277 V instead of 208Y/120 V. When the building is very large, the electric utility may establish spot-network substations on intermediate floors in the building as well as at the basement level.

When a commercial building is small enough to be supplied from a single transformer station, the recommended practice is to allow the utility to install the transformer and then purchase power at utilization voltage. Commercial building personnel are often not qualified to operate and maintain medium-voltage equipment, and any option to provide the transformer in return for a reduction in the rate should justify the expense and risk involved in owning the transformer. When the building is too large to be supplied from a single transformer station located at a point suitable to the supplying utility, power may be purchased at the utility distribution voltage and taken through or around the building to supply the transformers stepping down to utilization voltage. The NEC [6] and utility policy, with some exceptions, provide for only one service to a building; utilities, as a general rule, will not provide transformers that are suitable for installation indoors unless the transformers are installed in utility-approved vaults. In cases where commercial buildings have more than one tenant, some utilities will furnish the medium-voltage system and transformers in return for the right to sell power direct to the tenants, and for buildings supplied from a utility network.

Five basic circuit arrangements are used for medium- and low-voltage distribution in commercial facilities: radial-circuit, primary-selective, secondary-selective, secondary-network, and loop-circuit. The reader should recognize that the medium-voltage circuits and substations may be owned by either the utility company or the building owner, depending upon the electric tariffs, rates, local practices, and requirements of the particular electric utility serving the specific building site.

In the remainder of this chapter, where circuit breakers are shown in the figures, fused equipment may be the design choice. In this case, proper design considerations, including fault protection, safety interlocking, automatic or manual control, training, experience, availability, and capabilities of operating and maintenance personnel, should be fully evaluated when developing a safe and reliable system. See Chapter 9 for a discussion of electrical protection.

4.8.1 Radial Feeders. When power is brought into a commercial building at the utilization voltage, the simplest and the lowest cost way of distributing the power is to use a radial-circuit arrangement. Since the majority of commercial buildings are served at utilization voltage, the radial-circuit arrangement is used in the great majority of commercial buildings. The low-voltage, service entrance circuit comes into the building through service entrance equipment and terminates at a main switchgear assembly, switchboard, or panelboard. Feeder circuits are provided to the loads or to other switchboards, distribution cabinets, or panelboards.

When power is purchased at a medium voltage, one or more transformers may be located to serve low-voltage radial circuits. Circuit breakers or fused switches are required on both the medium- and low-voltage circuits in this arrangement except when the NEC [6] permits the medium-voltage device to serve for the secondary protection.

Figure 20 shows the two forms of radial-circuit arrangements most frequently used in commercial buildings. Under normal operating conditions, the entire load is served through the single incoming supply circuit, and, in the case of medium-voltage service, through the transformer. A fault in the supply circuit, the transformer, or the main bus will cause an interruption of service to all loads. A fault on one of the feeder or branch circuits should be isolated from the rest of the system by utilizing selectively coordinated main, feeder, and branch-circuit protective devices. Under this condition, continuity of service is maintained for all loads except those served from the faulted branch circuit.

Continuity of service to the loads in commercial buildings is very important from a safety standpoint as well as with regard to the normal activities of the occupants of the building. The safety aspect becomes more critical as the height of the building and number of people in the building increase. This requirement for continuity of service often requires multiple paths of power supply as opposed to a single path of power supply in the radial-circuit arrangement. However, modern distribution equipment has demonstrated sufficient reliability to justify the use of the radial-circuit arrangement in many commercial buildings. When the risk is slight and the consequence of service loss is unimportant, branch circuits and feeders are almost invariably radial feeders. As the demand or the size of the building, or both, increase, several smaller secondary substations rather than one large secondary substation may be required to maintain adequate voltage at the utilization equipment. Each of the smaller substations may be located close to the center of the load area that it is to serve. This arrangement, shown in Figs 21 and 22, will provide better voltage conditions, lower system losses, and offer a less expensive installation cost than the arrangement using relatively long high-amperage, low-voltage feeder circuits.

**Fig 20
Radial-Circuit Arrangements
in Commercial Buildings**

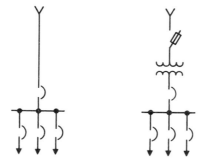

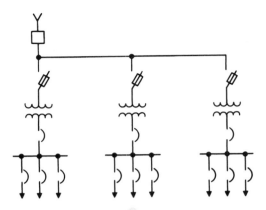

Fig 21
Radial-Circuit Arrangement—Common
Primary Feeder to Secondary-Unit Substations

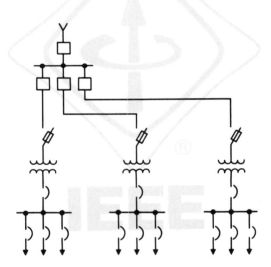

Fig 22
Radial-Circuit Arrangement—
Individual Primary Feeders
to Secondary-Unit Substations

The relative economics of radial-circuit arrangements using low- or medium-voltage feeders will vary with building size, demand, cost of floor space, and utility tariffs. Medium-voltage systems require investment in transformers, medium-voltage protective devices, medium-voltage cable, and, possibly, some rentable floor space for substation locations. On the other hand, the investment in feeder and riser circuits for a low-voltage system of the same capacity may become excessive when voltage-drop limitations are to be met.

A fault in a primary feeder, as shown in the arrangement in Fig 21, will cause the main protective device to operate and interrupt service to all loads. If the fault were in a transformer, service could be restored to all loads except those served from that transformer. If the fault were in a primary feeder, service could not be restored to any loads until the source of the trouble has been eliminated and repairs completed. Since it is to be expected that more faults will occur on the feeders than in the transformers, it becomes logical to consider providing individual circuit protection on the primary feeders as shown in Fig 22. This arrangement has the advantage of limiting outages, due to a feeder or transformer fault, to the loads associated with the faulted equipment. The cost of the arrangement in Fig 22 will usually exceed the cost of the arrangement in Fig 21.

4.8.2 Primary-Selective Feeders. The circuit arrangements of Fig 23 provide a means of reducing both the extent and duration of an outage caused by a primary feeder fault. This operating feature is provided through the use of duplicate primary feeder circuits and load interrupter switches that permit connection of each secondary substation transformer to either of the two primary feeder circuits. Each primary feeder circuit should have sufficient capacity to carry the total load in the building. Suitable safety interlocks for each pair of fused switches or circuit breakers are usually required to avoid closing both switches at the same time. Under normal operating conditions, the appropriate switches are closed in an attempt to divide the load equally between the two primary feeder circuits. Should a primary feeder fault occur, there will be an interruption of service to only half of the load. Service can be restored to all loads by switching the de-energized transformers to the other primary feeder circuit. The primary-selective switches are usually manually operated and outage time for half the load is determined by the time it takes to accomplish the necessary switching. An automatic throwover switching arrangement can be used to reduce the duration of interruption of service to half of the load. The additional cost of this automatic feature may be justified in many applications. If a fault occurs in a secondary substation transformer, service can be restored to all loads except those served from the faulted transformer.

The higher degree of service continuity afforded by the primary-selective arrangement is realized at a cost that is usually 10%–20% above the cost of the circuit arrangement of Fig 21 because an additional primary circuit and the primary switching equipment at each secondary substation is needed. The cost of the primary-selective arrangement, using manual switching, will sometimes be less than the radial-circuit arrangement.

A variation of the circuit arrangements shown in Fig 23 utilizes three primary-selective feeders and one standby feeder. Each feeder is sized between one-half and two-thirds of the total load and supplies one-third of the total load under normal conditions. Under emergency conditions, with a primary cable fault, the load on the faulted cable can be transferred to the standby feeder. Depending on the capacity of the standby feeder, the load can be transferred to the remaining normal feeder or left on the standby feeder until the cause of the failure is corrected.

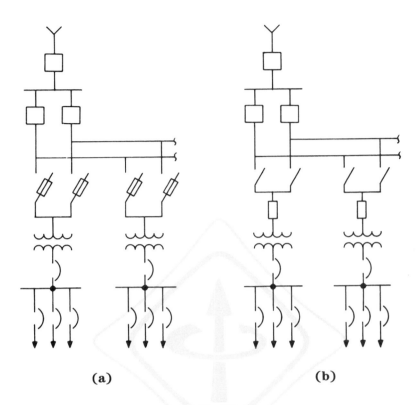

Fig 23
Primary-Selective Circuit Arrangements
(a) Dual Fused Switches
(b) Duplex Load Interrupter Switches with Transformer Primary Fuse

4.8.3 Secondary-Selective Feeders. Under normal conditions, the secondary-selective circuit arrangement in Fig 24 is operated as two separate radial systems. The secondary bus-tie circuit breaker or switch in the double-ended substation is normally open.

The load served from a secondary substation should be divided equally between the two bus sections. If a fault occurs on a primary feeder or in a transformer, service is interrupted to all loads associated with the faulted feeder or transformer. Service may be restored to all secondary buses by first opening the main secondary switch or circuit breaker associated with the faulted transformer and primary feeder, and then closing the bus-tie device in such a manner that all three cannot be in the closed position simultaneously. This prevents parallel operation of the two transformers and thereby minimizes the service interruptions to all loads on the bus when a fault occurs in either a primary feeder or a transformer. To prevent closing the tie on a faulted switchgear bus, a main tie/main safety interlock scheme may be provided to lock out the tie device whenever a secondary main has interrupted a downstream fault.

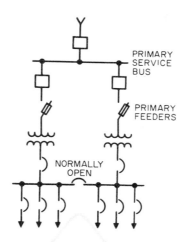

Fig 24
Secondary-Selective Circuit Arrangement
(Double-Ended Substation with Single Tie)

The cost of the secondary-selective circuit arrangement will depend upon the spare capacity in the transformers and primary feeders. The minimum transformer and primary feeder capacity will be determined by essential loads that should be served under standby operating conditions. If service is to be provided for all loads under standby conditions, then each primary feeder should have sufficient capacity to carry the total load, and each transformer should be capable of carrying the total load on both substation buses.

This type of circuit arrangement will be more expensive than either the radial or primary-selective circuit arrangement; but it makes restoration of service to all essential loads possible in the event of either a primary feeder or transformer fault. The higher cost results from the duplication of transformer capacity in each secondary substation. This cost may be reduced by load-shedding nonessential feeders.

A modification of the secondary-selective circuit arrangement is shown in Fig 25. In this arrangement, there is only one transformer in each secondary substation; but adjacent substations are interconnected in pairs by a normally open low-voltage tie circuit. When the primary feeder or transformer supplying one secondary substation bus is out of service, essential loads on that substation bus can be supplied over the tie circuit. The operating aspects of this system are somewhat complicated if the two substations are separated by distance. It may not be a desirable choice in a new building because a multiple-key interlock system would be required if it became necessary to avoid tying the two substations together while they were energized.

4.8.4 Secondary Network. High-rise and institutional buildings that have concentrated loads that require a power source with high reliability are often supplied from secondary systems. In a modern, large commercial building with heavy electronic and computer loads, the time it takes to operate a mechanical transfer

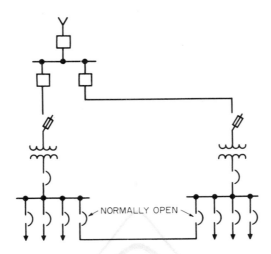

Fig 25
Secondary-Selective Circuit Arrangement
(Individual Substations with Interconnecting Ties)

switch or the time required for personnel to close a tie feeder is normally unacceptable. A secondary network is formed when two or more transformers having the same characteristics are supplied from separate feeders, and are connected to a common bus through network protectors. The distributed network and the spot network are the two basic types of secondary-network systems. The distributed network shown in Fig 26 is a widely dispersed system that has multiple transformer/network protector units connected to a cable grid. The grid is tapped to provide takeoffs to utility customers at commercial buildings. The spot network shown in Fig 27 is a localized distribution center consisting of two or more transformer/network protector units connected to a common bus called a "collector bus."

A typical commercial building spot network is illustrated in Fig 28. Feeders originating at the 13.8 kV service entrance substation are extended throughout the commercial building complex to supply spot networks at four locations. The feeders terminate at the transformer primary disconnecting device. In this particular design, the device is a fused load interrupter with a grounding switch located within the same enclosure. Although the grounding switch has a fault closing rating, it cannot be operated until the safety requirements of a key interlock scheme have been satisfied. The key interlocks prevent closing the grounding switch until all possible sources of supply to the feeder have been isolated.

In network system design, protection for the transformer primary is usually provided by the substation feeder breaker overcurrent relays. Since the substation breaker often feeds a group of transformers, the protection should be set high enough to prevent tripping on the sum of the individual transformer currents during contingency loading and inrush conditions. The application of primary fusing, as shown in Fig 28, offers a significant improvement over the limited protection

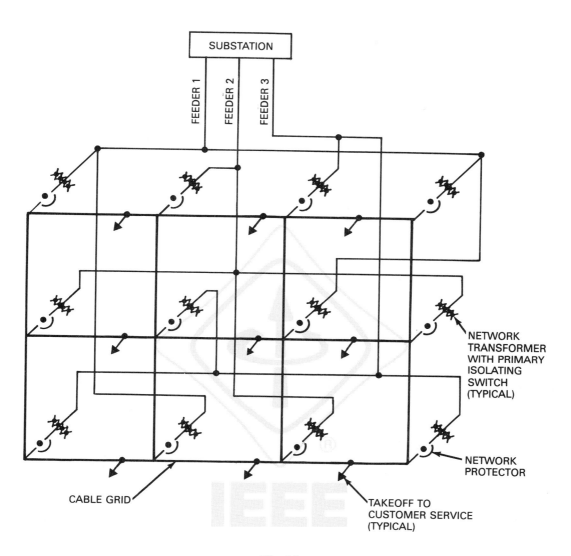

Fig 26
Distributed Secondary Network

provided by the substation breaker overcurrent relays alone. The requirements of
the NEC, Article 450 [6] call for primary protection by a circuit breaker to be set no
higher than 400% of transformer rated current. In this case, the requirement would
effectively limit the maximum setting of the primary protector to an average of
100% of transformer rating, which is impractical for operational and system coordination purposes. The circuit breaker overcurrent protection would have to be set
much higher than the 400% value to prevent unnecessary tripping. Consequently,
compliance with the NEC, Article 450 [6] could not be met.

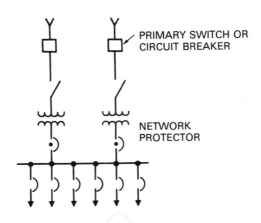

Fig 27
Basic Spot Network

Primary fusing provides a more sensitive level of protection, especially to liquid filled transformers in which tank rupture and serious fire are possible. In the case of liquid filled transformers, current-limiting fuses are recommended. There are limitations on the use of current-limiting fuses on large transformers because of their high cost and need for parallel fusing.

Transformers of the illustrated design in Fig 28 are the ventilated-dry-type with forced-air cooling. The transformers are each rated 1500 kVA on a self-cooled basis and collectively supply a peak load of 4500 kVA. Under normal peak loading conditions, the transformers are loaded to 75% of capacity. The system that is designed for first contingency operation provides full system load capacity with one transformer out of service. Under this condition, the remaining three transformers with cooling fans operating have a 33% increased capacity that maintains the transformer loading at 75% of its forced-air rating. Had liquid filled transformers been used in this design, the loading would have been at 75% of capacity with all transformers in service and at 87% capacity with cooling under first contingency operation. For 65 °C (149 °F) rise liquid filled transformers of this size, fan cooling provides a 15% increase above the self-cooled rating. For liquid filled transformers of this size with a dual rating of 55 °C (131 °F)/65 °C (149 °F) rise with fan cooling, a 28% increase above the self-cooled rating is available.

The electric characteristics of network transformers are essentially the same as secondary substation transformers. The one characteristic difference is the preferred impedance voltage rating. Secondary substation transformers have a typical rating of 5.75%, while network transformers similar to the 1500 kVA units in the illustrated system have a typical impedance rating of 7%. The higher rating provides a reduction in short-circuit current under fault conditions.

Conventional network protectors are self-contained units consisting of an electrically operated circuit breaker, special network relays, control transformers,

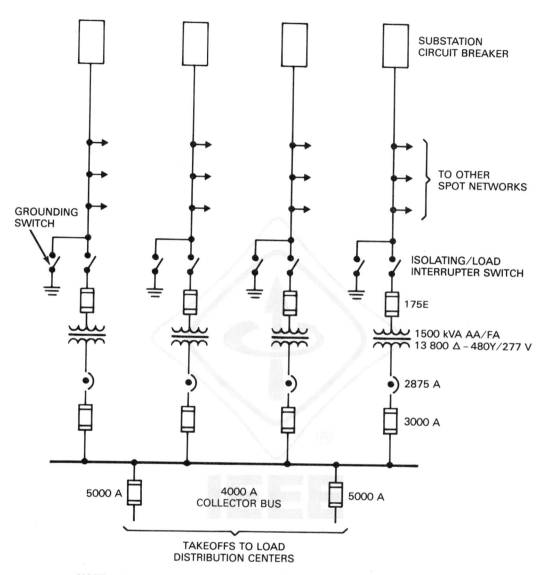

NOTE: All devices not shown; simplified diagram for illustration purposes only.

**Fig 28
Four-Unit Spot Network**

instrument transformers, and open-type fuse links. The protector will automatically close when the oncoming transformer voltage is greater than the collector bus voltage and will open when reverse current flows from the collector bus into the transformer. Reverse current flow can be the result of a fault beyond the line side of the protector, supplying load current back into the primary distribution system when the collector bus voltage is higher than the individual transformer voltage, or the opening of the transformer primary feeder breaker, which causes the collector bus to supply transformer magnetizing current via the transformer secondary winding.

Network protectors are not designed to provide overcurrent protection in accordance with the NEC [6], and, therefore, do not meet the requirements for customer-owned services, unless supplementary protection is added.

Fuses of a special alloy are included in the protector package. Their primary function is to protect the transformer under severe short-circuit conditions in the event the protector fails to open. The application of separately mounted current-limiting fuses offer several improvements over standard network protector fuse links. The fuse links have a less inverse time current characteristic, which allows fault currents to persist for an extended length of time. Conversely, the Class L current-limiting fuse has an extremely inverse characteristic, which provides a much faster clearing time for moderate level faults and current limitation when the prospective fault current is above a specific value.

Unlike open-type fuse links, the interruption of fault current by the Class L fuse takes place within an insulated tube where the released thermal energy and arcing are contained. With open-type fuse links, the protector's internal components are subjected to the effects of the rupturing element, which may cause flashover and result in serious damage to the protector. Additionally, mounting the fuse outside the network protector enclosure removes a significant heat source, which allows the protector to operate at a lower temperature.

Low-voltage power circuit breakers may be applied with separate network protection relays and used as network protectors. When so employed, circuit breakers offer some advantage. Racking-in and withdrawal procedures do not require physical contact with energized components. Greater fault interrupting capacity is provided, and integral overcurrent protection, which can meet the NEC [6] requirements for customer-owned services, is available. Ground-fault protection is also available; however, careful study is advised prior to application. Integral ground-fault protection applied singularly and not in conjunction with downstream protection may compromise the intended reliability of network system design.

In comparison to network protectors, power circuit breakers do have one disadvantage. The number of permissible mechanical operations for a breaker is far fewer than the number allowed for a protector. This limitation should be especially noted when large frame size breakers are considered for network application.

Instead of using cable in the distributed-type network, collector buses in modern commercial building spot-network design are usually metal-enclosed busway or a specially designed high-integrity bus structure preferred by utilities. Protection for the utility preferred bus is provided through the physical design, which utilizes an open-type construction with insulated bus bars widely spaced between phases and mounted overhead on insulated supports. The physical construction, which varies among utilities, affords excellent protection against electrical faults and the mechanical stresses imposed by short-circuit currents as high as 200 000 A.

When a metal-enclosed bus is used, several options are possible. Greater electrical integrity may be provided by the use of a bus manufactured to 5 kV design standards. Construction in the 5 kV design mandates a larger physical spacing between phases and a higher grade of bus insulation. Regardless of the type when any form of metal-enclosed bus is specified, the application of ground-fault protection is recommended.

Relay protection is the most common method of ground-fault protection. The fault current may be sensed by the ground return method, by the residual method, or by the zero-sequence method. Each of the methods have proved successful where appropriately applied; but they share a common limitation in that they cannot distinguish between in-zone and thru-zone ground faults unless incorporated in a complex protection scheme.

One particular method of ground-fault detection that is not prone to unnecessary tripping is enclosure monitoring. This method offers the distinct advantage of not requiring coordination with other protective devices. Enclosure monitoring is a simple concept that has been employed on various electric system components, such as motors, transformers, switchgear, and busways. When the collector bus is metal enclosed, it may be protected against arcing ground-fault damage using the scheme in Fig 29. As shown, the enclosure is grounded by a conductor, which is monitored by a current transformer. The enclosure is insulated at all termination points that are connected to other enclosures that are not in the same zone of protection. Additional protection can be provided by the use of thermal protectors above the busway and switchgear.

Secondary-network systems were originally designed to operate so that faults at the grid were allowed to burn clear rather than incur a disruption of service. This design philosophy may be acceptable for 216 V cable grids; however, when the common connection for the network protector/transformer units is a 480 V collector bus in a commercial building with an available short-circuit current in excess of 100 000 A, additional protection is advised. With emergency and standby generators and uninterruptible power supplies included in commercial building power system design, accepting the risk of serious electrical faults with an extended period of system downtime is not justified. Properly applied protection will significantly reduce fault damage and allow system restoration in minimal time.

Protection for spot-network systems is a subject that has received increased attention in recent years. Many engineers now believe that it is unwise to apply the same design concepts for commercial building spot networks as traditionally provided for utility-type distributed secondary-network systems. In response to a number of serious spot-network burndowns, a variety of protective schemes and devices not normally applied to secondary distributed networks have been developed and employed in spot-network systems. They include

(1) Transformer primary protection
(2) Network protector current-limiting fuses
(3) Ground-fault relaying
(4) Enclosure fault current monitoring
(5) High-integrity bus structures
(6) Thermal sensors
(7) Ultraviolet light detectors
(8) Infrared detectors

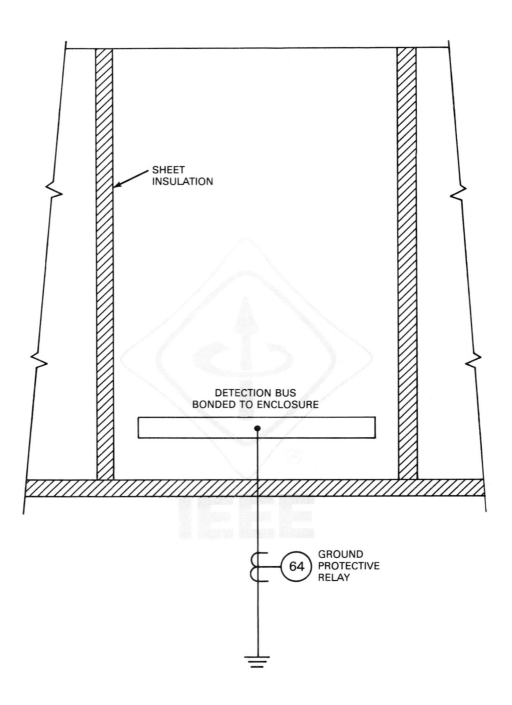

Fig 29
Enclosure Monitor

(9) Smoke detectors
(10) Radio-frequency interference and audible noise detectors
(11) Current-limiting cable limiters installed on each end at each cable where three or more cables per phase are utilized

While the application of the listed protective schemes and devices are not suggested for all spot networks, there is a minimum level of spot-network protection that is recommended for commercial buildings.

Spot networks are employed to provide a reliable source of power to important electrical loads. To ensure service continuity in the event a utility feeder is lost, spare capacity is built in to allow for at least one contingency. Planning for service continuity should be extended beyond the consideration of losing a utility feeder. The consequences of severe equipment damage, including the resulting system downtime, should also be considered. Spot-network systems, which incorporate transformer primary protection, improved network protector protection, and ground-fault protection judiciously applied, will enhance system reliability and, therefore, are recommended.

4.8.5 Looped Primary System. The looped primary system (see Fig 30) is basically a two-circuit radial system with the ends connected together to form a continuous loop. Early versions of the closed-loop system, as shown in Fig 30(a), were designed to be operated with all loop isolating switches closed. Although it is relatively inexpensive, this system has fallen into disfavor because its apparent reliability advantages are offset by the interruption of all service by a fault occurring anywhere in the loop, by the difficulty of locating primary faults, and by safety problems associated with the nonload break, or "dead break," isolating switches.

Newer open-loop versions shown in Fig 30(b), which are designed for modern underground commercial and residential distribution systems, utilize fully rated air, oil, and vacuum interrupters. Equipment is available in voltages up to 34.5 kV with interrupting ratings for both continuous load and fault currents to meet most system requirements. Certain equipment can close in and latch on fault currents, equal to the equipment interrupting values, and still be operational without maintenance.

With the elimination of the major disadvantages of the older closed-loop systems and the present demand for decentralized systems with low profile pad-mounted equipment and greater reliability than the simple radial system, the open-loop primary system has become a viable distribution solution.

The major advantages of the open-loop primary system over the simple radial system is the isolation of cable or transformer faults, or both, while maintaining continuity of service for the remaining loads. With coordinated transformer fusing provided in the loop-tap position, transformer faults can be isolated without any interruption of primary service. Primary cable faults will temporarily drop service to half of the connected loads until the fault is located; then, by selective switching, the unfaulted sections can be restored to service, which leaves only the faulted section to be repaired.

Disadvantages of the loop system are the increased costs to fully size cables, protective devices, and interrupters to total capacity of the load (entire load on one feeder), and the time delay necessary to locate the fault, isolate the section, and

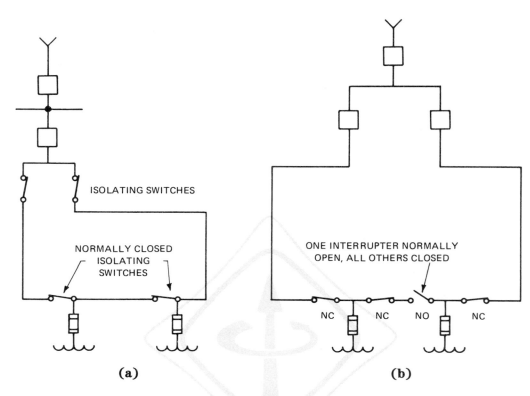

ISOLATING SWITCHES

NORMALLY CLOSED
ISOLATING
SWITCHES

ONE INTERRUPTER NORMALLY
OPEN, ALL OTHERS CLOSED

NC   NC   NO   NC

(a)   (b)

**Fig 30
Looped Primary-Circuit Arrangement
(a) Closed Loop (Obsolete)
(b) Open Loop**

restore service. The safety considerations in maintaining a loop system are more complex than for a radial or a primary-selective system.

The development of load break cable terminators and the regulatory requirements for total underground utility distribution have led to the use of loop-loop primary distribution circuit arrangements.

For systems that have a large quantity of small capacity transformers, the loop-loop design has the lowest cost of the loop-circuit arrangements. It has the same disadvantages as the old designs because it is still possible to connect a cable terminator into a cable fault.

Figure 31 shows a loop-loop system in which pad-mounted loop manually operated load break sectionalizing switches are provided in the main loop and load break cable terminators are provided in the secondary loops. The main loop is designed to carry the maximum system load, whereas secondary loops are fused to handle load concentrations smaller than the total system capacity. For additional discussion on looped systems, see Chapter 7.

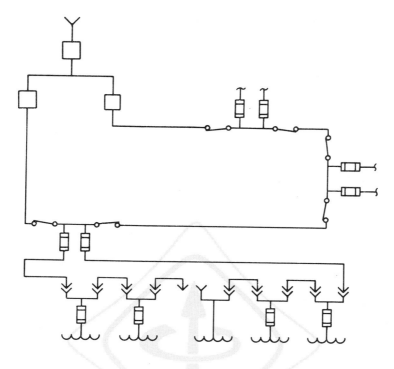

Fig 31
Loop-Loop Primary-Circuit Arrangement

4.9 Emergency and Standby Power Systems. Emergency electric services are required for protection of life, property, or business where loss might be the result of an interruption of the electric service. The extent of the emergency services required depends on the type of occupancy, the consequences of a power interruption, and the frequency and duration of expected power interruptions.

Municipal, state, and federal codes define minimum requirements for emergency systems for some types of public buildings and institutions. These shall be adhered to; but economics or other advantages may result in making provisions beyond these minimums (see the NEC, Articles 517, 700, 701, and 702 [6]). The following presents some of the basic information on emergency and standby power systems. For additional information, design details, and maintenance requirements, see IEEE Std 446-1987, IEEE Recommended Practice for Emergency and Standby Power Systems for Industrial and Commercial Applications (ANSI) [14], ANSI/NFPA 110-1988, Emergency and Standby Power Systems [9], and ANSI/NFPA 110A-1989, Stored Energy Systems [10], and Reference [28].

Emergency power systems should be separated from the normal distribution systems by using separate raceways and panelboards. The NEC [6] requires that each item of emergency equipment be clearly marked as to its purpose. In large public buildings, physical separation of the emergency system from the normal system elements would enhance the reliability of the emergency system in the event of fire or other contingencies.

4.9.1 Lighting. Exit and emergency lights that are sufficient to permit safe exit from buildings where the public may congregate should be supplied from an emergency power source (i.e., auditoriums, theaters, hotels, large stores, sports arenas, etc.). Local regulations should always be referred to for more specific requirements. When the emergency lighting units are not used under normal conditions, power should be immediately available to them upon loss of the normal power supply. When the emergency lights are normally in service and served from the normal power supply, provisions should be made to transfer them automatically to the emergency power source when the normal power supply fails.

Sufficient lighting should be provided in stairs, exits, corridors, and halls so that the failure of any one unit will not leave any area dark or endanger persons leaving the building. Adequate lighting and rapid automatic transfer to prevent a period of darkness is important in public areas. Public safety is improved and the chance of pilfering or damage to property is minimized.

ANSI/NFPA 101-1988, Life Safety Code [8] requires that emergency power sources for lighting be capable of carrying their connected loads for at least 90 minutes. There are cases in which provisions should be made for providing emergency service for much longer periods of time, such as in health care facilities, communications, police, fire fighting, and emergency services. A 2-3 hour capacity is more practical and, in many installations, a 5-6 hour capacity is provided. During a severe storm or catastrophe, the demands on hospitals, communications, police, fire fighting, and emergency services facilities will be increased. A third source of power to achieve the desired lighting reliability may be required.

When installation of a separate emergency power supply is not warranted but some added degree of continuity of service for exit lights is desired, they may be served from circuits connected ahead of the main service entrance switch for some occupancies. This assures that load switching and tripping due to faults in the building's electric system will not cause loss of the exit lights. However, this arrangement does not protect against failures in the electric utility system.

4.9.1.1 Illumination of Means of Egress. In its occupancy chapter, ANSI/NFPA 101-1988 [8] has illumination requirements for building egress, which includes stating the type of emergency lighting required (see Reference [31]).

Primary or normal illumination is required to be continuous during the time "the conditions of occupancy" require that the means of egress be available for use. ANSI/NFPA 101-1988 [8] specifies the illuminances and equipment for providing this type of lighting.

Emergency power sources listed in the NEC, Article 700 [6] include

(1) Storage batteries (rechargeable type) to supply the load for 90 minutes without the voltage at the load decreasing to 87.5% of normal
(2) Generator sets that will accept the emergency lighting load within 10 seconds, unless an auxiliary lighting source is available
(3) Uninterruptible power supplies

(4) Separate electric utility service, which is widely separated electrically and physically from the normal service

(5) Unit equipment (permanently installed) consisting of a rechargeable storage battery, automatic charger, lamp(s), and automatic transfer relay.

Refer to the ANSI/NFPA 101-1988, Sections 5-8 and 5-9 [8], "Illumination of Means of Egress" and "Emergency Lighting," respectively.

4.9.2 Power Loads. An emergency source for supplying power loads is required when loss of such a load could cause extreme inconvenience or hazard to personnel, loss of product or material, or contamination of property. The type and size of the emergency system should be determined through consideration of the health and convenience factors involved and whether the utilization affects health care facilities, communication systems, alarm systems, police, fire fighting, and emergency services facilities. The installation should comply with any applicable codes and standards and be acceptable to the authority that has jurisdiction. For example, health care facilities may require conformance to ANSI/NFPA 99-1990, Health Care Facilities [7] and the NEC, Article 517 [6]. Fire pump installations may require conformance to ANSI/NFPA 20-1990, Centrifugal Fire Pumps [4].

In laboratories where continuous processes are involved or where chemical, biological, or nuclear experimentation is conducted, requirements are very demanding insofar as power and ventilating system requirements are concerned. Loss of adequate power for ventilation could permit the spread of poisonous gases, biological contamination, or radioactive contamination throughout the building and can even cause loss of life. A building contaminated from radioactive waste could be a total loss or require expensive cleanup measures. Many processes or experiments cannot tolerate a power loss that could interrupt cooling, heating, agitation, etc.

Emergency power for fire pumps should be provided when water requirements cannot be met from other sources. Emergency power for elevators should also be considered when elevators are necessary to evacuate buildings or the cost seems warranted to avoid inconvenience to the public. This does not mean that the emergency power supply should have the full capacity for the demand of all elevators simultaneously.

4.9.3 Power Sources. Sources for emergency power may include batteries, local generation, or a separate source over separate lines from the electric utility. The quality of service required, the amount of load to be served, and the characteristics of the load will determine which type of emergency supply is required.

4.9.3.1 Batteries. Batteries offer an extremely reliable source of energy but also require regular maintenance. Their capacity in Ah is also limited. Inspection and tests of individual cells should be made at regular intervals to ensure that electrolyte levels and correct charges are maintained. When lead-acid batteries are used, they should be of the sealed-glass-jar type. Ample space should be provided together with adequate ventilation of the battery room. Batteries of the nickel-lead-alkaline or the nickel-cadmium type may be used provided that the characteristics of each battery type and each load are considered; these batteries are more suitable for standby service and require less attention than the lead-acid type.

Battery charging equipment will be determined by the battery characteristics and the type of load being served. The capacity of the equipment will depend on the size of the load and the length of time such load should be supplied. For low charging rates, electronically controlled chargers are generally used. Heavier loads may require motor-generator sets or heavy-duty silicon controlled rectifiers, the latter is the more advantageous choice from the standpont of efficiency and convenience.

For a simple and effective lighting installation that will permit building evacuation, small package units containing light, battery, charger, and relay are suggested. See the NEC, Article 700 [6] for installation requirements. Normal supply voltage also actuates a relay to de-energize the light on the emergency unit. When the normal supply fails, the relay drops out and the light is energized from the self-contained battery. Provision is made for testing the unit.

4.9.3.2 Local Generation. Local generation is advisable when service is absolutely essential for lighting or power loads, or both, and when these loads are relatively large and distributed over large areas. Several choices are available in the type of prime mover, voltage of the generator, and method of connection to the system. Various alternates should be considered. The prime mover supply may be steam, gas, gasoline, diesel fuel, or liquified petroleum gases (LPG).

For generators over 500 kW, gas turbine driven units may be a favorable choice. This type of unit has acceptable efficiency at full load; but it is much less efficient than other types of drives at partial load. Gas turbine driven units do not start as rapidly as other drives but are reliable and require a minimum of attention. (They generally will not meet the NEC [6] requirements for emergency systems. Generator sets requiring more than 10 seconds to develop power require that an auxiliary system supply power until the generator can pick up load.)

Fuel storage requirements should be determined after considering the frequency and duration of power outages, the types of emergency loads to be served, and the ease of replenishing fuel supplies. Some installations may require a supply sufficient for 3 months be maintained, while a 1 day supply may be adequate for others. Code requirements (see ANSI/NFPA 37-1990, Stationary Combustion Engines and Gas Turbines [5]) severely limit the amount of fuel that can be stored in buildings, so that fuel may have to be piped to a small local tank adjacent to the generator. The NEC [6] and other codes (e.g., EGSA 109C-1984, "Codes for Emergency Power by States and Major Cities" [12][42]) require an on-site fuel supply capable of operating the prime mover at full demand load for at least 2 hours.

Generator selection can only be made after a careful study of the system to which it is connected and the loads to be carried by it. The voltage, frequency, and phase relationships of the generator should be the same as in the normal system. The size of the generator will be determined by the load to be carried, with consideration given to the size of the individual motors to be started. Another consideration is the distortion created by the loads that the system will be supplying. The speed and voltage regulation required will determine the accuracy and sensitivity of regulating devices. When a generator is required to carry emergency loads only during power outages and should not operate in parallel with the normal system, the simplest type of regulating equipment is usually adequate. For parallel opera-

---

[42]EGSA publications are available from the Electrical Generating Systems Association, 10251 West Sample Road, Suite D, P.O. Box 9257, Coral Springs, FL 33075-9257.

tion, good quality voltage regulators and governors are needed to ensure proper active and reactive power loading of the generator. When the generator is small in relation to the system, it is usually preferable to have a large drooping characteristic in the governor and considerable compensation in the voltage regulator so that the local generator will follow the larger system rather than try to regulate it. Automatic synchronizing packages for paralleling generators are available that may include all the protective features required for paralleling generators. The design of this equipment should be coordinated with the characteristics of the generator.

4.9.3.3 Dual-Service Connection. When the local utility can provide two or more service connections over separate lines and from separate generation points so that system disturbances or storms are not apt to affect both supplies simultaneously, local generation or batteries may not be justified. A second line for emergency power should not be relied upon, however, unless total loss of power can be tolerated on rare occasions. The alternate feeder can either serve as a standby with primary switching or have its own transformer with secondary switching.

4.9.4 Transfer Methods. Figure 32(a) shows a typical switching arrangement in which a local emergency generator is used to supply the entire load upon loss of normal power supply. All emergency loads are normally supplied through device A. Device B is open and the generator is at rest. When the normal supply fails, the transfer switch undervoltage relay is de-energized and, after a predetermined time delay, closes its engine starting contacts. The time delay is introduced so that the

Fig 32
Typical Transfer Switching Methods
(a) Total Transfer
(b) Critical Load Transfer

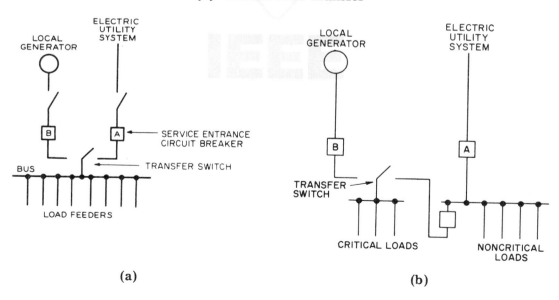

(a)

(b)

generator will not be started unnecessarily during transient voltage dips and momentary outages. When the alternate source is a generator, sufficient time or speed monitoring should be allowed to permit the generator to reach acceptable speed before transfer and application of load. It should be noted that the arrangement shown in Fig 32(a) does not provide complete protection against power disruption within the building.

Figure 32(b) shows a typical switching arrangement in which only the critical loads are transferred to the emergency source, in this case, a generator. For maximum protection, the transfer switch is located close to the critical loads. For further discussion on the application of automatic transfer switches, see Chapter 5.

Figure 33 shows two separate sources of power external to the buildings that have the necessary reliability to satisfy the need for emergency reliability and the need for emergency power. Relaying is provided to transfer the load automatically to either source if the other one fails. The control is arranged so that a transfer will not take place unless one source (alternate or normal) is energized. If the alternate supply is not able to carry the entire load, provisions should be made to drop noncritical loads when the transfer takes place.

Figure 34 shows a secondary-selective substation in which each incoming service feeds its respective bus and the loads connected to the bus. Should one service fail, the de-energized bus can be connected to the other bus via the tie device.

Normally, the two main overcurrent devices are closed and the tie circuit breaker is interlocked to permit the two mains or one main and the tie overcurrent devices to be closed, but prevents the three overcurrent devices from being closed simultaneously. Many other arrangements of switching and relaying are possible; but the four cases shown illustrate the basic principles. A careful study of each system should be made to determine exact power needs and critical features, and to select

Fig 33
Service Transfer Using Circuit Breakers

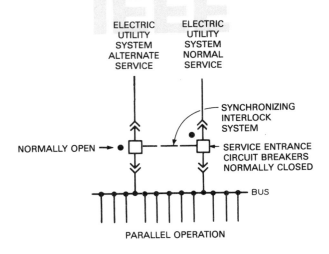

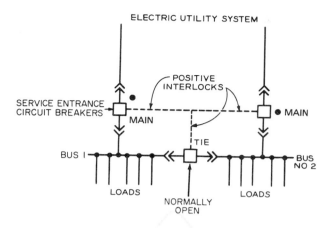

Fig 34
Secondary Transfer Using Circuit Breakers

an arrangement that meets these requirements, which is consistent with sound economics and applicable codes.

4.9.5 Special Precautions. Transfer of resistance and most power and lighting loads can be made as rapidly as desired, but is dependent upon the available switching equipment. Note that there will be a momentary outage on the order of a fraction of a second when transfer takes place with an immediate available alternate source. The outage time may be up to 10 seconds or longer if an idle generator set has to be started. Therefore, all equipment connected to the emergency source should be reviewed to determine the effect of the momentary outage and what precautions should be taken. The short outage is hardly noticeable with incandescent and fluorescent lights. However, mercury, ordinary sodium, and metal-halide lamps will drop out, so they should not be used for emergency lighting. Motors with undervoltage releases that are not equipped with built-in time delay and relay controls will also drop out. Computers may shut down, and some data may be lost. When outages caused by a transfer operation are a problem, consideration should be given to closed transition transfer switches, or to static transfer switches that operate within 4.17 ms.

Application of motor loads on limited power sources, such as standby and emergency generators, presents certain problems to system designers that are usually nonexistent on large utility-furnished services. The most critical problem is the sizing of the generator system so that it is capable of running all normally connected lighting, power, and motor loads plus starting the largest motor load. Information required in making the determination of generator size includes the kVA and kW of each load plus the starting kVA and kW of the largest motor. When the largest motor is the predominate load, the generator may be twice the size of the motor. Consultation with the manufacturer of generator systems is advised when large motor loads are contemplated. Other problems include

(1) Regeneration loads developed by operation of elevators

(2) Nuisance breaker tripping and possible motor damage incurred when switching between two out-of-phase energized power sources
(3) Means and methods of temporarily shedding motor loads prior to transfer to avoid overloading the limited power source
(4) Proper grounding of the generator neutral to avoid extraneous ground-fault tripping of the normal source breaker
(5) SCR-type loads that may produce harmonics that interfere with the operation of unprotected exciter-regulators

4.9.6 Transfer of Power. When local generation is operated in parallel with the utility system, there will generally be a momentary voltage dip as the local generator starts to supply the utility line until the reverse power relays operate to disconnect the utility line. The voltage should not go all the way to zero, as is the case with a transfer switch, but it can go low enough to cause equipment to drop out; so its effect should be considered. Faults on the utility system and in the building can also cause voltage dips. The relay system to protect both the utility supply and the building will require utility approval. Requirements for such relay systems vary from one utility to another. Furthermore, the degree of relay protection differs depending upon whether the two sources are just momentarily paralleled (as required for closed transition switching), or remain paralleled (as required for peak shaving and co-generation) (see Reference [22] and IEEE Std 1001-1988, IEEE Guide for Interfacing Dispersed Storage and Generating Facilities with Electric Utility Systems (ANSI) [17]).

Normally, the transfer from a normal to an alternate or emergency supply is accomplished automatically. The return to the normal supply can be automatic or manual. Manual return should be considered when the unexpected restoration of the normal supply would cause equipment to restart and endanger either the equipment or personnel. For example, elevators should not be permitted to restart automatically because rescue operations may be in progress to free trapped passengers.

4.10 Uninterruptible Power Supply (UPS) Systems. A UPS is a device or system that provides quality and continuity of an ac power source. Every UPS should maintain some specified degree of continuity of load for a specified stored energy time upon ac input failure (see NEMA PE1-1990, Uninterruptible Power Systems [20][43]). The term "UPS" commonly includes equipment, backup power source(s), environmental equipment (enclosure, heating, and ventilating equipment), switchgear, and controls, which, together, provide a reliable continuous quality electric power system.

The following definitions are given for clarification (see also Reference [26]):
(1) Critical Load — Is that part of the load that requires continuous quality electric power for its successful operation.
(2) Uninterruptible Power Supply (UPS) System — Consists of one or more UPS modules, an energy storage battery (per module or a common battery), and

---

[43] NEMA publications are available from the National Electrical Manufacturers Association, 2101 L Street, N.W., Washington, DC 20037.

accessories (as required) to provide a reliable and high-quality power supply. The UPS isolates the load from the primary and emergency sources, and, in the event of a power interruption, provides regulated power to the critical load for a specified period depending upon battery capacity. (The battery is normally sized to provide a capacity of 15 minutes when operating at full load.)

(3) UPS Module—Is the power conversion portion of the UPS system. A UPS module may be made entirely of solid-state electronic construction, or a hybrid combining rotary equipment (motor-generator) and solid-state electronic equipment.

A solid-state electronic UPS consists of a rectifier, an inverter, and associated controls along with synchronizing, protective, and auxiliary devices. UPS modules may be designed to operate either individually or in parallel.

A rotary UPS consists of a pony motor, a motor-generator or, alternatively, a synchronous machine in which the synchronous motor and generator have been combined in a single unit. This comprises a stator whose slots carry alternate motor and generator windings, and a rotor with dc excitation, a rectifier, an inverter, a solid-state transfer switch, and associated controls along with synchronizing, protective, and auxiliary devices.

(4) Nonredundant UPS Configuration (see Fig 35)—Consists of one or more UPS modules operating in parallel with a bypass circuit transfer switch and a battery. The rating and number of UPS modules are chosen to supply the critical load with no intentional excess capacity. Upon the failure of any UPS module, the bypass circuit automatically transfers the critical load to the bypass source without an interruption. The solid-state electronic UPS configuration relies upon a static transfer switch for transfer within 4.17 ms.

The rotary UPS configuration relies upon the stored energy of the flywheel to propel the generator and maintain normal voltage and frequency for the time that the electromechanical circuit breakers are transferring the critical load to the alternate source. All operational transfers are "make before break."

(5) "Cold" Standby Redundant UPS Configuration (see Fig 36) — Consists of two independent, nonredundant modules with either individual module batteries or a common battery. One UPS module operates on the line, and the other UPS module is turned off. Should the operating UPS module fail, its static bypass circuit will automatically transfer the critical load to the bypass source without an interruption to the critical load. The second UPS module is then manually energized and placed on the bypass mode of operation. To transfer the critical load, external "make before break" nonautomatic circuit breakers are operated to place the load on the second UPS bypass circuit. Finally, the critical load is returned from the bypass to the second UPS module via the bypass transfer switch. The two UPS modules cannot operate in parallel; therefore, a safety interlock circuit should be provided to prevent this condition. This configuration is rarely used.

(6) Parallel Redundant UPS Configuration (see Fig 37)—Consists of two or more UPS modules with static inverter turnoff(s), a system control cabinet,

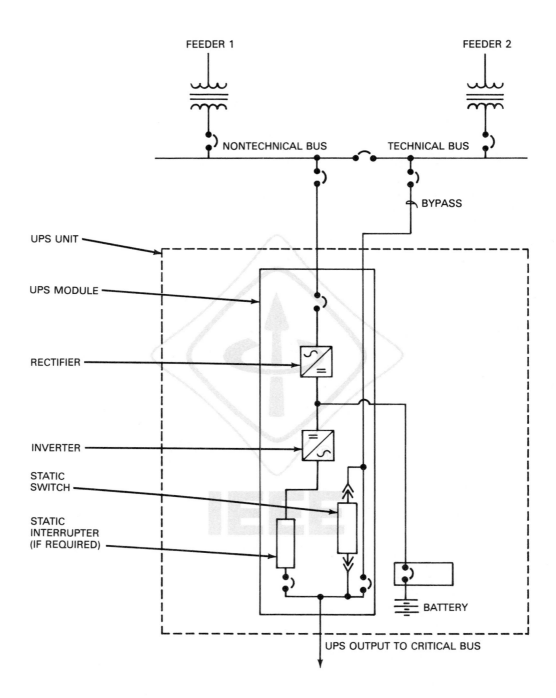

**Fig 35
Nonredundant UPS System Configuration**

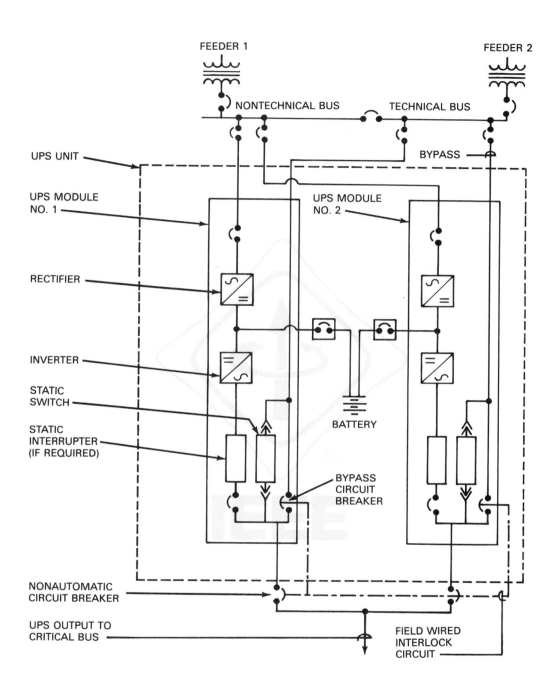

**Fig 36
"Cold" Standby Redundant UPS System**

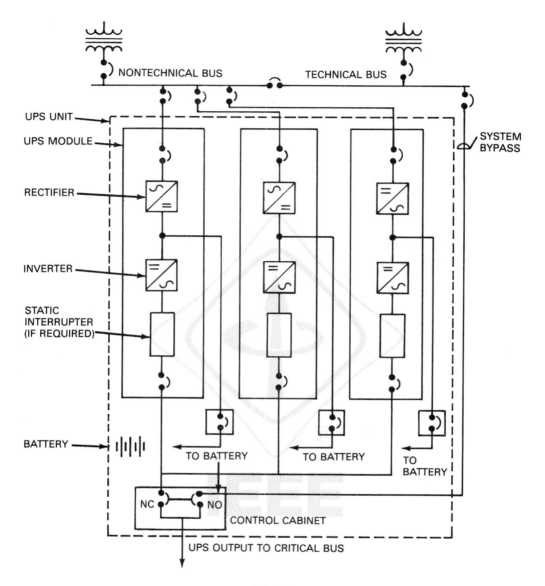

NONTECHNICAL BUS   TECHNICAL BUS

UPS UNIT

UPS MODULE

SYSTEM
BYPASS

RECTIFIER

INVERTER

STATIC
INTERRUPTER
(IF REQUIRED)

BATTERY

TO BATTERY   TO BATTERY   TO
BATTERY

NC   NO

CONTROL CABINET

UPS OUTPUT TO CRITICAL BUS

Fig 37
Parallel Redundant UPS System

and either individual module batteries or a common battery. The UPS
modules operate in parallel and normally share the load, and the system is
capable of supplying the rated critical load upon failure of any one UPS
module. A static interrupter will disconnect the failed UPS module from the
other UPS modules without an interruption to the critical load. A system
bypass is usually included to permit system maintenance.

(7) Isolated Redundant UPS Configuration (see Fig 38) — Uses a combination of automatic transfer switches and a reserve system to serve as the bypass source for any of the active systems (in this case, a system consists of a single module with its own system switchgear). The use of this configuration requires each active system to serve an isolated/independent load. The advantage of this type of configuration minimizes single-point failure modes, i.e., systems do not communicate via logic connections with each other; the systems operate independently of one another. The disadvantage of this type of system is that each system requires its own separate feeder to its dedicated load.

4.10.1 Application of UPS. (See Reference [26].)

(1) The nonredundant UPS may be satisfactory for many critical load applications.

(2) Parallel Redundant UPS System — When the criticality of the load demands the greatest protection and the load cannot be divided into suitable blocks, then installation of a parallel redundant UPS system is justified.

4.10.2 Power System Configuration for 60 Hz Distribution. In 60 Hz power distribution systems, the following basic concepts are used:

**Fig 38
Isolated Redundant UPS System**

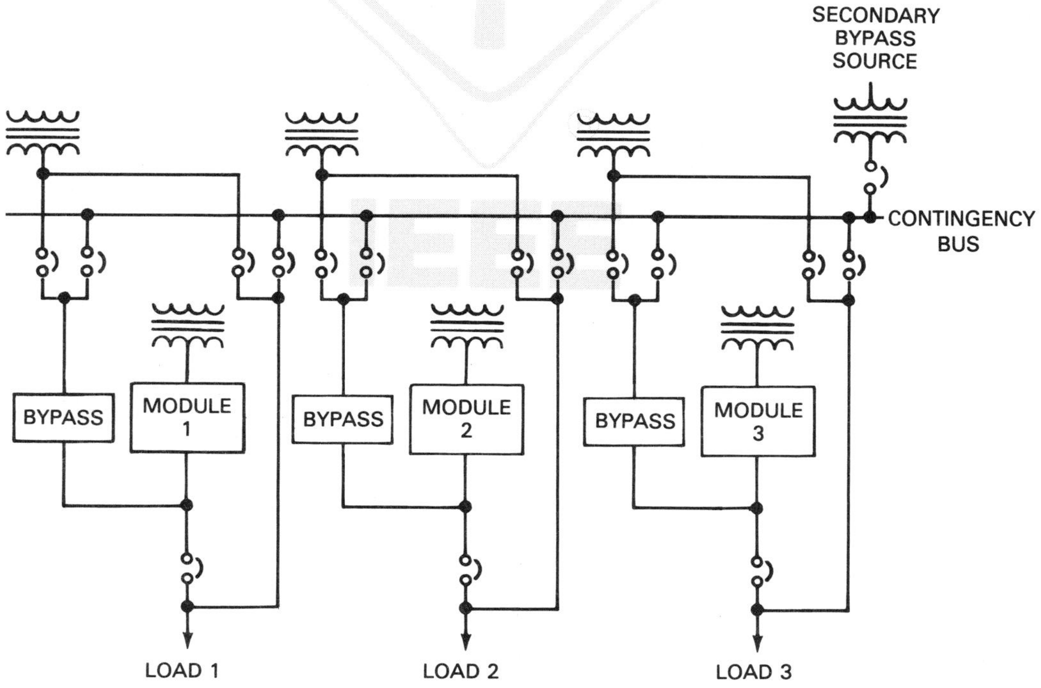

(1) Single-Module UPS System (see Fig 39) — Is a single unit that is capable of supplying power to the total load. In the event of an overload or if the unit fails, the critical bus is transferred to the bypass source via the bypass transfer switch. Transfer is uninterrupted.

(2) Parallel Capacity UPS System (see Fig 40) — Is two or more units capable of supplying power to the total load. In the event of overload, or if either unit fails, the critical load bus is transferred to the bypass source via the bypass transfer switch. Transfer is uninterrupted. Battery may be common or separate.

(3) Parallel Redundant UPS System (see Fig 41) — Is two or more units with more capacity than is required by the total load. If any unit fails, the remaining units should be capable of carrying the total load. If more than one unit fails, the critical bus will be transferred to the bypass source via the bypass transfer switch. Battery may be common or separate per module.

Fig 39
Single-Module UPS System

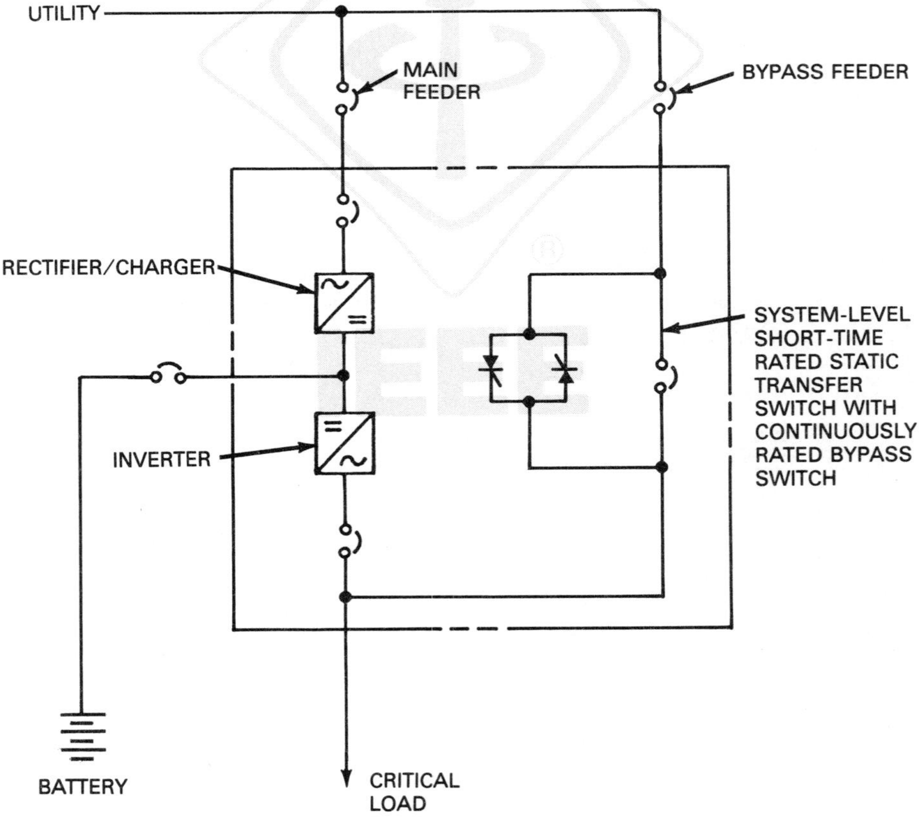

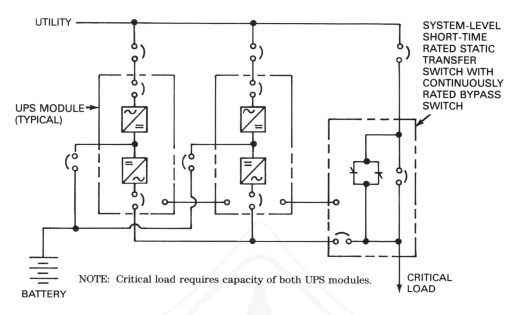

Fig 40
Parallel Capacity UPS System

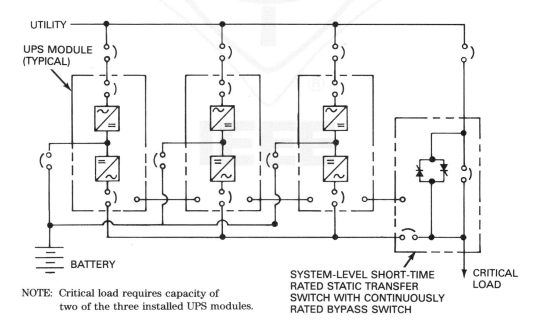

Fig 41
Parallel Redundant UPS System

(4) Dual Redundant UPS System (see Fig 42)—One UPS module is standing by, running unloaded. If the loaded module fails, the load is transferred to the standby module. Each rating is limited to the size of largest available module.

(5) Isolated Redundant UPS System (see Fig 43)—Multiple UPS modules, usually three, are individually supplied from transformer sources. Each UPS module supplies a critical load and is available to supply a common contingency bus. The common contingency bus supplies the bypass circuit for each UPS module. In addition to being supplied from the common contingency bus, the bypass switch of each module is supplied from an individual transformer source. Furthermore, the common contingency bus is also supplied from a separate standby transformer called a "secondary bypass source." The arrangement includes one UPS module in reserve as a "hot" standby. When a primary UPS module fails, the reserve UPS module is transferred to the load.

(6) Parallel Tandem UPS System (see Fig 44)—The tandem configuration is a special case of two modules in parallel redundancy. In this arrangement, both modules have rectifier/chargers, dc links, and inverters, also one of the

Fig 42
Dual Redundant UPS System

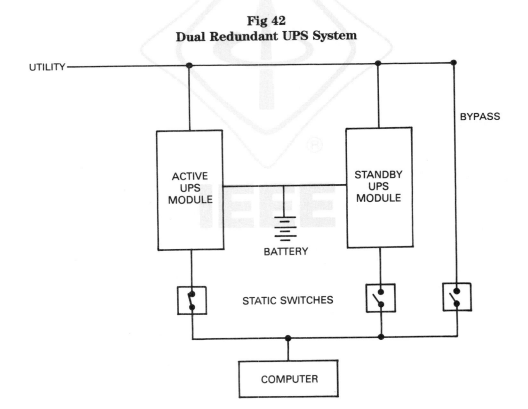

NOTE: Each module is capable of supplying the load.

modules houses the system-level static transfer switch. Either module can support full system load while the other has scheduled or corrective maintenance performed.

(7) Hot Tied Bus UPS System (see Fig 45)—The UPS tied bus arrangement consists of two individual UPS systems (single module, parallel capacity, or redundant), with each one supplying a critical load bus. The two critical load buses can be paralleled via a tie breaker (normally open) while remaining on inverter power, which allows greater user flexibility for scheduled maintenance or damage control due to various failures.

(8) Super-Redundant Parallel System — Hot Tied Bus UPS System (see Fig 46) — The super-redundant UPS arrangement consists of n UPS modules (limited by a 4000 A bus). Each UPS module is supplied from dual sources (either/or) to supply two critical "paralleling buses." Each paralleling bus is connected via a breaker to a "common bus" in parallel with the output feeder of one of the system static bypass switches. This junction is connected via a breaker to a "system critical load bus." A tie enables the two "system critical load buses" to be paralleled. Bypass sources for each system supply their own respective static bypass switches and maintenance bypasses.

Fig 43
Isolated Redundant UPS System

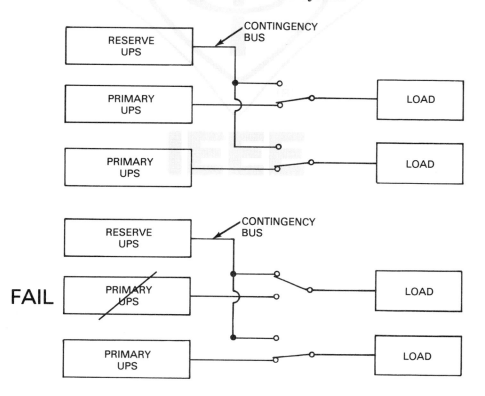

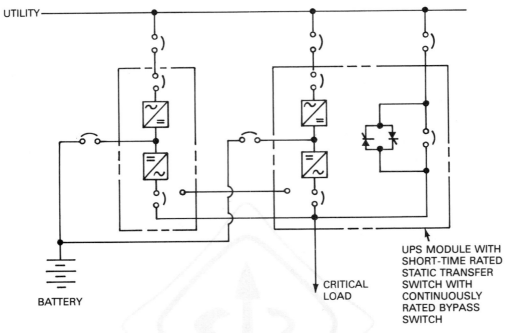

UTILITY

BATTERY

CRITICAL
LOAD

UPS MODULE WITH
SHORT-TIME RATED
STATIC TRANSFER
SWITCH WITH
CONTINUOUSLY
RATED BYPASS
SWITCH

NOTE: System can be for redundancy or capacity.

**Fig 44
Parallel Tandem UPS System**

The super-redundant UPS arrangement normally operates with the tie breaker open between the two "system critical load buses." When all UPS modules are supplying one paralleling bus, then the tie breaker is closed.

All operations are preselected, automatic, and allow the user to do module- and system-level reconfiguration without submitting either critical load to utility power.

(9) Uninterruptible Power with Dual Utility Sources and Static Transfer Switches (see Fig 47) — Essentially, uninterruptible electric power to the critical load may be achieved by the installation of dual utility sources, preferably from two separate substations, supplying secondary buses via step-down transformers as required.

Feeders from each of the two source buses are connected to static transfer switches as source 1 and source 2. A feeder from the load connection of the static transfer switch supplies a power line conditioner, if needed. The power line conditioner filters transients and provides voltage regulation. Filtered and regulated power is then supplied from the power line conditioner to critical load distribution switchgear. This system eliminates the need for energy storage batteries, emergency generators, and other equipment. The reliability of this system is dependent upon the two utility sources and power conditioners.

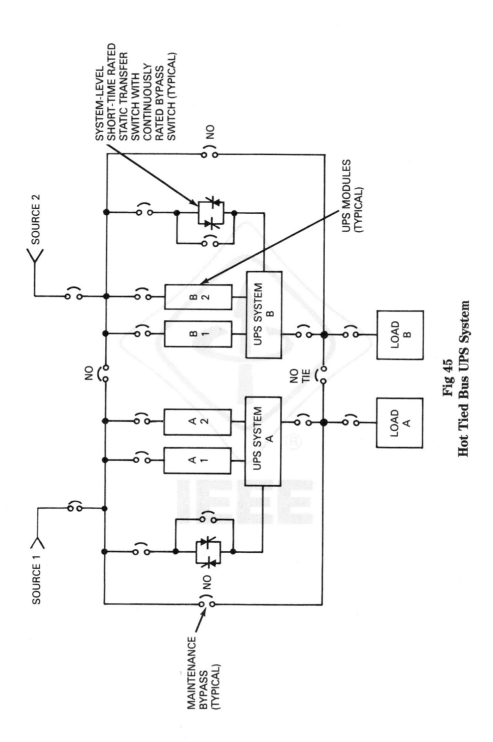

**Fig 45
Hot Tied Bus UPS System**

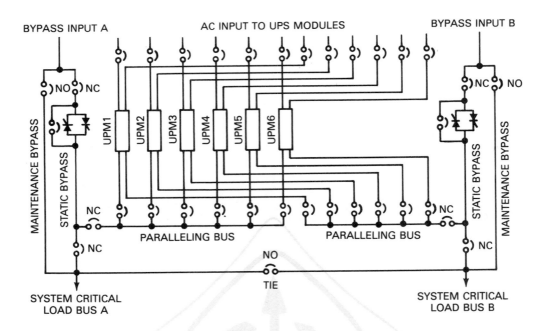

Fig 46
Super-Redundant Parallel System — Hot Tied Bus UPS System

4.10.2.1 Power System Configuration with 60 Hz UPS

(1) Electric Service and Bypass Connectors — Two separate electric sources, one to the UPS rectifier circuit and the other to the UPS bypass circuit, should be provided. When possible, they should emanate from two separate buses with the UPS bypass connected to the noncyclical load bus (also called the "technical bus"). This connection provides for the isolation of sensitive technical loads from the effects of UPS rectifier harmonic distortion and motor start-up current inrush.

(2) Maintenance Bypass Provisions — To provide for the maintenance of equipment, bypass provisions are necessary to isolate each UPS module or system.

4.10.2.2 UPS Distribution Systems. The UPS serves critical loads only. Non-critical loads are served by separate distribution systems that are supplied from either the noncyclical load bus ("technical bus") or the cyclical load bus ("non-technical bus"), as appropriate.

(1) Critical Load Protection — Critical load overcurrent devices equipped with fast-acting fuses to shorten the transient effects of undervoltage caused by short circuits will result in a reliable system. Solid-state transient suppressors (metal-oxide type) should also be supplied to lessen the overvoltage transients caused by reactive load switching.

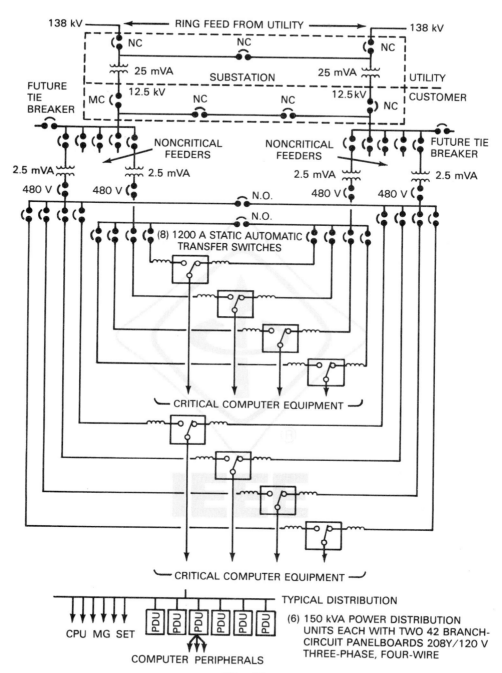

Fig 47
Uninterruptible Power with Dual Utility Sources and Static Transfer Switches

"Uninterruptible Power Supplies," Fig 14.26, by D. C. Griffith.
Reprinted and adapted with permission of the author.

(2) Critical Motor Loads—Due to the energy losses and the starting current inrush inherent in motors, the connection of motors to the UPS bus should be limited to frequency conversion applications, that is, motor-generator sets. Generally, due to the current inrush, motor-generator sets are started on the UPS bypass circuit. Motor-generator sets may be started on the rectifier/inverter mode of operation under the following conditions:

    (a) When the rating of the motor-generator set is less than 10% of the UPS rating

    (b) When reduced voltage and peak current starters, such as the wye-delta closed-transition type, are used for each motor load (see Chapter 6)

    (c) When more than one motor-generator set is connected to the critical bus, each set should be energized sequentially rather than simultaneously

Refer all applications requiring connection of induction and synchronous motor loads to the UPS manufacturer. Application rules differ depending upon the design and rating of the UPS.

4.10.3 Power System Configuration for 400 Hz Distribution. In 400 Hz power distribution systems, the following basic concepts are used:

(1) Direct-Utility Supply to Dual-Rotary Frequency Converters Parallel at the Output Critical Load Bus—Each frequency converter is sized for 100% load or the arrangement has redundant capacity. The frequency converters may be equipped with an inverter/charger and battery upon utility failure. Transfer from the utility line to the inverter occurs by synchronizing the inverter to the residual voltage of the motor.

(2) Dual-Utility Supply—Dual-utility feeders supply an automatic transfer switch. The automatic transfer switch supplies multiple-rotary frequency converters (flywheel equipped). The frequency converters are parallel at the critical load bus. Transfer from one utility line to another occurs within the ride-through capability time of the rotary frequency converters.

(3) UPS—A static or rotary UPS supplies multiple-frequency converters and other 60 Hz loads.

(4) UPS with Local Generation Backup—Both the utility feeder (connected to the normal terminals) and the feeder from the backup generation (connected to the emergency terminals) supply the automatic transfer switch. The automatic transfer switch in turn supplies the UPS. Critical load distribution is as described above.

(5) Parallel 400 Hz Single CPU Configuration—Two or more 60–400 Hz frequency converters are normally connected in a redundant configuration to supply the critical load. There is no static switch or bypass breaker. Note that, on static converters, it is possible to use a 400 Hz motor-generator as a bypass source.

(6) Common UPS for Single Mainframe Computer Site—Two 60–400 Hz frequency converters are normally connected in a redundant configuration supplying the mainframe computer, while a 60 Hz UPS supplies the peripherals.

(7) Alternative Combination UPS for Single Mainframe Computer Site — A 60 Hz UPS supplies a critical load bus that, in turn, supplies the peripherals plus the input to a motor-generator set frequency converter (60–400 Hz).

(8) Combination UPS for Multiple-Mainframe Computer Site — A utility source supplies a redundant 400 Hz UPS system. This paralleled system supplies a 400 Hz critical load distribution bus. Feeders from the 400 Hz distribution bus, equipped with line-drop compensators (LDCs) to reactive voltage drop, supply computer mainframes. A utility source also supplies a parallel redundant 60 Hz UPS system. This system supplies the critical peripheral load.

(9) Remote Redundant 400 Hz UPS — A 60 Hz UPS and a downstream parallel redundant 400 Hz motor-generator frequency conversion system with paralleling and distribution switchgear and line-drop compensators, which are all installed in the facility power equipment room with 60 and 400 Hz feeders distributed into the computer room.

(10) Point-of-Use Redundant 400 Hz UPS — A 60 Hz UPS and a parallel redundant frequency conversion system as in item (9) above except that the motor-generators are equipped with silencing enclosures and are installed in the computer room near the mainframes.

(11) Point-of-Use 400 Hz UPS — A 60 Hz UPS and a nonparalleled point-of-use static or rotary 400 Hz frequency converter installed in the computer room adjacent to each mainframe.

(12) Remote 400 Hz UPS — A 60 Hz UPS and a separate parallel redundant 400 Hz UPS installed in the power equipment room, which is similar to item (8) above.

(13) Wiring — For 400 Hz circuits, the reactance of circuit conductors may produce unacceptable voltage drops. Multiple-conductor cables and use of conductors in parallel, if necessary, should be installed in accordance with the NEC, Article 310-4 [1].

4.11 Voltage Regulation and Power Factor Correction. Power factor correction and voltage regulation are closely related. In many cases, the desired voltage regulation is costly to obtain. Larger or paralleled conductors to reduce voltage drop under load are, in many cases, the proper solution. However, power factor correction may also be justified for four reasons

(1) To improve voltage

(2) To lower the cost of electric energy, when the electric utility rates vary with the power factor at the metering point

(3) To reduce the energy losses in conductors

(4) To utilize the full capacity of transformers, switches, overcurrent devices, buses, and conductors for active power only, thereby lowering the capital investment and annual costs

4.11.1 Voltage Regulation. The goal of good voltage regulation is to control the voltage of the system so that it will stay within a practical and safe range of voltage tolerances under all design loads. Voltage at any utilization equipment should be within the guaranteed operative range of the equipment. The type and size of wires or cables, types of raceways, reactances of transformers and cables, selection of motor-starting means, circuit design, power factor correction, and the means and degree of loading will all affect voltage regulation.

Voltage regulation in any circuit, expressed in percentage, is

$$\frac{(\text{no-load voltage} - \text{full-load voltage}) \times 100}{\text{no-load voltage}} \qquad \text{(Eq 6)}$$

When it is not economical to control voltage drop through conductor sizing, circuit design, or other means, voltage regulators may be needed. Several types of voltage regulators, either automatic or manual, are available for all types and sizes of loads from individual electronic devices to the equipment for an entire laboratory or department store. Voltage regulators are frequently used by electric utility companies in their distribution system feeders and are seldom needed within commercial buildings, except for use with electronic equipment. Normally, the power and light distribution system within large commercial buildings can be designed economically and adequately without the use of large voltage regulators.

4.11.2 Power Factor Correction. When the type of load to be installed in the commercial building will result in a poor power factor, then an evaluation should be made to determine if installing capacitors is justified, either to stay within the power factor range specified by the electric utility in order to avoid penalty payments or to obtain a reduction in the electric bill. Care should be taken when adding capacitors to ensure that no resonant conditions could exist with the fundamental or harmonic frequencies. This is particularly true if SCR drives are used.

When large machines (like blowers, or refrigeration or air compressors) are to be installed, a study should be made to determine whether it would be economical to install a synchronous motor and utilize it for power factor correction. The cost of the synchronous motor with its controller should be compared with the cost of a squirrel-cage motor with its simpler controls plus separate static capacitors.

4.12 System Reliability Analysis. One of the questions often raised during the design of the power distribution system is how to make a quantitative comparison of the failure rate and the forced downtime in hours per year for different circuit arrangements, including radial, primary-selective, simple spot-network, and secondary-network circuits. This quantitative comparison could be used to make trade-off decisions involving the initial cost versus the failure rate and the forced downtime per year. The estimated cost of power outages at the various distribution points could be considered in deciding which type of circuit arrangement to use. Decisions could thus be based upon total life-cycle cost over the useful life of the equipment rather than on first cost.

4.12.1 Reliability Data for Electrical Equipment. In order to calculate the failure rate and the forced downtime per year of the power distribution system, it is necessary to have reliability data on the electric utility supply and each piece of electrical equipment used in the power distribution system. One of the best sources for this type of data is the extensive survey of equipment reliability included in IEEE Std 493-1990 (ANSI) [15], which presents average data for all equipment manufacturers and a variety of applications.

4.12.2 Reliability Analysis and Total Owning Cost. Statistical analysis methods involving the probability of power failure may be used to make calculations of the failure rate and the forced downtime for the power distribution system. The methods and formulas used in these calculations are given in IEEE Std 493-1990 (ANSI) [15]. Another source of this information would be MIL-HDBK-217, Reliability for Electric and Electronic Equipment Handbook [19].[44]

4.13 References. The following references shall be used in conjunction with this chapter:

[1] ANSI C2-1990, National Electrical Safety Code.

[2] ANSI C12.1-1988, Code for Electricity Metering.

[3] ANSI C84.1-1989, Voltage Ratings for Electric Power Systems and Equipment (60 Hz).

[4] ANSI/NFPA 20-1990, Centrifugal Fire Pumps.

[5] ANSI/NFPA 37-1990, Stationary Combustion Engines and Gas Turbines.

[6] ANSI/NFPA 70-1990, National Electrical Code.

[7] ANSI/NFPA 99-1990, Health Care Facilities.

[8] ANSI/NFPA 101-1988, Life Safety Code.

[9] ANSI/NFPA 110-1988, Emergency and Standby Power Systems.

[10] ANSI/NFPA 110A-1989, Stored Energy Systems.

[11] ASHRAE/IES 90.1-1989, Energy Efficient Design of New Buildings Except New Low-Rise Residential Buildings.

[12] EGSA 109C-1984, "Codes for Emergency Power by States and Major Cities," Coral Springs, FL: Electrical Generating Systems Association (EGSA).

[13] IEEE Std 142-1982, IEEE Recommended Practice for Grounding of Industrial and Commercial Power Systems (ANSI).

[14] IEEE Std 446-1987, IEEE Recommended Practice for Emergency and Standby Power Systems for Industrial and Commercial Applications (ANSI).

[15] IEEE Std 493-1990, IEEE Recommended Practice for the Design of Reliable Industrial and Commercial Power Systems (ANSI).

[16] IEEE Std 944-1986, IEEE Application and Testing of Uninterruptible Power Supplies for Power Generating Stations (ANSI).

[17] IEEE Std 1001-1988, IEEE Guide for Interfacing Dispersed Storage and Generating Facilities with Electric Utility Systems (ANSI).

---

[44]MIL publications are available from U.S. Navy Publications and Forms, 5801 Tabor Avenue, Philadelphia, PA 19120.

[18] IEEE Std 1109-1990, IEEE Guide for the Interconnection of User-Owned Substations to Electric Utilities.

[19] MIL-HDBK-217, Reliability for Electric and Electronic Equipment Handbook.

[20] NEMA PE1-1990, Uninterruptible Power Systems.

[21] UL 1778-1989, Uninterruptible Power Supplies.

[22] Castenschiold, R. "Closed-Transition Switching of Essential Loads," *IEEE Transactions on Industry Applications*, vol. 25, no. 3, May/Jun. 1989, pp. 403–407.

[23] Caywood, R. E. "Electric Utility Rate Economics," New York: McGraw-Hill, 1972.

[24] "Computer Support Systems," Brochure CGI007, Torrance, CA: Teledyne Inet, Jan. 1990.

[25] "Diesel Continuous Power Supply Systems," Leatherhead, Surrey, England: Holec Limited, Jun. 1990.

[26] "Electrical Engineering—Preliminary Design Considerations, Design Manual 4-1," U.S. Department of the Navy, Naval Facilities Engineering Command, Alexandria, VA.

[27] Griffith, D. C. "Uninterruptible Power Supplies," New York: Marcel Decker, 1989.

[28] "On-Site Power Generation," Johnson, G. (editor), Coral Springs, FL: Electrical Generating Systems Association (EGSA), 1989.

[29] "Power Producers' Interconnection Handbook," San Francisco, CA: Pacific Gas and Electric Company, Jun. 1986.

[30] "Requirements for Electric Service Installations (blue book)," New York: Consolidated Edison Company, May 1986.

[31] Stevens, R. E. "Designing Buildings for Fire Safety," NFPA publication number SPP-24.

[32] "System Configurations," Raleigh, NC: Exide Electronics.

[33] "Three-Phase Vacuum Contactors and Circuit Breakers," Campbell, CA: Ross Engineering Company, Brochure B-1005, Sep. 1974.

[34] "415 Hz Power Systems," Middletown, NY: KW Controls, Inc., 1989.

Chapter 5
Power Distribution Apparatus

5.1 General Discussion. Electric systems for commercial installations encompass a wide variety of electrical apparatus. There are numerous choices to be made between similar equipment, which either have overlapping functions or which are direct substitutes with varying advantages or degrees of acceptability to a particular application. The engineer making these basic decisions should consider all facets of the actual project including, but not limited to, protection; coordination; initial cost including installation, operational personnel and cost; maintenance facilities and cost; availability and cost of space; and the procurement time to meet objectives. Equipment connecting directly to the serving electric utility should be compatible with the utility's requirements.

General descriptions of apparatus frequently used in these electric systems follow in this order:

(1) Transformers
(2) Medium- and high-voltage fuses
(3) Metal-enclosed 5–34.5 kV load interrupter switchgear
(4) Metal-clad 5–34.5 kV circuit breaker switchgear
(5) Metal-enclosed, low-voltage 600 V power switchgear and circuit breakers
(6) Metal-enclosed distribution switchboards
(7) Primary-unit substations
(8) Secondary-unit substations
(9) Panelboards
(10) Molded-case circuit breakers
(11) Low-voltage fuses
(12) Service protectors
(13) Enclosed switches
(14) Bolted pressure switches
(15) High-pressure contact switches
(16) Network protectors
(17) Lightning and transient protection
(18) Load transfer devices
(19) Interlock systems

A brief explanation of equipment ratings is provided in 5.2.1. ANSI and NEMA Standards and other publications referred to in the text are listed in 5.22 and a bibliography is included in 5.23. Other equipment related to power conditioning (e.g., voltage regulators, power line conditioners, uninterruptible power supplies, adjustable frequency drives, etc.) are discussed elsewhere in this book.

The safety of high-voltage installations should also be considered. ANSI/NFPA 70-1990, National Electrical Code (NEC) [9][45] and ANSI C2-1990, National Electrical Safety Code (NESC) [1][46] are guidelines in this area. In addition, independent, nationally recognized testing laboratories (i.e., UL, Factory Mutual, etc.) publish standards on certain electrical apparatus. The code-enforcing agency will have final approval as to the acceptability of equipment; see Chapter 1, 1.6 for a discussion of the NEC, OSHA, equipment labeling, identification, and "approval by the code-enforcing agency" requirements for power system apparatus.

5.2 Transformers. Transformers in commercial installations are normally used to change a voltage level from a utility distribution voltage to a voltage that is usable within the building, and are also used to reduce building distribution voltage to a level that can be utilized by specific equipment. Applicable standards are the ANSI C57 Series and NEMA TR and ST Series.

5.2.1 Transformer Types. The following types of transformers are normally used in commercial buildings:

(1) Substation
(2) Primary-unit substation
(3) Secondary-unit substation (power center)
(4) Network
(5) Pad-mounted
(6) Indoor distribution

Many other types of transformers are manufactured for special applications, such as welding, constant voltage supply, and high-impedance requirements. Discussion of the special transformers and their uses is beyond the scope of this recommended practice.

(1) Substation Transformers — Used with outdoor substations, they are rated 750–5000 kVA for single-phase units and 750–25 000 kVA for three-phase units. The primary voltage range is 2400 V and up. Taps are usually manually operated while de-energized; but automatic load tap changing may be obtained. The secondary voltage range is 480–13 800 V. Primaries are usually delta connected, and secondaries are usually wye connected because of the ease of grounding the secondary neutral. The insulation and cooling medium is usually liquid. High-voltage connections are on cover-mounted bushings. Low-voltage connections may be cover-mounted bushings or an air terminal chamber.

---

[45]The numbers in brackets correspond to those in the references at the end of this chapter. ANSI publications are available from the Sales Department of the American National Standards Institute, 11 West 42nd Street, 13th Floor, New York, NY 10036. NFPA publications are available from Publications Sales, National Fire Protection Association, 1 Batterymarch Park, P.O. Box 9101, Quincy, MA 02269-9101.

[46]ANSI publications are available from the Sales Department of the American National Standards Institute, 11 West 42nd Street, 13th Floor, New York, NY 10036.

(2) Primary-Unit Substation Transformers—Used with their secondaries connected to medium-voltage switchgear, they are rated 1000–10 000 kVA and are three-phase units. The primary voltage range is 6900–138 000 V. The secondary voltage range is 2400–34 500 V. Taps are usually manually changed while de-energized; but automatic load tap changing may be obtained. Primaries are usually delta connected. The type may be oil, less-flammable liquid, air, dry, cast-coil, or gas. The high-voltage connections may be cover bushings, an air terminal chamber, or throat. The low-voltage connection is a throat.

(3) Secondary-Unit Substation Transformers—Used with their secondaries connected to low-voltage switchgear or switchboards, they are rated 112.5–2500 kVA and are three-phase units. The primary voltage range is 2400–34 500 V. The taps are manually changed while de-energized. The secondary voltage range is 120–480 V. The primaries are usually delta-connected, and secondaries are usually wye connected. The type may be oil, less-flammable liquid, air, dry, cast-coil, or gas. The high-voltage connections may be cover bushings, an air terminal chamber, or throat. The low-voltage connection is a throat.

(4) Network Transformers—Used with secondary-network systems, they are rated 300–2500 kVA. The primary voltage range is 4160–34 500 V. The taps are manually operated while de-energized. The secondary voltages are 208Y/120 V and 480Y/277 V. The type may be oil, less-flammable liquid, air, dry, cast-coil, or gas. The primary is delta connected, and the secondary is wye-connected. The high-voltage connection is generally a network switch (on-off-ground) or an interrupter-type switch with or without a ground position. The secondary connection is generally an appropriate network protector, or a low-voltage power air circuit breaker designed to provide the functional equivalent of a network protector.

ANSI C57.12.40-1990, Requirements for Secondary Network Transformers, Subway and Vault Types (Liquid Immersed) [6] applies to liquid immersed, subway- and vault-type network units. A subway-type unit is suitable for frequent or continuous operation while submerged in water; a vault-type unit is suitable for occasional submerged operation.

(5) Pad-Mounted Transformers—Used outside buildings where conventional unit substations might not be appropriate, and are either single-phase or three-phase units. Because they are of tamper-resistant construction, they do not require fencing. Primary and secondary connections are made in compartments that are adjacent to each other but separated by barriers from the transformer and each other. Access is through padlocked hinged doors designed so that unauthorized personnel cannot enter either compartment. Where ventilating openings are provided, tamper-resistant grills are used. Gauges and accessories are in the low-voltage compartment.

These units are rated 75–2500 kVA. The primary voltage range is 2400–34 500 V. Taps are manually changed while de-energized. The secondary voltage range is 120–480 V. Primaries are almost always delta connected or special construction wye connected, and secondaries are usually wye connected. A delta-connected tertiary is not acceptable with a three-legged core unless an upstream device opens all three phases for a single-phase fault. The type may be oil, less-flammable liquid, air, dry, cast-coil, or gas. The high-voltage connection is in an air terminal chamber that may contain just pressure- or disconnecting-type connectors or may have a disconnecting device, either fused or unfused. The connections may be for either single or loop feed. The low-voltage connection is usually by cable at the bottom; but it may also be by bus duct.

The dry-type, pad-mounted transformer does not have the inherent fire hazards of the oil filled, pad-mounted transformer and frequently the dry-type, pad-mounted transformer is mounted on the roofs of buildings so that it will be as near to the load center as possible.

ANSI C57.12.22-1989 [5] applies to oil immersed units with primary voltages of 16 340 V and below.

(6) Indoor Distribution Transformers—Used with panelboards and separately mounted, they are rated 1–333 kVA for single-phase units and 3–500 kVA for three-phase units. Both primaries and secondaries are 600 V and below (the most common ratio is 480–208Y/120 V). The cooling medium is air (ventilated or nonventilated). Smaller units have been furnished in encapsulated form. High- and low-voltage connections are pressure-type connections for cables. Impedances of distribution transformers are usually lower than those of substation or secondary-unit substation transformers.

Indoor and outdoor distribution transformers are also available at primary voltages of up to 34 500 V and 150 kV basic impulse insulation level (BIL).

The majority of transformers for distributing power at 480 V in a commercial building are usually referred to as "general-purpose transformers" and secondaries are typically rated at 208Y/120 V. These transformers are mostly of the dry-type, and some of the smaller sized ones are encapsulated. General-purpose transformers are used for serving 120 V lighting, appliances, and receptacles.

Virtually all power transformers used in commercial buildings are of the two-winding type, which may be referred to as *isolating* or *insulating* transformers, and are distinct from the one-winding type known as the *autotransformer*. The two-winding-type transformer provides a positive isolation between the primary and secondary circuits; which is desirable for safety, circuit isolation, reduction of fault levels, coordination, and reduction of electrical interference.

There are also a number of "specialty transformers" used for applications, such as x-ray machines, laboratories, electronic equipment, and special machinery applications. The health care applications are described in detail in IEEE Std 602-1986, IEEE Recommended Practice for Electric Systems in Health Care Facilities (ANSI) [30].[47] Specialty transformers used in applications where the least amount of leakage current could cause an arc and ignite the atmosphere (such as

---

[47] IEEE publications are available from the Institute of Electrical and Electronics Engineers, IEEE Service Center, 445 Hoes Lane, P.O. Box 1331, Piscataway, NJ 08855-1331.

in an oxygenated environment) or cause personal injury (such as in open heart surgery) will require an ungrounded secondary. In the most sensitive applications, the leakage current may be monitored and is controlled by introducing a grounded shield between the primary and secondary coils. Such a shield also reduces electromagnetic interference (EMI), which may be present in the primary.

5.2.2 Transformer Specifications. The following factors should be considered when specifying transformers:

(1) Kilovoltampere (kVA) Rating — Table 33 gives the preferred kVA ratings of both single-phase and three-phase transformers according to IEEE C57.12.00-1987, IEEE Standard General Requirements for Liquid-Immersed Distribution, Power, and Regulating Transformers (ANSI) [23].

(2) Voltage Ratings, Ratio, and Method of Connection (Delta or Wye) — All the preferred kVA ratings in Table 33 are obviously not available as standard at all voltage ratings and ratios. In general, the smaller sizes apply to lower voltages and the larger sizes to higher voltages. Voltage ratings and ratios should be selected in accordance with available standard equipment that is indicated in manufacturers' catalogs. This is recommended, if at all possible, both from the viewpoint of cost and time for initial procurement and for ready replacement, if necessary.

In most small size commercial projects, the 208Y/120 V secondary voltage is used because the majority of load is lighting and small appliances. A secondary voltage of 480Y/277 V, in addition to the 208Y/120 V circuits, may be required when loads are electric motors or have large lighting requirements.

Generally, a three-phase transformer secondary voltage should be selected at 480Y/277 V. This has become standard and is compatible with three-

Table 33
Preferred Kilovoltampere Ratings

Single-Phase	Three-Phase
3	9
5	15
10	30
15	45
25	75
37.5	112.5
50	150
75	225
100	300
167	500
250	750
333	1000
500	1500
833	2000
1250	2500
1667	3750
2500	5000
3333	7500
5000	10 000

phase motors, which are now rated 460 V standard. Under normal circumstances, a 460 V rating for the transformer secondary should not be selected unless the load is predominantly older motors rated 440 V and located close to the transformer. Phase-to-neutral 277 V circuits can serve fluorescent and high-intensity discharge (HID) lighting.

(3) Voltage Taps—Taps are used to change the ratio between the high- and low-voltage windings. Manual de-energized tap changing is usually used to compensate for differences between the transformer ratio and the system nominal voltage. The tap selected in the transformer should be based upon maximum no-load voltage conditions. For example, a standard transformer rated 13 200 V to 480 V may have four 2.5% taps in the 13 200 V winding (two above and two below 13 200 V). If this transformer is connected to a system whose maximum voltage is 13 530 V, then the 13 530 V to 480 V tap could be used to provide a maximum of 480 V at no-load.

Tap changers are classified as follows:

(a) Underload—Taps can be changed when the transformer is energized and loaded. These taps are used to compensate for excessive variations in the supply voltage. They are infrequently associated with commercial building transformers except as part of outdoor substations over 5000 kVA. Load tap changers can be controlled automatically or manually.

(b) No-Load—Taps can be changed only when the transformer is de-energized. Tap leads are brought to an externally operated tap changer with a handle capable of being locked in any tap position. This is a standard accessory on most liquid filled and sealed-type transformers. On very small liquid filled transformers and most ventilated-dry-type transformers, the taps are changed by moving internal links that are made accessible by a removable panel on the enclosure.

Manually adjustable (handle- or link-operable) taps are suitable for correcting long-term voltage conditions. They are not suitable for correcting short-term (hourly, daily, or weekly) voltage variations. Automatic tap changing or voltage regulating transformers are relatively expensive so that one of the following solutions might be more appropriate:

(i) Request improvement of the utility power supply regulation.

(ii) Segregate the circuits so that heavy variable loads are separated from more sensitive loads. When a source transformer constitutes a significant part of the impedance to a sensitive load, use a separate transformer (or secondary-unit substation) for such loads.

(iii) Use voltage regulating supplies for just the sensitive loads.

(4) Typical Impedance Values for Power Transformers—Typical impedance values for power transformers are given in Table 34. These values are at the self-cooled transformer kVA ratings and are subject to a tolerance of ±7.5%, as set forth in IEEE C57.12.00-1987 (ANSI) [23]. Nonstandard impedances may be specified at a nominally higher cost: higher impedances to reduce available fault currents or lower impedances to reduce voltage drop under

heavy-current, low-power factor surge conditions. Consult manufacturers' bulletins for impedances of small transformers because they can vary considerably.

(5) Insulation Temperature Ratings—Transformers are manufactured with various insulation material systems (as shown in Table 35). Performance data with reference to conductor loss and impedances should be referenced to a temperature of 40 °C over the rated average conductor temperature rise as measured by resistance. While Table 35 represents the limiting standard

Table 34
Transformer Approximate Impedance Values

High-Voltage Rating (volts)	Design Impedance (percent)	
	Low Voltage, Rated 480 V	Low Voltage, Rated 2400 V or Higher
Power Transformers		
2400 to 22 900	5.75	5.5
26 400, 34 400	6.0	6.0
43 800	6.5	6.5
67 000		7.0

Rated kVA	Design Impedance (percent)
Secondary-Unit Substation Transformers	
112½ through 225	Not less than 2
300 through 500	Not less than 4.5
Above 500	5.75
Network Transformers	
1000 and smaller	5.0
Above 1000	7.0

Table 35
Insulation Temperature Ratings in °C

Average Conductor Temperature Rise* (°C)	Maximum Ambient Temperature (°C)	Hot-Spot Temperature Differential* (°C)	Total Permissible Ultimate Temperature* (°C)	Class of Insulation System (°C)
55	40	10	105	105
65	40	15	120	120
80	40	30	150	150
115	40	30	185	185
150†	40	30	220	220

*Maximum at continuous rated load.
†Dry-type transformers using a 220 °C insulation system can be designed for lower temperature rises (115 °C or 80 °C) to conserve energy, increase life expectancy, and provide some continuous overload capability.

requirements, transformers with lower conductor losses and corresponding lower temperature rises are available, when longer life expectancy and reduced operating costs are desired. A Class 105 insulation system allows for a 55 °C rise with a total ultimate temperature of 105 °C. A Class 120 insulation system allows a 65 °C rise with a total permissible ultimate temperature of 120 °C. An 80 °C rise is allowed for a Class 150, a 115 °C rise is allowed for a Class 185, and a 150 °C rise is allowed for a Class 220. Materials or combinations of materials that may be included in each insulation material class are specified in IEEE C57.12.00-1987 (ANSI) [23].

(6) Insulation Classes — Voltage insulation classes and BILs are listed in Table 36.

(7) Sound Levels — Permissible sound levels are listed in Tables 37 and 38. Transformer sound levels can be a problem in commercial building interiors, especially where relative quiet is required, such as in conference rooms and certain office areas. Technical specifications can require transformer sound levels to be below those specified in these two tables. The effects of transformer sound levels can be minimized by placing the transformers in separate rooms, avoiding direct attachment of transformers to structural members, use of sound isolating pads or vibration dampers for mounting, and avoiding the mounting of transformers near plenums or stairwells where the sound will be directed into work areas. For large units, providing flexible connections from the transformer to long busway runs will reduce the transmission of vibrations.

(8) Effects of Transformer Failures — Transformer failures are rare; however, in high-rise buildings and in other buildings where the conditions for evacuation are limited, the effects of the failure of larger transformers can be serious. Air from transformer vaults should be exhausted directly outdoors. Dry-type transformers will usually be preferred to liquid filled transformers

Table 36
Voltage Insulation Classes and Dielectric Tests

Nominal System Voltage (kV)	Insulation Class	Dry Transformers		Oil Immersed Distribution Transformers		Oil Immersed Power Transformers	
		Basic Impulse Level (kV)	Low-Frequency Test (kV)	Basic Impulse Level (kV)	Low-Frequency Test (kV)	Basic Impulse Level (kV)	Low-Frequency Test (kV)
1.2	1.2	10	4	30	10	45	10
2.4	2.5	20	10	45	15	60	15
4.8	5.0	30	12	60	19	75	19
8.32	8.7	45	19	75	26	95	26
14.4	15.0	60	31	95	34	110	34
23.0	25.0	110	37	125	40	150	50
34.5	34.5	150	50	150	50	200	70

NOTE: Ventilated-dry-type transformers and cast-coil transformers can be built to match the BIL of the oil immersed distribution transformers.

Table 37
Sound Levels for Dry-Type Transformers in dB

Equivalent Two-Winding kVA	Self-Cooled Ventilated 1	Self-Cooled Sealed 2	Forced-Air-Cooled Ventilated* 3
0—9	45	45	
10—50	50	50	
51—150	55	55	
151—300	58	57	
301—500	60	59	
501—700	62	61	
701—1000	64	63	
1001—1500	65	64	
1501—2000	66	65	
2001—3000	68	66	
3001—4000	70	68	
4001—5000	71	69	
5001—6000	72	70	
6001—7500	73	71	
0—1167			67
1168—1667			68
1668—2000			69
2001—3333			71
3334—5000			73
5001—6667			74
6668—8333			75
8334—10 000			76

NOTES: *Does not apply to sealed-type transformers.

(1) Columns 1 and 2 — Class AA rating, column 3 — Class FA and AFA rating.

(2) As given in ANSI/NEMA ST20-1988, Dry-Type Transformers for General Applications, Part IV, Table 4-4, page 26 [B3][48] sound levels for dry-type units rated 1.2 kV and less differ from those given here.

Table 38
Sound Levels for Single-Phase and Three-Phase Oil Cooled Transformers in dB

Equivalent Two-Winding kVA	Without Fans	With Fans
0—300	56	
301—500	58	
501—700	60	70
701—1000	62	70
1001—1500	63	70
1501—2000	64	70
2001—3000	65	71
3001—4000	66	71
4001—5000	67	72
5001—6000	68	73
6001—7500	69	73
7501—10 000	70	74

---

[48]The numbers in brackets preceded by a B refer to the bibliographic references that are at the end of this chapter.

(even the less-flammable, liquid insulated types) where fire and smoke considerations are critical. Well-designed transformer protection can minimize the extent of damage to any type of transformer. Dry-type transformers, including the cast-coil-type, if subjected to faults for an extended period, can burn and generate smoke; liquid filled transformers can burst, burn, and generate smoke. Provisions can be made for dealing with these rare but still possible failure modes for large transformers in critical areas.

(9) Harmonic Content of Load — Very recent developments have indicated failures of certain types of transformers due to nonlinear loads, which cause third and higher harmonics to flow through the windings. When these harmonics are present, due to loads like computers, variable speed drives, electronic ballasts, HID lighting, arc furnaces, rapid mode switching devices, and similar electrical loads, consideration should be given to specifying a special transformer that is designed to withstand these harmonic currents and the fluxes they produce in the cores.

(10) When a transformer is able to be paralleled with another transformer, specifying $\%IR$, $\%IX$, and $\%IZ$ is required.

5.2.3 Transformer Construction. Transformers are constructed in several different types, which are discussed below. This section is generally applicable to transformers of the liquid filled, ventilated dry, or gas filled dry types. Liquid insulated and gas filled transformers have their windings brought out to bushings or to junction boxes on the ends or the top of the transformers. Ventilated-dry-type transformers usually have their windings terminated within the enclosure of the transformer to either stand-off insulators or bus bar terminals.

(1) Liquid Filled Transformers — Are constructed with the windings encased in a liquid-tight tank filled with insulating liquid. Liquid filled transformers should be avoided inside commercial buildings unless nonflammable or less-flammable liquids are used or unless proper precautions are taken by building a transformer vault that meets the requirements of the NEC [9], and then only if all applicable jurisdictional and insurance carrier requirements have been met. The liquid provides insulation between the various sections of the windings and between the windings and the tank, and serves as a cooling medium, absorbing heat from the windings and transferring it to the outside of the tank. To increase the transfer of heat to the air, tanks are provided with cooling fins (to increase the area of the radiating surface) or with external cooling tubes or radiators. The hot liquid circulates through the radiators, transferring the heat picked up in the transformer windings to the radiator and then to the surrounding air.

Fans are sometimes installed to force air over the radiators in order to increase the full load rating by approximately 15% on transformers rated 750–2000 kVA and 25% on transformers rated 2500–10 000 kVA.

It is essential that the liquid in the transformer be maintained, clean, and free from moisture. Moisture can enter the transformer through leaks in the tank covers or when moisture-laden air is drawn into the transformer. Transformers can draw air into the tanks through breathing action that results from changes in the volume of liquid, and air in the tank that occurs with changes in temperature. Most modern transformers are tightly sealed and do not breathe if they are free from leaks.

Insulating liquid, through the normal aging process, develops a small amount of acid that, if allowed to increase above well-established limits, can cause damage to insulation in the transformer. Yearly testing to determine the dielectric breakdown voltage of the liquid (a low dielectric test indicates the presence of water or other foreign material) and neutralization number (a high neutralization number indicates the presence of acid in the liquid) by a competent testing laboratory will greatly prolong the life of the transformer. Liquid samples should be withdrawn under carefully controlled conditions as directed by the group making the liquid test. In some areas, this service is available from the electric utility.

The classification and handling of existing liquid filled transformers with regard to PCB contamination is subject to strict control by environmental agencies. Information on the handling and maintenance of liquid filled transformers can be obtained from the *Federal Register*, manufacturers, local EPA offices, and, usually, the local electric utility. It is important that any existing liquid filled transformers that have not been "evaluated," tagged, or otherwise classified be properly handled. Liquid filled transformers, which contain from 50–500 parts per million (ppm) of PCB, have successfully been brought into the 0–50 ppm range, which is within the limits of non-PCB contamination.

(2) Ventilated-Dry-Type Transformers—Are constructed in much the same manner as liquid filled transformers, except that the insulating liquid is replaced with air, and larger clearances and different insulating materials are used to compensate for the lower dielectric strength of air (see IEEE C57.12.01-1989, IEEE Standard General Requirements for Dry-Type Distribution and Power Transformers Including Those with Solid-Cast and/or Resin-Encapsulated Windings (ANSI) [24]). Both ventilated-dry-type and sealed-dry-type transformers use a UL component recognized insulation system that is suitable for operation at an ultimate temperature of 220 °C. The normal temperature rise of the windings is 150 °C by resistance. If transformers are purchased with a 220 °C insulation system, but are rated for full load use at a lower temperature (115 °C or 80 °C rise), then an improvement in efficiency, overload capability, and life can be expected. Units rated over 600 V are listed under UL 1562-1990, Transformers, Distribution, Dry-Type—Over 600 V [38].[49] In addition, IEEE PC57.12.58, Guide for Conducting a Transient Voltage Analysis of a Dry-Type Transformer Coil [25] is a guide to making a transient analysis of the high-voltage winding to assure that the insulating system can withstand the repetitive transients prevalent in today's electric system due to vacuum switches and similar devices. IEEE PC57.12.58 [25] also indicates that a transient surge can be

---

[49]UL publications are available from Underwriters Laboratories, 333 Pfingsten Road, Northbrook, IL 60062.

doubled upon entering the high-voltage winding. IEEE C57.12.01-1989 (ANSI) [24] also applies.

Consideration should also be given to nonlinear harmonic loads, such as SCRs, UPS, rectifiers, and variable speed drive applications, since these higher harmonics can cause appreciably higher eddy and stray loss heating in the windings as well as very high currents in the neutrals of these transformers.

Very often special designs for nonlinear load applications are preferable to just oversizing the unit because of the skin effect at the higher frequencies.

The ventilated-dry-type transformer is provided with a sheet metal enclosure that surrounds the winding for mechanical protection of the windings and the safety of personnel. Ventilating louvers are installed in the enclosure to permit thermal circulation of air directly over the winding for cooling. Fans are sometimes installed to force air directly over the windings in order to increase the full load rating by approximately 33%. These types of transformers are normally installed indoors and require the periodic cleaning of the complete core and coil assembly and an adequate supply of clean ventilating air. These transformers are gaining acceptance in the 15 kV and 34.5 kV class, and can be built to match the BIL of liquid immersed transformers and with special enclosures for use outdoors. Meggering before energizing is recommended after a lengthy shutdown or lengthy periods when the insulation has been subjected to moisture.

(3) Sealed-Dry-Type Transformers — Sealed-dry-type transformers are constructed in essentially the same way as ventilated-dry-type transformers. The enclosing tank is sealed and operated under positive pressures. It may be filled with nitrogen or other dielectric gas. Heat is transferred from the winding to the gas within the transformer housing and from there to the tank and to the surrounding air.

The sealed-dry-type (gas filled) transformer can be installed both outdoors and indoors and in areas where a corrosive or dirty atmosphere would make it impossible to use a ventilated-dry-type transformer.

(4) Cast-Coil, Dry-Type Transformers — Are constructed with primary and secondary windings encapsulated (cast) in reinforced epoxy resin. Because of the cast-coil construction, they are ideal in applications where moisture or airborne contaminants, or both, are a major concern. This type of construction is available with primary voltage ratings through the 34.5 kV class and BIL ratings through 200 kV. These transformers are ideal alternatives for liquid or gas filled units in indoor or rooftop applications. They may be forced air cooled to increase their self-cooled ratings by 50%.

(5) Totally Enclosed, Nonventilated-Dry-Type Transformers — Are constructed in essentially the same way as ventilated-dry-type transformers. The enclosure, while not sealed, contains air, so the units have the same BIL capabilities as ventilated-dry-type transformers. The totally enclosed, nonventilated-dry-type transformer can be installed both indoors and outdoors and in areas where a corrosive or dirty atmosphere would make it impossible to use a ventilated-dry-type transformer. These units are available with fan cooling for a minimum 25% increase in capacity.

(6) Winding Temperature Measurement and Controls — Various temperature measurement equipment and controls are available for determining the winding temperature and for activating cooling, tripping, or alarm devices. To make sure the ultimate temperature of the insulating system is not exceeded, imbedded detectors should be wound in each low-voltage winding.

5.3 Medium- and High-Voltage Fuses. Medium- and high-voltage fuses are a part of many commercial power distribution systems. Applicable standards are ANSI C37.46-1981 (Reaff. 1988), Specifications for Power Fuses and Fuse Disconnecting Switches [4] and NEMA SG2-1986, High-Voltage Fuses [36].[50]

Modern fuses that are suitable for the range of voltages encountered fall into the following two general categories:

(1) Distribution Fuse Cutouts — According to the ANSI definition, the distribution fuse cutout has the following characteristics:
  (a) Dielectric withstand (BIL) strengths at distribution levels
  (b) Application primarily on distribution feeders and circuits
  (c) Mechanical construction basically adapted to pole or crossarm mounting except for the distribution oil cutout
  (d) Operating voltages correspond to distribution system voltages
     Characteristically, a distribution fuse cutout consists of a mounting (insulating support) and a fuseholder. The fuseholder, normally a disconnecting type, engages contacts supported on the mounting and is fitted with a simple, inexpensive fuse link. The fuseholder is lined with an organic material, usually horn fiber. Interruption of an overcurrent takes place within the fuseholder by the action of de-ionizing gases, which are liberated when the liner is exposed to the heat of the arc that is established when the fuse link melts in response to the overcurrent.

(2) Power Fuses — According to the ANSI definition, the power fuse is identified by the following characteristics:
  (a) Dielectric withstand (BIL) strengths at power class levels
  (b) Application in stations, substations, distribution feeders, and in metal-enclosed switchgear
  (c) Mechanical construction is adapted to mountings for use in all applications.

Power fuses have other characteristics that differentiate them from distribution fuse cutouts in that they are available in higher voltage, current, and interrupting-current ratings, and in forms suitable for indoor and enclosure applications as well as all types of outdoor applications.

A power fuse consists of a mounting plus a fuseholder or end fittings, which accept, respectively, a refill unit or fuse unit, or fuse. Many power fuses are available with blown fuse indicators, which provide a visual indication that a fuse has operated. Indoor mountings for use with fuse units rated up to 29 kV maximum can be furnished with an integral hookstick operated, load current interrupting

---

[50]NEMA publications are available from the National Electrical Manufacturers Association, 2101 L Street, N.W., Washington, DC 20037.

device, thus providing for single-pole live switching in addition to the fault inter-rupting function provided by the fuse.

Power fuses are typically classified as either expulsion-type or current-limiting-type, depending on such factors as construction, interrupting medium, and the method used to interrupt overcurrents. However, new developments in the area of medium-voltage fuses may not readily fit within either class. Such a new development is the electronic fuse.

(1) Expulsion-Type Power Fuses—The earliest forms of power fuses, being outgrowths of distribution fuse cutouts, were fiber lined, and circuit inter-ruption was also like that of cutouts. However, such fuses had limited inter-rupting capacity and could not be used within buildings or in enclosures, and thus led in the '30s to the development of solid material, boric acid power fuses. These fuses utilize densely molded solid boric acid powder as a lining for the interrupting chamber. This solid material lining liberates non-combustible, highly de-ionized steam when subjected to the arc established by the melting of the fusible element. Solid material, boric acid power fuses have higher interrupting capacities than fiber lined power fuses of identical physical dimensions, produce less noise, need less clearance in the path of the exhaust gases, and, importantly, can be applied with normal electrical clearances indoors, or in enclosures when equipped with exhaust control devices. (Exhaust control devices provide for quiet operation and contain all arc interruption products.) These advantages, plus their availability in a wide range of current and interrupting ratings, and time current character-istics have led to the wide use of solid material, boric acid power fuses in utility, industrial, and commercial power distribution systems.

(2) Current-Limiting Power Fuses—Introduction of current-limiting fuses in the United States occurred almost simultaneously with the development of solid material, boric acid power fuses. Current-limiting power fuses operate without expulsion of gases because all the arc energy of operation is absorbed by the powder or sand filler surrounding the fusible element. They provide current limitation if the overcurrent value greatly exceeds the fuse ampere rating, thereby reducing the stresses and possible damage in the circuit up to the fault. But, for lower overcurrent values, current limitation is not achieved. These fuses can be applied indoors or in enclosures, and require only normal electrical clearances. In addition to the protection of transformers, certain current-limiting fuses are for use with high-voltage motor starters.

(3) Electronic Power Fuses—Recently, another type of power fuse, the elec-tronic power fuse, has been introduced. This latest technological develop-ment combines many of the features and benefits of fuses and relays to provide coordination and ratings that are not obtainable with other power fuses. Electronic power fuses generally consist of two separate components: an electronic control module that provides the time current characteristics and the energy to initiate tripping; and an interrupting module that inter-rupts the current when an overcurrent occurs. The electronic control module makes it possible to provide a variety of time current characteristics, such as

instantaneous tripping or time delay tripping. Only the interrupting module is replaced following fuse operation.

5.3.1 Fuse Ratings

(1) High-Voltage, Fiber Lined Power Fuses — This category has its principal usage in outdoor applications at the subtransmission voltage level. This fuse is available in current ratings and three-phase symmetrical short-circuit interrupting ratings as shown in Table 39.

(2) High-Voltage, Solid Material, Boric Acid Fuses — High-voltage, solid material, boric acid fuses are available in two styles.

    (a) The end fitting and fuse unit style, in which fusible element, interrupting element, and operating element are all combined in an insulating tube structure called the "fuse unit," which is the replaceable section.

    (b) The fuseholder and refill unit style, in which only the fusible element and interrupting element are combined in an epoxy tube called the "refill unit," which is the only section replaced following operation.

Solid material, boric acid fuses in the end fitting and fuse unit styles are used universally: outdoors at subtransmission and distribution voltages in poletop or station-style mountings, as well as indoors at distribution voltages in mountings installed in metal-enclosed interrupter switchgear, indoor vaults, and pad-mounted switchgear. Indoor mountings incorporate an exhaust control device that contains most of the arc interruption products and virtually eliminates the noise accompanying a fuse operation. These exhaust control devices do not require a reduction of the interrupting ratings of the fuse. Outdoor mountings with exhaust control devices are also becoming available at distribution voltages.

Solid material, boric acid fuses in the end fitting and fuse unit style are available with current and interrupting ratings as shown in Table 40.

The solid material, boric acid fuses in the fuseholder and refill unit style can be used either indoors or outdoors at medium- and high-voltage distri-

Table 39
Maximum Continuous Current and Interrupting
Ratings for Horn Fiber Lined, Expulsion-Type Fuses

Rated Maximum Voltage (kV)	Continuous Current Ratings A (Maximum)				Maximum Interrupting Rating* kA rms Symmetrical
8.3	100	200	300	400	12.5
15.5	100	200	300	400	16.0
25.8	100	200	300	400	20.0
38.0	100	200	300	400	20.0
48.3	100	200	300	400	25.0
72.5	100	200	300	400	20.0
121	100	200			16.0
145	100	200			12.5
169	100	200			12.5

*Applies to all continuous current ratings.

butions. Indoor mountings for use with fuseholders and refill units rated up to 29 kV maximum are also available with integral load current interrupting devices for single-pole live switching. The fuses are available in current and interrupting ratings as shown in Table 41.

(3) Current-Limiting Power Fuses — Current-limiting power fuses that are suitable for the protection of auxiliary power transformers, small power transformers, and capacitor banks are available with current and interrupting ratings as shown in Table 42.

Current-limiting fuses for the protection of medium-voltage transformers are available with interrupting ratings to 80 kA (symmetrical) at 5.5 kV, 120 kA at 15.5 kV, and 44 kA at 25.8 kV and 38 kV.

Current-limiting fuses that are suitable only for use with high-voltage motor starters are available with current and interrupting ratings as shown in Table 43.

Table 40
Maximum Continuous Current and Interrupting
Ratings for Solid Material, Boric Acid Fuses (Fuse Units)

Rated Maximum Voltage (kV)	Continuous Current Ratings A (Maximum)			Corresponding Maximum Interrupting Ratings kA rms Symmetrical		
5.5			400	25.0		
17		200	400	14.0	25.0	
27		200	400	12.5	20.0	
38	100	200	300	6.7	17.5	33.5
48.3	100	200	300	5.0	13.1	31.5
72.5	100	200	300	3.35	10.0	25.0
121	100	250		5.0	10.5	
145	100	250		4.2	8.75	

Table 41
Maximum Continuous Current and Interrupting
Ratings for Solid Material, Boric Acid Fuses (Refill Units)

Rated Maximum Voltage (kV)	Continuous Current Ratings A (Maximum)			Corresponding Maximum Interrupting Ratings kA rms Symmetrical		
2.75	200	400	720*	7.2	37.5	37.5
4.8	200	400	720*	17.2	37.5	37.5
8.25	200	400	720*	15.6	29.4	29.4
15.5	200	400	720*	14.0	34.0	25.0
25.8	200	300		12.5	21.0	
38	200	300		6.25	17.5	

*Parallel fuses

(4) Electronic Power Fuses — Electronic power fuses are suitable for service entrance protection and the coordination of commercial distribution circuits because they have high current-carrying capability and unique time current characteristics designed for coordination with source-side overcurrent relays and load-side feeder fuses. They are ideally suited for load feeder protection and coordination because of their high continuous and interrupting ratings. Electronic fuses are available in current and interrupting ratings as shown in Table 44.

5.3.2 Fuse Applications

(1) Power Supply — When a commercial project is served by a utility at medium or high voltage and a transformer substation provides in-plant service at utilization voltage or primary distribution voltage, power fuses can be used as an economical primary-side overcurrent protective device for transformer banks rated to 161 kV with a 15 000 kVA maximum rating.

With their high short-circuit interrupting capability and high-speed operation, power fuses will protect the circuit by clearing faults at the trans-

Table 42
Maximum Continuous Current and Interrupting
Ratings for Current-Limiting Fuses

Rated Maximum Voltage (kV)	Continuous Current Ratings A (Maximum)				Corresponding Maximum Interrupting Ratings kA rms Symmetrical			
2.75	225	450*	750*	1350*	50.0	50.0	40.0	40.0
2.75/4.76		450*				50.0		
5.5	225	400	750*	1350*	50.0	62.5	40.0	40.0
8.25		125	200*			50.0	50.0	
15.5	65	100	125*	200*	85.0	50.0	85.0	50.0
25.8		50	100*				35.0	35.0
38		50	100*				35.0	35.0

*Parallel fuses

Table 43
Maximum Continuous Current and Interrupting
Ratings for Current-Limiting Fuses (Motor Starters)

Rated Maximum Voltage (kV)	R Designation	Continuous Current Ratings A (Maximum)	Corresponding Maximum Interrupting Ratings kA rms Symmetrical
2.54	50 R	700	50.0
2.75/5.5	—	750	50.0
5.0	50 R	700	50.0
7.2	18 R	390	50.0
8.3	6 R	170	50.0

Table 44
Maximum Continuous Current and Interrupting
Ratings for Electronic Fuses

Rated Maximum Voltage (kV)	Continuous Current Ratings A (Maximum)		Corresponding Maximum Interrupting Ratings kA rms Symmetrical	
5.5	—	600	—	40.0
17.0	400	600	14.0	40.0
29	200	600	12.5	40.0

former. In addition, power fuses can provide backup protection in the event of a transformer secondary overcurrent protective device malfunction.

In addition to providing overcurrent protection to the main power transformers, power fuses are used to provide protection for instrument transformers and for capacitor banks.

(2) Power Distribution — The principal functions of overcurrent protective devices at these primary voltages are

(a) To interrupt high values of overcurrent

(b) To act as backup protection in the event of a malfunction of the next downstream protective device

(c) To open circuits under overcurrent conditions

(d) To coordinate with the next upstream and downstream protective device

Modern medium-voltage power fuses can be used to provide this protection and coordination for virtually all types and sizes of distribution systems. Such fuses used with properly coordinated and designed load interrupter switches may be applied outdoors in vaults, or in metal-enclosed interrupter switchgear.

5.4 Metal-Enclosed 5–34.5 kV Interrupter Switchgear. Metal-enclosed interrupter switchgear can be used to provide switching capability and overcurrent protection through the use of interrupter switches and power fuses. An interrupter switch is an air switch equipped with an interrupter that makes or breaks specified currents. Interrupter switches depend on high operating speed to divert the arc from the main contacts during opening onto enclosing materials within the interrupters, which confine the arc and evolve gases to suppress it. Interrupter switchgear can also be used for ground-fault protection of resistance-grounded systems, if properly applied. Rated maximum voltages are 4.8 kV, 8.25 kV, 15.0 kV, 15.5 kV, 17.0 kV, 25.8 kV, 29.0 kV, and 38.0 kV with main bus ratings of 600 A, 1200 A, or 2000 A. Interrupting ratings are determined by the power fuses, for which maximum ratings are given in Tables 40–44. Power fuses are available in a wide range of current ratings and are offered in a selection of time current characteristics to provide proper coordination with other protective devices and with the thermal characteristics of the power transformer.

The interrupter switches, which may be manually or automatically operated, are rated 200 A, 600 A, or 1200 A, continuous and interrupting. Interrupter switches are also available with a vacuum or SF_6 gas as the interrupting medium. Generally, these have limited fault ratings and are used primarily for switching. They are compatible with automatic or remote control schemes, but may have the disadvantage of lacking a visible break as is available with most air-type interrupter switches. SF_6 interrupter switches are available for all ranges of medium-voltage applications, and vacuum switches are available up to 35 kV.

Interrupter switches of all types can be applied in combination with power fuses (including current-limiting fuses) to achieve greater ratings than may be possible when the interrupter switch is used alone. An applicable standard for metal-enclosed interrupter switchgear is IEEE C37.20.3-1987, IEEE Standard for Metal-Enclosed Interrupter Switchgear (ANSI) [20] and NEMA SG6-1990, Power Switching Equipment [37].

Metal-enclosed interrupter switchgear does not incorporate a reclosing feature because reclosing is rarely desirable in power systems for commercial buildings where the conductors are commonly arranged in cable trays or enclosed in raceways or busways. The rare faults that do occur in such installations require significant repair before re-energization.

Metal-enclosed interrupter switchgear can be used in high-continuity distribution circuits, such as the conventional (two-switch) and the split bus (three-switch) primary-selective systems. Furthermore, the switches can be manually operated or power operated (with either automatic or remote operation), depending on system operating requirements.

Interrupter switchgear is usually less expensive than metal-clad power switchgear (see 5.5). This permits the engineer to improve service continuity by providing more radial feeders per dollar of equipment cost with the use of interrupter switchgear.

5.4.1 Automatic Control Devices. Automatic control devices can be incorporated in metal-enclosed interrupter switchgear, in conjunction with motor-powered switch operators, to provide high service continuity through primary-selective systems by initiating the automatic transfer of sources that provide service to the main bus (or buses) in the event of a fault or outage on one of the sources. Optional features include provisions for manual or automatic back transfer (with open or closed transition), time delay on transfer, and lockout on faults.

Switch operators can typically be disconnected from the associated switches to permit the checking of the automatic transfer scheme without requiring a power interruption to the load.

Interrupter switch manufacturers can also provide an open phase or overcurrent relay system, which initiates circuit interruption to protect loads from single phasing that may occur as a result of broken conductors or fuse operations in the source-side circuit. These relays can also be applied to protect against single phasing due to load circuit fuse operations.

5.4.2 Auxiliary Equipment and Features. Metal-enclosed interrupter switchgear may include (in addition to interrupter switches and power fuses) instrument transformers, voltage and current sensors, meters, and other auxiliary devices, including motor powered switch operators for remote operation of the interrupter switches (or operation of the switches in an automatic transfer scheme, when used in conjunction with an automatic control device). The power fuses may be equipped with blown fuse indicators (for positive visual checking of fuses while in their mountings).

5.4.3 Capability Required. Metal-enclosed interrupter switchgear should comply with the NEC, Article 710-21(e) [9], which requires that interrupter switches, when used in combination with fuses or circuit breakers, safely withstand the effects of closing, carrying, or interrupting all possible currents up to the assigned maximum short-circuit rating. (See also IEEE C37.20.3-1987, 6.4.8 (ANSI) [20].) Fault interrupting ratings are not required for interrupter switches because the associated fuses should be selected to interrupt any faults that may occur.

5.5 Metal-Clad 5–34.5 kV Circuit Breaker Switchgear. Metal-clad switchgear is available with voltage ratings of 4.16–34.5 kV and with circuit breakers having interrupting ratings from 8.8 kA at 4.16 kV to 40 kA at 34.5 kV as standard. Continuous current ratings are 1200 A, 2000 A, 3000 A, and 3750 A. Applicable standards include IEEE C37.20.2-1987, IEEE Standard for Metal-Clad and Station-Type Cubicle Switchgear (ANSI) [19], IEEE C37.04-1979 (Reaff. 1988), IEEE Standard Rating Structure for AC High-Voltage Circuit Breakers Rated on a Symmetrical Current Basis (ANSI) [16], IEEE C37.09-1979, IEEE Test Procedure for AC High-Voltage Circuit Breakers Rated on a Symmetrical Current Basis (ANSI) [B4], IEEE C37.010-1979, IEEE Application Guide for AC High-Voltage Circuit Breakers Rated on a Symmetrical Current Basis (ANSI) (Includes Supplement IEEE C37.010d-1984 [ANSI]) [B5], IEEE C37.011-1979, IEEE Application Guide for Transient Recovery Voltage for AC High-Voltage Circuit Breakers Rated on a Symmetrical Current Basis (ANSI) [B6], IEEE C37.012-1979, IEEE Application Guide for Capacitance Current Switching of AC High-Voltage Circuit Breakers Rated on a Symmetrical Current Basis (ANSI) [B7], IEEE C37.1-1987, IEEE Standard Definition, Specification, and Analysis of Systems Used for Supervisory Control, Data Acquisition, and Automatic Control (ANSI) [B8], IEEE C37.2-1979, IEEE Standard Electrical Power System Device Function Numbers (ANSI) [B9], and IEEE C37.100-1981 (Reaff. 1989), IEEE Standard Definitions for Power Switchgear (ANSI) [22] for power circuit breakers.

Metal-clad switchgear has a circuit breaker as the main circuit interrupting and protective device. Major parts of the primary circuit, such as circuit switching or interrupting devices, buses, potential transformers, and control power transformers, are completely enclosed by grounded metal barriers. Circuit instruments, protective relays, and control switches are mounted on a hinged control panel or occasionally on a separate switchboard remote from the switchgear. The power circuit breaker is readily removable and has self-coupling disconnecting primary and secondary contacts. Potential transformers and control power transformer fuses may be provided in drawout assemblies to permit the safe changing of fuses.

Automatic shutters to shield the stationary primary contacts when the circuit breaker is removed are provided, as well as other necessary interlocking features to ensure a proper sequence of operation. The drawout feature facilitates inspection and maintenance of the circuit breaker. In addition, it permits the quick replacement of any circuit breaker with a spare and, therefore, provisions for bypassing it during circuit breaker maintenance periods are generally not required. The circuit breaker compartments have separable main and secondary (or control) disconnect contacts to achieve connected, test, and disconnect positions. The test position provides a feature whereby the circuit breaker may be electrically exercised while disconnected from the main power circuit. The disconnect position allows the circuit breaker to be disconnected from the main power and control supply, locked, and stored in its cubicle.

Metal-clad switchgear can provide the switching, isolation, protection, and instrumentation of all the incoming, bus tie, and feeder circuits. All parts are housed within grounded metal enclosures, thereby providing a high degree of safety for both personnel and equipment. All line conductors are opened simultaneously in the event of circuit breaker tripping. A wide variety of parameters can be programmed into the tripping function.

The insulation used in the vital points of the metal-clad switchgear is of the potential tracking-resistant-type and may be flame-retardant. Thus, the equipment presents a very minimum fire hazard and is suitable for indoor installations without being placed in a vault. For outdoor equipment, a weatherproof enclosure is provided over the same switchgear components as is used for the indoor switchgear assemblies. Protected aisle construction, which permits maintenance in inclement weather, can also be provided.

5.5.1 Circuit Breakers. Medium-voltage power circuit breakers may be of the following types:

(1) Minimum-oil-type circuit breaker, which are no longer manufactured for medium-voltage applications
(2) Air-type circuit breaker, which was the standard for medium voltage until recently and, therefore, constitutes the greatest number in use today. But, it now has limited availability.
(3) SF_6-type circuit breaker
(4) Vacuum-type circuit breaker

The two latter types are readily available in metal-clad switchgear through 15 kV. Manufacturers can provide current information on the availability of the vacuum-type and SF_6-type at all other medium voltages.

The air-type circuit breaker has, in certain ratings, the disadvantage of having very large and heavy arc extinguishing "chutes," which enclose the contacts. The SF_6-type and vacuum-type are typically lighter than the air-type of the same rating. Both the vacuum-type and SF_6-type contain the arcs, which do not permit the arc products to exhaust to the atmosphere. The failure rate of the vacuum interrupter has been so low that it is not normally considered an operating problem; mechanical indicators associated with the vacuum interrupters indicate when contact wear requires replacement. The SF_6-type circuit breaker, in frequent usage, may require periodic service of the gas system, which should be performed by properly trained specialists because the arcing products sealed in gas chambers may be toxic and also because the gas should not become contaminated.

Any device that interrupts a reactive load at high speed (and almost all fault currents reflect a significant X/R ratio) can introduce transient overvoltages into a circuit. These transients may be dangerous to insulation, may increase as "traveling waves," may cause restrike within the interrupting device, and may damage or cause interference with sensitive electronic equipment. Very high speed circuit interruption by current-limiting devices (e.g., current-limiting fuses, circuit breakers, static switches) may introduce such transients. Vacuum switches and circuit breakers have, in the past, been a source of high-speed interruption. However, with newer contact design, the problem has been somewhat reduced. The design engineer should evaluate the need for the protection of system insulation (particularly solid-state equipment, motors, and dry-type transformers) by properly selecting insulation levels (BIL), by inserting surge capacitors and suppressors (where required), and by selecting interrupting devices that will avoid damaging transients.

Vacuum-type and SF_6-type circuit breakers offer the advantage of faster clearing time than air-magnetic-type breakers. The SF_6-type does this without the potential transient voltage surge effects of vacuum breakers. For a tabulation of standard ratings of circuit breakers for metal-clad switchgear, see ANSI C37.06-1987, Preferred Ratings and Related Required Capabilities for AC High-Voltage Circuit Breakers Rated on a Symmetrical Current Basis [2].

5.5.2 Instrument Transformers and Protective Relaying. All of these circuit breakers utilize relays, which are operated by current and voltage transformers. This combination provides a wide range of protection that is field adjustable. With protective relaying, full tripping selectivity can usually be obtained between all of the circuit breakers in the equipment in case of faults.

5.5.3 Control. Circuit breakers are electrically operated devices and should be provided with a source of control power. Control power can be obtained from a battery or from a control power transformer located within the switchgear.

5.5.4 Main Bus Current Selection. Main bus continuous current and momentary ratings are available to match the ratings of the associated circuit breakers. By the proper physical arrangement of the source and load circuit breakers or bus taps, it is possible to engineer the lowest bus current requirements consistent with the system capacity. For example, it may be necessary to have a 2000 A source circuit breaker (or breakers), yet only require a 1200 A main bus. Regardless of the lower bus capacity at different points, the bus is designed and rated for the present and future current capacity at the maximum point. It would not be tapered for reducing current capacity. The bus should also be properly braced to withstand system momentary requirements.

5.5.5 Ground and Test Devices. Ground and test devices are drawout-modified circuit breakers, which temporarily replace a normal circuit breaker for grounding the load (and sometimes the line) circuits for safety purposes while they are maintained. These devices also permit the insertion of probes for measuring voltage, assuring that the circuit is not energized, fault location, and cable testing. Ground and test devices should be purchased for any major power circuit breaker installation.

5.6 Metal-Enclosed, Low-Voltage 600 V Power Switchgear and Circuit Breakers

5.6.1 Drawout Switchgear. Metal-enclosed, drawout switchgear using air-type circuit breakers is available for the protection and control of low-voltage circuits. Rigid ANSI Standards dictate the design, construction, and testing of switchgear to assure reliability to the user. Industry standards are IEEE C37.20.1-1987, IEEE Standard for Metal-Enclosed Low-Voltage Power Circuit Breaker Switchgear (ANSI) [18] and IEEE C37.13-1981, IEEE Standard for Low-Voltage AC Power Circuit Breakers Used in Enclosures (ANSI) [17].

Unlike distribution switchboards where a broad variety of protective devices or panelboards can be incorporated, the main, tie, and feeder positions in low-voltage power switchgear are limited to drawout circuit breakers. Drawout switchgear is more adaptable and procurable with complex control circuitry, such as sequential interlocking, automatic transfer, or complex metering. This type of switchgear is often used in multiple-bus arrangements, such as the double-ended substation consisting of two buses, each with a feeder breaker and a tie breaker; so that in the event of a feeder failure, one feeder can automatically be switched to serve both buses. Various arrangements are discussed in Chapter 4.

This class of switchgear is available in both indoor and outdoor construction. The latter usually is constructed to provide a sheltered aisle with an overhead circuit breaker removal device. An integral roof-mounted circuit breaker removal device is also available for indoor construction.

The individual air-type circuit breakers are in compartments isolated from each other and from the bus area. Compartments accommodate circuit breakers in ANSI sizes of 225 A, 600 A, 1600 A, 2000 A, 3000 A, and 4000 A, arranged in multiple high construction. Some manufacturers offer 800 A, 2500 A, and 3200 A, instead of 600 A, 2000 A, and 3000 A ratings. The air-type circuit breakers can be electrically or manually operated and equipped with added devices, such as shunt trip, undervoltage, auxiliary switches, etc. They are available either with electro-magnetic overcurrent direct-acting tripping devices or static tripping devices.

The drawout circuit breakers and compartments have separable main and secondary disconnect contacts to achieve connected, test, disconnect, and fully withdrawn positions. The test position provides a feature whereby the circuit breaker may be exercised while disconnected from the main power circuit. The disconnect position allows the circuit breaker to be disconnected from the main power and control supply, and then locked and stored in its compartment. In the fully withdrawn position, the circuit breaker is exposed for inspection and adjustments and may be removed from the switchboard for replacement or inspection.

Separate compartments are provided for required meters, relays, instruments, etc. Potential and control power transformers are usually mounted in these compartments so that they will be front accessible. Current transformers may be mounted around the stationary power primary leads within the circuit breaker compartment (front accessible) or in the rear bus area.

The rear section of the switchboard is isolated from the front circuit breaker section and accommodates the main bus, feeder terminations, small wiring, and terminal blocks. Bus work is usually aluminum, designed for an allowable temperature rise of 65 °C above an average 40 °C ambient. A copper bus is available at an added cost. Circuit breaker terminals are accessible from the rear of the switch-board. Cable lugs or busway risers are provided for top or bottom exits from the switchgear. Control wiring from the separable control contacts of the circuit breaker is extended to terminal blocks mounted in the rear section. These blocks accommodate remote control and intercompartment and frame wiring by the manufacturer.

5.6.2 Low-Voltage Power Air-Type Circuit Breakers. Low-voltage power air-type circuit breakers are long-life, quick-make (via a stored energy manual or electrical closing mechanism), quick-break switching devices with integral inverse time overload or instantaneous trip units. These circuit breakers also have a short-time (30 Hz) rating, which permits the substitution of short-time tripping devices in place of the instantaneous tripping feature. Interrupting ratings for each circuit breaker depend on the voltage of the system to which it is applied (that is, 240 V, 480 V, 600 V, alternating-current, 60 Hz) and whether it is equipped with an instantaneous or short-time tripping feature as part of the circuit breaker assembly or equivalent panel-mounted protective relays. It is this short-time rating of the circuit breakers that permits the designer to develop selective systems. These circuit breakers are open construction assemblies on metal frames, with all parts designed for accessible maintenance, repair, and ease of replacement. They are intended for service in switchgear compartments or other enclosures of deadfront construction at 100% of their rating in a 40 °C ambient without compensation or de-rating. Tripping units are field-adjustable over a wide range and are completely interchangeable within their frame sizes.

Static-type tripping units are available from most manufacturers. Static trip units may provide an additional degree or number of steps in selectivity when only a small margin of spread exists between optimum protective settings for connected loads downstream and utility or other existing protective device settings upstream. Static devices readily permit the inclusion of ground-fault protection as part of the circuit breaker assembly.

A low-voltage power circuit breaker can be used by itself or with integral current-limiting fuses in drawout construction or separately mounted fuses to meet interrupting current requirements up to 200 000 A symmetrical rms. When part of the circuit breaker, the fuses are combined with an integral mounted blown fuse indicator and breaker trip device to open all three phases.

Air-type circuit breakers may be used for the control and protection of large low-voltage motors. They can be equipped to provide disconnect, running overload, and short-circuit protection, and are generally not suitable when operation is highly repetitive. (See Chapter 6 for more information.)

5.6.3 Selection of Circuit Breaker Tripping Characteristics. The degree of service continuity available from a low-voltage distribution system depends on the degree of coordination between circuit breaker tripping characteristics. The method of tripping coordination will be a factor in determining the degree of service continuity and of initial cost.

All circuit breakers should have adequate interrupting capacity for the fault current at the point of application. It may not be possible, because of cost or other limitations, to obtain full selectivity; however, a fully selective system should be the design goal. In a selective system, the main circuit breaker is equipped with over-current trip devices that have long- and short-time delay functions. The feeder circuit breakers are equipped with overcurrent trip devices that have long-time delay and instantaneous functions, unless they are required to be selective with other protective devices nearer the load. In this case, the feeders are equipped with trip devices that have both long- and short-time delay.

In a selective system, only the circuit breaker nearest the fault trips. Service continuity is thus maintained through all other circuit breakers. The selective system offers a maximum of service continuity, with a slightly higher initial cost for the short-time functions instead of the standard instantaneous function.

5.7 Metal-Enclosed Distribution Switchboards. Metal-enclosed distribution switchboards are frequently used in commercial buildings at 600 V and below for service entrance, power, or lighting distribution, and as the secondary sections of unit substations. A wide range of protective devices and single- or multiple-section assemblies are available for large services from 40–4000 A. While 4000 A equipment is available, the use of smaller services is recommended. NEMA PB2-1989, Deadfront Distribution Switchboards [35] is applicable.

Equipment ground-fault protection is recommended when the switchboard is applied on grounded wye systems. It is required on electrical services of more than 150 V to ground for any service disconnecting mean rated 1000 A or more. See the NEC, Article 230-95 [9] for minimum requirements.

Automatic transfer between main and emergency sources is generally provided as a complete package with all of the power and control features built into the assembly by the manufacturer in accordance with applicable standards (see 5.18).

5.7.1 Components. The following components are available:
(1) Service protectors
(2) Molded-case circuit breakers, group or individually mounted
(3) Fusible switches
(4) Motor starters
(5) Low-voltage ac power circuit breaker (generally limited to main or tie position)
(6) Bolted pressure and high-pressure switches
(7) Transfer devices or switches
(8) Instrumentation, metering, and relaying—Instrumentation and metering include the utility company metering equipment, voltmeters, ammeters, wattmeters, voltage and current transformers, etc.

5.7.2 Construction Features
(1) Front Accessible—Front Connected
    (a) Designed to be installed against a wall.
    (b) All mechanical and electrical connections are made from the front.
    (c) Multiple-section switchboards have backs lined up.
    (d) Switchboards are enclosed on all sides except the bottom.
    (e) Maximum rating of 2000 A.
    (f) Drawout low-voltage ac power circuit breakers are not available as branch devices.

    (g) Load-side risers are not available.
  (2) Rear Accessible — Front Connected
    (a) Designed to be free-standing.
    (b) Designed for rear accessibility.
    (c) All main connections are made from the rear.
    (d) All normal maintenance in the main bus is performed from the rear.
    (e) All line and load connections for branch devices are made from the front.
    (f) Cross bus is located behind the branch devices and is accessible only from the rear.
    (g) Multiple-section switchboards have fronts lined up.
    (h) Capable of accepting all components.
  (3) Rear Accessible — Rear Connected
    (a) Designed to be free-standing.
    (b) Designed for rear accessibility.
    (c) All main connections are made from the rear.
    (d) All normal maintenance to the main bus is performed from the rear.
    (e) All line and load connections for branch devices are made from the rear.
    (f) All cross bus and line and load connections for branch devices are accessible only from the rear.
    (g) Multiple-section switchboards have fronts lined up.
    (h) Capable of accepting all components.

5.8 Primary-Unit Substations. Primary-unit substations are best described by their function, that is, to transform power from high or medium voltages down to a voltage above 1000 V, and to provide protection and control for the lower voltage feeder circuits. Primary-unit substations are most often used today in commercial buildings to convert any 13.2 – 34.5 kV service to 4160 V or 2400 V for large motors. They may be used, however, to provide service at any medium voltage when power is being purchased at a higher voltage.

These unit substations are physically and electrically coordinated, indoor or outdoor, combinations of primary-unit-substation-type transformers and metal-enclosed interrupter switchgear or power circuit breaker switchgear.

The incoming and secondary sections of primary-unit substations are available in arrangements to suit the many variations of power distribution circuits as described in 5.9.

For detailed information on the transformer section, refer to 5.2. Similarly, for detailed information on the switchgear section, refer to 5.3, 5.4, and 5.5.

5.9 Secondary-Unit Substations. Secondary-unit substations are best described by their function; i.e., to transform power from the 2300 – 35 000 V range down to 600 V or less and to provide protection and control for low-voltage feeder circuits. Secondary-unit substations consist of coordinated incoming line, transformer, and low-voltage sections. Each of these major sections is available in several forms for both indoor and outdoor application and to suit the many variations of power distribution circuit arrangements.

5.9.1 Basic Circuits. Four basic circuits are most widely used in the following order:

(1) Simple radial system
(2) Secondary-selective system
(3) Primary-selective system
(4) Secondary-network system

5.9.2 Incoming Line Section. For use with the simple radial, secondary-selective, or secondary-network system, this section will generally consist of one fuse and a two-position (open-close) 5 kV or 15 kV interrupter switch. This is the same device that is discussed in 5.4. Interrupter switches, or simply an air filled terminal chamber, may satisfy the application.

Primary-selective systems using interrupter switches, as discussed in Chapter 4, normally involve the use of two interrupter switches serving a common bus, or two interrupter switches serving two buses with an interrupter tie switch. In either case, one primary supply can serve the entire load if the other is not available. Transfer can be made manually or automatically (stored energy). When manual or automatic transfer is used, electrical or mechanical interlocks may be used to prevent inadvertently connecting the two sources together.

5.9.3 Transformer Section. This section transforms the incoming power from the higher primary to the lower secondary voltage. Ratings, voltages, and connections are as covered in 5.2.1. The transformer is mechanically and electrically coordinated to the incoming line (primary) section and to the low-voltage switchgear section.

5.9.4 Low-Voltage Switchgear Section. This section provides the protection and control for the low-voltage feeder circuits. It may consist of a drawout circuit breaker switchgear assembly, a metal-enclosed distribution switchboard, a panelboard mounted in or on the transformer section, or a single secondary protective device. Aluminum bus work has become the standard furnished by most manufacturers. Copper is also available, but at an additional cost.

For detailed information on the low-voltage switchgear section, refer to 5.6 and 5.7.

5.10 Panelboards. Electric systems in commercial buildings usually include panelboards, which utilize fusible or circuit breaker devices, or both. They are generally classified into two categories

(1) Lighting and appliance panels
(2) Power distribution panels

Panelboard mounting of motor starter units may also be involved. NEMA PB1-1990, Panelboards [34] and ANSI/UL 67-1988, Panelboards [10] are applicable.

5.10.1 Lighting and Appliance Panelboards. These panels have more than 10% of the overcurrent devices rated 30 A or less, for which neutral connections are provided. The number of overcurrent devices (branch-circuit poles) is limited to a maximum of 42 in any one box. When the 42 poles are exceeded, two or more separate boxes are required. A common front for multiple boxes is usually available.

Narrow width box constructions are used to fit into a 10 inch or 8 inch structural wide flange beam where mounting of a panelboard on a building column is appropriate. Column extensions and pull boxes are also available for this application.

Ratings of these panels are single-phase, two-wire 120 V or three-wire 120/240 V; 120/208 V, three-phase, three-wire 208 V, 240 V, or 480 V; and three-phase, four-wire 208Y/120 V or 480Y/277 V.

5.10.2 Power Distribution Panelboards. This type includes all other panelboards not defined as lighting and appliance panelboards. The 42 overcurrent protective device limitation does not apply. However, care should be exercised not to exceed practical physical limitations, such as the standard box heights and widths available. Common fronts for two or more boxes are often impractical from a weight and installation standpoint due to the size of this type of panelboard.

Ratings are single-phase, two- or three-wire; three-phase, three- or four-wire; 120/240 V through 600 V_{ac}, 250 V_{dc}; 50–1600 A, 1200 A maximum branch.

5.10.3 Motor Starter Panelboards. Rather than use an individual mounting, a small number of motor starters can be grouped into a panelboard. Motor starter panelboards consist of combination units utilizing either molded-case or motor circuit protector fusible disconnects. The combination starters are factory wired and assembled. Class A provides no wiring external to the combination starter; Class B provides control wiring to terminal blocks furnished near the side of each unit. When a large number of motors are to be controlled from one location or additional wiring between starters and to master terminal blocks is required, conventional motor control centers (MCCs) are most commonly used. See Chapter 6 for a discussion of MCCs.

5.10.4 Multiple-Section Panelboards. Both lighting and appliance panelboards or power distribution panelboards requiring more than one box are called "multiple-section" panelboards. Unless a main overcurrent device is provided in each section, each section should be furnished with a main bus and terminals of the same rating for connection to one feeder. The three methods commonly used for interconnecting multiple-section panelboards are as follows:

(1) Gutter Tapping—Increased gutter width may be required. Tap devices are not furnished with the panelboard.

(2) Subfeeding—A second set of main lugs (subfeed) are provided directly beside the main lugs of each panelboard section, except the last in the lineup.

(3) Throughfeeding—A second set of main lugs (throughfeed) are provided on the main bus at the opposite end from the main lugs of each section, except the last in the lineup. This method has the undesirable feature of allowing the current of the second panelboard section to flow through the main bus of the first section.

5.10.5 Panelboard Data. To assist the engineer planning an installation, manufacturers' catalogs provide a wide choice of panelboards for specific applications. Some very important rules governing the application of panelboards are described in the NEC [9].

(1) Six Circuit Rule—The NEC, Article 230-71 [9] specifies that a device may be suitable for service entrance equipment when not more than six main disconnecting means are provided (except two mains maximum in lighting and

appliance branch-circuit panelboards). In addition, a disconnecting means (which need not be a switch) shall be provided for the ground conductor as specified in the NEC, Article 230-75 [9].

(2) Thirty Conductor Rule—The NEC, Article 362-5 [9] states that wireways shall not contain more than 30 conductors at any cross section, unless the conductors are for signaling or motor control. It further states that the total cross sectional areas of all the conductors shall not exceed 20% of the internal cross section of the wireway. Column panels or panels fed by a single wireway are limited to three main conductors and 27 branch and neutral conductors (12 circuit panelboard, single-phase, three-wire). When the neutral bar is mounted in a column panel pullbox, this will be changed to two main conductors and 28 branch circuits (28 circuit panelboard).

(3) Gutter Tap Rule—The NEC, Article 240-21 [9] states that overcurrent devices shall be located at the point where the conductor to be protected receives its supply. But many exceptions apply to this rule. For example, exception number 2 to this paragraph permits omission of the main overcurrent device if the tap conductor (a) is not over 10 feet long, (b) is enclosed in a raceway, (c) does not extend beyond the panelboard it supplies, and (d) has an ampacity not less than the combined computed loads supplied and not less than the ampere rating of the switchboard, panelboard, or control device supplied. Gutter taps are permitted under this ruling.

5.11 Molded-Case Circuit Breakers. Standard designs of molded-case circuit breakers (MCCB) are quick-make and quick-break switching devices with both inverse time and instantaneous trip action. They are encased within rigid, non-metallic housings and vary greatly in size and rating. Standard frames are available with 30–4000 A current, and 120–600 V_{ac} and 125–250 V_{dc} ratings.

The smaller breakers are built in one-, two-, or three-pole construction and are sealed units without adjustable instantaneous trips. The larger ratings are usually available in three- or four-pole frames only and have interchangeable and adjustable instantaneous trip units. With modifications and new developments, the manufacturers' catalogs should be consulted to obtain the MCCB best suited for user requirements.

The current domestic standards are NEMA AB1-1986, Molded-Case Circuit Breakers [31] and ANSI/UL 489-1985, Molded-Case Circuit Breakers and Circuit Breaker Enclosures [12].

(1) Requirements—With few exceptions, the manufacturing, ratings, and performance requirements are the same for both standards. Typically, MCCBs are submitted for UL witness testing, which is repeated periodically for certification. The required switching tests are conducted sequentially with a set of MCCBs according to a listed schedule. The test samples undergo all tests, which include overload, endurance, and short circuit.

(2) Accessories—MCCBs are usually operated manually; but solenoids are available for remote tripping and electrical motor operators are available for remote operation with the larger frames. Other attachments are auxiliary contacts for signaling and undervoltage devices to trip the MCCB on

reduced system potential. All MCCB designs employ a trip-free mechanism, which prevents injury to an operator who closes a breaker into a fault. The larger frames have ground-fault designs utilizing external current transformers and relays to energize a shunt trip within the MCCB.

(3) Application — Ambient temperature and system frequency should be considered for all MCCBs. Unlike the power air-type circuit breakers, MCCBs usually require a 20% current de-rating when installed in enclosures. Several manufacturers offer 100% rated MCCBs with 600 A frames and larger. With few exceptions, conventional MCCB designs cannot be coordinated for selectivity. These breakers employ rapid mechanisms that have little inertia, and interrupting times at maximum fault levels are usually one cycle or less.

MCCBs employing electronic trip units and current transformers can be applied for selective coordination and their short-time ratings vary with each design. Many of these modern designs have internal ground-fault detection, which improves system protection.

5.11.1 Types of Molded-Case Circuit Breakers. These devices are available in the following general types:

(1) Thermal Magnetic — Employ temperature-sensitive bimetals, which provide inverse or time delayed tripping on overloads, and coils or magnet and armature designs for instantaneous tripping.

(2) Magnetic Only — Employ only instantaneous tripping and are used in welding or motor circuit application. The NEC [9] recognizes adjustable magnetic types only for motor circuit applications.

(3) Integrally Fused — Specially designed current-limiting fuses are housed within the molded case for extended short-circuit application in systems with 100 or 200 kA available, and interlocks are provided to ensure that the MCCB trips when any fuse operates.

(4) Current Limiting — Employs electromagnetic principles to effectively reduce the let-through magnitudes of current and energy (I^2t). Their ratings and number of effective operations are available in manufacturers' literature, and the designs are UL listed. However, some designs may be larger in size than typical circuit breakers.

(5) High Interrupting Capacity — Many manufacturers offer this type for application in systems having high fault currents. They employ stronger, high-temperature molded material, but retain the standard circuit breaker dimensions.

5.11.2 Use of Molded-Case Circuit Breakers. Molded-case circuit breakers are suitable in various equipment and installations.

(1) Individual enclosures
   (a) Wall-mounted, dust-resistant NEMA Types 1A and 12 (See ANSI/NEMA ICS6-1988, Enclosures for Industrial Control and Systems [8].)
   (b) Outdoor, raintight NEMA Type 3 (See ANSI/NEMA ICS6-1988 [8].)
   (c) Hazardous NEMA Types 4, 5, 7, and 9 (See ANSI/NEMA ICS6-1988 [8].)
(2) In panelboards and distribution switchboards
(3) In switchgear having rear-connected, bolt-on, plug-in, or drawout features

(4) In combination starters and motor control centers

(5) In automatic transfer switches

In case (5), molded-case circuit breakers may be used as a part of the automatic transfer switches to serve as service or feeder disconnects and to provide overcurrent protection. They may also be used as part of the automatic transfer switch when found suitable for this particular task and when operated by appropriate mechanisms in response to initiating signals, such as loss of voltage, etc. If MCCBs are used to combine both functions, an external manual operator should be provided for independent disconnections of both the normal and alternate supplies. Particularly in the larger sizes (current ratings), consideration should be given to the anticipated number of operations to which the equipment will be subjected because MCCBs are not designed for highly repetitive duty.

5.12 Low-Voltage Fuses. A fuse may be defined as "an overcurrent protective device with a circuit opening, fusible element part that is heated and severed by the passage of overcurrent through it." (See ANSI/NEMA FU1-1986, Low-Voltage Cartridge Fuses [7].) The fusible element opens in a time that varies approximately inversely with the square of the magnitude of current that flows through the fuse. The time current characteristic depends upon the rating and type of fuse. Non-time delay fuses are fuses that have no intentional built-in time delay. They are generally employed in other than motor circuits or in combination with circuit breakers where the circuit breaker provides protection in the overload current range and the fuse provides protection in the short-circuit current range.

Time delay fuses have intentional built-in time delays in the overload range. This time delay characteristic often permits the selection of fuse ratings that are closer to full load currents.

Time delay fuses are widely used because they have adequate time delay to permit their use as motor overcurrent running protection. Dual-element time delay fuses provide protection for both motors and circuits and make it possible to use a fuse whose current rating is not far above the full load current of the circuit. The fuse will permit starting inrush current of a motor, but stands ready to open the circuit on long continued overcurrent.

5.12.1 Fuse Ratings. Low-voltage fuses have current, voltage, and interrupting ratings, which should not be exceeded in practical application. In addition, some fuses are also rated according to their current-limiting capability as established by UL Standards and are so designated by a class marking on the fuse label (Classes J, K1, K5, L, RK1, RK5, etc.). Current-limiting capabilities are established by UL Standards according to the maximum peak current let-through and the maximum I^2t let-through of the fuse upon clearing a fault.

(1) Current Rating—Current rating of a fuse is the maximum dc or ac current in amperes at rated frequency, which it will carry without exceeding specified limits of temperature rise. Current ratings that are available range from milliamperes up to 6000 A.

(2) Voltage Rating—Voltage rating of a fuse is the maximum ac or dc voltage at which the fuse is designated to operate. Low-voltage fuses are usually given a voltage rating of 600 V, 300 V, 250 V, or 125 V_{ac} or V_{dc}, or both.

(3) Interrupting Rating — Interrupting rating of a fuse is the assigned maximum short-circuit current (usually ac) at rated voltage which the fuse will safely interrupt. Low-voltage fuses may have interrupting ratings of 10 000 A, 50 000 A, 100 000 A, or 200 000 A symmetrical rms.

5.12.2 Current Limitation. Current-limiting fuses allow less than the available current to flow into a fault for a time interval of less than a half-cycle, thereby reducing the current magnitude and the duration of the fault. It is designed so that, in the current-limiting range, a high enough arc voltage is developed as a fusible element melts to prevent the current from reaching the magnitude it otherwise would reach. The action is so fast that the current does not reach peak value in the first half-cycle (see Fig 48).

The current-limiting action limits the total energy flowing into a fault and thus minimizes mechanical and thermal stresses in the elements of the faulted circuit.

5.12.3 NEC Categories of Fuses. The NEC [9] recognizes two principal categories of fuses: plug fuses and cartridge fuses. In addition, the NEC [9] mentions the following fuses: time delay fuses, current-limiting fuses, noncurrent-limiting fuses, fuses over 600 V, and primary fuses.

(1) Plug Fuses — Are rated 125 V and are available with current ratings up to 30 A. Their use is limited to circuits rated 125 V or less, and they are usually employed in circuits supplied from a system having a grounded neutral and no conductor in those circuits operating at more than 150 V to ground. The NEC [9] requires Type S plug fuses in all new installations of plug fuses because they are tamper-resistant. A nonremovable adapter that screws into a standard Edison screw base limits the size of the Type S plug fuse, which can be inserted.

(2) Cartridge Fuses — Are constructed with cylindrical copper ends known as "ferrules" for ratings 60 A and below; and with knifeblade contacts for ratings above 60 A. For ratings above 600 A, the fuses are designed with holes for bolting into position. Table 45 shows cartridge fuses and fuseholder case sizes according to current and voltage. All fuses recognized by

**Fig 48
Current-Limiting Action of Fuses**

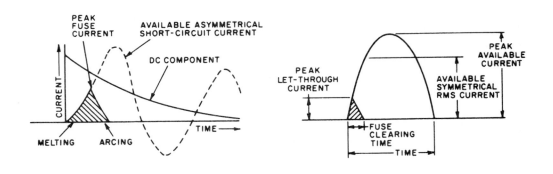

the NEC [9], which have interrupting ratings exceeding 10 000 A, should be marked on the fuse label with the designated interrupting rating. Fuses rated 10 000 A may also be so designated.

5.12.4 UL Listing Requirements. The UL Standard covering fuses requires the following:

(1) Fuses should carry 110% of their rating continuously when installed in the test circuit specified in the standard.

(2) Fuses of 0–60 A rating should open within 1 hour and fuses of 61–600 A rating within 2 hours when carrying 135% of rating in the specified test circuit. Fuses rated above 600 A should open within 4 hours when carrying 150% of rated current in the test circuit.

(3) Different current and voltage ratings of fuses should have specified physical dimensions, which prevent interchangeability.

(4) Fuses listed as having an interrupting rating in excess of 10 000 A should have their interrupting rating shown on the fuse.

5.12.5 Fuses Carrying Class Letter. UL (in conjunction with NEMA) has established standards for the classification of fuses by letter rather than by type. The class letter may designate interrupting rating, physical dimensions, degree of current limitation (maximum peak let-through current), and maximum clearing energy (A^2 seconds) under specific test conditions, or combinations of these characteristics. The descriptions of these classes are as follows:

(1) Class G Fuses, 0–60 A — Class G fuses are miniature fuses rated 300 V, primarily developed for use on 480Y/277 V systems for connections phase-to-ground. These fuses are available in ratings up to 60 A and carry an interrupting rating of 100 000 A symmetrical rms. Case sizes for 15 A, 20 A,

Table 45
Fuse Classification

(Cartridge fuses and fuseholders should be classified as listed here.)

Not Over 290 Volts	Not Over 300 Volts	Not Over 600 Volts
0—30	0—30	0—30
31—60	31—60	31—60
61—100	61—100	61—100
101—200	101—200	101—200
201—400	201—400	201—400
401—600	401—600	401—600
601—800	601—800	601—800
801—1200	801—1200	801—1200
1201—1600	1201—1600	1201—1600
1601—2000	1601—2000	1601—2000
2001—2500	2001—2500	2001—2500
2501—3000	2501—3000	2501—3000
3001—4000	3001—4000	3001—4000
4001—5000	4001—5000	4001—5000
5001—6000	5001—6000	5001—6000

NOTE: Fuses shall be permitted to be used for voltages at or below their voltage ratings.

30 A, and 60 A are each of a different length. Fuseholders designed for a specific case size will reject a larger fuse.

Class G fuses are considered to be time delay fuses according to UL if they have a minimum time delay of 12 seconds at 200% of their current rating.

(2) Class H Fuses, 0–600 A — Class H fuses have dimensions previously listed in the NEC [9]. These fuses are often referred to as "code fuses." Although these fuses are not marked with an interrupting rating, they are tested by UL on circuits that deliver 10 000 A_{ac} and may be marked 10 000 A_{ic}. They are rated 250 V or 600 V. The two fuses that are recognized as Class H fuses are

(a) One-time fuses (nonrenewable)

(b) Renewable fuses

The ordinary one-time cartridge fuse is the oldest type of cartridge fuse in use today. It utilizes a zinc or copper link and has limited interrupting capabilities. The use of the one-time fuse is decreasing due to its limited interrupting rating and lack of intentional time delay.

Renewable fuses are similar to one-time fuses, except that they can be taken apart after interrupting a circuit and the fusible element replaced. Renewal links are usually made of zinc. Their ends are clamped or bolted to the fuse terminals.

(3) Class J Fuses, 0–600 A — Class J fuses have specific physical dimensions that are smaller than the 600 V Class H fuses. Class H fuses cannot be installed in fuseholders that are designed for Class J fuses. Class J fuses are current-limiting and carry an interrupting rating of 200 000 A symmetrical rms. UL has also established maximum allowable limits for peak let-through current and let-through energy I^2t, which are slightly less than those for Class K1 fuses of the same current rating.

Time delay standards have been established for Class J fuses. To be UL listed as time delay, Class J fuses should have a minimum time delay of 10 seconds at 500% of rated current.

(4) Class K Fuses, 0–600 A — Class K designates a specific degree of peak let-through current and maximum clearing I^2t. Present Class K fuses have the same dimensions as Class H fuses, but have interrupting ratings higher than 10 000 A, i.e., 50 000 A, 100 000 A, or 200 000 A symmetrical rms. UL has established two Class K levels, K1 and K5, with Class K1 having the greatest current-limiting ability and K5, the least.

To be listed as time delay fuses, Class K fuses are required by UL to have a minimum time delay of 10 seconds at 500% of rated current (8 seconds for 250 V, 30 A fuses).

(5) Class R Fuses, 0–600 A — Class R designates a Class K fuse with a rejection feature on one end. All Class R fuses have a 200 000 A_{ic}. Class K5 fuses become Class RK5, and Class K1 fuses become Class RK1 fuses when rejection features are added.

(6) Class L Fuses, 601–6000 A — Class L fuses have specific physical dimensions and bolt-type terminals. They are rated 600 V and carry an interrupting rating of 200 000 A symmetrical rms. Class L fuses are current limiting and UL has specified maximum values of peak let-through current and I^2t for

each rating. UL has not established standards for time delay characteristics in the overload range for Class L fuses. However, Class L fuses may be labeled "time delay," and most of the available Class L fuses have a minimum time delay in the overload range of approximately 4 seconds at 500% of rated current. Time delay standards have been established for Class L fuses. To be UL listed as time delay, Class L fuses should have a minimum time delay of 10 seconds at 500% of rated current.

(7) Supplementary Fuses—There are other fuses with special characteristics and dimensions designed for supplementary overcurrent protection, some of which conform to UL Standards.

5.12.6 Cable Limiters (Protectors). Cable limiters are available for use in multiple-cable circuits to provide short-circuit protection for cables. Cable limiters are rated up to 600 V with interrupting ratings as high as 200 000 A symmetrical rms. They are rated according to cable size, that is, 4/0, 500 kcmil, etc., and have numerous types of terminations.

These limiters are designed to provide short-circuit protection for cables. They are used primarily in low-voltage networks or in service entrance circuits where more than two cables per phase are brought into a switchboard. A typical single-line diagram representing a cable limiter installation is shown in Fig 49. (Note that, for the isolation of faulted cable, the limiters should be located at each end of each cable.) The limiter does not provide overload protection as described in the NEC, Article 240 [9]. It does not have the characteristics associated with fuses, but will limit the extent of the fault while preserving service to the balance of the system.

In the event of the failure of a single cable, which is cleared by the limiters, the remaining cables would carry the load current continuously and could be damaged over a long period. If ground-fault protection is provided, it is likely that the entire feeder will be removed before the limiters operate; however, maintenance programs should include consideration of "blown limiters."

**Fig 49
Typical Circuit for Cable Limiter Application**

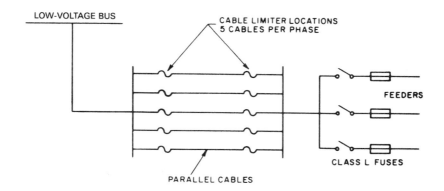

5.13 Service Protectors. A service protector is a nonautomatic circuit-breaker-type switching and protective device with an integral current-limiting fuse. Stored energy operation provides for manual or electrical closing. Switching under normal or abnormal current conditions, up to at least 12 times continuous current ratings of the service protector is permissible. It is capable of closing and latching against fault currents up to 200 000 A symmetrical rms. During fault interruption, the service protector will withstand the stresses created by the let-through current of the fuses. Therefore, for all operating conditions, including normal load, overload, and fault switching up to the maximum interrupting capacity, this dual-protective device will adequately open the circuit.

Downstream equipment is subject only to the let-through current of the fuses. Protection against single phasing is included in the design of service protectors.

Service protectors are generally available at continuous current ratings of 800 A, 1200 A, 1600 A, 2000 A, 3000 A, 4000 A, 5000 A, and 6000 A for use on 240 V_{ac} and 480 V_{ac}, in both two- and three-pole construction. They are used in both wall-mounted and free-standing compartments as well as in switchboards. Service protectors are often used with ground-fault protective equipment since their circuit breaker type of construction gives a total fault clearing time of under 3 Hz after shunt tripping by the ground-fault detector.

Manufacturers' catalogs should be consulted for complete ranges of equipment features and specific applications. IEEE C37.29-1981 (Reaff. 1985), IEEE Standard for Low-Voltage AC Power Circuit Protectors Used in Enclosures (ANSI) [21] is applicable.

5.14 Enclosed Switches. Enclosed switches are switches with or without fuse-holders, completely enclosed in metal, operable without opening the enclosure, and with provisions for padlocking in the off position. See NEMA KS1-1990, Enclosed Switches [32] and ANSI/UL 98-1986, Enclosed and Dead-Front Switches [11] for more information.

5.14.1 NEMA Requirements. NEMA requirements are as follows:

(1) General-Duty (Type GD) — General-duty switches are available in 30–600 A ratings and are intended for light service when usual load conditions prevail in systems not exceeding 240 V_{ac} and are for use with Class H fuses.

   They are capable of interrupting 600% of full load current 50 times at rated voltage. When properly designed and coordinated, these switches have been used in conjunction with certain types of fuses to achieve interrupting ratings through 200 000 A.

(2) Heavy-Duty (Type HD) — Heavy-duty switches are available from 30–1200 A ratings and are intended for systems not exceeding 600 V_{ac}. They may also be suitable for 600 V_{dc}. Various designs accommodate Class H, J, L, or R fuses, and approved kits are available to convert these switches for use with different fuse types.

   The ac interrupting ratings are based on their equivalent hp ratings. All HD switches that are approved for dc motors should successfully interrupt 400% of the full load current 50 times at rated voltage.

(3) UL has three switch categories:
  (a) General use without a hp rating
  (b) General use with a hp rating
  (c) Fuse motor circuit type
(4) UL interrupting requirements are as follows:
  (a) General-use switches should operate 50 times at 150% of nominal current.
  (b) Horsepower-rated switches have requirements similar to those of NEMA for ratings above 100 hp. Ratings less than 100 hp should interrupt load currents 50 times at approximately 160% of nominal current.

5.14.2 Application

(1) Current — Switches should have a current rating of at least 125% of the expected continuous load current.
(2) Frequency — Unless otherwise noted, all ac rated switches are approved for 60 Hz systems only.
(3) Temperature — Both NEMA and UL stipulate a maximum temperature limit of 30 °C (86 °F) rise throughout the conductor path when operated without fuses and carrying rated current, except for switches for use with 400 A and 600 A Class J fuses and all switches used with Class L fuses, which are permitted a maximum temperature rise of 60 °C (140 °F) when carrying 80% of their nameplate rating.
(4) Fused Switches — Switches approved with fuses are short-circuit tested at various magnitudes of fault current to determine the capability of the switch to either withstand let-through currents of the fuses or interrupt those current values that do not cause instantaneous fuse melting.
(5) Ground Fault — Switches approved as disconnects when used with ground-fault detectors employing a solenoid to open the switch should be carefully coordinated with fuses to ensure that the switch operates only within its interrupting capacity.

UL-listed switches with ratings from 30–1200 A for this application are tested according to ANSI/UL 1053-1982 (Reaff. 1988), Ground-Fault Sensing and Relaying Equipment [14] and are classified as follows:

(1) Class 1 service requires switches to be capable of interrupting at least 12 times their nameplate ratings.
(2) Class 2 service requires provisions to prevent opening of the switch on fault currents exceeding the normal ratings mentioned previously in this section.

5.15 Bolted Pressure Switches and High-Pressure Contact Switches

5.15.1 Manual Operations. A bolted pressure switch consists of movable blades and stationary contacts with arcing contacts and a simple toggle mechanism for applying pressure to both the hinge and jaw contacts in a manner similar to a bolted bus joint when the switch is closed. The operating mechanism consists of a spring that is compressed by the operating handle and released at the end of the operating stroke to provide quick-make and quick-break switching action.

A high-pressure contact switch has an over-center toggle mechanism with high-energy springs to achieve higher acceleration of parting (or closing) contacts. This provides for a higher interrupting capability.

5.15.2 Electrical Trip. The electrical trip, bolted pressure switch, or high-pressure contact switch, is basically the same as the manually operated switch, except that a stored energy latch mechanism and a solenoid trip release are added to provide simple and economical automatic electrical opening. These switches are designed for tripping from a remote location or for use with ground-fault protection equipment. Some switch designs have an integral ground-fault sensing scheme. The contact interrupting rating is 12 times continuous rating. Electrical trip, bolted pressure switches are capable of tripping at 55% of normal voltage and the opening time is approximately 6 Hz.

Both manually operated and electrical trip switches are designed for use with Class L current-limiting fuses. They are available in ratings of 800 A, 1200 A, 1600 A, 2000 A, 2500 A, 3000 A, 4000 A, and 6000 A, 600 V_{ac}, will carry 100% of rating, and are suitable for use on circuits having available fault currents of 200 000 A symmetrical rms. Both manually operated and electrical trip switches are available for switchboard mounting or in individual wall-mounted and free-standing enclosures.

Bolted pressure switches and high-pressure contact switches are covered by ANSI/UL 977-1984, Fused Power-Circuit Devices [13] and CSA Std C22.2-1980, Canadian Electrical Code, Part 2: Safety Standards for Electrical Equipment, Electrical Signs [15].[51] Manufacturers' catalogs should be consulted for the complete range of equipment features available and application information.

5.16 Network Protectors. The descriptions in the paragraphs below refer to packaged network protectors. These protectors, commonly used by electric utilities, cannot be used alone in a network system and still meet NEC [9] requirements. The most satisfactory method of application is to use standard, fully rated drawout circuit breakers (fused, if required) to accomplish all of the functions listed below. Coordinated forward protection can then be included, which is not a feature of the packaged network protector. The standard network relays or equivalent relays can then be included in the protection package. Maintenance safety considerations will be enhanced by the use of switchgear to serve as network protection. If standard circuit breakers are to be used, consult the manufacturer to make certain that the frequency of operation in the network application will not shorten the maintenance interval.

The network protector is a heavy-duty power air-type circuit breaker with special relaying designed to permit paralleling the outputs of a number of transformers, fed from different primary feeders, to a collector bus. Protectors are used in spot-network substations or secondary networks (see 4.8.4). The network protector serves to prevent backfeeding from the collector bus through the protector

---

[51] In the U.S., Canadian Standards Association (CSA) Standards are available from the Sales Department, American National Standards Institute (ANSI), 11 West 42nd Street, 13th Floor, New York, NY 10036. In Canada, they are available at the Canadian Standards Association (Standards Sales), 178 Rexdale Boulevard, Rexdale, Ontario, Canada M9W 1R3.

and through the transformer into the primary feeder. Such a backfeed could result from a fault in the high-voltage feeder, from another load on the primary line at a time when the line is disconnected from the utility power station, or even from the excitation current of the transformer when the utility feeder circuit breaker opens. When proper voltage is restored to a feeder, the network protector will close, permitting the re-energized feeder to accept its share of the load.

The network protector has no forward overcurrent protection other than fuses that are designed to open slowly under extremely heavy short-circuit currents. Originally, the concept was to permit faults in network cables to burn themselves clear and to allow all overcurrent devices that are downstream plenty of time to operate. The modern approach is to install cable limiters, as described in 5.12.5, at each end of each cable to isolate cable faults. The network protector fuses are intended to operate only to remove a protector and transformer from the secondary bus in the event of a relay or protector trip mechanism malfunction. This is to prevent backfeeding a faulted primary feeder or a network transformer.

The network protector has two plug-in relays, the master and phasing relays, which trip the protector circuit breaker if the power flow is from the collector to a transformer, and reclose the circuit breaker when the transformer secondary voltage is slightly above and leading the collector bus voltage. The settings of these relays involve a 360° vector diagram (power in or out, lead or lag) as well as magnitude settings to allow for differences in collector and transformer voltages. Although it may not always be desirable, these relays can be set to open the circuit breaker on reverse magnetizing current of the transformer. Adjustable desensitizing time relays may be used to avoid nuisance tripping.

The network protector is withdrawable and, in certain ratings, is available in a drawout design that may require special safety precautions (beyond those required for metal-clad switchgear) when withdrawn from an energized circuit. Other ratings require internal disconnection by maintenance personnel to withdraw the circuit breaker element.

An external handle can be used to lock the protector open, which is essential in preventing backfeeds during maintenance of the high-voltage feeders. Protectors are available as wall-, switchboard-, or transformer-mounted units that are bused directly from the transformer. Dustproof, dust-tight, dripproof, and submersible enclosures permit protector location in any available part of the building.

Conventional network systems are optimally fed from "balanced feeders," almost physically identical and dedicated to network service. The use of undedicated feeders requires special design precautions and is beyond the scope of this recommended practice.

Network protectors can be fitted with external control to trip and lockout in response to overcurrent, ground, or heat sensing relaying. Reclosing relaying of the protector should always be under the ultimate control of the protector master relay.

Occasionally, in a commercial building, elevators or other loads capable of regenerating into the system could cause the protectors to open because of reverse power. A relay can be added that desensitizes the system to reverse currents of this type.

Network protectors have essentially no overload capacity, while the transformer associated with the protector has heavy overload capabilities. Therefore, the protector will usually be rated on a current basis higher than the transformer full-load rating. Protectors are rated at 125 V, 240 V, 480 V, or 575 V with a maximum current of 5000 A.

5.17 Lightning and System Transient Protection. The insulation level of overhead lines is necessarily considerably higher than the isulation level of terminal apparatus, such as transformers, switchgear, potheads, etc., which comprise the service entrance to buildings. Such overhead lines are vulnerable to overvoltage, principally from direct or induced lightning voltages and switching surges. These overvoltages can have values varying from several times the impulse and low-frequency withstand strength of the terminal apparatus down to very low values.

It is a fundamental characteristic of traveling voltage waves that they tend to increase in voltage when they arrive at equipment having a surge impedance higher than that of the incoming line. The magnitude of such incoming waves will approximately double at the terminals of a transformer or at any open point in the circuit, such as an open circuit breaker. Because of this characteristic, equipment connected by cable to overhead circuits generally requires arrester protection at each end of the cable to guard against the possibility of transient overvoltages.

Protection against direct strokes is usually provided at outdoor substation installations in the form of grounded masts or overhead ground wires stretched above the installation to intercept lightning strokes that might otherwise terminate on the lines or apparatus.

5.17.1 Surge Arresters and Capacitors. In a modern power system, control of system transients to protect equipment from overvoltages and high rates of change in voltage are important. These can be caused by lightning (a direct stroke is rare, but induced transients are common), heavy current switching, switching highly inductive loads, and system fault related transients. The new higher speed circuit breakers (especially vacuum circuit breakers) and current-limiting fuses may generate high transients because of the high rates of current change that occur on a circuit opening. These transients can damage equipment, and, when a length of feeder line is involved, may, in effect, be amplified in traveling over that line. Most equipment has a BIL rating to indicate the ability of the equipment to withstand these impulses.

Surge arresters and capacitors will limit these overvoltage by attenuating the excess energy during the period that the transient occurs. Surge capacitors act as a short circuit to these transients (since their component frequencies are so high). The surge device should be placed close to the protected device if the effects of traveling wave voltage amplification are to be avoided. The modern protector typically used is of the gapless-metal-oxide type, which has excellent nonlinear characteristics for handling high transient overvoltages. Solid-state electronic equipment, motors, and, to a lesser extent, dry-type transformers are particularly susceptible to these overvoltages and frequently require protection. The technical specifications can require that the arresters be furnished as part of an integral assembly for certain equipment (e.g., dry-type transformers, metal-clad switchgear, interrupter switches, etc.).

Surge arresters are available in three types: station, intermediate, and distribution, in decreasing ability to handle the amounts of surge energy required to be dissipated. In the absence of an engineering study of insulation levels, arrester characteristics, and the extent of exposure to lightning and surges, the manufacturer should be consulted to determine the protection recommended for each piece of equipment for the application. For individual pieces of plug-in utilization equipment, small surge protection units can be used as plug-in devices to interface between the outlet and appliance. (See NEMA LA1-1986, Surge Arresters [33]; IEEE C62.1-1989, IEEE Standard for Gapped Silicon-Carbide Surge Arresters for AC Power Circuits (ANSI) [26]; and IEEE C62.2-1987, IEEE Guide for Application for Gapped Silicon-Carbide Surge Arresters for AC Systems (ANSI) [27].)

5.17.2 Apparatus and Electromagnetic Interference (EMI). Electronic systems, including computer and communication systems, are inherently sensitive to EMI from the power supply, from interconnecting control and data cables, and from radiated electric and magnetic fields from external equipment. This section is concerned with the installation of apparatus that can reduce these problems. The specification of equipment with controlled emission levels and the methods of interwiring and installation are as important as the correction of defects in the quality of power to sensitive equipment.

The desirability of separating the feeders, panelboard, and even unit substations from which power is taken for sensitive equipment from electrical noise, surge, or other power source problem areas, has already been discussed. Appliances may have specifications for noise emission radiation. The specifications of equipment with low levels of noise generation is an effective approach. Where electronic power control, such as chopping of phase angle control, is to be purchased, it is often possible to obtain the equipment with filters or noise suppressors built in. The use of surge arresters or suppressors is valuable in protecting sensitive equipment from transient damage but of limited value in reducing the effects of poor quality power.

When very sensitive equipment is involved, the specification of optoelectronic isolation (interface) in low-level input circuits, and of fiber optics for extended data and communication circuits is an extremely effective and practical approach. Less effective techniques in control and communication wiring include: metallic conduit (very effective, steel conduit is an excellent magnetic shield), shielding (and double shielding), twisted pairs (including twisting of nearby power cables), and, most importantly, effective grounding.

The following are items of equipment, when required, that can materially reduce the effect of poor quality power, which includes excessive transients, harmonics, dips, and poor voltage regulation:

(1) Isolation transformers can reduce common-mode interference. If the transformer is equipped with a grounded isolation shield, the effective coupling capacitance between the input and output circuits is substantially reduced.

(2) Motor-generator sets almost completely isolate the input and output circuits, can provide controlled voltage regulation, and are protective against momentary supply voltage dips from system switching or faults.

(3) Power conditioners are combinations of noise and harmonic filters, voltage regulators, surge suppressors, capacitors, and other protective devices that can assure the supply of high-quality power to appliances.

(4) The uninterruptible power supply (UPS), while designed for preventing disturbances during loss of utility power by use of rectifiers, storage batteries, and inverters, often employs many of the devices of the power conditioner to improve the quality of the output. If the input uses a quality isolation transformer, then the system output will be relatively resistant to incoming line disturbances. (See IEEE Std 518-1982, IEEE Guide for the Installation of Electrical Equipment to Minimize Electrical Noise Inputs to Controllers from External Sources (ANSI) [29] and References [41] and [42]).

5.18 Load Transfer Devices. This section is confined to commercial building circuitry of low and medium voltages. These devices are used for the transfer of critical loads from normal (such as purchased power) to emergency (such as another incoming feeder or feeders, or standby mechanically driven alternator or alternators) in the event of the failure of the normal source. To ensure continuity of power at the points of utilization, one or more such devices should be considered according to the various reliability classifications of utilization equipment and their logical economical circuit groupings.

The loads to be transferred should be selected according to the importance of the reliability classification of the utilization equipment or groups of equipment involved, considering the duration and frequency of interruptions allowable. Certain configurations for various levels are shown in Fig 50. The degree of reliability requirements may be classified as follows:

(1) Level 1 — Emergency critical loads involving safety to life and property and where emergency power is legally required.

(2) Level 2 — Loads of less critical nature than emergency but where standby power is legally required.

(3) Level 3 — Optional standby loads where power outage may cause discomfort or damage to a product, process, or building facility.

Examples of the above level classifications are as follows:

(1) Health care facilities, egress lighting, fire detection, fire pumps, ventilation, elevators, military systems, aeronautical safety, and certain communication systems

(2) Some transportation systems, critical controls for heating, ventilating, and cooling dump water, and refrigeration and sewage disposal

(3) Computer data processing systems and manufacturing processes

The selection of the load transfer devices should be based upon the required reliability classification of the utilization equipment. It should also be based upon an intimate knowledge of the various characteristics and limitations of the devices for the particular application, considering the characteristics of each utilization component to be served. The electrical engineer should give particular consideration to motor loads to be sure that both the transfer device and the standby power source have enough capacity for the large low-power factor currents that the

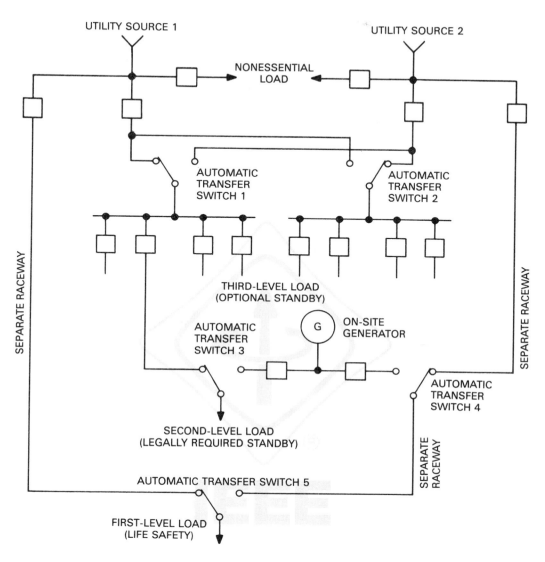

**Fig 50
Diagram Illustrating Multiple Automatic Double-Throw Transfer
Switches Providing Varying Degrees of Emergency and Standby Power**

motor loads may impose following the transfer. Critical loads should utilize automatic load transfer devices with adequate monitoring and control relays.

The typical characteristics of load transfer devices should include the capability to successfully and repeatedly make and break the load current at their various make and break power factors, to carry rated current continuously when closed (with due regard to possible deterioration of contacts under arcing conditions), to close on faults (if required by applicable standards) while remaining operable with the ability to subsequently carry and break load current and to withstand through-fault currents successfully while other circuit protective devices are clearing the faults. Also, emphasis should be placed on accessibility and ease of thorough inspection of contact elements after repeated subjections to such operations and load and fault currents. The ability of the device to withstand repeated operations that are required for the application is a basic requirement.

Load transfer devices are available in the following forms:

(1) Automatic or manual transfer switches available in ratings to 4000 A in low-voltage class, and to 1200 A in medium-voltage class. These switches may be fusible or nonfusible.

(2) Automatic power circuit breakers consisting of two or more power circuit breakers, which are mechanically or electrically interlocked, or both, rated 600–3000 A, in both low- and medium-voltage classes.

(3) Manually or electrically operated bolted pressure switches (600 V), which are fusible or nonfusible and are available from 800–6000 A.

(4) Static transfer switches, which effectively have no time loss between transfers, have been designed for special applications such as UPS systems. These switches transfer fast enough to avoid possible loss of synchronism, to prevent loss of illumination from discharge-type lighting, and to avoid loss of power.

5.18.1 Automatic Transfer Switches. Automatic transfer switches are primarily used for emergency and standby power generation systems. These transfer switches may incorporate overcurrent protection and are designed and applied in accordance with the NEC, Articles 230, 517, 700, 701, 702 and 710 [9], and other applicable standards. To comply with codes and standard requirements for reliability, automatic transfer switches are mechanically held and are electrically operated from the power source to which the load is to be transferred.

An automatic transfer switch is usually located at the main or secondary distribution bus, which feeds the branch circuits. Because of its location in the system, the capabilities that should be designed into the transfer switch are varied. For example, special consideration should be given to the following characteristics of an automatic transfer device:

(1) Its ability to close against high inrush currents

(2) Its ability to carry full rated current continuously from the normal and emergency sources

(3) Its ability to withstand fault currents

(4) Its ability to interrupt full load currents the appropriate number of times as specified in the applicable standard

(5) Additional electrical spacing and insulation, as needed, for two unsynchronized power sources

The arrangements shown in Fig 49 use low-voltage switches of double-throw construction that provide protection against the loss of one of the utility sources. In addition to loss of power from the utility sources, continuity of power to critical loads can also be disrupted by

(1) An open circuit within the building area on the load side of the incoming service
(2) Overload or fault conditions
(3) Electrical or mechanical failure of the electric power distribution system within the building

Therefore, the location of transfer switches and of the overcurrent protective devices should be given careful consideration. Many engineers advocate the use of multiple-transfer switches of lower current rating located near the load as well as one large transfer switch at the point of incoming service. A typical transfer scheme using multiple-transfer switches is shown in Fig 49.

5.18.2 Automatic Transfer Circuit Breakers. Circuit breakers may or may not require the energy from an electric storage battery for operation. While batteries make the transfer equipment independent of ac power, they require periodic maintenance. Extreme care should be exercised to assure that transfer control and operation do not in any way detract from either overcurrent protection or readily accessible disconnect means.

Magnetically operated transfer switches operate very rapidly because of their double-throw feature. Solenoid operated circuit breakers also operate very quickly. Motor operated circuit breakers are slower.

If power circuit breakers or molded-case circuit breakers serve both functions of load transfer and service entrance devices, they should include overcurrent protection and a readily accessible disconnect means. Frequently, remote manual and automatic trip means are justified for fire or other dangerous situations.

5.18.3 Automatic Load Transfer Devices. Automatic load transfer devices operate rapidly with a total operating time of usually less than 0.5 second, depending upon the rating of the transfer switch and the operating mechanisms. Therefore, transferring motor loads may require special consideration in that the residual voltage of the motor may be out of phase with that of the power source to which the motor is being transferred. Upon transfer, this phase differential may cause serious damage to the motor, and excessive current drawn by the motor may trip the overcurrent protective device. Motor loads above 50 hp with relatively low load inertia in relation to torque requirements, such as pumps and compressors, may require special controls. Automatic transfer devices can be provided with accessory controls that disconnect motors prior to transfer and reconnect them after transfer when the residual voltage has been substantially reduced. Problems may arise if the power source is applied too quickly while a significant voltage is still being self-generated by the motor. The rule of thumb is to delay reclosure time until motor residual voltage decays to 25% of rated voltage. The open circuit time constant of the motor and driven equipment should be obtained. Automatic transfer devices can also be provided with in-phase monitors that prevent retransfer to the normal source until both sources are synchronized. Another approach is to use a three-position transfer switch with accessory controls that allow the switch to pause in the neutral position, while the residual voltage decays substantially, before completing the transfer. Closed transition transfer occurs without any power interruption when both sources are present. Such arrangements are being used more frequently to reduce stress on the electrical equipment that is located on the load side of the transfer device. (See Reference [38].)

Other accessories include time delays of 0.5-6 seconds to ignore harmless momentary power dips and adjustable time delays of 2-30 minutes on retransfer to allow the normal source voltage to stabilize before assuming the load. Furthermore, consideration should be given to the minimum voltage at which the load will operate satisfactorily to determine if the automatic transfer device should be provided with close differential voltage protection. Additional accessories may include time delay on transfer to emergency, test switches, auxiliary contacts, remote annunciators, lockout relays, and switching neutral contacts as may be needed for ground-fault sensing.

Bypass isolation switches are also available to bypass the transfer device by connecting the load directly to the power source. This permits the isolation of the transfer switch for maintenance. (See IEEE Std 446-1987, IEEE Recommended Practice for Emergency and Standby Power Systems for Industrial and Commercial Applications (ANSI) [28] for more information.)

Automatic transfer of power sources may be bidirectional or unidirectional with manual reset from emergency to normal positions. The fully automatic bidirectional system should include a long time delay for the transfer back from emergency to normal sources to ensure recovery of the normal source to a stable situation.

A transfer device may include override time delay, or other means, to avoid transfer during a short-circuit condition in the transferable branch. In many systems, a short circuit on the utilization circuit to be transferred can be made to automatically lock out the transfer from normal to emergency source and from emergency to normal source until reset by hand.

Loss of potential from the normal power supply should start the transfer timer or the emergency alternator unit or units. A loss-of-potential alarm may be provided on an emergency utility supply to initiate action in order to restore the emergency supply. An alarm should be provided to indicate a transfer operation so that action may be initiated to restore the normal supply.

A transfer operation causes a momentary outage on the equipment transferred; consequently, the circuitry should be reviewed with regard to permitting automatic restart. In some instances, manual restart may be preferred. If the normal and emergency sources can be paralleled momentarily, the transfer switch may be equipped with a closed transition mode so it can be operated without such an outage.

In order to properly ground the neutral of the service source and alternate sources, it may be necessary to switch the neutral along with the phase conductors. Switching of the neutral conductor may also simplify ground-fault sensing.

5.19 Interlock Systems. Mechanical interlock systems prevent the closing (or opening) of switches, circuit breakers, contactors, or access features, such as doors, panels, or screens, unless certain actions are taken beforehand or at the same time. For example, having different switches on a common shaft or a bar preventing the closing of two adjacent breakers are positive mechanical interlocks. The controlled use of padlocks for equipment or access locks is also a form of limiting operation.

Key interlocks are mechanical interlocks in which the use of keys enforces the staged operation of equipment or access control. Examples of this are the use of keys to prevent closing of two feeds together in a manually operated double-ended substation; to prevent the opening of a screen or door in front of a medium-voltage fuse unless the associated interrupter switch is open; and to stop the closing of a switch served from a feeder unless the feeder service breaker has been opened and the ground-and-test devices inserted. Key interlock assemblies are available, which require the insertion and removal of several keys into a multiple-lock assembly block before any action can be taken. In any key interlock system, provisions for the control of keys is essential; duplicate keys in the hands of operators is a safety hazard.

5.20 Remote Control Contactors

5.20.1 Remote Control Lighting Contactors.
Remote control lighting contactors are used for controlling preselected branch circuits or complete lighting panelboards. They are generally used in sizes from 20–225 A and are mounted in panelboards or separate enclosures. The 20 A size is typically prescribed for branch circuits. Multiple-pole contactors can switch up to 12 circuits each. The most common control voltages are 24 V_{dc}, 24 V_{ac}, 120 V_{ac}, and 277 V_{ac}. The 24 V control voltage is commonly used in computerized energy management systems. Remote control makes it possible to turn blocks of light fixtures on or off from various local locations or from one central location. Figure 51 shows a typical control circuit for an electrically operated, mechanically held lighting contactor with multiple control stations. Figure 52 illustrates how the lighting contactors might be controlled by a central (remote) control in addition to local control. In addition to convenience of control, installation savings can be realized by reducing the length of power cable runs.

Fig 51
Mechanically Held, Electrically Operated Lighting Contactor
Controlled by Multiple Momentary Toggle-Type Control Stations

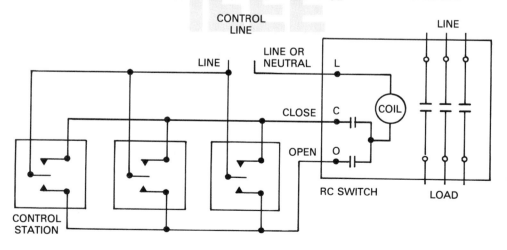

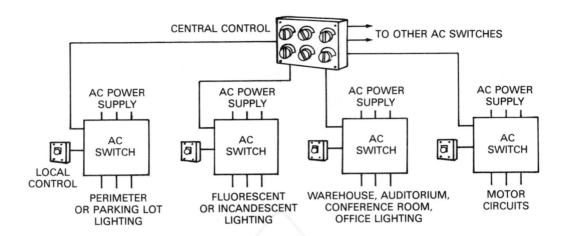

Fig 52
Remote Control Lighting Contactors
Controlled from Local and Remote Locations

Lighting contacts are actuated electromagnetically and are either magnetically or mechanically held. Magnetically held lighting contactors are usually controlled by an on-off single-pole, single-throw toggle switch and will drop open upon loss of control voltage.

Mechanically held lighting contactors will not change contact position upon drop or loss of control voltage. The operating coil is energized only during the opening or closing operation, thereby eliminating coil hum and power drain. In addition to toggle-type and rotary switches, a mechanically held lighting contactor can be controlled from computerized energy management systems, occupancy sensors, photoelectric cells and time clocks, as shown in Fig 53.

Auxiliary relays and optional interface control options may be used with lighting contactors to accommodate long runs between the lighting contactor and the control switch for two-wire control, low-voltage control, and for control by pilot contact devices. Interface control options can permit integration into larger control networks with relative ease and a high degree of flexibility. Reference [39] provides a more in-depth discussion of control options.

5.20.2 Remote Control Switches for Power Loads. Remote control switches provide for the convenient and accessible control of power circuits from any number of control stations. They are mechanically held and, therefore, will not change contact position upon loss of control voltage. Remote control switches for power loads are available in sizes from 30–4000 A, suitable for 600 V_{ac} service and are designed primarily for inductive loads. They may be used for lighting or non-inductive loads that exceed the capacities of smaller mechanically held lighting contactors. Standard control voltages are 120 V_{ac}, 240 V_{ac}, 277 V_{ac}, and 480 V_{ac}.

The simplicity and reliability of these switches are mainly due to the unique operating mechanism. Without the use of hooks, latches, or semipermanent magnets,

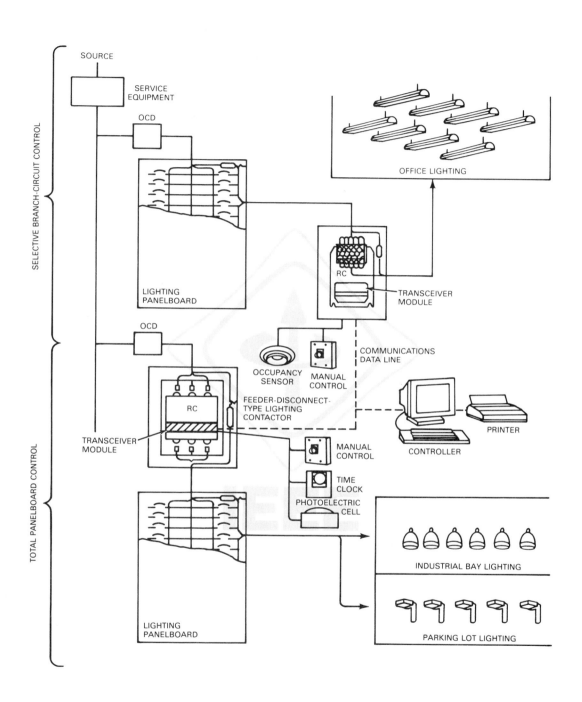

**Fig 53
Various Control Means for Remote Control Lighting Contactors**

the contacts are positively locked in position. The solenoid coil in the operating mechanism is energized only during the instant of operation. Auxiliary contacts in the switch automatically disconnect the coil when the switch has operated, thus eliminating continuous energization of the operating coil. The same operating power is used to open or close the switches, and controlling stations do not break any load current.

Remote control switches that are suitable for all classes of load are capable of carrying rated current continuously without contact deterioration or overheating. They are capable of closing against high inrush currents without contact welding or excessive contact erosion. They can interrupt locked-rotor motor currents or 600% overload at 0.40-0.50 power factor.

Remote control switches are often installed in panelboards that are required to withstand fault current in excess of 10 000 A. As a component of the panelboard, the remote control switch should be capable of withstanding the magnetic stress and thermal effects of the maximum available fault current.

Remote control switches are used when disconnection of circuits is a matter of safety to life or property. Wherever electrical power is being distributed over a wide area, remote control switches also provide economy and convenience. With their use, the electrical layout can be designed without regard to the accessibility of the disconnect switches, thus simplifying the distribution system and making it more flexible for future expansion.

Distribution panels can be located to provide direct feeders and short branch circuits resulting in minimum line voltage drops. Small conductors can be used for the control stations, and an unlimited number of stations can be used for each remote control switch, providing additional convenience and economy. In addition to pushbutton control stations, remote control switches can be operated by time switches, photoelectric cells, central control stations, break-glass stations, energy management systems, and auxiliary relays.

5.21 Equipment Ratings. All power equipment is almost always assigned nominal ratings for voltage, current, phases, and frequency. "Nominal" means the value at which the equipment is designed to be applied, and it is in reality a band or range of application. Unless otherwise indicated, these voltage and current ratings are based on rms values. It can be a mistake to use the basic nominal values without referring to the appropriate engineering criteria.

The conditions of application, which include ambient temperature and altitude, affect equipment ratings. The appropriate standards and manufacturers' data specify the limits of application and de-rating (or, in some cases, uprating), which should be applied. The NEC [9] establishes the de-rating values of certain types of equipment (particularly cables, circuit breakers, and switches) when operated under other than normal conditions.

Some standards, such as those for medium-voltage circuit breakers, include detailed factors affecting the application, such as system X/R ratios, which can only be determined as part of a system study. Some switchgear have a nominal MVA rating, which may differ markedly from the actual interrupting value as determined by an engineering study.

There are ratings for circuit breakers and switches, either implied (as in the case of many items of low-voltage equipment) or listed in detail, such as fault-make rating (ability to safely close in on a fault), short-time rating (overcurrent withstand), and interrupting rating (ability to clear current). The ratings of a piece of equipment, such as low-voltage switches and starters, may differ depending on whether the equipment is mounted in enclosures or in the "open" (unfortunately, the open rating is too often given when the equipment is typically enclosed). The NEC [9] often requires that equipment be de-rated (usually to 80%) unless it is specifically approved for 100% of rating. Low-voltage equipment usually has power terminals rated 60 °C (140 °F) or 75 °C (167 °F). Cable temperatures rated at 90 °C (194 °F) cannot be fully loaded when directly connected to these terminals without the danger of exceeding the temperature limitation of the terminals.

Typically, power, distribution, and general-purpose transformers can withstand overloads that are twice the normal rating for a very short time. On the other hand, circuit breakers and fuses are limited to overloads of perhaps 10%. Chapter 3 contains information on the performance of equipment under various voltage conditions including phase voltage unbalance. Insulation coordination at medium voltage involves a study of BIL ratings to assure system protection under surge conditions and the application of proper surge protection, if necessary, as described above.

5.22 References. The following references shall be used in conjunction with this chapter:

[1] ANSI C2-1990, National Electrical Safety Code.

[2] ANSI C37.06-1987, Preferred Ratings and Related Required Capabilities for AC High-Voltage Circuit Breakers Rated on a Symmetrical Current Basis.

[3] ANSI C37.12-1981, Guide Specifications for AC High-Voltage Circuit Breakers Rated on a Symmetrical Current Basis and a Total Current Basis.

[4] ANSI C37.46-1981 (Reaff. 1988), Specifications for Power Fuses and Fuse Disconnecting Switches.

[5] ANSI C57.12.22-1989, Requirements for Pad-Mounted Compartmental-Type, Self-Cooled, Three-Phase Distribution Transformers with High-Voltage Bushings, 2500 kVA and Smaller: High Voltage, 34 500 GrdY/19 920 V and Below; Low Voltage, 480 V and Below.

[6] ANSI C57.12.40-1990, Requirements for Secondary Network Transformers, Subway and Vault Types (Liquid Immersed).

[7] ANSI/NEMA FU1-1986, Low-Voltage Cartridge Fuses.

[8] ANSI/NEMA ICS6-1988, Enclosures for Industrial Control and Systems.

[9] ANSI/NFPA 70-1990, National Electrical Code.

[10] ANSI/UL 67-1988, Panelboards.

[11] ANSI/UL 98-1986, Enclosed and Dead-Front Switches.

[12] ANSI/UL 489-1985, Molded-Case Circuit Breakers and Circuit Breaker Enclosures.

[13] ANSI/UL 977-1984, Fused Power-Circuit Devices.

[14] ANSI/UL 1053-1982 (Reaff. 1988), Ground-Fault Sensing and Relaying Equipment.

[15] CSA Std C22.2-1990, Canadian Electrical Code, Part 2: Safety Standards for Electrical Equipment, Electric Signs.

[16] IEEE C37.04-1979 (Reaff. 1988), IEEE Standard Rating Structure for AC High-Voltage Circuit Breakers Rated on a Symmetrical Current Basis (ANSI).

[17] IEEE C37.13-1981, IEEE Standard for Low-Voltage AC Power Circuit Breakers Used in Enclosures (ANSI).

[18] IEEE C37.20.1-1987, IEEE Standard for Metal-Enclosed Low-Voltage Power Circuit Breaker Switchgear (ANSI).

[19] IEEE C37.20.2-1987, IEEE Standard for Metal-Clad and Station-Type Cubicle Switchgear (ANSI).

[20] IEEE C37.20.3-1987, IEEE Standard for Metal-Enclosed Interrupter Switchgear (ANSI).

[21] IEEE C37.19-1981 (Reaff. 1985), IEEE Standard for Low-Voltage AC Power Circuit Protectors Used in Enclosures (ANSI).

[22] IEEE C37.100-1981 (Reaff. 1989), IEEE Standard Definitions for Power Switchgear (ANSI).

[23] IEEE C57.12.00-1987, IEEE Standard General Requirements for Liquid-Immersed Distribution, Power, and Regulating Transformers (ANSI).

[24] IEEE C57.12.01-1989, IEEE Standard General Requirements for Dry-Type Distribution and Power Transformers Including Those with Solid-Cast and/or Resin-Encapsulated Windings (ANSI).

[25] IEEE PC57.12.58, Guide for Conducting a Transient Voltage Analysis of a Dry-Type Transformer Coil.

NOTE: When IEEE PC57.12.58 is completed and published by the IEEE, it will become IEEE Standards Board approved IEEE C57.12.58-199x.

[26] IEEE C62.1-1989, IEEE Standard for Gapped Silicon-Carbide Surge Arresters for AC Power Circuits (ANSI).

[27] IEEE C62.2-1987, IEEE Guide for Application of Gapped Silicon-Carbide Surge Arresters for AC Systems (ANSI).

[28] IEEE Std 446-1987, IEEE Recommended Practice for Emergency and Standby Power Systems for Industrial and Commercial Applications (ANSI).

[29] IEEE Std 518-1982, IEEE Guide for the Installation of Electrical Equipment to Minimize Electrical Noise Inputs to Controllers from External Sources (ANSI).

[30] IEEE Std 602-1986, IEEE Recommended Practice for Electric Systems in Health Care Facilities (ANSI).

[31] NEMA AB1-1986, Molded-Case Circuit Breakers.

[32] NEMA KS1-1990, Enclosed Switches.

[33] NEMA LA1-1986, Surge Arresters.

[34] NEMA PB1-1990, Panelboards.

[35] NEMA PB2-1989, Deadfront Distribution Switchboards.

[36] NEMA SG2-1986, High-Voltage Fuses.

[37] NEMA SG6-1990, Power Switching Equipment.

[38] UL 1562-1990, Transformers, Distribution, Dry-Type — Over 600 V.

[39] Castenschiold, R. "Closed Transition Switching of Essential Loads," *IEEE Transactions on Industry Applications*, vol. 25, no. 3, May/Jun. 1989.

[40] Chen, K. and Castenschiold, R. "Selecting Lighting Controls for Optimum Energy Savings," *1985 Conference Record*, 1985 Industry Applications Society Annual Meeting.

[41] Key, T. S. "Diagnosing Power Quality Related Computer Problems," *IEEE Transactions on Industry Applications*, Jul./Aug. 1979, p. 381.

[42] White, Atkinson, and Osborn. "Taming EMI in Microprocessor Systems," *IEEE Spectrum*, vol. 22, no. 12, Dec. 1985, p. 30.

5.23 Bibliography. The references in this bibliography are listed for informational purposes only.

[B1] ANSI C37.11-1979, Standard Requirements for Electrical Control for AC High-Voltage Circuit Breakers Rated on a Symmetrical Current Basis or a Total Current Basis.

[B2] ANSI C37.42-1981, Specifications for Distributions Cutouts and Fuse Links.

[B3] ANSI/NEMA ST20-1988, Dry-Type Transformers for General Applications.

[B4] IEEE C37.09-1979, IEEE Test Procedure for AC High-Voltage Circuit Breakers Rated on a Symmetrical Current Basis (ANSI).

[B5] IEEE C37.010-1979, IEEE Application Guide for AC High-Voltage Circuit Breakers Rated on a Symmetrical Current Basis (ANSI) (Includes Supplement IEEE C37.010d-1984 [ANSI]).

[B6] IEEE C37.011-1979, IEEE Application Guide for Transient Recovery Voltage for AC High-Voltage Circuit Breakers Rated on a Symmetrical Current Basis (ANSI).

[B7] IEEE C37.012-1979, IEEE Application Guide for Capacitance Current Switching of AC High-Voltage Circuit Breakers Rated on a Symmetrical Current Basis (ANSI).

[B8] IEEE C37.1-1987, IEEE Standard Definition, Specification, and Analysis of Systems Used for Supervisory Control, Data Acquisition, and Automatic Control (ANSI).

[B9] IEEE C37.2-1979, IEEE Standard Electrical Power System Device Function Numbers (ANSI).

[B10] IEEE C37.30-1971, IEEE Definitions and Requirements for High-Voltage Air Switches, Insulators, and Bus Supports (ANSI) (Includes Supplement IEEE C37.30a-1975 [ANSI]).

[B11] IEEE Std 80-1986, IEEE Guide for Safety in AC Substation Grounding (ANSI).

[B12] IEEE Std 142-1982, IEEE Recommended Practice for Grounding of Industrial and Commercial Power Systems (ANSI).

Chapter 6
Controllers

6.1 General Discussion. Controls play an important and growing role in commercial buildings. They are used in heating, lighting, ventilation, air conditioning, elevators, etc. Most commercial buildings require some form of automatic or programmed control. Controls cover so many fields that it is nearly impossible to separate them from the discussions of the systems that they control. However, this chapter covers those controls primarily associated with motors.

Heating, ventilating, air conditioning, refrigeration, pumping, elevators, and conveyors require the use of motors. They can be operated manually or automatically to respond and then perform the function for which they are intended. Furthermore, protection should be afforded the motor and the electric supply system. A motor controller causes the motor to respond to a signal from a pilot device and provides the required protection.

Most integral horsepower motors used in commercial buildings are of squirrel-cage design and are powered from three-phase, ac, low-voltage distribution systems. Controllers to be applied on distribution systems up to 600 V are generally given horsepower and current ratings by the manufacturer. The National Electrical Manufacturers Association (NEMA) publishes standardized ratings for such devices. These ratings range from 2 hp for size 00 to 1600 hp for size 9, based upon use in a 480 V system. (See ANSI/NEMA ICS2-1988, Industrial Control Devices, Controllers, and Assemblies [2].[52]) Unless special provisions are made to interrupt higher current, standard controllers are tested for interruption of current equal to 10 times the full-load current of their maximum horsepower rating.

Medium-voltage starters are standardized from 2500–7200 V. The interrupting rating is standardized for unfused Class E1 controllers from 25–75 MVA and for fused Class E2 controllers for 160–570 MVA. For special applications, voltages between 600 and 2500 V are utilized; for example, 830 V for pump panels, 1050 V for mining equipment, and 1500 V for special pumps. Standards for controllers between 600 and 1000 V are in preparation.

---

[52]The numbers in brackets correspond to those in the references at the end of this chapter. ANSI publications are available from the Sales Department of the American National Standards Institute, 11 West 42nd Street, 13th Floor, New York, NY 10036. NEMA publications are available from the National Electrical Manufacturers Association, 2101 L Street, N.W., Washington, DC 20037.

Controllers exist for special purposes, especially in the air-conditioning and heating industry. Other special controllers include lighting contactors, transfer switches, etc.

During the past decade, IEC-type controllers have gained increased use in North America. These controllers differ significantly from traditional NEMA-type controllers. Each type has relative advantages and disadvantages.

6.2 Starting. The primary function of a motor controller is starting, stopping, and protecting the motor to which it is connected.

A magnetically operated contactor connects the motor to the power source. This contactor is designed for a large number of repetitive operations in contrast with the typical circuit breaker application. Energizing its operating coil with a small amount of control power causes it to close its contacts, connecting each line of the motor to the power supply. If the controller is to be the reversing type, two contactors are used to connect the motor with the necessary phase relation for the desired shaft rotation.

Full voltage starting of the motor requires only that the contactor connect the motor terminals directly to the distribution system. Starting a squirrel-cage motor from standstill by connecting it directly across the line may allow inrush currents of approximately 500%–600% of rated current at a lagging power factor of 35%–50%. The inrush current of motors rated 5 hp and below usually exceeds 600% of the rated current. Small motors, for example, 0.5 hp, may have inrush currents of 10 times full-load motor current. Energy-efficient motors may even draw higher currents. For applications, such as ventilating fans or small pumps, this type of starting is not objectionable. As a result, most of these controllers are full voltage types. However, some applications, such as large compressors for air-conditioning and pumping installations, may require motors as large as several thousand horsepower. For many of the larger motors, the starting inrush current may be great enough to cause voltage dips, which may adversely affect the building's lighting system.

Electric utilities also have restrictions on starting currents, so that voltage fluctuations can be held to prescribed limits. Before applying large motors, starting limitations should be checked with the utility. Some type of starting that limits the current may be necessary. Some couplings or driven equipment have limitations on torque that may be safely applied. Such maximum torque limits may require reduced voltage starting.

Many kinds of reduced voltage starters are in common use. Figures 54–56 show the principles of the most common reduced voltage starters for squirrel-cage motors. In addition, the contactor sequence and control diagrams show the speed versus torque and voltage versus current characteristics of reduced voltage starters.

6.2.1 Part-Winding Starters. Part-winding starting of motors reduces inrush current drawn from the line to about 65% of locked-rotor current and reduces torque to about 42% of full voltage starting torque. This type of starting requires connecting part of the winding to the supply lines for the first step and connecting the balance in an additional step to complete the acceleration. Although special motors can be designed with any division of winding that is practicable, the typical motor used with part-winding starting has two equal windings.

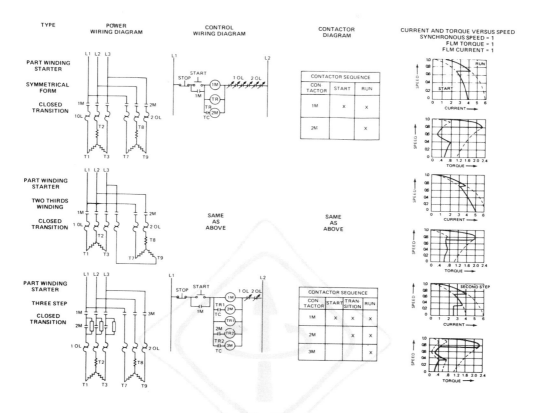

**Fig 54
Principles of the Most Common
Reduced Voltage Starters for Squirrel-Cage Motors, Part 1**
(The contactor sequence and control diagram show the
speed versus torque and voltage versus current characteristics.)

The total starting time should be set for about 2–4 seconds. Due to severe torque dip during the transfer, the transition time should be short and at approximately half-speed. The branch-circuit protection is usually set at 200% of each winding current. Part-winding starters are comparatively low cost but are only used for light starting loads, such as high-speed fans or compressors with relief or unloading valves.

6.2.2 Resistor or Reactor Starters. The simplest reduced voltage starting is obtained through a primary reactor or resistor. The voltage impressed across the motor terminals is reduced by the voltage drop across the reactor or resistor, and the inrush current is reduced proportionately. When the motor has accelerated for a predetermined interval, a timer initiates the closing of a second contactor to short the primary resistor, or reactor, and connect the motor to the full line voltage. The transition from starting to running is smooth since the motor is not disconnected during this transition.

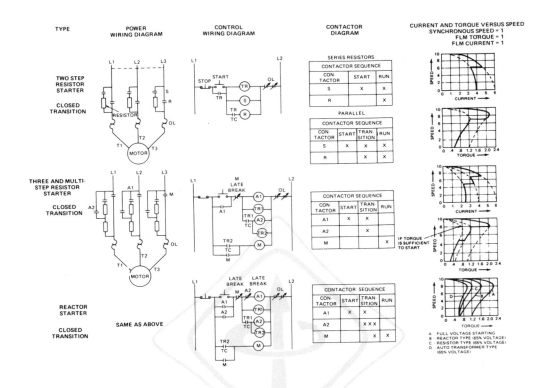

**Fig 55
Principles of the Most Common
Reduced Voltage Starters for Squirrel-Cage Motors, Part 2**
(The contactor sequence and control diagram show the
speed versus torque and voltage versus current characteristics.)

Also, the impressed voltage on the motor is a function of the speed at which it is running. Since the current decreases as the motor accelerates, this decreases the drop across the resistor or reactor. The starting torque of the motor is a function of the square of the applied voltage. Therefore, if the initial voltage is reduced to 50%, the starting torque of the motor will be 25% of its full voltage starting torque. If the drive has high inertia, such as centrifugal air-conditioning compressors, a compromise should be made between the starting torque necessary to start the compressor in a reasonable time and the inrush current that may be drawn from the system.

Resistor- and reactor-type reduced voltage starters provide closed transition and can be used with standard motors. The resistors are usually selected for 5 seconds on and 75 seconds off. Other conditions require specially selected resistors. A three-step resistor-type starter usually does not start rotating the motor until the end of the first step. At the second step, the starting torque is 45%–50% of normal starting torque. The time setting is also usually between 3 and 4 seconds. The

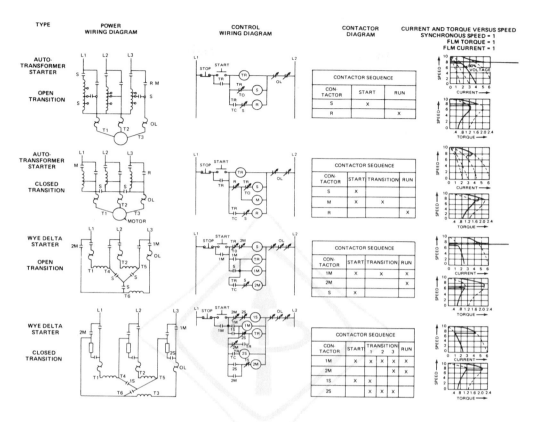

Fig 56
Principles of the Most Common
Reduced Voltage Starters for Squirrel-Cage Motors, Part 3
(The contactor sequence and control diagram show the
speed versus torque and voltage versus current characteristics.)

branch-circuit protection is the same as for full voltage starters. This is also true when reactor-type reduced voltage starters are used. They are difficult to adjust, however, and generally are only used for larger medium-voltage motors.

Reactor-type reduced voltage starters have somewhat better torque speed characteristics than resistor-type starters; but resistor-type starters are less expensive and, therefore, are used more frequently. Resistor-type starters have the disadvantage that the wattage dissipation during start up can be costly for large motors that are started frequently.

6.2.3 Autotransformer Starters. An autotransformer starter has characteristics that are similar, but at the same time more efficient, than the resistor-reactor starter. Since an autotransformer controller reduces the voltage by transformation, the starting torque of the motor will vary directly as the line current, even though the motor current is reduced directly with the voltage impressed on the motor. The formula generally used to calculate the starting current drawn from the line with an autotransformer is: the product of the motor locked-rotor current in A at full voltage times the square of the fraction of the autotransformer tap, plus one-fourth of the full-load current of the motor. The reason for this is that the magnetizing current of the autotransformers usually does not exceed 25% of the full-load motor current. Based on this formula, a motor with 100 A full-load current and 600 A locked-rotor current, when started on the 50% tap of an autotransformer, would only draw 175 A inrush current from the line. This is in contrast to the 300 A drawn from the line in a reactor-type or primary-resistor-type starter. If the voltage is reduced to 25% on starting, the torques will be identical on the reactor, primary resistor, and autotransformer starters. Autotransformer starters usually have taps for 65% and 80% voltage for motor up to 50 hp and taps for 50%, 65%, and 80% voltage for larger motors.

However, on the autotransformer starter, the torque of the motor does not increase with acceleration but remains essentially constant until the transfer is made from starting to running voltage. Also, with an autotransformer-type starter using a five-pole start contactor and a three-pole run contactor, the motor is momentarily disconnected from the line on transfer from the start to the run connection. This open transition may result in some voltage disturbance.

To overcome the objection of the open-circuit transition, a circuit known as the "Korndorfer connection" is in common use. This type of controller requires a two-pole and a three-pole start contactor instead of the five-pole. The two-pole contactor opens first on the transition from start to run, opening the connections to the neutral of the autotransformer. The windings of the transformer are then momentarily used as series reactors during the transfer. This allows a closed-circuit transition without losing the advantages of the autotransformer type of starter. Although it is somewhat more complicated, this type of starter is frequently used on high-inertia centrifugal compressors to obtain the advantages of low-line current surges and closed-circuit transition. Standard motors can be used with autotransformer starters. The time setting should be 3–4 seconds and 4–5 seconds, respectively, for open- and closed-transition autotransformer starters. Most newer autotransformer starters are of the closed transition type.

6.2.4 Wye-Delta (Y–Δ) Starters. Contactors 1M and 2M (as shown in the above circuits) for wye-delta starters carry 58% of the motor load; whereas contactors 1S and 2S carry 33.3% of the motor load. The NEMA rating of a wye-delta starter is higher than that of a full voltage starter that has the same contactor. In closed transition, contactor 2S is usually one size smaller than 1S. An overload relay is included in each phase and set at 58% of the full-load motor current. The time setting should be set somewhat longer than for part-winding starters; that is, 3–4 seconds on open transition and 3–5 seconds on closed transition autotransformer and wye-delta starters.

The branch-circuit protection has to be selected very carefully for open transition starters. The magnetic trip unit should not trip below 15 times full-load motor current or even higher to avoid tripping on the severe current peak at the transition. The current peak is especially high on autotransformer starters; but it could also be 13–14 times full-load motor current on open transition wye-delta starters.

On closed transition wye-delta and autotransformer starters, the standard branch-circuit protective device is selected in the same manner as are full voltage starters. Autotransformer starters are mostly used in the United States for ventilators, conveyors, machine tools, pumps, and compressors without relief valves. Wye-delta starters are extensively utilized in Europe, and in the United States, particularly for large air-conditioning units. Wye-delta starters can only be used if the motor has two terminals for each phase.

6.2.5 Series-Parallel Starters. Series-parallel starters are available, which initially connect the two windings of each phase in series in a conventional wye arrangement. Since this is maximum impedance, the inrush current is about 25% of full-voltage, locked-rotor current, and the torque is 25% of maximum starting torque. The second step removes one winding from each phase and allows the motor to run on the other winding, the same as in the first step of a part-winding controller. The third and final step connects the balance of the winding to the supply lines to affect the parallel connections for normal operation.

6.2.6 Solid-State Starters. Solid-state or electronic reduced voltage starters provide a smooth, stepless method of acceleration for standard squirrel-cage motors. Three methods of acceleration are available:

(1) Constant current acceleration, in which the motor is accelerated to full speed at a field-selectable, preset current level.
(2) Current ramp acceleration, in which the voltage is gradually increased to provide smooth stepless acceleration under varying loads.
(3) Linear timed acceleration, in which the motor is accelerated at a linear rate that is field-adjustable.

A tachometer feedback circuit is required for the latter type of acceleration. A solid-state control circuit provides control for the silicon controlled rectifiers, which are used to provide the variable voltage to the motor. A schematic diagram of the power circuit is shown in Fig 57. Contactors are often used in the power circuit to provide isolation between the motor and the load.

A typical enclosed solid-state, reduced voltage starter with fusible disconnect is shown in Fig 58. Solid-state starters are particularly suitable for applications that require extremely fast or a large number of operations, or both (several million under load). In addition to starting motors, solid-state controllers are also used for speed control of ac and dc motors. Speed control of ac motors is further discussed in 6.14.

6.2.7 Cost Comparison. Table 46 shows a relative cost comparison of some of the more commonly used reduced voltage starters. The cost of solid-state controllers varies considerably, depending on ratings and features. As shown, they are generally more expensive than electromagnetic controllers.

6.3 Protection. Protection of motor branch circuits is divided between protection on running overload and short-circuit protection. Running overloads are overloads up to locked-rotor current, which are usually six times, sometimes eight to ten times, full-load motor current.

Since 1972, NEMA Standards have recommended three overload relays (one per phase) for running overload protection. Most overload relays have thermal ele-

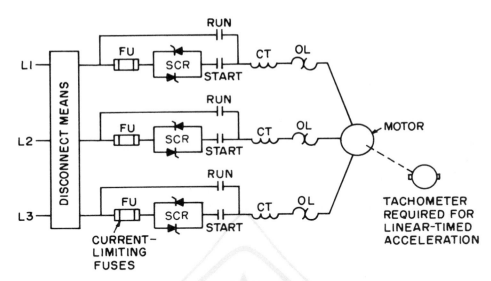

Fig 57
Simplified Wiring Diagram of a Solid-State Controller

ments that are heated by interchangeable heaters in series with the motor or fed through current transformers. The rating of these heaters are determined by the full-load motor current. Most IEC-type relays do not have interchangeable heaters but have bimetals that are heated by conduction from the heater plus current passing through the bimetal itself. The movement of the bimetal can be adjusted to utilize the relay for a certain current range (usually ±20% or 1:1.4).

NEMA Standards divide overload relays into three classes: Class 30, 20, and 10. Each class is defined by the maximum time in seconds in which the relay should function on six times its ultimate trip current (≤1.25 full-load motor current for motors having a service factor of ≥1.15, and ≤1.15 full-load motor current for motors having a service factor of 1.0).

All thermal responsive elements have an inverse time characteristic. This means that, for small overcurrents, considerable time elapses before tripping occurs. However, at high overcurrent (locked-rotor current), tripping occurs in a shorter time period. Class 20 overload relays are used for protection of T-frame motors and Class 30 for the older U-frame motors. No minimum trip time is standard. Overload relays should have sufficient thermal capacity to allow the motor to start.

Ambient compensated and noncompensated overload relays are available. There are basically two types of applications for ambient compensated relays. First, there is the case in which the ambient compensation mechanism is utilized to provide for appropriate trip characteristics over a wide range of ambient temperature. These are intended for general use. The other type of application is where the motor is in a controlled environment, and the starter is in a noncontrolled environment (for example, submersible pumps). In applications where the motor and controller are in the same environment, it is not advisable to use ambient compensated overload relays because the protected device (motor, cable) is not ambient compensated.

**Fig 58
Typical Solid-State Controller
with Fusible Disconnect**

Table 46
Cost Comparison of Reduced Voltage Starters

(All costs as multiples of the lowest item 20 hp
240 V part-winding starter [relative costs = 1].)

hp	240 V							
	(1)	(2)	(3)	(4)	(5)	(6)	(7)	(8)
20	1	1.4	2	2	2.75	2.2	6*	6.35*
50	1.4	2	2.7	4.3	5.5	4.3	6	6.35
100	6.5	7.3	9	7.5	8.8	7.4	9.5	9.85
200	13.5	15.8	20	13.8	—	13.8	12.5	12.85
500	21	21	27.5	30	—	22.6	—	—
	480 V							
20	1	1.3	2	1.6	2.5	1.8	6*	6.35*
50	1.5	2	3	2.2	3.2	2.4	6	6.35
100	3.1	4	5	4.3	5.5	4.5	7	7.35
200	6.3	7.5	9.5	8	9.3	8	9.5	9.85
500	13.5	16	20	22	—	22	20	20.35

NOTES: (1) Part-winding starter. The actual hp ratings are different. For 240 V, they are 20, 50, 150, 300, and 450 hp; for 480 V, they are 40, 75, 150, 350, and 600 hp.
  (2) Wye-delta starters, open transition. The 200 hp, 240 V device can be used up to 250 hp.
  (3) Wye-delta starter, closed transition.
  (4) Resistor-type, reduced voltage starter, two steps.
  (5) Resistor-type, reduced voltage starter, three steps.
  (6) Autotransformer starter, closed transition.
  (7) Solid-state, reduced voltage starter, constant current acceleration.
  (8) Solid-state, reduced voltage starter current ramp and linear-timed acceleration.

*The smallest device is not offered as a standard reduced voltage starter, but may be available as a nonstandardized item at a lower cost.

Most adjustable IEC-type overload relays include single-phase protection. These relays have a differential mechanism that enables this type of overload relay to provide appropriate protection when a three-phase motor runs without one phase.

Short-circuit protection for the total motor branch circuit (cable, starter, motor, disconnect device) is provided by the short-circuit protective device (SCPD). This device can be a fuse or a circuit breaker. In many cases, the crossover point between the SCPD and the overload relay characteristic is beyond its limit of self-protection of the relay, also often beyond the interrupting capacity of the contactor. However, the SCPD is selected to clear the fault while limiting damage to the other components and preventing a safety hazard.

In addition to protection of motors, protection of starters should also be considered as required by ANSI/NFPA 70-1990, National Electrical Code (NEC) [5].[53] With IEC-type starters, there are two levels of starter protection, depending upon

---

[53] ANSI publications are available from the Sales Department of the American National Standards Institute, 11 West 42nd Street, 13th Floor, New York, NY 10036. NFPA publications are available from Publications Sales, National Fire Protection Association, 1 Batterymarch Park, P.O. Box 9101, Quincy, MA 02269-9101.

the degree of damage permitted. Class J and Class RK-1 fuses should be considered when a high degree of protection is required (see Reference [9]).

In many cases, motors have inherent protectors that are placed in the winding of the motor. These protectors are sensitive to the motor winding temperature itself, to the rate-of-rise of the temperature, or to a combination of rate-of-rise and heat. Usually more than one sensor should be placed in the motor. Protectors are especially advantageous if a motor is used for intermittent duty. Since these devices are built into the motor, they should be furnished with the motor. Such protectors do not protect the motor branch circuit in case of a fault on the line side of the motor. Modern inherent motor protection is obtained with thermistors. Small thermistors in the form of disks, rods, or beads change their resistance at the switching point (temperature) from less than $500-1000 \ \Omega$ to $4000-10 \ 000 \ \Omega$. This change, upon amplification, is sensed by a thermistor relay, which then de-energizes the motor contactor.

If severe undervoltage occurs on the distribution system, the motor controllers will normally disconnect the motor from the line. Refer to the latest OSHA rules for the acceptability of automatically restarting motors upon voltage restoration. If the motor is under the control of a single contact pilot device, such as a thermostat or pressure switch, it may be allowed to restart when the normal voltage is restored. This type of control is called "undervoltage release." If there are a large number of motors on the feeder, the simultaneous restarting of all of these motors on return of the voltage could draw an unacceptably large inrush current from the line. Motors under the manual control of an operator are generally not allowed to restart until the operator pushes the button to energize the controller for each individual motor. Control that remains de-energized until actively restarted is referred to as "undervoltage protection." On some occasions, it may be desirable to measure the duration of the voltage dip and, if the undervoltage lasts less than some predetermined time, the motor is not disconnected. This feature is called "time delay undervoltage protection."

When motors are transferred from one source to another, that is, from an emergency generator to the utility, both the motor controller and motor are often momentarily de-energized. If the utility voltage is out of phase with the motor residual voltage, the motor will often draw excessive current and generate high-transient torques that may cause nuisance tripping of breakers and possibly cause damage to the motor or its load. To overcome this problem, automatic transfer switches often include in-phase monitors that prevent transfer until the residual voltage of the motor and the utility voltage are nearly synchronized, or delay transfer until the residual voltage has decayed to a safe level.

Many other special protection means for motors exist. Small voltage unbalances can cause very high currents in the rotor circuit. The condition is even more severe if the motor runs on one phase. After a three-phase motor is stopped, it usually cannot restart if one phase is interrupted. This could be especially dangerous for elevators. Phase failure relays and single-phase protection can be provided. In addition, ground-fault relays are sometimes used for disconnecting large motors in case of a ground fault. For applications with extremely long starting times, the overload relay may be bypassed during the starting period.

In recent years, solid-state overload relays have been developed in which motor damage curves are more closely matched. They lend themselves to special motor branch circuit protection applications. At present, they are mostly used for larger motors; but since their costs are steadily decreasing, it is expected that there will be an increased use of them in smaller motors. Current is sensed, transformed, rectified, and fed to an analog unit or microprocessor. Features vary significantly from one device to another; but devices are available that react to motor overload, phase loss, phase reversal, mechanical jam, ground fault, current unbalance, voltage unbalance, low-voltage conditions, and more. The response time can be changed to fit the application. Solid-state devices can be combined with inherent motor protectors, such as thermistors. Motor characteristics can be approximated, and memory can be provided if motor starts or overloads happen frequently. For very large motors, it is not too costly to use programmable motor protectors, in which data may be entered in a protective module so that the user can read pertinent motor data. Various functions, such as motor current, line voltage, and allowable acceleration, can be displayed. There are now many solid-state protective functions that either could not be obtained previously, or that were too costly and unreliable by electromechanical means.

6.4 Special Features. Many features and additional components can be added to motor controllers to make the motors and drives perform particular functions or to provide additional protection to the motors and systems. Unloading devices on compressors are frequently interlocked with motor control to assure that the motor will not attempt to start if the compressor is loaded, and will unload the compressor if the load exceeds the available motor torque. Vanes and dampers in air systems, and valves in liquid systems, may be similarly interlocked with the motor controller. If the building requires conveying systems to handle materials, elaborate interlocking between motor controllers can assure the proper sequencing of the starting and stopping of the conveyor drives.

Mechanically held contactors may be added to perform additional functions, such a motor feeder disconnect, and controlling other associated loads. The contacts are power driven, for example, by a single solenoid mechanism, into either the open or closed position and positively locked so loss of control voltage cannot cause them to change position.

Combination of switches, relays, and contactors may be used to automatically or manually alternate operation of multiple-pump motors or to operate them simultaneously to equalize the running time on each and still provide sufficient capacity for maximum load conditions.

6.5 Control Systems. There are different arrangements available for control systems in commercial buildings. The physical configuration varies depending upon the complexity of the system as indicated below.

6.5.1 Panelboard-Type Constructions. The fusible disconnect or circuit breaker for each branch circuit is placed in one enclosure. The handle is either attached to the door or to the disconnect device, and the door cannot be opened if the disconnect device is closed. The handle can be padlocked in the off position, and the disconnect device cannot be closed if the door is open. The starter and, if required, the control transformer, pushbuttons, and pilot devices are often placed in other enclosures on the panel. The two enclosures are interlocked so that the starter is only accessible if the disconnect device is in the off position. The disconnect device is usually connected to the bus bar system of a switch-type panelboard, and the two enclosures are connected by conduit or cable. Instead of separate enclosures, a unitized combination starter in one enclosure is also used and connected directly to the bus bar system.

6.5.2 Separate Enclosures. The disconnect devices may be put in separate enclosures and the starters, including pilot devices, etc., in the other enclosure. These devices are connected by conduit or cable. The starter and the specific type of disconnect device should be tested and approved as a combination. Each device should be approved by itself, except if the disconnect device is a fusible switch and the available short-circuit current is 5000 A or less for size 1 and 2 starters or 10 000 A for size 3 or 4 starters.

6.5.3 Combination Starters. The disconnect device, pilot device, starter, and control transformer are all placed in one enclosure and mounted on the wall or machine. Usually, the handle is attached to the disconnect device. The handle interlock is the same as in 6.5.1. Each combination starter should be approved for the available short-circuit current. All protective devices and combination reversing, reduced voltage, and multiple-speed starters are available. Combination starters capable of withstanding up to 100 kA of short-circuit current capacity at 480 V are also available.

6.5.4 Simplified Control Centers. Some manufacturers furnish standard combination starters in a modular construction, thus, width and length are multiples of a basic dimension. These enclosures are placed in front of a steel construction and wired to a bus system that is usually on the top or bottom of the steel structure. The height of the structure including the bus system is 90 inches.

6.5.5 Complex Panels. Complex panels are often placed in special large enclosures. Sometimes, these panels have group protection, which means that one branch-circuit protective device is used for several motors of one machine. Such panels are especially suitable for special equipment, such as programmable controllers, relays, electronic devices, timers, resistors, etc. This construction is frequently used for control of large machines or production lines rather than in commercial buildings.

6.5.6 Motor Control Centers. Motor control centers are preferred for applications involving central control of multiple motors. Applicable standards include ANSI/UL 845-1987, Motor Control Centers (with 9/7/88 Revision) [6][54] and ANSI/ NEMA ICS2-1988 [2]. The motor control center consists of a number of basic vertical structures. Each vertical structure has a vertical bus system connected to a horizontal bus system. The horizontal bus system is either behind, on the top, or on the bottom of the vertical bus system. The total height of the motor control center

---

[54]ANSI publications are available from the Sales Department of the American National Standards Institute, 11 West 42nd Street, 13th Floor, New York, NY 10036. UL publications are available from Underwriters Laboratories, 333 Pfingsten Road, Northbrook, IL 60062.

is 90 inches. Each vertical section has a number of basic units, which consist of combination or unitized combination starter with or without control transformers, combination reversing, or multiple-speed, reduced voltage starters, etc. These units are prefabricated and can be plugged into the vertical bus structure. This structure is braced for withstanding high fault current (42 kA and, in exceptional cases, 100 kA). Combination starters larger than size 4 are often bolted to the vertical bus structure. All compartments are multiples of a basic dimension (usually 3 inches or 6 inches). The total width of each vertical section is not standardized; although most manufacturers use a basic width of 20 inches with wider sections available.

Motor control centers generally consist of a factory-fabricated structural metal frame that houses the buses and the various controllers and their auxiliaries. Larger motor control centers are shipped to the job site in *shipping sections*, thus requiring the structural sections to be field-bolted and the buses to be connected. In order to ensure proper alignment of the structural frame and the buses, it may be necessary to install leveling channels at the front and the rear of the motor control centers. These steel channels are placed so that the webs are essentially flush with the surface of the floor and the legs embedded in the concrete. Thus, the channels serve as a level base for sections of the motor control centers that would not be available from the usual concrete floor.

Incoming and outgoing conduits require special consideration when specifying or selecting motor control centers. For example, consider the following:

(1) When space and aesthetics permit, conduits may be run horizontally on a ceiling (or other elevated structure) above the motor control centers. The conduits are then elbowed or bent downward to their point(s) of entry into the top of the motor control centers. Removable top plates are usually provided for convenient conduit entrances.

(2) In some instances, conduits may be run horizontally and directly into an upper section of the motor control centers. In this case, it may be necessary to include a box-like metal structure above, and attached to, and the standard motor control centers. Here again, plates are required to permit convenient conduit entry.

(3) In some cases, conduits are embedded in floors. Bending the conduit from the embedded points into the motor control centers may be impractical, and a waterproofed trench below the motor control centers may be necessary. Conduits enter the trench, and the conductors are formed (or bent) to proper position.

Metering is often installed in motor control centers. Currently, there are no standards for the number or types of meters. However, consideration should be given to voltmeters and respective voltmeter switches; ammeters for each motor starter or one ammeter, in the main bus, or both; kilowatt meters; power factor meters; and running time meters. These functions are also available on programmable, solid-state, multiple-function metering devices with digital readouts.

In addition to the above, when preparing specifications for motor control centers, consideration should be given to the following:

(1) Provide identification labels for each motor starter indicating the motor or device controlled; device, such as meters, circuit breakers, and fuses; and

each compartment, the enclosing devices, or the operational parts.

(2) Provide additional floor space for anticipated growth or maintenance, or both. Unless projected growth is within a relatively short time span (5 years or less), it is usually not economical to provide space within the motor control center for future equipment because of frequent product redesign, obsoleting the availability of the equipment when it is needed.

(3) Provide spare parts, such as entire starters, circuit breakers, or parts thereof, that are subject to periodic replacement.

(4) Provide spare fuses when fuses are specified as protective devices. At least one complete set of properly identified fuses should be included. Further, in the case of cartridge-type fuses, a suitable fuse puller may be desirable.

(5) Provide lockout tags/padlocks to ensure safe disconnection and service of motors and other equipment controlled by the motor starters. Highly visible tags and, sometimes, padlocks are necessary to prevent closure of the disconnect device while personnel are servicing the motors or equipment.

(6) Enclosure types should be specified depending upon location and service. The National Electrical Manufacturers Association (NEMA) has designated various types of structural enclosures, e.g., NEMA 1 for indoor (essentially dust-free and nonhazardous areas), NEMA 5 for areas where dust collects, etc. (See ANSI/NEMA ICS6-1988, Enclosures for Industrial Control and Systems [3] and NEMA 250-1985, Enclosures for Electrical Equipment (1000 Volts Maximum) [7].[55])

(7) Buses should be specified as either copper or aluminum, depending upon the most advantageous bus for the particular purpose. The proper conductor terminating or connecting devices, or both, should also be specified.

NEMA classifies that motor control centers as having either Class 1 or Class 2 assemblies. With either class, the user may specify the physical arrangement of units within the motor control center, subject to the design parameters of the manufacturer.

(1) Class 1 motor control centers are independent units and consist of mechanical groupings of mechanical motor control units, feeder-tap units, and other electrical devices arranged in convenient assembly. The manufacturers' drawings include overall dimensions of the motor control center, identification of units and their location in the motor control center, locations of incoming line terminals, mounting dimensions, available conduit entrance areas, and the location of the master terminal board, if required (Type C wiring only). The manufacturers' standard diagrams for individual units and master terminal boards identify electrical devices, electrical connections, and terminal numbering designations.

(2) Class 2 motor control centers are interconnected units and are the same as Class 1 motor control centers, except with the addition of manufacturer furnished electrical interlocking and wiring between units as specifically described in overall control system diagrams supplied by the purchaser. In

---

[55]NEMA publications are available from the National Electrical Manufacturers Association, 2101 L Street, N.W., Washington, DC 20037.

addition to the drawings furnished for Class 1 motor control centers, the manufacturer furnishes drawings that indicate factory interconnections within the motor control center.

(3) Class 1-S and 2-S are motor control centers with custom drawing requirements. They are the same as Class 1 and 2 motor control centers except for the fact that custom drawings are provided in lieu of standard drawings as specified by the user. Typical custom drawings include special identifications for electrical devices, special terminal numbering designations, and special sizes of drawings. The drawings supplied by the manufacturer convey the same information as drawings provided with Class 1 and 2 motor control centers, except that they are additionally modified as specified by the user.

To comply with NEMA ICS2-1988 [4], all circuit components within each unit should be factory wired. There are three types of wiring: Type A, Type B, and Type C.

(1) Type A user field wiring is connected directly to device terminals internal to the unit. Such wiring is only provided on Class 1 motor control centers.

(2) Type B user field wiring is for combination motor control units size 3 or smaller and is designated a B-D or B-T. B-D pertains to load wiring connections directly to the device terminals, which are located immediately adjacent, and readily accessible to, the vertical wireway. B-T pertains to load wiring connections directly to a load terminal block in, or adjacent to, the unit. Type B control wiring is to be connected to unit terminal blocks located in, or adjacent to, each combination motor control unit.

(3) Type C user field control wiring is connected directly to master terminal blocks mounted at the top or on the bottom of those vertical sections that contain combination motor control units or control assemblies. Combination motor control units and control assemblies are factory wired to their master terminal blocks. User field load wiring (for combination motor control units size 3 or smaller) is connected directly to master terminal blocks mounted at the top or on the bottom of vertical sections. Motor control unit load wiring is factory wired to the master terminal blocks. User field load wiring (for combination motor control units larger than size 3 and for feeder tap units) is connected directly to unit device terminals.

A control center can be located on each floor of a building to accommodate the motors on that floor, or all motor controllers may be grouped in a central location. The control center is self-supporting and may have units mounted on the front and back. Others may have units mounted only on the front. The motor control center may be mounted on the wall. Control centers that are approximately 20 inches or more in depth are usually more stable if a building is exposed to earthquakes.

The short-circuit capability of a motor control center is determined by both its structure, individual controllers, and other current-carrying components. Current-limiting reactors are available as part of the motor control center lineup to reduce available short-circuit currents so that economical combination starters can be specified.

The lowest component capability is the short-circuit rating of the entire center.

Under a bolted fault condition, the unit enclosures surrounding any SCPD (and any other equipment within this same enclosure) and the equipment so enclosed are allowed by UL and NEMA Standards to have specified damage as long as

(1) The fault current has been interrupted.
(2) A dielectric test on the line side of the unit is passed.
(3) The operating handle can open the unit door.
(4) The line connections are undamaged.
(5) The door is not blown open.

The short-circuit rating should be equal to or greater than the available fault current on the line terminals of the motor control center including the motor contributions. If current-limiting means (e.g., reactors, current-limiting circuit breakers or fuses) are used, only the components ahead of these current-limiting means should be capable of withstanding the available short-circuit current. (For further information, see Chapter 9.)

6.6 Low-Voltage Starters and Controllers. The most common controller is the across-the-line magnetic-type starter. An electromagnet, energized by either the line voltage or a lower voltage from a control transformer, closes the contacts of the contactor. The control voltage is usually 120 V, though other voltages between 24–600 V are used. Low control voltage has the disadvantage of needing larger control wires due to the increase of the control circuit current. Also, at voltages of 24 V and below, continuity may be a problem. Therefore, at low control voltage, it is often advisable to use two parallel contacts at the auxiliary device or sliding-type contacts to secure continuity. On the other hand, high control voltage requires greater insulation integrity and more precautions for the safety of personnel. Therefore, the most common control voltage is 120 V.

An overload relay is placed on the load side of the contactor that has overload protection in each phase. Overload relays may be nonambient or ambient compensated. Some overload relays have separate indicating contacts to which a light or other alarm can be wired to indicate tripping. Overload relays are trip-free; most of them should be reset manually, but some are available for automatic reset.

Automatic reset is not acceptable on machines where automatic restart of the motor could be hazardous. This should always be considered in the application of such devices.

The starters and contactors in Table 47 are standardized by NEMA. In addition, there are three classes of overload relays standardized by NEMA.

Overload relays should be selected to trip at 125% full-load current for 40 °C rise motors and 115% full-load motor current for all other motors; and in 10 (Class 10), 20 (Class 20), or 30 (Class 30) seconds or faster at six times the ultimate trip current with the overload sensing elements starting cold.

NEMA-rated or NEMA-type controllers are designed to provide a high level of performance over a wide set of application conditions. They are the form most commonly found in general use. Another form, IEC-type controllers, are rated based upon their performance in the laboratory. Varying application conditions constitute a greater consideration in their application. Typically, they are considerably smaller and provide a lower level of performance than similarly rated NEMA-

Table 47
NEMA Standardized Starters and Contactors

Size	00	0	1	2	3	4	5	6	7	8	9
Rated Current Closed A	9	18	27	45	90	135	270	540	810	1215	2250
Rated Current Open A	10	20	30	50	100	150	300	600	900	1350	2500
Hp at 480 V		2	5	10	25	50	100	200	400	600	900	1600
Hp at 240 V		1.5	3	7.5	15	30	50	100	200	300	450	800

type devices. Figure 59 illustrates this size difference. Both the NEMA-type and the IEC-type starters shown are rated for use at 10 hp on a 480 V system.

Low horsepower rated, IEC-type controllers appeal to European and some U.S. manufacturers because of marketing considerations (lower costs and smaller size). However, NEMA-type motor controllers are generally favored by U.S. users and specifiers for their longer life, higher short-circuit withstand capability, replaceable contacts, and broad application capability. NEMA-type motor controllers are the logical choice when the motor service factor, duty cycle, contactor life requirement, or short-circuit protective device are not known, or where unfamiliar terminal markings may cause a problem (see Reference [8]).

The definite-purpose contactor is another form of controller that exists for a specific type of application. Originally developed for use with hermetic compressors in air-conditioning equipment, their use has expanded to other application areas where heavy-duty, general-purpose controller characteristics are not required, such as resistance heating, crop drying equipment, commercial deep-fat fryers, etc. They are mostly current, rather than horsepower, rated and should fulfill special test requirements. For example, a contactor for air conditioning should be able to interrupt 6000 times for the following: 600% rated current at 240 V, 500% rated current at 480 V, and 400% rated current at 600 V.

There are also manually operated controllers that are generally limited to use in up to size 1 maximum. They are similar to circuit breakers but usually cannot interrupt the short-circuit current. They should have a much longer life than circuit breakers because they have to switch rated motor currents more frequently.

Solid-state contactors are used in applications that require an extremely high number of operations or where high shock or vibration resistance is required. The market share of solid-state contactors and starters will increase with their decreasing costs. Even today, solid-state devices are competitive costwise for many adjustable frequency drives and certain reduced voltage starting applications.

6.7 Multiple-Speed Controllers. Magnetic motor starters can be connected to multiple-speed motors to obtain different motor velocities.

    (1) The windings of each phase can be connected in two different ways so that the stator winding has half the number of poles (twice the speed) in one position compared to the other. The poles can be further connected in different ways to obtain:

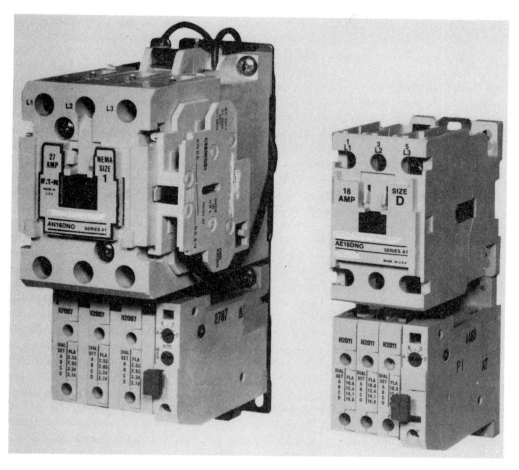

**Fig 59
NEMA versus IEC Size Comparison**

    (a) Equal power at the two speeds (double-parallel star connection at low speed; series delta at high speed)

    (b) Equal torque (delta at low speed; double-parallel star at high speed)

    (c) Variable torque (double torque at double speed; series wye at low speed; double-parallel wye at high speed)

    Attention should be paid so that a high torque and high overcurrent do not occur if the motor winding is switched suddenly from high speed to low speed. Therefore, the high-speed winding should first be disconnected.

 (2) If the required speed ratio is different than 2:1, two-speed motors should have two separate windings.

 (3) By combining items (1) and (2), motors having three and four speeds can be developed. Controllers are available for these motors.

(4) Pole amplitude modification (PAM) motors are available for arbitrarily selected speed ratios (for example, 3.2). Conventional two-speed motor controllers are used with these motors.

6.8 Fire Pump Controllers. Fire pump controllers are frequently installed in commercial buildings to control the operation of fire pumps so as to maintain water pressure in standpipes and sprinkler systems under the heavy water demand of a fire. Such fire pumps are driven either by electric motors or diesel engines. The degree of reliability of electric power and environmental conditions usually determine when a diesel engine driver is used. It is not unusual for a large installation, such as a warehouse or high-rise building, to have multiple fire pumps, either zoned to serve specific areas, or in parallel for standby service.

Fire pump controllers differ from the usual combination motor controller in that there is only limited overcurrent protection built into the controller, generally providing only locked-rotor overcurrent protection for the fire pump motor. Further, short-circuit protection is usually provided with circuit breakers or integrally fused circuit breakers; but, where fuses are acceptable, they should be capable of carrying locked-rotor current continuously. The standard to which fire pump controllers are generally required to conform is ANSI/NFPA 20-1990, Installation for Centrifugal Fire Pumps, Chapter 7 [4].

Although these controllers may provide either manual operation (nonautomatic) or combined manual and automatic operation, the manual-only type is rarely installed. The manual controller starts and stops the fire pump by means of a start-stop station on the controller. Provisions are available for remote start stations; but remote stop stations are not permitted. In addition, an emergency mechanical operator, externally operable, is provided to mechanically close and hold in the contactor. The combined manual and automatic controller, in addition to the means just described, starts and stops the fire pump from a pressure switch within the controller, starting when the water pressure at the fire pump discharge header drops to the predetermined setting and stopping when the pressure rises to the higher predetermined setting, but only after a minimum running time. A variation of the above, requiring manual stopping after an automatic start, is becoming increasingly common. Various starting means, such as across-the-line, reduced voltage, part, winding, and wye-delta, are available.

Power for these controllers is usually obtained from either a tap ahead of the main service disconnect for the facility or as a separate service. Where two power sources are required to be available for fire pump operation, a separate dedicated transfer switch or a fire pump controller/automatic transfer switch assembly is used.

These controllers are generally wired to the power source near the service entrance or, in some cases, are provided with a separate service. ANSI/NFPA 20-1990 [4] mandates that fire pump controllers have a withstand rating at least equal to the maximum short-circuit current that can flow to the controller.

6.9 Medium-Voltage Starters and Controllers. It may be practical to operate larger motors (100 hp and above) in commercial buildings at a medium motor voltage of 2300/2500 V, 4000/4800 V, 6600/7200 V, and, in exceptional cases, up to 13 800 V. Medium-voltage starters are available in current ratings of 400 A and 800 A. The resulting lower currents result in less line disturbance upon starting, and since the motors are not on the same distribution network as the lighting and other low-voltage devices, any reflected voltage drop due to starting will be minimal.

Controllers for medium-voltage motors are divided into two NEMA classifications. Class E1 controllers employ their contacs for both starting and stopping the motor and interrupting short circuits or faults exceeding operating overloads. Class E2 controllers employ their contacts for starting and stopping the motor and employ fuses for interrupting short circuits or faults exceeding operating overloads. Class E1 controllers should interrupt up to 50 MVA fault currents, and Class E2 controllers use medium-voltage Class R fuses for protection.

These controllers function similarly to low-voltage controllers, though they are quite different in mechanical design. They are available as full voltage or reduced voltage, induction, or synchronous motor starters, and as multiple-speed starters. The contactors may be of the air-break, oil-immersed, vacuum, or SF_6 type. For motors operating at 13.8 kV, vacuum, SF_6, or power air breakers are used as switching devices (see Reference [10]).

A disconnecting switch or multiple-disconnecting-type fuses can be used to provide a motor disconnecting means. Control circuits are usually 120 V and supplied through control transformers. The extent of relaying can vary greatly, depending upon the degree and type of protection required. Compact, fused disconnect, interlocked drawout starters with built-in protective relaying are available for motors rated up to 5000 hp. Medium-voltage controllers should withstand high basic impulse insulation levels (BIL), for example, 60 kV for 5 kV controllers, since they are often installed close to the service entrance point. However, standards allow the controllers to be designed to an appreciably lower BIL level if surge arresters are installed.

6.10 Synchronous Motor Starters. To start a synchronous motor, it should be brought up to synchronous speed, or nearly so, with the dc field de-energized, and, at or near synchronism, the field should be energized to pull the motor into step. A small induction motor may be mounted on the shaft of the synchronous motor for bringing it up to speed. The induction motor should have fewer poles than the synchronous motor, so that it may reach the required speed. If the exciter that supplies the field is mounted on the motor shaft, it may be used as a motor for starting, provided that a separate dc supply is available to energize it. However, since most synchronous motors are polyphase and provided with a damper winding, the common practice is to start them as squirrel-cage induction motors, with the torque supplied by the induced current in the damper winding. Like squirrel-cage motors, they may be connected directly to the line or started on reduced voltage. When they are started on reduce voltage from an autotransformer, the usual practice is to close the starting contactor first, connecting the stator to the reduced voltage; then, at a speed near synchronism, to open the starting contactor and close the running contactor, connecting the stator to full line voltage. A short time later, the field contactor is closed, connecting the field to its supply lines. The field may be energized before the running contactor has closed, which will result in a little less line disturbance; but the pull-in torque will be lower.

Instead of an autotransformer to supply the reduced voltage, any of the methods for starting squirrel-cage motors may be used. These include starting resistance in the stator circuit, starting reactance in the stator circuit, and combinations of reactance and autotransformer. The Korndorfer system of autotransformer connection can also be used.

6.11 DC Motor Controls. DC motors are started by either full voltage or reduced voltage starting methods. Generally, full voltage starting is limited to motors of 2 hp or less because of very high starting currents. Reduced voltage starting is accomplished by inserting a resistance in series with the armature winding. As counter-electromotive force builds up in the armature, the external starting resistance can be gradually reduced and then removed as the motor comes up to speed, either by a current relay or by timers in steps. All resistance should be removed from the circuit as soon as the motor reaches full speed. Motor characteristics and the resistors are different for series and for shunt motors. Speed control of dc motors can be accomplished by varying resistance in the shunt or series fields or in the armature circuit. Reversing is accomplished by reversing the flow of current through either the armature or the field.

An increasing number of solid-state dc motor drives are used for adjustable speed applications. In many cases, single- or three-phase ac power is converted to dc because dc motors are easier to regulate than ac motors.

6.12 Pilot Devices. There are manual and automatic pilot devices that initiate the control of motors.

6.12.1 Manually Operated Devices. These are pushbuttons (on–off) with normally open and normally closed contacts; selector switches having two, three, or four positions; or master switches (mostly cam switches). Selector switches open and close the coil circuit of the contactor and often connect to an automatic switch (see 6.12.2) in a third position. Selector switches that have spring return keep a certain operational mode only as long as the switch is held in a given position. Most pushbuttons can change the state of the circuit only as long as they are *pushed*; though there are other pushbuttons (infrequently used) that have maintained contacts. A contactor can hold itself in after the on pushbutton closes a *normally open* contact by means of an auxiliary contact *(normally open)* on the starter, which closes a circuit parallel to the *normally open* contact of the pushbutton. Various pushbuttons and indicating lights are included in pushbutton stations and control panels.

There are three standardized pushbutton lines: standard duty (only for limited varieties), heavy duty, and, the most universal line, oil tight.

There are two lines of standardized oil-tight pushbuttons. In each line, the diameter of the hole in the cover is identical for all operators. The oil-tight pushbutton has all possible variations available.

The most popular line of oil-tight pushbuttons has a 30 mm diameter mounting hole; whereas, the smaller line has a 22.5 mm diameter hole. There are a great number of operators available, such as mushroom-head buttons, key-operated switches, various handles, push-to-test pilot lights, pilot lights for full voltage or with transformers on low voltage, etc.

Some typical pushbutton elementary circuit connections for ac full voltage starters are shown in Fig 60.

Fig 60
Typical Pushbutton Control Circuits

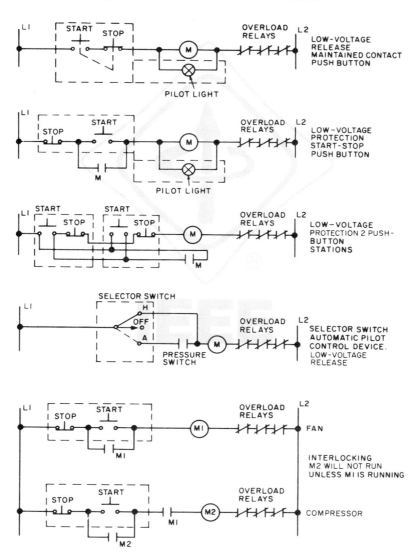

6.12.2 Automatically Operated Devices. A great number of automatic pilot devices are available to switch a coil circuit depending on the actuating medium.

(1) Conventional limit switches convert a mechanical motion into an electrical control signal. The moving object comes into direct contact with the limit switch actuator. These limit switches have various actuators depending on what kind of movement controls the state of the limit-switch contacts. Limit switches should have an extremely long life and should be oil tight. They have normally open and normally closed contacts.

(2) Proximity limit switches are becoming more popular. They operate when an object approaches a sensor. The advantage of this is that physical contact with the object is not necessary; therefore, longer life is possible. The system can be triggered by changing the magnetic field, especially if the approaching object has an iron contact or section. Another method is to change the inductance, capacitance, or resistance when the object approaches the sensor. Photoelectric means are also used.

(3) Float switches are a type of limit switch actuated by the level of a liquid.

(4) Pressure switches respond to pressure changes of a gas or liquid. If the maximum pressure is 500 lbs./in^2 and the medium is not harmful to a diaphragm, the switches are diaphragm operated. If the required pressure approaches 2500 lbs./in^2, the pressure switches are usually bellows operated. If the maximum pressure is still higher, piston-type pressure switches are used.

(5) Sail switches are air pressure actuated devices that are inserted in air ducts. For example, consider a motor-driven fan providing essential flow of air within a duct. Should the motor continue to run, though the fan is mechanically disconnected from the motor by a broken fan belt, the sail switch would sense the change in air flow and provide a signal to actuate an alarm and stop the motor. Differential pressure switches may also be used to sense the failure of fan or blower systems.

(6) Temperature switches use a sensor or bulb to sense the temperature of the surrounding medium. A change in pressure in the sensing element actuates the contacts.

(7) If the control circuit is more complex and requires logic, many possibilities exist to develop a logic diagram.

    (a) The simplest way to develop a logic diagram is by using electro-mechanical relays. Electrically operated, electrically held control relays typically have 4, 8, and up to 12 normally open or normally closed contacts. These contacts are very often field-convertible from normally open to normally closed.

    (b) Mechanically held relays only need to be momentarily energized to change position. They have up to 12 normally open or normally closed contacts, and maintain their position even after control power is removed.

    (c) Timing relays or timers are relays that have delayed contacts. Timers are available with adjustable delay time from a few cycles to 3 minutes and more. They are sometimes field-convertible from delay on to delay off.

In addition to these simple timers, there are also timers for repeat cycling, which repeat a certain on-off pattern (interval timer). These timers begin a timing period after an initial switch is actuated. After a selected time, the output switch is restored to its initial position.

Timers operate on various principles, such as motor-driven, dashpot, thermal, solid-state, or pneumatic. Pneumatic timers generally do not have a very high repeat accuracy ($\pm 10\% - \pm 20\%$); but they have a moderately long life (approximately 1 million operations) and are low in cost. The actuating time element can be varied by a needle valve or the length of a groove (linear timer). Solid-state timers have higher accuracy and longer life.

(8) Complex systems that have a large number of logic elements frequently use solid-state elements.

    (a) Solid-state relays are usually similar to solid-state contactors but have only pilot duty rating.

    (b) Hard-wired, solid-state logic systems require a different wiring technique than is used for relay logic. A relay usually has one input (coil) and several outputs (normally open, normally closed, with or without time delay contacts), whereas solid-state logic can have several inputs and one output, or combinations thereof.

The basic elements obtained by transistor logic are AND, OR, NOR, NAND, memory, time delay, retention memory, etc. The advantage of this is that solid-state devices do not wear out. Very complicated control systems will become simpler with solid-state devices than with relays.

6.12.3 Programmable Controllers. Programmable controllers are microcomputer-based, solid-state devices that are often programmed in a format similar to the familiar relay logic ladder diagram. These controllers utilize digital logic, which consists of input and output interfaces, the central processor, the memory, the program, and the necessary power supply. The advantage of this is that they can be easily reprogrammed in case of a change in the sequence. Further, many functions that are difficult to obtain with standard relays can be added, such as counting, arithmetic functions, and various sequential and timing functions. Programmable controllers are generally less expensive than relays if the logic is complex. To energize valves, contactors, starters, etc., small solid-state output switches are generally used for interfacing.

The central processing unit (CPU) checks by scanning perforated tapes or switch positions or input/output (I/O) cards and the control plan that is stored in memory. There is read-only memory (ROM), read and write memory, and random access memory (RAM). Certain types of ROMs may be changed by ultraviolet light or other electronic means. They can also be programmable and are called "programmable read-only memory (PROM)." Read and write memories are usually RAMs that can be programmed and changed. However, unless backed up by a battery supply, they will lose their state if power is lost. The complexity of the programs determines the size of memory required.

The most commonly used CPU today is the microprocessor. With sufficient memory, the new programmable controllers are actually small computers that also play an important role in the energy management of commercial buildings.

Besides relay logic ladder diagrams (still the most common program language), Boolean algebra, and other logic formats (AND, OR, etc.), and various high-level computer languages can be used.

The program can be studied and checked on computer terminals. Troubleshooting is made easier with a diagnostic device to check the status of the controller elements. In addition, most personal computers have self-diagnostic routines to detect errors.

Electromechanical control devices or transducers can also be used as input sources for programmable controllers. Transducers measure properties and convert those properties (pressure, temperature, speed, power, etc.) to an electrical output that is applied to the input of the controller.

The input and output devices usually work on 120 V_{ac}, but supply a low-level output (approximately 5 V_{dc} to the controller). Transducers made by different manufacturers are usually interchangeable because they have standard outputs. Printed-circuit cards have indicating lights to show if they are in an on or off state. Each output circuit can have a separate fuse.

6.13 Speed Control of DC Motors. The shunt motor is excited by a field with a constant or adjustable voltage. In a series motor, both armature and field circuits are in series and, therefore, have identical currents. The compound motor contains both a shunt and series field. The speed of the motor is proportional to the counterelectromotive force and inversely proportional to field strength. The torque is proportional to the field strength and the armature current. Thus, the speed of the motor can be regulated by either changing the armature current or field current. Both currents can be controlled by voltage adjustment, which can be accomplished for the dc motor by resistance or by solid-state voltage controls. Since

$$CEMF = V - I_a R_a \qquad\qquad (Eq\ 7)$$

where

$CEMF$ = Counter-generated voltage.
V = Applied armature voltage.
I_a = Armature current.
R_a = Armature resistance.

it can be seen that the motor speed will always vary to provide a counterelectromotive force to match this equation.

The speed of the shunt motor can be increased by changing the field strength, which means adding resistance to the field circuit. Braking of shunt motors can be obtained by switching a resistor into the isolated armature circuit, which causes the motor to act as a generator, or by applying current in the opposite direction through various control means. The speed of series motors can be regulated by adding resistors in the armature and field circuit or, as is most often done, by adding resistors in parallel to the armature to dc.

Numerous circuits have been utilized for dc motor control. Motor-generator (Ward-Leonard) control was widely used until a few years ago. The output voltage of a shunt-connected generator that applies power to the drive motor is adjusted by field control and, thereby, changes the motor speed. Today, the speed of dc motors is mainly adjusted by the phase control of silicon-controlled rectifier (SCR) converters that operate directly off the ac input lines. The SCR is a controlled-rectifier-type PNPN semiconductor with a gate on the second positive layer (P). After being triggered, the SCR stays in the on state until the main circuit is either interrupted or reverse voltage is applied.

In order to keep the ripple to a minimum, full wave, bridge-type rectification with two SCRs and two diodes is used for unidirectional armature control for single-phase ac. For three-phase ac, three SCRs or six SCRs and three diodes are used. In three-phase controls, triggering occurs by utilizing three independent trigger circuits, which are powered from the secondary windings of one three-phase transformer in Y, thus producing the same delay on all three phases. The dc voltage is dependent on $\cos x$, where x is the delay angle to trigger the SCR. If $x = 180°$, the voltage is zero; if $x = 0$, the output voltage is maximum. The motor can be reversed in different ways

(1) By reversing the polarity of the field windings, using either electromagnetic contactors or solid-state devices
(2) By reversing the polarity of the armature voltage with contactors
(3) By reversing the polarity of the armature voltage using two SCR converters, one connected to produce reversed voltage polarity
(4) By reversing the generator field voltage in a motor-generator (Ward-Leonard) system

6.14 Speed Control of AC Motors. The speed of ac induction motors is dependent upon synchronous speed, the number of poles, and the slip. Synchronous speed is controlled by the supply frequency; whereas slip is dependent on voltage and current regulation.

Motors with stator windings for different numbers of poles are generally used for two speed (high and low) requirements. The controller is designed for two steps, and speed cannot be varied in a wide stepless range.

The simplest stepless speed control is possible with wound-rotor motors. Since the speed of a wound-rotor motor at a given output torque depends on its secondary resistance, an adjustable resistance in the rotor circuit provides adjustable shaft speed.

AC adjustable frequency drives provide an attractive means of utilizing standard induction motors, energy savings, and a wide range of speed and torque control. The most common types of these drives used in industry are variable voltage input (VVI) and pulse width modulated (PWM). In these types, the ac-dc-ac or dc link control, and the one SCR converter turns three-phase or three-phase ac into adjustable frequency and adjustable voltage ac by an SCR inverter. The frequency controls the speed of the motor, and the voltage is controlled to maintain a constant ratio of voltage to frequency (V/f), as required, to prevent overexcitation of the motor. Both braking and reversal are accomplished by reversing the phase rotation of the output ac controls. The ac-dc-ac controls are used with both induction and synchronous motors, and can provide very wide speed ranges, including speeds above the synchronous speed of the motor at 60 Hz.

The other popular type of adjustable frequency controller is the cycloconverter, in which an adjustable frequency and voltage ac output for the motor is synthesized directly from the input ac waveforms, with no intervening dc step. Cycloconverter drives are more economical than dc link drives when very low speeds (approximately 10% of 60 Hz synchronous speed) are needed.

6.15 Power System Harmonics from Adjustable Speed Motor Controls. Both dc and ac adjustable speed drives using solid-state techniques (SCR converters, inverters, or cycloconverters) have nonsinusoidal, square-edged ac input current waveforms. These currents may be considered to contain harmonic frequency components (that is, current components at multiples of power frequency) that propagate through the power system feeding the drive. The effects of this are usually harmless, but can be troublesome. For instance, a harmonic component can excite a resonant condition between a power factor correcting capacitor bank and the inductance of the power system, causing damaging overvoltages to appear at or near the capacitors. In the rare cases when such harmonics problems occur, they can be readily eliminated by such a simple method as changing a capacitor bank rating. Electrical consultants and equipment suppliers can provide valuable advice on the prevention or cure of harmonics problems.

6.16 References. The following references shall be used in conjunction with this chapter:

[1] ANSI/NEMA ICS1-1988, General Standards for Industrial Control and Systems.

[2] ANSI/NEMA ICS2-1988, Industrial Control Devices, Controllers, and Assemblies.

[3] ANSI/NEMA ICS6-1988, Enclosures for Industrial Control and Systems.

[4] ANSI/NFPA 20-1990, Installation of Centrifugal Fire Pumps.

[5] ANSI/NFPA 70-1990, National Electrical Code.

[6] ANSI/UL 845-1987 (with 9/7/88 Revision), Motor Control Center.

[7] NEMA 250-1985, Enclosures for Electrical Equipment (1000 Volts Maximum).

[8] "IEC and NEMA Motor Control Application Considerations," *Consulting/Specifying Engineer*, Oct. 1989.

[9] Brozek, James P. "Type 2 Protection of IEC Starters—A Recommended Method," *IEEE Conference Record of 1990 Industry Applications Society Annual Meeting.*

[10] Kussy, F. W. and Warren, J. L. *Design Fundamentals for Low-Voltage Distribution and Control*, New York: Marcel Dekker, Inc., 1987.

Chapter 7
Services, Vaults, and Electrical Equipment Rooms

7.1 Incoming Lines and Service Laterals. Incoming lines and service laterals are an extension of a building's electric distribution system, connecting the facility to the serving utility's service point. The operating voltage, ownership, maintenance, and burden of installation cost for this portion of the electric system will vary from region to region.

When planning, it is important to route the incoming circuits to avoid clearance conflicts with existing or future underground or overhead structures. Poles located in areas subject to vehicular traffic may require curbs or barriers for protection. When open-wire lines pass near buildings, adequate clearances should be provided to avoid accidental contact by occupants, maintenance, or inspection personnel, and firemen. Some considerations for medium-voltage services are different than those for utilization voltage services, such as qualifications for operating and maintenance personnel, and utility and code requirements.

In many states, because of the potential hazard of electric conductors to the general public, services are required by law to meet minimum construction standards. Underground and overhead services are normally built in accordance with local electric utility standards and the requirements of ANSI C2-1990, National Electrical Safety Code (NESC) [1][56] or state codes (i.e., General Order 95 in California). The utility may assign responsibility for this construction to the commercial customer.

7.1.1 Overhead Service. For small buildings supplied at utilization voltages, the overhead service lateral is generally terminated at a bracket on the building at sufficient height to provide the required ground clearance given in provisions of the NESC [1]. Larger buildings may be served by open-wire lines terminating at a transformer stepdown substation outside the building, or at a cable terminal pole where the stepdown substation is inside the building. An alternative method is to use aerial cable (insulated cables, shielded where applicable, supported by a grounded messenger) attached to poles.

---

[56]The numbers in brackets correspond to those in the references at the end of this chapter. ANSI publications are available from the Sales Department of the American National Standards Institute, 11 West 42nd Street, 13th Floor, New York, NY 10036.

Open-wire lines may consist of copper, copperweld, aluminum, or aluminum conductor, steel-reinforced (ACSR) conductors attached to insulators, supported on pins mounted on wood or epoxiglass crossarms or on pole brackets attached to wood poles set in the ground. Metal or reinforced concrete structures are also sometimes used. The conductors may be suspended from crossarms on suspension-type insulators or clamped to horizontally mounted post type insulators bolted to poles. The NESC [1] spells out all clearances. This construction should meet the applicable voltage and BIL levels of the service and is subject to acceptance by the serving utility.

The design of an open-wire line depends upon the following factors:

(1) Safety
  (a) Safety to the public, providing the necessary clearances from the line to buildings, railroad tracks, driveways, etc.
  (b) Safety to personnel who may operate and maintain the line, involving adequate climbing and working space on the pole, spacing between conductors on the crossarm, and interphase spacing between items of equipment on the pole. Spacing requirements should consider the effects of wind-induced galloping of conductors.
  (c) Mechanical strength, involving consideration of wind and ice loads, diameter of the pole, the size and strength of wire, etc. Margins of safety for power lines are given in the NESC [1] for construction grades B, C, and N. Grade B is the strongest.

(2) Insulation
  (a) Protection against lightning surges. This is handled by shielding the line from direct strokes and induced surges through the use of surge arresters, one or more shield wires installed above the power conductors, and by greater insulation. System and pole grounds drain lightning currents from the line following a lightning stroke. Proper grounding will help minimize lightning damage.
  (b) Protection against voltage surges caused by power switching. The above mentioned preventive measures also apply for switching surges.

For the details on designing a line, refer to various electrical handbooks, to the local electric utility company, and to experts proficient in this professional work. Flashover characteristics of insulators can be obtained from manufacturers' catalogs; however, this does not necessarily provide the coordination of the insulation level required.

Right-of-way grants may be required from the owner for lines on private property, and permits are required from the responsible governmental authority for lines on public property.

7.1.2 Lines Over and On Buildings. Although the NESC [1] provides clearance requirements, the installation of open-wire lines over buildings should be avoided because it is poor practice; they interfere with the activities of fire fighters and maintenance or security personnel, and present a safety hazard. Use of fully insulated cable, with a grounded metallic sheath for medium-voltage cable, is an alternative to open-wire construction over or near buildings. When attached to a building's exterior, the service conductors should be in grounded metallic conduit.

Where this conduit is mounted on the roof or other flammable building material, the conduit should be encased in 2 inches of concrete. The concrete encasement can be omitted for voltages under 300 V. If open wire is installed over buildings, minimum clearances for personnel should be maintained over all areas accessible to personnel. Clearances should meet provisions of the NESC [1] or rules of the local code enforcement authority. "Tree" coverings applied to bare overhead conductors, for voltages up to and including 15 kV, do not have adequate insulation values to prevent injury from accidental contact. These coverings are used to reduce interruptions caused by momentary tree branch contact with the wire. All conductors operating above 2000 V to ground should be fully insulated and shielded or considered as bare conductors. Bare conductors should be guarded by height or barriers meeting requirements of the NESC [1] or the ANSI/NFPA 70-1990, National Electrical Code (NEC) [3].[57]

7.1.3 Weather and Environmental Considerations. In designing any outdoor structure, weather forces should be considered. A building is designed to withstand wind on its walls and a snow or water load on its roof. Similarly, an overhead electric line should be designed to withstand a wind load on the poles and conductor, a well as an ice load on the conductor. The severity of the weather factor varies by location throughout the United States, and reference may be made to the "General Loading Map" in the NESC [1]. In damp, foggy, or polluted atmospheres, contamination of insulator surfaces becomes a problem, and special insulators having an unusually long leakage distance should be used to prevent leakage currents across the surface of the insulators. Resistance grounded insulators may be used to control the effect of atmospheric contamination on insulator performance.

7.1.4 Underground Service. Certain conditions may require underground construction. Examples of the conditions include: conflicts with overhead structures that cannot be bypassed with aerial construction, load density, local ordinances, or regulatory requirements governing construction in new residential subdivisions. Aesthetics may also be a factor to consider.

An underground system is relatively free from any of the problems associated with an overhead system. However, in case of failure, the repair time and expense of an underground system is considerably greater. Underground systems almost always cost substantially more than equivalent overhead systems. This is especially true of conduits and manhole underground systems. The use of direct-burial-type underground systems, such as underground residential distribution (URD) and commercial and industrial park underground distribution (CIPUD) yields a considerable cost saving for new developments over the cost of an equivalent conduit-and-manhole system.

(1) URD is a direct burial, single-phase distribution system used by utilities for new residential developments. Organic insulated and jacketed cables are

---

[57] ANSI publications are available from the Sales Department of the American National Standards Institute, 11 West 42nd Street, 13th Floor, New York, NY 10036. NFPA publications are available from Publications Sales, National Fire Protection Association, 1 Batterymarch Park, P.O. Box 9101, Quincy, MA 02269-9101.

used together with premolded or encapsulated splices and termination devices. Pad-mounted transformers may be utilized, or transformers and switching devices may be installed in prefabricated fiberglass or expoxiresin "box pads" or in cast manholes.

(2) CIPUD is a direct burial, three-phase system used by utilities for commercial/ industrial park distribution. The cable system is looped through pad-mounted transformers or switchgear installed above ground, or installed in suitable below grade boxes or vaults. CIPUD and URD cable systems are generally run behind curbing in grassy areas to minimize paving costs as well as to provide accessibility. Conduit or duct sleeves are generally installed under paved crossings to eliminate the need for breaking and restoring paving when installing or removing cable or making repairs.

(3) Fault indicators assist in enabling rapid determination and repair of faults on CIPUD and URD cable systems. Installed at cable termination locations, the fault indicator displays a signal whenever fault current has passed through its sensor. The device either resets automatically after the system is re-energized or is reset manually.

(4) Maintenance and operation of CIPUD and URD systems should be entrusted to the utility or to specially trained contractors or facility personnel for both safety and operational reasons. These systems are not suitable for the high-density loads of urban centers because of the direct burial aspect and the limited load and fault handling capability of the equipment.

7.1.5 Service Entrance Conductors Within a Building. Regardless of voltage, when service entrance conductors have to pass through the building to the service equipment, a safety hazard presents itself because that part of the circuit in the building is not generally protected against short circuits, overloads, or arcing faults. However, when the distance is as short as possible, the hazard is considered to be minimal. Longer circuits from the point of entrance through the building wall to the service equipment should be installed in a raceway encased in at least 2 inches of concrete. Greater additional protection is provided by the use of metallic conduit suitably encased in concrete, which is considered by the NEC [3] as "outside the building." This protects the building by confining any fire or arcing (because of a short circuit) within the concrete envelope.

The safety of property and life is always enhanced by encasing the service entrance conductors in concrete inside the building. Concrete encased raceway may be installed along ceilings, under the basement floor, or on the roof. Although busway is sometimes used for service entrance conductors, it is very difficult to provide protection for arcing faults. Neutral grounds normally employed at both the main switchboard and the service transformer(s) make fault detection complicated. From services fed by spot networks, primary sensing is generally ineffective. Overheat detectors that are spotted frequently over the busway have proven valuable; but cooperation from the serving utility to interrupt the utility protection devices at their equipment is necessary. Cable limiters are often installed at both ends of all phase cables whenever three or more cables per phase are utilized. These limiters isolate faulted cables, allowing for a maximum continuity of service.

Service entrance conductors within buildings should be installed to meet or exceed the minimum requirements of the NEC [3] as modified by state laws and local ordinances. Any questions concerning the application of these rules should be taken up with the local code-enforcing authority that has jurisdiction. It should be recognized that most of these codes cover minimum requirements and are not intended to be recommended design criteria.

Cable systems should be routed to avoid high ambient temperature that is caused by steam lines, boiler rooms, etc. Also, avoid running raceways over the roof where conductors are subject to direct sunlight, or re-radiated or convective heat. Where cable is lead-sheathed, duct runs through cinder beds should be avoided unless the duct system is encased in a sufficiently thick envelope of concrete to make it impervious to the acid condition that is prevalent in cinder beds. The lead sheath should also be jacketed to protect against corrosion. Precaution should be taken with polyethylene, crosslinked polyethylene, and other organic jacketed cables to prevent chemical degradation of the jacketing when hydrocarbons may be present, such as in fueling areas, marsh land, landfill areas, and similar locations. Cable systems should be protected from oils and chemicals that are used as preservatives in wood poles by suitable barriers on the riser pole, and enclosed in a raceway at a suitable distance from the pole.

Manholes and pull boxes are recommended in long duct runs to facilitate pulling and splicing the cables. Spare ducts should be considered to provide for the contingency in which a faulted cable becomes frozen in a duct and cannot be removed for replacement. This also simplifies installation of future cables that may be required for load growth. A duct system should not be laid in the same trench with gas or sewer service.

7.2 Service Entrance Installations. Service entrance conductors comprise that portion of the system between the client-owned service equipment and the utility's service drop or lateral. Service equipment includes the main service control or disconnect for the electric supply and consists of one or more circuit breakers, or switches and fuses, and accessories as well as the metering equipment.

Service entrance conductors and service equipment are generally paid for and owned by the client. Design features are frequently influenced or controlled by the electric utility company. Billing metering instruments are nearly always owned and maintained by the electric utility company. Current and voltage transformers used exclusively for billing metering purposes may be furnished by the electric utility company at either the customer's or the electric utility company's expense. Refer to Chapter 4 for details on utility metering.

The relationship of service entrance equipment design and characteristics to the incoming lines or feeders, and to the distribution switchgear or switchboard are of vital importance to the customer and to the electric utility company. Therefore, it is important that the engineer serve both the client and the electric utility company by developing a design that satisfies client requirements without interfering with the quality of the electric service to other customers of the electric utility.

7.2.1 Number of Services. The number of services supplied to a building or a group of buildings will depend upon several factors.

(1) The degree of reliability required for the installation as related to the reliability of the power source — When service reliability is important, multiple services or standby service, with load transfer arrangements between various parts of the building distribution system, may be indicated. In some cases, economic considerations may indicate acceptance of reduced service availability and the interruption of nonessential loads during an emergency. If more than one service is required by a client, an additional charge may be assessed by the utility.

(2) The magnitude of the total load — Since the capacity of an individual service is limited by the utility to a maximum current value, additional services may be provided, as required, to meet building demands.

(3) The availability of more than one system voltage from the utility — If more than one voltage is available, the utility may, for example, supply 208Y/120 V for lighting and receptacles at one or more service entrance points, and 480Y/277 V for power.

(4) The physical size of the building or the distances separating buildings comprising a single facility — Tall buildings, occupying a large ground area, and widely separated smaller buildings will often be supplied from multiple services.

(5) The NEC [3] and local code requirements, including items such as fire walls.

(6) Additional capacity to serve future loads — The initial design should consider requirements for future services and feeders. If service capacity is determined by the NEC [3] (partially illustrated in Chapter 2), service capacity will generally be more than adequate to serve all but major load additions. Service equipment design, however, should be such that additional feeder protective devices may be added.

7.2.2 Physical Arrangement. The physical arrangement of the service entrance will vary considerably, depending upon the type of distribution system employed by the utility and the type of building being served. In some cases, the utility will supply service from one or more transformer vaults located directly outside the building with bus stabs through the basement wall. This is the usual arrangement for buildings of moderate height in heavily loaded areas of many large cities. Transformer vaults are sometimes located within the building itself — in the basement and on the upper floors of tall buildings. Underground service, by means of cable from a manhole or a pole in the street, is sometimes provided; while, in other cases, overhead services may be available.

In all cases, service entrance equipment rooms should be easily accessible to qualified persons, be dry, well lighted, and should comply in all respects with the requirements of the electric utility and local code authorities that have jurisdiction. Plans should be made for possible future replacement of equipment. Provision for smoke exhaust should be considered in the event of an electrical fire.

7.2.3 Low- and Medium-Voltage Circuits. Service entrance equipment is one of the most important parts of the electric supply system for buildings because it is through this equipment that the entire load of the building is served. The service entrance equipment that is installed initially should either be adequate for all future loads or be designed such that it can be supplemented or replaced without interfering with the normal operation of the building that it serves.

Because the service entrance is part of the building and involves equipment that is important to the utility company, the choice of service equipment and service voltage should be a cooperative decision between the building's electrical design engineer and the local electric utility. This should be accomplished early in the design phase in order to allow the building designer to adapt the design to the present and future supply plans of the utility and to enable the utility to supply power to the building in a manner that considers both present and future requirements. The electrical design engineer should furnish load and other data to the electric utility company to assist it in determining the effect of the building load on its system and, where necessary, plan for the expansion of its facilities. The utility will, at this time, inform the design engineer as to the type of services available, their voltages, and the options of overhead or underground electric service. The building engineer should assist the electrical designer in determining, with the help of the utility, the service point and its termination. Other data pertinent to the system design, such as short-circuit current or the kVA available at the service entrance; service reliability; costs; space requirements for poles, substations, transformer vaults, metering equipment, inrush current limitations, and design standards; and similar information should be obtained from the utility.

An increasing number of utilities are offering medium-voltage services to moderate and large commercial facilities. Although the higher voltages pose unfamiliar problems to the commercial facility's electric distribution system, there are many benefits to both the facility and the serving utility. Planning a medium-voltage distribution system within a building requires more time and effort, but affords the design engineer a greater degree of flexibility in selecting equipment and designing electrical facilities within the building. The utility saves considerable capital investment in switching and transformation equipment, which should be reflected in a lower electric rate structure for services at these voltages.

The following is a checklist of items to be considered in connection with electric service:

(1) Complete characteristics of loads to be served
    (a) Kilovoltampere demand, both initial and future, at various utilization voltages, and method of calculation
    (b) Service continuity requirements
    (c) Voltage requirements and limitations of voltage variations
    (d) Special loads, such as x-ray machines and computers
    (e) Superimposing of carrier current onto the electric system for signals, clocks, or communications
    (f) Largest motor inrush current
    (g) Significant low power factor loads
    (h) Nonlinear loads that generate harmonics (such as solid-state, adjustable speed motor drives)
(2) Complete characteristics of all types of service available from the utility under its electrical rate structure
    (a) Voltages available and voltage ranges

    (b) Billing demand clauses of the rates

    (c) Rates and special clauses, such as exclusive service, standby service, power factor penalty, fuel cost adjustments, all-electric service, electric heating/air conditioning, and interruptible service

    (d) Possible need for equipment to transform, regulate, or otherwise modify the characteristics of the available electric service to meet the requirements of the building

(3) Physical and mechanical requirements of the service entrance

    (a) Number of locations at which service may be supplied

    (b) Type of service — overhead or underground cable, or bus

    (c) Points of service termination, including information as to which parts of the service installation will be owned, installed, and maintained by the utility

    (d) Location and type of metering equipment, including provisions for totalizing demand and for submetering, where permitted, and provisions for mounting and wiring the electric utility's meters and metering transformers. The utility may have requirements for remote metering/ monitoring.

    (e) Space and other requirements for utility vaults, poles, and similar equipment, and access provisions for its installation, maintenance, testing, and meter reading

    (f) Avoidance of structural interferences (particularly critical when using busway)

    (g) Equipment — Construction may have to meet special requirements of the utility for the isolation of and blocking for utility system maintenance.

    (h) Service cable should meet the utility's specification for the ability to handle return ground-fault currents. Sizing conductors to match the utility's standard sizes may prove beneficial in facilitating emergency replacement of the cables or terminations.

(4) Electrical requirements for service entrance

    (a) Equipment voltage and BIL levels and coordination of surge protection

    (b) System capacity and fault capability, both present and future (future may be defined as "foreseeable")

    (c) Requirements for the coordination of overcurrent protective devices. Types, sizes, and settings should be acceptable to the serving utility.

    (d) Utility-approved types of service and metering equipment, utility grounding methods, and requirements for the coordination of ground-fault protection for grounded service systems

(5) Schedule data

    (a) The date service and preliminary construction schedule will be required

    (b) The dates when full estimated initial load and full load will be required

    (c) Temporary construction service requirements

(6) Medium-voltage system design flexibility — Most utilities that provide medium-voltage services will only require the service equipment to meet their requirements for interconnection. This allows the design engineer to

have complete control over the design of medium- and low-voltage distribution systems throughout the building. Local codes, the NEC [3], and, in some cases, the NESC [1] will govern the design of the medium-voltage installation. The advantages of this design flexibility include

(a) Medium-voltage rises can be designed as feeders. Fully protected at the service equipment, concrete encasement is no longer required. Standard wiring practices can be used. All medium-voltage conductors and raceways should be adequately marked to indicate their operating voltage.

(b) Unit substations can be installed in electrical rooms rather than utility vaults. The design of the rooms can be more flexible with the selection of proper equipment. Transformer insulation types can be specified to suit the design of the electrical rooms, reducing fire resistance construction features.

(c) Transformers can be selected and located to optimize kVA size and floor space, which is valuable to the facility's owner. The designer has complete control of the transformer specifications to control voltage, connection configuration, and impedance to optimize voltage drop and the short-circuit contribution to the low-voltage overcurrent devices. The design engineer should, however, review transformer primary-voltage ratings, voltage taps, and transformer connections with the utility company for compatibility with the utility's system voltage and variations. The configuration of the transformers can also be controlled to allow for redundancy in the system with primary and/or secondary transfer between multiple medium-voltage supplies.

(7) Medium-voltage service costs — Although medium-voltage services provide benefits to both the utility and the facility, the facility should bear the cost of installation, operation, and maintenance of the medium-voltage service equipment, which usually includes the cost of transformer losses. These additional costs should be offset by the rates offered by the serving utility. An economic analysis should be performed to substantiate the initial investment. In some cases, the electrical requirements of the facility may not meet the low-voltage service characteristics of the serving utility, forcing the facility to a medium-voltage service to take advantage of the flexibility that it affords.

Maintenance and operation of the system will require special training of the facility's maintenance personnel. A contractor experienced with medium-voltage systems should be trained in the event of equipment malfunction, replacement, or system expansion.

7.2.4 Load Current and Short-Circuit Capacity. The design of service equipment depends not only upon the continuous current requirements of the circuit and, hence, the current capacity of the service equipment, but also on the short-circuit current available at the service bus. Low-voltage equipment will, therefore, be divided into three categories

(1) Low-Capacity Circuits — For service entrances with current ratings less than 600 A, which are fed by individual transformers, service equipment problems may be minor. However, engineers should check the electric utility for short-circuit duty because some transformers have very low impedance, which results in high short-circuit current. When several services are supplied from one transformer bank, short-circuit duty may be 15 000–100 000 A (depending upon the backup system and transformer impedance). When the available fault current is less than 10 000 A, a wide choice of equipment exists. If initial investment is an important factor, the simplest fused disconnect switch or molded-case circuit breaker may be used. Ground-fault protection in lower current installations is at the option of the design engineer, and coordination problems are usually not serious.

(2) Medium-Capacity Circuits — While there is no standard definition, typical medium-capacity services may have short-circuit duties ranging from 50 000–100 000 A. The available fault current can be determined only by the utility. The impedance of power transformers may run on the order of 5%; while that of distribution transformers, so common in this type of installation, may be 2% or less. At the latter impedances, the short-circuit current may exceed 50 times the normal load rating.

At these fault duties, a number of fused-type devices are available, as well as certain high-interrupting capacity unfused and fused circuit breakers. While high-interrupting capacity breakers, breakers with current limiters, and switches with current-limiting fuses may protect the basic device, it is imperative that the engineer closely examine the device's limiting characteristics to assure that the limiting effect will protect downstream interrupting devices. Unless test curves or manufacturers' data, or both, assure such protection, it is incumbent on the engineer to include additional current-limiting devices downstream in the system.

For grounded-wye electrical services of more than 150 V to ground, but not exceeding 600 V phase-to-phase, ground-fault protection is required by the NEC [3] for any disconnecting means rated 1000 A or more. The design engineer may elect to use it at lower levels.

(3) High-Capacity Circuits — All service entrances that have an available short-circuit capability in excess of 65 000 A can be considered high-capacity service entrances. Buildings that are fed from ac secondary networks or very large buildings that are fed from a number of parallel transformers are in this class. In these installations, it is imperative that breakers (with or without current limiters) or switches with current-limiting fuses that are suitable for the available fault current be utilized when possible. Multiple sources may prove effective in lowering available fault current. Reactors, current-limiting buses, and current-limiting cable systems have been used to reduce fault currents that are available at the protecting devices but are becoming less common as device-interrupting capacities have been increased. Where they are used, it is important to analyze the resultant X/R ratio of the circuit in the short-circuit calculations to be able to assure the protection of system components.

7.2.5 Limiting Fault Current

(1) Multiple Sources — The simplest method to eliminate excessively large available short-circuit currents is to divide the electric circuits of the building into several independent parts, where practicable and where permitted by the utility, with each part fed by one three-phase transformer or by a group of transformers. If the entire load of a large building takes six parallel transformers to supply it, it may be possible, by integrating the design of the building's electric system and the utility company's supply system, to divide the six parallel transformers into two groups of three transformers each or into six single-transformer loads. This method is less effective for smaller buildings fed directly from the utility ac secondary network because serving the building at more than one entrance may not reduce the available short-circuit current on any entrance where network ties exist.

In order to reduce the available short-circuit current at the load side of a large service switch or circuit breaker, the service can be divided into six or fewer smaller service disconnecting switches or circuit breakers equipped with current-limiting fuses. While each of these devices is capable of handling the available short-circuit current at the service point, the let-through current is greatly reduced by the use of smaller current-limiting fuses than would be the case for a single large service with current-limiting fuses. This arrangement permits the use of equipment that has lower interrupting ratings further downstream, and may be considered where permitted by local codes.

A variation of the "divide-and-conquer" technique of fault current reduction consists of designing additional reactance into the cable circuits connecting each network transformer to a common bus or in designing additional reactance into the cable circuits from each individual transformer bus to separate service entrance circuit breakers or switches within the building. Reactance of these cable circuits is controlled by spacing between the phases, which are individually put into nonmetallic conduit buried in concrete. Normally, open tie circuit breakers or switches between the various service entrance buses are interlocked with the main service entrance circuit breakers or switches. This arrangement permits serving a very heavy load with only moderate amounts of available fault current and good flexibility in case of outages of transformers or primary feeders. However, local codes and the utility company should first be checked to determine if this arrangement will be acceptable.

Where high-capacity, low-voltage service is the only choice provided by the utility or is existing, consideration should be given to the use of step-up indoor transformers to limit short-circuit current while facilitating distribution.

(2) Reactors (see also Chapter 9) — With utility agreement, reactors can be put in the main service connection to reduce the fault current to the rating of the service entrance main and feeder circuit breakers or switches. These reactors are often of high continuous current rating, which makes them quite large. Reactors in the main supply can reduce fault currents to about

60 000 A. This is within the rating of available circuit breakers or switches. It is unnecessary and usually not economical to reduce the fault current further with reactors. Use of a reactor will decrease the short-circuit power factor. The peak current as a result of the lower power factor should not exceed the peak current at which the protective devices have been tested.

Smaller reactors can also be used to feed a smaller group of overcurrent devices that may be part of the service entrance equipment to reduce the short-circuit current. Small, enclosed reactors are available in current ratings up to 800 A, which may be economical when applied to feed groups of smaller devices. The use of these reactors permits better selectivity in the coordination of circuit breakers for tripping under fault conditions and permits the use of protective devices of reduced interrupting capacity. Enclosures or guards should be considered for reactors that are accessible to personnel or exposed to physical damage.

Where reactors, or reactance of any type, are used, consideration should be given to the voltage drop introduced into the circuit.

(3) Current-Limiting Busway — Current-limiting busway is designed with all the bars of one phase installed side by side instead of with interlacing bars of different phases as with the usual low-reactance bus. Current-limiting busway is effective in reducing very large short-circuit currents, such as 200 000 A, down to levels of approximately 100 000 A, but is usually less effective in reducing the short-circuit currents to much lower values. Where the local code regulations permit the installation of the service entrance equipment at a point some distance from the entrance into the building, the use of current-limiting busway presents a solution to the problem of reducing the available short-circuit current, provided that transition units for phase transposition are used frequently enough to balance the reactances and resultant voltages.

However, most codes require that the service disconnect switch or circuit breaker be applied closest to the point where the utility company enters the building. In this case, the service entrance switch or circuit breaker should be adequate for the available short-circuit duty. Current-limiting busway is not feasible unless the runs are long enough to provide the desired reactance. Again, the design engineer should review the resulting X/R ratio of the circuit in the short-circuit calculations to ensure protection of system components. Busway can sometimes be used between the fully rated main service switch or circuit breaker and downstream switchboards if the run is of sufficient length. (See the example of this in Chapter 9.)

(4) Current Limiting by Cables — Current-limiting effects can also be achieved by means of cable in which the spacing between phases is controlled by design. If cables are used ahead of the service disconnect switch or circuit breaker, local electrical codes may require that they be installed in conduit that is buried in concrete. If permitted by local codes and the utility, separation of the cables for each phase in three equally spaced nonmetallic conduits provides a relatively high current-limiting reactance, while also reducing the chances of a phase-to-phase cable fault to nearly zero. Equal

(delta) spacing is necessary to keep the reactances balanced, while non-metallic conduit is required to prevent destructive heating by currents that are induced in metal conduit.

Local codes may require the use of a separate vault for service entrance current-limiting busway where no separate protective device exists ahead of the current-limiting busway.

(5) Current-Limiting Fuses — Current-limiting fuses start to limit current at approximately 20–30 times their current rating and to clear the circuit before the current has reached its peak value on the first half-cycle. Their current-limiting action results from interrupting the fault current before it can increase to its maximum asymmetrical value during the first half-cycle. The fault current that flows through the fuse while it is melting and interrupting the circuit is called the "let-through current." Its peak magnitude depends directly on both the continuous rating of the fuse and the fault current available in the system if the fuse were not in the circuit. The higher the continuous rating of the fuse, the more current it will let through.

The curves in Chapter 9 illustrate the typical peak let-through based upon available symmetrical fault current for various current-limiting fuse sizes. To convert these peak let-through values to a symmetrical downstream interrupting rating that will match a given circuit breaker, it is necessary to use a dividing factor of 1.8–2.3, depending on the X/R ratio of the distribution system. When the equipment's symmetrical and asymmetrical let-through rating is not known, the design engineer should request recommendations from the manufacturer based upon actual test values obtained in high power laboratories.

The use of large current-limiting fuses with the service entrance switch or circuit breaker may protect the service equipment, but does not necessarily protect the lower rated equipment beyond the service entrance, such as motor control centers, distribution switchboards, or panelboards. These circuits can be protected by smaller downstream current-limiting fuses to reduce the let-through, or by the use of current-limiting busway, cable runs, or reactors as previously discussed.

The current-limiting fuse has one of the disadvantages of the instantaneous trip circuit breaker. Selectivity between mains and feeders is possible over the entire range of fault current if the ratio of fuse ratings is more than 2:1 or 3:1. Unbalanced faults may blow only one fuse in a fused switch, leaving the entire load operating in single phase. With the large amount of motor load in modern buildings, this can have serious consequences unless the motors are adequately protected from overload and single-phasing damage. To prevent this damage, blown fuse indicators are available that can signal the tripping device of the switch to open it, or the fuse may be mounted in a built-in circuit breaker or service protector and fuse combination with inherent anti-phase protection. Sensitive anti-single-phase protection may also be obtained for motors by the appropriate selection of motor overload devices, including bimetallic anti-single-phase types and electronic types. It may also be obtained by the proper application of dual-element fuses.

7.2.6 Ground-Fault Protection. The NEC, Article 230-95 [3], covers ground-fault protection requirements for equipment. Arcing ground faults, if not interrupted quickly, can cause extensive destruction of equipment, particularly on high-capacity 480Y/277 V circuits. Fully coordinated ground-fault protection schemes are recommended on such systems. The achievement of proper ground-fault coordination may require a combination of fixed time delay, inverse time delay, and zone-selective interlocking, etc.; between the main service and feeder overcurrent protective devices and also between the feeder and subfeeder overcurrent protective device. The design engineer should check the effectiveness of coordination in all such cases.

7.3 Vaults and Pads for Service Equipment

7.3.1 Vaults. Service transformers and associated switching and protective equipment are often located in vaults. Special precautions should be taken to remove the heat given off by the transformers. Equipment vaults should be located so that they can be ventilated to the outside atmosphere without the use of flues and ducts, where practical. Natural ventilation is considered to be the most reliable means of ventilation. The total net area of the ventilator should not be less than 3 in^2/kVA of installed transformer capacity. Additional ventilation may be required by local codes or electric utilities. Where the load peaks in the summer and where the average outdoor temperature during 24 hour periods in the summer exceed 30 °C, the ventilator area should be increased, or an auxiliary means of removing the heat from the vaults, such as fans, should be used.

When long vertical ventilating shafts from the vault to the top of the building should be used, it is necessary to have a larger vent area to compensate for the added resistance to the flow of air. The long vertical shaft should also have a divider (air-in/air-out) with the air-in portion carried down to just above the floor in the vault to better promote circulation. For such shafts to overcome air friction, a fan should be installed with its discharge directed toward the shaft air-out opening to increase the velocity of the air through the ventilation shaft. The fan should have a cord and plug to facilitate its replacement, and it may be either single-phase or three-phase. A signal light at the entrance door should indicate fan failure. Controls for the fan should not be permitted in the vault and should be accessible only to authorized personnel.

Suitable screening should be used to prevent birds, insects, vermin, or rodents from entering the shaft. The ceiling, walls, and floor should be of fire-resistant construction. Reinforced concrete is preferred. When oil insulated transformers are used and persons occupy the area adjacent to the vault wall, or when an explosion may otherwise damage a building wall, the vault walls should be sufficiently strong to withstand an explosion. The hazard may also be reduced by limiting the ratio of vault volume in cubic feet to net ventilated area in square feet. Tests indicate that where this ratio does not exceed 50:1, an 8 inch reinforced concrete wall will suffice.

Any opening from a vault into a building should be provided with a tight fitting UL-approved 3 hour fire door. The vault should be free of all foreign pipes or duct systems. A sump, with protective cover or grate, should be provided in the vault floor, to catch and hold any oil or liquid spillage. The floor should be pitched to the sump. The floor should be sealed with an adequate coating before installing equipment. Door sills should be of sufficient height to retain all of the oil from the largest transformer. Fire dampers may be required at air duct openings.

Grade-level gratings are suitable for underground vaults and will also suffice for a combination access hatchway and ventilation well when the vault is in the basement of a building and adjacent to an outside wall. Gratings for sidewalk service vaults should be made strong enough to support the wheels of trucks and should satisfy local code requirements. Gratings of net free-air area equal to 63%–70% of gross grate area are available commercially to meet various loading requirements. Grating for roadway service should comply with Class H20 highway loadings.

When multiple banks of transformers supplied from different sources are used, they should be installed in separate compartments to prevent fire in one compartment from affecting adjacent transformers. Switchgear associated with the transformers should also be separately enclosed so that a failing transformer can be isolated without entering the transformer compartment and to prevent transformer trouble from involving the switchgear. Consult local codes, the NEC [3], insurance underwriters, and the local utility for specific vault construction requirements.

7.3.2 Outdoor Pads. Pad-mounted, three-phase transformers and switching equipment are now being installed in many applications. Designed for installation on surface pads, pad-mounted components are an economical and safe means for providing service. Cables enter and leave via the bottom of the component, hence presenting no energized parts to create a hazard. While no supplementary enclosure is required, an enclosure may be provided for unusual or aesthetic purposes. Traffic protecting posts should be provided in areas that are accessible to vehicles. If placed in a vault, provision for the insertion and withdrawal of the pad-mounted unit by crane should be allowed. Landscaping or architectural fencing may be used for concealment. Additional space may be required for operation and maintenace by the utility.

7.3.3 Safety and Environment. Except for outdoor, pad-mounted equipment, which meets the requirements of the NESC, Article 380 [1], outdoor substations should be enclosed by walls or fences. Adequate aisles should be provided for safe operation and maintenance. Proper clearances, both vertically and horizontally, should be maintained. Fence safety clearances should be maintained as required by IEEE Std 1119-1988, IEEE Guide for Fence Safety Clearances in Electric-Supply Stations (ANSI) [8].[58]

All equipment, operating handles, fences, etc., should be adequately grounded. See IEEE Std 142-1982, IEEE Recommended Practice for Grounding of Industrial and Commercial Power Systems (ANSI) [4] for a complete discussion of grounding requirements and methods. High-voltage warning signs should be prominently displayed. Enclosures, equipment, operating handles, etc.; should be locked.

---

[58] IEEE publications are available from the Institute of Electrical and Electronics Engineers, IEEE Service Center, 445 Hoes Lane, Piscataway, NJ 08855-1331.

Substations should not be located near windows or roofs where live parts may be reached, or where a fire in the substation could be transmitted to the building (see IEEE Std 979-1984 (Reaff. 1988), IEEE Guide for Substation Fire Protection (ANSI) [6] for fire precautions). Local utilities, authorities, and insurance underwriters may require liquid filled transformers to be located not less than certain minimum distances from building openings unless suitably baffled (see IEEE Std 980-1987, and IEEE Std 640-1985, IEEE Guide for Power Station Noise Control [5]). Indoor substations should have the same general safety considerations as an outdoor substation, even though they are usually metal clad or enclosed. They should have a separate enclosure or should be placed in separate locked rooms, and should be accessible to authorized personnel only.

Multiple-escape means to the outdoors or to other parts of the building should be provided from vaults, and located in front of and to the rear of switchgear and the rows of transformers. These emergency escape means should be hinged doors with panic bars on the inside of the doors for quick direct escape in the event of trouble.

7.4 Network Vaults for High-Rise Buildings. The electric demands generated by large modern metropolitan office or commercial high-rise buildings almost invariably require the installation of a multiple number of transformers in close proximity to the structure. Each transformer is connected to a common low-voltage bus through a network protector. Many new buildings have power supplied at two or more locations, one beneath the sidewalk and others in the building or perhaps on the roof. Typically, these installations could provide up to 6000 kVA at 208 V or up to 15 000 kVA at 480 V at one point of service (see 7.2.5).

The design of major network installations divides naturally into two parts. First, it is necessary to establish a utilization voltage, the number of transformers, and the number of service points. It is then necessary to match utility standards with clients' building designs. The design should satisfy client and utility requirements and also meet municipal regulations, all within a framework of economics. The ability to install, maintain, or replace a component of the supply system without interruption of service is the backbone of network design.

7.4.1 Network Principles. To more fully appreciate the subject of specifying and designing network installations, it is first necessary to understand the principles of a network system. The network is designed to meet power demands on a contingency basis, which is to say that, with a predetermined number of components (for example, transformers) out of service, full-load capability is maintained. This is accomplished by operating the remaining equipment above its nameplate rating and allowing slightly reduced service voltage levels. Networks are generally designed as first or second contingency systems. First contingency networks generally utilize two or three primary feeders. Second contingency networks may utilize three, four, six, or more primary feeders. If full-load capability can be maintained with two sets of components (that is, primary feeders) out of service, the system is defined as a "second contingency network." If full-load capability can be maintained with only one set of components out of service, the system is defined as a "first contingency network."

True contingency design also requires that the primary feeder supply system as well as the substations and switching stations ahead of them be built and operated with the distribution system in mind. The last implication in contingency design is that all network equipment, including the associated high- or low-voltage cable ties, is sufficiently isolated. As a result of this the failure or destruction of a single component in the system will cause only first contingency operation until repairs or replacements can be effected.

7.4.2 Preliminary Vault Design. The initial step is to prepare a simple sketch of the proposed installation by means of a standard vault equipment arrangement showing any adaptations required in the building structure or any interferences with existing obstructions located beneath the sidewalk. The standard designs are similar for subsidewalk or in-building locations. Designs should include the following considerations:

(1) Sufficient space should be available with reasonable proximity to customer load centers.
(2) Subsurface conditions should be favorable. It is desirable to avoid the added expense of pilings or footings.
(3) Installations should be designed so that environmental factors (for example, water) present no serious problems. As an example, underground transformers may be cooled by natural convection with an all-welded construction and corrosion-resistant finish. Interior transformers may be of the ventilated-dry-type with natural and forced air cooled ratings, providing for the safety of nonexplosive, nonflammable equipment.
(4) Conformance to municipal regulations should include
  (a) General structural design with sidewalk loads that may be in the order of 600 lbs./ft^2 or highway loads
  (b) Location and size of ventilation and access panels—These factors provide only for the basic adequacy of an installation at a particular location.

7.4.3 Detailed Vault Design. Many other specific considerations are involved in the safe and reliable design of network installations. Major considerations for properly designed vaults follow:

(1) Ventilation—Ventilation should be directly to the atmosphere and sized at not less than 3 inches of net open area/1 kVA of transformer capacity. The electric utility and local codes should be consulted for more stringent requirements. Forced ventilation, if required, should be a minimum of 3 ft^3/minute per kVA of transformer capacity, unless a higher rating is required by local codes or the utility.
(2) Construction—Below grade vaults shall be reinforced concrete for strength and explosion confinement. Vaults are constructed to be as water-tight as possible; but drainage (when permitted and practical) is also provided to eliminate stagnant or casual water accumulation. Concrete floors should be coated with a suitable sealant to prevent concrete dust from being convected into the core and coils of dry-type transformers and onto circuit breaker parts.

(3) Ventilation Ratio—The ventilation ratio relates to the construction and ventilation discussion above. To avoid excessive pressure in the event of a secondary explosion in a vault containing oil filled equipment, the ratio of vault volume to net ventilation area should be as small as practical. Such a ratio should be less than 50 ft^3/ft^2 of open ventilation area, typically 30 ft^3/ft^2.

(4) Access—Direct, rapid access is required at any time for maintenance or emergency operating personnel. An acceptable access route should be included for replacing transformers. Removable slabs or walls are sometimes considered acceptable.

(5) Isolation and Protection—The effects of equipment failure can be reduced by either isolation or protective relaying, or both. In the extreme, the following effects may result from such failure:

(a) Oil filled equipment—Explosion, tank rupture, fire, smoke, danger of secondary explosion (re-explosion of volatile vapors generated by destruction of Class A and B materials)

(b) High-flame point liquid (when permitted by applicable codes)—Violent tank rupture, a form of explosion

(c) Ventilated-dry-type—Open—Smoke with very limited fire possibilities. Very limited possibility of secondary explosion.

(d) Sealed-dry-type—Normally not considered hazardous. While such failures are rare, the possibilities cannot be neglected. Of course, the location of the equipment in relation to people will determine the overall degree of hazard. For example, a transformer failing in a sidewalk vault may be relatively innocuous compared to a similar failure in an electrical room adjacent to a public area.

Some utilities depend almost entirely upon isolation for safety. At the higher voltages, utility practice may preclude the use of protective relaying, which would detect low-level faults. In these instances, strong masonry vaults that are vented to the outside are depended upon to contain the effects of a fault. In larger installations, transformers are placed in individual vaults, and the network protectors and collector buses may be similarly isolated. When the vaults or network area are part of the building distribution system rather than utility owned, protective relaying that will cover all zones is required. Such protection will include, as a minimum, overcurrent and ground-fault protection. It may, in addition, include differential protection, transformer liquid level, liquid temperature, winding hot-spot temperature, pressure/vacuum and sudden pressure trip, or alarm. Heat- or arc-sensing, or smoke detection devices in the room or vault, or even inside larger pieces of equipment, may be provided. None of these, however, can relieve the need for physical isolation, which may be required to protect the building's occupants and the public from the effects of such failure. The use of machine room floors or other heavy equipment areas for the location of the electrical rooms or vaults further enhances the afforded protection. Great care should be exercised to ensure that no smoke or fumes can, under any condition, enter the normal building ventilation system.

(6) Apparatus Arrangement — It is important to provide adequate working space around electrical equipment for the operation and maintenance of the following:

(a) Drawout of switchgear

(b) Replacement of fuses, cable limiters, or cables

(c) Access to equipment accessories

(d) Cleaning

(e) Air circulation

(f) Access to equipment for replacement purposes without disturbing other equipment

(g) Cable pulling and installation

(7) Miscellaneous

(a) Heavy-duty roof structures for vehicular traffic (Class H20) loading

(b) Interference by curb cuts or driveways

(c) Future street widenings or grade changes

(d) Improved drainage

(e) Spare conduits within buildings

(f) Duct arrangement for separation of primary and secondary feeders

(g) Effects of unnecessarily long cable ties on voltage regulation

(h) Balanced equipment loading

(i) Access for heavy test sets to the vault switchgear

(j) Normal illumination for routine inspection and maintenance with power supply receptacles for additional lighting and test equipment use. Consideration should be given to having part or all of this supply on the building emergency source during outage conditions.

(k) Transformer noise reduction should be considered in the design of the vault by

(i) Avoiding room dimensions that are half wavelengths of transformer noise frequencies in all directions

(ii) Damping treatment in the room if the above dimensions cannot be avoided

(iii) Isolation of the transformer from the ground by use of sound-absorbing pads

(iv) Use of flexible connections

(v) Placement of ventilation ducts so they do not transmit or amplify sounds

Published sound levels of transformers are usually based on meaurements taken in large rooms. Actual sound levels measured in smaller rooms will be higher.

7.5 Service Rooms and Electrical Closets. Service and distribution equipment is generally located in electrical rooms, while subdistribution equipment is generally located in electrical closets. These areas should be as close to the areas they serve as is practical.

The rooms should be sized so that there is sufficient access and working space around all electrical equipment to permit its ready and safe operation and maintenance. The doors should be of sufficient size to permit easy installation or removal of the electrical equipment contained therein.

7.5.1 Space Requirements. To provide flexibility for future expansion and growth, the electrical rooms and closets should be sized somewhat larger than the minimum criteria dictated by the NEC [3]. The minimum clear working space in front of electrical equipment is clearly spelled out in the NEC [3]. Additional utility working space may be required. Particular attention should be given to the space and clearance requirements of busway equipment, such as bus plugs and large fixed switches and circuit breakers.

7.5.2 Illumination. Adequate illumination should be provided for all such areas in accordance with the NEC [3], the NESC [1], and ANSI/IES RP7-1983, Practice for Industrial Lighting [2].[59]

7.5.3 Ventilation. Ventilation should be provided to limit the ambient temperature of the room to 40 °C (104 °F). When a transformer other than a signal-type transformer is installed in an electrical closet or room, some local codes require that a system of mechanical ventilation be provided. Refer to 7.3.1 and 7.4.3 for mechanical ventilation requirements.

7.5.4 Foreign Facilities. Electrical rooms and closets should only contain the facilities necessary for the electrical installation's operation and maintenance. Electrical closets or rooms are not to be used for storage or other purposes. Certain local codes prohibit a raceway, wiring panel, or device of a telephone system from being installed in this area. The same codes would even be more stringent on the running of water, gas, or other nonelectrical pipes or ventilating ducts through electrical rooms or closets. When local codes are not specific about this, the dictates of good judgment or practice should apply. Condensation can drip from cold water pipes and ventilating ducts. Sleeves and slots or other openings should be provided for cable and busway entrances. Those that are not in use should be sealed with pipe caps, plugs, or barriers. All openings with cables should be sealed with approved duct sealer or other materials. Fire stops should be provided in accordance with code requirements where busways or wiring troughs pass between floors or fire-rated walls. Sills or elevated sleeve openings may be used to prevent seepage of liquids around cables or busways. When facilities are used for other purposes, unqualified personnel who enter become exposed to an unfamiliar environment. This could result in an electrical injury or in accidental or malicious tampering with the electrical equipment.

7.6 References. The following references shall be used in conjunction with this chapter:

[1] ANSI C2-1990, National Electrical Safety Code.

[2] ANSI/IES RP7-1983, Practice for Industrial Lighting.

---

[59]ANSI publications are available from the Sales Department of the American National Standards Institute, 11 West 42nd Street, 13th Floor, New York, NY 10036. IES publications are available from the Illuminating Engineering Society, 345 East 47th Street, New York, NY 10017.

[3] ANSI/NFPA 70-1990, National Electrical Code.

[4] IEEE Std 142-1982, IEEE Recommended Practice for Grounding of Industrial and Commercial Power Systems (ANSI).

[5] IEEE Std 640-1985, IEEE Guide for Power Station Noise Control.

[6] IEEE Std 979-1984 (Reaff. 1988), IEEE Guide for Substation Fire Protection (ANSI).

[7] IEEE Std 980-1987, IEEE Guide for Substation Fire Protection (ANSI)

[8] IEEE Std 1119-1988, IEEE Guide for Fence Safety Clearances in Electric-Supply Stations (ANSI).

Chapter 8
Wiring Systems

8.1 Introduction. Wiring systems in commercial buildings use cable and busway systems. A typical building will have both, as each has advantages for particular applications. The first part of this chapter, 8.1 through 8.14, discusses cable; the latter part, 8.15 through 8.26, discusses busway.

8.2 Cable Systems. The primary function of cables is to carry energy reliably between source and utilization equipment. In carrying this energy, there are heat losses generated in the cable that should be dissipated. The ability to dissipate these losses depends on how the cables are installed, and this affects their ratings.

Cables may be installed in raceway, cable trays, underground in duct or direct buried, messenger supported, in cable bus, or as open runs of cable.

The selection of conductor size requires consideration of the load current to be carried and the loading cycle, emergency overloading requirements and duration, fault clearing time and interrupting capacity of the cable overcurrent protection or source capacity, voltage drop, ambient temperatures that may exist for 3 hours or more, circuit length through hot ambient temperature, and system frequency, e.g., 400 Hz, for the particular installation conditions. Caution should be exercised when locating conductors in high ambient heat areas so that the operating temperature will not exceed that designated for the type of insulated conductor involved.

Insulations can be classified in broad categories as solid insulations, taped insulations, and special-purpose insulations. Cables incorporating these insulations cover a range of maximum and normal operating temperatures and exhibit varying degrees of flexibility, fire resistance, and mechanical and environmental protection.

The installation of cables requires care in order to avoid excessive pulling tensions that could stretch the conductor or insulation shield, or rupture the cable jacket when pulled around bends. The minimum bending radius of the cable or conductors should not be exceeded during pulling around bends, at splices, and particularly at terminations to avoid damage to the conductors. The engineer should also check each run to ensure that the conductor jamming ratio is correct and the maximum allowable sidewall pressure is not exceeded.

Provisions should be made for the proper terminating, splicing, and grounding of cables. Minimum clearances should be maintained between phases and between phase and ground for the various voltage levels. The terminating compartments should be heated to prevent condensation from forming. Condensation or contamination on medium-voltage terminations could result in tracking over the terminal surface with possible flashover.

Many users test cables after installation and periodically test important circuits. Test voltages are usually dc of a level recommended by the cable manufacturer for the specific cable. Usually, this test level is well below the dc strength of the cable; but it is possible for accidental flashovers to weaken or rupture the cable insulation due to the higher transient overvoltages that can occur from reflections of the voltage wave. IEEE Std 400-1980 (Reaff. 1987), IEEE Guide for Making High-Direct-Voltage Tests on Power Cable Systems in the Field (ANSI) [18][60] provides a detailed discussion on cable testing.

The application and sizing of all cables rated up to 35 kV is governed by ANSI/ NFPA 70-1990, National Electrical Code (NEC) [4].[61] Cable use may also be covered under state and local regulations recognized by the local electrical inspection authority having jurisdiction in a particular area.

The various tables in this chapter are intended to assist the electrical engineer in laying out and understanding, in general terms, requirements for the cable system under consideration.

8.3 Cable Construction

8.3.1 Conductors. The two conductor materials in common use are copper and aluminum. Copper has historically been used for conductors of insulated cables due primarily to its desirable electrical and mechanical properties. The use of aluminum is based mainly on its favorable conductivity-to-weight ratio (the highest of the electrical conductor materials), its ready availability, and the lower cost of the primary metal.

The need for mechanical flexibility usually determines whether a solid or a stranded conductor is used, and the degree of flexibility is a function of the total number of strands. The NEC [4] requires conductors of No. 8 AWG and larger to be stranded. A single insulated or bare conductor is defined as a "conductor;" whereas an assembly of two or more insulated conductors, with or without an overall covering, is defined as a "cable."

Stranded conductors are available in various configurations, such as standard concentric, compressed, compact, rope, and bunched, with the latter two generally specified for flexible service. Bunch-stranded conductors (not shown in Fig 61) consist of a number of individual strand members of the same size that are twisted together to make the required area in circular mils for the intended service. Unlike

---

[60]The numbers in brackets correspond to those in the references at the end of each chapter. IEEE publications are available from the Institute of Electrical and Electronics Engineers, IEEE Service Center, 445 Hoes Lane, P.O. Box 1331, Piscataway, NJ 08855-1331. ANSI publications are available from the Sales Department of the American National Standards Institute, 11 West 42nd Street, 13th Floor, New York, NY 10036.

[61]ANSI publications are available from the Sales Department of the American National Standards Institute, 11 West 42nd Street, 13th Floor, New York, NY 10036. NFPA publications are available from Publications Sales, National Fire Protection Association, 1 Batterymarch Park, P.O. Box 9101, Quincy, MA 02269-9101.

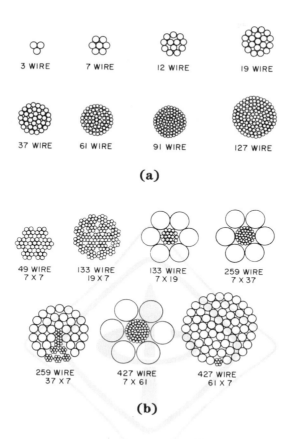

**Fig 61
Conductor Stranding
(a) Concentric Lay Strands
(b) Concentric Rope-Lay Strands**

the individual strands in a concentric-stranded conductor, the strands in a bunch-stranded conductor are not controlled with respect to one another, as shown in Fig 61. This type of conductor is usually found in portable cords.

8.3.2 Comparison Between Copper and Aluminum. Aluminum requires larger conductor sizes to carry the same current as copper. For equivalent ampacity, aluminum cable is lighter in weight and larger in diameter than copper cable. The properties of these two metals are given in Table 48.

The 36% difference in thermal coefficients of expansion and the different electrical natures of their oxide films require consideration in connector designs. An aluminum-oxide film forms immediately upon the exposure of the fresh aluminum surface to air. Under normal conditions, it slowly builds up to a thickness in the range of 3–6 nanometers and stabilizes at this thickness. The oxide film is essentially an insulating film or dielectric material and provides aluminum with its corro-

Table 48
Properties of Copper and Aluminum

	Copper Electrolytic	Aluminum EC Grade
Conductivity, % IACS* at 20 °C	100.0	61.0
Resistivity, $\Omega \cdot$ cmil/ft at 20 °C	10.371	17.002
Specific gravity at 20 °C	8.89	2.703
Melting point, °C	1083	660
Thermal conductivity at 20 °C, $(cal \cdot cm)/(cm^2 \cdot °C \cdot s)$†	0.941	0.58
Specific heat, $cal/(g \cdot °C)$†		
for equal weights	0.092	0.23
for equal direct-current resistance	0.184	0.23
Thermal expansion, in; equal to constant $\times 10^{-6} \times$ length in inches $\times$ °F	9.4	12.8
steel = 6.1		
18-8 stainless = 10.2		
brass = 10.5		
bronze = 15		
Relative weight for equal direct-current resistance and length	1.0	0.50
Modulus of elasticity, $(lb/in^2) \times 10^6$	16	10

*International annealed copper standard.
†In this table, cal denotes the gram calorie.

sion resistance. Copper produces its oxide rather slowly under normal conditions, and the film is relatively conducting, presenting no real problem at connections.

Approved connector designs for aluminum conductors provide increased contact areas and lower unit stresses than are used for copper cable connectors. These terminals possess adequate strength to ensure that the compression of the aluminum strands exceeds their yield strength and that a brushing action takes place that destroys the oxide film to form an intimate aluminum contact area yielding a low-resistance connection. Recently developed aluminum alloys provide improved terminating and handling as compared to electrical conductor (EC) grades.

Water should be kept from entering the strand space in aluminum conductors at all times. Any moisture within a conductor, either copper or aluminum, is likely to cause corrosion of the conductor metal or impair insulation effectiveness.

8.3.3 Insulation. Basic insulating materials are either organic or inorganic, and there are a wide variety of insulations classified as organic. Mineral insulated cable employs the one inorganic insulation, magnesium oxide (MgO), that is generally available.

Insulations in common use are:

(1) Thermosetting compounds, solid dielectric
(2) Thermoplastic compounds, solid dielectric

(3) Paper laminated tapes
(3) Varnished cloth, laminated tapes
(5) Mineral insulation, solid dielectric granular

Most of the basic materials listed in Table 49 should be modified by compounding or mixing them with other materials to produce desirable and necessary properties for manufacturing, handling, and end use. The thermosetting or rubber-like materials are mixed with curing agents, accelerators, fillers, and antioxidants in varying proportions. Crosslinked polyethylene (XLPE) is included in this class. Generally, smaller amounts of materials are added to the thermoplastics in the form of fillers, antioxidants, stabilizers, plasticizers, and pigments.

(1) Insulation Comparison — The aging factors of heat, moisture, and ozone are among the most destructive to organic-based insulations, so the following comparisons are a gauge of the resistance and classification of these insulations:

(a) Relative Heat Resistance — The comparison in Fig 62 illustrates the effect of a relatively short period of exposure at various temperatures on the hardness characteristic of the material at that temperature. The basic differences between thermoplastic and thermosetting insulation, excluding aging effect, are evident.

(b) Heat Aging — The effect on elongation of an insulation (or jacket) when subjected to aging in a circulating air oven is an acceptable measure of heat resistance. The air oven test at 121 °C, which is contained in some

Table 49
Commonly Used Insulating Materials

Common Name	Chemical Composition	Properties of Insulation	
		Electrical	Physical
Thermosetting			
Crosslinked polyethylene	Polyethylene	Excellent	Excellent
EPR	Ethylene propylene rubber (copolymer and terpolymer)	Excellent	Excellent
Butyl	Isobutylene isoprene	Excellent	Good
SBR	Styrene butadiene rubber	Excellent	Good
Oil base	Complex rubberlike compound	Excellent	Good
Silicone	Methyl chlorosilane	Good	Good
TFE*	Tetrafluoroethylene	Excellent	Good
ETFE†	Ethylene tetrafluorethylene	Excellent	Good
Neoprene	Chloroprene	Fair	Good
Class CP rubber‡	Chlorosulfonated polyethylene	Good	Good
Thermoplastic			
Polyethylene	Polyethylene	Excellent	Good
Polyvinyl chloride	Polyvinyl chloride	Good	Good
Nylon	Polyamide	Fair	Excellent

*For example, Teflon® or Halon®.
†For example, Tefzel®.
‡For example, Hypalon®.

Teflon, Halon, Tefzel, and Hypalon are registered trademarks of E.I. duPont de Nemours and Company, Inc.

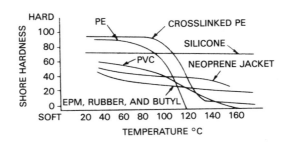

Fig 62
Typical Values for Hardness versus Temperature

specifications, is severe, but provides a relatively quick method of grading materials for possible use at elevated conductor temperatures or in hot-spot areas. The 150 °C oven aging is many times more severe and is used to compare materials with superior heat resistance. The temperature ratings of insulations in general use are shown in Table 50. Depending upon the operating conditions, the maximum shield temperature should also be considered (see ICEA P-45-482-1979, Short-Circuit Performance of Metallic Shields and Sheaths of Insulated Cable [9].)[62]

(c) Ozone and Corona Resistance—Exposure to accelerated conditions, such as higher concentrations of ozone (as standardized by NEMA WC5-1973 (Reaff. 1979 and 1985), Thermoplastic-Insulated Wire and Cable for the Transmission and Distribution of Electrical Energy (ICEA S-61-402, Third Edition [22])[63] for butyl, 0.03% ozone for 3 hours at room temperature), or air oven tests followed by exposure to ozone, or exposure to ozone at elevated temperatures, aid in measuring the ultimate ozone resistance of the material. Insulations exhibiting superior ozone resistance under accelerated conditions are silicone, rubber, polyethylene, crosslinked polyethylene (XLPE), ethylene propylene rubber (EPR), and polyvinyl chloride (PVC). In fact, these materials are, for all practical purposes, inert in the presence of ozone. However, that is not the case with corona discharge.

The phenomenon of corona discharge produces concentrated and destructive thermal effects along with the formation of ozone and other ionized gases. Although corona resistance is a property associated with cable over 600 V, in a properly designed and manufactured cable, damaging corona is not expected to be present at operating voltage. Materials exhibiting less susceptibility than polyethylene and XLPE to such discharge activity are the ethylene propylene rubbers (EPR).

---

[62] ICEA publications are available from the Insulated Cable Engineers Association, P.O. Box 440, South Yarmouth, MA 02664.

[63] NEMA publications are available from the National Electrical Manufacturers Association, 2101 L Street, N.W., Washington, DC 20037.

Table 50
Rated Conductor Temperature Ratings

Insulation Type	Maximum Voltage Class (kV)	Maximum Operating Temperature (°C)	Maximum Overload Temperature* (°C)	Maximum Short-Circuit Temperature (°C)
Paper (solid-type) multiconductor and single conductor, shielded	9	95	115	200
	29	90	110	200
	49	80	100	200
	69	65	80	200
Varnished cambric	5	85	100	200
	15	77	85	200
	28	70	72	200
SBR	2	75	95	200
Butyl rubber	5	90	105	200
	35	85	100	200
Oil-base rubber	35	70	85	200
Polyethylene (crosslinked)†	35	90	130	250
EPR†	35	90	130	250
Chlorosulfonated polyethylene	2	90	130	250
Polyvinyl chloride	2	60	85	150
	2	75	95	150
	2	90	105	150
Silicone rubber	5	125	150	250
Ethylene tetra fluoroethylene	2	150	200	250

*Operation at these overload temperatures shall not exceed 100 hours/year. Such 100 hour overload periods shall not exceed five.
†Cables are also available in 69 kV and higher ratings.

(d) Moisture Resistance — Insulations such as XLPE, polyethylene, and EPR, exhibit excellent resistance to moisture as measured by standard industry tests, such as the ICEA Accelerated Water Absorption Test — Electrical Method (EM-60) (see NEMA WC3-1980 (Reaff. 1986), Rubber-Insulated Wire and Cable for the Transmission and Distribution of Electrical Energy (ICEA S-19-81, Sixth Edition) [21], NEMA WC5-1973 (Reaff. 1979 and 1985), Thermoplastic-Insulated Wire and Cable for the Transmission and Distribution of Electrical Energy (ICEA S-61-402, Third Edition) [22], NEMA WC7-1988, Cross-Linked-Thermosetting-Polyethylene-Insulated Wire and Cable for the Transmission and Distribution of Electrical Energy [23], and NEMA WC8-1988, Ethylene-Propylene-Rubber-Insulated Wire and Cable for the Transmission and Distribution of Electrical Energy [24]). The electrical stability of these insulations in water as measured by capacitance and power factor is impressive. A degradation phenomenon called "treeing" has been found to be aggravated by the presence of water. This phenomenon appears to occur in solid dielectric insulations and is more prevalent in polyethylene and

XLPE than in EPR. The capacitance and power factor of natural polyethylene and some crosslinked polyethylenes are lower than those of
EPR or other elastomeric power cable insulations.

(2) Insulations in General Use — Insulations in general use for 2 kV and above
are shown in Table 50. Solid dielectrics, both thermoplastic and thermosetting, are used much more frequently, while the laminated constructions,
such as paper and lead cables, are declining in popularity in commercial
building service.

The generic names given for these insulations cover a broad spectrum of
actual materials, and the history of performance on any one type may not
properly be related to another in the same generic family.

8.3.4 Cable Design. The selection of power cable for particular circuits or feeders should be based on the following considerations:

(1) Electrical — Dictates conductor size, type, and thickness of insulation, correct
materials for low- and medium-voltage designs, consideration of dielectric
strength, insulation resistance, specific inductive capacitance (dielectric constant), and power factor.

(2) Thermal — Compatible with ambient and overload conditions, expansion,
and thermal resistance.

(3) Mechanical — Involves toughness and flexibility, consideration of jacketing
or armoring, and resistance to impact, crushing, abrasion, and moisture.

(4) Chemical — Stability of materials on exposure to oils, flame, ozone, sunlight,
acids, and alkalies.

(5) Flame Resistance — Cables installed in cable tray should be listed by a nationally recognized testing laboratory as being flame-retardant and marked for
installation in cable tray. The marking may be "Type TC, for use in cable
trays," or "for CT use," depending on the voltage and construction.

(6) Low Smoke — The NEC [4] authorized the addition of the suffix "LS" to the
cable marking on any cable construction that was flame-retardant and had
limited smoke characteristics. The criteria for "Limited Smoke" was being
developed at the time this recommended practice was published. While the
NEC [4] does not specifically require the use of LS constructions in any area,
this requirement might be considered for occupancies with large populations or high-rise occupancies.

(7) Toxicity — All electrical wire and cable installed or terminated in any building in the State of New York after December 16, 1987, should have the toxicity
level and certain other data for the product on file with the New York Secretary of State.

The installation of cable in conformance with the NEC [4] and state and local
codes under the jurisdiction of a local electrical inspection authority requires evidence of listing for use in the intended application and occupancy by a nationally
recognized testing laboratory, such as Underwriters Laboratories Inc. (UL). Some
of the more common commercial types listed in the NEC [4] are discussed below.

8.3.4.1 Low-Voltage Cables. Low-voltage power cables are generally rated at
600 V, regardless of the voltage used, whether 120 V, 208 V, 240 V, 277 V, 480 V, or
600 V.

The selection of 600 V power cable is oriented more toward physical rather than electrical service requirements. Resistance to forces, such as crush, impact, and abrasion, becomes a predominant factor, although good electrical properties for wet locations are also needed.

The 600 V compounds of crosslinked polyethylene (XLPE) are usually filled with carbon black or mineral fillers to further enhance the relatively good toughness of conventional polyethylene. The combination of crosslinking the polyethylene molecules through vulcanization plus fillers produces superior mechanical properties. Vulcanization eliminates polyethylene's main drawback of a relatively low melting point of 105 °C (121 °F). The 600 V construction consists of a copper or aluminum conductor with a single extrusion of insulation in the specified thickness.

Rubber-like insulations, such as ethylene propylene rubber (EPR) and styrene butadiene rubber (SBR), require outer jackets for mechanical protection, usually of polyvinyl chloride (PVC), neoprene, or CP rubber. However, the newer EPR insulations have improved physical properties that do not require an outer jacket for mechanical protection. A list of the more commonly used 600 V conductors and cables is provided below. Cables are classified by conductor operating temperatures and insulation thicknesses in accordance with the NEC [4].

(1) EPR or XLPE Insulated, With or Without a Jacket—Type RHW for 75 °C (167 °F) maximum operating temperature in wet or dry locations, Type RHH for 90 °C (194 °F) in dry locations only, and Type RHW-2 for 90 °C (194 °F) maximum operating temperature in wet and dry locations.

(2) XLPE or EPR Insulated, Without a Jacket —Type XHHW for 75 °C (167 °F) maximum operating temperature in wet locations and 90 °C (194 °F) in dry locations only, and Type XHHW-2 for 90 °C (194 °F) maximum operating temperature in wet and dry locations.

(3) PVC Insulated, Nylon Jacketed—Type THWN for 75 °C (167 °F) maximum operating temperature in wet or dry locations, and Type THHN for 90 °C (194 °F) in dry locations only.

(4) PVC Insulated, Without Jacket—Type THW for 75 °C (167 °F) maximum operating temperature in wet or dry locations.

The preceding conductors are suitable for installation in conduit, duct, or other raceway, and, when specifically approved for the purpose, may be installed in cable tray (1/0 AWG and larger) or direct buried, provided NEC [4] requirements are satisfied.

Cables in items (2) and (4) are usually restricted to conduit or duct. Single conductors may be furnished paralleled or multiplexed, as multiconductor cables with an overall nonmetallic jacket, or as aerial cable on a messenger.

(5) Metal Clad Cable, Type MC—Is a multiconductor cable employing either an interlocking tape armor or a continuous metallic sheath (corrugated or smooth), with or without an overall jacket. The maximum temperature rating of the cable is based upon the temperature rating of the individual insulated conductors used, which are usually Type XHHW, XHHW-2, RHH/ RHW, or RHW-2. Type MC cable may be installed in any raceway, in cable tray, as open runs of cable, direct buried, or as aerial cable on a messenger.

(6) Power and Control Tray Cable, Type TC — Is a multiconductor cable with an overall flame-retardant nonmetallic jacket. The individual conductors may be any of the above, and the cable has the same maximum temperature rating as the conductors used. Type TC may be installed in cable trays, raceways, or, where supported in outdoor locations, by a messenger wire.

Note that the temperatures listed are the maximum rated operating temperatures as specified in the NEC [4].

8.3.4.2 Power Limited Circuit Cables. When the power in the circuit is limited to the levels defined in the NEC, Article 725 [4] for remote control, signaling, and power limited circuits, then Class 2 (CL2) or Class 3 (CL3) power limited circuit cable or power limited tray cable (Type PLTC) may be utilized as the wiring method. These cables, which are rated 300 V, include both copper conductors for electrical circuits and thermocouple alloys for thermocouple extension wire.

Cables installed in ducts, plenums, and other spaces used for environmental air should be plenum cable Type CL2P or CL3P. Cables installed in vertical runs and penetrating more than one floor, or cables installed in vertical runs in a shaft should be riser cable Type CL2R or CL3R. Limited-use Type CL2X or CL3X cables may be installed in dwellings or in raceway in buildings. Cables installed in cable tray should be Type PLTC.

If the circuit is not Class 2 or Class 3 power limited, then 600 V branch-circuit conductors or cable should be used.

Similarly, power limited, fire protective signaling circuit cable may be used on circuits that comply with the power limitations in Article 760 of the NEC [4]. Type FPLP cable is required for plenums, Type FPLR cable for risers, and Type FPL cable for general-purpose fire alarm use. If the circuit is not power limited, then 600 V cables should be used. Type NPLFP cable is required for plenums, Type NPLFR cable for risers, and Type NPLF cable for general-purpose fire alarm use.

8.3.4.3 Medium-Voltage Cables. Type MV (medium-voltage) power cables have solid extruded dielectric insulation and are rated from 2001–35 000 V. These single- and multiconductor cables are available with nominal voltage ratings of 5 kV, 8 kV, 15 kV, 25 kV, and 35 kV. Solid dielectric 69 kV and 138 kV transmission cables are also available; however, they are not listed in the NEC [4].

EPR and XLPE are the usual insulating compounds for Type MV cables; however, polyethylene and butyl rubber are also available. The maximum operating temperatures are 90 °C (194 °F) for EPR and XLPE, 85 °C (185 °F) for butyl rubber, and 75 °C (167 °F) for polyethylene.

Type MV cables may be installed in raceways in wet or dry locations. The cable should be specifically listed for installation in cable tray, direct buried, exposure to sunlight, exposure to oils, or for messenger-supported wiring.

Multiconductor Type MV cables that also comply with the requirements for Type MC (metal-clad) cables may be labeled as Type MV or MC and may be installed as open runs of cable.

8.3.4.4 Shielding of Medium-Voltage Cable. For operating voltages below 2 kV, nonshielded constructions are normally used. Above 2 kV, cables are required to be shielded to comply with the NEC [4] and ICEA Standards. The NEC [4] does permit the use of nonshielded cables up to 8 kV provided the conductors are listed by a nationally recognized testing laboratory and are approved for the purpose. Where nonshielded conductors are used in wet locations, the insulated conductor(s) should have an overall nonmetallic jacket or a continuous metallic sheath, or both. Refer to the NEC [4] for specific insulation thicknesses for wet and dry locations.

Since shielded cable is usually more expensive than nonshielded cable, and the more complex terminations require a larger terminal box, nonshielded cable has been used extensively at 2400 V and 4160 V and, occasionally, at 7200 V. However, any of the following conditions may dictate the use of shielded cable:

(1) Personnel safety
(2) Single conductors in wet locations
(3) Direct earth burial
(4) Where the cable surface may collect unusual amounts of conducting materials (e.g., salt, soot, conductive pulling compounds)

Shielding of an electric power cable is defined as the "practice of confining the electric field of the cable to the insulation surrounding the conductor by means of conducting or semiconducting layers, or both, which are in intimate contact or bonded to the inner and outer surfaces of the insulation." In other words, the outer insulation shield confines the electric field to the space between the conductor and the shield. The inner or strand stress relief layer is at or near the conductor potential. The outer or insulation shield is designed to carry the charging currents and, in many cases, fault currents. The conductivity of the shield is determined by its cross sectional area and the resistivity of the metal tapes or wires employed in conjunction with the semiconducting layer.

The metallic shield, which is available in several forms, is an electrostatic shield and is not designed to carry fault currents. The most common is the tape shield consisting of a copper tape, 3–5 mils thick, which is helically applied over the insulation shield.

A modification of the tape shield consists of a corrugated copper tape applied longitudinally over the insulation shield. This permits full electrical use of the tape as a current-carrying conductor, and it is capable of carrying a much greater fault current than a helically wrapped tape.

Another type is a wire shield, where copper wires are helically applied over the insulation screen with a long lay. Typically, a wire shield will have 15%–20% less cross sectional area than a tape shield.

A modification of the wire shielding system consists of six corrugated copper drain wires embedded in an extruded black conducting chlorinated polyethylene (CPE) combination insulation shield and jacket.

An extruded lead sheath may also be used as a combination shield and mechanical covering. The thickness of the lead can be varied to provide the desired cross sectional area to carry the required fault current. The lead also provides an excellent moisture barrier for direct burial applications.

The stress control layer at the inner and outer insulation surfaces, by its close bonding to the insulation surface, presents a smooth surface to reduce stress concentrations and to minimize void formation. Ionization of the air in such voids can progressively damage insulating materials and eventually cause failure.

Insulation shields have several purposes

(1) To confine the electric field within the cable.

(2) To equalize voltage stress within the insulation, minimizing surface discharges.

(3) To protect cable from induced potentials.

(4) To limit electromagnetic or electrostatic interference to communication receivers, e.g., radio, television.

(5) To reduce shock hazard (when properly grounded).

Figure 63 illustrates the electrostatic field of a shielded cable.

The voltage distribution between a nonshielded cable and a grounded plane is illustrated in Fig 64. Here, it is assumed that the air is the same, electrically, as the insulation, so that the cable is in a uniform dielectric above the ground plane to permit a simpler illustration of the voltage distribution and field associated with the cable.

In a shielded cable (see Fig 63), the equipotential surfaces are concentric cylinders between conductor and shield. The voltage distribution follows a simple logarithmic variation, and the electrostatic field is confined entirely within the insulation. The lines of force and stress are uniform and radial, and cross the equipotential surfaces at right angles, which eliminates any tangential or longitudinal stresses within the insulation or on its surface.

The equipotential surfaces for the nonshielded system (see Fig 65) are cylindrical but not concentric with the conductor, and cross the cable surface at many different potentials. The tangential creepage stress to ground at points along the cable may be several times the normal recommended stress for creepage distance at terminations in dry locations for nonshielded cable operating on 4160 V systems.

Fig 63
Electrical Field of Shielded Cable

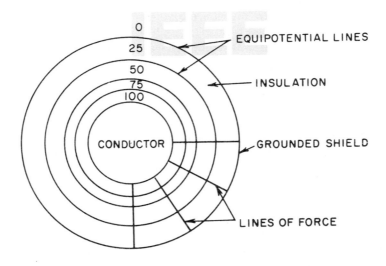

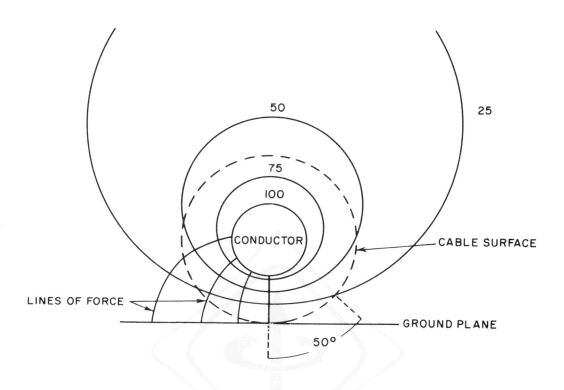

Fig 64
Electrical Field of Conductor on Ground Plane in Uniform Dielectric

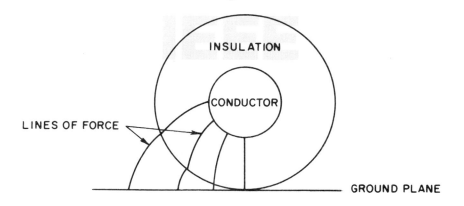

Fig 65
Unshielded Cable on Ground Plane

Surface tracking, burning, and destructive discharges to ground could occur under these conditions. However, properly designed nonshielded cables as described in the NEC [4] limit the surface energies available, which could protect the cable from these effects.

Typical cables supplied for shielded and nonshielded applications are illustrated in Fig 66.

8.4 Cable Outer Finishes. Cable outer finishes or outer coverings are used to protect the underlying cable components from the environmental and installation conditions associated with the intended service. The choice of a cable outer finish for a particular application is based on the same performance criteria as used for insulations, namely electrical, thermal, mechanical, and chemical. A combination of metallic and nonmetallic coverings are available to provide the total protection needed for the particular installation and operating conditions. Specific industry requirements for these coverings are defined in IEEE, UL, ICEA, and ASTM Standards.

8.4.1 Nonmetallic Finishes

(1) Extruded Jackets — There are outer coverings, either thermoplastic or vulcanized, which may be extruded directly over the insulation, or over electrical shielding systems of metal sheaths or tapes, copper braid, or semi-conducting layers with copper drain wires or spiraled copper concentric wires; or over multiconductor constructions. Commonly used materials include polyvinyl chloride (PVC), chlorinated polyethylene (CPE), nitrile butadiene/polyvinyl chloride (NBR/PVC), crosslinked polyethylene (XLPE), polychloroprene (neoprene), and chlorosulfonated polyethylene (hypalon). While the detailed characteristics may vary due to individual manufacturers' compounding, these materials provide a high degree of moisture, chemical, and weathering protection, are reasonably flexible, provide some degree of electrical isolation, and are of sufficient mechanical strength to protect the insulating and shielding components from normal service and installation damage. Materials are available for service temperatures from −55 °C (−67 °F) to +115 °C (239 °F).

(2) Fiber Braids — This category includes braided, wrapped, or served synthetic or natural fiber materials selected by the cable manufacturer to best meet the intended service. While asbestos fiber has been the most common material used in the past, fiberglass is now used extensively for employee health reasons. Some special industrial applications may require synthetic or cotton fibers applied in braid form. All fiber braids require saturants or coating and impregnating materials to provide some degree of moisture and solvent resistance as well as abrasion and weather resistance.

Glass braid is used on cable to minimize flame propagation, smoking, and other hazardous or damaging products of combustion.

8.4.2 Metallic Finishes. This category of materials is widely used where a high degree of mechanical, chemical, or short-time thermal protection of the underlying cable components is required by the application. Commonly used materials are

(a) Single-Conductor Cable (600 V or 5 kV Nonshielded)

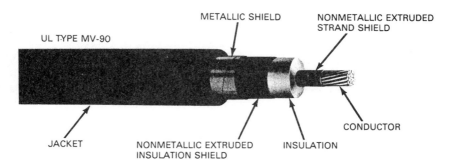

(b) Type MV Medium-Voltage Single Conductor Cable (5–35 kV)

(c) Type TC Power and Control Tray Cable (600 V)

(d) Type MC Metal-Clad Power and Control Cable (600 V–35 kV)

(e) Messenger-Supported Wiring (600 V–35 kV)

Fig 66
Commonly Used Shielded and Nonshielded Constructions

interlocked galvanized steel, aluminum, or bronze armor; extruded lead or aluminum; longitudinally applied, welded, and corrugated aluminum or copper sheath; and helically applied round or flat armor wires. The use of any of these materials, alone or in combination with others, does reduce the flexibility of the overall cable.

Installation and operating conditions may involve localized compressive loadings, occasional impact from external sources, vibration and possible abrasion, heat shock from external sources, extended exposure to corrosive chemicals, and condensation.

(1) Interlocked Armor—Provides mechanical protection with minimum reduction in flexibility. While not entirely impervious to moisture or corrosive agents, interlocked armor does provide mechanical protection against impact and abrasion and protection from thermal shock by acting as a heat sink for short periods of localized exposure.

When moisture protection is required, an inner jacket over the cable core and under the armor is required. If an inner jacket is not used, 600 V cable in wet locations can only be rated for 75 °C (167 °F) unless the newer Types RHW-2 or XHHW-2 conductors are used; in which case, the cable can then be rated 90 °C (194 °F) wet or dry.

When corrosion resistance is required, for environmental conditions, direct burial, or embedment in concrete, an overall jacket is required.

The use of interlocked galvanized steel armor should be avoided on single-conductor ac power circuits due to high hysteresis and eddy current losses. This effect, however, is minimized by using three-conductor cables with overall armor or with aluminum armor on single-conductor cables.

Commonly used interlocked armor materials are galvanized steel, aluminum (for less weight and corrosion resistance), marine bronze, and other alloys for highly corrosive atmospheres.

(2) Corrugated Metal Sheath—Longitudinally welded and corrugated metallic sheaths (corrugations formed perpendicular to the cable axis) have been used for many years in direct buried communication cables, but only since 1960 has this method of cable core protection been applied to power and control cable. The sheath material may be of aluminum, copper, copper alloy, or a bimetallic composition with the choice of material selected to best meet the intended service.

The corrugated metal sheath offers mechanical protection equal to or greater than interlocked armor but at a lower weight. The aluminum or copper sheath may also be used as the equipment grounding conductor, either alone or in parallel with a grounding conductor within the cable.

The sheath is made from a metal strip that is longitudinally formed around the cable, welded into a continuous, impervious metal cylinder, and corrugated for pliability and increased radial strength. This sheath offers maximum protection from moisture and liquid or gaseous contaminants. An extruded nonmetallic jacket should be used over the metal sheath for direct burial, embedment in concrete, or in areas that are corrosive to the metal sheath. This cable construction is always rated 90 °C (194 °F) in wet or dry locations.

(3) Lead—Pure or alloy lead is occasionally used in power cable sheaths for moisture protection in underground manholes and tunnels, or underground duct distribution systems subject to flooding. While not as resistant to crushing loads as interlocked armor or a corrugated metal sheath, its very high degree of corrosion and moisture resistance makes lead attractive in these applications. Protection from installation damage can be provided by an outer jacket of extruded material.

Pure lead is subject to work hardening and should not be used in applications where flexing may be involved. Copper- or antimony-bearing lead alloys are not as susceptible to work hardening as pure lead, and may be used in applications involving limited flexing. Lead or its alloys should *never* be used for repeated flexing service.

One problem encountered with the use of lead sheathed cable is in the area of splicing and terminating. Installation personnel experienced in the art of wiping lead sheath joints are not as numerous as they were many years ago, which poses an installation problem for many potential users. However, many insulation systems do not require lead sleeves at splices and treat the lead like any other metallic sheath.

(4) Aluminum or Copper—Extruded aluminum or copper, or die-drawn aluminum or copper sheaths are used in certain applications for weight reduction and moisture penetration protection. While more crush-resistant than lead, aluminum sheaths are subject to electrolytic attack when installed underground. Under these conditions, aluminum sheathed cable should be protected with an extruded outer jacket.

Mechanical splicing sleeves are available for use with aluminum sheathed cables, and sheath joints can be made by inert gas welding, provided that the underlying components can withstand the heat of welding without deterioration. Specifically designed hardware is available for terminating the sheath at junction boxes and enclosures.

(5) Wire Armor—Significant mechanical protection and particularly longitudinal strength can be obtained with the use of spirally wrapped or braided round steel armor wire. This type of outer covering is frequently used in submarine cable and vertical riser cable for mechanical protection and support. As noted for steel interlocked armor, this form of protection should be used only on three-conductor power cables to minimize sheath losses.

8.4.3 Single-Conductor and Multiconductor Constructions. Single-conductor cables are usually easier to handle and can be furnished in longer lengths as compared to multiconductor cables. The multiconductor constructions have smaller overall dimensions than the same number of single-conductor cables, which can be an advantage when space is important.

Sometimes, the outer finish can influence whether the cable should be supplied as a single-conductor or multiconductor cable. For example, as mentioned previously, the use of steel interlocked or steel wire armor on ac cables is practical on multiconductor constructions, but should be avoided on single-conductor cables. It is also more economical to apply a metallic sheath or armor over multiconductor constructions rather than over each of the single-conductor cables.

8.4.4 Physical Properties of Materials for Outer Coverings. Depending on the environment and application, the selection of outer finishes to provide the degree of protection needed can be complex. For a general appraisal, Table 51 lists the relative properties of some commonly used materials.

8.5 Cable Ratings

8.5.1 Voltage Rating. The selection of the cable insulation (voltage) rating is made on the basis of the phase-to-phase voltage of the system in which the cable is to be applied, whether the system is grounded or ungrounded, and the time in which a ground fault on the system is cleared by protective equipment. It is possible to operate cables on ungrounded systems for long periods of time with one phase grounded due to a fault. This results in line-to-line voltage stress across the insulation of the two ungrounded conductors. Therefore, such a cable should have greater insulation thickness than a cable used on a grounded system where it is impossible to impose full line-to-line potential on the other two unfaulted phases for an extended period of time.

Therefore, 100% insulation level cables are applicable to grounded systems provided that the protection devices will clear ground faults within 1 minute. On ungrounded systems where the clearing time of the 100% level category cannot be met, and yet there is adequate assurance that the faulted section will be cleared within 1 hour, 133% insulation level cables are required. On systems where the time required to de-energize a grounded section is indefinite, a 173% insulation level is used.

8.5.2 Conductor Selection. The selection of conductor size is based on the following considerations:

(1) Load current criteria as related to loadings, NEC [4] requirements, thermal effects of the load current, mutual heating, losses produced by magnetic induction, and dielectric losses

(2) Emergency overload criteria

Table 51
Properties of Jackets and Braids

Material	Abrasion Resistance	Flexibility	Low Temperature	Heat Resistance	Fire Resistance
Neoprene	Good	Good	Good	Good	Good
CP rubber	Good	Good	Fair	Excellent	Good
Crosslinked polyethylene	Good	Poor	Poor	Excellent	Poor
Polyvinyl chloride	Fair	Good	Fair	Good	Fair
Polyurethane	Excellent	Good	Good	Good	Poor
Glass braid	Fair	Good	Good	Excellent	Excellent
Nylon	Excellent	Fair	Good	Good	Fair
ETFE	Excellent	Poor	Excellent	Good	Fair

NOTE: Chemical resistance and barrier properties depend on the particular chemicals involved, and any questions should be referred to the cable manufacturer.

(3) Voltage-drop limitations

(4) Fault current criteria

(5) Frequency criteria

(6) Hot-spot temperature criteria

(7) Length of cable in elevated ambient temperature areas

(8) Equipment termination requirements

8.5.3 Load Current Criteria. The ampacity tables in the NEC [4] for low- and medium-voltage cables should be used where the NEC [4] applies. These tables are derived from IEEE S-135, Power Cable Ampacities (IPCEA) [13].

All ampacity tables show the minimum conductor size required; but conservative engineering practice, future load growth considerations, voltage drop, and short-circuit heating may make the use of larger conductors necessary.

Large groups of cables should be carefully considered, as de-rating due to mutual heating may be limiting. Conductor sizes over 500 kcmil require the consideration of paralleling two or more smaller size cables because the current-carrying capacity per circular mil of conductor decreases for ac circuits due to the skin and proximity effects. The reduced ratio of surface to cross sectional area of the larger conductors is a factor in the reduced ability of the larger conductor to dissipate heat. When multiple cables are used, consideration should be given to phase placement of the cables to minimize the effects of the uneven distribution of current in the cables, which will also reduce ampacity. Although the material cost of cables may be less for two smaller conductors, this cost saving may be offset by increased installation costs.

The use of load factor in underground runs takes into account the heat capacity of the duct bank and surrounding soil that responds to average heat losses. The temperatures in the underground section will follow the average loss, thus permitting higher short period loadings. The load factor is the ratio of average load to peak load. The average load is usually measured on a daily basis; the peak load is the average of a 30 minute to 1 hour period of the maximum loading that occurs in 24 hours.

For direct buried cables, the average cable surface temperature is limited to 60 °C (140 °F) or 70 °C (158 °F), depending on soil conditions, to prevent moisture migration and thermal runaway.

Cables should be de-rated when in proximity to other loaded cables or heat sources, or when the ambient temperature exceeds the ambient temperature on which the ampacity (current-carrying capacity) tables are based.

The normal ambient temperature of a cable installation is the temperature the cable would assume at the installed location with no load being carried on the cable. A thorough understanding of this temperature is required for a proper determination of the cable size required for a given load. For example, the ambient temperature for a cable exposed in the air and isolated from other cables is the temperature of that cable before load is applied, assuming, of course, that this temperature is measured at the same time of day and with all other conditions exactly the same as they will be when the required load is actually being carried. It is also assumed that, for cables in air, the space around the cable is large enough so that the heat generated by the cable can be dissipated without raising the temperature of the room as a whole. Unless exact conditions are specified, the following ambients are commonly used in the calculation of current-carrying capacity:

(1) Indoors — The ampacity tables in the NEC [4] are based upon an ambient temperature of 30 °C (86 °F) for low-voltage cables. In most parts of the United States, 30 °C (86 °F) is too low for summer months, at least for some parts of a building. The NEC [4] Type MV cable ampacity tables are based upon a 40 °C (104 °F) ambient air temperature. In any installation, where the conditions are accurately known, the measured temperature should be used; otherwise, use 40 °C (104 °F). Refer to the NEC, Article 318 [4] for cables installed in cable tray.

Sources of heat adjacent to the cables under the most adverse conditions should be taken into consideration when calculating current-carrying capacity. This is usually done by correcting the ambient temperature for localized hot spots. These may be caused by steam lines or other heat sources that are adjacent to the cable, or they may be due to sections of the cable running through boiler rooms or other hot locations. Rerouting may be necessary to avoid this problem.

(2) Outdoors — An ambient temperature of 40 °C (104 °F) is commonly used as the maximum for cables installed in the shade and 50 °C (122 °F) for cables installed in the sun. In using these ambient temperatures, it is assumed that the maximum load occurs during the time when the ambient temperature will be as specified. Some circuits probably do not carry their full load during the hottest part of the day or when the sun is at its brightest, so that an ambient temperature of 40 °C (104 °F) for outdoor cables is probably reasonably safe for certain selected circuits, otherwise, use 50 °C (122 °F). Refer to the NEC, Article 310 [4] ampacity tables and associated notes for the calculations to be used for outdoor installations and Article 318 for cables installed in cable tray.

(3) Underground — The ambient temperature used for underground cables varies in different sections of the country. For the northern sections, an ambient temperature of 20 °C (68 °F) is commonly used. For the central part of the country, 25 °C (77 °F) is commonly used, while, for the extreme south and southwest, an ambient temperature of 30 °C (86 °F) may be necessary. The exact geological boundaries for these ambient temperatures cannot be defined, and the maximum ambient temperature should be measured in the earth at a point away from any sources of heat at the depth at which the cable will be buried. Changes in the earth-ambient temperature will lag changes in the air-ambient temperature by several weeks.

The thermal characteristics of the medium surrounding the cable are of primary importance in determining the current-carrying capacity of the cable. The type of soil in which the cable or duct bank is buried has a major effect on the current-carrying capacity of cables. Porous soils, such as gravel and cinder fill, usually result in a temperature increase and lower ampacities than normal sandy or clay soil. The type of soil and its thermal resistivity should be known before the size of the conductor is calculated.

The moisture content of the soil has a major effect on the current-carrying capacity of cables. In dry sections of the country, cables may have to be de-rated or other precautions taken to compensate for the increase in thermal resistance that is due to the lack of moisture. On the other hand, in ground that is continuously wet or under tidewater conditions, cables may safely carry higher than normal currents. Shielding is necessary for even 2400 V circuits in continuously wet or alternately wet and dry conditions. When the cable passes from a dry area to a wet area, which provides natural shielding, there will be an abrupt voltage gradient stress, just as at the end of shielded cables terminated without a stress cone. Nonshielded cables specifically designed for this service are available.

Alternate wet and dry conditions have also been found to accelerate the progress of water treeing in solid dielectric insulations.

Ampacities in the NEC tables take into account the grouping of adjacent circuits. For ambient temperatures different from those specified in the tables, more than three conductors in a cable or raceway, or other installation conditions, the de-rating factors to be applied are contained in Tables 310-16 through 310-19, "Notes to Ampacity Tables of 0 to 2000 V," and "Notes to Tables 310-69 through 310-84."

8.5.4 Emergency Overload Criteria. The normal loading limits of insulated wire and cable are based on many years of practical experience and represent a rate of deterioration that results in the most economical and useful life of such cable systems. The rate of deterioration is expected to result in a useful life of 20–30 years. The life of cable insulation is about halved, and the average rate of thermally caused service failures about doubled for each 5 °C–15 °C (41 °F–59 °F) increase in normal daily load temperature. Additionally, sustained operation over and above maximum rated operating temperature or ampacities is not a very effective or economical expedient because the temperature rise is directly proportional to the conductor loss, which increases as the square of the current. The greater voltage drop might also increase the risks to equipment and service continuity.

As a practical guide, the ICEA has established maximum emergency overload temperatures for various insulations. Operation at these emergency overload temperatures should not exceed 100 hours/year, and such 100 hour overload periods should not exceed five during the life of the cable. Table 52 provides uprating factors for short-time overloads for various types of insulated cables. The uprating factor, when multiplied by the nominal current rating for the cable in a particular installation, will give the emergency or overload current rating for the particular insulation.

A more detailed discussion on emergency overload and cable protection is contained in IEEE Std 242-1986, IEEE Recommended Practice for Protection and Coordination of Industrial and Commercial Power Systems (ANSI), Chapter 11 [16].

8.5.5 Voltage-Drop Criteria. The supply conductor, if not of sufficient size, will cause excessive voltage drop in the circuit, and the drop will be in direct proportion to the circuit length. Proper starting and running of motors, lighting equipment, and other loads that have heavy inrush currents should be considered. The NEC [4] recommends that the steady-state voltage drop in power, heating, or lighting feeders be no more than 3%, and the total drop, including feeders and branch circuits, be no more than 5% overall.

Table 52
Uprating for Short-Time Overloads*

Insulation Type	Voltage Class (kV)	Conductor Operating Temperature (°C)	Conductor Overload Temperature (°C)	Uprating Factors for Ambient Temperature							
				20 °C		30 °C		40 °C		50 °C	
				Cu	Al	Cu	Al	Cu	Al	Cu	Al
Paper (solid type)	9	95	115	1.09	1.09	1.11	1.11	1.13	1.13	1.17	1.17
	29	90	110	1.10	1.10	1.12	1.12	1.15	1.15	1.19	1.19
	49	80	100	1.12	1.12	1.15	1.15	1.19	1.19	1.25	1.25
	69	65	80	1.13	1.13	1.17	1.17	1.23	1.23	1.38	1.38
Varnished cambric	5	85	100	1.09	1.08	1.10	1.10	1.13	1.13	1.17	1.17
	15	77	85	1.05	1.05	1.07	1.07	1.09	1.09	1.13	1.13
	28	70	72								
SBR	0.6	75	95	1.13	1.13	1.17	1.17	1.22	1.22	1.30	1.30
	5	90	105	1.08	1.08	1.09	1.09	1.11	1.11	1.14	1.14
Butyl RHH	15	85	100	1.09	1.08	1.10	1.10	1.13	1.13	1.17	1.17
	35	80	95	1.09	1.09	1.11	1.11	1.14	1.14	1.20	1.20
Oil-base rubber	35	70	85	1.11	1.11	1.14	1.14	1.20	1.20	1.29	1.29
Polyethylene (crosslinked)	35	90	130	1.18	1.18	1.22	1.22	1.26	1.26	1.33	1.33
Silicone rubber	5	125	150	1.08	1.08	1.09	1.09	1.10	1.10	1.12	1.11
EPR	35	90	130	1.18	1.18	1.22	1.22	1.26	1.26	1.33	1.33
Chlorosulfonated polyethylene	0.6	75	95	1.13	1.13	1.17	1.17	1.22	1.22	1.30	1.30
Polyvinyl chloride	0.6	60	85	1.22	1.22	1.30	1.30	1.44	1.44	1.80	1.79
	0.6	75	95	1.13	1.13	1.17	1.17	1.22	1.22	1.30	1.30

*To be applied to normal rating determined for such installation conditions.

8.5.6 Fault Current Criteria. Under short-circuit conditions, the temperature of the conductor rises rapidly. Then, depending upon the thermal characteristics of the insulation, sheath, surrounding materials, etc., it cools off slowly after the short-circuit condition is removed. For each insulation, the ICEA recommends a transient temperature limit for short-circuit duration times not in excess of 10 seconds.

Failure to check the conductor size for short-circuit heating could result in permanent damage to the cable insulation due to disintegration of the insulation material, which may be accompanied by smoke and the generation of combustible vapors. These vapors will, if sufficiently heated, ignite and possibly start a fire. Less seriously, the insulation or sheath of the cable may be expanded to produce voids, leading to subsequent failure. This becomes especially important in cables rated at 5 kV and higher.

In addition to the thermal stresses, mechanical stresses are set up in the cable through expansion when heated. As the heating is usually very rapid, these stresses may result in undesirable cable movement. However, on modern cables, reinforcing binders and sheaths considerably reduce the effect of such stresses. Within the range of temperatures expected with coordinated selection and application, the mechanical aspects can normally be discounted except with very old or lead sheathed cables.

During short-circuit or heavy pulsing currents, single-conductor cables will be subjected to forces that tend to either attract or repel the individual conductors with respect to each other. Therefore, cables installed in cable trays, racks, switchgear, motor control centers, or switchboard cable compartments should be secured to prevent damage caused by such movements.

The minimum conductor size requirements for various rms short-circuit currents and clearing times are shown in Table 53. The initial and final conductor temperatures from ICEA P-32-382-1989, Short-Circuit Characteristics of Insulated Cable [8] are shown for the various insulations. Table 50 provides conductor temperatures (maximum operating, maximum overload, and maximum short-circuit current) for various insulated cables.

The shield can be damaged if exposed to excessive fault currents. ICEA P-45-482-1979 [9] recommends that the ground-fault current not exceed 2000 A for one-half of a second. Some lighter duty shield constructions may have a lower current limit; check with the cable manufacturer. To limit ground-fault shield conductor exposure, the recommended practice is to utilize current-limiting overcurrent protective devices or employ low-resistance grounded supply systems for a maximum ground-fault current of 400–2000 A with suitably sensitive relaying. Without such limiting, it is likely that the occurrence of a ground fault could require the replacement of substantial lengths of cable. Grounding of the shield at all splice and termination points will direct fault currents into multiple paths and reduce shield damage. A more detailed discussion of fault current and cable protection is contained in IEEE Std 242-1986 (ANSI) [16].

8.5.7 Frequency Criteria. In general, three-phase, 400 Hz power systems are designed in the same way as 60 Hz systems; however, the specifier should be aware that the higher frequency will increase the skin and proximity effects on the conductors, thereby increasing the effective copper resistance. For a given current, this

Table 53
Minimum Conductor Sizes, in AWG or kcmil, for Indicated Fault Current and Clearing Times

Total RMS Current (amperes)	Polyethylene and Polyvinyl Chloride, 75—150 °C				Oil Base and SBR, 75—200 °C				XLPE and EPR, 90—250 °C			
	1/2 Cycle (0.0083 s)		10 Cycles (0.166 s)		1/2 Cycle (0.0083 s)		10 Cycles (0.166 s)		1/2 Cycle (0.0083 s)		10 Cycles (0.166 s)	
	Cu	Al	Cu	Al	Cu	Al	Cu	Al	Cu	Al	Cu	Al
5000	10	8	4	2	10	8	4	3	12	10	4	3
15 000	6	4	2/0	4/0	6	4	1/0	3/0	6	4	1	3/0
25 000	3	2	4/0	350	4	2	3/0	250	4	3	3/0	250
50 000	1/0	2/0	400	700	1	2/0	350	500	2	1/0	300	500
75 000	2/0	4/0	600	1000	1/0	3/0	500	750	1/0	3/0	500	700
100 000	4/0	300	800	1250	3/0	250	700	1000	2/0	4/0	600	1000

increase in resistance results in increased heating and may require a larger conductor. The higher frequency will also increase the reactance, and this, combined with the increased resistance, will increase the voltage drop. The higher frequency will also increase the effect of magnetic materials upon cable reactance and heating. For this reason, the cables should not be installed in steel or magnetic conduit, steel wireway, or run along magnetic structural members within the building.

The curves in Fig 67 show the ac/dc resistance ratio that exists on a 400 Hz system and the resulting reduction in current rating that is necessary from a heating standpoint to counteract the effect of the increased frequency.

The reactance can be taken as directly proportional to the frequency without introducing any appreciable errors. This method of determining reactance does not take into account the reduction due to proximity effect; but this change is not large and the error introduced by neglecting it is small.

The curves are applicable to any 600 V cable in the same nonmagnetic conduit, or to any Type MC cable with an aluminum or bronze sheath or interlocking armor.

When voltage drop is the limiting factor, then paralleling smaller conductors should be considered.

8.5.8 Elevated Ambient Temperature. The ambient temperature of the area where cables are installed should be considered in determining the allowable ampacity of the circuit.

Cables and insulated conductors rated 2000 V or less that are installed in areas where the ambient temperature is higher than that permitted in the NEC, Tables 310-16 through 310-19 [4] should have the allowable ampacity reduced by the ampacity correction factors listed in the appropriate table.

The ampacity of cables and insulated conductors rated over 2000 V that are installed in areas where the ambient temperature is either higher or lower than the temperatures specified in the NEC, Tables 310-69 through 310-84 [4] may be determined by using the formula contained in Note 1 of "Notes to Tables 310-69 through 310-84."

8.5.9 Hot-Spot Temperature Criteria. The allowable ampacity of a cable or insulated conductor should be reduced whenever more than 6 feet of the run is in a higher ambient temperature area. Refer to 8.5.8 for the applicable correction factors.

8.5.10 Termination Criteria. Equipment termination requirements should be considered, e.g., the manufacturer of a circuit breaker may specify a minimum conductor size for a particular breaker rating. Also, on 600 V terminations, the rating of the termination may require the cable to be operated at a lower temperature, such as 60 °C (140 °F) or 75 °C (167 °F).

8.6 Installation. There are a variety of ways to install power distribution cables in commercial installations. The engineer's responsibility is to select the method most suitable for each particular application. Each method has characteristics that make it more suitable for certain conditions than others; that is, each method will transmit power with a unique combination of reliability, safety, economy, and quality for a specific set of conditions. These conditions include the quantity and characteristics of the power being transmitted, the distance of transmission, and the degree of exposure to adverse mechanical and environmental conditions.

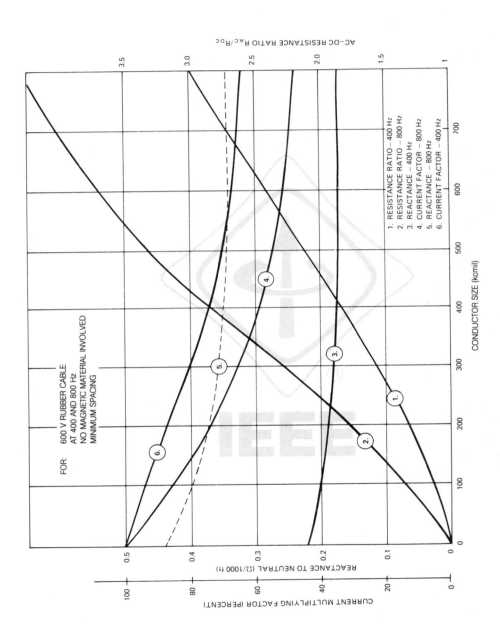

**Fig 67
Curves Showing the AC/DC Resistance Ratio That Exists on a 400 Hz System**

340

8.6.1 Layout. The first consideration in wiring systems layout is to keep the distance between the source and the load as short as possible. This consideration should be tempered by many other important factors in order to arrive at the lowest cost system that will operate within the reliability, safety, economy, and performance factors that are required. Some other factors that should be considered for various routings are the cost of additional cable and raceway versus the cost of additional supports, inherent mechanical protection provided in one alternative versus additional protection required in another, clearance for and from other facilities, and the need for future revision.

8.6.2 Open Wire. This method was used extensively in the past. Although it has now been replaced in most applications, it is still quite often used for primary power distribution over large areas where conditions are suitable.

Open-wire construction consists of single conductors on insulators that are mounted on poles or structures. The conductors may be bare or have a covering or jacket for protection against corrosion or abrasion.

The attractive features of this method are its low initial cost and the fact that damage can be detected and repaired quickly. On the other hand, noninsulated conductors are a safety hazard and are also very susceptible to mechanical damage and electrical outage from birds, animals, lightning, etc. There is an increased safety hazard where crane or boom truck use may be involved. In some areas, contamination on insulators and conductor corrosion can result in increased maintenance costs. Also, electrical leakage currents have been known to cause fires on wood crossarms and poles.

Due to the large conductor spacing, open-wire circuits have a higher reactance than circuits with more closely spaced conductors, producing a larger voltage drop. This problem is reduced by operating at a higher voltage and power factor.

Exposed open-wire circuits are more susceptible to outages from lightning than other installation methods. This problem may be minimized through the use of overhead ground wires, surge arresters, or special insulators.

8.6.3 Aerial Cable. Aerial cable is usually used for incoming or service distribution between commercial buildings. As a logical replacement for open wiring, it provides greater safety and reliability and requires less space. Properly protected cables are not a safety hazard and are not easily damaged by casual contact. They are, however, open to the same objections as open wire in regard to vertical and horizontal clearances. Aerial cables are frequently used in place of more expensive conduit systems, where the mechanical protection of the conduit is not required. They are also generally more economical for long runs of one or two cables than are cable tray installations. It is cautioned that aerial cable having a portion of the run in conduit should be de-rated to the ampacity in conduit for this condition.

Aerial cables may be either self-supporting or messenger-supported. They may be attached to pole lines or structures. Self-supporting aerial cables have high tensile strength conductors for this application.

Multiple single conductors, Types MV, RHH and RHW without outer braids, and THW; or multiconductor cables, Types MI, MC, SE, UF, TC, MV, or other factory-assembled multiconductor control, signal, or power cable that are identified for the use, may be messenger-supported. See the NEC, Article 321 [4] for specific requirements.

Cables may be messenger-supported either by spirally wrapping a steel band around the cables and the messenger or by pulling the cable into rings suspended from the messenger. The spiral wrap method is used for factory-assembled cable; both methods are used for field assembly. A variety of spinning heads are available for the application of spiral wire banding in the field. The messenger used on factory-assembled, messenger-supported wiring is required to be copper covered steel or a combination of copper covered steel and copper, and the assembly should be secured to the messenger by a flat copper binding strip. Single insulated conductors should be cabled together.

Factory-preassembled aerial cables are particularly susceptible to installation damage from high stress at support sheaves while being pulled in.

Self-supporting cable is suitable for only relatively short spans. Messenger-supported cable can span longer distances, depending on the weight of the cable and the tensile strength of the messenger. The supporting messenger provides the strength to withstand climatic rigors and mechanical shock. The messenger should be grounded in accordance with the NEC [4].

A convenient feature available in one form of factory-assembled aerial cable makes it possible to form a slack loop to connect a circuit tap without cutting the cable conductors. This is done by reversing the direction of the spiral of the conductor cabling every 10–20 feet.

Spacer cable is an electric distribution line construction that consists of an assembly of one or more covered conductors separated from each other and supported from a messenger by insulating spacers. This is another economical means of transmitting power overhead between buildings. Available for use in three-phase 5–15 kV grounded or ungrounded systems, the insulated nonshielded phase conductors provide protection from accidental discharge through contact with ground level equipment, such as aerial ladders or crane booms. Uniform line electrical characteristics are obtained through the balanced geometric positioning of the conductors with respect to each other by the use of plastic or ceramic spacers located at regular intervals along the line. Low terminating costs are obtained because the conductors are nonshielded.

8.6.4 Open Runs. This is a low-cost method in which adequate support surfaces are available between the source and the load. It is most useful in combination with other methods, such as branch runs from cable trays, and when adding new circuits to existing installations.

This method employs multiconductor cable attached to surfaces, such as structural beams and columns. Type MC cable is permitted to be installed in this manner in commercial buildings as well as power limited control and telephone circuits. For architectural reasons, it is usually limited to service areas, above hung ceilings, and electrical shafts.

8.6.5 Cable Tray. A cable tray is defined as "a unit or assembly of units or sections, and associated fittings, made of metal or other noncombustible material forming a continuous rigid structure used to support cables." These supports include ladders, troughs, and channels, and have been increasing in popularity in commercial electric systems for the following reasons: low installed cost, system flexibility, improved reliability, accessibility for repair or addition of cables, and space saving when compared with conduit where a larger number of circuits with common routing are involved.

Cable trays are available in a number of styles, materials, and mechanical load carrying capability. Special coatings or materials for corrosion protection are also available.

Initial planning of a cable tray system should consider occupancy requirements as given in the NEC [4] and also allow additional space for future system expansion.

Covers, either ventilated or nonventilated, may be used when additional mechanical protection is required or for additional electrical shielding when communication circuits are involved. When cable trays are continuously covered for more than 6 feet with solid, unventilated covers, the cable ampacity rating should be de-rated as required by the NEC, Section 318 [4].

A solid fixed barrier is required for separation of cables rated over 600 V from those rated 600 V or less. Barrier strips are not required when the cables over 600 V are Type MC.

Seals or fire stops may be required when passing through walls, partitions, or elsewhere to minimize flame propagation.

In stacked tray installations, it is good practice to separate voltages, locating the lowest voltage cables in the bottom tray and increasingly higher voltage cables in an ascending order of trays. In a multiple-phase system, all phase conductors should be installed closely grouped in the same tray.

Cable tray provides a convenient economical support method when more than three cables are being routed in the same direction. Single conductors are not permitted in cable tray in commercial occupancies. Type MC cable can be installed in cable tray and, when only one or two cables have to be routed to a separate location, the cable can then be installed as open runs of cable. Type TC cable requires the use of a raceway between the cable tray and the termination point.

The steel or aluminum metal in a cable tray can also be used as an equipment grounding conductor when the tray sections are listed by a nationally recognized testing laboratory as having adequate cross sectional area, and are bonded using mechanical connectors or bonding jumpers. Refer to the NEC, Section 318-7 [4] for complete requirements.

8.6.6 Cable Bus. Cable bus is used for transmitting large amounts of power over relatively short distances. It is a more economical replacement for conduit or busway systems, but more expensive than cable tray. It also offers better reliability, safety, and lower maintenance than open-wire or bus systems.

Cable bus is a hybrid between cable tray and busway. It uses insulated conductors in an enclosure that is similar to cable tray with covers. The conductors are supported at maintained spacings by nonmetallic spacer blocks. Cable buses are furnished either as components for field assembly or as completely assembled sections. The use of completely assembled sections is recommended when the run is short enough so that splices may be avoided. Multiple sections requiring joining may preferably employ continuous conductors.

The conductors are generally spaced one cable diameter apart so that the rating in air may be attained. This spacing is also close enough to provide low reactance, resulting in minimum voltage drop.

8.6.7 Conduit. Among conduit systems, rigid steel provides the greatest degree of mechanical protection available in above ground conduit systems. Unfortunately, this is also a relatively high cost system. For this reason, it is being replaced, where possible, by other types of conduit and wiring systems. Where applicable, rigid aluminum, rigid nonmetallic conduit (NMC), electrical metallic tubing (EMT), intermediate metal conduit (IMC), electrical nonmetallic tubing (ENMT), and plastic, fiberglass, and cement ducts may be used. Cable trays and open runs of Type MC cable are also being utilized.

Conduit systems offer some degree of flexibility in permitting replacement of existing conductors with new ones. However, in case of fire or short-circuit current faults, it may be impossible to remove the conductors. In this case, it is necessary to replace both conduit and wire at great cost and delay. Also, during fires, conduits may transmit corrosive fumes into equipment where these gases can do a lot of damage. To keep flammable gases out of such areas, seals should be installed.

With magnetic conduits, an equal number of conductors of each phase should be installed in each conduit; otherwise, losses and heating will be excessive. A single conductor, for example, should not be installed in steel conduit.

Refer to the NEC [4] for regulations on conduit use.

Underground ducts are used when it is necessary to provide good mechanical protection, for example, when overhead conduits are subject to extreme mechanical abuse or when the cost of going underground is less than providing overhead supports. In the latter case, direct burial (without conduit) may be satisfactory under certain circumstances.

Underground ducts use rigid steel, plastic, or fiberglass conduits encased in concrete, or precast multiple-hole concrete duct banks with close fitting joints. When the added mechanical protection of concrete is not required, heavy wall versions of fiberglass conduits are direct buried as are rigid steel and plastic conduits. Medium-voltage, low-voltage, signal, and communication systems should not be installed in the same manhole. Manholes intended for cable splices or for drain provisions on long length cables should have adequate provisions for grounding.

Cables used in underground conduits should be suitable for use in wet areas. Some cost savings can be realized by using flexible plastic conduits with factory-installed conductors.

When a relatively long distance between the point of service entrance into a building and the service entrance protective device is unavoidable, the requirements of the NEC, Section 230-6 [4] apply. The conductors should be placed under at least 2 inches of concrete beneath the building; or they should be placed in conduit or duct and enclosed by concrete or brick not less than 2 inches thick. They are then considered outside the building.

8.6.8 Direct Burial. Cables may be buried directly in the ground, where permitted by the NEC [4], when the need for future maintenance along the cable run is not anticipated nor the protection of conduit required. The cables used should be suitable for this purpose, that is, resistant to moisture, crushing, soil contaminants, and insect and rodent damage. Direct buried cables rated over 600 V shall be shielded and should also provide an exterior ground path for personnel safety in the event of accidental dig-in. Multiconductor nonshielded Type MC cables rated up to 5000 V are also permitted to be direct buried. Refer to the NEC, Tables 300-5 and 710-3(b) [4] for minimum depth requirements.

The cost savings of this method over duct banks can vary from very little to a considerable amount. Cable trenching or burying machines, when appropriate, can significantly reduce the installation cost of direct buried cable, particularly in open field construction, such as shopping centers. While this system cannot be readily added to or maintained, the current-carrying capacity for a cable of a given size is usually greater than that for cables in ducts. Buried cable should have selected backfill for suitable heat dissipation. It should be used only when the chances of its being disturbed are minimal or it should be suitably protected. Relatively recent advances in the design and operating characteristics of cable fault location equipment and subsequent repair methods and material have diminished the maintenance mean time to repair.

8.6.9 Hazardous (Classified) Locations. Wire and cable installed in locations where fire or explosion hazards may exist should comply with the NEC, Articles 500 through 517 [4]. The authorized wiring methods are dependent upon the class and division of the specific area (see Table 54). The wiring method should be approved for the class and division, but is not dependent upon the group, which defines the hazardous substance.

Equipment and the associated wiring system approved as "intrinsically safe" is permitted in any hazardous location for which it has been approved. However, the installation should prevent the passage of gases or vapors from one area to another. Intrinsically safe equipment and wiring is not capable of releasing sufficient electrical or thermal energy under normal or abnormal conditions to cause ignition of

Table 54
Wiring Methods for Hazardous Locations

(Based on ANSI/NFPA 70-1990 [4])

Wiring Method	Class 1 Division		Class 2 Division		Class 3 Division 1 or 2
	1	2	1	2	
Threaded rigid metal conduit	X	X	X	X	X
Threaded steel intermediate metal conduit	X	X	X	X	X
Rigid metal conduit				X	X
Intermediate metal conduit (IMC)				X	X
Electrical metallic tubing (EMT)				X	X
Rigid nonmetallic conduit					X
Type MI mineral insulated cable	X	X	X	X	X
Type MC metal-clad cable		X		X	X
Type SNM shielded nonmetallic cable		X		X	X
Type MV medium-voltage cable		X			
Type TC power and control tray cable		X		X	
Type PLTC power limited tray cable		X		X	
Enclosed gasketed busways or wireways		X			
Dust-tight wireways				X	X

a specific flammable or combustible atmospheric mixture in its most easily ignitable concentration.

Seals should be provided in the wiring system to prevent the passage of the hazardous atmosphere along the wiring system from one division to another or from a Division I or II hazardous location to a nonhazardous area. The sealing requirements are defined in the NEC, Articles 501 through 503 [4]. The use of multiconductor cables with a gas-/vapor-tight continuous outer sheath, either metallic or nonmetallic, can significantly reduce the sealing requirement in Class 1, Division II hazardous locations.

8.6.10 Installation Procedures. Care should be taken in the installation of raceways to ensure that no sharp edges exist to cut or abrade the cable as it is pulled in. Another important consideration is not to exceed the maximum allowable tensile strength or manufacturers' recommendations for the maximum sidewall pressure of a cable. These forces are directly related to the force exerted on the cable while it is being pulled in. These forces can be decreased by shortening the length of each pull and reducing the number of bends. The force required for pulling a given length can be reduced by the application of a pulling compound on cables in conduit and the use of rollers in cable trays.

When the cable is to be pulled by the conductors, the maximum tension in pounds is limited to 0.008 times the area of the conductors, in circular mils, within the construction. The allowable tension should be reduced by 20%–40% when several conductors are being pulled simultaneously since the tension is not always evenly distributed among the conductors. This allowable tension should be further reduced when the cable is pulled by a grip placed over the outer covering. A reasonable figure for most jacketed constructions would be 1000 pounds/grip; but the calculated conductor tension should not be exceeded. Pulling eyes connected to each conductor provide the maximum allowable pulling tension. Reusable pulling eyes are available.

Sidewall pressures on most single conductors limit pulling tensions to approximately 450 pounds times the cable diameter (inches) times the radius of the bend (feet). Triplexed and paralleled cables would use their single-conductor diameters and a factor of 225 pounds and 675 pounds, respectively, instead of the 450 pound factor for a single conductor.

For duct installations involving many bends, it is preferable to feed the cable into the end closest to the majority of the bends (since the friction through the longer duct portion without the bends is not yet a factor) and pull from the other end. Each bend gives a multiplying factor to the tension it sees; therefore, the shorter runs to the bends will keep this increase in pulling tensions to a minimum. However, it is best to calculate pulling tensions for installation from both ends of the run and install from the end requiring the least tension.

The minimum bending radii is 8 times the overall cable diameter for nonshielded single-conductor and multiconductor cables and 12 times for metal tape shielded or lead covered cables. The minimum bending radius for nonshielded Type MC cable with interlocking armor or a corrugated sheath is 7 times the overall diameter of the metallic sheath; for shielded cables, the minimum bending radius is 12 times the overall diameter of one of the individual conductors or 7 times the overall diameter of the multiconductor cable, whichever is greater. Type MC cable with a smooth metallic sheath requires a greater minimum bending radius; refer to the NEC, Section 334-11 [4]. The minimum bending radius is applicable to bends of even a fraction of an inch in length, not just the average of a long length of cable being bent.

When installing cables in wet underground locations, the cable ends should be sealed to prevent moisture entry into the conductor strands. These seals should be left intact or remade after pulling if disrupted, until splicing, terminating, or testing is to be done. This practice is recommended to avoid unnecessary corrosion of the conductors and to safeguard against moisture entry into the conductor strands that would generate steam under overload, emergency loadings, or short-circuit conditions after the cable is energized.

8.7 Connectors

8.7.1 Types Available. Connectors are classified as "thermal" or "pressure," depending upon the method used to attach them to the conductor.

Thermal connectors use heat to make soldered, silver soldered, brazed, welded, or cast-on terminals. Soldered connections have been used with copper conductors for many years, and their use is well understood. Aluminum connections may also be soldered satisfactorily with the proper materials and technique. However, soldered joints are not commonly used with aluminum. Shielded arc welding of aluminum terminals to aluminum cable makes a satisfactory termination for cable sizes larger than 4/0 AWG. Torch brazing and silver soldering of copper cable connections are used, particularly for underground connections with bare conductors, such as in ground mats. Exothermic welding kits utilizing carbon molds are also used for making connections with bare copper or bimetallic (copperweld) cables for ground mats and for junctions that will be below grade. These are satisfactory as long as the conductors to be joined are dry and the welding charge and tool are proper. The exothermic welding process has also proved satisfactory for attaching connectors to insulated power cables.

Mechanical and compression pressure connectors are used for making joints in electric conductors. Mechanical connectors obtain the pressure to attach the connector to the electric conductor from an integral screw, cone, or other mechanical parts. Thus, a mechanical connector applies force and distributes it suitably through the use of bolts or screws and properly designed sections. The bolt diameter and number of bolts are selected to produce the clamping and contact pressures required for the most satisfactory design. The sections are made heavy enough to carry rated current and withstand the mechanical operating conditions. These are frequently not satisfactory with aluminum, since only a portion of the strands are distorted by this connector.

Compression connectors are those in which the pressure to attach the connector to the electric conductor is applied externally, changing the size and shape of the connector and conductor.

The compression connector is basically a tube with the inside diameter slightly larger than the outer diameter of the conductor. The wall thickness of the tube is designed to carry the current, withstand the installation stresses, and withstand the mechanical stresses resulting from thermal expansion of the conductor. A joint is made by compressing the conductor and tube into another shape by means of a specially designed tool and die. The final shape may be indented, cup, hexagon, circular, or oval. All methods have in common the reduction in cross sectional area by an amount sufficient to assure intimate and lasting contact between the connector and the conductor. Small connectors can be applied with a small hand tool. Larger connectors are applied with a hydraulic compression tool.

A properly crimped joint deforms the conductor strands sufficiently to have good electrical conductivity and mechanical strength, but not so much that the crimping action excessively compresses the strands, thus weakening the joint.

Mechanical and compression connectors are available as tap connectors. Many connectors have an independent insulating cover. After a connection is made, the cover is assembled over the joint to insulate, and, in some cases, to seal against the environment.

8.7.2 Connectors for Aluminum. Aluminum conductors are different from copper in several ways, and these property differences should be considered in specifying and using connectors for aluminum conductors (see Table 48). The normal oxide coating on aluminum has a relatively high electrical resistance. Aluminum has a coefficient of thermal expansion that is greater than copper. The ultimate and the yield strength properties and the resistance to creep of aluminum are different from the corresponding properties of copper. Corrosion is possible under some conditions because aluminum is anodic to other commonly used metals, including copper, when electrolytes from even humid air are present.

(1) Mechanical Properties and Resistance to Creep — Creep has been defined as the "continued deformation of material under stress." The effect of excessive creep resulting from the use of an inadequate connector that applies excessive stress could be the relaxation of contact pressure within the connector, and a resulting deterioration and failure of the electrical connection. In mechanical connectors for aluminum, as for copper, proper design can limit residual unit bearing loads to reasonable values, with a resulting minimum plastic deformation and creep subsequent to that initially experienced upon installation. Connectors for aluminum wire can accommodate a range of conductor sizes, provided that the design takes into account the residual pressure on both minimum and maximum conductors.

(2) Oxide Film — The surface oxide film on aluminum, though very thin and quite brittle, has a high electrical resistance and, therefore, should be removed or penetrated to ensure a satisfactory electric joint. This film can be removed by abrading with a wire brush, steel wool, emery cloth, or similar abrasive tool or material. A plated surface, whether on the connector or bus, should never be abraded; it can be cleaned with a solvent or other means that will not remove the plating.

Some aluminum fittings are factory filled with a connection aid compound, usually containing particles that aid in obtaining low contact resistance. These compounds act to seal connections against oxidation and corrosion by preventing air and moisture from reaching contact surfaces. Connection to the inner strands of a conductor requires deformation of these strands in the presence of the sealing compound to prevent the formation of an oxide film.

(3) Thermal Expansion — The linear coefficient of thermal expansion of aluminum is greater than that of copper and is important in the design of connectors for aluminum conductors. Unless provided for in the design of the connector, the use of metals with coefficients of expansion that are less than that of aluminum can result in high stresses in the aluminum during heat cycles, causing additional plastic deformation and significant creep. Stresses can be significant, not only because of the differences in coefficients of expansion, but also because the connector may operate at an appreciably lower temperature than the conductor. This condition will be aggravated by the use of bolts that are of a dissimilar metal or have different thermal expansion characteristics from those of the terminal.

(4) Corrosion — Direct corrosion from chemical agents affects aluminum no more severely than it does copper and, in most cases, less. However, since aluminum is more anodic than other common conductor metals, the opportunity exists for galvanic corrosion in the presence of moisture and a more cathodic metal. For this to occur, a wetted path should exist between external surfaces of the two metals in contact to set up an electric cell through the electrolyte (moisture), resulting in corrosion of the more anodic of the two, which, in this instance, is the aluminum.

Galvanic corrosion can be minimized by the proper use of a joint compound to keep moisture away from the points of contact between dissimilar metals. The use of relatively large aluminum anodic areas and masses minimizes the effects of galvanic corrosion. Plated aluminum connectors should be protected by taping or other sealing means.

(5) Types of Connectors for Aluminum Conductors — UL has listed connectors approved for use on aluminum, which have successfully withstood UL performance tests that are contained in ANSI/UL 486B-1982, Wire Connectors for Aluminum Conductors [6].[64] Both mechanical and compression connectors are available. The most satisfactory connectors are specifically designed for aluminum conductors to prevent any possible troubles from creep, the presence of oxide film, and the differences of coefficients of expansion between aluminum and other metals. These connectors are usually satisfactory for use on copper conductors in noncorrosive locations. The connection of an aluminum connector to a copper or aluminum pad is similar to the connection of bus bars. When both the pad and the connector are plated and the connection is made indoors, few precautions are necessary. The contact surfaces should be clean; if not, a solvent should be used. Abrasive cleaners are undesirable since the plating may be removed. In normal application, steel, aluminum, or copper alloy bolts, nuts, and flat washers may be used. A light film of a joint compound is acceptable, but not mandatory. When either

---

[64] ANSI publications are available from the Sales Department of the American National Standards Institute, 11 West 42nd Street, 13th Floor, New York, NY 10036. UL publications are available from Underwriters Laboratories, 333 Pfingsten Road, Northbrook, IL 60062.

of the contact surfaces is not plated, the bare surface should be cleaned by wire brushing and then coated with a joint compound. Belleville washers are suggested for heavy-duty applications where cold flow or creep may occur, or where bare contact surfaces are involved. Flat washers should be used wherever Belleville washers or other load concentrating elements are employed. The flat washer should be located between the aluminum lug, pad, or bolt and the outside edge of the Belleville washer with the neck or crown of the Belleville against the bolting nut to obtain satisfactory operation. In outdoor or corrosive atmospheres, the above applies with the additional requirement that the joint be protected. A nonplated aluminum-to-aluminum connection can be protected by the liberal use of a non-oxide compound.

In an aluminum-to-copper connection, a large aluminum volume compared to the copper is important as is the placement of the aluminum above the copper. Again, coating with a joint compound is the minimum protection; painting with a zinc chromate primer or thoroughly sealing with a mastic or tape is even more desirable. Plated aluminum should be completely sealed against the elements.

(6) Welded Aluminum Terminals—For aluminum cables 250 kcmil and larger, which carry large currents, excellent terminations can be made by welding special terminals to the cable. This is best done by the inert gas shielded metal arc method. The use of inert gas eliminates the need for any flux to be used in making the weld. The welded terminal is shorter than a compression terminal because the barrel for holding the cable can be very short. It has the advantage of requiring less room in junction or equipment terminal boxes. Another advantage is the reduced resistance of the connection. Each strand of the cable is bonded to the terminal, resulting in a continuous metal path for the current from every strand of the cable to the terminal.

Welding of these terminals to the conductors may also be achieved by using the tungsten electrode type of ac welding equipment. The tungsten arc method is slower, but, for small work, gives somewhat better control.

The tongues or pads of the welded terminals, such as the large compression connectors, are available with bolt holes to conform to NEMA Standards for terminals to be used on equipment.

(7) Procedure for Connecting Aluminum Conductors (see Fig 68)

   (a) When cutting cable, avoid nicking the strands. Nicking makes the cable subject to easy breakage [see Fig 68(a)].
   (b) Contact surfaces should be cleaned. The abrasion of contact surfaces is helpful even with new surfaces, and is essential with weathered surfaces. Do not abrade plated surfaces [see Fig 68(b)].
   (c) Apply joint compound to the conductor if the connector does not already have it [see Fig 68(c)].
   (d) Only use connectors specifically tested and approved for use on aluminum conductors.
   (e) On mechanical connectors, tighten the connector with a screwdriver or wrench to the required torque. Remove excess compound [see Fig 68(d)].

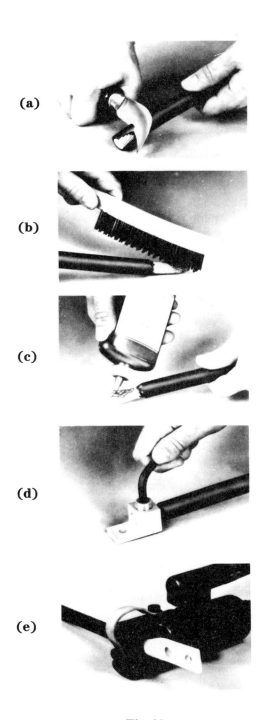

(a)

(b)

(c)

(d)

(e)

**Fig 68
Procedures for Connecting Aluminum Conductors**

    (f) On compression connectors, crimp the connector using the proper tool and die. Remove excess compound [see Fig 68(e)].

    (g) Always use a joint compound that is compatible with the insulation and as recommended by the manufacturer. The oxide film's penetrating or removing properties of some compounds aids in obtaining good initial conductivity. The corrosion inhibiting and sealing properties of some compounds help ensure the maintenance of continued good conductivity and the prevention of corrosion.

    (h) When making an aluminum-to-copper connection that is exposed to moisture, place the aluminum conductor above the copper. This prevents soluble copper salts from reaching the aluminum conductor, which could result in corrosion. If there is no exposure to moisture, the relative position of the two metals is not important.

    (i) When using insulated conductors outdoors, extend the conductor insulation or covering as close to the connector as possible to minimize weathering of the joint. Outdoors, whenever possible, joints should be completely protected by tape or other means. When outdoor joints are covered or protected, the protection should completely exclude moisture, as the retention of moisture could lead to severe corrosion.

8.7.3 Connectors for Cables of Various Voltages. Standard mechanical or compression connectors are recommended for all primary voltages, provided that the bus is noninsulated. Welded connectors may also be used for conductors sized in circular mils. Up to 600 V, standard connector designs present no problem for insulated or noninsulated conductors. The standard compression connectors are suitable for use on nonshielded conductors up to 5 kV. Above 5 kV and on shielded 5 kV conductors, stress considerations make it desirable to use tapered end compression connectors or semiconducting tape construction to provide the same effect.

8.7.4 Performance Requirements. Electric connectors for commercial buildings are designed to meet the requirements of the NEC [4]. They are evaluated on the basis of their ability to pass secureness, heating, heat cycling, and pullout tests as specified in ANSI/UL 486A-1982, Wire Connectors and Soldering Lugs for Use with Copper Conductors [5] and ANSI/UL 486B-1982, Wire Connectors for Aluminum Conductors [6]. These standards were revised to incorporate more stringent requirements for aluminum terminating devices. The reader is cautioned to specify and use only those lugs meeting the requirements of current UL Standards.

8.7.5 Electrical and Mechanical Operating Requirements. Electrically, the connectors shall carry the current without exceeding the temperature rise of the conductors being joined. Joint resistance that is not appreciably greater than that of an equal length of the conductor being joined is recommended to assure continuous and satisfactory operation of the joint. In addition, the connector should be able to withstand momentary overloads or short-circuit currents to the same degree as the conductor itself. Mechanically, a connector should be able to withstand the effects of the environment within which it is operating. When installed outdoors, it should withstand temperature extremes, wind, vibration, rain, ice, sleet, gases, chemical attack, etc. When used indoors, any vibration from rotating machinery, corrosion caused by plating or manufacturing processes, elevated temperatures from furnaces, etc., should not materially affect the performance of the joint.

8.8 Terminations

8.8.1 Purpose.
A termination for an insulated power cable should provide certain basic electrical and mechanical functions. These essential requirements include the following:

(1) Electrically connect the insulated cable conductor to electrical equipment, bus, or noninsulated conductor.

(2) Physically protect and support the end of the cable conductor, insulation, shielding system, and overall jacket, sheath, or armor of the cable.

(3) Effectively control electrical stresses to provide both internal and external dielectric strength to meet desired insulation levels for the cable system.

The current-carrying requirements are the controlling factors in the selection of the proper type and size of the connector or lug to be used. Variations in these components are related to the base material used for the conductor within the cable, the type of termination used, and the requirements of the electric system.

The physical protection offered by the termination will vary considerably, depending on the requirements of the cable system, the environment, and the type of termination used. The termination should provide an insulating cover at the cable end to protect the cable components (conductor, insulation, and shielding system) from damage by any contaminants that may be present, including gases, moisture, and weathering.

Shielded medium-voltage cables are subject to unusual electrical stresses where the cable shield system is ended just short of the point of termination. The creepage distance that should be provided between the end of the cable shield, which is at ground potential, and the cable conductor, which is at line potential, will vary with the magnitude of the voltage, the type of terminating device used, and, to some degree, the kind of cable used. The net result is the introduction of both radial and longitudinal voltage gradients that impose dielectric stress of varying magnitudes at the end of the cable. The termination provides a means of reducing and controlling these stresses within the working limits of the cable insulation and the materials used in the terminating device.

8.8.2 Definitions.
The definitions for cable terminations are obtained from IEEE Std 48-1990, IEEE Standard Test Procedures and Requirements for High-Voltage Alternating-Current Cable Terminations (ANSI) [14].

A Class 1 medium-voltage cable termination, or more simply, a Class 1 termination, provides:

(1) Some form of electrical stress control for the cable insulation shield termination.

(2) Complete external leakage insulation between the medium-voltage conductor(s) and ground.

(3) A seal to prevent the entrance of the external environment into the cable and to maintain the pressure, if any, within the cable system. This classification encompasses what was formerly referred to as a "pothead."

A Class 2 termination provides only items (1) and (2), some form of electrical stress control for the cable insulation shield termination, and complete external leakage insulation, but no seal against external elements. Terminations within this classification would be stress cones with rain shields or special outdoor insulation added to give complete leakage insulation, and the newer slip-on terminations for cables having extruded insulation that do not provide a seal as in Class 1.

A Class 3 termination provides only item (1), some form of electrical stress control for the cable insulation shield termination. This class of termination is used primarily indoors. Typically, this would include handwrapped stress cones (tapes or pennants) and the slip-on stress cones.

8.8.3 Cable Terminations. The requirements imposed by the installation location dictate the termination design class. The least critical is an indoor installation within a building or inside a sealed protective housing. Here, the termination is subjected to a minimum exposure to the elements, i.e., sunlight, moisture, and contamination. IEEE Std 48-1990 (ANSI) [14] refers to what is now called a "Class 3 termination" as an "indoor termination."

Outdoor installations expose the termination to a broad range of elements and require that features be included in its construction to withstand this exposure. The present Class 1 termination defined in IEEE Std 48-1990 (ANSI) [14] was previously called an "outdoor termination." In some areas, the air can be expected to carry a significant amount of gaseous contaminants and liquid or solid particles that may be conducting, either alone or in the presence of moisture. These environments impose an even greater demand on the termination to protect the cable end, prevent damaging contaminants from entering the cable, and for the termination itself to withstand exposure to the contaminants. The termination may be required to perform its intended function while partially or fully immersed in a liquid or gaseous dielectric. These exposures impose upon the termination the necessity of complete compatibility between the liquids and exposed parts of the termination, including any gasket sealing material. Cork gaskets have been used in the past; but the newer materials, such as tetrafluoroethylene (TFE) and silicone, provide superior gasketing characteristics. The gaseous dielectrics may be nitrogen or any of the electronegative gases, such as sulfur hexafluoride, that are used to fill electrical equipment.

8.8.3.1 Nonshielded Cable. Cables have a copper or aluminum conductor with thermosetting or thermoplastic insulation and no shield. Terminations for these cables generally consist of a lug and may be taped. The lug is fastened to the cable by one of the methods described in 8.7, and tape is applied over the lower portion of the barrel of the lug and down onto the cable insulation. Tapes used for this purpose are selected on the basis of compatibility with the cable insulation and suitability for application in the environmental exposure anticipated.

8.8.3.2 Shielded Cable. Cables rated over 2000 V have either a copper or aluminum conductor with an extruded solid dielectric insulation, such as ethylene propylene rubber (EPR) or crosslinked polyethylene (XLPE), or a laminated insulating system, such as oil impregnated paper tapes or varnished cloth tapes. A shielding system should be used on solid dielectric cables rated 5 kV and higher unless the cable is specifically listed or approved for nonshielded use (see 8.3.4.4).

When terminating shielded cable, the shielding is terminated far enough back from the conductor to provide the necessary creepage distance between the conductor and the shield. This abrupt ending of the shield introduces longitudinal stress over the surface of the exposed cable insulation. The resultant combination of radial and longitudinal electrical stress at the termination of the cable results in maximum stress occurring at this point. However, these stresses can be controlled and reduced to values within the safe working limits of the materials used for the termination. The most common method of reducing these stresses is to gradually increase the total thickness of insulation at the termination by adding, over the insulation, a premolded rubber cone or insulating tapes to form a cone. The cable shielding is carried up the cone surface and terminated at a point approximately 1/8 inch behind the largest diameter of the cone. A typical tape construction is illustrated in Fig 69. This form is commonly referred to as a "stress-relief cone" or "geometric stress cone." This function can also be accomplished by using a high dielectric constant material, as compared to that of the cable insulation, either in tape form or in a premolded tube, applied over the insulation in this area. This method results in a low-stress profile and is referred to as "capacitive stress control."

It is advisable to consult individual manufacturers of cable, and terminating and splicing materials for their recommendations on terminating and splicing shielded cables.

8.8.3.3 Termination Classes. A Class 1 termination is designed to handle the electrical functions as defined in 8.8.2. A Class 1 termination is used in areas that may have exposure to moisture or contaminants, or both. As pointed out in 8.8.3, the least severe requirements are those for a completely weather-protected area within a building or in a sealed protective housing. In this case, a track-resistant insulation, such as a silicone rubber tape or tube, would be used to provide the external leakage insulation function. The track-resistant surface would not necessarily need the skirts (also called "fins" or "rain shields"). The design of the termination to provide stress control and cable conductor seal can be the same for a weather-protected, low-contamination area as for a high-contamination area. When a Class 1 termination is installed outdoors, the design of the termination will vary according to the external leakage insulation function that will be in the form of silicone rubber, EPDM rubber, or porcelain insulation with rain shields. Of these forms, porcelain has the better resistance to long-term exposure in highly contaminated areas and to electrical stress with arc tracking. Because of these features, they are usually chosen for use in coastal areas where the atmosphere is salty. The choice in other weather-exposed areas is usually based on such factors as ease of installation, time of installation, overall long-term corrosion resistance of components, device cost, and past history. Typical Class 1 terminations are shown in Figs 70 and 71.

A Class 2 termination is different from a Class 1 termination only in that it does not seal the cable end to prevent the entrance of the external environment into the cable or maintain the pressure, if any, within the cable. Therefore, a Class 2 termination should not be used where moisture can enter into the cable. For a nonpressurized cable, typical of most industrial power cable systems using solid dielectric

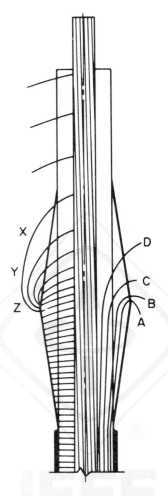

X, Y, Z — Electric Stress Lines
A, B, C, D — Equipotential Lines

**Fig 69
Stress-Relief Cone**

insulation, this seal is usually very easy to make. In the case of a poured porcelain terminator (commonly known as a "pothead"), the seal is normally built into the device. For a tape or slip-on terminator, the seal against external elements can be obtained by using tape (usually silicone rubber) to seal the conductor between the insulation and connector, assuming that the connector itself has a closed end.

The Class 3 termination only provides some form of stress control. Formerly known as an "indoor termination," it is recommended for use only in weather-protected areas. Before selecting a Class 3 termination, consideration should be given to the fact that, while it is not directly exposed to the elements, there is no

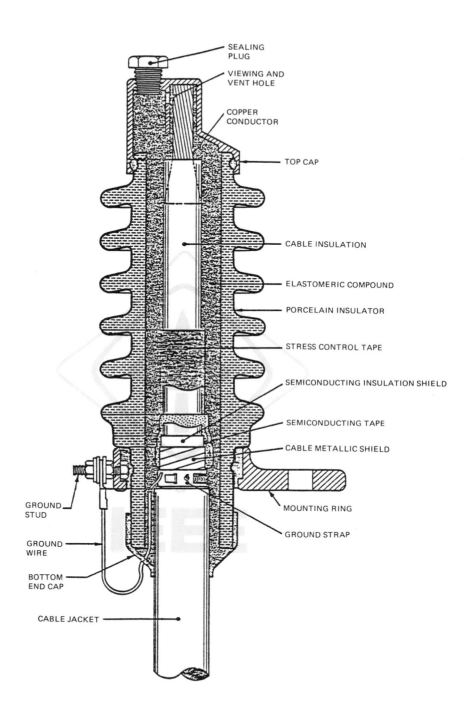

**Fig 70
Typical Class 1 Porcelain Terminator (for Solid Dielectric Cables)**

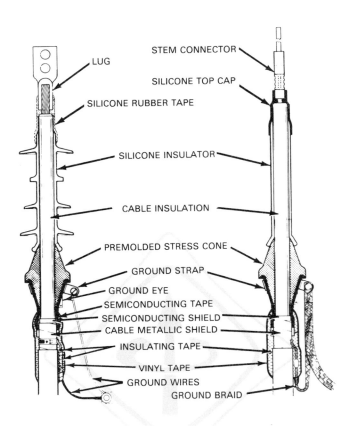

STEM CONNECTOR

LUG

SILICONE TOP CAP

SILICONE RUBBER TAPE

SILICONE INSULATOR

CABLE INSULATION

PREMOLDED STRESS CONE

GROUND STRAP

GROUND EYE

SEMICONDUCTING TAPE

SEMICONDUCTING SHIELD

CABLE METALLIC SHIELD

INSULATING TAPE

VINYL TAPE

GROUND WIRES

GROUND BRAID

Fig 71
Typical Class 1 Molded-Rubber Terminator (for Solid Dielectric Cables)

guarantee of the complete absence of some moisture or contamination. As a result, the lack of external leakage insulation between the medium-voltage conductor(s) and ground (or track-resistant material), and the seal to prevent the moisture from entering the cable, can result in shortened life of the termination. In general, this practice should be avoided. A typical Class 3 termination is shown in Fig 72.

8.8.3.4 Other Termination Design Considerations. Termination methods and devices are available in ratings of 5 kV and above for either single-conductor or three-conductor installations and for indoor, outdoor, or liquid immersed applications. Mounting variations include bracket, plate, flanged, and free-hanging types.

Both the cable construction and the application should be considered in the selection of a termination method or device. Voltage rating, desired basic impulse insulation level (BIL), conductor size, and current requirements are also considerations in the selection of the termination device or method. Cable construction is the controlling factor in the selection of the proper entrance sealing method and the stress-relief materials or filling compound.

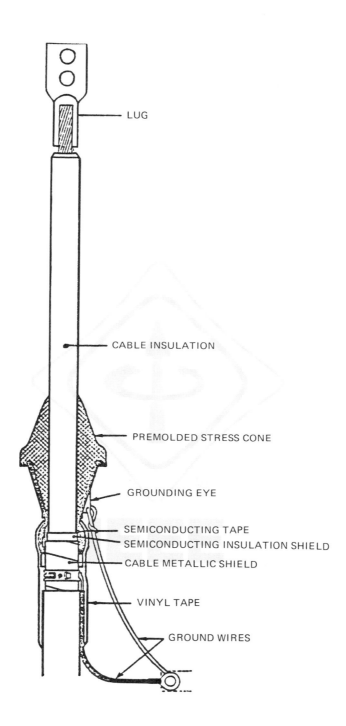

Fig 72
Typical Class 3 Molded-Rubber Terminator (for Solid Dielectric Cables)

Application, in turn, is the prime consideration for selecting the termination device or method, mounting requirements, and desired aerial connectors. Cable systems may be categorized into two general groups: nonpressurized and pressurized. Most power cable distribution systems are nonpressurized and utilize solid dielectric insulation.

8.8.3.5 Termination Devices and Methods. The termination hardware used on a pressurized cable system, which can also be used on a nonpressurized system, includes a hermetically sealed feature used to enclose and protect the cable end. A typical design consists of a metallic body with one or more porcelain insulators with fins (also called "skirts" or "rain shields"). The body is designed to accept a variety of optional cable entrance fittings, while the porcelain bushings, in turn, are designed to accommodate a number of cable sizes and aerial connections. These parts are assembled in the field onto the prepared cable end, with stress-relief cones required for shielded cables, and the assembled unit is filled with an insulating compound. Considerable skill is required for proper installation of this Class 1 termination, particularly in filling and cooling, in order to avoid shrinkage and the formation of voids in the fill material. Similar devices are available that incorporate high dielectric filling compounds, such as oil and thermosetting polyurethane resin, which do not require heating.

Advances in terminations for single-conductor cables include units designed to reduce the required cable end preparation and installation time, and to eliminate the hot-fill-with-compound step. One termination, applicable only to solid dielectric cables, is offered with or without a metal porcelain housing and requires the elastomeric materials to be applied directly to the cable end. Another termination consists of a metal porcelain housing filled with a gelatin-like substance designed to be partially displaced as the termination is installed on the cable. This latter unit may be used on any compatible nonpressurized cable.

The advantages of the preassembled terminations include simplified installation procedures, reduced installation time, and consistency in the overall quality and integrity of the installed system.

Preassembled Class 1 terminations are available in ratings of 5 kV and above for most applications. The porcelain housing units include flanged mounting arrangements for equipment mounting and liquid immersed applications. Selection of preassembled termination devices is essentially the same as for poured compound devices with the exception that those units using solid elastomeric materials generally should be sized, with close tolerance, to the cable diameters to ensure proper fit.

Another category of termination devices incorporates preformed stress-relief cones (see Fig 71). The most common preformed stress cone is a two-part elastomeric assembly consisting of a semiconducting lower section formed in the shape of a stress-relief cone and an insulating upper section. With the addition of medium-voltage insulation protection from the stress cone to the termination lug (a track-resistant silicone tape or tube, or silicone insulators or fins for weather-exposed areas) and by sealing the end of the cable, the resultant termination is a Class 2 termination, for use in areas exposed to moisture and contamination, but is not required to hold pressure.

Taped terminations, although generally more time-consuming to apply, are very versatile. Generally, taped terminations are used at 15 kV and below; however, there have been instances where they were used on cables up to 69 kV. On non-shielded cables, the termination is made with only a lug and a seal, usually tape. Termination of shielded cables requires the use of a stress-relief cone and cover tapes in addition to the lug. The size and location of the stress cone is controlled primarily by the operating voltage and whether the termination is exposed to or protected from the weather.

A creepage distance of 1 inch/kV of nominal system voltage is commonly used for protected areas, and a 2–3 inch distance allowed for exposed areas. Additional creepage distance may be gained by using a nonwetting insulation, fins, skirts, or rain hoods between the stress cone and conductor lug. For weather-exposed areas, this insulation is usually a track-resistant material, such as silicone rubber or porcelain.

Insulating tapes for the stress-relief cone are selected to be compatible with the cable insulation, and tinned copper braid and semiconducting tape are used as conducting materials for the cone. A solid copper strap or solder blocked braid should be used for the ground connection to prevent water wicking along the braid.

Some of the newer terminations do not require a stress cone. They utilize a stress-relief or grading tape or tube. The stress-relief or grading tape or tube is then covered with another tape or a heat shrinkable tube for protection against the environment. The exterior tape or tube may also provide a track-resistant surface for greater protection in contaminated atmospheres.

8.8.4 Jacketed and Armored Cable Connectors. Outer coverings for these cables may be nonmetallic, such as neoprene, polyethylene, or polyvinyl chloride; or metallic, such as lead, aluminum, or galvanized steel; or both, depending upon the installation environment. The latter two metallic coverings are generally furnished in an aluminum or galvanized steel tape helically applied and interlocked over the cable core or a continuously welded and corrugated aluminum sheath. The terminations available for use with these cables provide a means of securing the outer covering and may include conductor terminations. The techniques for applying them vary with cable construction, voltage rating, and the requirements for this installation.

The outer covering of multiconductor cables should be secured at the point of termination using cable connectors that are approved both for the cable and the installation conditions.

Type MC metal-clad cables with a continuously welded and corrugated sheath or an interlocking tape armor require, in addition to cable terminators, an arrangement to secure and ground the armor. Fittings available for this purpose are generally referred to as "armored cable connectors." These armored cable connectors provide mechanical termination and electrically ground the armor. This is particularly important on the continuous corrugated aluminum sheath because the sheath is the grounding conductor. In addition, the connector may provide a water-tight seal for the cable entrance to a box, compartment, pothead, or other piece of electrical equipment. These connectors are sized to fit the cable armor and are designed for use on the cable alone, with brackets or with locking nuts or adapters for application to other pieces of equipment.

8.8.5 Separable Insulated Connectors. These are two-part devices used in conjunction with medium-voltage electrical apparatus. A bushing assembly is attached to the medium-voltage apparatus (transformer, switch, fusing device, etc.) and a molded plug-in connector is used to terminate the insulated cable and connect the cable system to the bushing. The deadfront feature is obtained by fully shielding the plug-in connector assembly.

Two types of separable insulated connectors, for application at 15 kV and 25 kV, are available: load break and nonload break. Both utilize a molded construction design for use on solid dielectric insulated cables (rubber, crosslinked polyethylene, etc.) and are suitable for submersible applications. The connector section of the device has an elbow (90 °C [194 °F]) configuration to facilitate installation, improve separation, and save space (see IEEE Std 386-1985, IEEE Standard for Separable Insulated Connectors for Power Distribution Systems Above 600 V (ANSI) [17]).

Electrical apparatus may be furnished with only a universal bushing wall for the future installation of bushings for either the load-break or nonload-break dead-front assemblies. Shielded elbow connectors may be furnished with a voltage detection tap to provide a means for determining whether or not the circuit is energized.

8.8.6 Performance Requirements. Design test criteria have been established for terminations in IEEE Std 48-1990 (ANSI) [14], which specifies the short-time ac 60 Hz and impulse withstand requirements. Also listed in this design standard are maximum dc fieldproof test voltages. Individual terminations may safely withstand higher test voltages, and the manufacturer should be contacted for such information. All devices employed to terminate insulated power cables should meet these basic requirements. Additional performance requirements may include thermal load cycle capabilities of the current-carrying components, the environmental performance of completed units, and the long-term overvoltage withstand capabilities of the device.

8.9 Splicing Devices and Techniques. Splicing devices are subjected to a somewhat different set of voltage gradients and dielectric stress than that of a cable termination. In a splice, as in the cable itself, the greatest stresses are around the conductor and connector area and at the end of the shield. Splicing design should recognize this fundamental consideration and provide the means to control these stresses to values within the working limits of the materials used to make up the splice.

In addition, on shielded cables, the splice is in the direct line of the cable system and should be capable of handling any ground or fault currents that may pass through the cable shielding.

The connectors used to join the cable conductors together should be electrically capable of carrying the full rated load, emergency overload, and fault currents without overheating, as well as being mechanically strong enough to prevent accidental conductor pullout or separation.

Finally, the splice housing or protective cover should provide adequate protection to the splice, giving full consideration to the nature of the application and its environmental exposure.

(1) 600 V and Below — An insulating tape is applied over the conductor connection to electrically and physically seal the joint. The same taping technique is employed in the higher voltages, but with more refinement to cable end preparation and tape applications.

Insulated connectors are used when several relatively large cables have to be joined together. These terminators, called "moles" or "crabs," are, fundamentally, insulated buses with a provision for making a number of tap connections that can be very easily taped or covered with an insulating sleeve. Connectors of this type enable a completely insulated multiple connection to be made without the skilled labor normally required for careful crotch taping or the expense of special junction boxes. One widely used connector is a preinsulated multiple joint in which the cable connections are made mechanically by compression cones and clamping nuts. Another type is a more compact preinsulated multiple joint in which the cable connections are made by standard compression tooling that indents the conductor to the tubular cable sockets. Also available are tap connectors that accommodate a range of conductor sizes and have an independent insulating cover. After the connection is made, the cover is snapped shut to insulate the joint.

Insulated connectors lend themselves particulary well to underground services and commercial building wiring where a large number of multiple-connections have to be made.

(2) Over 600 V — Splicing of nonshielded cables up to 8 kV consists of assembling a connector, usually soldered or pressed onto the cable conductors, and applying insulating tapes to build up the insulation wall to a thickness of 1.5–2 times that of the original insulation on the cable. Care should be exercised in applying the connector and insulating tapes to the cables; but it is not as critical with nonshielded cables as with shielded cables.

Aluminum conductor cables require a moistureproof joint to prevent moisture entry into the stranding of the aluminum conductors.

Splices on solid dielectric cables are made with uncured tapes that will fuse together after application and provide a waterproof assembly. It is necessary, however, to use a moistureproof adhesive between the cable insulation and the first layer of insulating tapes. Additional protection may be obtained through the use of a moistureproof cover over the insulated splice. This cover may consist of additional moistureproof tapes and paint or a sealed weatherproof housing of some type.

8.9.1 Taped Splices. Taped splices (see Fig 73) for shielded cables have been used quite successfully for many years. Basic considerations are essentially the same as for nonshielded cables. Insulating tapes are selected not only on the basis of dielectric properties but also for compatibility with the cable insulation. The characteristics of the insulating tapes should also be suitable for the application of the splice. This latter consideration should include such details as providing a moisture seal for splices subjected to water immersion or direct burial, thermal

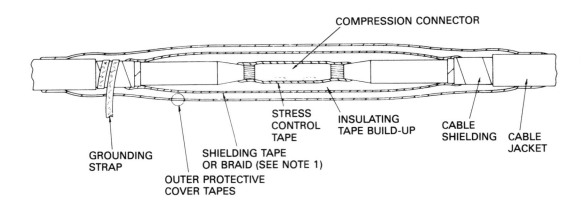

COMPRESSION CONNECTOR

STRESS CONTROL TAPE

INSULATING TAPE BUILD-UP

CABLE SHIELDING

CABLE JACKET

GROUNDING STRAP

OUTER PROTECTIVE COVER TAPES

SHIELDING TAPE OR BRAID (SEE NOTE 1)

NOTES: (1) Heavy braid jumper or perforated strip should be used across the splice to carry possible ground-fault current. Stress control tape should cover strands completely, lapping slightly onto insulation taper.
(2) Consult individual cable supplier for recommended installation procedures and materials.

Fig 73
Typical Taped Splice in Shielded Cable

stability of tapes for splices subjected to elevated ambient and operating temperatures, and ease of handling for applications of tape on wye or tee splices.

Connector surfaces should be smooth and free from any sharp protrusions or edges. The connector ends are tapered, and indentations or distortion caused by pressing tools are filled and shaped to provide a round, smooth surface. Semiconducting tapes are recommended for covering the connector and the exposed conductor stranding to provide a uniform surface over which insulating tapes can be applied. Cables with a solid dielectric insulation are tapered and those with a tape insulation are stepped to provide a gradual transition between the conductor/connector diameter and the cable insulation diameter prior to application of the insulating tape. This is done to control the voltage gradients and resultant voltage stress to values within the working limits of the insulating materials. The splice should not be overinsulated to provide additional protection since this could restrict heat dissipation at the splice area and risk splice failure.

A tinned or coated copper braid is used to continue the shielding function over the splice area. Grounding straps are applied to at least one end of the splice for grounding purposes, and a heavy braid jumper is applied across the splice to carry the available ground-fault current. Refer to 8.10.1 for single-point grounding to reduce sheath losses.

Final cover tapes or weather barriers are applied over the splice to seal it against moisture entry. A splice on a cable with a lead sheath is generally housed in a lead sleeve that is solder wiped to the lead cable sheath at each end of the splice. These lead sleeves are filled with compound in much the same way as potheads.

Hand taped splices may be made between lengths of dissimilar cables if proper precautions are taken to ensure the integrity of the insulating system of each cable and that the tapes used are compatible with both cables. One example of this would be a splice between a rubber insulated cable and an oil impregnated, paper insulated cable. Such a splice should have an oil barrier to prevent the oil impregnated in the paper cable from coming in contact with the insulation on the rubber cable. In addition, the assembled splice should be made completely moistureproof. This requirement is usually accomplished by housing the splice in a lead sleeve with wiped joints at both ends. A close fitting lead nipple is placed on the rubber cable and sealed to the jacket of the cable with tape or epoxy. The solder wipe is made to this lead sleeve.

Three-way wye and tee splices and the several other special hand taped splices that can be made all require special design considerations. In addition, a high degree of skill on the part of the installer is a prime requirement for proper installation and service reliability.

8.9.2 Preassembled Splices. Similar to the preassembled terminators, there are several variations of factory-made splices. The most basic is an elastomeric unit consisting of a molded housing sized to fit the cables involved, a connector for joining the conductors, and tape seals for sealing the ends of the molded housing to the cable jacket. Other versions of elastomeric units include an overall protective metallic housing that completely encloses the splice. These preassembled elastomeric splices are available in two- and three-way tees and multiple configurations for applications up to 35 kV. They can be used on most cables having an extruded solid dielectric insulation.

The preassembled splice provides a moistureproof seal to the cable jacket and is suitable for submersible, direct burial, and other applications where the splice housing should provide protection for the splice to the same degree that the cable jacket provides protection to the cable insulation and shielding system. An advantage of these preassembled splices is the reduction in time required to complete the splice after cable end preparation. However, the solid elastomeric materials used for the splice are required to be sized, with close tolerance, to the cable diameters in order to ensure a proper fit.

8.10 Grounding of Cable Systems. For safety and for reliable operation, the shields and metallic sheaths of power cables should be grounded. Without grounding, shields would operate at a potential considerably above ground. Thus, they would be hazardous to touch and would cause rapid degradation of the jacket or other material intervening between the shield and ground. This is caused by the capacitive charging current of the cable insulation that is on the order of 1 mA/feet of conductor length. This current normally flows, at power frequency, between the conductor and the earth electrode of the cable, normally the shield. In addition, the shield or metallic sheath provides a fault return path in the event of insulation failure, permitting rapid operation of the protection devices.

The grounding conductor and its attachment to the shield or metallic sheath, normally at a termination or splice, should have an ampacity not less than that of the shield. In the case of lead sheath, the ampacity of the grounding conductor should be adequate to carry the available fault current without overheating until it is interrupted. Attachment to the shield or sheath is frequently achieved with solder, which has a low melting point; thus an adequate area of attachment is required.

There is much disagreement as to whether the cable shield should be grounded at both ends or at only one end. If grounded at only one end, any possible fault current should traverse the length from the fault to the grounded end, imposing high current on the usually very light shield conductor. Such a current could readily damage or destroy the shield and require replacement of the entire cable rather than only the faulted section. With both ends grounded, the fault current would divide and flow to both ends, reducing the duty on the shield with consequently less chance of damage. There are modifications to both systems. In one, single-ended grounding may be attained by insulating the shields at each splice or sectionalizing point, and grounding only the source end of each section. This limits possible shield damage to only the faulted section. Multiple grounding, rather than just grounding at both ends, is simply the grounding of the cable shield or sheath at all access points, such as manholes or pull boxes. This also limits possible shield damage to only the faulted section.

8.10.1 Sheath Losses. Currents are induced in the multiple-grounded shields and sheaths of cables by the current flowing in the power conductor. These currents increase with the separation of the power conductors and increase with decreasing shield or sheath resistance. This sheath current is negligible with three-conductor cables, but with single-conductor cables separated in direct burial or separate ducts, it can be appreciable. For example, with three single-conductor 500 kcmil cables, laid parallel on 8 inch centers with twenty spiral No. 16 AWG copper shield wires, the ampacity is reduced by approximately 20% by this shield current. With single-conductor, lead sheathed cables in separate ducts, this current is important enough that single-ended grounding is mandatory. As an alternative, the shields are insulated at each splice (at approximately 500 foot intervals) and crossbonded to provide sheath transposition. This neutralizes the sheath currents, but still provides double-ended grounding. Of course, these sheaths and the bonding jumpers should be insulated; their voltage differential from ground may be in the 30–50 V range. For details on calculating sheath losses in cable systems, consult ANSI C2-1990, National Electrical Safety Code (NESC) [3].

Difficulties may arise from current attempting to flow via the cable shield, unrelated to cable insulation failures. To prevent this, all points served by a multiple-grounded shielded cable need to be interconnected with an ample grounding system. The insulation between shield sections at splices of single-end grounded shield systems should have sufficient dielectric strength to withstand possible abnormal voltages as well. This system requires interconnecting grounding conductors of suitably low impedance so that lightning, fault, and stray currents will follow this path rather than the cable shield. Cable shield ground connections should be made to this system, which should also be connected to the grounded element of the source supplying the energy to the cable. Duct runs, or direct burial routes, generally include a heavy grounding conductor to ensure such interconnection. For further details, refer to IEEE Std 142-1982 (ANSI) [15] and the NESC [4].

8.11 Protection from Transient Overvoltage. Cables rated up to 35 kV that are used in power distribution systems have insulation strengths well above that of other electrical equipment of similar voltage ratings. This is to compensate for installation handling and possibly a deterioration rate greater than that for insulation that is exposed to less severe ambient conditions. This high-insulation strength may or may not exist in splices or terminations, depending on their design and construction. Except for deteriorated points in the cable itself, the splices or terminations are most affected by overvoltages of lightning and switching transients. The terminations of cable systems that are not provided with surge protection may flashover due to switching transients. In this event, the cable would be subjected to possible wave reflections of even higher levels, possibly damaging the cable insulation; however, this is a remote possibility in medium-voltage cables.

Like other electrical equipment, the means employed for protection from these overvoltages is usually surge arresters. These may be for protection of associated equipment as well as the cable. Distribution or intermediate class arresters are used, applied at the junctions of open-wire lines and cables, and at terminals where switches may be open. Surge arresters are not required at intermediate positions along the cable run in contrast to open-wire lines.

It is recommended that surge arresters be connected between the conductor and the cable shielding system with short leads to maximize the effectiveness of the arrester. Similarly recommended is the direct connection of the shields and arrester ground wires to a substantial grounding system to prevent surge current propagation through the shield.

Aerial, messenger-supported, fully insulated aerial cables that are messenger-supported and spacer cables are subject to direct lightning strokes, and a number of such cases are on record. The incidence rate is, however, rather low, and, in most cases, no protection is provided. Where, for reliability, such incidents should be guarded against, a grounded shield wire, similar to that used for bare aerial circuits, should be installed on the poles a few feet above the cable. Grounding conductors down the pole need to be carried past the cable messenger with a lateral offset of approximately 18 inches to guard against side flashes from the direct strokes. Metal bayonets, when used to support the grounded shielding wire, should also be kept no less than 18 inches clear of the cables or messengers.

8.12 Testing

8.12.1 Application and Utility. Testing, particularly of elastomeric and plastic (solid) insulations, is a useful method of checking the ability of a cable to withstand service conditions for a reasonable future period. Failure to pass the test will either cause breakdown of the cable during testing or otherwise indicate the need for its immediate replacement.

Whether or not to routinely test cables is a decision each user has to make. The following factors should be taken into consideration:

(1) If there is no alternate source for the load supplied, testing should be done when the load equipment is not in operation.

(2) The costs of possible service outages due to cable failures should be weighed against the cost of testing. With solid dielectric insulation, failures of cables in service may be reduced approximately 90% by dc maintenance testing.

(3) Personnel with adequate technical capability should be available to do the testing, make observations, and evaluate the results.

The procedures discussed in this chapter are intended to be used as a recommended practice, and many variations are possible. At the same time, variations made without a sound technical basis can negate the usefulness of the test or even damage equipment.

With solid dielectric cable (elastomeric and plastic), the principal failure mechanism results from progressive degradation due to ac corona cutting during service at the locations of the manufacturing defects, installation damage, or accessory workmanship shortcomings. Initial tests reveal only gross damage, improper splicing or terminating, or cable imperfections. Subsequent use on ac usually causes progressive enlargement of such defects in proportion to their severity.

Oil impregnated paper (laminated) cable with a lead sheath (PILC) usually fails from water entrance at a perforation in the sheath, generally within 3–6 months after the perforation occurs. Periodic testing, unless very frequent, is therefore likely to miss many of these cases, making this testing method less effective with PILC cable.

Testing is not useful in detecting possible failure from moisture induced tracking across termination surfaces, since this develops principally during periods of precipitation, condensation, or leakage failure of the enclosure or housing. However, terminals should be examined regularly for signs of tracking and the condition corrected whenever it is detected.

8.12.2 Alternating-Current versus Direct-Current. Cable insulation can, without damage, sustain application of dc potential equal to the system basic impulse insulation level (BIL) for very long periods. In contrast, most cable insulations will sustain degradation from ac overpotential, proportional to the overvoltage, time of exposure, and the frequency of the applications. Therefore, it is desirable to utilize dc for any testing that will be repetitive. While the manufacturers use ac for the original factory test, it is almost universal practice to employ dc for any subsequent testing. All discussion of field testing hereafter applies to dc high-voltage testing.

8.12.3 Factory Tests. All cable is tested by the manufacturer before shipment, normally with ac voltage for a 5 minute period. Nonshielded cable is immersed in water (ground) for this test; shielded cable is tested using the shield as the ground return. Test voltages are specified by the manufacturer, by the applicable specification of the ICEA, or by other specifications such as those published by the Association of Edison Illuminating Companies (AEIC); refer to AEIC CS5-1987, Specifications for Thermoplastic and Crosslinked Polyethylene Insulated Shielded Power Cables Rated 5 kV through 69 kV [1] and AEIC CS6-1987, Specifications for Ethylene Propylene Rubber Insulated Shielded Power Cable Rated 5 kV through 69 kV [2].[65] In addition, a test may be made using dc voltage or two to three times

---

[65] AIEC publications are available from the Association of Edison Illuminating Companies, 51 East 42nd St., New York, NY 10017.

the rms value used in the dc test. On cables rated over 2 kV, corona tests may also be made.

8.12.4 Field Tests. As well as having no deteriorating effect on good insulation, dc high voltage is the most convenient to use for field testing since the test power sources or test sets are relatively light and portable. However, it should be recognized that a correlation between dc test results and cable life expectancy has never been established.

The primary benefit of dc high-voltage testing is to detect conducting particles left on the creepage surface during splicing or termination. Voltages for such testing should not be so high as to damage sound cable or component insulation but should be high enough to indicate incipient failure of unsound insulation that may fail in service before the next scheduled test.

Test voltages and intervals require coordination to attain suitable performance. One large industrial company with more than 25 years of cable testing experience has reached over a 90% reduction of cable system service failures through the use of voltages specified by ICEA. These test voltages are applied at installation, after approximately 3 years of service, and every 5–6 years thereafter. The majority of test failures occur at the first two tests; test (or service) failures after 8 years of satisfactory service are less frequent. The importance of uninterrupted service should also influence the test frequency for specific cables. Tables 55 and 56 specify cable field test voltages.

The AEIC has specified test values (AEIC CS5-1987 [1] and AEIC CS6-1987 [2]) for 1968 and newer cables approximately 20% higher than the ICEA values.

IEEE Std 400-1980 (Reaff. 1987), IEEE Guide for Making High-Direct-Voltage Tests on Power Cable Systems in the Field (ANSI) [18] specifies much higher voltages than either the ICEA or the AEIC. These much more severe test voltages, as shown in Table 57, are intended to reduce cable failures during operation by overstressing the cable during shutdown testing and causing weak cable to fail at that time. These test voltages should not be used without the concurrence of the cable manufacturer; otherwise, the cable warranty will be voided.

Cables that are to be tested should have their ends free of equipment and clear from ground. All conductors not under test should be grounded. Since equipment

Table 55
ICEA DC Test Voltages (kV) After Installation
Pre-1968 Cable

| Insulation Type | Grounding | 5 kV | Maintenance Test Rated Cable Voltage | | 35 kV |
			15 kV	25 kV	
Elastomeric:	Grounded	27	47	—	—
butyl, oil base, EP	Ungrounded	—	67	—	—
Polyethylene,	Grounded	22	40	67	88
including cross-linked polyethylene	Ungrounded	—	52	—	—

Table 56
ICEA DC Test Voltages (kV) After Installation
1968 and Later Cable*

Insulation Type	Insulation Level (%)	5 kV 1	2	15 kV 1	2	25 kV 1	2	35 kV 1	2
Elastomeric:	100	25	19	55	51	80	60	—	—
butyl and oil base	133	25	19	65	49	—	—	—	—
Elastomeric:	100	25	19	55	41	80	60	100	75
EP	133	25	19	65	49	100	75	—	—
Polyethylene,	100	25	19	55	41	80	60	100	75
including cross-linked polyethylene	133	25	19	65	49	100	75	—	—

NOTES: Columns 1 — Installation tests, made after installation, before service; columns 2 — maintenance tests, made after cable has been in service.

*These test values are lower than for pre-1968 cables because the insulation is thinner. Hence, the ac test voltage is lower. The dc test voltage is specified as three times the ac test voltage, so it is also lower than that of older cables.

Table 57
Cable DC Test Voltages (kV) for Installation and Maintenance
(See IEEE Std 400-1980 (ANSI) [18])

System Voltage (kV)	BIL (kV)	Test Voltage (kV) 100% Insulation Level	133% Insulation Level
2.5	60	40	50
5	75	50	65
8.7	95	65	85
15	110	75	100
23	150	105	140
28	170	120	
34.5	200	140	

NOTE: These test voltages should not be used without the cable manufacturers' concurrence because the cable warranty will be voided.

to which cable is customarily connected may not withstand the test voltages allowable for cable, either the cable has to be disconnected from this equipment, or the test voltage has to be limited to levels that the equipment can tolerate. The latter constitutes a relatively mild test on the cable condition, and the predominant leakage current measured is likely to be that of the attached equipment. Essentially, this tests the equipment, not the cable. It should also be recognized that some preassembled or premolded cable accessories may have a lower BIL than the cable itself, and this should be considered when establishing the test criteria.

In field testing, in contrast to the go-no-go nature of factory testing, the leakage current of the cable system should be closely watched and recorded for signs of approaching failure. The test voltage may be raised continuously and slowly from zero to the maximum value, or it may be raised in steps, pausing for 1 minute or more at each step. Potential differences between steps are on the order of the ac rms rated voltage of the cable. As the voltage is raised, current will flow at a relatively high rate to charge the capacitance, and, to a much lesser extent, to supply the dielectric absorption characteristics of the cable, as well as to supply the leakage current. The capacitance charging current subsides within a second or so, the absorption current subsides much more slowly and will continue to decrease for 10 minutes or more, ultimately leaving only the leakage current flowing.

At each step, and for the 5–15 minute duration of the maximum voltage, the current meter (normally a microammeter) is closely watched. Except when the voltage is first increased at each step, if the current starts to increase, slowly at first, then more rapidly, the last remnants of insulation at a weak point are failing, and total failure will occur shortly thereafter unless the voltage is reduced. This is characteristic of approximately 80% of all elastomeric insulation test failures.

In contrast to this avalanche current increase to failure, sudden failure (flashover) can occur if the insulation is already completely or nearly punctured. In the latter case, voltage increases until it reaches the sparkover potential of the air gap length, then flashover occurs. Polyethylene cables exhibit this characteristic for all failure modes. Conducting leakage paths, such as at terminations or through the body of the insulation, exhibit a constant leakage resistance independent of time or voltage.

One advantage of step testing is that a 1 minute absorption stabilized current may be read at the end of each voltage step. The calculated resistance of these steps may be compared as the test progresses to the next voltage step. At any step where the calculated leakage resistance decreases markedly (approximately 50% of that of the next lower voltage level), the cable could be near failure and the test should be discontinued short of failure as it may be desirable to retain the cable in serviceable condition until a replacement cable is available. On any test in which the cable will not withstand the prescribed test voltage for the full test period (usually 5 minutes) without current increase, the cable is considered to have failed the test and is subject to replacement as soon as possible.

The polarization index is the ratio of the current after 1 minute to the current after 5 minutes of maximum test voltage, and, on good cable, it will be between 1.25 and 2. Anything less than 1 minute should be considered a failure, and between 1 and 1.25 only a marginal pass.

After completion of the 5 minute maximum test voltage step, the supply voltage control dial should be returned to zero and the charge in the cable allowed to drain off through the leakage of the test set and voltmeter circuits. If this takes too long, a bleeder resistor of 1 MΩ/10 kV of test potential can be added to the drainage path, discharging the circuit in a few seconds. After the remaining potential drops below 10% of the original value, the cable conductor may be solidly grounded. All conductors should be grounded when not on test, during the testing of other conductors, and for at least 30 minutes after the removal of the dc test potential. They may be touched only while the ground is connected to them; otherwise, the release of absorption current by the dielectric may again raise their potential to a dangerous level.

8.12.5 Procedure. Load is removed from the cables either by diverting the load to an alternate supply, or by shutdown of the load served. The cables are de-energized by switching, tested to ensure voltage removal, grounded, and then disconnected from the attached switching equipment. (In case they are left connected, lower test potentials are required.) Surge arresters, potential transformers, and capacitors should also be disconnected.

All conductors and shields should be grounded. The test set is checked for operation, and, after its power has been turned off, the test lead is attached to the conductor to be tested. At this time, and not before, the ground should be removed from that conductor, and the bag or jar (see 8.12.6) applied over all of the terminals, by covering all noninsulated parts at both ends of the run. The test voltage is then applied slowly, either continuously or in steps as outlined in 8.12.4. Upon completion of the maximum voltage test duration, the charge is drained off, the conductor grounded, and the test lead removed for connection to the next conductor. This procedure is repeated for each conductor to be tested. Grounds should be left on each tested conductor for not less than 30 minutes.

8.12.6 DC Corona and Its Suppression. Starting at approximately 10–15 kV and increasing at a high power of the incremental voltage, the air surrounding all bare conductor portions of the cable circuit becomes ionized from the test potential on the conductor and draws current from the conductor. This ionizing current indication is not separable from that of the normal leakage current, and reduces the apparent leakage resistance value of the cable. Wind and other air currents tend to blow the ionized air away from the terminals, dissipating the space charge and allowing ionization of the new air, thus increasing what is known as the "direct corona current."

Enclosing the bare portions of both end terminations in plastic or glass jars, or plastic bags, prevents the escape of this ionized air, thus it becomes a captive space charge. Once formed, it requires no further current, so the direct corona current disappears. Testing up to approximately 100 kV is possible with this treatment. Above 100 kV, larger bags or a small bag inside a larger one are required. In order to be effective, the bags should be blown up so that no part of the bag touches the conductor.

An alternative method to minimize corona is to completely tape all bare conductor surfaces with standard electrical insulating tape. This method is superior to the bag method for corona suppression; but it requires more time to adequately tape all the exposed ends.

8.12.7 Line Voltage Fluctuations. The very large capacitance of the cable circuit makes the microammeter extremely sensitive to even minor variations in the 120 V, 60 Hz supply to the test set. Normally, it is possible to read only average current values or the near steady current values. A low harmonic content, constant voltage transformer improves this condition moderately. Complete isolation and stability are attainable only by using a storage battery and a 120 V, 60 Hz inverter to supply the test set.

8.12.8 Resistance Evaluation. Medium-voltage cable exhibits extremely high insulation resistance, frequently many thousands of megohms. While insulation resistance alone is not a primary indication of the condition of the cable insulation, the comparison of the insulation resistance of the three-phase conductors is useful. On circuits less than 1000 feet long, a ratio in excess of 5:1 between any two conductors is indicative of some questionable condition. On longer circuits, a ratio of 3:1 should be regarded as a maximum. Comparison of insulation resistance values with previous tests may be informative; but insulation resistance varies inversely with temperature, with winter insulation resistance measurements much higher than those obtained under summer conditions. An abnormally low insulation resistance is frequently indicative of a faulty splice, termination, or a weak spot in the insulation. Test voltages greater than standard values have been found practical in locating a weak spot by causing a test failure where the standard voltage would not cause breakdown. Fault location methods may also be used to locate the failure.

8.12.9 Megohmmeter Test. Since the insulation resistance of a sound medium-voltage cable circuit is generally on the order of thousands to hundreds of thousands of megohms, a megohmmeter test will reveal only grossly deteriorated insulation conditions of medium-voltage cable. For low-voltage cable, however, the megohmmeter tester is quite useful, and is probably the only practical test. Sound 600 V cable insulation will normally withstand 20 000 V or higher dc. Thus, a 1000 V or 2500 V megohmmeter is preferable to the lower 500 V testers for such cable testing.

For low-voltage cables, temperature corrected comparisons of insulation resistances with other phases of the same circuit, with previous readings on the same conductor, and with other similar circuits are useful criteria for adequacy. Continued reduction in the insulation resistance of a cable over a period of several tests is indicative of degrading insulation; however, a megohmmeter will rarely initiate the final breakdown of such insulation.

8.13 Locating Cable Faults. In electric power distribution systems, a wide variety of cable faults can occur. The problem may be in a communication circuit or in a power circuit, either in the low- or medium-voltage class. Circuit interruption may have resulted, or operation may continue with some objectionable characteristic. Regardless of the class of equipment or the type of fault involved, the one common problem is to determine the location of the fault so that repairs can be made.

The vast majority of cable faults encountered in an electric power distribution system occurs between conductor and ground. Most fault locating techniques are made with the circuit de-energized. In ungrounded or high resistance grounded, low-voltage systems, however, the occurrence of a single line-to-ground fault will not result in automatic circuit interruption, and, therefore, the process of locating the fault may be carried out by special procedures with the circuit energized.

8.13.1 Influence of Ground-Fault Resistance. Once a line-to-ground fault has occurred, the resistance of the fault path can range from almost zero up to millions of ohms. The fault resistance has a bearing on the method used to locate the failure. In general, a low-resistance fault can be located more readily than one of high resistance. In some cases, the fault resistance can be reduced by the application of voltage that is sufficiently high to cause the fault to break down as the excessive current causes the insulation to carbonize. The equipment required to do this is quite large and expensive, and its success is dependent, to a large degree, on the insulation involved. Large users indicate that this method is useful with paper and elastomeric cables, but generally of little use with thermoplastic insulation.

The fault resistance that exists after the occurrence of the original fault depends on the cable insulation and construction, the location of the fault, and the cause of the failure. A fault that is immersed in water will generally exhibit a variable fault resistance and will not consistently arc over at a constant voltage. Damp faults behave in a similar manner until the moisture has been vaporized. In contrast, a dry fault will normally be much more stable and, consequently, can be more readily located.

For failures that have occurred in service, the method of system grounding and available fault current, as well as the speed of relay protection, will be the influencing factors. Because of the greater carbonization and conductor vaporization, a fault resulting from an in-service failure can generally be expected to be of a lower resistance than one resulting from overpotential testing.

8.13.2 Equipment and Methods. A wide variety of commercially available equipment and a number of different approaches can be used to locate cable faults. The safety considerations outlined in 8.12 should be observed.

The method used to locate a cable fault depends on the

(1) Nature of the fault
(2) Type and voltage rating of the cable
(3) Value of rapid location of faults
(4) Frequency of faults
(5) Experience and capability of personnel

8.13.2.1 Physical Evidence of the Fault. Observation of a flash, a sound, or smoke accompanying the discharge of current through the faulted insulation will usually locate a fault. This is more probable with an overhead circuit than with underground construction. The discharge may be from the original fault or may be intentionally caused by the application of test voltages. The burned or disrupted appearance of the cable will also serve to indicate the faulted section.

8.13.2.2 Megohmmeter Instrument Test. When the fault resistance is sufficiently low that it can be detected with a megohmmeter, the cable can be sectionalized and each section tested to determine which one contains the fault. This procedure may require that the cable be opened in a number of locations before the fault is isolated to one replaceable section. This could, therefore, involve considerable time and expense, and might result in additional splices. Since splices are often the weakest part of a cable circuit, this method of fault locating may introduce additional failures at a subsequent time.

8.13.2.3 Conductor Resistance Measurement. This method consists of measuring the resistance of the conductor from the test location to the point of fault by using either the Varley or the Murray loop test. Once the resistance of the conductor to the point of fault has been measured, it can be translated into distance by using handbook values of resistance per unit length for the size and conductor material involved, correcting for temperature as required. Both of these methods give good results that are independent of fault resistance, provided the fault resistance is low enough that sufficient current for readable galvanometer deflection can be produced with the available test voltage. Normally, a low-voltage bridge is used for this resistance measurement. For distribution systems using cable insulated with organic materials, relatively low-resistance faults are normally encountered. The conductor resistance measurement method has its major application on such systems. Loop tests on large conductor sizes may not be sensitive enough to narrow down the location of the fault.

High-voltage bridges are available for higher resistance faults but have the disadvantage of increased cost and size as well as requiring a high-voltage dc power supply. High-voltage bridges are generally capable of locating faults with a resistance to ground of up to 1 or 2 MΩ, while a low-voltage bridge is limited to the application in which the resistance is several kilohms or less.

8.13.2.4 Capacitor Discharge. This method consists of applying a high-voltage and high-current impulse to the faulted cable. A high-voltage capacitor is charged by a relatively low-current capacity source, such as that used for high-potential testing. The capacitor is then discharged across an air gap or by a timed closing contact into the cable. The repeated discharging of the capacitor provides a periodic pulsing of the faulted cable. The maximum impulse voltage should not exceed 50% of the allowable dc cable test voltage since voltage doubling can occur at open circuit ends. Where the cable is accessible, or the fault is located at an accessible position, the fault may be located simply by sound. Where the cable is not accessible, such as in duct or directly buried, the discharge at the fault may not be audible. In such cases, detectors are available to trace the signal to the location of the fault. The detector generally consists of a magnetic pickup coil, an amplifier, and a meter to display the relative magnitude and direction of the signal. The direction indication changes as the detector passes beyond the fault. Acoustic detectors are also employed, particularly in situations where no appreciable magnetic field external to the cable is generated by the tracing signal.

In applications where relatively high-resistance faults are anticipated, such as with solid dielectric cables or through compound in splices and terminations, the impulse method is the most practical method presently available and is the one most commonly used.

8.13.2.5 Tone Signal. A tone signal may be used on energized circuits. A fixed frequency signal, generally in the audio frequency range, is imposed on the faulted cable. The cable route is then traced by means of a detector, which consists of a pickup coil, receiver, and a headset or visual display, to the point where the signal leaves the conductor and enters the ground return path. This class of equipment has its primary application in the low-voltage field and is frequently used for fault location on energized ungrounded circuits. On systems over 600 V, the use of a tone signal for fault location is generally unsatisfactory because of the relatively large capacitance of the cable circuit.

8.13.2.6 Radar System. A short duration low-energy pulse is imposed on the faulted cable and the time required for propagation to and return from the point of fault is monitored on an oscilloscope. The time is then translated into distance to locate the fault. Although this equipment has been available for a number of years, its major application in the power field has been on long distance high-voltage lines. In older test equipment, the propagation time is such that it cannot be displayed with good resolution for relatively short cables. However, recent equipment advances have largely overcome this deficiency. The major limitation of this method is its inability to adequately determine the difference between faults and splices on multiple-tapped circuits. An important feature of this method is that it will locate an opening in an otherwise unfaulted circuit.

8.13.3 Selection. The methods already listed represent some of the methods available to locate cable faults. They range from very simple to relatively complex. Some require no equipment, others require equipment that is inexpensive and can be used for other purposes, while still others require special equipment. As the complexity of the means used to locate a fault increases, so does the cost of the equipment, and also the training and experience required for those who are to use it.

In determining which approach is most practical for any particular facility, the size of the installation and the amount of circuit redundancy that it contains should be considered. The importance of minimizing the outage time of any particular circuit should be evaluated. The cable installation and maintenance practices and the number and time of anticipated faults will determine the expenditure that can be justified for test equipment. Equipment that requires considerable experience and operator interpretation for accurate results may be satisfactory for an application with frequent cable faults but ineffective when the number of faults is so small that adequate experience cannot be obtained. Because of these factors, many companies employ firms that offer the service of cable fault locating. Such firms are usually located in large cities and cover a large area with mobile test equipment.

While the capacitor discharge method is most widely used, no single method of cable fault location can be considered to be most suitable for all applications. The final decision on which method or methods to use depends upon the evaluation of the advantage and disadvantage of each in relation to the particular circumstances of the facility in question. As a last resort, opening splices in manholes and testing the cable between manholes can be used to locate the faulted cable.

8.14 Cable Specification. Once the correct cable has been determined, it can be described in a cable specification. Cable specifications generally start with the conductor and progress radially through the insulation and coverings. The following is a checklist that can be used in preparing a cable specification:

(1) Number of conductors in cable and phase identification required
(2) Conductor size (AWG, kcmil) and material
(3) Insulation (rubber, polyvinyl chloride, XLPE, EPR, etc.)
(4) Voltage rating and whether system requires 100%, 133%, or 173% insulation level
(5) Shielding system; required on cable systems rated 8 kV and above, and may be required on systems rated over 2001–8000 V.
(6) Outer finishes

(7) Installation approvals required (for use in cable tray, direct burial, messenger supported, wet location, exposure to sunlight or oil, etc.)
(8) Applicable UL listing
(9) Test voltage and partial discharge voltage
(10) Ground-fault current value and time duration
(11) Cable accessories, if any, to be supplied by cable manufacturer

An alternative method of specifying cable is to furnish the ampacity of the circuit (amperes), the voltage (phase-to-phase, phase-to-ground, grounded, or ungrounded), and the frequency, along with any other pertinent system data. Also required is the installation method and the installation conditions (ambient temperature, load factor, etc.). For either method, the total number of linear feet of conductors required, the quantity desired shipped in one length, any requirement for pulling eyes, and whether it is desired to have several single-conductor cables paralleled or triplexed on a reel should also be given.

8.15 Busway. Busways originated as a result of a request from the automotive industry in Detroit in the late '20s for an overhead wiring system that would simplify electrical connections for electric motor driven machines and permit a convenient arrangement of these machines in production lines. From this beginning, busways have grown to become an integral part of the low-voltage electric distribution system for industrial plants at 600 V and below.

Busways are particularly advantageous when numerous current taps are required. Plugs with circuit breakers or fusible switches may be installed and wired without de-energizing the busway.

Power circuits over 1000 A are usually more economical and require less space with busways than with conduit and wire. Busways may be dismantled and reinstalled in whole or in part to accommodate changes in the electric distribution system layout.

8.16 Busway Construction. Originally, a busway consisted of bare copper conductors supported on inorganic insulators, such as porcelain, mounted within a nonventilated steel housing. This type of construction was adequate for the current ratings of 225–600 A then used. As the use of busways expanded and increased loads demanded higher current ratings, the housing was ventilated to provide better cooling at higher capacities. The bus bars were covered with insulation for safety and to permit closer spacing of bars of opposite polarity in order to achieve lower reactance and voltage drop.

In the late '50s, busways were introduced that utilized conduction for heat transfer by placing the insulated conductor in thermal contact with the enclosure. By utilizing conduction, current densities are achieved for totally enclosed busways that are comparable to those previously attained with ventilated busways. Totally enclosed busways of this type have the same current rating regardless of mounting position. A stack of one bus bar per phase is used where each bus bar is up to approximately 7 inches wide (1600 A). Higher ratings will use two (3000 A) or three stacks (5000 A). Each stack will contain all three phases and the neutral to minimize circuit reactance.

Early busway designs required multiple nuts, bolts, and washers to electrically join adjacent sections. The most recent designs use a single bolt for each stack (with bars up to 7 inches wide). All hardware is captive to the busway section when shipped from the factory. Installation labor is greatly reduced with corresponding savings in installation costs.

Busways are available with either copper or aluminum conductors. Compared to copper, aluminum has lower electrical conductivity, less mechanical strength, and, upon exposure to the atmosphere, quickly forms an insulating film on the surface. For equal current-carrying ability, aluminum is lighter in weight and less costly.

For these reasons, aluminum conductors will have electroplated contact surfaces (tin or silver) and use Belleville springs at electrical joints, and bolting practices that accommodate aluminum's mechanical properties. Copper busway will be physically smaller (cross section) while aluminum busway is lighter in weight and lower in cost. Copper plug-in busway is more tolerant of cycling loads such as welding.

Busway is usually supplied in 10 foot sections. Since the busway should conform to the building, all possible combinations of elbows, tees, and crosses are available. Feed and tap fittings to other electrical equipment, such as switchboards, transformers, motor control centers, etc., are provided. Plugs for plug-in busway use fusible switches and molded-case circuit breakers. Standard busway current ratings are 20–5000 A for single-phase and three-phase service. Neutral conductors may be supplied, if required. Busway including plug-in devices, can incorporate a ground bar, if specified.

Four types of busways are available, complete with fittings and accessories, providing a unified and continuous system of enclosed conductors (see Fig 74):

(1) Feeder busway for low-impedance transmission of power
(2) Plug-in busway for easy connection or rearrangement of loads
(3) Lighting busway to provide electric power and mechanical support to fluorescent, high-intensity discharge, and incandescent fixtures
(4) Trolley busway for mobile power tapoffs to electric hoists, cranes, portable tools, etc.

8.17 Feeder Busway. Feeder busway is used to transmit large blocks of power. It has a very low and balanced circuit reactance for the control of voltage at the utilization equipment (see Fig 75).

Feeder busway is frequently used between the source of power, such as a distribution transformer or service drop, and the service entrance equipment. Industrial plants use feeder busway from the service equipment to supply large loads directly and to supply smaller current ratings of feeder and plug-in busway, which in turn supply loads through power takeoffs or plug-in units.

Available current ratings range from 600–5000 A, 600 V_{ac}. The manufacturer should be consulted for dc ratings. Feeder busway is available in single-phase and three-phase service with 50% and 100% neutral conductor. A ground bus is available with all ratings and types. Available short-circuit current ratings are 50 000–200 000 A symmetrical rms (see 8.22.2). The voltage drop of low-impedance feeder busway with the entire load at the end of the run ranges from 1–3 V/100 feet,

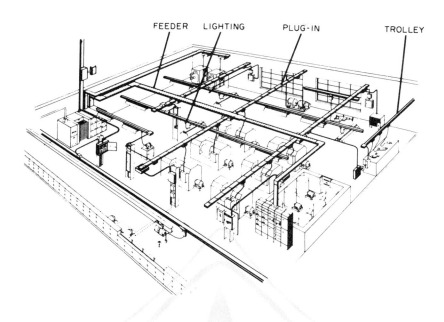

**Fig 74
Illustration of Versatility of Busways,
Showing Use of Feeder, Plug-in, Lighting, and Trolley Types**

line-to-line, depending upon the type of construction and the current rating used (see 8.22.3).

Feeder busway is available in indoor and weatherproof (outdoor) construction. Weatherproof construction is designed to shed water (rain). It should be used indoors where the busway may be subjected to water or other liquids. If NEMA "3R" equipment is suitable (see NEMA BU1-1988, Busways [19]), weatherproof busway should be used. Busway of any type is not suitable for immersion in water.

8.18 Plug-in Busway. Plug-in busway is used in industrial plants as an overhead system to supply power to utilization equipment. It serves as an elongated switchboard or panelboard running through the area with covered plug-in openings provided at closely spaced intervals to accommodate the plug-in devices placed on the busway near the loads that they supply.

Plug-in tapoff rearrangement is greatly facilitated by the use of flexible bus drop cable. The plug may be removed from the busway together with the bus drop cable and reinstalled with the machine in a minimum of time (see Fig 76).

Available plug-in devices include fusible switches, circuit breakers, static voltage protectors, ground indicators, combination starters, lighting contactors, and capacitor plugs.

Most plug-in busway is totally enclosed with current ratings from 100–4000 A. Usually plug-in and feeder busway sections, of the same manufacturer, above 600 A have compatible joints, so that they are interchangeable in a run. Plug-in busway

Fig 75
Feeder Busway

may be inserted in a feeder run when a tapoff is desired. Plug-in tapoffs are generally limited to maximum ratings of 800 A.

Short-circuit current ratings vary from 15 000–150 000 A symmetrical rms (see 8.22.2). The voltage drop ranges from 1–3 V/100 feet, line-to-line, for evenly distributed loading. If the entire load is concentrated at the end of the run, these values double (see 8.22.3).

A neutral bar may be provided for single-phase loads, such as lighting. Neutral bars vary from 25%–100% of the capacity of the phase bars.

A ground bar is often added for greater system protection and coordination under ground-fault conditions. The ground bar provides a low-impedance ground path and also reduces the possibility of arcing at the joint under high-level ground faults. (See 8.22.2 for additional details.)

8.19 Lighting Busway. Lighting busway is rated at a maximum of 60 A, 300 V to ground, with two, three, or four conductors. It may be used on 480Y/277 V or 208Y/120 V systems and is specifically designed for use with fluorescent and high-intensity discharge lighting (see Fig 77).

Lighting busways provide power to the lighting fixture and also serve as the mechanical support for the fixture. Auxiliary supporting means called "strength

**Fig 76
Installation View of Small Plug-in
Busway Showing Individual Circuit Breaker
Power Tapoff and Flexible Bus Drop Cable**

**Fig 77
Lighting Busway Supporting and Supplying
Power to High-Intensity Discharge Fixture**

beams" are available. Strength beams may be supported at maximum intervals of 16 feet. This permits the strength beam supports to conform to building column spacing. The strength beams provide support for the lighting busway as required by the NEC [4].

Fluorescent lighting fixtures may be suspended from the busway or they may be ordered with plugs and hangers attached for close coupling of the fixture to the busway. The busway may also be recessed in or surface mounted to dropped ceilings.

Lighting busway is also used to provide power for light industrial applications.

8.20 Trolley Busway. Trolley busway is constructed to receive stationary or movable takeoff devices. It is used on a moving production line to supply electric power to a motor or a portable tool moving with a production line, or where operators move back and forth over a range of 10–20 feet to perform their specific operations.

8.21 Standards. Busways are designed to conform to the following standards:
  (1) The NEC, Article 364 [4]
  (2) ANSI/UL 857-1989, Busways and Associated Fittings [7]
  (3) NEMA BU1-1988 [19]
ANSI/UL 857-1989 [7] and NEMA BU1-1988 [19] are primarily manufacturing and testing standards. The NEMA Standard is generally an extension of the UL Standard to areas that UL does not cover. The most important busway parameters are resistance R, reactance X, impedance Z, and short-circuit testing and rating.

The NEC [4] governs busway installation and some of the requirements are as follows:
  (1) Busway may be installed only where it is located in the open and is visible. Installation behind panels is permitted if access is provided and the following conditions are met:
      (a) No overcurrent devices are installed on the busway other than for an individual fixture.
      (b) The space behind the panels is not used for air-handling purposes.
      (c) The busway is the totally enclosed nonventilating type.
      (d) The busway is so installed that the joints between sections and fittings are accessible for maintenance purposes.
  (2) Busway may not be installed where it will subject to physical damage, corrosive vapors, or in hoistways.
  (3) When specifically approved for the purpose, busway may be installed in a hazardous location, outdoors, or in wet or damp locations.
  (4) Busway should be supported at intervals not exceeding 5 feet unless otherwise approved. Where specifically approved for the purpose, horizontal busway may be supported at intervals up to 10 feet, and vertical busway may be supported at intervals up to 16 feet.
  (5) Busway should be totally enclosed where passing through floors and for a minimum distance of 6 feet above the floor to provide adequate protection from physical damage. It may extend through walls provided any joints are outside the walls.

State and local electrical codes may have specific requirements in addition to ANSI/UL 857-1989 [7] and the NEC [4]. Appropriate code-enforcing authorities and manufacturers should be contacted to ensure that requirements are met.

8.22 Selection and Application of Busways. To properly apply busways in an electric power distribution system, some of the more important items to consider are the following.

8.22.1 Current-Carrying Capacity. Busways should be rated on a temperature rise basis to provide safe operation, long life, and reliable service.

Conductor size (cross sectional area) should not be used as the sole criterion for specifying busway. Busway may appear to have adequate cross sectional area and yet have a dangerously high temperature rise. The UL requirement for temperature rise (55 °C) (see ANSI/UL 857-1989 [7]) should be used to specify the maximum temperature rise permitted. Larger cross sectional areas can be used to provide lower voltage drop and temperature rise.

Although the temperature rise will not vary significantly with changes in ambient temperature, it may be a significant factor in the life of the busway. The limiting factor in most busway designs is the insulation life, and there is a wide range of types of insulating materials used by various manufacturers. If the ambient temperature exceeds 40 °C (104 °F) or a total temperature in excess of 95 °C (203 °F) is expected, then the manufacturer should be consulted.

8.22.2 Short-Circuit Current Rating. The bus bars in busways may be subjected to electromagnetic forces of considerable magnitude by a short-circuit current. The generated force per unit length of bus bar is directly proportional to the square of the short-circuit current and inversely proportional to the spacing between bus bars. Short-circuit current ratings are generally assigned and tested in accordance with NEMA BU1-1988 [19]. The ratings are based on the use of an adequately rated protective device ahead of the busway that will clear a short circuit in 3 cycles, and application in a system with short-circuit power factor that is not less than that given in Table 58.

If the system on which the busway is to be applied has a lower short-circuit power factor (larger X/R ratio) or a protective device with a longer clearing time, the short-circuit current rating may have to be reduced. The manufacturer should then be consulted.

The required short-circuit current rating should be determined by calculating the available short-circuit current and X/R ratio at the point where the input end

Table 58
Busway Ratings as a Function of Short-Circuit Power Factor

Busway Rating (symmetrical rms amperes)	Power Factor	X/R Ratio*
10 000 or less	0.50	1.7
10 001—20 000	0.30	3.2
Above 20 000	0.20	4.9

*X/R is load reactance X divided by load resistance R.

of the busway is to be connected. The short-circuit current rating of the busway should equal or exceed the available short-circuit current.

The short-circuit current may be reduced by using a current-limiting fuse at the supply end of the busway to cut it off before it reaches maximum value.

Short-circuit current ratings are dependent on many factors, such as bus bar center line spacing, size, and strength of bus bars and mechanical supports.

Since the ratings are different for each design of bus bar, the manufacturer should be consulted for specific ratings. Short-circuit current ratings should include the ability of the ground return path (housing and ground bar, if provided) to carry the rated short-circuit current. Failure of the ground return path to adequately carry this current can result in arcing at joints with attendant fire hazard. The ground-fault current can also be reduced to the point that the overcurrent protective device does not operate.

8.22.3 Voltage Drop. Line-to-neutral voltage drop V_D in busways may be calculated by the following formulas. The exact formulas for concentrated loads at the end of the line are, with V_R known,

$$V_D = \sqrt{(V_R \cos \phi + IR)^2 + (V_R \sin \phi + IX)^2} - V_R \tag{Eq 8}$$

and with V_S known,

$$V_D = V_S + IR \cos \phi + IX \sin \phi - \sqrt{V_S^2 - (IX \cos \phi - IR \sin \phi)^2} \tag{Eq 9}$$

where

$$V_R = V_S \frac{Z_L}{Z_S}, \quad V_D = V_S - V_R \tag{Eq 10}$$

NOTE: Multiply the line-to-neutral voltage drop by $\sqrt{3}$ to obtain the line-to-line voltage drop in a three-phase systems. Multipy the line-to-neutral voltage drop by 2 to obtain the line-to-line voltage drop in single-phase systems.

The approximate formulas for concentrated loads at the end of the line are

$$V_D = I(R \cos \phi + X \sin \phi) \tag{Eq 11}$$

$$V_{pr} = \frac{S(R \cos \phi + X \sin \phi)}{10 \, V_k^2} \tag{Eq 12}$$

The approximate formula for distributed load on a line is

$$V_{pr} = \frac{S(R \cos \phi + X \sin \phi) L}{10 \, V_k^2} \left(1 - \frac{L_1}{2L}\right) \tag{Eq 13}$$

where

V_D = Voltage drop, in volts.
V_{pr} = Voltage drop, in percent of voltage at sending end.
V_S = Line-to-neutral voltage at sending end, in volts.
V_R = Line-to-neutral voltage at receiving end, in volts.
ϕ = Angle whose cosine is the load power factor.

R = Resistance of circuit, in ohms per phase.
X = Reactance of circuit, in ohms per phase.
I = Load current, in amperes.
Z_L = Load impedance, in ohms.
Z_S = Circuit impedance, in ohms, plus load impedance, in ohms, added vectorially.
S = Three-phase apparent power for three-phase circuits or single-phase apparent power for single-phase circuits, in kilovoltamperes.
V_k = Line-to-line voltage, in kilovolts.
L_1 = Distance from source to desired point, in feet.
L = Total length of line, in feet.

The preceding formulas for concentrated loads may be verified by the trigonometric analysis shown in Fig 78. From this figure, it can be seen that the approximate formulas are sufficiently accurate for practical purposes. In practical cases, the angle between V_R and V_S will be small (much smaller than in Fig 78, which has been exaggerated for illustrative purposes). The error in the approximate formulas diminishes as the angle between V_R and V_S decreases and is zero, if that angle is zero. This latter condition will exist when the X/R ratio (power factor) of the load is equal to the X/R ratio (power factor) of the circuit through which the load current is flowing.

In actual practice, loads may be concentrated at various locations along the feeders, uniformly distributed along the feeder, or in any combination of the two. A comparison of the approximate formulas for concentrated end loading and uniform loading will show that a uniformly loaded line will have one-half the voltage drop as that due to the same total load concentrated at the end of the line. This aspect of the approximate formula is mathematically exact and entails no approximation. Therefore, in calculations of composite loading involving approximately uniformly loaded sections and concentrated loads, the uniformly loaded sections may be treated as end-loaded sections having one-half the normal voltage drop of

Fig 78
Diagram Illustrating Voltage Drop and Indicating
Error When Approximate Voltage-Drop Formulas Are Used

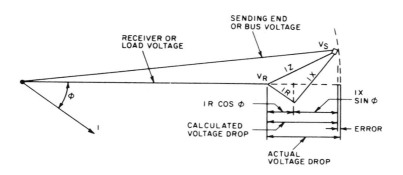

the same total load. Thus, the load can be divided into a number of concentrated loads distributed at various distances along the line. The voltage drop in each section may then be calculated for the load that it carries.

Three-phase voltage drops may be determined with reasonable accuracy by using Tables 59 and 60. These are typical values for the particular types of busway shown. The voltage drops will be different for other types of busway and will vary slightly by manufacturer within each type. The voltage drop shown is for three-phase, line-to-line, per 100 feet, at rated load on a concentrated loading basis for feeder, plug-in, and trolley busway. Lighting busway values are single-phase, distributed loading. For other loading and distances, use the formula

$$\text{voltage-drop } V_D = \text{table } V_D \left(\frac{\text{actual load}}{\text{rated load}} \right) \left(\frac{\text{actual distance (feet)}}{100 \text{ feet}} \right) \qquad \text{(Eq 14)}$$

The voltage drop for a single-phase load connected to a three-phase busway is 15.5% higher than the value shown in the tables. Typical values of resistance and reactance are shown in Table 61. Resistance is based on normal room temperature (25 °C [77 °F]). This value should be used in calculating the short-circuit current available in systems because short circuits can occur when busway is lightly loaded or initially energized. To calculate the voltage drop when fully loaded (75 °C [167 °F]), the resistance of copper and aluminum should be multiplied by 1.19.

8.22.4 Thermal Expansion. As load is increased, the bus bar temperature will increase and the bus bars will expand. The lengthwise expansion between no load and full load will range from 0.5 to 1 inch per 100 feet. The amount of expansion will depend on the total load, the size and location of the tapoffs, and the size and duration of varying loads. To accommodate the expansion, the busway should be mounted by using hangers that permit it to move. It may be necessary to insert expansion lengths in the busway run. To locate expansion lengths, the method of support, the location of power takeoffs, the degree of movement permissible at each end of the run, and the orientation of the busway should be known. The manufacturer can then make recommendations as to the location and number of expansion lengths.

8.22.5 Building Expansion Joints. Busway, when crossing a building expansion joint, should include provisions for accommodating the movement of the building structure. Fittings providing for 6 inches of movement are available.

8.22.6 Welding Loads. The busway and the plug-in device should be properly sized when plug-in busway is used to supply power to welding loads. The plug sliding contacts (stabs) and protective device (circuit breaker or fused switch) should have sufficient rating to carry both the continuous and peak welding load. This is normally done by determining the equivalent continuous current of the welder based on the maximum peak welder current, the duration of the welder current, and the duty cycle. Values may be obtained from the welder manufacturer. Loads 600 A and greater require special attention including consideration of bolted taps. As previously stated, copper busway is more tolerant of cycling loads than aluminum busway. When aluminum busway is used, cycling loads should be referred to the manufacturer.

Table 59
Voltage-Drop Values for Three-Phase Busways with Copper Bus Bars, in Volts per 100 Feet, Line-to-Line, at Rated Current with Entire Load at End*

Current Rating (amperes)	Load Power Factor (Percent, lagging)									
	20	30	40	50	60	70	80	90	95	100
Totally Enclosed Feeder Busway										
600	2.28	2.51	2.73	2.93	3.09	3.23	3.31	3.31	3.23	2.83
800	1.75	1.93	2.08	2.23	2.35	2.44	2.49	2.48	2.42	2.10
1000	1.51	1.81	2.11	2.39	2.66	2.92	3.15	3.33	3.39	3.29
1350	1.60	1.87	2.13	2.37	2.60	2.80	2.98	3.11	3.13	2.96
1600	1.90	2.10	2.27	2.43	2.56	2.67	2.73	2.72	2.66	2.31
2000	1.82	2.00	2.16	2.30	2.43	2.52	2.57	2.55	2.49	2.15
2500	1.75	1.91	2.06	2.18	2.29	2.36	2.40	2.37	2.30	1.96
3000	1.96	2.14	2.30	2.43	2.55	2.63	2.67	2.63	2.55	2.17
4000	1.84	2.01	2.16	2.29	2.40	2.49	2.53	2.49	2.42	2.07
5000	1.67	1.83	1.98	2.11	2.22	2.30	2.35	2.33	2.27	1.96
Totally Enclosed Plug-In Busway										
225	1.92	2.08	2.22	2.36	2.46	2.54	2.56	2.52	2.42	2.04
400	2.26	2.40	2.52	2.60	2.66	2.70	2.66	2.54	2.40	1.90
600	4.91	5.03	5.10	5.11	5.04	4.89	4.62	4.11	3.67	2.38
800	5.75	5.91	6.00	6.02	5.96	5.80	5.50	4.92	4.42	2.92
1000	4.77	4.91	4.98	5.02	4.98	4.84	4.60	4.12	3.70	2.46
1350	3.72	3.84	3.92	3.94	3.94	3.84	3.68	3.32	3.01	2.06
1600	3.58	3.70	3.78	3.82	3.80	3.72	3.54	3.22	2.92	2.00
2000	4.67	4.79	4.86	4.86	4.82	4.68	4.42	3.94	3.52	2.30
2500	4.08	4.20	4.26	4.30	4.26	4.14	3.94	3.54	3.18	2.12
3000	3.76	3.87	3.92	3.94	3.90	3.80	3.60	3.24	2.90	1.92
4000	4.64	4.74	4.80	4.79	4.73	4.57	4.30	3.81	3.38	2.15
5000	3.66	3.75	3.78	3.78	3.78	3.62	3.40	3.02	2.70	1.76
Lighting, Single Phase, Distributed Loading										
30	0.84	1.11	1.38	1.65	1.89	2.13	2.40	2.51	2.20	2.75
60	1.08	1.38	1.62	1.98	2.22	2.46	2.70	2.88	3.00	3.00
Trolley										
100	1.16	1.38	1.56	1.74	1.90	2.06	2.20	2.28	2.30	2.18

NOTES: Voltage-drop values are based on bus bar resistance at 75 °C (room ambient temperature 25 °C plus average conductor temperature at full load of 50 °C rise).

*Divide values by 2 for distributed loading.

8.23 Layout. Busway should be tailored to the building in which it is installed. Once the basic engineering work has been completed and the busway type, current rating, number of poles, etc., are determined, a layout should be made for all but the simplest straight runs. The initial step in the layout is to identify and locate the building structure (walls, ceilings, columns, etc.) and other equipment that is in the busway route. A layout of the busway to conform to this route is then made. Although the preliminary layout (drawings for approval) can be made from architectural drawings, it is essential that field measurements be taken to verify building and busway dimensions prior to the release of the busway for manufacture. When dimensions are critical, it is recommended that a section be held for field check of dimensions and manufactured after the remainder of the run has been installed. Manufacturers will provide quick delivery on limited numbers of these field-check sections.

Table 60
Voltage-Drop Values for Three-Phase Busways with Aluminum Bus Bars, in Volts per 100 Feet, Line-to-Line, at Rated Current with Entire Load at End*

Current Rating	Load Power Factor (Percent, lagging)									
(amperes)	20	30	40	50	60	70	80	90	95	100
Totally Enclosed Feeder Busway										
600	1.64	1.93	2.21	2.48	2.73	2.96	3.16	3.30	3.34	3.17
800	1.69	1.95	2.21	2.44	2.66	2.86	3.03	3.14	3.15	2.94
1000	1.51	1.81	2.11	2.39	2.66	2.92	3.15	3.33	3.39	3.29
1350	1.60	1.87	2.13	2.37	2.60	2.80	2.98	3.11	3.13	2.96
1600	1.70	1.97	2.22	2.45	2.67	2.87	3.04	3.14	3.15	2.94
2000	1.57	1 81	2.03	2.23	2.42	2.59	2.73	2.81	2.81	2.60
2500	1.56	1.78	1.98	2.18	2.35	2.51	2.63	2.70	2.69	2.48
3000	1.64	1.94	2.14	2.37	2.58	2.78	2.94	3.04	3.05	2.85
4000	1.60	1.83	2.04	2.24	2.42	2.59	2.71	2.79	2.78	2.56
Totally Enclosed Plug-In Busway										
100	2.05	2.63	3.20	3.76	4.30	4.83	5.33	5.79	5.98	6.01
225	1.94	2.22	2.49	2.73	2.96	3.15	3.31	3.41	3.40	3.13
400	3.47	3.66	3.81	3.92	3.99	3.99	3.92	3.69	3.45	2.64
600	4.62	4.89	5.12	5.30	5.41	5.45	5.37	5.10	4.80	3.76
800	4.09	4.34	4.54	4.70	4.81	4.84	4.78	4.54	4.28	3.36
1000	3.22	3.43	3.61	3.75	3.85	3.89	3.86	3.70	3.50	2.79
1350	2.92	3.10	3.12	3.36	3.44	3.48	3.44	3.28	3.08	2.44
1600	3.98	4.20	4.38	4.51	4.59	4.61	4.52	4.27	3.99	3.07
2000	3.48	3.68	3.85	3.99	4.07	4.09	4.04	3.83	3.60	2.81
2500	2.83	3.00	3.13	3.24	3.30	3.32	3.27	3.10	2.92	2.27
3000	3.68	3.85	3.99	4.09	4.14	4.12	4.01	3.74	3.47	2.60
4000	3.11	3.27	3.40	3.50	3.55	3.55	3.47	3.26	3.04	2.31

NOTES: Voltage-drop values are based on bus bar resistance at 75 °C (room ambient temperature 25 °C plus average conductor temperature at full load of 50 °C rise).

*Divide values by 2 for distributed loading.

Busway has great physical and electrical flexibility. It may be tailored to almost any layout requirement. However, some users find it a good practice to limit their busway installations to a minimum number of current ratings and maintain as many 10 foot lengths as possible. This enables them to reuse the busway components to maximum advantage when production line changes, etc., require relocation of the busway.

Another important consideration when laying out busway is coordination with other trades. Since there is a finite time lapse between job measurement and actual installation, other trades may use the busway clear area if coordination is lacking. Again, standard components can help since they are more readily available (sometimes from stock). By reducing the time between final measurement and installation, in addition to proper coordination, the chances of interference from other trades can be reduced to a minimum.

Finally, terminations are a significant part of busway layout considerations. For ratings 600 A and above, direct bus connections to the switchboard, motor control center, etc., can reduce installation time and problems. For ratings up to 600 A, direct bus terminations are generally not practical nor economical. These lower current ratings of busway are usually fed by short cable runs.

Table 61
Typical Busway Parameters, Line-to-Neutral, in Milliohms per 100 Feet, 25 °C

Current Rating (amperes)	Feeder Busway Aluminum		Copper		Plug-In Busway Aluminum		Copper	
	R	X	R	X	R	X	R	X
100	—	—	—	—	29.1	5.0	—	—
225	—	—	—	—	6.74	3.45	4.44	3.94
400	—	—	—	—	3.20	4.33	2.31	2.76
600	2.56	0.99	2.28	1.68	3.03	3.80	1.92	4.35
800	1.78	0.81	1.27	0.98	2.03	2.52	1.78	3.80
1000	1.59	0.50	1.05	0.82	1.35	1.57	1.20	2.52
1350	1.06	0.44	0.76	0.65	0.88	1.06	0.75	1.44
1600	0.89	0.41	0.70	0.53	0.93	1.24	0.61	1.17
2000	0.63	0.31	0.52	0.41	0.68	0.86	0.56	1.24
2500	0.48	0.25	0.38	0.32	0.44	0.56	0.42	0.86
3000	0.46	0.21	0.35	0.30	0.43	0.62	0.32	0.66
4000	0.31	0.16	0.25	0.21	0.28	0.39	0.26	0.62
5000	—	—	0.19	0.15	—	—	0.17	0.39
Lighting								
30	—	—	—	—	—	—	79.0	3.0
60	—	—	—	—	—	—	51.0	3.0
Trolley								
100	—	—	12.6	4.3	—	—	—	—

NOTE: Resistance values increase as temperature increases. Reactance values are not affected by temperature. The above values are based on conductor temperature of 25 °C (normal room temperature) since short circuits may occur when busway is initially energized or lightly loaded. To calculate voltage drop when fully loaded (75 °C), multiply resistance of copper and aluminum by 1.19.

8.24 Installation. Busway installs quickly and easily. When compared with other distribution methods, the reduced installation time for busway can result in direct savings on installation costs. In order to ensure maximum safety, reliability, and long life from a busway system, proper installation is essential. The guidelines below can serve as an outline from which to develop a complete installation procedure and timetable.

8.24.1 Procedure Prior to Installation

(1) Manufacturers supply installation drawings on all but the simplest of busway layouts. Study these drawings carefully. When drawings are not supplied, make your own.

(2) Verify actual components on hand against those shown on the installation drawing to be sure that there are no missing items. Drawings identify components by catalog number and location in the installation. Catalog numbers appear on section nameplate and carton label. Location on the installation (item number) will also be on each section.

(3) During storage (prior to installation), all components, even the weatherproof type, should be stored in a clean, dry area and protected from physical damage.

(4) Always read manufacturers' instructions for installation of individual components. If you are still in doubt, ask for more information; never guess.

(5) Electrical testing of individual components prior to installation is recommended. Identification of defective pieces prior to installation will save considerable time and money.

(6) Finally, preposition hanger supports (drop rods, etc.) and hangers if of the type that can be prepositioned. You are now ready to begin the actual installation of busway components.

8.24.2 Procedure During Installation

(1) Almost all busway components are built with two dissimilar ends that are commonly called "bolt end" and "slot end." Refer to the installation drawing to properly orient the bolt and slot ends of each component. This is important because it is not possible to properly connect two slot ends or bolt ends.

(2) Lift individual components into position and attach to hangers. It is generally best to begin this process at the end of the busway run that is most rigidly fixed (for example, the switchboards).

(3) Pay particular attention to "TOP" labels and other orientation marks, when applicable.

(4) As each new component is installed in position, tighten the joint bolt to proper torque per manufacturers' instructions. Also install any additional joint hardware that may be required.

(5) On plug-in busway installations, attach plug-in units in accordance with manufacturers' instructions and proceed with wiring.

(6) Outdoor busway may require removal of weep hole screws and the addition of joint shields. Pay particular attention to installation instructions to ensure that all steps are followed.

8.24.3 Procedure After Installation. Be sure to recheck all steps to ensure that you have not forgotten anything. Be particularly sure that all joint bolts have been properly tightened. At this point, the busway installation should be almost complete. However, before energizing, the complete installation should be properly tested.

8.25 Field Testing. The completely installed busway run should be electrically tested prior to being energized. The testing procedure should first verify that the proper phase relationships exist between the busway and associated equipment. This phasing and continuity test can be performed in the same manner as similar tests on other pieces of electrical equipment.

All busway installations should be tested with a megohmmeter or high-potential voltage to be sure that excessive leakage paths between phases and ground do not exist. Megohmmeter values depend on the busway construction, type of insulation, size and length of busway, and atmospheric conditions. Acceptable values for a particular busway should be obtained from the manufacturer.

If a megohmmeter is used, it should be rated 1000 V_{dc}. Normal high-potential test voltages are twice rated voltage plus 1000 V for 1 minute. Since this may be above the corona starting voltage of some busway, frequent testing is undesirable. For additional details, see NEMA BU1.1-1986 [20].

8.26 Busways Over 600 V (Metal-Enclosed Bus). Busway over 600 V is referred to as "metal-enclosed bus" and consists of three types: isolated phase, segregated phase, and nonsegregated phase. Isolated phase and segregated phase are utility-type busways used in power generation stations. Industrial plants outside of power generation areas use nonsegregated phase bus for the connection of transformers and switchgear and the interconnection of switchgear lineups. The advantage of metal-enclosed bus over cable is a simpler connection to equipment (no potheads required). It is rarely used to feed individual loads.

8.26.1 Standards. Metal-enclosed bus was included first in the 1975 NEC. The NEC [4] requires that the metal-enclosed bus nameplate specify its rated

    (1) Voltage

    (2) Continuous current

    (3) Frequency

    (4) 60 Hz withstand voltage

    (5) Momentary current

The NEC [4] further requires that metal-enclosed bus be constructed and tested in accordance with IEEE C37.20.1-1987, IEEE Standard for Metal-Enclosed Low-Voltage Power Circuit Breaker Switchgear (ANSI) [10], IEEE C37.20.2-1987, IEEE Standard for Metal-Clad and Station-Type Cubicle Switchgear (ANSI) [11], and IEEE C37.20.3-1987, IEEE Standard for Metal-Enclosed Interrupter Switchgear (ANSI) [12].

8.26.2 Ratings. IEEE C37.20.1-1987 (ANSI) [10], IEEE C37.20.2-1987 (ANSI) [11], and IEEE C37.20.3-1987 (ANSI) [12] specify the voltage, insulation, and the continuous and momentary current levels for metal-enclosed bus (see Table 62). The ratings are equal to the corresponding values for metal enclosed switchgear.

8.26.3 Construction. Metal enclosed (nonsegregated phase) bus consists of aluminum or copper conductors with bus supports usually of glass polyester or porcelain. Bus bars are insulated with sleeves or by the fluid bed process. After installation, joints are covered with boots or tape. Metal enclosed bus is totally enclosed. The enclosure is fabricated from steel in lower continuous current ratings and aluminum or stainless steel in higher ratings. Normal lengths are 8–10 feet with a cross section of approximately 16 inches × 26–36 inches, depending on conductor

Table 62
Voltage, Insulation, Continuous Current, and
Momentary Current Ratings of Nonsegregated Phase Metal-Enclosed Bus

Voltage (kV, rms)		Continuous Current (A)	Insulation, Withstand Level (kV)			Momentary Current (kA, asymmetrical)
Nominal	Rated Maximum		Power Frequency (rms), 1 min	DC Withstand, 1 min	Impulse	
4.16	4.76	1200	19.0	27.0	60	19–78
13.8	15.00	2000	36.0	50.0	95	19–78
23.0	25.80	3000	60.0	—	125	58
34.5	38.00	—	80.0	—	150	58

size and spacing. Electrical connection points are electroplated with either silver or tin. Indoor and outdoor (weatherproof) constructions are available.

8.26.4 Field Testing. After installation, the metal-enclosed bus should be electrically tested prior to being energized. Phasing and continuity tests can be performed with other associated electrical equipment on the job. Megohmmeter tests can be made that are similar to those described for busway under 600 V. High-potential tests should be conducted at 75% of the values shown in Table 62.

Continuous current ratings are based on a maximum temperature rise of 65 °C (149 °F) of the bus (30 °C [86 °F] if joints are not electroplated). Insulation temperature limits vary with the class of insulating material. Maximum total temperature limits for metal-enclosed bus are based on 40 °C (104 °F) ambient temperature. If the ambient temperature will exceed 40 °C (104 °F), the manufacturer should be consulted.

The momentary current rating is the maximum rms total current (including direct-current component) that the metal-enclosed bus can carry for 10 cycles without electrical, thermal, or mechanical damage.

8.27 References. The following references shall be used in conjunction with this chapter:

[1] AEIC CS5-1987, Specifications for Thermoplastic and Crosslinked Polyethylene Insulated Shielded Power Cables Rated 5 kV through 69 kV.

[2] AEIC CS6-1987, Specifications for Ethylene Propylene Rubber Insulated Shielded Power Cables Rated 5 kV through 69 kV.

[3] ANSI C2-1990, National Electrical Safety Code.

[4] ANSI/NFPA 70-1990, National Electrical Code.

[5] ANSI/UL 486A-1982, Wire Connectors and Soldering Lugs for Use with Copper Conductors.

[6] ANSI/UL 486B-1982, Wire Connectors for Use with Aluminum Conductors.

[7] ANSI/UL 857-1989, Busways and Associated Fittings.

[8] ICEA P-32-382-1969, Short-Circuit Characteristics of Insulated Cable.

[9] ICEA P-45-482-1979, Short-Circuit Performance of Metallic Shields and Sheaths of Insulated Cable.

[10] IEEE C37.20.1-1987, IEEE Standard for Metal-Enclosed Low-Voltage Power Circuit Breaker Switchgear (ANSI).

[11] IEEE C37.20.2-1987, IEEE Standard for Metal-Clad and Station-Type Cubicle Switchgear (ANSI).

[12] IEEE C37.20.3-1987, IEEE Standard for Metal-Enclosed Interrupter Switchgear (ANSI).

[13] IEEE S-135, Power Cable Ampacities (IPCEA).

[14] IEEE Std 48-1990, IEEE Standard Test Procedures and Requirements for High-Voltage Alternating-Current Cable Terminations (ANSI).

[15] IEEE Std 142-1982, IEEE Recommended Practice for Grounding of Industrial and Commercial Power Systems (ANSI).

[16] IEEE Std 242-1986, IEEE Recommended Practice for Protection and Coordination of Industrial and Commercial Power Systems (ANSI).

[17] IEEE Std 386-1985, IEEE Standard for Separable Insulated Connectors for Power Distribution Systems Above 600 V (ANSI).

[18] IEEE Std 400-1980 (Reaff. 1987), IEEE Guide for Making High-Direct-Voltage Tests on Power Cable Systems in the Field (ANSI).

[19] NEMA BU1-1988, Busways.

[20] NEMA BU1.1-1986, Instructions for Handling Installation, Operation, and Maintenance of Busway Rated 600 Volts or Less.

[21] NEMA WC3-1980 (Reaff. 1986), Rubber-Insulated Wire and Cable for the Transmission and Distribution of Electrical Energy (ICEA S-19-81, Sixth Edition).

[22] NEMA WC5-1973 (Reaff. 1979 and 1985), Thermoplastic-Insulated Wire and Cable for the Transmission and Distribution of Electrical Energy (ICEA S-61-402, Third Edition).

[23] NEMA WC7-1988, Cross-Linked-Thermosetting-Polyethylene-Insulated Wire and Cable for the Transmission and Distribution of Electrical Energy (ICEA S-66-524).

[24] NEMA WC8-1988, Ethylene-Propylene-Rubber-Insulated Wire and Cable for the Transmission and Distribution of Electrical Energy (ICEA S-68-516).

8.28 Bibliography. The references in this bibliography are listed for informational purposes only.

[B1] ANSI/UL 44-1985, Rubber-Insulated Wires and Cables.

[B2] ANSI/UL 62-1985, Flexible Cord and Fixture Wire.

[B3] ANSI/UL 83-1985, Thermoplastic-Insulated Wires and Cables.

[B4] ANSI/UL 493-1988, Thermoplastic-Insulated Underground Feeder and Branch-Circuit Cables.

[B5] ANSI/UL 854-1986, Service-Entrance Cables.

[B6] ANSI/UL 1569-1985, Metal-Clad Cables.

[B7] ANSI/UL 1581-1985, Reference Standard for Electrical Wires, Cables, and Flexible Cords.

[B8] IEEE C37.95-1989, IEEE Guide for Protective Relaying of Utility-Consumer Interconnections (ANSI).

[B9] IEEE Std 100-1988, IEEE Standard Dictionary of Electrical and Electronics Terms, Fourth Edition (ANSI).

[B10] IEEE Std 120-1989, IEEE Master Test Guide for Electrical Measurements in Power Circuits.

[B11] IEEE Std 404-1986, IEEE Standard for Cable Joints for Use with Extruded Dielectric Cable Rated 5000 through 46 000 Volts, and Cable Joints for Use with Laminated Dielectric Cable Rated 2500 through 500 000 Volts (ANSI).

[B12] IEEE Std 446-1987, IEEE Recommended Practice for Emergency and Standby Power Systems for Industrial and Commercial Applications (ANSI).

[B13] IEEE Std 525-1987, IEEE Guide for the Design and Installation of Cable Systems in Substations (ANSI).

[B14] IEEE Std 532-1982, IEEE Guide for Selecting and Testing Jackets for Cables (ANSI).

[B15] IEEE Std 575-1988, IEEE Guide for the Application of Sheath-Bonding Methods for Single-Conductor Cables and the Calculation of Induced Voltages and Currents in Cable Sheaths (ANSI).

[B16] IEEE Std 592-1977, IEEE Standard for Exposed Semiconducting Shields on Premolded High-Voltage Cable Joints and Separable Insulated Connectors.

[B17] IEEE Std 816-1987, IEEE Guide for Determining the Smoke Generation of Solid Materials Used for Insulations and Coverings of Electric Wire and Cable.

[B18] NEMA HP100-1985, High Temperature Instrumentation and Control Cables.

[B19] NEMA HP100.1-1985, High Temperature Instrumentation and Control Cables Insulated and Jacketed with FEP Fluorocarbons.

[B20] NEMA HP100.2-1985, High Temperature Instrumentation and Control Cables Insulated and Jacketed with ETFE Fluoropolymers.

[B21] NEMA HP100.3-1987, High Temperature Instrumentation and Control Cables Insulated and Jacketed with Cross-Linked (Thermoset) Polyolefin (XLPO).

[B22] NEMA HP100.4-1985, High Temperature Instrumentation and Control Cables Insulated and Jacketed with ECTFE Fluoropolymers.

[B23] NEMA WC50-1976 (Reaff. 1982 and 1988), Ampacities, Including Effect of Shield Losses for Single-Conductor Solid-Dielectric Power Cable 15 kV Through 69 kV (ICEA P-53-426, Second Edition).

[B24] NEMA WC55-1986, Instrumentation Cables and Thermocouple Wire (ICEA S-82-552).

[B25] NEMA WC57-1990, Standard for Control Cables (ICEA S-73-532).

[B26] UL 13-1990, Power Limited Circuit Cables.

[B27] UL 910-1991, Test Method for Fire and Smoke Characteristics of Electrical and Optical-Fiber Cables Used in Air-Handling Spaces.

[B28] UL 1072-1986, Medium-Voltage Power Cables.

[B29] UL 1277-1989, Electrical Power and Control Tray Cables with Optional Optical-Fiber Members.

[B30] *Underground Systems Reference Book*, Chapter 10, New York: Association of Illuminating Companies, 1957.

Chapter 9
System Protection and Coordination

9.1 General Discussion. Electric power systems in commercial and institutional buildings should be designed to serve loads in a safe and reliable manner. One of the major considerations in the design of a power system is the adequate control of phase-to-ground, phase-to-phase, and three-phase short-circuit faults. Short-circuit current is an overcurrent resulting from a fault of negligible impedance between live conductors having a difference in potential under normal operating conditions. The fault path may include the path from active conductors via earth to the neutral. Uncontrolled short circuits can cause service outages with accompanying lost time and associated inconvenience, interruption of essential facilities or vital services, extensive equipment damage, fire damage, and possibly personnel injury or fatality.

Electric power systems should be as fault-free as possible through careful system and equipment design and should be properly installed and maintained. However, even with these precautions, faults do occur. Some precipitating causes are loose connections; voltage surges; deterioration of insulation; accumulation of moisture, vermin, or rodents; seepage from concrete; contaminants; concrete and/or cement dust; intrusion of metallic or conducting objects, such as fish tapes, tools, core drills, jackhammers, or construction equipment; and undetermined phenomena.

When a short circuit occurs on a power system, undesirable things happen.

(1) Electrical arcs, flashes, and burning can occur at the fault location with consequent smoke generation from the fuel load of combustibles.

(2) Increased current will flow from the various sources to the fault location. All components carrying the fault currents are subject to increased thermal and mechanical stress. This mechanical stress varies as a function of the peak current squared and the thermal stress varies as a function of both rms current squared and of the duration of current flow (I^2t).

(3) Voltages decrease throughout the system for the duration of the fault, voltage drops in proportion to the magnitude of the current; maximum voltage drop will occur at the fault location (to zero voltage for bolted fault).

(4) Enclosures that are in contact with live conductors can be subject to elevated voltages and can increase the hazard of electric shock.

The fault should be quickly removed from the power system to minimize the effects of these undesirable conditions, including arcing and burning. This is the job of the circuit protective devices, circuit breakers, and fuses. The protective device should have the ability to interrupt the maximum short-circuit current, which can flow for a bolted fault at the device location.

All conductive components should have the capability to carry the short-circuit current until it is successfully interrupted. Equipment grounding should be adequate to limit voltage on faulted enclosures to safe values.

The bolted fault value of short-circuit current results when the fault offers no impedance to the flow of short-circuit current and the magnitude of current is limited only by the impedance of the circuit elements. This condition results in a maximum short-circuit current and is frequently referred to as the available short-circuit current. Bolted short circuits are very rare, however, and the fault usually involves arcing and burning. Under these conditions, fault currents may be much lower than bolted fault values and may present special problems of detection and isolation.

When the fault involves ground, as it very often will, the protective enclosure may experience elevated potential, which can increase the exposure of personnel to shock hazard. The likelihood of injury and death increases as a function of shock voltage and duration. It is important to maintain adequate equipment grounds to minimize exposure voltage and to rapidly detect and isolate the fault to reduce the duration of exposure.

For a simple example, consider Fig 79(a). The impedance that determines the flow of load current is the 20 Ω impedance of the motor. If a short circuit occurs at F, the only impedance limiting the flow of short-circuit current is the transformer impedance (0.1 Ω compared with 20 Ω for the motor). Therefore, the short-circuit current is 1000 A, or 200 times as great as the load current. Consequently, the circuit protective device should have the ability to safely interrupt 1000 A.

If the load grows and a larger transformer is substituted for the original unit, then the short circuit at F_1 (see Fig 79(b)) becomes limited by 0.01 Ω, which is the impedance of the larger transformer. Although the load current is still 5 A, the short-circuit current increases to 10 000 A, which the circuit protective device should be able to interrupt.

9.1.1 Single- and Multiple-Pole Interrupters. Circuit breakers and fuses can be designed for single- or multiple-pole use. The protective function of circuit breakers automatically actuates the switching function. Generally, fused switches including safety switches, service protectors, and bolted pressure switches have separate protective and switching functions. If one pole of the multiple-pole circuit is actuated by the protective sensors, all poles are usually opened simultaneously. This same feature can be specified for service protectors, bolted pressure switches, and certain other types of fused switches.

Single or unbalanced phase voltage conditions that result from loss of voltage on one line conductor of a multiple-phase or three-wire single-phase system may arise from the failure of the utility supply, system defects, or operation of single-pole interrupters. Under such conditions, part of the system lighting remains energized, while some portions of the system may be subjected to undervoltage; unbalanced voltages; backfeeds through loads, including voltages from rotating equipment; or prolonged faults. The designer should evaluate the extent of protection required to

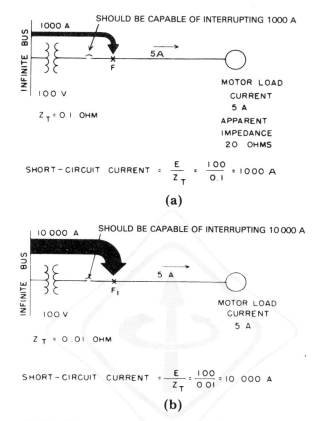

SHORT-CIRCUIT CURRENT $= \dfrac{E}{Z_T} = \dfrac{100}{0.1} = 1000$ A

(a)

SHORT-CIRCUIT CURRENT $= \dfrac{E}{Z_T} = \dfrac{100}{0.01} = 10\,000$ A

(b)

NOTE: Values were chosen for illustrative purposes only.

**Fig 79
Short Circuit on Load Side of Main Switch**

provide an effective system including undervoltage protection, ground-fault protection, and their relationships to the type of circuit operating device used.

9.1.2 Sources of Short-Circuit Currents. When determining the magnitude of short-circuit currents, it is extremely important that all sources of short circuit be considered and that the impedance characteristics of these sources be known.

There are four basic sources or short-circuit current

(1) Local system generators
(2) Synchronous motors
(3) Induction motors
(4) Electric utility systems (remote generation)

All of these can feed current into a fault (see Fig 80).

9.1.3 Rotating Machine Reactance. The impedance of a rotating machine consists primarily of reactance and is not one simple value (as for a transformer or a

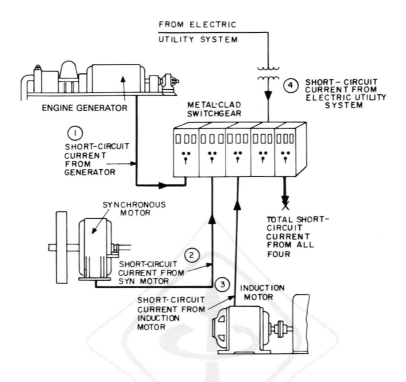

Fig 80
Total Short-Circuit Current Equals Sum of Sources

piece of cable). It is also complex and variable with time. For example, if a short circuit is applied to the terminals of a generator, the short circuit starts out at a high value and decays to a steady-state value after some time has elapsed from its inception. Since the field excitation voltage and speed have remained relatively constant within the short interval of time after inception of the fault, the reactance of the machine may be assumed to have changed with time, which explains the change in the current value.

Expression of such a variable reactance at any instant requires a complicated formula involving time as one of the variables. Therefore, for the sake of simplification, three values of reactance are assigned to rotating machines (motors and generators) for the purpose of calculating short-circuit currents at specified times. These values are called the subtransient reactance, transient reactance, and synchronous reactance. They are described as follows:

(1) The subtransient reactance X''_d is the apparent reactance of the stator winding at the instant the short circuit occurs, and it determines the current flow during the first few cycles after the short circuit.

(2) The transient reactance X'_d determines the current during the period following that when the subtransient reactance is the controlling value; it is effective up to 0.5 second or longer, depending upon the design of the machine.

(3) The synchronous reactance X_d is the reactance that determines the current flow when a steady-state condition is reached. It is not effective until several seconds after the short circuit occurs; consequently, it is not generally used in short-circuit current calculations.

A synchronous motor (or generator) has the same kinds of reactance as an induction motor but usually has different values. Induction motors have no field coils; but the rotor bars act like the amortisseur winding in a generator. Therefore, induction motors are said to have subtransient reactance only.

9.1.4 Utility Source. The available utility three-phase short-circuit current and three-phase short-circuit X/R ratio plus single line-to-ground short-circuit current and single line-to-ground short-circuit X/R ratio should be obtained from the serving utility. If the data furnished are at the primary voltage, it should be modified by the transformer impedance and voltage ratio.

9.1.5 Symmetrical and Asymmetrical Currents. The word "symmetrical" describes the displacement of the ac waves from the zero axis. If the envelopes of the peaks of the current waves are symmetrical around the zero axis, they are called "symmetrical current envelopes" (see Fig 81). If the envelopes are not symmetrical around the zero axis, they are called "asymmetrical current envelopes" (see Fig 82). The envelope is a line drawn through the peaks of the waves. The magnitude of the dc component of an asymmetrical current at any instant is the value of the offset between the axis of symmetry of the asymmetrical current and the zero axis (see Figs 87 and 88).

Most short-circuit currents are asymmetrical during the first few cycles after the short circuit occurs. The asymmetrical current is at a maximum during the first cycle after the short-circuit occurs and, in a few cycles, gradually becomes symmetrical. An oscillogram of a typical short-circuit current is shown in Fig 83 (see also Figs 87 and 88).

9.1.6 Why Are Short-Circuit Currents Asymmetrical? In ordinary power systems, the applied or generated voltage wave shapes are sinusoidal. When a short

Fig 81
Symmetrical AC Wave

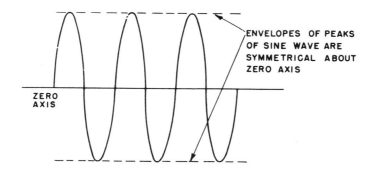

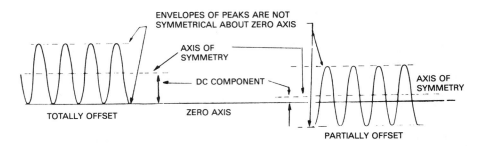

Fig 82
Asymmetrical AC Wave

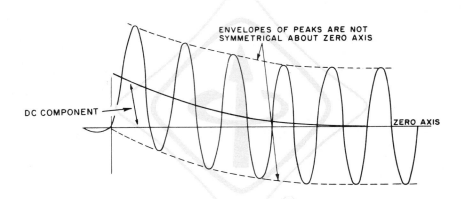

Fig 83
Typical Short Circuit

circuit occurs, approximately sinusoidal currents result. The following discussion assumes sinusoidal wave voltages and currents.

The power factor of a current circuit is determined by the series resistance and reactance of the circuit (from the fault back to and including the source or sources of the short-circuit currents). For example, in Fig 84, the reactance equals 19%, the resistance equals 1.4%, and the short-circuit power factor equals 7.3%, which is determined by the formula

$$\text{power factor} = \frac{R}{\sqrt{R^2 + X^2}} \qquad \text{(Eq 15)}$$

The relationship of the resistance and reactance of a circuit is sometimes expressed in terms of the X/R ratio of the circuit, which is 13.6 (see Fig 84).

In high-voltage power circuits, the resistance of the circuit back to and including the power source is low compared with the reactance of the circuit. Therefore, the

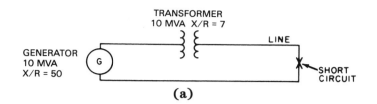

(a)

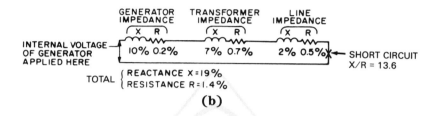

(b)

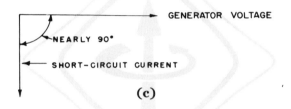

(c)

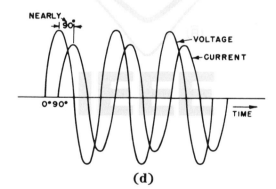

(d)

Fig 84
Phase Relations of Voltage and Short-Circuit Currents
(Medium-Voltage Generator Feeding a Distribution Line)
(a) Circuit Diagram
(b) Impedance Diagram
(c) Vector Diagram
(d) Sine Waves Corresponding to Vector Diagram (c) for Circuit (a)

short-circuit current lags the source voltage by almost 90° (see Fig 84). Low-voltage power circuits (below 600 V) tend to have a larger percentage of resistance, and the current will lag behind the voltage by less than 90°.

If a short circuit occurs at the peak of the voltage wave in a circuit containing only reactance, the short-circuit current will start at zero and trace a sine wave, which will be symmetrical about the zero axis (see Fig 85). If a short circuit occurs at the zero point of the voltage wave, the current will start at zero; but it cannot follow a sine wave symmetrically about the zero axis because the current should lag behind the voltage by 90°. This can only happen if the current is displaced from the zero axis as shown in Fig 86.

The two cases shown in Figs 85 and 86 are extremes. Figure 85 shows a totally symmetrical current, and Fig 86 shows a completely asymmetrical current. If the fault occurs at any point between zero voltage and peak voltage, the current will be asymmetrical to a degree dependent upon the point at which the short circuit occurs on the voltage wave.

To produce maximum asymmetry, when a circuit contains resistance, the short circuit should always occur at the zero point on the voltage wave. However, the point on the voltage wave at which the short circuit should occur to produce a symmetrical short-circuit current wave depends on the ratio of reactance to resistance (X/R ratio). The actual point on the voltage wave at which a short circuit should be initiated to produce a symmetrical current is the angle whose tangent equals the X/R ratio of the circuit.

For example, when $X/R = 6.6$ (15% pf), the angle on the voltage wave = arctan 6.6 = 81.384°; when $X/R = 3.0$, the angle on the voltage wave = arctan 3.0 = 71.565°.

9.1.7 The DC Component of Asymmetrical Short-Circuit Currents. Asymmetrical currents are analyzed in terms of two components, a symmetrical current and a dc component, as shown in Fig 87. As previously discussed, the symmetrical component is at a maximum at the inception of the short circuit and decays to a steady-

**Fig 85
Symmetrical Current and Voltage
in a Zero Power Factor Circuit**

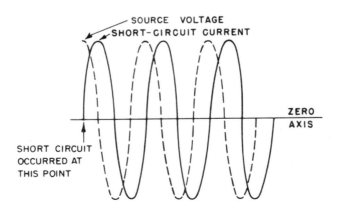

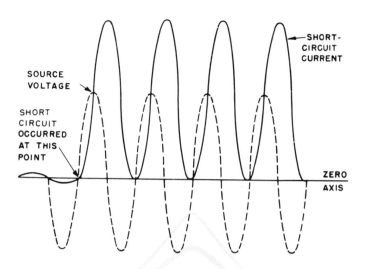

**Fig 86
Asymmetrical Current and Voltage
in a Zero Power Factor Circuit**

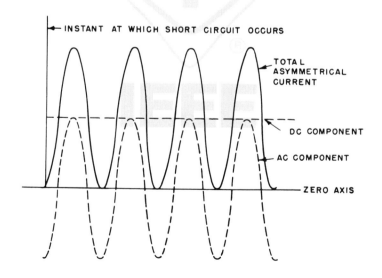

**Fig 87
Components of Current Shown in Fig 86**

state value due to the apparent change in machine reactance. In all practical circuits (that is, those containing resistance), the dc component will also decay (to zero) as the energy represented by the dc component is dissipated as I^2R loss in the resistance of the circuit. Figure 88 illustrates the decay of the dc component.

The rate of decay of the dc component is a function of the resistance and reactance of the circuit. In practical low-voltage circuits, the dc component decays to zero in from one to six cycles.

9.1.8 Total Short-Circuit Current. The total symmetrical short-circuit current usually has several sources, as illustrated in Fig 89. The first source is the utility, the second is local generation, and synchronous motors, if any, are a third source. Induction motors, a fourth source, are located in every building. Because rotating machine currents usually decay over time due to the reduction of flux in the machine after a short circuit, the total short-circuit current decays with time (see Fig 89). Considering only the symmetrical part of the short-circuit current, the magnitude is highest at the first half-cycle after a short circuit and is of lower value a few cycles later. Note that the induction motor component will almost entirely disappear after one or two cycles, except for very large motors where it may be present longer than four cycles.

The magnitude during the first few cycles is further increased by the dc component. This component also decays with time, accentuating the difference in magnitude of a short-circuit current at the first cycle after a short circuit occurs a few cycles later. The maximum asymmetrical current is available on only one phase of a three-phase system due to a three-phase fault.

Fig 88
Decay of DC Component and
Effect of Asymmetry of Current

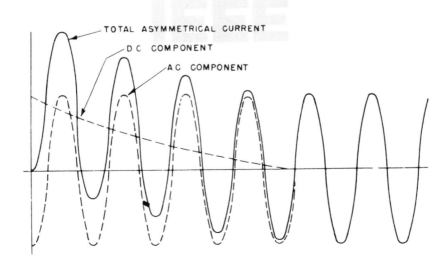

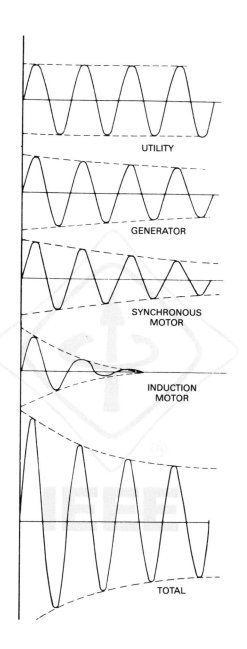

**Fig 89
Symmetrical Short-Circuit Currents
from Four Sources Combined into Total**

9.2 Short-Circuit Calculations. The calculation of the precise value of an asymmetrical current at a given time after the inception of a fault is rather complex. Consequently, simplified methods have been developed that yield the short-circuit currents that are required to match the assigned ratings of various system protective devices and equipment.

The value of the symmetrical short-circuit current is determined through the use of the proper impedance in the following equation:

$$I = E/Z \qquad\qquad \text{(Eq 16)}$$

where

E = The system driving voltage.

Z (or X) = The proper system impedance (or reactance) of the power system back to and including the source or sources of the short-circuit current.

The value of the proper impedance is determined with regard to the basis of rating for the device or equipment under consideration.

9.2.1 Type of Power System Faults. Faults or short circuits on a three-phase power system can be of several types. The protective device or equipment should have the ability to interrupt or withstand the fault current, and conductive components should have the ability to withstand the resulting mechanical and thermal stresses for any type of fault that can occur. The basic types of faults will be described; but it should be noted that the basic fault calculation for the selection of equipment is the three-phase bolted fault.

9.2.2 Three-Phase Bolted Faults. A three-phase bolted fault describes the condition where the three phase conductors are physically held together with zero impedance between them, just as if they were bolted together. This type of fault condition is not the most frequent in occurrence; however, it generally results in maximum short-circuit values and, for this reason, is the basic fault calculation in commercial power systems.

9.2.3 Line-to-Line Bolted Faults. In most three-phase power systems, the line-to-line bolted fault currents are approximately 87% of the three-phase bolted fault currents. A detailed calculation is seldom required.

9.2.4 Line-to-Ground Bolted Faults. In solidly grounded systems, the line-to-ground bolted fault current value is usually about equal to the three-phase bolted fault current value for the location being examined. Under certain conditions, such as a bolted line-to-ground fault at the secondary terminal of the connected transformer, the line-to-ground bolted fault current value can theoretically exceed the three-phase bolted current value (however, tests show that, in practical systems, the ground-fault current is less than the bolted three-phase fault current). Most often, the ground-fault current will be significantly lower than the three-phase bolted fault current due to the relatively high impedance of the ground return circuit (i.e., conduit, busway enclosure, grounding conductor, etc.).

In resistance grounded high-voltage systems, the resistor is generally selected to limit the ground-fault current to a value ranging between 1–2000 A. Line-to-ground fault magnitudes on these systems are limited primarily by the resistor itself, and a complicated line-to-ground short-circuit current calculation is generally not required.

9.2.5 Arcing Faults. Power system faults may also be arcing in nature. Arcing faults can display a much lower level of short-circuit current than a bolted fault at the same location. These lower levels of current are due in part to the impedance of the arc. While system components should be capable of interrupting and withstanding the thermal and mechanical stresses of bolted short-circuit currents, arcing faults usually present different problems. Arcing faults may be difficult to detect because of the smaller currents. Sustained arcs can present safety hazards to people and also cause extensive damage because of the burning and welding effect of the arc as well as from the conductive products of ionization. Table 63 presents multipliers that can be applied to bolted fault currents at point of fault to estimate approximate values of arcing fault currents compared to bolted fault values.

9.3 Selection of Equipment. To provide for personnel safety, to minimize equipment damage, and to maintain a high degree of service continuity, equipment should be selected to detect faults quickly and accurately and to remove them in the shortest possible period of time.

Most protective devices employ the detection of current for operation. Fuses and certain types of circuit breakers are inherently current sensitive. A wide variety of protective relays is available to detect abnormal conditions of voltage, frequency, or real or reactive power. Relays can be used to determine current or power direction, and differential relays can be used to compare current magnitude and direction at two or more locations. Relays are only detecting devices and should be used in conjunction with circuit breakers or motorized switches to remove the detected faults.

The circuit protective devices should be selected to successfully detect and interrupt the fault condition rapidly enough so that any circuit element is not subjected to conditions beyond its rating. Proper selection is dependent upon a knowledge of the magnitudes of short-circuit current that can be expected for the various types of faults that may be experienced. Short-circuit calculations are the method by which these values are predicted.

Table 63
Approximate Minimum Values of
Arcing Fault Currents in Per Unit of Bolted Values

Type of Fault	Nominal System Voltage		
	600 V	480 V	208 V
Three-phase	0.94	0.89	0.12
Single-phase, line-to-line	0.85	0.74	0.02
Single-phase, line-to-ground	0.40	0.38	0
Three-phase, one transformer primary fuse open	0.88	0.80	0

9.3.1 Equipment Rating. To provide for personnel safety and to minimize equipment damage, it is absolutely essential to use equipment with short-circuit ratings equal to or greater than the available short-circuit current to which the equipment can be subjected. ANSI/NFPA 70-1990, National Electrical Code (NEC), Section 110-9 [7][66] states that "devices intended to break current shall have an interrupting rating sufficient for the voltage employed and for the current which must be interrupted."

For any given location, there may be a choice of one of several types of protective devices. Selection of a specific device then depends on factors such as protective characteristics, economics, component protection, maintainability, user preference, etc.

Equipment can be applied at a location where the available short-circuit current is higher than the short-circuit rating of the device, provided that current-limiting fuses or circuit breakers "upstream" from the device limit the "let-through" current to a level the "downstream" equipment can withstand.

The calculated available short-circuit current may be found on the line side of the device. For a fault on the load side of the device, the actual current that the device does interrupt may be less than the available current due to the impedance of the device, the impedance of the arc on contact parting, and the ability of the device to limit current as in the case of a current-limiting fuse or circuit breaker. The basic concept is that the device should have the ability, when applied at a location with a given available short-circuit current, to safely interrupt a fault at its load terminals.

It is also necessary to identify the short-circuit rating of circuit conducting components, such as busway, bus structures within switchgear and panelboards, and insulated conductors. The short-circuit rating refers to the ability of the equipment to withstand the available short-circuit current under specified test conditions. Short-circuit ratings for noninterrupting circuit components should exceed the let-through currents of overcurrent protective devices in the circuit (if it is lower than the calculated available short-circuit current at the point of the fault).

9.4 Basis of Short-Circuit Current Calculations. The basis for rating interrupting devices and the time after the inception of the fault at which the devices operate determines the type of short-circuit calculation required.

Most equipment is rated on a symmetrical basis. However, unusual circuit configurations may result in a device being applied within its symmetrical rating; but it will be subjected to asymmetrical currents beyond its capabilities. To avoid this possibility, all short-circuit calculations should include a consideration of the X/R ratio at the point of fault. The X/R ratio establishes the power factor of the short-circuit current. Data are available to allow estimation of asymmetrical multiplying factors based upon the X/R ratio and circuit power factor.

---

[66]The numbers in brackets correspond to those in the references at the end of each chapter. ANSI publications are available from the Sales Department of the American National Standards Institute, 11 West 42nd Street, 13th Floor, New York, NY 10036. NFPA publications are available from Publications Sales, National Fire Protection Association, 1 Batterymarch Park, P.O. Box 9101, Quincy, MA 02269-9101.

Since the short-circuit current may change during the time following the inception of a fault, the speed of operation and the basis for rating the devices establishes the circuit impedances to be used in the basic equation $I = E/Z$.

9.4.1 Total Current Basis of Rating. ANSI C37.6-1971, Schedule of Preferred Ratings for AC High-Voltage Circuit Breakers Rated on a Total Current Basis [4] lists high-voltage circuit breakers. (Some standards still refer to voltages above 600 V or 1000 V as "high voltage." ANSI C84.1-1989, Electric Power Systems and Equipment — Voltage Ratings (60 Hz) [B1][67] lists 1000–100 000 V as "medium voltage." For reasons of convenience in this chapter, we retain the term "high voltage.") IEEE C37.5-1953, Determining the Rms Value of a Sinusoidal Current Wave and a Normal-Frequency Recovery Voltage and for Simplified Calculation of Fault Currents [2] previously described the calculation of short-circuit duties to apply these circuit breakers. It was superseded by ANSI C37.5-1969, Methods for Determining Values of a Sinusoidal Current Wave, a Normal-Frequency Recovery Voltage, and a Guide for Calculation of Fault Currents for Application of AC High-Voltage Circuit Breakers Rated on a Total Current Basis [3], which describes a revised calculation for obtaining short-circuit duties to apply to total current rated circuit breakers. Both IEEE C37.5-1979, IEEE Guide for Calculation of Fault Currents for Application of AC High-Voltage Circuit Breakers Rated on a Total Current Basis [12][68] and C37.6-1971 [4] have been withdrawn because all modern high-voltage circuit breakers are rated on the basis of symmetrical current.

The first-cycle duty (momentary) was determined by ANSI C37.5-1953 [2] as follows:

(1) A symmetrical short-circuit current value was calculated using the subtransient reactance X_d'' for all sources of short-circuit current in the equivalent circuit of the power system.

(2) Multiplying factors were applied to this calculated symmetrical value to determine asymmetrical short-circuit duty. In the revised calculation procedure in ANSI C37.5-1969 [3], the first-cycle duty (momentary) calculation is very similar. Differences occur in modified reactance values for small and medium-sized induction motors.

According to ANSI C37.5-1953 [2], the interrupting duty was determined using an equivalent circuit with the subtransient reactance X_d'' for synchronous generators, the transient reactance X_d' for synchronous motors, and ignoring the contribution of induction motors. The short-circuit interrupting current calculated from the circuit was then multiplied by a factor that depends on the circuit breaker rated interrupting time and on the power system operating conditions.

The contact parting time short-circuit (interrupting) duty calculated by the ANSI C37.5-1969 [3] method used subtransient reactance X_d'' for synchronous generators; 1.5 times subtransient reactance, 1.5 X_d'', for synchronous motors; and modified subtransient reactances for induction motors that are divided into three

---

[67] The numbers in brackets preceded by a B refer to the bibliographic references that are at the end of this chapter.

[68] IEEE publications are available from the Institute of Electrical and Electronics Engineers, IEEE Service Center, 445 Hoes Lane, P.O. Box 1331, Piscataway, NJ 08855-1331.

categories, each with a different reactance multiplier in the power system reactance network equivalent circuit. The circuits then reduced to an equivalent X (reactance) value, and an E/X symmetrical short-circuit current was calculated. Then a multiplying factor obtained from curves in ANSI C37.5-1969 [3] was applied to obtain the total short-circuit duty to be compared with the capability of a total current rated circuit breaker. The multiplying factor depended on the circuit breaker contact parting time, the fault point X/R ratio, and the proximity of generation. IEEE C37.5-1979 (ANSI) [12] described the fault point X/R ratio calculation utilizing a resistance network corresponding to the reactance network.

Low-voltage protective devices and equipment, including power circuit breakers, molded-case circuit breakers, motor control centers, motor controllers, fuses, and busway are rated on the basis of the maximum available symmetrical current at some specified power factor (X/R ratio). Their short-circuit ratings are based on current during the first cycle only. Therefore, the subtransient reactance X_d'' is used for all sources of short-circuit current.

9.4.2 Symmetrical Current Basis of Rating. ANSI C37.06-1987, Preferred Ratings and Related Required Capabilities for AC High-Voltage Circuit Breakers Rated on a Symmetrical Current Basis [1] lists the high-voltage circuit breakers. The rated symmetrical short-circuit current listed for a circuit breaker in this standard applies only at rated maximum voltage. The short-circuit capability at a lower actual operating voltage is higher and is found by multiplying the rated short-circuit current by the voltage ratio (maximum rated voltage/operating voltage) for voltage between rated maximum voltage and $1/K$ times rated maximum voltage [K = rated voltage range factor]).

The calculation method used to apply symmetrically rated circuit breakers is described in IEEE C37.010-1979 (Reaff. 1988), IEEE Application Guide for AC High-Voltage Circuit Breakers Rated on a Symmetrical Current Basis (ANSI) [11]. The first-cycle duty calculation by this standard is exactly the same as in IEEE C37.5-1979 (ANSI) [12]. The result is an asymmetrical first-cycle duty that is compared with the asymmetrical closing and latching capabilities of the symmetrically rated circuit breaker.

The contact parting time short-circuit (interrupting) duty calculation, as described in IEEE C37.010-1979 (ANSI) [11], uses the same reactance network as the calculation described in IEEE C37.5-1979 (ANSI) [12] and the same E/X calculation current value. A different multiplying factor is applied to E/X to establish the duty to be compared with the symmetrical short-circuit interrupting capability of a symmetrically rated circuit breaker.

As long as the X/R ratio for each network element or the fault point X/R ratio is 15 or less, the multiplying factor is 1.0. When the X/R ratio is 15 or less, the asymmetrical short-circuit duty never exceeds the symmetrical short-circuit duty by a margin greater than that by which the circuit breaker's asymmetrical short-circuit capability, as required by the standards, exceeds its symmetrical short-circuit capability.

When the X/R ratio exceeds 15, the multiplier usually exceeds 1.0. Multiplying factors are determined from curves in IEEE C37.010-1979 (ANSI) [11] and depend on the contact parting (interrupting) time of the circuit breaker. The fault point

X/R ratio calculation from IEEE C37.010-1979 (ANSI) [11] is the same as the calculation in IEEE C37.5-1979 (ANSI) [12].

9.4.3 Comparison of Duty Calculation Methods. Calculation methods in IEEE C37.5-1979 (ANSI) [12] (for total current basis rated circuit breakers) and IEEE C37.010-1979 (ANSI) [11] (for symmetrical current basis rated circuit breakers) differ from ANSI C37.5-1953 [2] principally in data collection (not only reactance values, but also *X/R* ratios or resistance values are needed for system components) and in the treatment of reactances.

The first-cycle (momentary) duty calculated by present methods will not generally differ from that calculated by the earlier method. The interrupting duty calculated by the present method is often higher because of the increased motor contributions recognized. For a further description of these procedures, see IEEE Std 141-1986, IEEE Recommended Practice for Electric Power Distribution for Industrial Plants (ANSI) [19].

9.5 Details of Short-Circuit Current Calculations. The general nature of ac short-circuit currents has been discussed in 9.2, and it was determined that the basic equation for the calculation of short-circuit currents is $I = E/Z$, where

E = The system driving voltage.
Z (or X) = The proper impedance (or reactance) of the power system back to and including the source or sources of the short-circuit current.

The proper value of impedance depends on the basis of the short-circuit rating for the device or equipment under consideration.

In this section, details of the short-circuit current calculations will be presented. Much of the work of such a study involves the representation of the proper system impedances from the point of fault back to and including the source(s) of short-circuit current.

9.5.1 Step-by-Step Procedure. The following steps identify the basic considerations in making short-circuit current calculations. In the simpler systems, several steps may be combined; for example, a combined single-line and impedance diagram may be used.

(1) Prepare a system single-line diagram, which is fundamental to short-circuit analysis. It should include all significant equipment and components and show their interconnections. Figure 90 illustrates a typical system single-line diagram.

(2) Decide on fault locations and the type of short-circuit current calculations required, based on the type of equipment being applied. Consider the variations of system operating conditions that are required to display the most severe duties. Assign bus numbers of suitable identification to the fault locations.

(3) Prepare an impedance diagram. For systems above 600 V, two diagrams are usually required to calculate interrupting and momentary duty for high-voltage circuit breakers. Determine the type of short-circuit current rating required for various kinds of equipment as well as the machine reactances

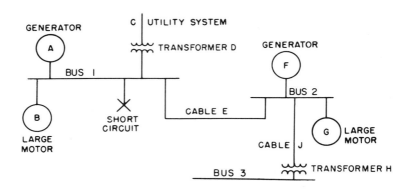

Fig 90
Typical System Single-Line Diagram

to use in the impedance diagram. Select suitable kVA and voltage bases for the study when the per unit system is used.

In order to develop accurate fault currents, it is necessary to know the subtransient and transient reactances of synchronous machines and the subtransient reactances of induction machines. In calculating the short-circuit currents of low-voltage systems, a realistic approximation involving a mix of synchronous and induction machines assumes a contribution at the machine terminals under bolted conditions of four times rated full-load current. This implies a source reactance of approximately 25%.

(4) For the designated fault locations and system conditions, resolve the impedance network and calculate the required symmetrical currents (E/Z or E/X). When calculations are made on a computer, submit impedance data in the proper form as required by the specific program.

9.5.2 System Conditions for Most Severe Duty. Sometimes, several of the intended or possible system conditions should be investigated to reveal the most severe duties for various components. Severe duties are those that are most likely to tax the capabilities of components.

Future growth and change in the system can modify short-circuit currents. For example, the initial utility available short-circuit duty for an in-building system being investigated may be 150 MVA; but future growth plans may call for an increase in available duty to 750 MVA several years later. This increase could substantially raise the short-circuit duties on the in-building equipment. Therefore, the increase should be included in present calculations so that adequate in-building equipment can be selected. In a similar manner, future in-building expansions very often will raise short-circuit duties in various parts of the power system, so that future expansions should also be considered.

The most severe duty usually occurs when the maximum concentration of machinery is in operation and all interconnections are closed. To determine the conditions that will most likely influence the critical duty, the following questions should be answered:

(1) Which machines and circuits are to be considered in actual operations?

(2) Which switching units are to be open, which closed?

(3) What future expansions or system changes will affect in-building short-circuit currents?

9.5.3 Preparing Impedance Diagrams. The impedance diagram displays the interconnected circuit impedance that controls the magnitude of short-circuit currents. The diagram is derived from the system single-line diagram, showing an impedance for every system component that exerts a significant effect on short-circuit magnitude. Not only should the impedance be interconnected to reproduce actual circuit conditions; but it would be helpful to preserve the same arrangement pattern used in the single-line diagram (see Fig 91).

9.5.4 Component Impedance Values. As they are collected, component impedance values may be expressed in terms of any of the following units:

(1) Ohms per phase (actually line-to-neutral single-phase impedance)

(2) Percent on rated kVA or a reference kVA base

(3) Per unit on reference kVA base

In formulating the impedance diagram, all impedance values should be expressed in the same units, either in Ω per phase or per unit on a reference kVA base. (Percent is a form of per unit — percent = per unit $\times$ 100.)

9.5.5 Combining Impedances. An impedance Z containing resistance R and reactance X is a complex quantity analyzed as a vector. It is frequently expressed in the form $R + jX$ as illustrated in Fig 92.

When combining impedances in series, the magnitudes of the impedance Z cannot be added directly. The resistance R and reactance X should be added separately, and then Z can be computed ($Z = \sqrt{R^2 + X^2}$). Figure 93 illustrates the addition of impedance in series.

**Fig 91
Equivalent Impedance Diagram
for System in Fig 90**

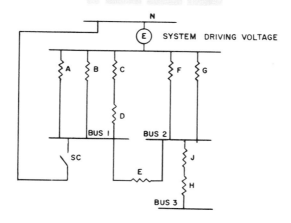

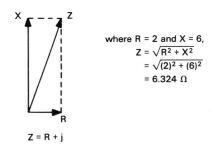

$$Z = R + j$$

**Fig 92
Impedance Vectors**

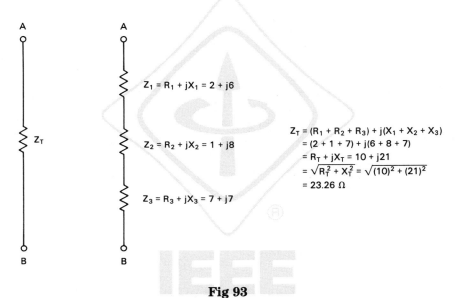

**Fig 93
How Series Impedances Are Added**

When combining several impedances in parallel, the equivalent impedance is found by taking the reciprocal of the sum of several impedance reciprocals, using the expression

$$1/Z_T = 1/Z_1 + 1/Z_2$$

The following formulas are used to find impedance reciprocals where $Z = R + jX$ and $1/Z \; (= Y) = G + jB$.

Components of $1/Z$ found from components of Z are

$$G = R/(R^2 + X^2)$$
$$-B = X/(R^2 + X^2)$$

(Eq 17)

Components of Z found from components of $1/Z$ are

$$R = G /(G^2 + B^2)$$
$$X = -B /(G^2 + B^2)$$

(Eq 18)

Figure 94 illustrates the addition of reciprocals to find an equivalent impedance, using a table for recording the calculation steps.

9.5.6 Use of Per Unit, Percent, or Ohms. Short-circuit current calculations can be made with impedances represented in per unit, percent, or ohms. All representations yield identical results. A single system should be used throughout any calculation and, at the outset, this decision should be made.

In general, if the system being studied has several different voltage levels or is a high-voltage system, per unit impedance representation will often provide the easier, more straightforward calculation. The per unit system is ideal for studying multiple-voltage systems. Also, data for most of the component included in high-voltage networks (machines, transformers, and utility systems) are given in per unit or percent values, making further conversion simple.

Percent impedance representation is only a variation of the per unit system

percent impedance = per unit impedance (100)

(Eq 19)

**Fig 94
Combining Impedances in Parallel**

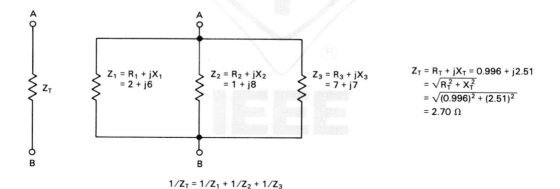

$1/Z_T = 1/Z_1 + 1/Z_2 + 1/Z_3$

Element	Z R	X	1/Z G	−B	Sum of Squares*
Z_1	2.0	6.0 →	0.05	0.15	40.0
Z_2	1.0	8.0 →	0.0154	0.1231	65.0
Z_3	7.0	7.0 →	0.0714	0.0714	98.0
Z_T	0.996	2.51 ←	0.1368	0.3445	0.1374

*$R^2 + X^2$ when finding $1/Z$ from Z and $G^2 + B^2$ when finding Z from $1/Z$.

For commercial building short-circuit calculations, the percent and per unit methods are equivalent, and both have the same advantages.

Where few or no voltage transformations are involved and for low-voltage systems in which many conductors are included in the impedance network, representation of system elements in Ω may provide easier, more straightforward calculations.

Characteristic impedance data is given for system components in Tables 64–66. Such data is commonly given as a percentage based on the equipment kVA rating or as an ohmic value. The conversion equations for the three systems are

$$\text{per unit impedance} = \frac{\text{percent impedance}}{100}$$

$$\text{per unit impedance (on chosen kVA base)} = \frac{\Omega \times \text{kVA base}}{1000 \times \text{kV}^2}$$

(Eq 20)

Table 64
Transformers

(a)
**Typical Per Unit R and X Values for Indoor, Open Dry-Type
150 °C Rise Transformers Rated from 15–2500 kVA, Three-Phase,
2.5–15 kV Primaries, 208, 240, 480, 600 V Wye or Delta Secondaries**

kVA	HV (kV)	LV (V)	%Z	X/R	R	X
15	2.5–15	208Y–600	3.00	0.5	0.027	0.013
30	2.5–15	208Y–600	5.00	1.0	0.035	0.035
45	2.5–15	208Y–600	5.00	1.0	0.035	0.036
75	2.5–15	208Y–600	5.50	2.0	0.025	0.049
112.5	2.5–15	208Y–600	4.50	1.5	0.025	0.037
150	2.5–15	208Y–600	4.50	2.0	0.020	0.040
225	2.5–15	208Y–600	5.00	2.5	0.019	0.046
300	2.5–15	208Y–600	5.00	2.8	0.017	0.047
500	2.5–15	208Y–600	5.00	4.0	0.012	0.049
750	2.5–15	208Y–600	5.75	2.0	0.026	0.051
1000	2.5–15	208Y–600	5.75	2.5	0.021	0.053
1000	2.5–15	480Y	8.00	3.8	0.021	0.077
1500	2.5–15	208Y–600	5.75	3.3	0.017	0.055
2000	2.5–15	208Y–600	5.75	4.0	0.014	0.056
2500	2.5–15	208Y–600	5.75	4.3	0.013	0.056

(b)
**Typical Per Unit R and X Values for Indoor, Open Dry-Type
150 °C Rise Transformers Rated from 25–500 kVA, Single-Phase,
5 and 15 kV Primaries, 120/240 V Wye or Delta Secondaries**

kVA	HV (kV)	LV (V)	%Z	X/R	R	X
25	5		4	2	0.018	0.036
to	to	120/240	to	to		
500	15		6	4	0.015	0.058

Table 64 *(Continued)*

(c)
Typical Range of Per Unit Values for Indoor, Open Dry-Type
150 °C Rise Transformers Rated from 15–500 kVA, Three-Phase,
480 V Primary, 208 V Wye Secondary

kVA	%Z	X/R	R	X
15	4.5	0.41	0.042	0.017
to	to	to		
500	5.9	2.09	0.025	0.053

(d)
Typical Range of Per Unit R and X Values for Indoor, Open Dry-Type
150 °C Rise Transformers Rated from 5–167 kVA, Single-Phase,
240×480 V, 480 V, 600 V Primaries, 120/240 V Secondaries

kVA	HV (kV)	LV (V)	%Z	X/R	R	X
5	240×480		3	0.6	0.026	0.015
to	to	120/240	to	to		
167	600		6	2.0	0.027	0.051

where

Ω = Line-to-neutral values (single conductor).
kVA base = Three-phase base kVA.
kV = Line-to-line voltage.

9.5.7 Per Unit Representation. The per unit system is a method of expressing numbers in a form that allows them to be easily compared. Impedances of circuit components are, therefore, a ratio on a chosen base number, i.e., the chosen kVA base. The kVA base chosen may be the kVA rating of one of the predominant pieces of system equipment, such as a generator or transformer. However, an arbitrary number, such as 10 000 kVA, may be selected as the kVA base. The number selected should be one that will result in component impedances that are not excessively large or small and can be easily handled in the calculations.

Component impedance may be given on bases other than the chosen kVA base. The conversion equation for one kVA base to another is

$$\text{per unit impedance on new base} = \text{per unit on old base} \times \left(\frac{\text{new kVA base}}{\text{old kVA base}} \right)$$

(Eq 21)

The procedure for making short-circuit calculations using the per unit system is given in 9.5.1 and involves converting the impedance of each circuit element to a per unit value on the common kVA base. The network is resolved to the point of fault to obtain a total per unit fault impedance.

Table 65
Approximate Impedance Data—Insulated Conductors—60 Hz
(Ω/1000 feet Each Conductor)

Size AWG or kCM	Resistance (25 °C)				Reactance—600 V—THHN			
	Copper		Aluminum		Several 1/C		1 Multicond.	
	Metal	Nonmet.	Metal	Nonmet.	Mag.	Nonmag.	Mag.	Nonmag.
14	2.5700	Same	4.2200	Same	0.0493	0.0394	0.0351	0.0305
12	1.6200	Same	2.6600	Same	0.0468	0.0374	0.0333	0.0290
10	1.0180	Same	1.6700	Same	0.0463	0.0371	0.0337	0.0293
8	0.6404	Same	1.0500	Same	0.0475	0.0380	0.0351	0.0305
6	0.4100	Same	0.6740	Same	0.0437	0.0349	0.0324	0.0282
4	0.2590	Same	0.4240	Same	0.0441	0.0353	0.0328	0.0285
2	0.1640	0.1620	0.2660	Same	0.0420	0.0336	0.0313	0.0273
1	0.1303	0.1290	0.2110	Same	0.0427	0.0342	0.0319	0.0277
1/0	0.1040	0.1020	0.1680	Same	0.0417	0.0334	0.0312	0.0272
2/0	0.0835	0.0812	0.1330	Same	0.0409	0.0327	0.0306	0.0266
3/0	0.0668	0.0643	0.1060	0.1050	0.0400	0.0320	0.0300	0.0261
4/0	0.0534	0.0511	0.0844	0.0838	0.0393	0.0314	0.0295	0.0257
250	0.0457	0.0433	0.0722	0.0709	0.0399	0.0319	0.0299	0.0261
300	0.0385	0.0362	0.0602	0.0592	0.0393	0.0314	0.0295	0.0257
350	0.0333	0.0311	0.0520	0.0507	0.0383	0.0311	0.0388	0.0311
400	0.0297	0.0273	0.0460	0.0444	0.0385	0.0308	0.0286	0.0252
500	0.0244	0.0220	0.0375	0.0356	0.0379	0.0303	0.0279	0.0250
600	0.0209	0.0185	0.0319	0.0298	0.0382	0.0305	0.0278	0.0249
750	0.0174	0.0185	0.0264	0.0301	0.0376	0.0301	0.0271	0.0247
1000	0.0140	0.0115	0.0211	0.0182	0.0370	0.0296	0.0260	0.0243

NOTE: Increased resistance of conductors in magnetic raceway is due to the effect of hysteresis losses. The increased resistance of conductors in metal nonmagnetic raceway is due to the effect of eddy current losses. The effect is essentially equal for steel and aluminum raceway. Resistance values are acceptable for 600 V, 5 kV, and 15 kV insulated conductors.

Size AWG or kCM	Reactance—5 kV				Reactance—15 kV			
	Several 1/C		1 Multicond.		Several 1/C		1 Multicond.	
	Mag.	Nonmag.	Mag.	Nonmag.	Mag.	Nonmag.	Mag.	Nonmag.
8	0.0733	0.0586	0.0479	0.0417				
6	0.0681	0.0545	0.0447	0.0389	0.0842	0.0674	0.0584	0.0508
4	0.0633	0.0507	0.0418	0.0364	0.0783	0.0626	0.0543	0.0472
2	0.0591	0.0472	0.0393	0.0364	0.0727	0.0582	0.0505	0.0439
1	0.0571	0.0457	0.0382	0.0332	0.0701	0.0561	0.0487	0.0424
1/0	0.0537	0.0430	0.0360	0.0313	0.0701	0.0561	0.0487	0.0424
2/0	0.0539	0.0431	0.0350	0.0305	0.0661	0.0529	0.0458	0.0399
3/0	0.0521	0.0417	0.0341	0.0297	0.0614	0.0491	0.0427	0.0372
4/0	0.0505	0.0404	0.0333	0.0290	0.0592	0.0474	0.0413	0.0359
250	0.0490	0.0392	0.0324	0.0282	0.0573	0.0458	0.0400	0.0348
300	0.0478	0.0383	0.0317	0.0277	0.0557	0.0446	0.0387	0.0339
350	0.0469	0.0375	0.0312	0.0274	0.0544	0.0436	0.0379	0.0332
400	0.0461	0.0369	0.0308	0.0270	0.0534	0.0427	0.0371	0.0326
500	0.0461	0.0369	0.0308	0.0270	0.0517	0.0414	0.0357	0.0317
600	0.0439	0.0351	0.0290	0.0261	0.0516	0.0413	0.0343	0.0309
750	0.0434	0.0347	0.0284	0.0260	0.0500	0.0400	0.0328	0.0301
1000	0.0421	0.0337	0.0272	0.0255	0.0482	0.0385	0.0311	0.0291

NOTE: These are only representative figures. Reactance is affected by cable insulation type, shielding, conductor outside diameter, conductor spacing in three-conductor cable, etc. In commercial buildings, medium-voltage impedances normally do not affect short-circuit calculations significantly.

Table 66
Busways

Busway Type	Ampere Rating	Resistance (R)	Reactance (X)	Impedance (Z)
600 V				
		Ω/100 Feet Line-to-Neutral		
		60 Hz Alternating Current		
Feeder with aluminum bus bars	600	0.00331	0.00228	0.00402
	800	0.00210	0.00081	0.00226
	1000	0.00163	0.00079	0.00181
	1350	0.00143	0.00052	0.00153
	1600	0.00108	0.00051	0.00119
	2000	0.00081	0.00037	0.00089
	2500	0.00064	0.00030	0.00071
	3000	0.00054	0.00024	0.00059
	4000	0.00041	0.00018	0.00045
	5000	0.00032	0.00013	0.00035
Feeder with copper bus bars	800	0.00200	0.00228	0.00304
	1000	0.00132	0.00081	0.00156
	1350	0.00099	0.00079	0.00126
	1600	0.00088	0.00052	0.00102
	2000	0.00066	0.00051	0.00083
	2500	0.00059	0.00037	0.00062
	3000	0.00040	0.00030	0.00050
	4000	0.00034	0.00024	0.00042
	5000	0.00025	0.00018	0.00031
Plug-in with aluminum bus bars	800	0.00210	0.00114	0.00238
	1000	0.00163	0.00110	0.00197
	1350	0.00143	0.00069	0.00159
	1600	0.00108	0.00066	0.00127
	2000	0.00081	0.00044	0.00092
	2500	0.00064	0.00035	0.00073
	3000	0.00054	0.00028	0.00061
	4000	0.00041	0.00021	0.00046
	5000	0.00032	0.00016	0.00036
Plug-in with copper bus bars	800	0.00200	0.00460	0.00500
	1000	0.00132	0.00114	0.00174
	1350	0.00099	0.00110	0.00148
	1600	0.00088	0.00069	0.00112
	2000	0.00066	0.00066	0.00093
	2500	0.00050	0.00044	0.00067
	3000	0.00040	0.00035	0.00053
	4000	0.00034	0.00028	0.00044
	5000	0.00025	0.00021	0.00032
CL with aluminum bus bars	1000	0.00220	0.0069	0.0072
	1350	0.00200	0.0064	0.0067
	1600	0.00148	0.0064	0.0066
	2000	0.00112	0.0058	0.0059
	2500	0.00090	0.0054	0.0055
	3000	0.00077	0.0050	0.0051
	4000	0.00059	0.0042	0.0042
CL with copper bus bars	1000	0.00177	0.0069	0.0071
	1350	0.00134	0.0069	0.0070
	1600	0.00121	0.0064	0.0065
	2000	0.00090	0.0064	0.0065
	2500	0.00070	0.0058	0.0058
	3000	0.00058	0.0054	0.0054
	4000	0.00041	0.0046	0.0046

Table 66 (Continued)

Metal-Clad Busways Ampere Rating		Bus Size	Resistance per 100 Feet at 50 °C	600 Hz Reactance per 100 Feet
		5 and 15 kV		
1200	Copper	1¼ × 4	0.00102	0.0049
2000	Copper	1⅜ × 6	0.00049	0.00415
3000	Copper	2⅜ × 6	0.00245	0.0045
4000	Copper	6 in. square tube	0.000135	0.0029
1200	Aluminum	1⅜ × 4	0.00118	0.0045
2000	Aluminum	1⅝ × 6	0.00059	0.0037
3000	Aluminum	2⅜ × 6	0.000295	0.0041
4000	Aluminum	6 in. round tube	0.00019	0.0036

The fault kVA is calculated by using the following equation:

$$\text{fault kVA} = \frac{\text{kVA base}}{\text{total per unit fault impedance}} \qquad \text{(Eq 22)}$$

The short-circuit current (for three-phase systems) can be calculated by using the following equation:

$$\text{short-circuit current} = \frac{\text{fault kVA}}{\sqrt{3} \times \text{rated kV at fault}} \qquad \text{(Eq 23)}$$

where

Rated kV = Line-to-line voltage.

9.5.8 Electric Utility System. The electric utility system is usually represented by a single equivalent reactance referred to the user's point of connection. The per unit reactance of the utility is, therefore, 1.0, which is based on the available short-circuit current from the utility. This value is obtained from the utility and may be expressed in several ways.

(1) Three-phase short-circuit current in kVA or MVA available at a given voltage
(2) Three-phase short-circuit current and X/R ratio plus single line-to-ground short-circuit current and X/R ratio available at a given voltage
(3) Percent or per unit reactance on a specified kVA and voltage base
(4) Reactance in Ω per phase (sometimes $R + jX$) referred to a given voltage

Example—Conversion to per unit on a 10 000 kVA base (kVA_b).
(1) Available three-phase short-circuit kVA = 500 000 kVA (500 MVA)

$$X_{\text{pu}} = \left(\frac{\text{kVA}_b}{\text{kVA}_{\text{sc}}}\right) = \left(\frac{10\,000}{500\,000}\right) = 0.02$$

(2) Available three-phase short-circuit current = 20 940 A at 13.8 kV

$$X_{\text{pu}} = \left(\frac{\text{kVA}_b}{\sqrt{3}\,I_{\text{sc}}\,\text{kV}}\right) = \left(\frac{10\,000}{\sqrt{3}\,(20\,940)(13.8)}\right) = 0.02$$

(3) Equivalent utility reactance = 0.2 pu on a 100 000 kVA base

$$X_{pu} = X_{pu(old)} \left(\frac{kVA_b}{kVA_{(old)}} \right) = 0.2 \times \left(\frac{10\,000}{100\,000} \right) = 0.02$$

(4) Equivalent utility reactance = 0.38 Ω per phase at 13.8 kV

$$X_{pu} = X \left(\frac{kVA_b}{kV^2\,(1000)} \right) = 0.38 \left(\frac{10\,000}{(13.8)^2\,1000} \right) = 0.02$$

(5) $X/R = 4$

$$R_{pu} = \left(\frac{0.02}{4} \right) = 0.005$$

9.5.9 Transformers. Transformer reactance (impedance) will most commonly be expressed as a percent value ($\%X_T$ or $\%Z_T$) on the transformer rated kVA. (Impedance values are usually expressed on the self-cooled kVA rating at rated temperature rise.)

Example—A 500 kVA transformer with an impedance of 5% on its kVA rating (assume impedance is all reactance).

Conversion to per unit on a 10 000 kVA base (kVA_b)

$$X_{pu} = \left(\frac{\%X_T}{100} \right) \left(\frac{kVA_b}{transformer\ kVA} \right) = \left(\frac{5}{100} \right) \left(\frac{10\,000}{500} \right) = 1.0$$

9.5.10 Busways, Cables, and Conductors. The resistance and reactance of busway, cables, and conductors will most frequently be available in terms of Ω per phase, per unit length.

Example—250 feet of a three-conductor 500 kcmil cable (600 V) installed in steel conduit on a 480 V system

Conversion to per unit on a 10 000 kVA base (kVA_b)

R = 0.0244 Ω/1000 feet

R = 0.0061 Ω/250 feet

X = 0.0279 Ω/1000 feet

X = 0.0070 Ω/250 feet

$$R_{pu} = R \left(\frac{kVA_b}{kV^2\,(1000)} \right) = 0.0061 \left(\frac{10\,000}{(0.48)^2\,(1000)} \right) = 0.2648$$

$$X_{pu} = X \left(\frac{kVA_b}{kV^2\,(1000)} \right) = 0.0070 \left(\frac{10\,000}{(0.48)^2\,(1000)} \right) = 0.3038$$

Z_{pu} = 0.2648 + j0.3038

Although they are small in magnitude, include the resistances of high- and medium-voltage elements since they significantly affect the X/R ratio needed for high- or medium-voltage interrupting duty calculations.

9.5.11 Rotating Machines. Machine reactances are usually expressed in terms of percent reactance $\%X_m$, or per unit reactance X_{pu}, on the normal rated kVA of the machine. Either the subtransient reactance X'' or the transient reactance X' should be selected, depending on the type of short-circuit current calculation required. Motor-rated kVA can be estimated, given the motor hp, as follows:

Type of Machine	Rated kVA
All	$\dfrac{V_{rated} I_{rated} \sqrt{3}}{1000}$ (exact)
Induction motors and 0.8 power factor synchronous motors	rated hp (approximate)
1.0 power factor synchronous motors	0.8 rated hp (approximate)

9.5.12 Motors Rated 600 V or Less. In systems of 600 V or less, the large motors (that is, motors over 50 hp) are usually few in number and represent only a small portion of the total connected hp. These large motors can be represented individually, or they can be combined with the smaller motors, representing the complete group as one equivalent motor in the impedance diagram. Small motors are turned off and on frequently, so it is practically impossible to predict which ones will be on line when a short circuit occurs. Therefore, all small motors are generally assumed to be running, and all are considered as one large motor.

Where more accurate data are not available, the following procedures may be used in representing the combined reactance of a group of miscellaneous motors:

(1) In all 208 V systems and 480 V commercial building systems, a substantial portion of the load consists of lighting; so assume that the running motors are grouped at the transformer secondary bus and have a reactance of 25% on a kVA base equal to 50% of the transformer kVA rating.

(2) Groups of small induction motors that are served by a motor control center can be represented by considering the group to have a reactance of 25% on a kVA rating equal to the connected motor hp.

Example

(1) Conversion to per unit on a 10 000 kVA base (kVA_b). A 500 hp, 0.8 pf synchronous motor has a subtransient reactance X''_d of 15%

$$X''_{pu} = \left(\frac{\%X''_d}{100}\right)\left(\frac{kVA_b}{\text{motor kVA}}\right) = \left(\frac{15}{100}\right)\left(\frac{10\,000}{500}\right) = 3.0$$

NOTE: Refrigeration chillers (such as those made by Carrier, Trane, and York) have motors rated in kW. IEC-type motors have kW ratings. To obtain an equivalent kVA, divide kW by an assumed power factor of about 0.85.

9.5.13 Other Circuit Impedances. There are other circuit impedances, such as those associated with circuit breakers, current transformers, bus structures, and connections, which are usually neglected in short-circuit current calculations. The

accuracy of the calculation is not generally affected because the effects of the impedances are small and omitting them provides conservative (higher) short-circuit
currents. The system designer may want to include these impedances in some
cases. Obtain the data for such calculations from the manufacturer.

9.5.14 Shunt Connected Impedances. In addition to the components already
mentioned, every system includes other components or loads that are represented
in a diagram as shunt connected impedances, e.g., lights, furnaces, and capacitors.
A technically accurate solution requires that these impedances be included in the
equivalent circuit used in calculating a short-circuit current; but practical considerations allow the general practice of omitting them. Such impedances have relatively high values, and their omission will not significantly affect the calculated
results.

9.5.15 System Driving Voltage. The system driving voltage E in the basic equation for short-circuit currents $I = E/Z$ can be represented by the use of a single
overall driving voltage as illustrated in Fig 91, rather than the array of individual
unequal generated voltages acting within individual rotating machines. This single
driving voltage is equal to the prefault voltage at the point of fault connection. The
equivalent circuit is a valid transformation accomplished by Thevenin's theorem
and permits an accurate determination of the short-circuit current for the assigned
values of system impedance. The prefault voltage referred to is ordinarily taken as
system nominal voltage at the point of fault, as this calculation leads to the full
value of short-circuit current that may be produced by the probable maximum
operating voltage.

In making a short-circuit current calculation on three-phase balanced systems, a
single-phase representation of a three-phase system and the system driving voltage
E is expressed in line-to-neutral volts. Line-to-neutral voltage is equal to line-to-line
voltage divided by $\sqrt{3}$.

When using the per unit system, if the system per unit impedances are established on voltage bases equal to system nominal voltages, the per unit driving
voltage is equal to 1.0. In the per unit system, both line-to-line and line-to-neutral
voltage have equal values, i.e., both would have values of 1.0.

When system impedance values are expressed in Ω per phase rather than per
unit, the system driving voltage is equal to the system line-to-neutral voltage, i.e.,
277 V for a 480Y/277 V system.

9.5.16 Determination of Short-Circuit Currents. After the impedance diagram
has been prepared, the short-circuit currents can be determined. This can be accomplished by longhand calculation, network analyzer, or digital computer techniques.

Simple radial systems, such as those used in most low-voltage systems, can be
easily resolved by longhand calculations, though digital computers can yield significant time savings, particularly when short-circuit duties at many system locations
are required and when resistance is being included in the calculation.

9.5.17 Longhand Solution. A longhand solution requires the combining of
impedances in series and parallel from the source driving voltage to the location of
the fault being calculated to determine the simple equivalent network impedance.
The calculation to derive the symmetrical short-circuit current is $I = E/Z$ (or E/X)
where

E = System driving voltage.

Z (or X) = Single equivalent network impedance (or reactance).

When calculations are made using per unit, the following formulas apply:

(1) Symmetrical three-phase short-circuit current in per unit

$$I_{pu} = \frac{E_{pu}{}^*}{Z_{pu}} \qquad \text{(Eq 24)}$$

(2) Symmetrical three-phase short-circuit current in A

$$I = I_b(I_{pu}) = \frac{I_b{}^*}{Z_{pu}} \qquad \text{(Eq 25)}$$

(3) Symmetrical three-phase short circuit per unit kVA

$$kVA_{pu} = E_{pu}(I_{pu}) \qquad \text{(Eq 26)}$$

(4) Symmetrical three-phase short circuit kVA

$$kVA_{pu} = kVA_b(kVA_{pu}) = \frac{kVA_b{}^*}{Z_{pu}} \qquad \text{(Eq 27)}$$

where

I_{pu} = Per unit current.

Z_{pu} = Equivalent network per unit impedance.

E_{pu} = Per unit voltage.

kVA_{pu} = Per unit kVA.

I_b = Base current in A.

kVA_b = Base kVA.

* = A simplified equality that applies only where $E_{pu} = 1.0$.

When calculations are made using Ω, the symmetrical three-phase short circuit in A is

$$I = \frac{E_{L-N}}{Z} \qquad \text{(Eq 28)}$$

where

E_{L-N} = Line-to-neutral voltage.

Z = Equivalent network impedance in Ω per phase.

A new combination of impedances to determine the single equivalent network impedance is required for each fault location.

The longhand solution for a radial system is fairly simple. For systems containing loops, simultaneous equations may be necessary, though delta-wye network transformations can usually be used to combine impedances. Electronic calculators can be excellent timesavers in making longhand calculations. An example of a longhand solution is included below.

Example—The Building Power System
(1) System Single-Line Diagram—Figure 95 is a single-line diagram of a building power system served from a utility spot network. The diagram includes
   (a) Utility short-circuit duty at the network bus
   (b) Conductor size, number, type, and length
   (c) Kilovoltamperes and impedance of 30 kVA and 150 kVA transformers
   (d) Lumped connected hp of induction motors
(2) Type and Locations of Short Circuits—Short-circuit currents are required at all buses where protective devices will be located (buses 1–18). Symmetrical three-phase short-circuit currents are required since all devices are rated 480 V and below. The most severe duty will occur with all circuit breakers closed, with a maximum short-circuit duty of 55 600 A, three-phase, symmetrical from the utility spot network.
(3) System Impedance Diagrams—The impedance diagram for this system is shown in Fig 96. Since most buses are at the 480 V level, the example uses system impedances in Ω rather than in per unit. All impedances are given in Ω per phase at 480 V. The impedance values as shown in the diagram resulting from the calculations shown in Table 67.
(4) Calculations—Longhand calculations to determine E/Z for bus 1 are shown in Table 68.

The calculations are usually done by computer, using methods outlined in 9.5.5. The number of longhand calculations becomes too cumbersome for practical application. Most commercial programs utilize the per unit system since it is simple to program once the system data has been organized on a consistent base. Commercial programs similar to the one shown in Table 69 for the illustrated system also show results using the appropriate asymmetrical multiplier for different types of interrupting devices.

Note that all the ohmic impedances shown on Fig 96 have been calculated on a 480 V base. If such ohmic values are used, the short-circuit current for 208 V buses should be determined using 480 V impedances and then multiplied by the transformer ratio squared (in this case, $TR^2 = 208^2/480^2 = 0.18777$). Using per unit impedances, the short-circuit current for any bus can be determined using the calculations in 9.5.7.

9.5.18 Determination of Line-to-Ground Fault Currents. The technique of symmetrical components will allow us to express the bolted line-to-ground fault current as follows:

$$I_{\text{L-G}} = \frac{3E}{Z_1 + Z_2 + Z_0 + 3R_0} \qquad \text{(Eq 29)}$$

where

E = Line-to-neutral voltage.
Z_1 = Positive-sequence impedance.
Z_2 = Negative-sequence impedance.
Z_0 = Zero-sequence impedances.
R_0 = Resistance of neutral grounding resistor, if any.

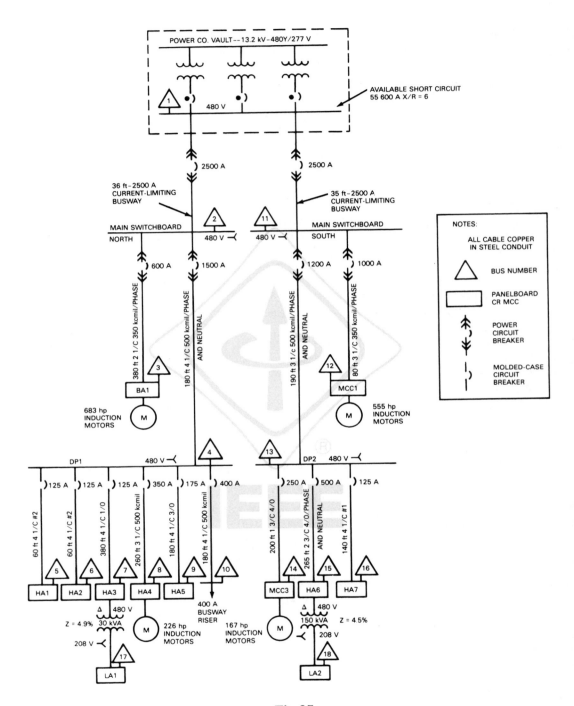

**Fig 95
Single-Line Diagram — Specific 480Y/277 V Network Served Building**

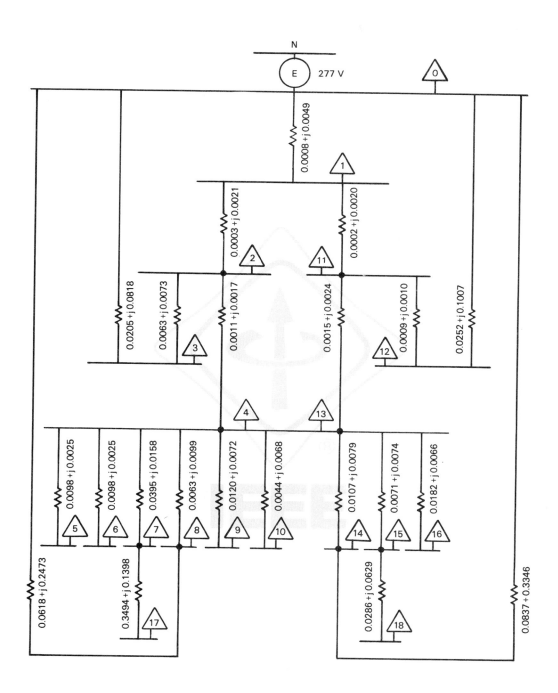

Fig 96
Impedance Diagram for the System in Fig 95
(All impedances in Ω at 480 V.)

Table 67
System Impedances in Ohms and Per Unit for Fig 96

Bus-to-Bus		Calculation	X/R	Ohms			Per Unit (1 MVA Base)		
				Z	R	X	Z	R	X
Utility									
0	1	277 V/55 600 A	6.00	0.0050	0.0008	0.0049			
0	1	$\dfrac{1 \text{ MVA}}{(55.6 \text{ kA})(0.48 \text{ kV})\sqrt{3}}$	6.00				0.0216	0.0036	0.0213
Motors									
0	3	$\dfrac{(25\%)(0.48 \text{ kV})^2(10)^3}{(100)(683 \text{ hp})}$	4.00	0.0843	0.0205	0.0818			
0	3	$\dfrac{(25\%)(1 \text{ MVA})(10)^3}{(100)(683 \text{ hp})}$	4.00				0.3660	0.0888	0.3551
0	8	$\dfrac{(25\%)(0.48 \text{ kV})^2(10)^3}{(100)(226 \text{ hp})}$	4.00	0.2549	0.0618	0.2473			
0	8	$\dfrac{(25\%)(1 \text{ MVA})(10)^3}{(100)(226 \text{ hp})}$	4.00				1.1062	0.2683	1.0732
0	12	$\dfrac{(25\%)(0.48 \text{ kV})^2(10)^3}{(100)(555 \text{ hp})}$	4.00	0.1038	0.0252	0.1007			
0	12	$\dfrac{(25\%)(1 \text{ MVA})(10)^3}{(100)(555 \text{ hp})}$	4.00				0.4505	0.1093	0.4370
0	14	$\dfrac{(25\%)(0.48 \text{ kV})^2(10)^3}{(100)(167 \text{ hp})}$	4.00	0.3449	0.0837	0.3346			
0	14	$\dfrac{(25\%)(1 \text{ MVA})(10)^3}{(100)(167 \text{ hp})}$	4.00				1.4970	0.3631	1.4523
Transformers									
15	18	$\dfrac{(4.5\%)(0.48 \text{ kV})^2(10)^3}{(100)(150 \text{ kVA})}$	2.20	0.0691	0.0286	0.0629			
15	18	$\dfrac{(4.5\%)(1 \text{ MVA})(10)^3}{(100)(150 \text{ kVA})}$	2.20				0.3000	0.1241	0.2731
7	17	$\dfrac{(4.9\%)(0.48 \text{ kV})^2(10)^3}{(100)(30 \text{ kVA})}$	0.40	0.3763	0.3494	0.1398			
7	17	$\dfrac{(4.9\%)(1 \text{ MVA})(10)^3}{(100)(30 \text{ kVA})}$	0.40				1.6333	1.5165	0.6066

Table 67 *(Continued)*

Bus-to-Bus		Calculation	X/R	Ohms			Per Unit (1 MVA Base)		
				Z	R	X	Z	R	X
Current-Limiting Busway									
1	2	$\dfrac{(36\ \text{ft})(0.0007+\text{j}\,0.0058)\,\Omega}{(100\ \text{ft})}$	8.29	0.0021	0.0003	0.0021			
1	2	$\dfrac{(36)(0.0007+\text{j}\,0.0058)(1\ \text{MVA})}{(0.48\ \text{kV})^2}$	8.29				0.0091	0.0011	0.0091
1	11	$\dfrac{(35\ \text{ft})(0.0007+\text{j}\,0.0058)\,\Omega}{(100\ \text{ft})}$	8.29	0.0020	0.0002	0.0020			
1	11	$\dfrac{(35)(0.0007+\text{j}\,0.0058)(1\ \text{MVA})}{(0.48\ \text{kV})^2}$	8.29				0.0089	0.0011	0.0088
Cable — All Copper in Steel Conduit									
500 kcmil — Single Conductor									
2	4	$\dfrac{(180\ \text{ft})(0.0244+\text{j}\,0.0379)\,\Omega}{(4/\text{ph})(1000\ \text{ft})}$	1.55	0.0020	0.0011	0.0017			
2	4	$\dfrac{(180)(0.0244+\text{j}\,0.0379)(1\ \text{MVA})}{4/\text{ph}(1000\ \text{ft})(0.48\ \text{kV})^2}$	1.55				0.0088	0.0048	0.0074
11	13	$\dfrac{(190\ \text{ft})(0.0244+\text{j}\,0.0379)\,\Omega}{(3/\text{ph})(1000\ \text{ft})}$	1.55	0.0029	0.0015	0.0024			
11	13	$\dfrac{(190)(0.0244+\text{j}\,0.0379)(1\ \text{MVA})}{3/\text{ph}(1000\ \text{ft})(0.48\ \text{kV})^2}$	1.55				0.0124	0.0067	0.0104
4	8	$\dfrac{(260\ \text{ft})(0.0244+\text{j}\,0.0379)\,\Omega}{(1000\ \text{ft})}$	1.55	0.0117	0.0063	0.0099			
4	8	$\dfrac{(260)(0.0244+\text{j}\,0.0379)(1\ \text{MVA})}{(1000\ \text{ft})(0.48\ \text{kV})^2}$	1.55				0.0509	0.0275	0.0428
4	10	$\dfrac{(180\ \text{ft})(0.0244+\text{j}\,0.0379)\,\Omega}{(1000\ \text{ft})}$	1.55	0.0081	0.0044	0.0068			
4	10	$\dfrac{(180)(0.0244+\text{j}\,0.0379)(1\ \text{MVA})}{(1000\ \text{ft})(0.48\ \text{kV})^2}$	1.55				0.0352	0.0191	0.0296
350 kcmil — Single Conductor									
2	3	$\dfrac{(380\ \text{ft})(0.0333+\text{j}\,0.0383)\,\Omega}{(2/\text{ph})(1000\ \text{ft})}$	1.15	0.0096	0.0063	0.0073			
2	3	$\dfrac{(380)(0.0333+\text{j}\,0.0383)(1\ \text{MVA})}{2/\text{ph}(1000\ \text{ft})(0.48\ \text{kV})^2}$	1.15				0.0419	0.0275	0.0316
11	12	$\dfrac{(80\ \text{ft})(0.0333+\text{j}\,0.0383)\,\Omega}{(3/\text{ph})(1000\ \text{ft})}$	1.15	0.0014	0.0009	0.0010			
11	12	$\dfrac{(80)(0.0333+\text{j}\,0.0383)(1\ \text{MVA})}{3/\text{ph}(1000\ \text{ft})(0.48\ \text{kV})^2}$	1.15				0.0059	0.0039	0.0044

Table 67 *(Continued)*

Bus-to-Bus		Calculation	X/R	Ohms			Per Unit (1 MVA Base)		
				Z	R	X	Z	R	X
Cable — All Copper in Steel Conduit									
4/0 — Three Conductor									
13	14	$\dfrac{(200\ \text{ft})(0.0534+j\,0.0295)\,\Omega}{(1000\ \text{ft})}$	0.55	0.0122	0.0107	0.0059			
13	14	$\dfrac{(200)(0.0534+j\,0.0295)(1\ \text{MVA})}{(1000\ \text{ft})(0.48\ \text{kV})^2}$	0.55				0.0530	0.0464	0.0256
13	15	$\dfrac{(265\ \text{ft})(0.0534+j\,0.0295)\,\Omega}{(2\ \text{ph})(1000\ \text{ft})}$	0.55	0.0081	0.0071	0.0039			
13	15	$\dfrac{(265)(0.0534+j\,0.0295)(1\ \text{MVA})}{2/\text{ph}\,(1000\ \text{ft})(0.48\ \text{kV})^2}$	0.55				0.0351	0.0307	0.0170
3/0 — Three Conductor									
4	9	$\dfrac{(180\ \text{ft})(0.0668+j\,0.0300)\,\Omega}{(1000\ \text{ft})}$	0.45	0.0132	0.0120	0.0054			
4	9	$\dfrac{(180)(0.0668+j\,0.0300)(1\ \text{MVA})}{(1000\ \text{ft})(0.48\ \text{kV})^2}$	0.45				0.0572	0.0522	0.0234
1/0 — Three Conductor									
4	7	$\dfrac{(380\ \text{ft})(0.104+j\,0.0312)\,\Omega}{(1000\ \text{ft})}$	0.30	0.0413	0.0395	0.0119			
4	7	$\dfrac{(380)(0.104+j\,0.0312)(1\ \text{MVA})}{(1000\ \text{ft})(0.48\ \text{kV})^2}$	0.30				0.1791	0.1715	0.0515
#1 — Three Conductor									
13	16	$\dfrac{(140\ \text{ft})(0.1303+j\,0.0319)\,\Omega}{(1000\ \text{ft})}$	0.24	0.0188	0.0182	0.0045			
13	16	$\dfrac{(140)(0.1303+j\,0.0319)(1\ \text{MVA})}{(1000\ \text{ft})(0.48\ \text{kV})^2}$	0.24				0.0815	0.0792	0.0194
#2 — Three Conductor									
4	5	$\dfrac{(60\ \text{ft})(0.1640+j\,0.0313)\,\Omega}{(1000\ \text{ft})}$	0.19	0.0100	0.0098	0.0019			
4	5	$\dfrac{(60)(0.1640+j\,0.0313)(1\ \text{MVA})}{(1000\ \text{ft})(0.48\ \text{kV})^2}$	0.19				0.0435	0.0427	0.0082
4	6	$\dfrac{(60\ \text{ft})(0.1640+j\,0.0313)\,\Omega}{(1000\ \text{ft})}$	0.18	0.0119	0.0118	0.0021			
4	6	$\dfrac{(60)(0.1640+j\,0.0313)(1\ \text{MVA})}{(1000\ \text{ft})(0.48\ \text{kV})^2}$	0.18				0.0519	0.0510	0.0093

Table 68
Short-Circuit Calculations for Bus 1 in Ohms for Fig 96

Series Impedances to Bus 2 and Bus 11 from Bus 0

Bus-to-Bus		R	X		Bus-to-Bus		R	X
0	8	0.0618	0.2473		0	14	0.0837	0.3346
8	4	0.0063	0.0099		14	13	0.0107	0.0079
4	2	0.0011	0.0017		13	11	0.0015	0.0024
0-8-4-2		0.0692	0.2589		0-14-13-11		0.0959	0.3449
0	3	0.0205	0.0818		0	12	0.0252	0.1007
3	2	0.0063	0.0073		12	11	0.0009	0.001
0-3-2		0.0063	0.0891		0-12-11		0.0261	0.1017

Parallel Impedance Formulae

$$R = \frac{G}{G^2 + B^2} \qquad X = \frac{B}{G^2 + B^2} \qquad G = \frac{R}{R^2 + X^2} \qquad B = \frac{X}{R^2 + X^2}$$

Parallel Impedances to Bus 2 and Bus 11 from Bus 0

				$R^2 + X^2$	G	B	$G^2 + B^2$
0-8-4-2		0.0692	0.2589 $\longrightarrow$	0.0718	0.9635	3.604953	
0-3-2		0.0063	0.0891 $\longrightarrow$	0.0080	0.7896	11.16751	
0	2	0.0079	0.0668 $\longleftarrow$		1.7532	14.77247	221.2994
0-14-13-11		0.0959	0.3429 $\longrightarrow$	0.1268	0.7564	2.704745	
0-12-11		0.0261	0.1017 $\longrightarrow$	0.0110	2.3675	9.225243	
0-11		0.0205	0.0784 $\longleftarrow$		3.1240	11.92999	152.0839

Series Impedances to Bus 1 from Bus 0 through Bus 2 and Bus 11

		R	X				R	X
0	2	0.0079	0.0668		0	11	0.0205	0.0785
2	1	0.0003	0.0021		11	1	0.0002	0.002
0-2-1		0.0082	0.0689		0-11-1		0.0207	0.0805

Parallel Impedances to Bus 1 from Bus 0

				$R^2 + X^2$	G	B	$G^2 + B^2$
0-2-1		0.0082	0.0689 $\longrightarrow$	0.0048	1.7100	14.3194	
0-11-1		0.0207	0.0805 $\longrightarrow$	0.0069	2.9977	11.64334	
0	1	0.0008	0.0049 $\longrightarrow$	2.465e-5	32.4544	198.7830	
0	1	0.0007	0.0043 $\longleftarrow$		37.1621	224.7457	51 891.67

$$Z = \sqrt{R^2 + X^2} = 0.0044$$
$$I = E/Z \qquad = 63100 \qquad \text{rms A symmetrical}$$

Table 69
Computer-Generated Short-Circuit Calculation Results
for All Buses That Are Similar to the System in Fig 96

NOTE: Figures used vary from figures derived and as shown in Tables 67 and 68.

```
                                                                    IEEE

CASE: 1-FCY



                    THREE PHASE SHORT CIRCUIT PROGRAM - VERSION 1.44
                FIRST CYCLE CALC. FOR BREAKER DUTIES PER ANSI C37.13-1981
            TOT. CURRENT & FLOWS FROM COMPLEX NETWORK, X/R FROM SEPARATE R & X
                    04-06-1990      1 MVA BASE       60 Hz

CUSTOMER ABC
XYZ, NY

CASE: 1-FCY
THREE-PHASE FAULT CALCULATIONS

BUS 1           E/Z= 62.547 kA(  52.00MVA)AT-79.72DEG.,X/R=  5.69,  0.480 kV
                Z=   0.003431 +J   0.018922

                MAX. LOW VOLTAGE FUSE DUTY =   64.56 SYM
                MAX. LOW VOLTAGE POWER CIRCUIT BREAKER DUTY  =      62.55
                MAX. LV MCCB OR ICCB (RATED >20kA INT.) DUTY =      64.56
                MAX. LV MCCB OR ICCB (RATED 10-20kA INT.)DUTY =     71.82

   CONTRIBUTIONS IN kA
      BUS  TO  BUS      MAG        ANG       BUS  TO  BUS      MAG        ANG
    REMOTE    1       55.596   -80.528    2        1        3.765   -72.396
      11      1        3.236   -74.347

BUS 2           E/Z= 47.736 kA(  39.69MVA)AT-80.02DEG.,X/R=  5.92,  0.480 kV
                Z=   0.004368 +J   0.024816

                MAX. LOW VOLTAGE FUSE DUTY =   49.66 SYM
                MAX. LOW VOLTAGE POWER CIRCUIT BREAKER  DUTY =      47.74
                MAX. LV MCCB OR ICCB (RATED >20kA INT.) DUTY =      49.66
                MAX. LV MCCB OR ICCB (RATED 10-20kA INT.)DUTY =     55.24

   CONTRIBUTIONS IN kA
      BUS  TO  BUS      MAG        ANG       BUS  TO  BUS      MAG        ANG
     4        2        0.997   -73.948    3        2        2.851   -71.561
     1        2       43.927   -80.701

BUS 3           E/Z= 20.046 kA(  16.67MVA)AT-61.24DEG.,X/R=  1.93,  0.480 kV
                Z=   0.028873 +J   0.052598

                MAX. LOW VOLTAGE FUSE DUTY =   20.05 SYM
                MAX. LOW VOLTAGE POWER CIRCUIT BREAKER  DUTY =      20.05
                MAX. LV MCCB OR ICCB (RATED >20kA INT.) DUTY =      20.05
                MAX. LV MCCB OR ICCB (RATED 10-20kA INT.)DUTY =     20.05

   CONTRIBUTIONS IN kA
      BUS  TO  BUS      MAG        ANG       BUS  TO  BUS      MAG        ANG
     2        3       16.980   -58.772    INDMOT    3        3.167   -74.558
```

Table 69 *(Continued)*

```
BUS 4        E/Z= 35.689 kA(  29.67MVA)AT-73.24DEG.,X/R=  3.37,  0.480 kV
             Z=   0.009718 +J   0.032272

             MAX. LOW VOLTAGE FUSE DUTY =    35.69 SYM
             MAX. LOW VOLTAGE POWER CIRCUIT BREAKER DUTY    =    35.69
             MAX. LV MCCB OR ICCB (RATED >20kA INT.) DUTY =    35.69
             MAX. LV MCCB OR ICCB (RATED 10-20kA INT.)DUTY =    36.24

        CONTRIBUTIONS IN kA
           BUS  TO  BUS     MAG      ANG      BUS  TO  BUS     MAG       ANG
            2       4      34.684  -73.216   8        4      1.005   -74.101
           10       4       0.000  -73.241   5        4     -0.000    -7.610
            6       4      -0.000   -7.610   7        4      0.000   -13.156
            9       4       0.000  -96.506

BUS 5        E/Z= 16.344 kA(  13.59MVA)AT-34.39DEG.,X/R=  0.69,  0.480 kV
             Z=   0.060728 +J   0.041572

             MAX. LOW VOLTAGE FUSE DUTY =    16.34 SYM
             MAX. LOW VOLTAGE POWER CIRCUIT BREAKER  DUTY  =    16.34
             MAX. LV MCCB OR ICCB (RATED >20kA INT.) DUTY =    16.34
             MAX. LV MCCB OR ICCB (RATED 10-20kA INT.)DUTY =    16.34

        CONTRIBUTIONS IN kA
           BUS  TO  BUS     MAG      ANG      BUS  TO  BUS     MAG       ANG
            4       5      16.344  -34.394

BUS 6        E/Z= 16.344 kA(  13.59MVA)AT-34.39DEG.,X/R=  0.69,  0.480 kV
             Z=   0.060728 +J   0.041572

             MAX. LOW VOLTAGE FUSE DUTY =    16.34 SYM
             MAX. LOW VOLTAGE POWER CIRCUIT BREAKER  DUTY  =    16.34
             MAX. LV MCCB OR ICCB (RATED >20kA INT.) DUTY =    16.34
             MAX. LV MCCB OR ICCB (RATED 10-20kA INT.)DUTY =    16.34

        CONTRIBUTIONS IN kA
           BUS  TO  BUS     MAG      ANG      BUS  TO  BUS     MAG       ANG
            4       6      16.344  -34.394

BUS 7        E/Z=  5.608 kA(   4.66MVA)AT-23.48DEG.,X/R=  0.43,  0.480 kV
             Z=   0.196708 +J   0.085462

             MAX. LOW VOLTAGE FUSE DUTY =     5.61 SYM
             MAX. LOW VOLTAGE POWER CIRCUIT BREAKER  DUTY  =     5.61
             MAX. LV MCCB OR ICCB (RATED >20kA INT.) DUTY =     5.61
             MAX. LV MCCB OR ICCB (RATED 10-20kA INT.)DUTY =     5.61

        CONTRIBUTIONS IN kA
           BUS  TO  BUS     MAG      ANG      BUS  TO  BUS     MAG       ANG
           17       7       0.000  -45.284   4        7      5.608   -23.483

BUS 8        E/Z= 14.498 kA(  12.05MVA)AT-62.57DEG.,X/R=  1.98,  0.480 kV
             Z=   0.038213 +J   0.073640

             MAX. LOW VOLTAGE FUSE DUTY =    14.50 SYM
             MAX. LOW VOLTAGE POWER CIRCUIT BREAKER  DUTY  =    14.50
             MAX. LV MCCB OR ICCB (RATED >20kA INT.) DUTY =    14.50
             MAX. LV MCCB OR ICCB (RATED 10-20kA INT.)DUTY =    14.50

        CONTRIBUTIONS IN kA
           BUS  TO  BUS     MAG      ANG      BUS  TO  BUS     MAG       ANG
            4       8      13.474  -61.610   INDMOT   8      1.051   -75.039
```

Table 69 *(Continued)*

```
BUS 9          E/Z= 13.178 kA(  10.96MVA)AT-39.69DEG.,X/R=  9.83,  0.480 kV
               Z=   0.070238 +J   0.058292

               MAX. LOW VOLTAGE FUSE DUTY =    13.18 SYM
               MAX. LOW VOLTAGE POWER CIRCUIT BREAKER DUTY   =    13.18
               MAX. LV MCCB OR ICCB (RATED >20kA INT.) DUTY  =    13.18
               MAX. LV MCCB OR ICCB (RATED 10-20kA INT.)DUTY =    13.18

     CONTRIBUTIONS IN kA
        BUS  TO  BUS     MAG      ANG      BUS  TO  BUS     MAG       ANG
        4         9     13.178  -39.690

BUS 10         E/Z= 18.669 kA(  15.52MVA)AT-60.05DEG.,X/R=  1.74,  0.480 kV
               Z=   0.032168 +J   0.055822

               MAX. LOW VOLTAGE FUSE DUTY =    18.67 SYM
               MAX. LOW VOLTAGE POWER CIRCUIT BREAKER  DUTY =    18.67
               MAX. LV MCCB OR ICCB (RATED >20kA INT.) DUTY =    18.67
               MAX. LV MCCB OR ICCB (RATED 10-20kA INT.)DUTY =   18.67

     CONTRIBUTIONS IN kA
        BUS  TO  BUS     MAG      ANG      BUS  TO  BUS     MAG       ANG
        4        10     18.669  -60.047

BUS 11         E/Z= 47.804 kA(  39.74MVA)AT-80.13DEG.,X/R=  5.95,  0.480 kV
               Z=   0.004312 +J   0.024789

               MAX. LOW VOLTAGE FUSE DUTY =    49.78 SYM
               MAX. LOW VOLTAGE POWER CIRCUIT BREAKER  DUTY =    47.80
               MAX. LV MCCB OR ICCB (RATED >20kA INT.) DUTY =    49.78
               MAX. LV MCCB OR ICCB (RATED 10-20kA INT.)DUTY =   55.38

     CONTRIBUTIONS IN kA
        BUS  TO  BUS     MAG      ANG      BUS  TO  BUS     MAG       ANG
       13        11      0.749  -73.122    12        11     2.546   -74.519
        1        11     44.528  -80.570

BUS 12         E/Z= 39.866 kA(  33.14MVA)AT-73.96DEG.,X/R=  3.51,  0.480 kV
               Z=   0.008338 +J   0.028996

               MAX. LOW VOLTAGE FUSE DUTY =    39.87 SYM
               MAX. LOW VOLTAGE POWER CIRCUIT BREAKER  DUTY =    39.87
               MAX. LV MCCB OR ICCB (RATED >20kA INT.) DUTY =    39.87
               MAX. LV MCCB OR ICCB (RATED 10-20kA INT.)DUTY =   40.93

     CONTRIBUTIONS IN kA
        BUS  TO  BUS     MAG      ANG      BUS  TO  BUS     MAG       ANG
       11        12     37.288  -73.893   INDMOT    12     2.578   -74.898

BUS 13         E/Z= 32.211 kA(  26.78MVA)AT-71.39DEG.,X/R=  3.00,  0.480 kV
               Z=   0.011917 +J   0.035389

               MAX. LOW VOLTAGE FUSE DUTY =    32.21 SYM
               MAX. LOW VOLTAGE POWER CIRCUIT BREAKER  DUTY =    32.21
               MAX. LV MCCB OR ICCB (RATED >20kA INT.) DUTY =    32.21
               MAX. LV MCCB OR ICCB (RATED 10-20kA INT.)DUTY =   32.21

     CONTRIBUTIONS IN kA
        BUS  TO  BUS     MAG      ANG      BUS  TO  BUS     MAG       ANG
       11        13     31.456  -71.345    14        13     0.755   -73.279
       15        13      0.000  -23.357    16        13     0.000    -8.864
```

Table 69 *(Continued)*

```
BUS 14        E/Z= 13.754 kA(  11.44MVA)AT-46.17DEG.,X/R=  1.10,  0.480 kV
              Z=  0.060562 +J   0.063084

              MAX. LOW VOLTAGE FUSE DUTY =    13.75 SYM
              MAX. LOW VOLTAGE POWER CIRCUIT BREAKER DUTY  =    13.75
              MAX. LV MCCB OR ICCB (RATED >20kA INT.) DUTY =    13.75
              MAX. LV MCCB OR ICCB (RATED 10-20kA INT.)DUTY =   13.75

     CONTRIBUTIONS IN kA
        BUS  TO  BUS     MAG      ANG        BUS  TO  BUS     MAG       ANG
        13       14     13.079  -44.536   INDMOT     14     0.776   -74.864


BUS 15        E/Z= 16.743 kA(  13.92MVA)AT-48.88DEG.,X/R=  1.15,  0.480 kV
              Z=  0.047247 +J   0.054119

              MAX. LOW VOLTAGE FUSE DUTY =    16.74 SYM
              MAX. LOW VOLTAGE POWER CIRCUIT BREAKER  DUTY =    16.74
              MAX. LV MCCB OR ICCB (RATED >20kA INT.) DUTY =    16.74
              MAX. LV MCCB OR ICCB (RATED 10-20kA INT.)DUTY =   16.74

     CONTRIBUTIONS IN kA
        BUS  TO  BUS     MAG      ANG        BUS  TO  BUS     MAG       ANG
        18       15     0.000  -48.879     13       15     16.743   -48.879


BUS 16        E/Z=  9.936 kA(   8.26MVA)AT-28.59DEG.,X/R=  0.55,  0.480 kV
              Z=  0.106297 +J   0.057939

              MAX. LOW VOLTAGE FUSE DUTY =     9.94 SYM
              MAX. LOW VOLTAGE POWER CIRCUIT BREAKER  DUTY =     9.94
              MAX. LV MCCB OR ICCB (RATED >20kA INT.) DUTY =     9.94
              MAX. LV MCCB OR ICCB (RATED 10-20kA INT.)DUTY =    9.94

     CONTRIBUTIONS IN kA
        BUS  TO  BUS     MAG      ANG        BUS  TO  BUS     MAG      ANG
        13       16     9.936  -28.594


BUS 17        E/Z=  1.502 kA(   0.54MVA)AT-22.00DEG.,X/R=  0.40,  0.208 kV
              Z=  1.713218 +J   0.692062

              MAX. LOW VOLTAGE FUSE DUTY =     1.50 SYM
              MAX. LOW VOLTAGE POWER CIRCUIT BREAKER  DUTY =     1.50
              MAX. LV MCCB OR ICCB (RATED >20kA INT.) DUTY =     1.50
              MAX. LV MCCB OR ICCB (RATED 10-20kA INT.)DUTY =    1.50

     CONTRIBUTIONS IN kA
        BUS  TO  BUS     MAG      ANG        BUS  TO  BUS     MAG      ANG
         7       17     1.502  -21.996


BUS 18        E/Z=  7.512 kA(   2.71MVA)AT-62.12DEG.,X/R=  1.89,  0.208 kV
              Z=  0.172787 +J   0.326589

              MAX. LOW VOLTAGE FUSE DUTY =     7.51 SYM
              MAX. LOW VOLTAGE POWER CIRCUIT BREAKER  DUTY =     7.51
              MAX. LV MCCB OR ICCB (RATED >20kA INT.) DUTY =     7.51
              MAX. LV MCCB OR ICCB (RATED 10-20kA INT.)DUTY =    7.51

     CONTRIBUTIONS IN kA
        BUS  TO  BUS     MAG      ANG        BUS  TO  BUS     MAG      ANG
        15       18     7.512  -62.118
```

In the case of a solidly grounded system, $3R_0 = 0$, and assuming that Z_2 is approximately equal to Z_1, the expression becomes

$$I_{\text{L-G}} = \frac{3E}{2Z_1 + Z_0}$$

From this expression, we can derive the following

$$I_{\text{L-G}} = \frac{E}{Z_1}\left(\frac{3}{Z_0/Z_1 + 2}\right)$$

This expression shows the line-to-ground fault current as a function of the three-phase bolted fault current (E/Z_1) and the ratio of the zero-sequence impedance and the positive-sequence impedance.

In the strictest sense, the quantity above can only be treated as a scalar, if Z_0 and Z_1 have the same phase angles. It has been shown, however, that useful results can be obtained using Z_0 and Z_1 as scalars. (See Reference [B7] for details.)

Practical circuit values of the Z_0/Z_1 ratio may range from 1–50, depending on the construction of the ground return circuit. Some typical values of the Z_0/Z_1 ratio are 2 for aluminum conduit (with or without internal ground conductor), 4–14 for steel conduit (with internal ground conductor, the ratio will generally not exceed 4) depending on the size of conduit, and 15–30 for cable in magnetic armor (see References [B4] and [B7]).

The type of ground return circuit should be known to calculate the bolted line-to-ground fault currents. A selection should then be made to determine the points where ground-fault current levels are required. Generally, a good indication will be those locations where the lower levels of three-phase bolted fault currents were found.

The selection and coordination of ground-fault protective device settings should consider the minimum arcing ground-fault currents. This type of fault can be particularly destructive. With a knowledge of the ground-fault current levels for a system, settings can be selected that will avoid excessive equipment damage.

Ground-fault calculations will be performed for the system shown in Fig 95. The following locations are selected for calculation:

Bus Number	Equipment Name
2	Main switchboard — north
11	Main switchboard — south
4	DP1
13	DP2
16	Panel HA 7
7	Panel HA 3

These calculations will be performed utilizing the following procedure for the series elements in the ground circuit:

Element Z_1		Z_0/Z_1 Ratio			Element Z_0
( )	×	( )	=		( )
( )	×	( )	=		( )
( )	×	( )	=		( )

$$(Z_1) \qquad\qquad\qquad\qquad\qquad\qquad (Z_0)$$

$$\text{Effective overall } Z_0/Z_1 = \frac{(Z_0)}{(Z_1)}$$

The effective overall Z_0/Z_1 ratio is then substituted in the expression

$$I_{\text{L-G}} = \left(\frac{E}{Z_1}\right)\left(\frac{3}{Z_0/Z_1 + 2}\right)$$

which is solved for the bolted line-to-ground fault current. The minimum arcing ground-fault current value is then obtained by utilizing the 0.38 multiplier from Table 63.

For these calculations, representative values of Z_0/Z_1 ratios were taken from ANSI C37.06-1987 [1] and ANSI C37.6-1971 [4]. The cable conductors are run in metallic conduit, and a representative Z_0/Z_1 ratio of 10 is used. A Z_0/Z_1 ratio of 5 is used with current-limiting busway. A good assumption is that the Z_0/Z_1 ratio for the utility source is approximately 1.0. Table 70 shows the calculations and results.

9.5.19 Effect of Low Available Utility kVA. Even when utility fault currents are held down to a low level, it is not always safe to specify protective devices with limited interrupting capacity. Overnight, the available fault kVA that the utility can deliver might be doubled or tripled. Since the destructive thermal and magnetic forces vary as the square of the current, any increase in fault level could result in a disastrous situation. The protective device selected should be one that takes system growth into consideration.

9.5.20 General Discussion of Short-Circuit Current Calculations

(1) Motor Contribution — Synchronous and induction motors will feed additional short-circuit current to a fault at their terminals at a value approximately equal to their locked-rotor rating. For this reason, they can be represented in equivalent circuits by their locked-rotor impedances that are fed by line voltage. The locked-rotor current rating usually is assumed to be four to five times the motor full-load current. This is a conservative figure and on the safe side. Actual contribution is normally somewhat less.

(2) Limiting Fault Current — The asymmetrical short-circuit current will continue to flow for several cycles depending upon the X/R ratio of the system. The asymmetrical fault current will eventually decay to the final symmetrical value of the current that was calculated in the examples. Since the asymmetrical current is always greater than the symmetrical, the largest amount of destructive energy flows during the first few cycles after the fault

Table 70
Line-to-Ground Fault Calculations in Ohms for Fig 96

Bus-to-Bus		Z_1	Z_0/Z_1	Z_0	E	$^*I_{\text{L-G}}$	$0.38 \times {}^*I_{\text{L-G}}$ Min. Arcing
Main Switchboard North							
0	1	0.005	1	0.005			
1	2	0.0021	5	0.0105			
		0.0071		0.0155	277	27 980	10 632
Main Switchboard South							
0	1	0.005	1	0.005			
1	11	0.002	5	0.01			
		0.007		0.015	277	28 655	10 889
DP1							
0	1	0.005	1	0.005			
1	2	0.0021	5	0.0105			
2	4	0.002	10	0.02			
		0.0091		0.0355	277	15 475	5880
DP2							
0	1	0.005	1	0.005			
1	11	0.002	5	0.01			
11	13	0.0029	10	0.029			
		0.0099		0.044	277	13 025	4950
HA3							
0	1	0.005	1	0.005			
1	2	0.0021	5	0.0105			
2	4	0.002	10	0.02			
4	7	0.0413	10	0.413			
		0.0504		0.4485	277	1513	575
HA7							
0	1	0.005	1	0.005			
1	11	0.002	5	0.01			
11	13	0.0029	10	0.029			
13	16	0.0188	10	0.188			
		0.0287		0.232	277	2871	1091

$$^*I_{\text{L-G}} = \frac{3}{Z_1} \times \frac{E}{\left(\dfrac{Z_0}{Z_1} + 2\right)}$$

is initiated. The amount of destructive energy is proportional to the square of the current and the time the fault persists. Therefore, it is very important to limit the current to the smallest value possible.

9.6 Method of Reducing the Available Short-Circuit Current. The available short-circuit current on a distribution system decreases from the source to the load because the circuit impedance increases. The rate of current decrease or impedance increase is a function of the circuit design. With the design and insertion of impedance in the circuit between the power source and the building protective equipment, the short-circuit values throughout the building may be appreciably decreased and, at some points, may be lowered enough to permit lower rated, less expensive equipment to be used.

9.6.1 Effect of Distribution Circuit Lengths on Short-Circuit Current. When design considerations of voltage regulation, space, and economics permit, the circuit impedance at any point may be increased by the proper selection of cables, busways, and, principally, the choice of the circuit arrangement.

Physically, the circuit length depends on the location of the service entrance switch, load or distribution centers, riser shafts and riser tapoffs, and is affected by the circuit type and installation method.

When the available short-circuit current is high, a small increase in the impedance of the service entrance feeders and parts of the network system, as can be accomplished by using increased spacing between phase conductors, is very effective in reducing the maximum fault currents.

9.6.2 Current-Limiting Fuses. The term "current limitation" is associated with short circuits only. Since overloads are generally lower in magnitude, it is permissible for the overcurrent device to take many seconds to open. In contrast, since the short-circuit current is greater in magnitude, it is necessary for the fuse to operate as quickly as possible. When a fuse operates in its current-limiting range, it will clear in less than a half-cycle (0.008 second). A current-limiting fuse is one which, when operating in its current-limiting range, limits the instantaneous peak current to a value much less than that to which the short-circuit current would rise if the fuse were not in the circuit and clears in a half-cycle or less. Total clearing time is designated as t_c. The fundamental purpose of current-limiting fuses is to limit this instantaneous peak current to as low a value as possible and for it to clear quickly in order to limit the amount of damaging energy let-through to the circuit and components that the fuse is protecting. Figure 97 illustrates the current-limiting ability of such a fuse. The degree of current limitation afforded by a fuse depends on several factors: the fuse type, the A rating of the fuse, and the available short-circuit current. Figure 98(a) and (b) shows the typical current-limiting curves for two classes of current-limiting fuses.

9.6.3 Current-Limiting Reactors. Reactors are useful devices for reducing the interrupting duty imposed on protective equipment. Where standard rated circuit breakers can be used, it is usually not economical to substitute a reactor and a lower rated circuit breaker. However, when a reactor can be used to reduce the rating of several circuit breakers or to reduce interrupting duty to within the capacity of standard circuit breakers, the installation may be economically justified.

When installing reactors, consideration should be given to power loss, space, and voltage drop. If they are to be installed in combination power and lighting circuits, lamp flicker problems, as well as motor starting torque requirements, should be

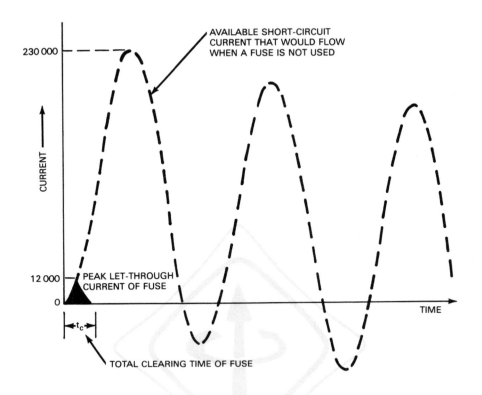

Fig 97
Current-Limiting Action of a Current-Limiting Fuse

investigated. The addition of reactors will increase the X/R ratios discussed previously, so special attention is required to check asymmetrical withstand or interrupting ratings.

9.6.4 Current-Limiting Busways. Current-limiting busways are another means of reducing short-circuit currents. They are available in ratings of approximately 1000–4000 A. Typical reactance values are shown in Table 66. At 0.70 to 0.90 lagging power factor, the voltage drop in the busway ranges from 1–2 V/10 feet (line-to-line) at full current rating. On large network systems with short-circuit currents up to 200 000 A symmetrical, the short run from the network bus to the switchboard is often sufficient to reduce the short-circuit currents to 100 000 A or less. As an example, the length of busway required to reduce a 180 000 A duty to 100 000 A is about 40 feet at 480 V or 20 feet at 240 V.

The impedance of the current-limiting busways is constant during all types of faults, such as low-level arcing faults.

A disadvantage of the current-limiting busway is the voltage drop in the busway. For each application of current-limiting busway, the voltage drop should be calculated. Although to obtain the desired reduction in short-circuit current the voltage

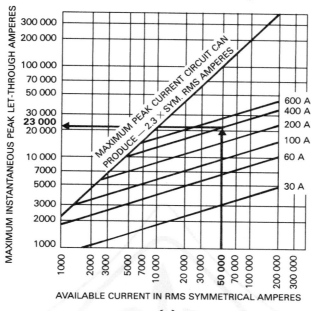

(a)

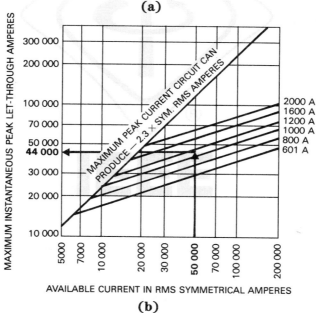

(b)

Fig 98
Typical Current-Limiting Fuse Peak Let-Through Curves
(a) (Class RK1) Peak Let-Through Current Data at 600 V_{ac}
(b) (Class L) Peak Let-Through Current Data at 600 V_{ac}

drop in the current-limiting busway is generally small, there are some applications in which the voltage drop is so high that the busway is not recommended and current-limmiting fuses should be used instead (current-limiting fuse application will be discussed later in this chapter). When the voltage drop in the current-limiting busway is too high, another possibility is to break the load into smaller parts. Dividing the load among four feeders reduces the voltage drop to 25% of its former value, provided that four current-limiting busways, each with the same impedance as the single busway, are utilized. To make a given reduction in short-circuit current, the percent voltage drop is the same on a 480 V system as on a 240 V system; but the length of busway required at 480 V is twice that at 240 V, and thus costs twice as much. On those current-limiting busway applications in which the voltage drop is satisfactory, the power loss in the busway is not a significant item.

When all factors are considered, including cost, it may be advisable to use current-limiting busways or reactors on some applications and combinations of current-limiting fuses and other equipment on other applications. Sometimes both are applicable on the same job. For example, current-limiting busways might be used from the transformer to the switchboard to reduce the duty to 75 000 A. Current-limiting fuses might be used in the switchboard in combination with circuit breakers rated less than 75 000 A interrupting. Current-limiting feeders might supply some equipment, such as a motor control center, to reduce the short-circuit current to 25 000 A. The first 20 or 30 feet of some busway feeders might be changed to current-limiting busways. The resulting combinations of current-limiting fuses and molded-case circuit breakers or other equipment might be used in other places.

9.6.5 Examples of Reducing Available Short-Circuit Current. Figure 99 illustrates the interrupting duty of protective equipment that might be required in various parts of a building for both a higher and lower voltage system. Cable conductors may be used in some circuits and either feeder of the current-limiting busway in others.

Use of a current-limiting device in the 2000 A switch will protect the downstream components. For example, in Fig 99(a), a current-limiting Class RK1 fuse will reduce the available fault current at the motor control center from 120 000 A to approximately 50 000 A (using one manufacturer's current-limiting curves); this current-limiting action will allow for less bracing requirements on the bus structure in the motor control center.

As an alternative, the building wiring may be designed to connect the service switch or circuit breaker directly to the bus takeoff or, if it is not feasible because of structural limitations or for other reasons, it may be located some distance away (as shown in Fig 99). The examples of Fig 99(a) and (c) assume that the service entrance switch or circuit breaker is directly at the end of the network bus takeoff. Figure 99(b) assumes that 37 feet of current-limiting busway is between the network bus and the service switch. This demonstrates the rapid reduction of available short-circuit current by lengthening the circuit when the available current value is high. In this case, the reduction is from 140 000 A to 80 000 A. Generally, it is advisable to have the service switch or circuit breaker as near to the bus takeoff as possible. The main interrupting device should be rated for the full available short-circuit current at the point of entrance.

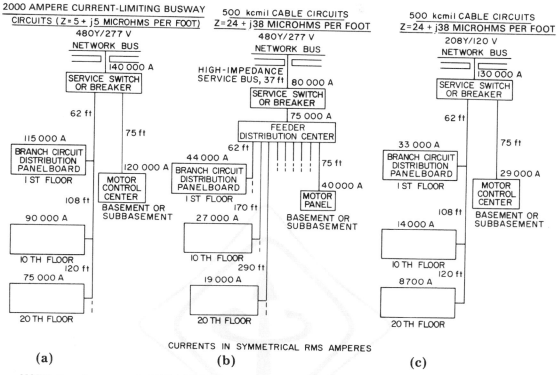

(a) **(b)** **(c)**

CURRENTS IN SYMMETRICAL RMS AMPERES

NOTE: Supply source six 2000 kVA 7% Z transformers for 460Y/265 V and six 1000 kVA 4% Z transformers for 208Y/120 V.

Fig 99
Typical Available Short-Circuit Currents on
Large Office Building Distribution Systems

Progressing away from the service entrance switch, the available fault current decreases, but not necessarily at a rate desirable for the protective equipment.

The effect of circuit length is illustrated in Figs 99 and 100, which depict the initial rapid decrease in available fault current with the increase in length of the circuit and show the diminishing rate of improvement as the circuit lengthens. The reduction of short-circuit current by lengthening the circuit is more effective at the higher current values than at the lower current values, where the relative improvement is much less and probably would not justify a lengthening of the circuit.

Figure 100 also illustrates general examples of the application of the methods and devices outlined above for controlling and limiting short-circuit currents in large building distribution circuits.

There are many other combinations of circuit elements that can be used in the layout of building wiring. In specific instances, the actual design depends on the type and magnitude of load, the service supply installation, building structure, local

445

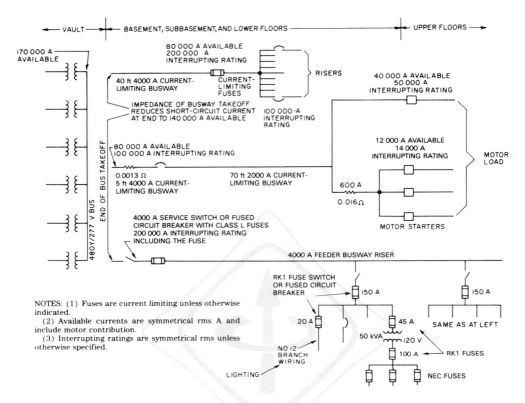

**Fig 100
Some Possible Arrangements to Limit and
Control Available Short-Circuit Current**

code requirements, reliability of service required, economic considerations, and the engineer's evaluation of these factors.

9.7 Selective Coordination. The major objective of the designer of an electric power system is to design a system so that faults will be removed in the shortest period of time possible, while maintaining a high degree of service continuity. The area of outage should be restricted as far as is practical. The goals of maximum protection and service continuity can most closely be realized by the proper selection and adjustment of high-speed protective devices. In order to properly select and adjust protective devices, a protective device coordination study is performed using the data from the short-circuit study and the time current curves of protected equipment withstand characteristics, as well as protective equipment operating characteristics. This chapter only explores a few of the aspects of a coordination study with particular emphasis on low-voltage systems. The procedures for conducting a coordination study are completely developed in IEEE Std 242-1986, IEEE Recommended Practice for Protection and Coordination of Industrial and Commercial Power Systems (ANSI) [20].

9.7.1 Coordination of Protective Devices. For a great many power systems, the optimum degree of protective device coordination consists of selective coordination in which only the protective device nearest the fault opens to remove a short circuit, and the other upstream protective devices remain closed.

On all power systems, the protective device should be selected and set to open *before* the thermal and mechanical limitations of the protected components are exceeded.

9.7.2 Preliminary Steps in a Coordination Study. Protective device coordination that balances device protection against the needs of service continuity is achieved and maintained only as a result of following a multiple-step procedure through to completion. When a short-circuit study has been performed as described earlier in this chapter, many of the preliminary coordination steps have already been undertaken.

Initially, a single-line diagram should be made of the system to be coordinated (see Fig 101). The diagram is used as a base on which pertinent data and information regarding relays, circuit breakers, fuses, current transformers, and operating equipment are recorded, while at the same time it provides a convenient representation of the relationships of the circuit protective devices with one another. The next step is to record all applicable impedances and ratings. Using these values, a short-circuit study is then made to determine the maximum and minimum short-circuit currents available at any particular point in the system.

Available fault current values can be noted on the system single-line diagram, and on the partial single-line diagrams used in coordination studies.

A further step is to ascertain the maximum load currents that will exist under normal operating conditions in each of the power system circuits, the transformer's magnetizing in-rush currents and starting currents, and the accelerating times of large motors. These values will determine the maximum current that circuit protective devices should carry without operating. The upper limit of current sensitivity will be determined by the smallest value resulting from the following considerations:

(1) Maximum available short-circuit current obtained by calculation
(2) Requirements of applicable codes and standards for the protection of equipment, such as cables, motors, and transformers
(3) Thermal and mechanical limitations of equipment

As a last preliminary step, the characteristic time current curves of all the protective devices to be coordinated should be obtained. These should be plotted on standard log-log coordination paper to facilitate the coordination study (see Figs 102 and 103).

9.7.3 Mechanics of Achieving Coordination. The process of achieving coordination among protective devices in series is essentially one of selecting individual units to match particular circuit or equpiment protection requirements, and of plotting the time current characteristic curves of these devices on a single overlay sheet of log-log coordination paper (5-cycle by 5-cycle), which is similar to Keuffel & Esser Company paper number 53599.

The achievement of coordination is a trial-and-error routine in which the various time current characteristic curves of the series array of devices are matched one against the other on a graph plot.

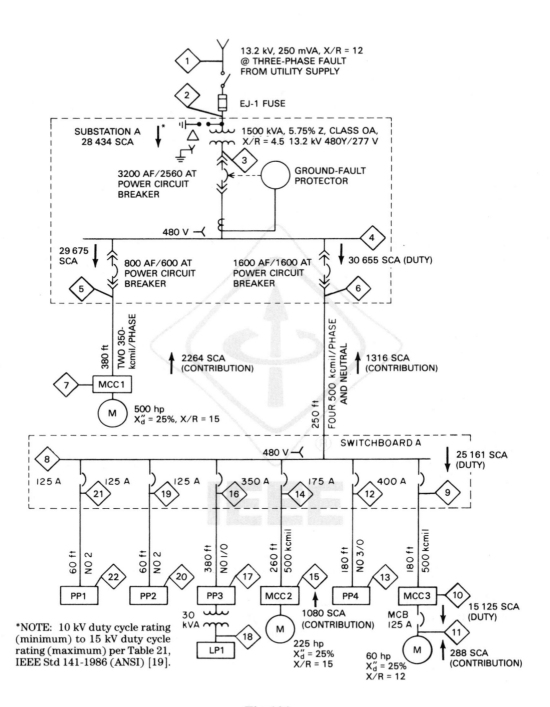

Fig 101
Typical Distribution Single-Line Diagram (Coordination Example)

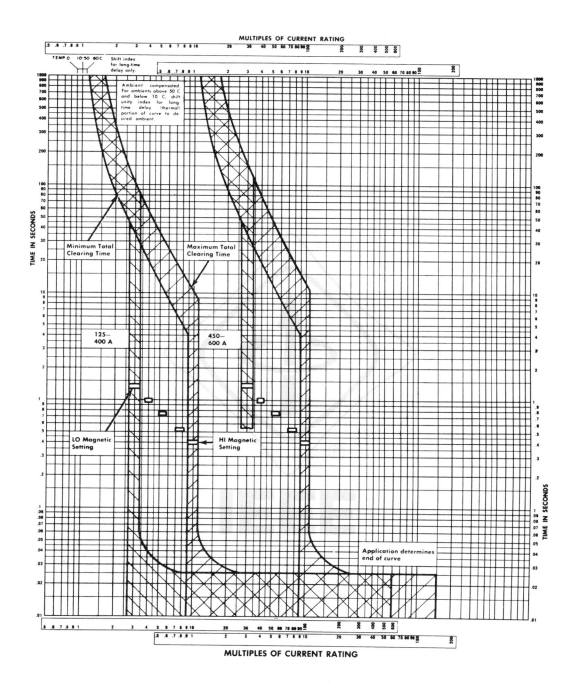

**Fig 102
Time Current Curves for 125–600 A Molded-Case Circuit Breakers**

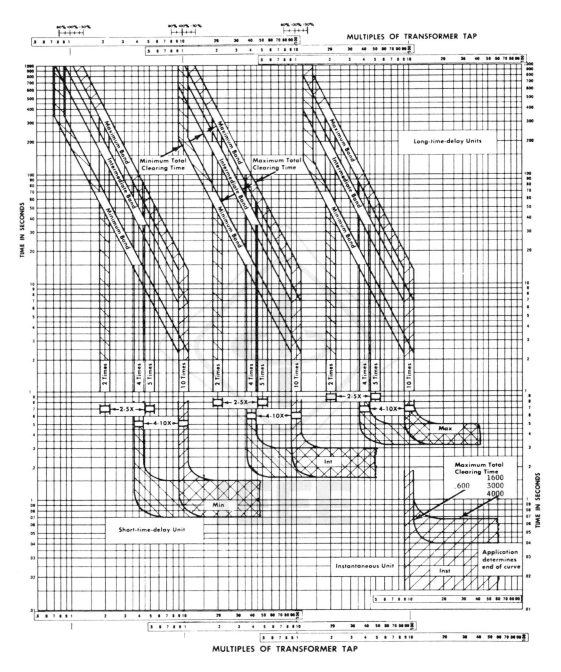

NOTE: Taps provided above the continuous current rating do not provide thermal self-protection.

**Fig 103
Time Current Curves for 600–4000 A Power Circuit Breakers**

When selecting protective devices, one should recognize ANSI, IEEE, NEMA, and NEC requirements and adhere to the limiting factors of coordination, such as load current, short-circuit current, and motor starting. The selected protective devices should operate within these boundaries, while providing selective coordination whenever possible.

Selective coordination (also called "discrimination") is usually obtained in low-voltage systems when the log-log plot of time current characteristics displays a clear space between the characteristics of the protective devices operating in series, i.e., no overlap should exist between any two time current characteristics if full selective coordination is to be obtained. An additional time allowance should be made for induction disk relay and for relay and fuse curve accuracy (see IEEE Std 242-1986 [20]). Quite often, the coordination study will not demonstrate complete selective coordination because a compromise has to be made between the competing objectives of maximum protection and maximum service continuity (see Fig 101).

9.7.4 Coordination Examples with Explanations. Now, we will examine the coordination of a portion of the power system that is shown in the single-line diagram in Fig 101.

The first level of coordination and protection to be considered is at the transformer primary. When selecting the primary protection, six factors should be taken into account

(1) Transformer full-load current
(2) The NEC, Section 450-3 [7]
(3) IEEE C57.109-1985 (ANSI) [18] (liquid filled transformers)
(4) Magnetizing inrush current of the transformer
(5) IEEE C57.12.59-1989 (ANSI) [16] (dry-type transformers)

NOTE: Use item (3) or (5) for transformer type, as appropriate.

(6) Hot load pickup current

In considering these factors, one can plot the withstandability of the transformer curves, and then select a fuse that meets the listed criteria (see Figs 105, 106, and 107).

To fully utilize the transformer, the protective device should carry transformer full-load current. Furthermore, the NEC [7] limits the maximum fuse size or device setting that can be utilized. For the liquid filled transformer shown in Figs 105, 106, and 107 (1500 kVA), a maximum fuse rating of three times or a relay pickup value of six times transformer full-load current is based on using a main secondary protective device. See the NEC, Section 450-3 [7] for more complete information.

Whatever primary device is used, it should be capable of withstanding transformer magnetizing inrush current. This point is usually selected as eight to twelve times transformer full-load current for a period of 0.1 second.

The transformer primary fuse should also be capable of withstanding the inrush current that occurs when a transformer that is carrying the load experiences a momentary loss of source voltage, followed by the re-energization of the transformer (such as when a source-side circuit breaker operates to clear a temporary upstream fault, and then automatically recloses). In this case, the inrush current is made up of two components: the magnetizing inrush current of the transformer, and the inrush current associated with the connected loads. The ability of the primary fuse to withstand combined magnetizing and load inrush current is referred to as "hot load pickup" capability.

As a result of the combined magnetizing and load inrush current, the integrated heating effect on the transformer primary fuse is equivalent to that of a current that has a magnitude of between 12 and 15 times the primary full-load current for a duration of 0.1 second.

These four factors (transformer full-load current, transformer mechanical and thermal withstand curve, transformer magnetizing current, and NEC [7] requirements) usually are shown on a single common base, time current curve along with the characteristics of an appropriately selected and rated protective device, considering the characteristics plotted.

(1) Transformer full-load current (at 480 V)

$$\frac{1500 \text{ kVA}}{(13.2 \text{ kV}) \sqrt{3}} = 66 \text{ A}$$

(2) NEC maximum fuse rating
   3 × full-load current with main secondary protective device

$$3 \times \frac{1500 \text{ kVA}}{(13.2 \text{ kV}) \sqrt{3}} = 197 \text{ A}$$

(3) Through-fault current duration — See the curve in Fig 104 for the Category II transformer, which plots current values at 58% of those shown to account for the delta-wye transformer connection. (Use the frequent fault curve from IEEE C57.109-1985, Fig 2 [18].)

(4) Magnetizing inrush current for 0.1 second (Also refer to IEEE C57.12.59-1989 [16] for dry-type transformers.)

$$8 \times \frac{1500 \text{ kVA}}{(13.2 \text{ kV}) \sqrt{3}} = 525 \text{ A}$$

Figure 105 shows the values for the calculated characteristics plus those of a suitably rated fuse that meets the criteria discussed. Figures 106 and 107 show similar curves for delta-delta and wye-wye connected transformers. See IEEE Std 242-1986, 10.8.3.2 and 10.6.3.3, pages 422–480 [20] for a more detailed description of the plotting procedure.

(5) Hot load pickup current for 0.1 second

$$12 \times \frac{1500 \text{ kVA}}{(13.2 \text{ kV}) \sqrt{3}} = 792 \text{ A}$$

Once the transformer parameters are plotted, rather than setting or selecting a primary main protective device, select the settings for the largest downstream load device. By starting at the load device first, the lower boundary of coordination is established.

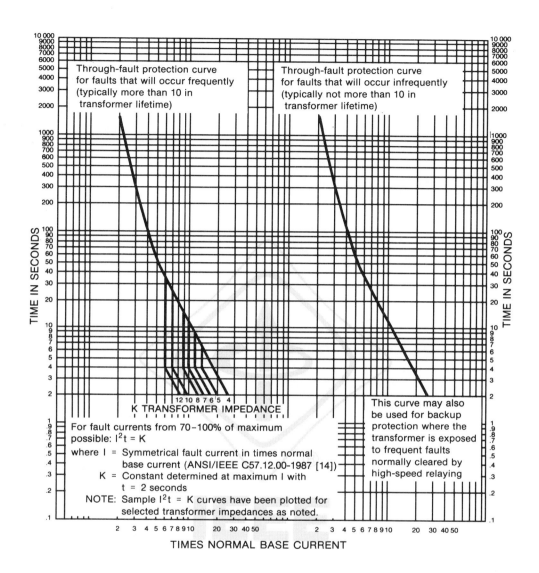

Fig 104
Through-Fault Protection Curves for
Liquid Immersed or Dry-Type Category II Transformers
(501–1667 kVA, Single-Phase; 501–5000 kVA, Three-Phase)

In the example under consideration, the switchboard "A" feeder is the largest downstream load feeder in substation "A" (1600 A). In switchboard "A," the motor control center (MCC) 3 feeder is the largest downstream feeder. In MCC 3, the largest load device is a 125 A molded-case circuit breaker feeding a 60 hp motor.

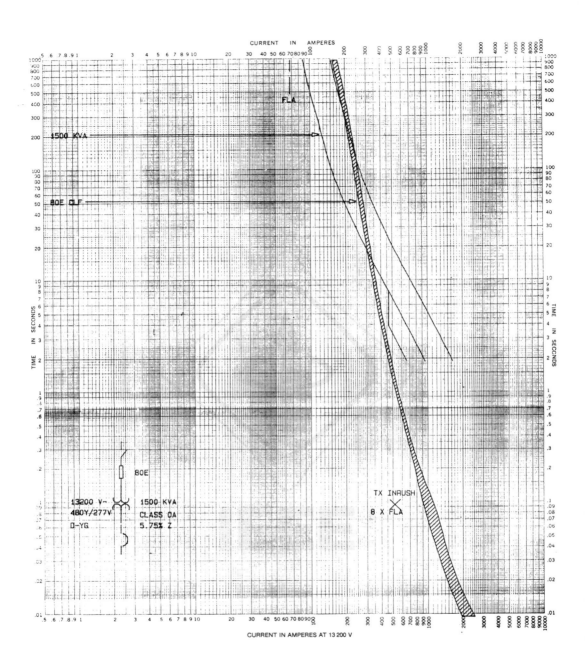

Fig 105
Time Current Curves Showing Primary Fuse
and Transformer Protection Criteria for a Delta-Wye,
Solidly Grounded Transformer for the System Shown in Fig 101

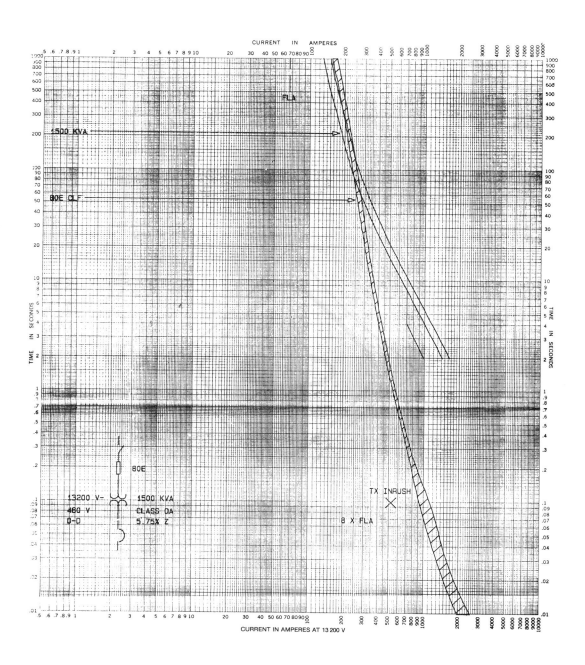

Fig 106
Time Current Curves Showing Primary Fuse and Transformer Protection
Criteria for a Delta-Delta Transformer for the System Shown in Fig 101

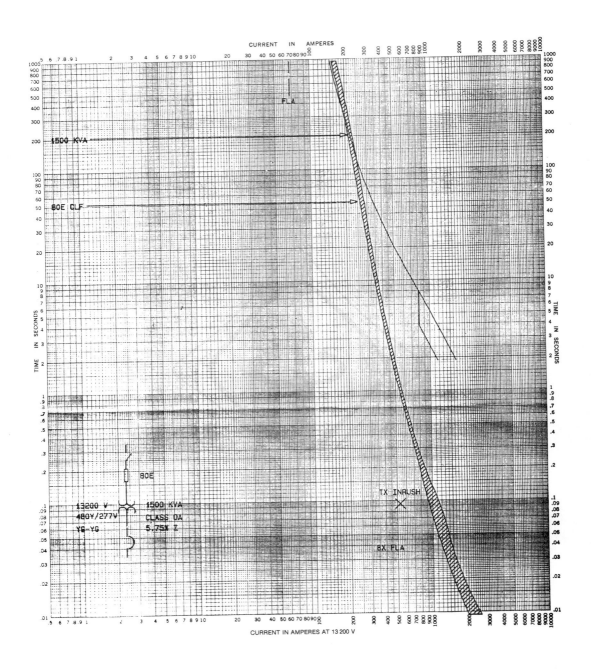

**Fig 107
Time Current Curves Showing Primary Fuse and Transformer
Protection Criteria for a Wye-Wye Transformer Solidly Grounded
on Both Primary and Secondary for the System Shown in Fig 101**

Using the coordination graph paper with the transformer parameters plotted on it, sketch in the motor current characteristics. When motor data are not available, it is usually assumed that the locked-rotor current is equal to six times the motor full-load current, and that the motor acceleration time is 10 seconds. Using these values for current and time, the straight-line characteristic is obtained. Motor running overcurrent protection that is provided by thermal overload relays or fuses should be shown. Once the motor starter and overload characteristics are plotted, the feeder device setting can be determined.

Normally, it is recommended that the setting of the motor feeder instantaneous element be at least twice the locked-rotor current. In this example, since the locked-rotor current is equal to 450 A, the instantaneous elements of the 125 A molded-case circuit breaker should be set at 900 A, or approximately 7.5 times the circuit breaker trip rating. By overlaying the previously drawn curves on the circuit breaker curve (see Fig 102), its characteristics are drawn in with an instantaneous setting as calculated ($7.5\ X$).

Once this device is drawn in, the lower limit of coordination is established. The remaining steps consist of overlaying the curves already drawn over the characteristic curves of each series upstream device in sequence, and selecting minimum settings for protection while obtaining coordination data.

The next upstream device in series is a 400 A molded-case circuit breaker. Overlay the curves already drawn on its characteristic curve, and select a setting that coordinates with the 125 A circuit breaker as set. The minimum instantaneous setting to coordinate will be $4\ X$. The 400 A circuit breaker characteristic is then drawn in using the $4\ X$ instantaneous setting. Most adjustable molded-case circuit breakers have only one adjustment, which is the instantaneous element.

It should be observed from the coordination between the 125 A and the 400 A circuit breakers that selective coordination between series instantaneous protective devices is seldom possible. In this example, any three-phase line-to-line or line-to-ground fault above 1500 A in magnitude results in a loss of service on the 400 A feeder.

When coordination of the instantaneous operating devices is required, the use of fuses in one or more protective devices may enable such coordination if adequate ratios between upstream and downstream fuses are maintained (see Table 71).

The next upstream device in the example system is a 1600 A low-voltage power circuit breaker equipped with an electrical trip device with long- and short-time adjustments. For the long-time portion of the curve (see Fig 104), there is a choice of a minimum, intermediate, or maximum long-time delay band. For the short-time portion, there is a choice of 2–5 times or 4–10 times the device pickup current value and a minimum, intermediate, or maximum time delay adjustment.

To select the settings for this device, again overlay the paper with the previously drawn curves on the characteristic curves on the device to be set. When this is done, it is found that a minimum long-time delay band and a short-time setting of $2\ X$ with a minimum time delay band coordinate well with the downstream devices, while providing maximum protection with minimum settings.

The final low-voltage device to be set in the series array being studied is the main secondary circuit breaker. This device is a 2500 A low-voltage power circuit breaker equipped with a static trip device with long- and short-time adjustments. The same choice of bands and ranges exists for this device as for the previously set 1600 A power circuit breaker.

Again, overlay the coordination paper with the downstream device curves on the characteristic curve of the device to be set. When this is done, it is found that a minimum long-time delay band, a short-time setting of 2 X, and a short-time delay band coordinate well with the downstream device and allow room between the overall curve and the ANSI curve from IEEE C57.109-1985, IEEE Guide for Transformer Through-Fault-Current Duration (ANSI) [18] to fit the primary protective device curve.

It should be noted that equipment protected by the 1600 A and 2500 A circuit breakers with short-time delay have a withstand rating capable of handling fault currents for the period of time that is needed to open the breaker, i.e., 0.19 second and 0.33 second, respectively.

The end point of the short-time delay band is cut off at the maximum available short-circuit current from the 1500 kVA transformer, in this case, 28 500 A.

Now that one complete series array of low-voltage protective devices has been set, the primary device can be selected. In this example, an EJ type is to be used as the primary protective device. A good point to start in selecting the fuse rating is 1.3 times the transformer full-load current, in this case, 85 A. When the coordination plot is placed over the 80E A fuse characteristic, it can be noted that coordination with the main secondary device is obtained.

To complete the system coordination study for phase overcurrent devices, the settings of the remaining series protective devices have to be selected. These settings are selected in the same manner as those described previously. Once the settings are all selected, one merely needs to set the devices as determined by the

Table 71
Typical Selectivity Schedule

Line Side	Class L Time Delay Fuse 601—6000A	Load Side Class L Fuse 601—6000A	Load Side Class K1 Fuse 0—600A	Load Side Class J Fuse 0—600A	Class K5 Dual-Element Fuse 0—600A
Class L time delay fuse 601—6000A	2:1	2.5:1	2:1	2:1	4:1
Class L fuse 601—6000A	2:1	2:1	2:1	2:1	4:1
Class K1 fuse 0—600A			2:1	2:1	8:1
Class J fuse 0—600A			2:1	2:1	8:1
Class K5 dual-element fuse 0—600A			1.5:1	1.5:1	2:1

study to carry out the coordination. Figure 108 shows the phase coordination for the system that is shown in Fig 101.

**Fig 108
Time Current Curves Showing Complete
Phase Coordination for the System Shown in Fig 101**

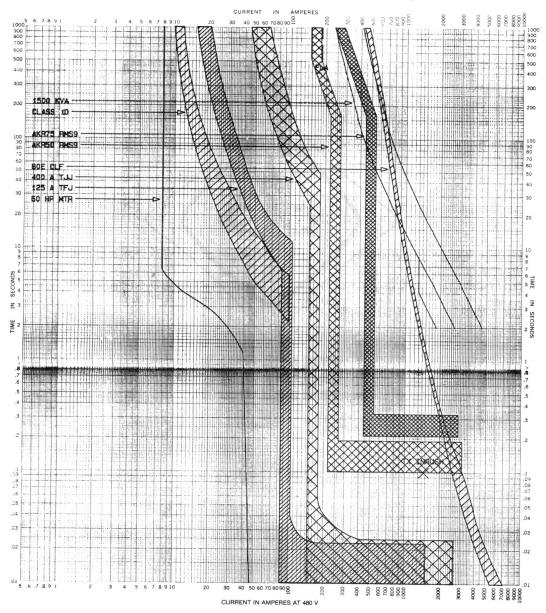

Calculations similar to those in 9.5.18 were performed, and it was determined that the ground-fault protection on the main circuit breaker should be set at 600 A pickup current and a time of 0.2 second. The ground-fault protection on the 600 A and 1600 A power circuit breakers was set at 400 A and a time of 0.1 second.

These settings were selected because they were the most consistent with the goals of protection and service continuity. The 400 A, 0.1 second pickup on the feeder circuit breakers gives good and fast protection, yet it will be selective with small downstream protective devices (20 A and 25 A molded-case circuit breakers).

The ground-fault protective device at the main circuit breaker is set higher both in time and current to provide selective coordination with the ground-fault protection devices on the feeder circuit breakers. Figure 109 shows the ground coordination for the system shown in Fig 101 overlaying the phase devices.

9.8 Fuses. The system coordination methods that have been described in previous sections cover general approaches. Fuse characteristics, such as current-limiting ability, I^2t coordination, and related material, are detailed in the following sections. By observing the principles previously stated and those in the following sections, an effectively coordinated system that involves fuses can be developed.

9.8.1 Fuse Coordination. Fuse time current curves that are plotted on standard log-log coordination paper are available from fuse manufacturers (see Fig 110). There are usually two sets of time current curves for each fuse. One curve shows the minimum melt characterisitc of the fuse, and the second shows the maximum total clearing time. In cases where only average melting curves are available, the manufacturers' recommendation to derive minimum melt and maximum total clearing times from these curves should be adhered to.

When coordinating fuses, the maximum total clearing time characteristic of the downstream fuse should fall below the minimum melt characteristic of the next upstream fuse.

Figure 111 illustrates how fuses selectively coordinate with one another for any value of short-circuit current. Note that, for selectivity, the total clearing energy of fuse B should be less than the melting energy of fuse A.

9.8.2 Fuse Selectivity Ratio Tables. The results of the phenomena displayed in Fig 111 for various types of fuses tested at rated voltage are presented in the form of ratio tables by various fuse manufacturers. Table 71 shows one manufacturer's selectivity schedule for various combinations of fuses.

An example of using the ratios in Table 71 is found in Fig 112 in which a 1600 A Class L fuse is to be selectively coordinated with a 400 A Class RK5 time-delay-type fuse.

Table 71 may be used as a simple checklist for selectivity, regardless of the short-circuit current involved. When closer fuse sizing is desired, check with the fuse manufacturers because the ratios may be reduced for lower values of short-circuit current. A coordination study may be desired (which is not required if ratios are adhered to) and can be accomplished by plotting time current characteristic curves on standard NEMA log-log graph paper. Since fuse ratios for high-voltage fuses to low-voltage fuses are not available, it is recommended that the fuses in question be plotted on log-log graph paper.

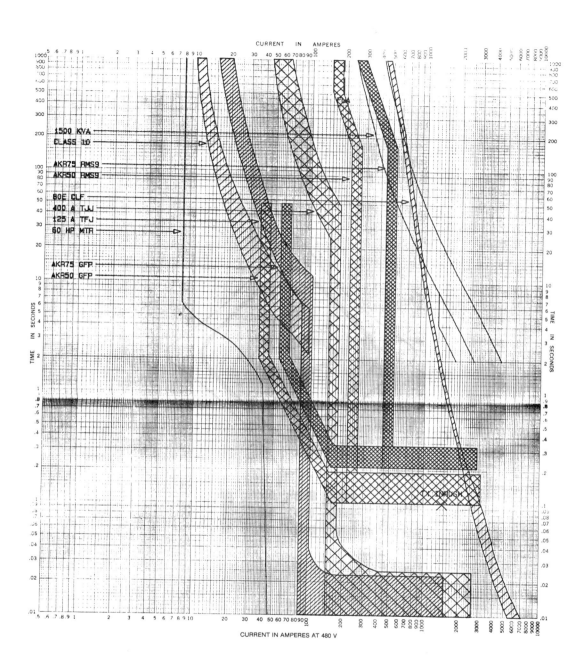

**Fig 109
Time Current Curves Showing Complete
Ground Coordination for the System Shown in Fig 101**

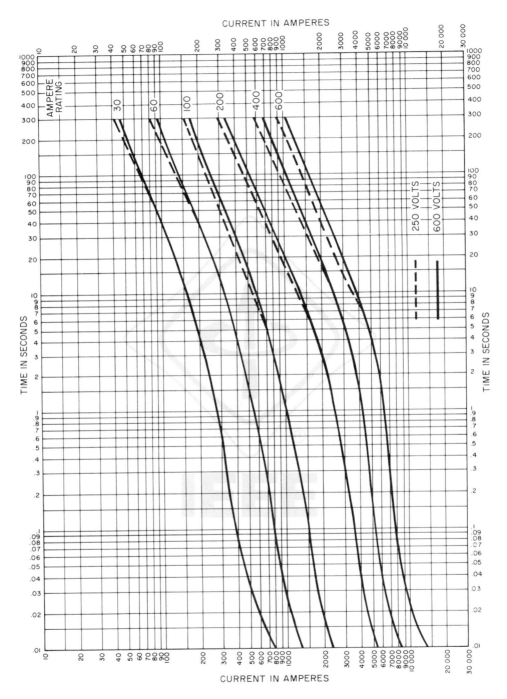

Fig 110
Typical Total Clearing Time versus Current Curves for Type RK5 Fuses

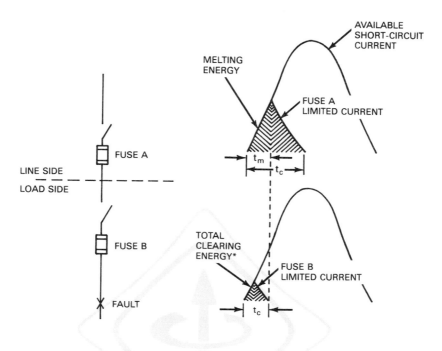

*INDICATES, BUT DOES NOT EQUAL, ENERGY.

Fig 111
Selectivity of Fuses
(Total clearing energy of fuse B should be less than melting energy of fuse A.)

9.8.3 Fuse Time Current Characteristic Curves. Fuse time current characteristic curves are available in the form of melt and total clearing time curves on transparent paper, which are easily adapted to tracing. A typical example of coordinating high- and low-voltage fuses using graphic analysis is shown in Fig 113. Note that the total clearing time curve of the 1200 A fuse is plotted against the minimum melt curve of the 125E 5 kV fuse. The curves are referred to as low voltage (240 V) for the study of secondary faults.

Care should be taken when coordinating the high- and low-side protection of a delta-wye transformer. For a line-to-ground fault on the wye side, one phase of the delta will see 16% more per unit current than the low side line. For a phase-to-ground fault on the secondary, the primary fuses will see only 58% of the phase-to-phase fault currents.

9.8.4 I^2t Values for Coordination. Depending on the class of fuses considered for application, there may be times when fuse I^2t values are required. Then one merely needs to compare the total clearing I^2t of the downstream fuse with the minimum melt I^2t of the next upstream fuse. When the downstream fuse's clearing

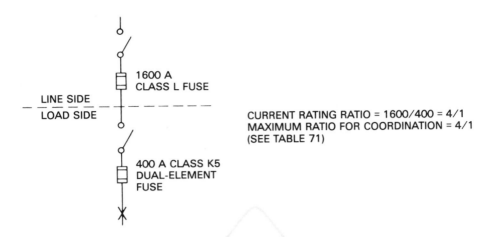

LINE SIDE
LOAD SIDE

1600 A
CLASS L FUSE

400 A CLASS K5
DUAL-ELEMENT
FUSE

CURRENT RATING RATIO = 1600/400 = 4/1
MAXIMUM RATIO FOR COORDINATION = 4/1
(SEE TABLE 71)

NOTE: If Class K1 time delay fuse used ratio is 2:1 (Class L 1200 A to Class K1 time delay 600 A).

**Fig 112
Typical Application Example Using the Data in Table 71**
(Selective coordination is apparent as fuses meet coordination/ratio requirements.)

I^2t is less than the upstream fuse's minimum melt I^2t, the fuses coordinate. All data used should be supplied by the manufacturer and will apply only to that manufacturer's fuse types.

9.8.5 Coordinating Fusible Unit Substation. The fusible unit substation shown in Fig 114 illustrates a 1000 kVA transformer that is supplied at 13.2 kV, serving a 480 V three-phase, three-wire switchboard. The primary 13.2 kV fuses are 80E power fuses and the 480 V secondary main fuses are 1200 A Class L fuses. The largest feeder is 400 A and is protected by 400 A Class RK1 time delay fuses and serves a 400 A motor control center.

The first step in coordinating this system is to follow the four factors on transformer protection that were given in 9.7.4. Then the minimum melt curve for the 80E power fuses is traced. This curve is referred to as 480 V for a study of secondary faults. The total clearing time curve for Class L 1200 A fuses is then traced on the graph in order to study the coordination between primary and secondary fuses.

The next step is to trace the minimum melt curve of the 1200 A Class L fuses and the total clearing time curve of the 400 A fuses. Noting complete coordination between the main and feeder fuses, the last step is to follow the above procedure to study the largest motor control center fuses and the 400 A feeder fuses. See Fig 115 for the completed coordination study.

The other procedure, which is quite often used to check coordination between low-voltage fuses, is to use a ratio chart that eliminates curve tracing (see 9.8.2). If a ratio chart analysis is used, the only curves that should be drawn are the primary and secondary fuse curves (as explained above). The 400 A Class RK1 time delay

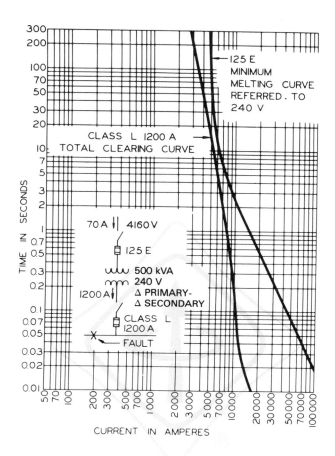

Fig 113
Coordination Study of Primary and
Secondary Fuses Showing Selective System

fuses can be installed in multiple switches equipped with ground-fault trip devices when full coordination with upstream fault protection is to be obtained.

9.8.6 Summary. Coordination is a multiple-step procedure consisting of the comparison and selection of protective devices and their ratings. The engineer who undertakes a coordination study should make all decisions concerning compromises between protection and selectivity.

9.8.7 Fuse Current-Limiting Characteristics. Due to the speed of response to short-circuit currents, fuses have the ability to cut off the current before it reaches dangerous proportions. Figure 116(a) illustrates the current-limiting ability of fuses. The available short-circuit current flows if there are no protective devices or if there is a delay as a result of the operation of a mechanical inertia-type device.

The large loop is one loop of a sine wave. This represents the first half-cycle of fault current available, which flows if no protective device is in the circuit. The

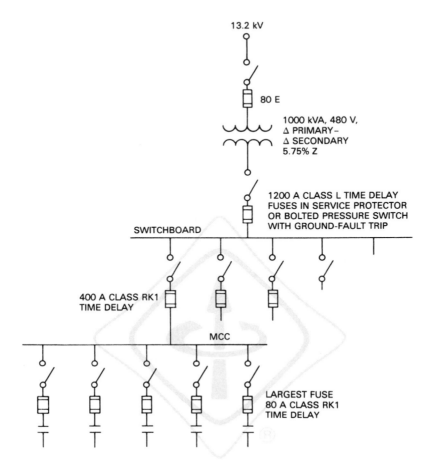

13.2 kV

80 E

1000 kVA, 480 V,
Δ PRIMARY –
Δ SECONDARY
5.75% Z

1200 A CLASS L TIME DELAY
FUSES IN SERVICE PROTECTOR
OR BOLTED PRESSURE SWITCH
WITH GROUND-FAULT TRIP

SWITCHBOARD

400 A CLASS RK1
TIME DELAY

MCC

LARGEST FUSE
80 A CLASS RK1
TIME DELAY

Fig 114
Fusible Unit Substation

current starts at zero in the circuit, rises to the peak of the loop, and returns to zero in a half-cycle of time (TIME). On a 60 Hz system, this happens 120 times each second. The peak of the wave represents the peak available current. See 9.1.6 for the effects of asymmetry on this process.

The effective value of the half-cycle of current, which is the value read on an ammeter, is $\sqrt{2}$ times the peak current. This is called the "root-mean-square value" and is not the same as the average value.

The small triangular wave in Fig 116(a) represents the performance of a current-limiting fuse on a fault current much higher than its rating. The fault current starts to rise; but melts the fuse element before the full available current can get through the fuse. The current through the fuse returns to zero, and the total elapsed time is represented by TIME. The peak of the triangular wave represents the peak current

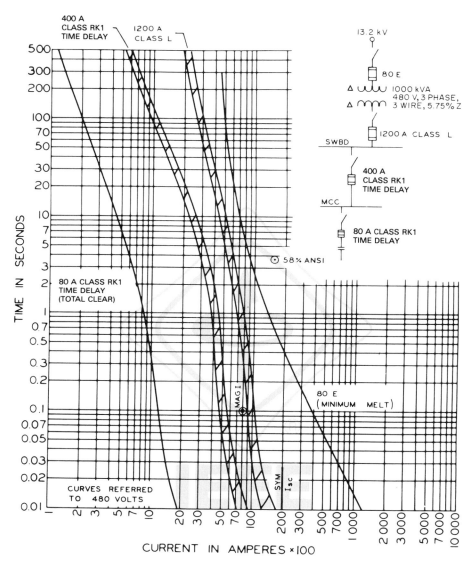

Fig 115
Completed Coodination Study of Low-Voltage
Fusible Substation Shows Complete Selectivity
(Class L and Class RK1 fuse curves are typical of one manufacturer's fuses.)

that the fuse lets through. This current can also be expressed in equivalent or apparent rms A (that is, the rms value of a symmetrical sinusoidal current that has the same peak current as the fuse let-through current).

It should be noted that current-limiting fuses limit both the fault current and the fault time.

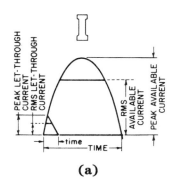

(a)

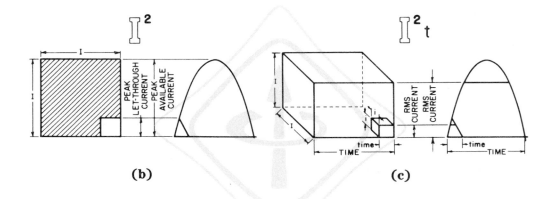

(b) **(c)**

Fig 116
Effects of I, I^2, and I^2t
(a) Current Limitation
(b) Mechanical (Electromagnetic) Force
(c) Heating Effect (Thermal Energy)

Figure 116(b) shows I_p^2, which is a measure of the mechanical force caused by peak short-circuit current where the force is proportional to the square of this current. This is associated with the electromagnetic force that mechanically stresses, and can damage, improperly designed bus structures, cable supports, etc. It immediately becomes apparent that squaring the peak available current can create a much larger square than squaring the peak let-through current of the current-limiting fuse. The difference in the size of the two squares is the difference between having and not having a current-limiting fuse in the circuit.

Figure 116(c) shows the heating effect I^2t, which is a measure of the thermal energy of a fault with and without a current-limiting fuse. In the case of I_p^2t, the rms current should be used instead of the peak current (as in the case of mechanical force I_p^2). The difference in size between the large and small cube-like figures represents the difference in energy between having and not having a current-limiting fuse in a circuit involving a high-magnitude available fault current. In extreme cases, where the effects of I^2t heating cannot be limited, points of failure typically involve pigtails and heater coils of motor starters and the possible welding of the contact in circuit-making devices.

For available fault currents greater than the "threshold current" of the fuse (value of the current in which the fuse becomes current limiting), a current-limiting fuse will limit the peak let-through current I_p to a value less than the available fault current and will clear the fault in less than one half-cycle, letting through only a portion of the available short-circuit energy. The degree of current limitation is usually represented in the form of peak let-through current charts.

Downstream equipment should be capable of withstanding voltage surges developed by a rapid drop in current or high $\mathrm{d}i/\mathrm{d}t$.

9.8.7.1 Peak Let-Through Current Charts. Peak let-through current charts, also referred to as "current-limiting effect curves," are useful for determining the degree of short-circuit protection that a fuse provides to the equipment located beyond it. These charts plot fuse instantaneous peak let-through current as a function of the available symmetrical rms current as shown in Fig 98, which is a typical manufacturers' curve. The straight line running from the lower left to the upper right shows a 2.4 (some manufacturers show 2.3, depending upon the power factor or X/R ratio of the test circuit) relationship between the instantaneous peak current that could occur without a current-limiting device in the circuit and the available symmetrical rms current. The following data can be determined from the peak let-through current charts:

(1) Peak-current let-through magnetic effect
(2) Apparent symmetrical rms let-through current heating effect

These data may then be compared to short-circuit ratings of static circuit elements, such as wire and bus.

Using the peak let-through current chart in Fig 117, we can enter at 100 000 A available symmetrical rms and read the following fuse let-through values:

(1) Peak let-through current — 10 000 A.
(2) Apparent symmetrical rms let-through current — 4000 A.

An example showing the application of the peak let-through current charts is represented in Fig 118 in which the component is protected by a 100 A current-limiting fuse, and the fuse let-through current values are needed with 100 000 A symmetrical rms available at the line side of the component.

This procedure will yield a value of symmetrical rms let-through current that can be compared with the rating of a downstream component, if the latter has been given a withstand time rating of one half-cycle or longer under a test power factor of 15%. When this method is used and the results are marginal, it is important that the manufacturer of the equipment (particularly in the case of molded-case circuit breakers) be consulted. I_p values should not exceed the component's peak withstand rating.

Knowing the short-circuit withstand capability of the component under consideration, a comparison can be made to establish short-circuit protection between maximum clearing I^2t and peak let-through current I_p.

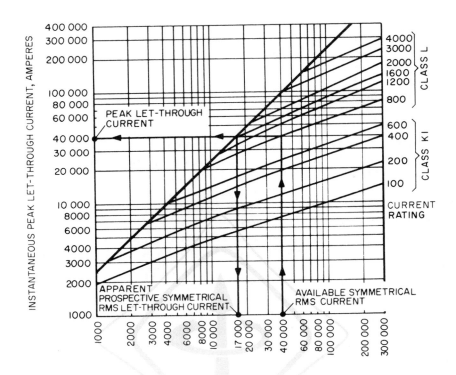

AVAILABLE SYMMETRICAL RMS SHORT-CIRCUIT CURRENT, AMPERES

**Fig 117
Peak Let-Through Current as a Function
of Available Symmetrical RMS Fault Current**

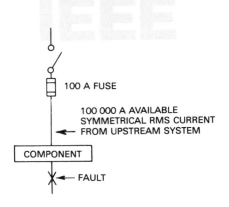

**Fig 118
Application Example of Fuse Let-Through Charts**

I^2t is a measure of the energy that a fuse lets through while clearing a fault. Every piece of electrical equipment is limited in its capability to withstand electrical destruction. When equipment is given an I^2t withstand rating, maximum clearing I^2t values for the fuses are available from manufacturers.

Magnetic forces can be substantial under short-circuit conditions and should also be examined. These forces vary with the square of the peak current I_p^2 and can be reduced considerably when current-limiting fuses are used. Some types of electrical equipment should be examined from the standpoint of peak circuit withstand as well as I^2t withstand.

9.8.8 Application of Fuses. Fuses that have 100 000 A or 200 000 A symmetrical rms interrupting ratings and are sized according to the NEC [7] requirements may provide adequate protection (both overload and short circuit) for the system components as well as provide increased interrupting capacity to handle future system growth. These fuses will also prevent unnecessary outages by isolating a faulted circuit if they are selected according to the selectivity ratios presented in Table 71.

An examination of fuse let-through charts for current-limiting fuses will reveal the adequacy of bus bracing requirements and wire protection when the withstand ratings are known.

Time delay fuses are most effectively applied in transformer and motor circuits because they can be sized close to the full-load rating without opening under transient conditions.

9.8.9 Bus Bracing Requirements. Reduced bus bracing requirements may be attained with current-limiting fuses. Figure 119 shows an 800 A motor control center protected by 800 A Class L fuses. The maximum available fault current to the motor control center (taking into consideration future growth) is 40 000 A symmetrical rms. To this available fault current from the upstream power system should be added the maximum fault contribution from the motors served from this motor control center (for example, with maximum motors in operation, drawing a rated full-load current of 700 A, the local motor contribution to the fault is 700 A

Fig 119
Example for Determining Bracing
Requirements for 800 A Motor Control Center

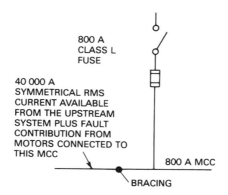

800 A
CLASS L
FUSE

40 000 A
SYMMETRICAL RMS
CURRENT AVAILABLE
FROM THE UPSTREAM
SYSTEM PLUS FAULT
CONTRIBUTION FROM
MOTORS CONNECTED TO
THIS MCC

800 A MCC

BRACING

× 4 = 2800 A). If a noncurrent-limiting device were used in front of the motor control center, the bracing requirement would be a minimum of 42 800 A symmetrical rms. Since current-limiting fuses are used, however, a substantial reduction in bracing may be possible. Entering the let-through chart of Fig 98(b) at 40 000 A, the apparent symmetrical rms let-through current for the 800 A fuse is 17 000 A. Thus, after adding the local motor fault current contribution of 2800 A, the total maximum available fault current at the motor control center main bus is 19 800 A. This would allow the standard bracing (see ANSI/NEMA ICS2-1990, Industrial Control Devices, Controllers, and Assemblies, Part ICS 2-322, p. 3 [5][69]) for an 800 A bus of 22 000 A symmetrical rms to be used. Depending on the fault current (considering future growth), other types and sizes of bus structures may be specified with reduced bracing.

NOTE: The bus bracing rating is stated as an rms symmetrical current for "x" cycles.

9.8.10 Circuit Breaker Protection. Circuit breakers may be applied in circuits where the available short-circuit current exceeds the interrupting rating of the circuit breakers when they are protected by current-limiting fuses properly applied in accordance with information from the circuit breaker manufacturer. Circuit breaker installations that were made several years ago may not meet present short-circuit current requirements because of changes to the electric system. These types of installations may also be protected from excessive short-circuit currents by the application of current-limiting fuses.

Reference should be made to circuit breaker manufacturers' literature for recommended circuit breaker fuse protection charts, which are the results of extensive testing. Fuse manufacturers now publish data that recommend fuse sizes and types for molded-case circuit breakers and should also be consulted.

9.8.11 Wire and Cable Protection. Sizing fuses for conductor protection according to the NEC [7] will assure short circuit as well as overload protection of conductors. Where noncurrent-limiting devices are used, short-circuit protection for small conductors may not be available, and reference should be made to ICEA wire damage charts (see IEEE C37.010-1979 (ANSI) [11]) for the short-circuit withstand capabilities of copper and aluminum cable.

Small conductors are protected from short-circuit currents by current-limiting fuses even though the fuse rating may be 300%–400% or higher of the conductor rating as allowed by the NEC [7] for motor branch-circuit protection.

9.8.12 Motor Starter Short-Circuit Protection. UL tests motor starters under short-circuit conditions. This short-circuit test may be used to establish a withstand rating for starters. Starters of 50 hp and less are tested with 5000 A available short-circuit current; starters over 50 hp are tested with 10 000 A (see IEEE C37.5-1979 [12]).

When applying starters in systems with high available fault currents, current-limiting fuses can reduce the let-through current to a value less than that established by the UL test procedures already described.

---

[69]NEMA publications are available from the National Electrical Manufacturers Association, 2101 L Street, N.W., Washington, DC 20037.

Figure 120 is a typical single-line diagram of a motor circuit in which the available short-circuit current has been calculated to be 40 000 A symmetrical rms at the motor control center, and the fuse is to be selected so that short-circuit protection as well as backup motor running protection is provided. When a Class RK1 time delay fuse sized at 125% of motor full-load current (17.5 A fuse) is chosen, the 40 000 A symmetrical rms will be limited by the fuse to let through current of less than 2000 A apparent symmetrical rms, and the fault will be cleared in less than one half-cycle. Since the apparent rms let-through current and clearing time are substantially less than the short-circuit withstand values established by the UL test for size 1 starters, this starter is considered to be protected from short-circuit damage. The apparent symmetrical rms let-through current can be determined from fuse manufacturers' let-through charts for 17.5 A time delay fuses.

IEC motor starters and contactors are widely used in the United States; but they present new problems in protection. Though they represent space and initial cost savings, the IEC starters have a lower withstand capability than their NEMA counterparts. In order to achieve the same level of protection for IEC devices as for NEMA devices, carefully select the motor starter fuse. For example, a 60 A Class RK5 fuse may let through 5900 A when the prospective short-circuit current is 50 000 A. Using a Class RK1 fuse reduces the let-through current to 3200 A, and using a Class J fuse reduces the let-through current to 2650 A.

9.8.13 Transformer Fuse Protection. Distribution transformers with low-voltage secondaries may be protected by fusing primary and secondary connections in accordance with the NEC, Section 450-3 [7]. Figure 121 shows low-voltage fuses for a 1000 kVA transformer that will provide overload protection.

**Fig 120
Selection of Fuses to Provide Short-Circuit
Protection and Backup Protection for Motor Starters**

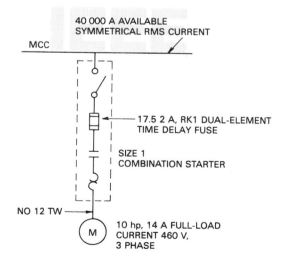

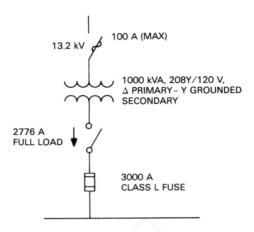

Fig 121
Typical Low-Voltage Distribution
Transformer Secondary Protection

Transformers are quite frequently used in low-voltage electric distribution systems to transform 480 V to 208Y/120 V. These transformers can be protected by using time delay fuses sized at 100%–125% of the primary full-load current. Some consideration should be given to the magnetizing inrush current since, for dry-type transformers, this current may be as high as 20–25 times rating. These inrush currents can easily be checked against the time delay fuse melting curve at 0.1 second (usually taken as the maximum duration of inrush current). Where dry-type and liquid filled transformers have inrush currents of about 12 times rating that last for 0.1 second, time delay fuses may be sized at 100%–125%. Figure 122 shows a 225 kVA lighting transformer properly protected with time delay fuses.

The NEC, Section 450-3 [7] covers overcurrent protection for transformers. It may be provided by protective devices in both primary and secondary circuits. However, the NEC, Section 450-3 [7] does spell out those conditions under which protection in only the primary is allowed. It also spells out the conditions under which secondary protection backed up with primary protection of the transformer is allowed (see the NEC, Table 450-3 (a) (2) [7]). With delta-wye transformations and under line-to-ground secondary fault conditions, the affected primary overcurrent devices will see only 58% of the comparable secondary short-circuit current.

9.8.14 Motor Running Overcurrent Protection. Single- and three-phase motors can be protected by the use of time delay fuses for motor running overcurrent protection sized according to the NEC [7]. These sizes vary from 100%–125% of motor full-load current, depending on service factor and temperature rise. When overload relays are used in motor starters, a larger size time delay fuse may be used to coordinate with the overload relays.

Combination fused motor starters that employ overload relays sized for motor running overcurrent protection (≤ 115% for 1.0 S.F. and ≤ 125% for ≥ 1.15 S.F.)

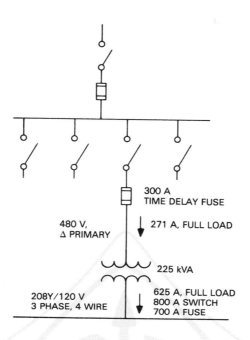

**Fig 122
Typical Protection for
225 kVA Lighting Transformer**

should incorporate time delay fuses sized at 115% for 1.0 S.F. and 125% for ≥ 1.15 S.F. or the next larger standard size to serve as backup protection. A combination motor starter with backup fuses will provide comprehensive protection. Figure 123 illustrates the protection for a motor circuit.

Three-phase motor single-phasing protection may be provided by time delay fuses that are sized at approximately 125% of the motor running current. Loss of one phase will result in an increase to 173%–200% of the line current to the motor. This will be sensed by the motor fuses because they are sized at 125%. Provided the fuses are sized to the actual motor running current, the single-phasing current will open the fuses before damage to the windings occurs. When the motors are running well under full load, anti-single phasing may be provided by sensitive anti-single-phasing-type motor overload relays.

9.8.15 Fuse Device Maintenance. Modern silver-sand and copper-link fuses require little if any maintenance. Occasionally, a visual and infrared inspection of the fuse retainers, or clips, is recommended to ensure that there is adequate pressure between contact making parts, and also so that overheating because of a bad connection is easily detected. Fuse characteristics do not change with age; hence, no maintenance is required for those fuses in storage.

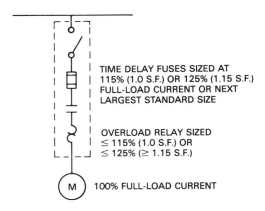

TIME DELAY FUSES SIZED AT
115% (1.0 S.F.) OR 125% (1.15 S.F.)
FULL-LOAD CURRENT OR NEXT
LARGEST STANDARD SIZE

OVERLOAD RELAY SIZED
$\leq$ 115% (1.0 S.F.) OR
$\leq$ 125% ($\geq$ 1.15 S.F.)

(M) 100% FULL-LOAD CURRENT

Fig 123
Protection for Typical Motor Circuit

9.9 Current-Limiting Circuit Breakers. When using circuit breakers, current-limiting characteristics can be obtained in the following ways:

(1) Auxiliary current-limiting fuses are internally mounted in molded-case circuit breakers. These are usually special-purpose fuses designed for breaker application.
(2) Current-limiting fuses are used with low-voltage power circuit breakers. The fuses are usually mounted on the drawout circuit breaker or within a separate drawout assembly, or on switchboards or switchgear.
(3) Nonfused current-limiting circuit breakers utilize very fast tripping speeds so that the potential high-magnitude fault current is limited during the first half-cycle of fault current, just as is achieved by some classes of current-limiting fuses. The unit operates as a conventional circuit breaker for overload and lower level fault currents.

Current-limiting circuit breakers, including those incorporating current-limiting fuses, are intended for applications needing the overload/overcurrent and switching functions of the circuit breaker in systems where available fault current exceeds the rated fault current capabilities of the circuit breaker, or other components of the power distribution system. When the current-limiting element used in conjunction with the circuit breaker is properly selected, the current limiter operates only in the event of a low-impedance fault (in a power system of high-fault current capacity) to provide protection against high peak fault current for the circuit components, including the circuit breaker and the downstream circuit components. The conventional elements of the circuit breaker will clear the overloads and low-magnitude fault currents, which are the most frequent causes of automatic operation of protective equipment in low-voltage systems. The current-limiting element handles the relatively infrequent high-magnitude fault currents, and operation of any fuse will trip the circuit breaker. This isolates the faulted portion of the system and will preclude the possibility of this causing single-phase operation of motors downstream from the device.

9.10 Ground-Fault Protection. The NEC, Sections 230-95 and 517-14 [7] requires knowledge of the levels of ground-fault currents to properly set and coordinate ground-fault protective devices. The NEC, Section 230-95 [7] states that "ground-fault protection of equipment shall be provided for grounded-wye electrical services of more than 150 V to ground, but not exceeding 600 V phase-to-phase for any service disconnecting means rated 1000 A or more." This ground-fault protection may consist of overcurrent devices and current transformers, or other equivalent protective equipment, which shall operate to cause the service disconnecting means to open all ungrounded conductors of the faulted circuit. The maximum setting of the ground-fault protection shall be 1200 A and the maximum time delay is 1 second for ground faults equal to or greater than 3000 A.

The NEC, Section 230-95, "Fine Print Notes" [7] explains that it may be desirable to include ground-fault protection for service disconnecting means rated less than 1000 A and also that additional installations of ground-fault protective equipment will be needed on feeders and branch circuits when maximum continuity of electrical service is necessary. In health care facilities, when ground-fault protection is provided on the service disconnecting means, the NEC, Section 517-14 [7] requires the additional step of ground-fault protection in the next level of feeder downstream toward the load.

The NEC, Article 240-13 [7] now requires ground-fault protection of equipment for each building or structure main disconnecting means rated at 1000 A or more.

Figure 124 shows methods of detecting the ground-fault current. If a transformer (see Fig 124(a)) is the source of supply and its ground return current can be measured, a simple current transformer may be used to detect the flow of ground-fault current back to the neutral connection of the transformer windings.

Fig 124
Detecting Ground-Fault Current
(a) With Current Transformer or Sensor
(b) With Current Transformer
(c) With Current Sensors

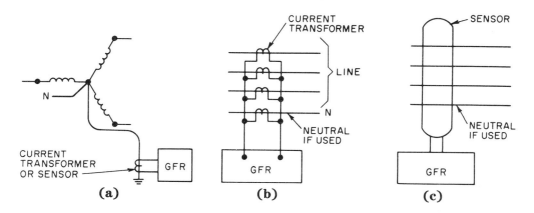

This method can also be used if the power system includes the neutral conductor (that is, loads may be connected line-to-neutral) provided the current transformer is located between the power transformer ground connection and the neutral conductor connection and also that the neutral conductor remains an insulated-isolated conductor (that is, no additional neutral conductor ground connections are made downstream).

Figures 124(b) and (c) depend on the principle that the phasor sum of all currents flowing from and returning to a source of power is zero. If there is any current flow through ground, then this current, when added to that flowing through the line and neutral conductors, should equal zero. Therefore, the unbalanced current or flux through the current transformers or sensor should equal that of the ground-fault current. The sensor usually consists of a single-window-type current transformer with an opening large enough to accept all of the phase and neutral conductors and is designed to handle only a limited burden ground-fault relay specially matched to it.

Figure 125 illustrates a typical single-line diagram with ground-fault relaying. The term "relay" includes electronic or solid-state relays as well as electromechanical devices. These relays may be specified with various pickup levels of current and with various time delay setting ranges. Full coordination with line or phase protective relaying is desirable. The simplest system involves time delay and current selectivity. A better system utilizes blocking signals or zone-selective interlock from the downstream device to delay the tripping of the upstream device to give the former a chance to clear the fault. A number of systems providing this kind of

**Fig 125
Typical Ground-Fault Relaying**

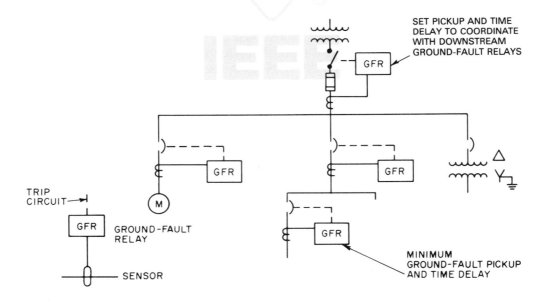

protection are available for protecting secondary-unit substations, double-ended substations, networks, and other sources, and information concerning such protection can be obtained from the switch or circuit breaker manufacturer. When intentional delaying is introduced, the equipment should be designed to handle the short-circuit current for that duration.

9.11 References. The following references shall be used in conjunction with this chapter:

[1] ANSI C37.06-1987, Preferred Ratings and Related Required Capabilities for AC High-Voltage Circuit Breakers Rated on a Symmetrical Current Basis.

[2] ANSI C37.5-1953, Determining the Rms Value of a Sinusoidal Current Wave and a Normal-Frequency Recovery Voltage and for Simplified Calculation of Fault Currents.

[3] ANSI C37.5-1969, Methods for Determining Values of a Sinusoidal Current Wave, a Normal-Frequency Recovery Voltage, and a Guide for Calculation of Fault Currents for Application of AC High-Voltage Circuit Breakers Rated on a Total Current Basis.

[4] ANSI C37.6-1971, Schedule of Preferred Ratings for AC High-Voltage Circuit Breakers Rated on a Total Current Basis.

[5] ANSI/NEMA ICS2-1988, Industrial Control Devices, Controllers, and Assemblies.

[6] ANSI/NFPA 20-1990, Centrifugal Fire Pumps.

[7] ANSI/NFPA 70-1990, National Electrical Code.

[8] ANSI/NFPA 110-1988, Emergency and Standby Power Systems.

[9] ANSI/NFPA 110A-1989, Stored Energy Systems.

[10] ICEA P-32-382-1969, Short-Circuit Characteristics of Insulated Cable.

[11] IEEE C37.010-1979 (Reaff. 1988), IEEE Application Guide for AC High-Voltage Circuit Breakers Rated on a Symmetrical Current Basis (Includes Supplement C37.010d) (ANSI).

[12] IEEE C37.5-1979, IEEE Guide for Calculation of Fault Currents for Application of AC High-Voltage Circuit Breakers Rated on a Total Current Basis.

[13] IEEE C37.13-1981, IEEE Standard for Low-Voltage AC Power Circuit Breakers Used in Enclosures (ANSI).

[14] IEEE C57.12.00-1987, IEEE Standard General Requirements for Liquid-Immersed Distribution, Power, and Regulating Transformers (ANSI).

[15] IEEE C57.12.01-1989, IEEE Standard General Requirements for Dry-Type Distribution and Power Transformers Including Those with Solid Cast and/or Resin-Encapsulated Windings.

[16] IEEE C57.12.59-1989, IEEE Guide for Dry-Type Transformer Through-Fault Current Duration.

[17] IEEE C57.94-1982 (Reaff. 1987), IEEE Recommended Practice for Installation, Application, Operation, and Maintenance of Dry-Type General Purpose Distribution and Power Transformers (ANSI).

[18] IEEE C57.109-1985, IEEE Guide for Transformer Through-Fault-Current Duration (ANSI).

[19] IEEE Std 141-1986, IEEE Recommended Practice for Electric Power Distribution for Industrial Plants (ANSI).

[20] IEEE Std 242-1986, IEEE Recommended Practice for Protection and Coordination of Industrial and Commercial Power Systems (ANSI).

[21] NEMA AB1-1986, Molded Case Circuit Breakers.

[22] NEMA BU1-1988, Busways.

[23] NEMA PB2.2-1988, Application Guide for Ground Fault Protection Devices for Equipment.

9.12 Bibliography. The references in this bibliography are listed for information purposes only.

[B1] ANSI C84.1-1989, Electric Power Systems and Equipment — Voltage Ratings (60 Hz).

[B2] Beeman, D. L., ed. *Industrial Power Systems Handbook*, New York: McGraw-Hill, 1955.

[B3] Freund, Arthur, *Overcurrent Protection, Electrical Construction and Maintenance*, New York: McGraw-Hill, 1980.

[B4] Gienger, J. A., Davidson, O. C., and Brendel, R. A. "Determination of Ground-Fault Current on Common AC Grounded Neutral Systems in Standard Steel or Aluminum Conduit," *AIEE Transactions, pt. II*, vol. 79, May 1960, pp. 84–90.

[B5] Huening, W. C., Jr. "Interpretation of New American National Standards for Power Circuit Breaker Application," *IEEE Transactions on Industry and General Applications*, vol. IGA-5, Sep./Oct. 1969, pp. 501–523.

[B6] *Industrial Control Equipment*, Underwriters Laboratories, Inc., Bulletin 508, paragraphs 121, 131, and 144.

[B7] Kaufman, R. H. "Let's Be More Specific about Equipment Grounding," *Proceedings of the American Power Conference*, 1972.

[B8] Reichenstein, Hermann W. *Applying Low-Voltage Fuses, Electrical Construction and Maintenance*, New York: McGraw-Hill, 1979.

[B9] *Short-Circuit Current Calculations*, General Electric Company Bulletin GET-3550D.

Chapter 10
Lighting

10.1 General Discussion. The era of electric lighting began a little more than a century ago with the invention of the incandescent lamp. Prior to that time, daylight was the principal illuminant in commercial buildings, with flame sources occasionally used to allow for earlier starting times or somewhat longer operations late in the day after daylight had faded.

Electric lighting has proved to be a high-technology industry, with manufacturers devoting effort to research and development. Consequently, in recent decades, a succession of new, more efficient light sources, auxiliary equipment, and luminaires have been introduced. Research in basic seeing factors has also been pursued for many years, and a succession of developments has provided greater knowledge of many of the fundamental aspects of the quality and quantity of lighting. Some of these developments make it possible to provide for visual task performance using considerably less lighting energy than in the past.

Today, energy conservation, cost, and availability, both present and future, should guide decisions on every energy using subsystem of a building. Despite the dramatic reduction in the energy required to produce effective illumination, lighting continues to account for 40% of commercial building energy use. This chapter will include ways to reduce the energy requirements for lighting, yet provide adequately for the well-being and needs of the occupants and the objectives of the owners.

Since there is much documentation elsewhere on lighting technology and design, reference will be made to the appropriate sources of such information. Application techniques and controls that save energy and costs will also be stressed in the material presented here. Chapter 17 of this book "Electrical Energy Management," also addresses the subject of lighting, specifically its relationship to energy conservation.

10.1.1 Lighting Objectives. Owner objectives for lighting may vary over a broad scale depending on whether fast and accurate visual performance in a business-like environment is desired or whether the creation of mood and atmosphere in a space is of paramount importance. Lighting has great flexibility in this regard, and designers can vary its distribution and color, use its effect on room surfaces and objects to achieve dramatic, sparkling, somber, relaxing, or attention getting effects, as desired.

481

In recent years, the psychology of lighting has had some in-depth study, and some guides are now available to aid designers and application engineers in using lighting to create the attributes in an environment that will result in the appropriate subjective reactions of the occupants of a space (see References [5]-[8][70]).

The desired objectives for lighting should be accomplished through an energy-efficient design.

10.1.2 Lighting Regulations. In 1976, the Federal Energy Agency (FEA) (the forerunner to the DoE) began an attempt to mandate energy conservation. This resulted in a document called "The Model Code for Energy Conservation in New Buildings." The FEA asked that the states adopt a mandatory lighting efficiency standard at least as stringent as ANSI/ASHRAE 90-80, Section 9. Subsequently, an ASHRAE/IES Committee was formed, and the original standard was revised. The resulting standard, ASHRAE/IES 90.1-1989, Energy Efficient Design of New Buildings Except New Low-Rise Residential Buildings [2],[71] is a useful guide to energy-efficient building design. As of this date, ANSI has not approved this standard.

On May 6, 1987, the DoE published, in the *Federal Register*, a note of a proposed interim rule entitled "Energy Conservation Voluntary Performance Standards for New Commercial and Multi-Family High-Rise Residential Buildings." When issued, the rule will become mandatory for all federal buildings and a voluntary recommendation for all other facilities. In addition, many states and cities have adopted their own standards. These standards may include both new and retrofit construction and may have very profound effects on the selection of light sources, ballasts, and application of controls. Some standards limit the amount of energy that may be utilized for lighting. Another approach is to offer "trade-offs" for the use of controls and other energy-conserving techniques. Therefore, it is incumbent on designers to become familiar with standards effecting their projects and to use efficient light sources, luminaires, and control techniques to achieve the lighting effect that is desired. Equally as important, it requires appropriate controls to turn off the lighting when it is not needed.

For existing buildings, a limit on fixed lighting load (expressed in W/ft^2) has been applied in some legal jurisdictions. Another approach used is to make an audit of the lighting energy consumed as of some appropriate base date and mandate an arbitrary percentage reduction from that figure. Credits may be available when approved energy conservation techniques, such as the use of electronic ballasts, occupancy sensors, or automated scheduling controls, are installed.

Additional credits may be granted for the use of daylighting techniques with controls to reduce lighting energy use. Some jurisdictions now use the annual energy budget (Btu/ft^2 per year) for a building on which all building subsystems should draw (e.g., lighting, heating, cooling, ventilation, hot water, etc.). Budgets will vary depending on building type and the climate where it is located. The owner/

---

[70]The numbers in brackets correspond to those in the references at the end of each chapter.

[71] ASHRAE publications are available from the American Society of Heating, Refrigerating, and Air-Conditioning Engineers, 1791 Tullie Circle, N.E., Atlanta, GA 30329. IES publications are available from the Illuminating Engineering Society, 345 East 47th Street, New York, NY 10017.

designer has a choice as to how to allot this energy budget among the subsystems of a particular building.

The United States General Services Administration (GSA) has issued a circular (see Reference [9]) that specifies certain illuminances (fc) for workstations and other areas of federal buildings.

Users of this recommended practice are advised to maintain an awareness of the regulations that may affect lighting power or energy use in a building for which they have responsibility. ASHRAE/IES 90.1-1989 [2] may be used as a guide; however, local regulations may be more restrictive. The designer would be well advised to use the most efficient light sources, luminaires, and control techniques for lighting the buildings that they are designing.

10.2 Lighting Terminology. The following are common lighting terms with commonly applied definitions:

ballast. An electrical device that is used with one or more discharge lamps to supply the appropriate voltage to a lamp for starting, to control lamp current while it is in operation, and, usually, to provide for power factor correction. Ballasts may be magnetic core and coil, electronic, or resistive.

brightness. The subjective attribute of any light sensation, including the entire scale of the qualities "bright," "light," "brilliant," "dim," and "dark." Brightness has been used in the past to refer to measurable photometric brightness. The preferable term for the latter is "luminance," which reserves the term "brightness" for the subjective sensation.

contrast. Indicates the degree of difference in light reflectance of the details of a task compared with its background. Contrast includes both specular and diffuse components of reflection.

coefficient of utilization (CU). For a specific room, the ratio of the average lumens delivered by a luminaire to a horizontal work plane to the lumens generated by the luminaire's lamps alone. The work plane is usually (but not necessarily) considered to be 30 inches above the floor.

efficacy. See lumens per watt (lm/W).

equivalent sphere illumination (ESI). The measure of the effectiveness with which a practical lighting system renders a task visible compared with the visibility of the same task that is lit inside a sphere of uniform luminance.

fixture. *See* luminaire.

footcandle (fc). A unit of illuminance (light incident upon a surface) that is equal to 1 lm/ft^2. In the international system, the unit of illuminance is lux (1 fc = 10.76 lux).

footlambert (fl). The unit of luminance that is defined as 1 lm uniformly emitted by an area of 1 ft^2. In the international system, the unit of luminance is candela per square meter (cd/m^2).

glare. The undesirable sensation produced by luminance within the visual field. It may cause annoyance (discomfort glare) or a temporary loss in visual performance (disabling glare).

high-intensity discharge (HID) lamps. A group of lamps filled with various gases that are generically known as mercury, metal halide, high-pressure sodium, and low-pressure sodium.

illuminance. The unit density of light flux (lm/unit area) that is incident on a surface. In the British system $1 \text{ lm/ft}^2 = 1$ fc; in the metric system, $1 \text{ lm/m}^2 = 1$ lux.

lamp. Generic term for a manmade source of light.

lumen (lm). The international unit of luminous flux or the time rate of the flow of light.

lumens per watt (lm/W). The ratio of lumens generated by a lamp to the watts consumed by the lamp. Traditionally, this term has not included the ballast watts for discharge lamps because of the many types of ballasts available. *See also* efficacy.

luminaire. A complete lighting unit that consists of parts designed to position a lamp (or lamps) in order to connect it to the power supply and to distribute its light.

luminaire efficiency. The ratio of lumens emitted by a luminaire and of the lumens generated by the lamp (or lamps) used.

luminance. The light emanating from a light source or the light reflected from a surface (the metric unit of measurement is cd/m^2).

lux. The metric measure of illuminance that is equal to 1 lumen uniformly incident upon 1 m^2 (1 lux = 0.0929 fc).

rated life of a ballast or lamp. The number of burning hours at which 50% of the units have burned out and 50% have survived.

reflectance. The ratio of the light reflected by a surface to the light incident. An approximation of a diffuse surface's reflectance may be obtained with a light meter. Surface specularity will greatly affect reflectance measurements.

relative visual performance (RVP). The potential task performance based upon the illuminance and contrast of the lighting system performance.

task-ambient lighting. A concept involving a component of light directed toward tasks from appropriate locations by luminaires located close to the task for energy efficiency. Ambient lighting is provided to fill in otherwise unlighted areas, reduce contrasts in the environment, and supply additional light on the tasks.

veiling reflections. Reflected light from a task that reduces visibility because the light is reflected specularly from shiny details of a task, which brightens those details and reduces contrast with the background.

visual comfort probability (VCP). A rating of a lighting system expressed as a percentage of people who, if seated at the center of the rear of a room, will find the lighting visually acceptable in relation to the perceived glare.

visual task. Work that requires illumination in order for it to be accomplished.

work plane. The plane in which visual tasks are located. For offices and schools, it is usually considered to be a horizontal plane 30 inches above the floor; however, it can be any plane (vertical, sloping, or horizontal) at any height.

10.3 Illumination Quality. Some quality of light factors are
   (1) Providing illumination without discomfort caused by glare
   (2) Providing the light so that veiling reflections in task details are minimized
   (3) Using a high color rendering source in which the appearance of people, food, appointments of a space, etc., are critical, or where the task itself has colors that should be discriminated
   (4) Selecting sources and luminaires that will provide sparkle and modeling on certain types of objects
   (5) Using sources, equipment, and techniques that will help provide the desired atmosphere in a space

10.3.1 Visual Comfort. In offices, schools, libraries, drafting rooms, and similar spaces, it is desirable to provide illumination without annoyance or discomfort due to luminaire or window brightness. Reference to visual comfort probability (VCP) data, which are available from luminaire manufacturers, is helpful in the selection of luminaires that will not produce discomfort. The Illuminating Engineering Society (IES) indicates that values above 70 VCP will generally result in satisfactory conditions; however, lower values may be satisfactory for many circumstances since the center of the rear of a room reference condition has the lowest VCP.

A great many types of shielding materials are available, including a variety of lenses, polarizers, and louvers. One material may differ significantly from another in its brightness properties, so it is necessary to have a manufacturer's VCP data to properly evaluate each material being considered.

VCP data may also be helpful in selecting lighting for a store. If a bright, stimulating store atmosphere is desired, a luminaire with a lower VCP than is desirable for an office may help. On the other hand, a store in which a subdued, relaxing atmosphere is desired should probably have a luminaire with a high VCP.

Windows with a direct view of the sun, clouds, sky, or bright buildings are sometimes a source of visual discomfort. For this reason, windows should have shades, vertical blinds, draperies, low-transmission glass, or other suitable shielding to reduce the brightness in the field of view. Compromises between high light transmission glass for daylighting and brightness control at the workstation are necessary.

10.3.2 Veiling Reflections. These reflections reduce task visibility by lowering the contrast between the details of the task (for example, a specular reflection from a graphite pencil stroke) and its background. Veiling reflections occur when a light source and the eye of a worker are at the mirror angle of reflection with the specular detail of a visual task. They are often difficult to eliminate by just shifting the viewing angle because luminaires (or windows) that produce these effects are substantial in area, and there are frequently many of them to produce reflections.

The most important single factor in minimizing veiling reflection effects is geometry. If the sources that light the task can be positioned out of the mirror angle of reflection with respect to the task and the worker's eyes, task visibility will be greatest. This is frequently practical in private offices where desk location is known. It is also possible with built-in workstation lighting, when lights are located on both sides to illuminate the task. Unfortunately, in many workstations, light sources are under a shelf or cabinet directly in front of the task, which is usually the worst possible location for a light.

In a general office or drafting room, it is best to position desks or drafting boards between rows of ceiling luminaires, with workers facing parallel with the rows so that more of the light on the tasks comes from the sides and not from luminaires on the ceiling immediately in front of the desks. Certain lighting distributions, such as polarizing lighting panels, bat-wing lenses, bat-wing reflectors on luminaires, and indirect lighting may also reduce the effect of veiling reflections.

Indirect lighting (without accompanying direct task lighting) works best in large rooms with low height furniture and is of little or no use with high screens or room dividers. Veiling effects are minimal when the ceiling is uniformly lit, and the tasks can be lit by a large area of ceiling. This begins to approach a reference condition known as "sphere lighting." However, even in large, open plan spaces in which workstation furniture stands 5 or 6 feet high, or many screens are used to partition the space, or both, the utilization of indirect lighting is greatly reduced and, therefore, the effect of veiling reflections is substantially increased.

Progress has been made in predicting and evaluating the effects of veiling reflections. The term "equivalent sphere illumination (ESI)" has resulted from research in this new technology (see 10.4).

10.3.3 Room Finishes. The reflectance of room surfaces is an important factor in the efficient utilization of light and, therefore, the efficient utilization of lighting energy. It is also important to visual comfort because luminances should be within certain well-established limits (ratios) in areas where demanding visual tasks are performed.

For the best utilization of the available light, the ceiling should be painted white. The walls, floor, and equipment finishes should be within the recommended reflectance range listed in Table 72.

To get even higher utilization of the available light, proposals are sometimes made to employ finishes on walls, floors, and desks, the reflectances of which are even higher than those listed in Table 72. Specifiers are cautioned against such experiments, as the recommended reflectance values have been well established over several decades of practice. Lighter finishes could create legitimate complaints of glare and upset the brightness relationships that are necessary for visual comfort.

Lighting engineers should include a specification for room reflectance as part of their design or ensure that they are consulted by those who will make the color specifications.

Table 72
Recommended Surface Reflectances for Offices

Surface	Center-point Tolerances	Equivalent Range (%)
Ceiling finishes*	0.80 + 15%	80—92
Walls	0.50 ± 20%	40—60
Furniture, machines, and equipment	0.35 ± 25%	26—44
Floors	0.30 ± 30%	21—39

*Reflectances for finish only. Overall average reflectance of textured acoustic materials may be somewhat lower.

Certain portions of walls, trim surfaces, or room appointments may have a higher or lower reflectance than the limits of the ranges in Table 72, if these areas are thought of as accents and restricted to no more than 10% of the total visual field.

In stores, restaurants, theaters, and similar commercial areas in which there is less need for balanced brightnesses, departures from the recommended reflectances may be made, when done with discretion. Lighting engineers should be aware of such departures and consider them in their computations, or the lighting result may be quite different from what has been anticipated.

10.3.4 Color. Color is a complex subject involving both physical parameters that can be expressed in mathematical terms and psychological factors that relate to individual interpretations of color.

Certain colors seem to be warm in character, while others are considered to be cool. Light sources have such characteristics, and their color may sometimes be a factor in source selection in order to complement a warm, cool, or neutral color scheme. Warmth or coolness in color scheme and light source may also be a factor in the preception of temperature by occupants of the space. This could have energy implications for space heating or cooling in the winter or summer.

Certain light sources may have high efficacy of light production with fair or poor color rendition. Others may have excellent color rendition with only moderate efficacy. In recent years, phosphor developments have resulted in fluorescent lamps with excellent color and good to excellent efficacy. These factors should be weighed along with many others in light source specification for particular applications.

There are two terms that can provide useful color information about lighting. One is *chromaticity*, or apparent color temperature, sometimes called "correlated color temperature;" the other is *color rendering index*, symbolized as Ra in color literature.

Chromaticity is the measure of the warmth or coolness of a light source, which is expressed in the Kelvin (K) temperature scale. It describes the appearance of the theoretical black body of physics, a perfect absorber and emitter of radiation, if it were to be heated to incandescence. At the first phase of incandescence, the object is a ruddy red. At higher temperatures, the color changes from a range of warm, yellowish white colors to white, and then to cool blue-white colors at still higher temperatures.

Some of the general service incandescent lamps and warm white fluorescent colors have a chromaticity of 3000 K. Cool white fluorescent lamps have a chromaticity of 4200 K. Chromaticities of sunlight and skylight vary over a broad range throughout the day.

Chromaticity provides no information about how well a light source will render various object colors. Daylight has excellent color rendition, though the appearance of colors will vary with the time of day, season, latitude, weather, and other atmospheric conditions. An incandescent lamp emits relatively small amounts of blue and green light relative to red, so it tends to mute or "gray" cool object colors, such as blue. Some discharge lamps are regarded as high color rendering types; others are not as good.

A measure of how well a light source renders colors is the color rendering index (Ra). This is a number that compares a specific light source of interest against a reference source on a 0 – 100 scale. The system is limited, and sometimes misunderstood, because a comparison of two sources is meaningful only if the two sources being compared have the same chromaticity. It would not be meaningful to compare the Ra of an incandescent lamp with that of a cool white fluorescent lamp because the chromaticity of an incandescent is 3000 K while, for a cool white fluorescent, it is 4200 K. A comparison could be made between cool white (Ra = 66) and deluxe cool white (Ra = 89) because both have the same chromaticity.

From a design viewpoint, if the appearance of colors is important, one approach might be to select a chromaticity whose warmth or coolness is suitable for a particular application, and then find a source with a high Ra in that chromaticity. Often an experienced designer or colorist is called upon to select the color scheme for a commercial area. The lighting engineer should ensure that the color specifications are reasonable for good visual comfort in areas where good seeing is critical, and that the assumptions made for ceiling, wall, and floor reflectances are realistic in order to ensure a satisfactory light design.

Theories of lighting and color perception are continually evolving. One lighting engineer has recently authored a new concept of color perception. While this theory will require time in order to validate or disprove it, see Reference [14] for more information.

10.3.5 Psychological Factors. There is a great deal that is subjective about how individuals react to a space. Nevertheless, in recent years, studies of the psychology of lighting have provided valuable data as to how statistically significant groups of people react to various kinds of lighting. For example, criteria have been developed that allow one to use lighting to create impressions of a public or a private space. These criteria can be extremely helpful in applying lighting in such areas as lobbies, private offices, cafeterias, conference rooms, libraries, and general offices (see References [5] – [8]).

10.4 Illumination Quantity. The Illuminating Engineering Society (IES) changed the basis for its recommended levels of illumination in 1979 (see *IES Lighting Handbook*, 1987 Edition (reference volume) [10] and *IES Lighting Handbook*, 1987

Edition (application volume) [11]). The previous system involved single number target values of footcandles (or ESI) for various tasks representing an averaging of assumptions about user eyesight, age, task demand, etc. The new system involves illuminance ranges that correlate with the recommendations in CIE Report no. 29, "Guide on Interior Lighting" [4],[72] which are summarized in Table 73. This approach can be considered to be an interim step that is based primarily on an international consensus. It is intended to replace this system with a scientifically based method at some future time; the pending research has yet to be completed. For additional discussion, see Reference [13].

To determine the nominal design illuminance from the range, Table 74 is consulted and weighting factors assessed. The selection of the weighting factor depends on the age of the workers, the reflectance of the task background, and the demand for speed and accuracy in performance of the task. All these factors are identified by recent research as significant variables that affect task performance.

The individual designer is required to make more specific decisions and to accept more responsibility for the performance of the lighting system with the new IES system than in the past.

There are subtleties and refinements in the new IES system that cannot be covered in detail in this brief summary. One very important one has to do with tasks that are subject to veiling reflections (which abound in commercial buildings), where use of the visibility metric equivalent sphere illumination (ESI) may be helpful in comparing various lighting systems. It should be cautioned that ESI values, whether measured or computed, cannot be directly compared with the illuminance values listed in Table 73. But an assessment of ESI for several lighting systems of interest can help to determine which one would be better in creating task visibility.

ESI is a relatively new metric measurement that involves both quality and quantity aspects of illumination design. It allows the comparison of actual or proposed lighting systems with a reference sphere lighting condition using a standardized task and an observer.

The sphere lighting condition is a convenient reference condition, not an ideal lighting situation. Imagine a visual task located in the center of a uniformly bright sphere interior (for flat paper tasks, a hemisphere would be satisfactory). If a small aperture is created in the sphere surface so that the task can be viewed at the reference viewing angle (25°), the contrast of the task can be measured with a visual task photometer (VTP) and the contrast rendition factor (CRF) computed. The VTP can also be used in the field to determine the CRF of the same task under an actual lighting system. When the CRFs are computed for the identical task, allow the effectiveness of the field lighting system in creating task visibility to be compared with that of the sphere reference condition or with other practical lighting systems.

In a room, ESI varies greatly from one location to another. It also varies greatly even at a single point in a room depending on viewing direction. For this reason, it

---

[72]CIE publications are available from the International Commission on Illumination, Kelegasse 27, P.O. Box 169, A-1030 Vienna, Austria.

Table 73
Illuminance Recommended for Use in
Selection of Values for Interior Lighting Design*

Category	Range of Illuminances[†] in Lux (Footcandles)	Type of Activity
A	20-30-50 [†] (2-3-5)[‡]	Public areas with dark surroundings
B	50-75-100[‡] (5-7.5-10)[‡]	Simple orientation for short temporary visits
C	100-150-200[‡] (10-15-20)[‡]	Working spaces where visual tasks are only occasionally performed
D	200-300-500[§] (20-30-50)[§]	Performance of visual tasks of high contrast or large size: for example, reading printed material, typed originals, handwriting in ink and good xerography, rough bench and machine work, ordinary inspection, rough assembly
E	500-750-1000[§] (50-75-100)[§]	Performance of visual tasks of medium contrast or small size: for example, reading medium-pencil handwriting, poorly printed or reproduced material, medium bench and machine work, difficult inspection, medium assembly
F	1000-1500-2000[§] (100-150-200)[§]	Performance of visual tasks of low contrast or very small size: for example, reading handwriting in hard pencil on poor quality paper and very poorly reproduced material, highly difficult inspection
G	2000-3000-5000[‖] (200-300-500)[‖]	Performance of visual tasks of low contrast and very small size over a prolonged period: for example, fine assembly, very difficult inspection, fine bench and machine work
H	5000-7500-10000[‖] (500-750-1000)[‖]	Performance of very prolonged and exacting visual tasks: for example, the most difficult inspection, extra fine bench and machine work, extra fine assembly
I	10000-15000-20000[‖] (1000-1500-2000)[‖]	Performance of very special visual tasks of extremely low contrast and small size: for example, surgical procedures.

*Adapted from Reference [4].
†Maintained in service.
‡General lighting throughout room
§Illuminance on task
‖Illuminance on task, obtained by a combination of general and local (supplementary) lighting

**Table 74
Weighting Factors to Be Considered in Selecting a Specific
Illuminance with the Ranges of Values for Each Category in Table 73***

Task and worker characteristics	Weight		
	−1	0	+1
Workers ages	Under 40	40—55	Over 55
Speed or accuracy or both	Not important	Important	Critical
Reflectance of task background	Greater than 70%	30%–70%	Less than 30%

*Weighting factors are to be determined based on worker and task information. When the algebraic sum of the weighting factors is −3 or −2, use the lowest value in the illuminance ranges D through I of Table 73; when −1 to +1, use the middle value; and, when +2 or +3, use the highest value.

is desirable to have some method of predicting in advance what the ESI will be at many locations in a space so as to know the best location for tasks that are subject to veiling reflections. A number of computer software service companies have programs available that can provide this information before a particular lighting system is installed.

This discussion of ESI is presented here to acquaint the user with this approach to illumination design. Consult References [10] and [11] for a complete discussion of this subject.

The design practice committee of the IES now has an approved method for evaluating the ESI in spaces where task locations are not known in advance of occupancy (see Reference [12]).

10.5 Light Sources. Electric light sources and daylight have a range of characteristics in terms of efficacy (lm/W), color, source size (optical implications), lumen maintenance, starting and restarting attributes, and economics.

Table 75 shows the lm/W efficacy (not including ballast watts) for all the major general lighting sources. Lamp efficacy, lumen maintenance, life, and optical control are the major factors affecting lighting economics. Economic comparisons of lamp/luminaire combinations are an important basis for selecting an appropriate lighting system. Computer programs that will evaluate these combinations are available from computer software service companies and from manufacturers.

10.5.1 Incandescent Lamps. Incandescent lamps have tungsten filaments and lamp efficacies generally ranging between 17–24 lm/W. This is the lowest efficacy of any of the light sources used. However, incandescent lamps, due to good optical control, may be energy efficient when used to light a small area from a distance, as with spotlighting in stores or theatrical lighting.

Incandescent lamps are not recommended for lighting sizable areas that have long operating hours. For athletic fields and infrequently accessed storage areas where operating hours are short, incandescent lighting should be considered.

Table 75
Appropriate Initial Efficacies for
the lm/W Range of Commonly Used Lamps

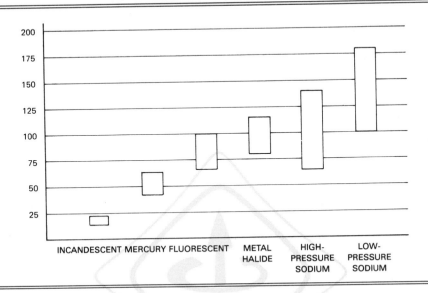

10.5.1.1 Life and Efficacy of Incandescent Lamps. The life and efficacy of incandescent lamps are inversely related. This is the only light source for which this is true. The lower the operating temperature of the filament, the lower the rate of tungsten evaporation and the longer the lamp life. However, the lower the filament temperature, the lower the lamp efficacy. The factors of life, efficacy, energy cost, and maintenance labor rates have been related in an economic equation to determine optimum lamp life for incandescent lamps. These criteria vary among users, accounting for some of the various life ratings found among incandescent lamps available today.

However, at today's higher energy costs, the more efficient incandescent lamp should be selected. Some of the extremely long-lived lamps waste considerable energy. For example, it is possible to select the next lower wattage while getting the same amount of light in standard incandescent lamps. This saves considerable energy and money compared with extremely long-lived incandescent lamps. Difficulty of access and the labor cost of lamp replacement also affect the choice of lamp life. An economic analysis is suggested.

Higher efficiency krypton filled incandescent lamps are available at wattages slightly below standard values and lives somewhat longer than those of general-service incandescent lamps. Their considerably higher initial cost may be paid back by their reduced operating cost and longer life.

10.5.1.2 Color of Incandescent Lamps. The rendition of incandescent lamp color is usually regarded as very good even though its spectrum is unbalanced in favor of warm colors and it is comparatively low in the cool colors. A number of colors are available in incandescent lamps by using filters applied to the lamp bulbs. Many more colors are available using separate color filters as in theatrical and display lighting equipment. These colors are less efficient than the several fluorescent lamp colors available, since considerable light is absorbed in the filter.

10.5.1.3 Tungsten Halogen Lamps. Tungsten halogen lamps are incandescent lamps that make use of the halogen cycle to prevent deposits of evaporated tungsten from collecting on the inner bulb surface. Consequently, their lumen output does not drop appreciably during life. Lamp life is about double that of standard service lamps.

10.5.1.4 Dimming. Incandescent lamps can be dimmed simply by reducing the voltage at the lamp socket. Variable autotransformers and solid-state devices are most often used for this effect. Dimming may be desirable for certain special effects when more than one level of illumination is necessary. However, it should be remembered that light output drops much more rapidly than wattage as incandescent lamps are dimmed and lamp efficacy is greatly reduced. Further information on dimming is included in 10.9.3.

10.5.2 Fluorescent Lamps. Fluorescent lamps are electric arc discharge sources that depend on a two-step process for generating light. The electric arc discharge through low-pressure mercury vapor generates ultraviolet radiation which, in turn, excites phosphors deposited on the bulb wall of the lamp to generate visible light. The phosphor is vitally important because it determines the efficacy, color, and lumen maintenance of the light produced. Phosphor composition also affects the cost of the lamp.

Fluorescent lamps generate light more efficiently than incandescent or mercury lamps, though there is great variation in fluorescent lamp efficacy depending on color and wattage. Fluorescent lamps also have characteristics of low brightness and diffusion, making them excellent for many lighting applications where high brightness could cause specular reflections in tasks, and where luminaire discomfort glare should be controlled.

10.5.2.1 Types of Fluorescent Lamps. Fluorescent lamps have tubular bulbs and are made in a variety of lengths for a number of operating currents. For example, 430 mA is the typical operating current of rapid-start and slimline fluorescent lamps. Ballasts can be selected that will operate these lamps at 200 mA or 300 mA for reduced wattage and similarly reduced light output.

High-output fluorescent lamps are operated at 800 mA and have about 45% higher light output per unit of length than the 430 mA lamps. There are also extra-high-output fluorescent lamps operated at 1500 mA that generate 60%–70% more light per unit length than even high-output lamps. The more highly loaded lamps have applications in which higher levels of illumination are needed or where higher mounting heights are involved and the number of lamps or fixtures, or both, can be reduced for economic reasons.

10.5.2.2 Reduced Wattage Fluorescent Lamps. Fluorescent lamps of specific lengths and tube diameters have individual electrical characteristics to which their ballasts should be designed. Consequently, substitutions of different lamp types in sockets designed for a particular lamp can rarely be accomplished satisfactorily, even if the lengths allow a physical fit. In recent years, due to rising energy costs, lamp manufacturers have made reduced wattage fluorescent lamps that fit in the existing sockets of particular lamps, and that will operate satisfactorily on the existing ballasts. These lamps reduce the lamp/ballast system power consumption by 10%–20%, depending on the lamp type and luminaire type. They are available for the 3 foot and 4 foot rapid-start lamps, and for slimline, high-output, and extra-high-output types.

The first generation of these reduced wattage lamps reduced light output in about the same proportion as wattage. The result was a 5 W reduction in the most popular F40 lamp. If less light was acceptable, this was satisfactory. However, if lighting maintenance procedures were improved, such as cleaning luminaires every year or two and replacing all lamps in groups every 3 or 4 years, the average lighting level maintained might be equal to, or even greater than, that maintained with the standard higher wattage lamps.

The latest generation of fluorescent lamps employ new, more efficient fluorescent phosphors and optimized bulb diameter. This has made it possible to provide a 5% increase in luminous efficacy over the standard 40 W lamp with an improved color rendering index (CRI) at 80 and a 20% increase in average lamp life to 24 000 hours using a T-8 bulb size. This lamp is particularly useful when the lighting W/ft^2 has been legislated. This new lamp provides light at the lowest cost and least energy use in new lighting installations as well as being suited for retrofit in existing systems where energy reductions are desired but reduced illumination is not.

Three new types of fluorescent lamps have been introduced. The SL and PL "compact" lamps may be adapted to socket use in place of incandescent lamps. Using electronic circuitry, the 18 SL lamp will replace a 75 W incandescent and has an average life of 10 000 hours. PL lamps use a twin or quad configuration with sockets to replace existing incandescent lamps or are supplied in ballasted fixtures designed specifically for them. A third type of fluorescent lamp is the "biaxial." This lamp type is similar in general configuration to the SL and PL types except it is not compact (12–18 inches long). For best performance, biaxial lamps should be used with compatible ballasts. All reduced wattage fluorescent lamps are more sensitive to temperature than standard lamps. Consequently, they are not recommended for use in areas where the ambient temperature is less than 15.6 °C (60 °F). (Standard lamps are satisfactory down to 10 °C [50 °F].) In addition, these lamps are not recommended for use on dimming systems or for use in emergency lighting units. Standard lamps should be used for these applications.

10.5.2.3 Fluorescent Lamp Colors. A great variety of fluorescent lamp colors can be obtained by simply changing phosphor components or their relative amounts. Specification of lamp color principally involves matters of efficiency and aesthetics.

Much of the early history of fluorescent lamp colors involved compromises between lamp efficacy and color rendition. Lamps high in lm/W efficacy were only fair in color rendition, while those that provided excellent color rendition were substantially lower in efficacy.

Now, however, due to improved phosphor technology, there are lamp colors that combine high color rendition with high efficacy. Such lamps are considerably higher in price than the higher color rendering lamps of lower efficacy; but economic analyses show that they are cost effective.

For applications of fluorescent lighting in some types of stores, restaurants, and homes, and where the good appearance of people, food, merchandise, or furnishings is essential, high color rendering fluorescent lamps should be considered.

Several saturated colors, such as pink, blue, red, gold, and green, are available in fluorescent lamps. These are obtained through the use of fluorescent phosphors in the lamp that generate the color of light desired. In a few cases, a colored filter is employed integrally with the glass tube to increase the color saturation.

10.5.2.4 Fluorescent Lamps and Temperature. The starting and operating characteristics of fluorescent lamps are significantly affected by temperature. See 10.13 for a discussion of this subject.

10.5.2.5 Dimming of Fluorescent Lamps. Equipment for dimming is now available for 30 W and 40 W rapid-start fluorescent lamps. This broadens their possible applications to auditoriums, restaurants, ballrooms, churches, studios, and theaters. It also provides an opportunity to integrate electric lighting systems with daylighting in order to maintain constant levels of task lighting indoors when daylight varies in quantity.

Various electronic systems have been devised for dimming fluorescent systems. Manufacturers should be consulted for information on the characteristics of their equipment, such as dimming range, starting reliability at various brightness levels, cost, etc.

Special ballast designs also make it possible to flash high-output fluorescent lamps with good life performance. The ballasts provide somewhat greater cathode heating during operation than does conventional lamp operation. A lead wire from this ballast to the sign contactor provides lamp control during on-off periods. Principal applications are for signs and attention getting displays in which the flashing uses less energy than continuous burning. Some flashing ballasts are designed to permit satisfactory outdoor operation during weather extremes.

10.5.3 High-Intensity Discharge (HID) Lamps. The mercury metal-halide, high-pressure sodium and low-pressure sodium lamps are in this lamp family. These are electric arc discharge lamps that require ballasts, but are distinguished from fluorescent lamps, in general, by the higher pressure (a little more than one atmosphere) in their arc tubes and a more intense, shorter discharge path.

10.5.3.1 Mercury Lamps. Mercury lamps have been widely used for indoor and outdoor applications for many years. However, since the advent of the substantially more efficient metal-halide and high-pressure sodium lamps, they are now rarely specified. Applications for mercury lamps would be in the 175 W and lower wattage range in which the more efficient metal-halide lamps have no equivalent. For example, in certain stores with lower ceilings and relatively low requirements for illumination level, mercury lamps can be an appropriate choice.

Mercury lamps have a very long life, usually in excess of 24 000 hours. However, they are also characterized by poor lumen maintenance especially on constant wattage (CW) and constant wattage autotransformer (CWA) ballasts. Practical economic life is generally more on the order of 12 000–16 000 hours. Used for longer hours, they waste more energy with their greatly reduced light output. The phosphors used in mercury lamps provide reasonably good color rendition. Most of the earlier lamp types with poorer color have been discontinued.

Self-ballasted mercury lamps are characterized by extremely long lives and with mean efficacies that are about the same as incandescent lamps over their rated lives. They use an incandescent tungsten filament in series with the mercury arc tube to replace the magnetic ballast during start-up and steady-state operations. The filament consumes about 60% of total lamp power during operation, substantially reducing the overall efficacy of this source. Furthermore, the extremely long life of the lamp works to the disadvantage of lighting system efficiency due to the fairly rapid degradation of light output from the mercury arc tube. At the end of life, efficacy is lower than that of typical incandescent lamps. Their principal virtue is long life and reduced labor cost for lamp replacement. However, conventional mercury lamps with their separate ballasts would be far more cost effective.

10.5.3.2 Metal-Halide Lamps. Metal-halide lamps are substantially more efficient than mercury lamps and should be employed when an HID lamp is appropriate and color is important. These lamps are equipped with either clear outer bulbs or with phosphor coated bulbs.

The phosphor coated bulb's color rendering is superior to that of the clear bulb, though that of the clear bulb is still good. The clear bulb will provide better optical control in a lighting fixture than a phosphor coated bulb. Therefore, the clear bulb is preferred for floodlighting equipment in which precise beam control is desired, projecting light over long distances, restricting light to a specific target area with minimum spill, etc.

Electrically, two types of metal-halide lamps are available. One type requires a ballast designed specifically for metal-halide lamps. The other type, available in 325 W, 400 W, and 1000 W ratings, is electrically interchangeable with mercury lamps on most of the commonly used mercury ballasts (about 80% of existing types). The 325 W lamp works on the same 400 W ballast as the 400 W interchangeable lamp. This makes it possible to upgrade existing mercury lighting systems in retrofit applications simply by changing lamps in existing fixtures without increasing the connected load or energy consumption. In the case of the 325 W lamp, energy use can be reduced and more light provided by retrofitting. If lighting is considered adequate before retrofitting, it may be possible to eliminate some fixtures.

Metal-halide lamp technology is still evolving rapidly, and there appear to be significant opportunities for developing lamps with much improved performance over that currently achieved. At present, however, there are certain limits placed on metal-halide lamps as to burning position, or significant differences in lamp performance in one position compared with another. Certain lamps may also have requirements for operation in enclosed luminaires only. Users are advised to consult published manufacturers' data for current information on lamp operating conditions.

10.5.3.2.1 Mercury and Metal-Halide Self-Extinguishing Lamps. Both mercury and metal-halide lamps produce considerable ultraviolet energy in their arc discharge. Some of this energy is very useful for phosphor coated outer bulbs, which react with the phosphor to generate light that improves the color rendition of the total light from the lamp. None of this ultraviolet light gets out of the lamp because the glass outer bulb will not transmit it.

However, in rare instances, the outer bulbs of mercury lamps have been broken during service and the arc tube continues to operate. If people are in the area for an extended period of time, they can experience a temporary reddening of the skin or irritation of the eyes due to the erythemal action of the ultraviolet light coming directly from the arc tube.

Consequently, lamp manufacturers produce lamp models with a disconnecting feature that will deactivate a lamp within a short time after the outer bulb has been broken. These are available for new or existing installations employing open fixtures where the lamps could be broken. Use of enclosed fixtures that do not permit foreign objects to break the lamps allow standard lamps to be used.

10.5.3.2.2 Starting Characteristics of Mercury and Metal-Halide Lamps. Both mercury and metal-halide lamps require 5–8 minutes of starting time (warm-up) before they reach full light output. This is because the vapor pressure of the light generating arc tube gases is quite low at the start, and it takes several minutes for the elements to vaporize into the arc stream.

If mercury or metal-halide lamps experience a momentary power interruption during normal operation or a low enough voltage dip, the arc tube will be extinguished and a 5–8 minute cooling period will be required because, with the lamp off and the arc tube hot, the vapor pressure in the arc tube is too high for the available voltage from the ballast to restart the lamp. As the lamp cools, the vapor pressure drops and the voltage required to start the arc decreases and, eventually, the available ballast voltage is sufficient. Higher ballast voltage would allow faster restart, but would also increase ballast cost. An incandescent lamp can be furnished in many HID luminaires to provide light during warm-up and for emergency purposes.

Higher voltage ballasts are required for lamps to start below 10 °C (50 °F). However, light output is not affected by ambient temperature.

10.5.3.3 High-Pressure Sodium Lamps. These lamps are characterized by a relatively high-pressure electric arc discharge (slightly over one atmosphere) in a special ceramic arc tube containing a small amount of sodium in an amalgam form. When the lamp is first started, there is very low pressure in the arc tube, and the sodium generates its characteristic monochromatic color. However, at operating pressure, the spectral output broadens, and all visible light wavelengths are present, though in different proportions compared with other familiar sources. The lm/W efficacy is quite high and, since source size is quite small, control of light distribution is good.

High-pressure sodium lamps have three characteristics that are quite different from other high-intensity discharge lamps: fast warm-up, fast restrike, and much better lumen maintenance. Warm-up to full light output generally occurs within 2 minutes, and restrike within 1 minute. Ballasts for high-pressure sodium lamps have a high-voltage, low-current starting circuit that generates a pulse of about 2500 V. This provides a fast restrike time, usually less than 1 minute after a power interruption. Ballasts are available (at higher cost) that will provide instant restrike for certain wattage high-pressure sodium lamps should that be necessary. High-pressure sodium lamps provide a mean lumen maintenance of almost 90%

over their approximately 24 000 hour rated life, which is substantially better than that provided by metal-halide and mercury lamps.

A high-pressure sodium lamp has also been developed to provide better color than standard lamps. This is accomplished by changes in the electrical characteristics of the lamp's arc tube. The improved color involves some sacrifice in lm/W efficacy and lamp life.

10.5.3.4 Low-Pressure Sodium Lamps. These lamps have the highest lm/W efficacy. Where color recognition is important, this type of lamp may not be acceptable. Most of the spectral energy output is concentrated between 589 and 590 nanometers and, hence, the light has a highly monochromatic color. The arc discharge takes place in a tube containing vaporized sodium in the free state. Low-pressure sodium lamps are more like fluorescent lamps in physical size than high-intensity discharge lamps.

Low-pressure sodium lamps have two different wattage types. One has the typical light depreciation wattage relationships of most lamps (light dropping off as operating hours increase). The other type has a wattage that may vary with hours of operation. Any wattage change will affect light output and also the capacity of the power distribution system. Suppliers should be contacted for wattage data throughout lamp life and light output.

If lamp color is not a factor in application, economic comparisons should be made between low-pressure sodium and other lamps, such as high-pressure sodium. Such comparisons will involve assumptions for utilization of light with typical luminaires in order to determine how effectively each lamp/fixture combination delivers light to tasks and their relative (or absolute) costs. Computer programs are available for such comparisons.

Care should be taken in disposing of these lamps as the free sodium in contact with water can create a fire hazard. For this reason, their use is not permitted in and around coal mines. The manufacturers' specific instructions should be followed when disposing of burned-out lamps.

10.6 Ballasts. All fluorescent and high-intensity discharge lamps should have ballasts to perform several functions. These include
  (1) To provide the appropriate voltage to start the lamp
  (2) To provide the appropriate voltage to maintain the lamp in operation
  (3) To provide power factor correction.
Ballasts consume from 3%–25% of a lighting system's energy. They also have an effect on the life, light output, and lumen maintenance of the lamps in the system. Hence, specification of an appropriate ballast is highly important to satisfactory performance, energy conservation, and the economy of a lighting system. Since there are many different types of ballasts that have a variety of characteristics, it is recommended that manufacturers' literature be consulted for specific details and operating data.

10.6.1 Fluorescent Lamp Ballasts. The rapid-start ballast is the predominant type in use today in fluorescent lamps. This ballast provides a low-voltage source of heat for fluorescent lamp cathodes and allows the lamp to start within 1 or 2 seconds after voltage is applied.

Rapid-start lamps are available to operate at 430 mA, 800 mA (high output), and 1500 mA (extra high output). The most popular and economical type of ballast available operates two rapid-start lamps in series. The power factor is corrected to be in excess of 90% leading. This slightly leading power factor can help improve the system power factor for a building, since other loads usually have a lagging power factor.

Solid-state electronic ballasts were first introduced in the early '80s. The specifier should state the percentage of harmonic current acceptable to the project. Caution should be exercised because some models generate increased harmonic currents that may cause overloading of neutral circuits. Electronic ballasts are available with a third harmonic content that is less than 3% and a total harmonic distortion (THD) that is less than 5%. Typically, input watts to a two-lamp electronic ballast operating at 277 V are

Lamp Type	Input Watts (Bare/Enclosed Fixture)
F40T12 (40 W lamps)	72/70
F40T12 (34 W lamps)	60.5/59
F032T8	65/63

Properly designed solid-state ballasts do increase luminaire efficiency by approximately 15%.

10.6.1.1 Grounding. ANSI/NFPA 70-1990, National Electrical Code (NEC) [1][73] requires that all fixtures and lighting equipment (including ballasts) should be grounded. Rapid-start ballasts require a starting aid consisting of a grounded metal strip running the full length of the lamp. When grounded, the metal of the fluorescent fixture housing normally acts as a starting aid.

10.6.1.2 Slimline Lamp. Slimline lamps are a lamp type that start instantly due to the high open-circuit voltage available from the ballast when the switch is closed. They may be operated at any of several currents (e.g., 200 mA, 300 mA, and 425 mA) by selecting the appropriate ballast. The most popular and economical ballast available for slimline lamps is the two-lamp series type. However, a lead-lag ballast is also available that operates a pair of lamps in parallel, one at leading and one at lagging power factor, so the net result is a high-power factor circuit. The lead-lag ballast is more costly than the series type.

10.6.1.3 Low-Loss Ballasts. Within recent years, manufacturers have developed fluorescent lamp ballasts that reduce ballast losses by almost half with respect to conventional ballasts. These ballasts run cooler due to lower watt loss and last considerably longer. Though somewhat higher in cost than conventional ballasts, they are cost effective and should be considered for all new lighting and as replacements for ballast failures in existing installations. Energy-conserving ballasts are available in both magnetic core and coil types, and electronic types. The

---

[73] ANSI publications are available from the Sales Department of the American National Standards Institute, 11 West 42nd Street, 13th Floor, New York, NY 10036. NFPA publications are available from Publications Sales, National Fire Protection Association, 1 Batterymarch Park, P.O. Box 9101, Quincy, MA 02269-9101.

electronic types operate the lamps at a higher frequency, which further increases system efficacy.

10.6.1.4 Voltage. Ballasts are available for the standard distribution system voltages, and should be operated at no more than 5% higher or 10% lower than the rating. Higher voltage will overheat ballasts and shorten life. Lower voltage will reduce lamp life, and lamps may fail to start.

10.6.1.5 Temperature. Extremes of hot or cold can be damaging to ballast life and performance. Ambient temperature in the areas where ballasts are installed will affect the ballast operating temperature. Fixture design will also have an effect. The ballast case hot-spot temperature should not exceed 90 °C (194 °F) during operation.

Thermally protected ballasts (Class P) will disconnect themselves from the circuit when their case temperature exceeds 90 °C (194 °F), or cause the lights to cycle off and on when the condition causing the excessive temperature persists. Ballasts without thermal cutouts will have their lives shortened when operated above 90 °C (194 °F).

Most ballasts designed for indoor operation of fluorescent lamps provide voltage for the satisfactory starting of lamps at 10 °C (50 °F) (15.6 °C [60 °F] for the reduced wattage fluorescent lamps). Low-temperature ballasts are available that can provide higher voltage to start lamps in ambient temperatures as low as –28.9 °C (–20 °F).

10.6.1.6 Lamp Burnouts. Ballasts may overheat when lamps flicker near the end of life or when one of a pair of lamps is removed from the lampholder. Flickering or burned-out lamps should be replaced promptly to prevent possible ballast damage.

10.6.1.7 Lamp Removal. When lamps are removed from fixtures that remain energized, a small amount of energy is consumed by the ballast at a very low power factor (except for slimline fixtures that have circuit-interrupting lampholders). This is due to the magnetized current flowing through the ballast primary. A series ballast for a pair of 4 foot rapid-start lamps will consume 6.5 W with the lamps removed. If the fixture will not be relamped within a short period of time, it may be disconnected from its power source by qualified personnel in order to eliminate this loss.

Electronic ballasts are available for use with high-pressure sodium lamps. They are built with a solid-state control circuit and a reactor that monitors lamp and line operating conditions and then establishes the proper value of ballast required to operate the lamp at its rated power. The same caution is suggested in selecting the electronic ballast for high-pressure sodium applications as was discussed for fluorescent applications.

10.6.1.8 Fusing. It is desirable to use an in-line fuseholder and time delay fuse with each ballast. This will prevent an entire area from being blacked out due to the failure of one ballast. It also provides a safe means of replacing ballasts without opening the branch-circuit breakers.

10.6.1.9 Switching. Circuit breakers that are used for frequent switching of fluorescent lamps should be UL listed as "SWD" for this duty.

10.6.1.10 Radio Interference. Radio interference from fluorescent lighting systems may be minimized by the use of appropriate lenses on the luminaire and by the installation of available filters in the circuit feeding the ballasts.

10.6.2 High-Intensity Discharge (HID) Lamp Ballasts. Most of the previous comments on ballasts in fluorescent lamps also apply to ballasts for HID lamps. The functions of HID lamps are the same and, since most are electromagnetic devices, their characteristics are also similar.

Ballasts for high-pressure sodium lamps differ from other high-intensity discharge lamp ballasts because they have a high-voltage pulse to aid in starting the lamp. This pulse also aids in restarting hot lamps within about 1 minute if a momentary voltage dip or power interruption occurs that extinguishes the lamp.

With older high-pressure sodium lamp ballasts, it is necessary to change burned-out lamps promptly and to not keep the ballast energized for several days without a functioning lamp because the pulsed starting aid would be damaged. At present, some ballasts do not have such limitations.

10.6.2.1 Grounding. HID ballasts should be grounded in compliance with the NEC [1] or local codes, when appropriate.

10.6.2.2 Voltage. Lag- and reactor-type ballasts should have a supply voltage within $\pm 5\%$ of the design voltage. For constant wattage autotransformer (CWA) ballasts, the voltage should be within $\pm 10\%$ of the design voltage. HID ballasts are available for a wide variety of utilization voltages.

10.6.2.3 Fusing. It may be desirable to use a line-side fuse with HID ballasts. This will prevent branch-circuit breakers from opening if there is a defective ballast on the circuit.

10.6.2.4 Radio Interference. A small amount of interference may be detected during lamp starting. There should be no objectional interference during operation.

10.7 Luminaires. Many factors should be considered when selecting luminaires for a space. Some of the principal ones are listed in 10.7.1. Some of these may receive more emphasis than others for a specific installation, but all will exert some influence.

10.7.1 General Considerations That Affect the Selection of a Lighting System
(1) Architectural character of the space to be lighted
    (a) Size and proportions
    (b) Layout of furnishings
    (c) Structural and mechanical features
(2) Designer's concept of how space should appear
    (a) Lighting patterns that emphasize structure or layout, or are design elements in and of themselves
    (b) Unobtrusive lighting patterns
(3) Styling of luminaires
    (a) Simple
    (b) Decorative

NOTE: Decorative luminaires are often inappropriate in providing illumination. Hence, they may be installed primarily for their decorative effect when the general lighting is provided by another system, such as cove lighting or downlighting systems.

  (4) Suitability for specific visual tasks or activities (e.g., office, store, warehouse, factory)
    (a) Light distribution of luminaires
       (i)   Direct, indirect, and intermediate types
       (ii)   Diffusion or directional qualities
       (iii) Creation of shadows
       (iv)   Veiling reflections
       (v)   Uniformity of illumination
    (b) Visual comfort
       (i)   Use of appropriate shielding and diffusing media
       (ii)   Opaque or luminous-sided luminaires
       (iii) Viewing orientation
       (iv)   Visual comfort probability (VCP)

  (5) Efficiency
    (a) Utilization of direct, indirect, and intermediate types
    (b) Power requirements

  (6) Flexibility
    (a) Movable office furniture with task lighting fixtures
    (b) Movable shelf-mounted fixtures
    (c) Movable free-standing indirect fixtures
    (d) Movable plug-in recessed troffers

  (7) Maintenance
    (a) Susceptibility to dirt collection
    (b) Ease of cleaning
    (c) Ease of relamping
    (d) Durability
    (e) Characteristics of plastics, paints, and metals used

  (8) Coordination with mechanical system
    (a) Luminaires for air supply
    (b) Luminaires for air return and their effect on air changes, fan horse-power, etc.
    (c) Lighting contribution to building heating and cooling loads
    (d) Heat redistribution systems
    (e) Heat storage/recovery systems

10.7.2 Special Lighting Distributions. Lighting control techniques are available for luminaires with certain distributions that are intended to minimize the effects of veiling reflections in visual tasks. These include the use of polarizing materials, batwing and radial batwing distributions, and indirect lighting. It should be pointed out that the geometric relationship between eye, task, and light source location is the biggest factor, by far, in reducing veiling reflections. If luminaires are not present in the offending zone (mirror angle for the eye with respect to task), veiling reflections are minimized. It may not be possible to achieve this geometry, however, especially in rooms with multiple occupancy. Hence, special lighting materials may be helpful.

A comparison of various lighting materials using ESI criteria (see 10.4) can be helpful in assessing the relative effectiveness of these materials in creating optimum task visibility in particular lighting situations.

10.7.3 Lighting and Other Building Subsystems. The integration of the lighting system with other environmental features, such as air conditioning, space partitioning, fire protection, and acoustical control, is receiving increased attention. There is opportunity in this area for advancing system design, in order to make lighting an integral part of the building structure and efficiently coordinating all the other control features that are needed in a modern commercial environment. Figure 126 shows a ceiling system that is coordinated with the space module and provides for lighting, air supply, air return, space partitioning, and acoustical treatment.

10.7.4 Visual Comfort Probability (VCP). Glare evaluation data have been developed through an empirically derived formula that assesses all the factors in a room that contribute to glare and confirmed by the testing of many people to establish their brightness tolerances on a statistical basis. This system is called "visual comfort probability (VCP)" and has already been discussed in 10.3.1. VCP tables for specific luminaires are available from luminaire manufacturers.

10.7.5 Luminaires and Air Conditioning. Lighting imposes a load on cooling systems in the summer and contributes to building heat in the winter. Certain types of luminaires have characteristics that may be useful in mechanical systems; for example, some luminaires are designed to supply conditioned air to spaces. Others may be used as air returns; while some can do both simultaneously. Neither air supply nor air return is essential in order to utilize lighting for heating in buildings.

Luminaires that are designed to supply air have an air supply path that is separate from the lamp compartment. Since the air temperature ranges from cold to hot during the seasons, this separation is necessary in order to ensure the good performance of fluorescent lamps, which are temperature-sensitive. One advantage of air supply luminaires is that they allow for a cleaner, simpler ceiling appearance without an obvious pattern of air diffusers. In addition, they may allow for greater flexibility in modular design by being able to supply air (and return it) within any desired module. Air supply luminaires are appropriate with uniform lighting layouts. They may not be appropriate for non-uniform layouts in which luminaires are expected to be relocated during a building's life-cycle.

Air return luminaires provide a path for air to be returned from occupied spaces to the mechanical equipment room by way of the luminaire lamp compartment (for maximum heat transfer) and the ceiling cavity. An alternative is to bypass the luminaire and return air directly to the ceiling cavity. In the latter case, the following benefits apply, but to a lesser degree than air return through the lamp compartment:

(1) Reduced heat gain in occupied space
(2) Reduced requirement for air exchanges in space due to item (1)
(3) Reduced duct size and fan horsepower due to item (1)

Fig 126
Ceiling System Providing Illumination, Air Supply,
Air Return, Space Partitioning, and Acoustical Treatment

(The ceiling configuration also improves visual comfort by shielding most of the luminaires from view.)

(4) Reduced luminaire operating temperature and heat radiation, which improves thermal comfort with the higher room air temperature that occurs in summer

(5) Reduced fluorescent lamp operating temperature and increased light output by about 10%

(6) Reduced ballast operating temperature and longer life for older conventional ballasts. (There is little effect on the life of low-watt loss ballasts, which run much cooler than the older type.)

Local codes should be consulted to determine if air return luminaires or the bypass technique, which has air moving through a ceiling cavity, should be employed.

10.8 Lighting Application Techniques. The purpose of commercial building lighting should be defined before there is any further description of lighting application techniques.

There are several different purposes for illuminance. The purpose and functional requirements of a lighting system should be clearly defined for each area of a building or its surroundings. Only after these needs are adequately defined can the illuminance system be intelligently planned and executed. Every legitimate need of lighting should be considered. Illuminance systems are purchased, installed, and operated to accomplish specific functions. Lighting is provided for its functional ability to provide for seeing, mood, direction, color, aesthetics, and for other considerations that exist in an occupied space.

(1) Seeing — It is necessary for objects to be visible in order to perform tasks with them. One cannot perform a drafting function unless there is appropriate light for the task. Tasks vary quite widely in their illuminance requirements, with respect to both the quantity and quality of the light. Quantity can be calculated and measured easily. Quality is somewhat more difficult to evaluate, though there is practical guidance in the form of recommended reflectances and brightness ratios for major room and work surfaces. The advent of systems such as visual comfort probability (VCP), contrast rendition factor (CRF), and equivalent sphere illuminance (ESI) have provided further insights into the ability of lighting systems to provide seeing light and not just quantity of light. These systems are also useful tools for comparing specific luminaires and lighting systems of interest.

(2) Mood — Illuminance may be used to provide a feeling of warmth, comfort, invitation, efficiency, excitement, or urgency. Many lighting installations should be evaluated in terms of the mood to be conveyed. Obvious violations come to mind: the use of cold, glaring fluorescent fixtures in a tavern; or warm, relaxing lighting for a bank teller work area. Generally, it is desirable to provide a business-like environment in the interest of efficiency in offices and schools, and for other daytime activities. A warm, relaxing atmosphere is frequently desired in specialty shops and for evening activities, such as dining in restaurants or attending the theater. Warmer colors can help in establishing mood.

(3) Direction and Information — Lighting can be used to give direction. As people pass through areas, they may be subtly and unconsciously drawn to the brighter area, unless the brighter area conveys a feeling of discomfort or danger, such as a glaring spotlight or floodlight. Exit lights, advertising, and directional signs are examples of lighting intended to convey information. In some cases, different colors of lighting can be used to delineate pathways in large, open-plan office areas.

(4) Aesthetic Lighting — Aesthetic lighting is designed to make objects and people look pleasing to the eye. This type of lighting is perhaps the most difficult to define and achieve. The lighting designer or architect should be acutely

aware of the aesthetic result desired in order to specify the best lighting systems. Architectural lighting frequently serves other needs, such as security, seeing, identification, and mood alterations.

There are several lighting techniques that may be evaluated for specific applications (see Table 76). These include:

(1) Uniform
(2) Non-uniform
(3) Task-ambient

Uniform and non-uniform lighting techniques are usually variations on ceiling installed lighting. Task-ambient lighting techniques are those in which some light sources may be integrated into, or mounted on, the furniture; others may be free-standing or movable, or mounted on the ceiling. These system classifications are defined, and their advantages, disadvantages, preferred and undesirable uses, cautions, and other application techniques are discussed below.

10.8.1 Uniform Lighting. By definition, uniform lighting illuminates spaces and areas on and around the immediate work or task area equally. The use of uniform lighting has been criticized because of the potential for wasted energy from lighting in both task and non-task areas uniformly. Uniform lighting is frequently applied to

Table 76
Relative Merit of Uniform, Non-Uniform, and Task-Ambient Lighting Systems

	Uniform	Non-Uniform	Task Ambient
Description	Even illuminance throughout the area at task lighting level	Most light on the task, with general and non-critical levels reduced	Direct lighting for task up close with ambient lighting from adjacent indirect or ceiling-mounted luminaires
Advantages	Where tasks not defined Where task areas not known Low levels of illuminance Good eye adaptation Uniform appearance	Energy efficient, lower initial cost Provides space interest, emphasizes work areas	Energy efficient Tax benefits Easily movable
Disadvantages	May be least energy efficient Monotonous May have higher initial cost	Must know tasks, task locations Must be moved as task changes	Veiling reflections Space confining Expensive fixtures Wiring and switching may be more difficult
Typical uses	Large homogenous areas Libraries, drafting rooms Clerical offices, supermarkets, cafeterias Gymnasiums, sportsfields Hallways and corridors	Large non-homogenous areas Smaller areas, private offices Low employee density areas Restrooms VDT and microfilm viewing areas Hallways and corridors	Small work areas Furniture systems Open office plans

areas in which the task or the task areas are not defined. Typical of these is 500 lux or 700 lux uniformly applied to speculative office space.

Actual tenant use of the space should dictate an area-by-area appraisal of the lighting system with its intended use taken into account.

The principal application for uniform lighting is in areas where the activity taking place occurs uniformly and continuously throughout the entire space and where task locations are quite close together, such as in classrooms or densely occupied office space (see Fig 127). It should not be installed as a substitute for proper planning when it is not really required. Fixtures may be kept on site but not installed until the specific locations of workstations are known. An alternative approach, considering the 50–60 year life-cycle of a building during which time tasks may be performed anywhere in the space, is to install luminaires capable of supplying uniform illumination, but with switching controls that would allow a non-uniform lighting result in the space (see Fig 128).

The disadvantages of uniform illuminance are
(1) Relatively high energy consumption with the whole space lit to the same value
(2) Monotonous in appearance
(3) Minimum visual stimulus in the area (However, an even illuminance tends to make small areas look and feel larger.)

Typical spaces where uniform illuminance can be used to best advantage include
(1) Densely occupied office space
(2) Data processing centers
(3) Classrooms
(4) Gymnasiums
(5) Mass merchandising stores
(6) Sports fields

In order to promote energy efficiency in uniform lighting installations, consideration should be given to multiple-level switching that uses two-level ballasts, switching one of a pair of ballasts in luminaires, switching of small areas of luminaires, and switching to lower lighting levels near windows, which can be utilized as a light source during daylight hours. In some states, multiple-level switching is mandatory. An alternative to it is to use occupancy sensors that automatically turn off the lights when the space is unoccupied.

Coefficients of utilization values that are published by luminaire manufacturers are used to calculate average illumination levels for uniform lighting. Actual illumination values in a real space will be higher than average in the center of the space and lower near the edges of it. In small rooms, illumination may be 30% higher than average in the center, varying to near average in very large rooms. Consequently, uniform illumination can be reduced if tasks are located near the center of small and medium-sized rooms. Conversely, work locations near walls should be avoided unless task lighting is provided.

10.8.2 Non-Uniform Lighting. Non-uniform lighting in task areas is achieved by putting more illuminance on the task and less on noncritical and general areas near the task. This concept has become popular because it has the potential for greater energy efficiency.

Fig 127
Near-Horizontal Drafting Boards Positioned So That
Draftsmens' Views Are Parallel with the Luminaires

(If boards are placed between rows of troffers, veiling reflections will be minimized.
The near-vertical boards are positioned at right angles to the rest; but veiling reflections [and shadows] are minimal for such boards, regardless of their orientation.)

Fig 128
Two U-Shaped Fluorescent Lamps, with Efficient,
Low-Brightness Anodized Aluminum Luminaire, Are
Employed in Each Unit to Supply Lighting from the Ceiling

(Flexibility in switching could allow non-uniform lighting distributions in the space; but the owner has the capability of providing task lighting anywhere in the space during the building's life-cycle.)

Non-uniform illuminance should be avoided in areas where non-uniformity might cause confusion or misdirection, and thus be a hazard to safety. With reasonable care and design skill, non-uniform illuminance may be applied successfully to most light situations. (see Fig 129). It is usually not feasible for areas such as classrooms, gymnasiums, and spaces, which may be used for a variety of tasks at the same or different times. Wall, ceiling, floor, and equipment reflectances are more critical with non-uniform lighting in order to minimize large changes in contrast. Dark walls may impart a feeling of being in a cave and may cause adaptation problems. The light reflected from walls (luminance) should not present a ratio in excess of 1:5 with the visual task as a worker glances up from his or her task. This requirement can necessitate a wall illuminance equal to that of the task with darkroom finishes. Consider a wall reflectance of 20% and a task reflectance that approaches 100%; the 5:1 illuminance ratio is met with the same illuminance on the wall and task. Therefore it is important to provide reasonably high wall, floor, and ceiling reflectances. Ideally, the values should not be less than those specified in Table 72. Note that the key word here is *reflected*; the light the eye sees is that which is reflected, and not the incident illuminance. Lighter finishes provide acceptable luminance ratios with non-uniform lighting. This method of energy reduction is far superior to arbitrary methods of illuminance reduction, lamp removal, or other methods that may detract from the illuminance performance criteria.

Non-uniform illuminance exhibits the following advantages:

(1) Energy efficiency
(2) Low initial cost
(3) No sacrifice in lighting quality with careful environmental design

The application of non-uniform lighting is appropriate when the task and task area are well defined (see Fig 130). This assumes a knowledge of what the task is and where it takes place. Greater time, effort, and design capability should be expended to provide an adequate system. System flexibility is required to provide for unforeseen and future contingencies during the life-cycle of a building.

The disadvantages of non-uniform illuminance include the following:

(1) Expenditure of more engineering time in the design stage
(2) Need to define task areas and tasks before actual occupancy in order to provide adequate information during the design phase
(3) Potential for confusion in large areas, although the change in patterns can be used to good effect in creating mood, indicating direction, and communicating information

Spaces where non-uniform illuminance can be used effectively include

(1) Offices
    (a) Small or private offices
    (b) Executive offices
    (c) Special-purpose areas, such as
        (i)   Microfilm viewing
        (ii) Reception areas
        (iii) General offices where employee density is not great
        (iv) Restrooms

Fig 129
Two Foot by Two Foot Luminaires Containing U-Shaped Fluorescent Lamps

(Luminaires are mounted on the appropriate modules of the "waffle-patterned" ceiling structure to supply non-uniform lighting. Units are mounted closer together over workstations, and farther apart in circulation areas.)

Fig 130
Secretarial Area with Uniform Lighting That Has Non-Uniform Possibilities

(Luminaires closest to desks are operated at full light output, while those over file areas are dimmed or switched to half-level. File cabinets, walls, and floors have high reflectance finishes to keep brightness balanced in work areas.)

  (2) Schools
    (a) Offices
    (b) Sewing classes
    (c) Cafeteria serving lines
    (d) Library checkout counters
    (e) VDT and microfilm viewing
    (f) Restrooms
  (3) Merchandising
    (a) Checkout counters
    (b) High-profit merchandise areas
    (c) Sale merchandise areas
    (d) Offices
    (e) Carpet sample areas
    (f) Advertising

10.8.3 Task-Ambient Lighting. Task-ambient lighting is a particular form of non-uniform illuminance that combines task illuminance and ambient (general) illuminance. One potential advantage of task-ambient illuminance is improved energy efficiency, as in the non-uniform system previously described (see Fig 131). There are also other advantages.

10.8.3.1 Advantages of Task-Ambient Lighting. Some forms of task-ambient lighting provide illumination that can be readily moved as the task location moves. This has the following advantages:

  (1) More light where it is needed, less in other areas
  (2) Potential for energy reduction, fewer fixtures, and individual control of fixtures
  (3) Potential reduced cost of lighting system
  (4) Possible tax advantage (lighting is classified as office furniture, which has an accelerated depreciation value)
  (5) Potential for reduced veiling reflections (higher ESI) where geometry between eye, task, and light source is optimized
  (6) Ease of cleaning and relamping
  (7) Uncluttered ceiling

10.8.3.2 Disadvantages of Task-Ambient Lighting. Task-ambient lighting has some of the following disadvantages:

  (1) Need for receptacles at all task lighting locations
  (2) Care to ensure that contrast ratios for visual comfort are not exceeded (this may require wall washing fixtures)
  (3) Need for higher ceiling heights or proper indirect luminaire optics to avoid light "puddles" on the ceiling (this can produce glare and reduce task visibility)
  (4) If the ambient portion is provided by indirect lighting, the luminaire faces upward and consequently requires more frequent cleaning.
  (5) Task-ambient fixtures may be purchased by persons who have the training to select and use them correctly. Installation may be by persons who do or do not understand them.

Fig 131
Open-Plan Office Area with Task-Ambient Lighting

(Energy efficiency is provided by close-up task lighting from fluorescent luminaires positioned to supply light from directions that minimize veiling reflections. The ambient lighting is distributed uniformly over the task and surrounding areas by indirect units that accommodate high-pressure sodium lamps.)

The task component of task-ambient lighting may take two forms: (1) furniture-mounted lighting built into a workstation (see Fig 132), or (2) floor-mounted fixtures that can be placed adjacent to a desk. Some fixtures provide both direct task lighting and indirect ambient lighting (see Fig 133). The direct contribution puts more light on the task than in the surrounding area. It is vital that the direct lighting luminaire be positioned so that it will not produce veiling reflections that will reduce visibility. The indirect contribution provides general lighting, which helps reduce the contrast between the task and the surrounding area. Other types of luminaires are also available that supply direct task lighting only. Their position with respect to workers' eyes and to the task is critical in avoiding veiling reflections.

Fig 132
Task-Ambient Lighting in a Workstation

(Two-thirds of the light on the task is supplied by fluorescent desk-mounted luminaires located at each side of the desk. This arrangement avoids veiling reflections.
About one-third of the light supplied on the task [and the total light in surrounding
areas] comes from indirect partition-mounted luminaires for high-pressure sodium
lamps.)

The ambient lighting component may be supplied in two ways: (1) by conventional luminaires on the ceiling, or (2) by indirect fixtures utilizing HID or fluorescent lamps with the output directed to the ceiling and adjacent walls. Some of
these units may also provide direct task lighting. Indirect luminaires (see Fig 134)
are available in a variety of forms including bollards, shelf- or partition-mounted

Fig 133
Task-Ambient Lighting with a Single Luminaire Providing Both Components

(The potential for veiling reflections with a light source directly in front of the worker is minimized due to the geometry of the relatively high mounting of the lighting unit and the narrowness of the desk from front to back. The top of the luminaire is covered with a parabolic wedge louver, inverted to reduce direct glare.)

units, and free-standing open units. Some fixtures have both metal-halide and high-pressure sodium lamps for higher efficiency than that produced by metal-halide lamps alone and improved color compared with high-pressure sodium alone. The basic fixture can be mounted on a floor stand, on shelving, or on display fixtures in stores. Application of this lighting technique can be seen in libraries, stores, schools, and offices. Asymmetrical reflectors are available for providing special light distributions for units that are located adjacent to walls.

When ceiling-mounted troffers are used for ambient lighting, a plug-in system of wiring should be considered so that luminaires can be relocated as task locations change.

Fig 134
Installation for Luminaires Supplying Ambient Lighting
(a) Luminaire Is Mounted on Top of a Room Divider Partition

(It is important that luminaire optics provide for a broad distribution of light across the ceiling to make it as uniformly bright as possible.)

In some applications, totally indirect lighting is employed. This works best in large rooms with uniformly low furniture. Large rooms utilize illumination much better than small rooms because less lighting energy is dissipated on walls and through doors and windows. When the furniture is uniformly low, tasks on desks and drafting boards can be lighted from a large expanse of uniformly bright ceiling, which reduces veiling reflections and improves task visibility. However, when furniture, partitions, screens, etc., are high, they block out much of the ceiling's contribution to lighting the task and, therefore, visibility suffers. Dark finishes on furniture, screens, etc., compound the problem.

Fig 134 *(Continued)*
Installation for Luminaires Supplying Ambient Lighting
(b) A Free-Standing Unit That May Be Moved

Typical task-ambient applications include the following:

(1) Offices
    (a) Open-plan office areas
    (b) Reception areas
    (c) VDT and microfilm viewing
    (d) Inspection areas
    (e) Isolated workstations
(2) Schools
    (a) Study carrells
    (b) VDT and microfilm viewing areas
    (c) Carpentry and mechanical trade shops
    (d) Food serving line
(3) Merchandising
    (a) Sales desks
    (b) Checkout stations
    (c) Merchandise gondolas

The application of task-ambient systems can be successful if the considerations of VCP, contrast rendition, ESI, and good lighting practice are carefully evaluated and compared with equivalent non-uniform and uniform systems. Engineering comparisons should include not only the illuminance parameters, but also the net life-cycle cost, tax advantages, energy consumption, and maintenance.

No single lighting system is right for all applications (see Fig 135). Many different requirements can be found in a single building (see Fig 136). Any large project will require that both uniform and non-uniform lighting techniques be applied, and comparisons should be made to choose the best system for the particular task, task area, and illuminance goal.

10.8.4 Special Lighting Considerations for Stores. Lighting is employed as a sales aid in modern merchandising (see Fig 137). To realize the full potential of lighting in successful merchandising requires more than just a general lighting system that provides illumination for the appraisal of merchandise. Lighting can draw attention to specific displays by lighting parts of the display to at least five times the level of the surrounding area. This is the purpose of spotlighting and of lighting units built into display fixtures, such as showcases, shelves, wall cases, etc. Effective spotlighting may sometimes be achieved with energy efficiency by employing 50 W, 12 V PAR spotlamps instead of 150 W R-lamps and PAR lamps. Lighting can also be a vital factor in creating an ambiance in a store through patterns of brightness and color.

There are many possibilities for lighting patterns to enhance design or provide distinctive appearance. They can vary from small, compact downlights with incandescent or HID lamps that occupy less than 1% of a ceiling area, to complete ceiling illumination.

The use of HID lamps for lighting stores is increasing (see Fig 138). Metal-halide lamps that have good color rendition and high efficiency are replacing some of the unshielded fluorescent strip lighting that has long been the trademark of the mass merchandising store. High-pressure sodium lamps are also being utilized in combination with metal-halide lamps for greater energy efficiency (see Fig 139).

**Fig 135
Parabolic Wedge Louver**

(The parabolic wedge louver provides exceptionally low luminaire brightness in this office. This type of lighting is recommended for rooms in which video display terminals (VDTs) are in use to avoid veiling reflections in terminal screens. Windows should also be covered if daytime brightness is likely to be reflected in terminal screens.)

**Fig 136
Building Retrofitted with Energy-Conserving Lamp Products Substantially
Reduce Operating Costs for the Building, Yet Do Not Sacrifice the Environment**

(1) Ellipsoidal reflector lamps of 75 W replaced 150 W reflector flood lamps in the deep, baffled downlights in all elevator lobbies, yet provide the same light as before.

(2) Reduced wattage fluorescent lamps replaced standard lamps in all office areas.

(3) High-pressure sodium lighting replaced mercury lighting in a multiple-level parking garage.

Fig 137
Special Lighting in a Store

The lighting in this store contains three elements to perform specific merchandising functions:
(1) A uniform level of general lighting is provided by fluorescent luminaires so that merchandise can be appraised anywhere in the space.
(2) Fluorescent lighting is concealed in the wall cases to define the perimeter of the store, contribute to a feeling of spaciousness, and enhance the appearance and attractiveness of the merchandise displayed in it.
(3) Accent lighting is provided on feature displays to attract attention.
The net effect of these lighting elements is an interesting, pleasant atmosphere.

**Fig 138
Illumination in a Supermarket by Metal-Halide Lamps in Recessed Luminaires**

(Some owners prefer this type of lighting to bare-strip fluorescent units because there is less glare directed toward the eye and more sparkle from the merchandise. Economic analyses indicate such lighting can be as energy efficient and economically competitive as unshielded fluorescent lighting.)

**Fig 139
Indirect Lighting in a Store**

(Luminaires are mounted on top of display fixtures. Each luminaire employs one metal-halide and one high-pressure sodium lamp.)

10.8.5 Electric Lighting and Daylighting. Daylighting is receiving increased attention in order to reduce building energy use. Wherever windows are employed, lighting units at the building perimeter should be switched separately so that they can be turned off when daylighting is adequate. Such control may be automatic or manual; however, in some jurisdictions, energy credits are available when automatic controls are used.

New possibilities for windows and daylighting are being explored in building design with varying degrees of success. Skylights may be installed to increase the use of daylighting and significantly reduce the electrical energy that is required for lighting. It should be remembered, however, that glazing transmits thermal energy at a far higher rate than a well insulated wall. Consequently, in order to determine whether windows are truly energy efficient in buildings, their heat gain in summer and heat loss in winter should be evaluated, as well as any conservation of electric lighting, to determine their net effect on total building energy (the same evaluation should also be made for skylights). When analyses show that windows are effective in the overall net use of energy for a particular building in a particular climate, they should be used. For northern climates, it may be energy efficient to use large windows with southern exposures and much smaller ones with northern exposures.

10.8.6 Outdoor and Sports Area Lighting. The lighting of parking areas around commercial buildings needs to be carefully designed in order to provide for the safety of people and the security of property. To achieve these objectives, adequate amounts of properly distributed light are needed throughout the environment, to reveal such hazards as curbs and steps, and to illuminate dark and potentially dangerous areas (see Figs 140 and 141). To effectively light these areas, luminaires should be selected to meet a specific light level and uniformity, and installed in such a manner as to minimize glare for pedestrians and drivers, and to avoid light spilling onto adjacent properties.

10.8.6.1 Guides for Good Floodlighting Results. Since there are many variables in floodlight production, such as pole placement, mounting heights, light level requirements, and the size and shape of the area, floodlights should be selected based upon the performance data that are supplied by the fixture manufacturer.

Floodlights are designated by the type and wattage of the lamp they use and by their light distribution or beam spread. Beam spreads can be determined by isofootcandle diagrams that are supplied for specific floodlights. The beam spread angle defines what are considered to be the outer limits of the luminaire's coverage. Overlapping of beams in multiple-floodlighting unit installations is desirable for uniformity and for safety in the event of outages.

To obtain uniform light, the distance between poles should not exceed four times the fixture mounting height. A floodlight will effectively light an area out to two mounting heights from the base of its mounting locations. Further separation of poles requires the aiming angle of the floodlight to be raised, which will result in a lower utilization of light and an increased fixture glare. With proper spacing, sufficient overlap between adjacent floodlights will ensure uniform lighting with minimal shadows.

Outdoor sports lighting is a specialized form of floodlighting (see Fig 142). Specific design consideration should be given to each sports lighting application to minimize

Fig 140
Parking Area Lighted with "High-Mast" 90 Foot Poles

(High poles minimize the number of poles and trenching for wiring needed in the lot; they also provide for more uniform light distribution, less direct glare, and simpler, less confusing patterns of brightness that a multiplicity of low poles sometimes creates. The light sources are 400 W high-pressure sodium lamps.)

Fig 141
Parking Garage Lighted with 150 W High-Pressure Sodium Lamps

(The luminaires are surface mounted on the concrete slab, and the structural beams provide natural shielding. The moderately high-reflectance concrete floor and ceiling help provide a reasonable distribution of illuminance through the area with interreflected light.)

fixture brightness or glare in the eyes of both players and spectators. Therefore, pole locations, mounting heights, and luminaire aiming should be selected judiciously for each sports lighting system. For example, in "aerial" sports, the lighting is designed to light the ball in play as well as the players and the playing surface.

10.8.6.2 Light Sources for Floodlighting. Economy of installation and operation is materially influenced by light source size, wattage, and the amount of light it produces. For these reasons, HID lamp usage has become more common in outdoor and sports lighting areas.

Fig 142
High School Football Field Lit with 1500 W Metal-Halide Lamps

(This high-wattage, high-efficiency source greatly reduces the number of lamps and floodlights required to meet the lighting objectives, compared with other sources of lower output and efficiency. They also reduce transformer capacity and the need for power distribution equipment. Though the life of this high light output source is only 3000 hours, this is quite adequate because of the relatively low annual operating hours of sports stadiums.)

The most appropriate HID light sources for outdoor applications are high-pressure sodium and metal-halide lamps. Almost without exception, the high-pressure sodium lamp will be the choice for greatest economy and least use of energy (see Fig 143). It also has a life rating that is as long as mercury lamps and provides reasonable color rendition. Mercury lamps are inefficient when compared with high-pressure sodium and metal-halide lamps and have little merit now.

Metal-halide lamps have shorter life ratings than either mercury or high-pressure sodium lamps. However, these lamps are the preferred choice when the color of landscaping, the appearance of athletic team and marching band uniforms, or racehorse colors are important; hence, metal-halide lamps are often chosen for outdoor sports lighting areas.

10.9 Control of Lighting. In its simplest form, electric lighting control is exercised manually by means of a switch located in a luminaire, a pull cord attached to the luminaire switch, or a fixed wall switch. This provides an on-off control for a particular luminaire or branch circuit. Often, more complex controls are desirable in order to control the level of illumination or provide light at a specific time. Such requirements can be met with manual or automatic controls. Circuit breakers in panelboards are sometimes used instead of switches; however, unless they are designed for switching duty, this practice should be avoided.

10.9.1 Switching for 480Y/277 V Distribution Systems. Wall switches that are approved for 300 V can be employed to switch the lighting fixtures on 277 V branch circuits. According to the NEC [1], these wall switches may be employed when the

Fig 143
Typical Data Available for the Evaluation of
High-Pressure Sodium Lamps Used for Floodlighting
(a) Isofootcandle Chart Shows a 400 W High-Pressure Sodium Lamp Floodlight
(b) Specific Footcandles for the Contours

(Manufacturers' data should be consulted for specific floodlights of interest.)

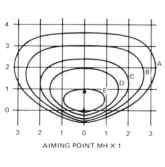

AIMING POINT MH X 1

(a)

INITIAL fc TABLE

	Mounting Height		
	30	40	50
A	0.18	0.1	0.06
B	0.35	0.2	0.13
C	0.89	0.5	0.32
D	1.8	1.0	0.64
E	3.6	2.0	1.3
F	8.9	5.0	3.2

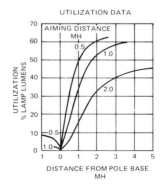

UTILIZATION DATA

DISTANCE FROM POLE BASE
MH

(b)

voltage between switches is limited to 300 V, using grounded barriers as necessary when the voltage exceeds 300 V, as when two phases of the 480 V system occupy the same enclosure.

10.9.2 Remote Control Switching Relays and Lighting Contactors. Low-voltage (usually 24 V) remote control switching systems can be used for branch-circuit and individual luminaire control. This type of control employs a low-voltage switch at the control point to actuate a relay in the branch circuit. Since the branch-circuit wiring goes only to the luminaire and relay and not to the control point, there may be some cost saving in the wiring. Substantial savings result with remote control in installations where considerable flexibility is desired and control is employed at several locations.

Lighting contactors are used for controlling large blocks of lighting and for multiple preselected control of branch circuits. They are generally available in 20–225 A sizes and are mounted in panelboards or separate enclosures. Larger contactors are not generally used for the control of lighting. The 20 A size is typically used to control branch circuits. Multiple-pole contactors can switch up to 12 circuits each. A typical multiple-pole contactor is shown in Fig 144(a). The larger sizes are intended for switching entire panelboards and are generally furnished in two- and three-pole arrangements (see Fig 144(b)). Standard control voltages are 24 V_{dc}, 24 V_{ac}, 120 V_{ac}, and 277 V_{ac}. The standard control voltages of 24 V_{ac} or 24 V_{dc} are commonly used with building management systems. Remote control makes it possible to turn blocks of light on or off from convenient locations or from one central location. In addition to the convenience of control, installation savings can be realized by using control wires, thereby reducing power cable runs.

Lighting contactors are actuated electromagnetically and are held either magnetically or mechanically. Magnetically held lighting contactors are usually controlled by an on-off single-pole, single-throw toggle switch and will change contact position upon loss of control voltage.

Mechanically held lighting contactors will not change contact position upon a drop in or loss of control voltage. The operating coil is only energized during the opening or closing operation, thereby eliminating coil hum and power drain. A mechanically held lighting contactor can be controlled from any number of control stations (as shown in Fig 144(c)), or from time switches, photoelectric cell relays, occupancy sensors, or computerized building management system (see Fig 144(d)).

Auxiliary relays and other interface control options may be used with lighting contactors to accommodate long runs between the lighting contactor and the control switch for two-wire control, for low-voltage control, and for control by pilot contact devices.

Due to energy considerations, the practice of switching large blocks of lighting by contactors is changing in favor of controlling much smaller numbers of luminaires. In some applications, energy codes may limit the area where lighting can be on a single switch, or may require that each individual office or workstation be switched independently.

Control systems are available that employ microprocessor logic to reduce the number of wires through multiplexing. Coded commands can be multiplexed over a control cable to the control points.

(a)
A 12-Pole, 20 A Lighting Contactor for Control of Branch Circuits

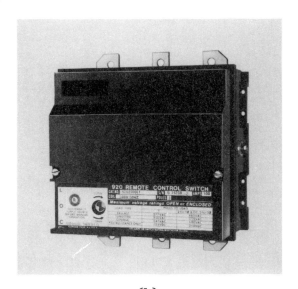

(b)
A Three-Pole, 100 A RC Switch for Panelboard Mounting

Fig 144
Typical Components for the Remote Control of Lighting

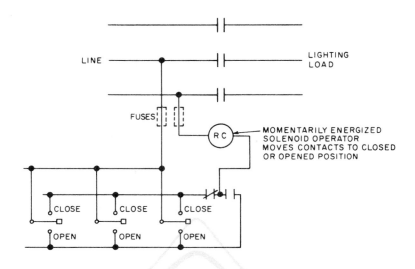

(c)
Mechanically Held Electrically Operating Lighting Contactor
Controlled by Multiple Momentary-Toggle-Type Control Stations

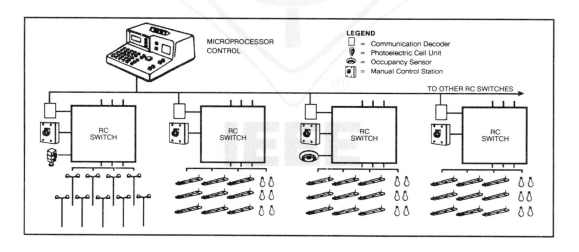

(d)
Various Control Means for Remote Control

Fig 144 *(Continued)*
Typical Components for the Remote Control of Lighting

Control points have receiver/switches; the latter component is usually a low-voltage relay. Logic functions can be preprogrammed into a control device to turn lighting on and off over a 24 hour or a weekly period. Overrides are available, and some systems can be accessed with touchtone telephones. Systems that control lighting in both time and space save considerable lighting energy when compared to past control practices (see Reference [3]).

Remotely controlled circuit breakers are available that provide for switching branch circuits directly, which obviates the use of lighting relays. These devices are rated 120 V or 277 V single-phase, 20 A and are capable of being controlled directly from the outputs of microprocessor-based controllers. The specifier should consult building management systems specialists and circuit breaker suppliers regarding the availability and application of these new devices.

10.9.3 Dimming and Flashing of Lamps. Dimming of incandescent lamps has been employed for many years because changing the voltage at the lamp socket provides a simple method of varying the light output. Several methods of voltage variation may be employed; but the availability of solid-state semiconductor dimmers has made them the most popular type available. Units controlling up to several hundred lamp watts are small and relatively inexpensive. Losses are generally less than 2% of the connected lamp load.

Other possible methods of dimming include

(1) Variable resistance (rheostat)
(2) Variable autotransformer
(3) Variable reactance
(4) Solid-state electronics

Dimming devices can be manually controlled by an operator. However, it is often desirable to provide the dimming control at locations that are distant from the load or at several stations. Geared motor drives can be employed for this purpose on low-voltage (typically 24 V) remote control systems.

Incandescent lamps can be flashed with a contactor, which switches the circuit on and off. Some commercial contactors are motor driven units that can flash lamps simultaneously or in a desired sequence. Small button contactors employing thermal elements can also be used in individual lamp sockets to create flashing. Since the lamps are inoperative during part of a flashing cycle, actual service time will be longer than the rated life of the lamps.

Practical dimming and flashing can be accomplished with certain fluorescent lamps (see 10.5.2.5).

10.9.4 Dimming of HID Lamps. Through the use of solid-state technology, the dimming of HID lamps has not only become practical, but has also materialized as an energy-saving method of lighting control. A dimming range from full to about 50% light output can be accomplished without adverse results to the life of the lamp.

A typical dimming system incorporates a centralized control panel with remote dimming controls. Dimming systems can be single-phase or three-phase, which control most standard size HID lamps. Dimming response times, although not instantaneous, are not significantly long, and, considering the reduced power consumption that dimming proportionately produces, should not be a major concern.

HID lamp dimming systems can be applied both indoors and outdoors. Indoor applications could include schools, hospitals, factories, stores, auditoriums, etc. Outdoor applications could include highways, tunnels, parking lots, shopping malls, etc.

One reason for dimming HID lamps is to maintain a constant level of illumination on tasks during the life of the lighting system. With new lamps and clean fixtures, the lamps are dimmed well below their maximum output, saving considerable energy. As lamps depreciate and fixtures collect dirt between cleanings, power is increased to keep task illumination constant.

Interfacing with time clocks, computers, occupancy sensors, or photocells can also be considered in a dimming system.

10.10 Lighting Maintenance. Light loss due to dirt, dust, and grime depends upon the type of lighting fixture used, the dirt conditions in the atmosphere, and the time between cleanings. Losses will range from 8%–10% in a "clean" environment to more than 50% under severe conditions. The longer lives of fluorescent and HID lamps reduce the frequency of relamping and the coincident cleaning of the lighting fixture. The planned effectiveness of a lighting installation can only be achieved by physical maintenance. With proper planning for maintenance during the design period, it is possible to significantly reduce the initial cost, the operating cost, and the energy consumption of a lighting system. Energy has always been a major cost component of any lighting system, and, when energy cost is reduced, the total life-cycle cost of the lighting system will be directly reduced.

Common practice in lighting system design has been to provide excess initial illumination to allow for the reduction in light as system components deteriorate due to dirt and age. The use of light loss factors in the planning of installations is a necessary admission that no amount of physical maintenance can keep the output of a system up to its initial level. The value of the light loss factor used indicates the amount of the uncontrollable depreciation expected, together with the results of the effort expended to overcome the controllable factors in depreciation.

The lumen maintenance of most lamps is published by manufacturers and provides a means of evaluating light output at various points in the life-cycle of a lamp. Some fluorescent lamps at rated life will only produce 80%–85% of their initial light output. Some mercury lamps produce only 40% of their initial light output at rated life. Obviously, mercury lamps should be replaced in groups well before they reach their rated life, or they will waste both energy and money. Planned lighting maintenance is the most efficient, economical approach to solving lighting system problems. A properly planned relamping program will arrest lumen depreciation and avoid burn-outs, thereby maintaining higher illumination levels without additional energy costs. The reduction of burn-outs gives an added advantage in saving labor, time, and expense, which would otherwise be involved in burn-out replacement. A properly planned periodic cleaning program will arrest luminaire dirt depreciation that is due to dirt accumulation on lamp and luminaire surfaces.

When most lamps in an area are of the same life and operated for the same length of time, the practice of group relamping and coordinated cleaning often reduces lighting maintenance costs substantially. This procedure may be utilized advantageously in incandescent, fluorescent, and HID lamp installations. The practice involves replacing all of the lamps in an area at the same time after they have already been operated the greater part of their useful life. There are several variables involved, such as the labor cost of individually replacing lamps compared to that of group replacement, and the number, type, and cost of the lamps. When lamps within the same area have different operating hours, group relamping may not be practical.

Due to the long life of fluorescent and HID lamps, these systems should have the lamps and luminaires periodically cleaned several times between relampings. Group relamping should be scheduled at the same time as the cleaning.

The timing of relamping and cleaning should be in accordance with the plans of the lighting system designer. When intervals between operations are too long, excessive loss of light results. When intervals are too short, labor, equipment, and lamps are wasted.

10.11 Voltage. The efficiency, light output, life, and power consumption of incandescent lamps are all substantially affected by their operating voltage. For this reason, they should be operated at or near their rated voltage to give the best value to the user. Incandescent lamps are manufactured and labeled for use with specific voltages at the socket, such as 115 V, 120 V, 125 V, etc. See Chapter 3 for a discussion on the effect of voltage variations on lamp life and efficiency.

The 120 V general-service incandescent lamps is considered to be the standard voltage incandescent lamp because a large majority of electric utilities provide 120 V service to their customers. Incandescent lamps are also available for operation at higher voltages, such as 230 V, 250 V, and 277 V. Higher voltage incandescent lamps are less efficient (except for tungsten-halogen-types) and not as rugged as 120 V lamps since the high-voltage tungsten filaments are smaller in diameter, are longer, and are more fragile than those of standard voltage lamps.

When lower voltage incandescent lamps are inadvertently inserted in higher supply socket voltages (for example, a 120 V lamp in a 277 V socket), they may shatter. Consequently, the application of higher voltage incandescent lamps is sometimes prohibited by certain users.

Incandescent lamp life is often considered the principal criterion of lamp performance. Actually, lamp efficiency is nearly always more important. These two factors are inversely related in incandescent lamps. Hence, lamps designed for longer life operate at reduced efficiency; those with a high lighting efficiency design (such as photo-flood lamps) have relatively short lives. Lamp life design is based on the total cost of the light, assuming typical operating conditions and costs that prevail among most users. In some instances, when electric energy rates may be very low or the labor cost of lamp replacements high, the economic picture is altered. For such applications, lamp manufacturers have a line of special-service incandescent lamps, the life of which is about 2.5 times that of general-service lamps. Special-service lamp efficiency is about 15% lower than that of the general-service type.

Some incandescent lamps are available with lives of 5000–10 000 hours or longer. These are so low in lm/W efficacy that they are uneconomical for use except when installed in difficult to access locations, when the labor cost to replace them is very high, or when special equipment may be necessary in order to change burned-out lamps. A cost analysis is recommended to determine the suitability of their use.

A magnetic coil fluorescent lamp ballast that is designed for 120 V primary supply can typically start and operate lamps at +5% or –10% of the design voltage. However, when operated for long periods at the extremes of these voltage limits, the lamps will not operate at their normal photometric, life, and power ratings, and the ballasts may be damaged. Ballast manufacturers suggest a somewhat more narrow voltage tolerance for sustained operating periods. For example, one manufacturer advises that the limits should be 100 V and 125 V for its 120 V ballasts. For a 277 V ballast, the indicated limits are 254 V and 289 V. Manufacturers' data should be utilized in determining recommended voltage limits on specific ballasts.

The life and light output ratings of fluorescent lamps are based on their use, with ballasts providing proper operating characteristics. Ballasts that do not provide proper electrical values may substantially reduce either lamp life or light output, or both. Ballasts certified as being built to the specifications adopted by certified ballast manufacturers do provide electrical values that meet or exceed minimum requirements. This certification assures the user, without individual testing, that lamps will operate at values close to their ratings. Ballasts for HID lamps are often designed with primary voltage taps. Connection should be made to the tap that corresponds most closely to the supply voltage.

Fluorescent and HID lamp ballasts are made for the higher branch-circuit voltage (277 V) employed in commercial buildings. There are also 480 V primary ballasts for HID lamps. When this voltage is available for lighting systems, sizable savings may be realized as a result of reduced wiring and distribution equipment costs. Fluorescent lamps and fixtures will be the same for 277 V lighting as for 120 V. Ballasts are approximately the same size and cost for either 120 V or 277 V lighting.

When the 480Y/277 V power supply is employed in a building distribution system, an effective and economical system is obtainable by connecting fluorescent luminaires line-to-neutral (see Fig 145). With 277 V panelboards, fewer circuits are needed (as shown in Fig 145 by comparing two areas that are similar in size and investigating the quantity of conduit, copper conductors, and branch circuits).

Wye-connected three-phase, four-wire supply circuits at 120 V or 277 V line-to-neutral provide a very economical system to supply large general lighting loads of fluorescent or HID lighting. However, the ballasts may draw a considerable third harmonic current component that flows in the neutral (or fourth) wire. For this reason, the NEC [1] requires that the neutral conductor be the same size wire as the other three circuit conductors when utilizing three-phase conductors and a common neutral in branch circuits between lighting loads and the serving branch-circuit panelboard. The neutral conductor cannot be reduced in size, which is permitted with incandescent and other resistive loads. The neutral counts as a fourth conductor for the purpose of calculating conductor ampacities.

When incandescent lighting is used in certain areas in addition to 277 V fluorescent lighting, 120 V is obtained by dry-type, step-down transformers that would serve a 120 V branch-circuit panelboard for both lighting and 120 V receptacles. More complete coverage of this subject is presented in Chapter 4.

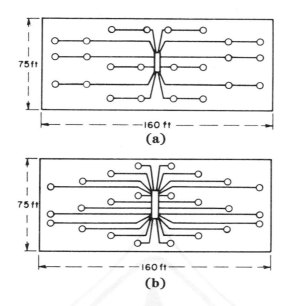

Fig 145
Comparison of Distribution Required for 277 V and 120 V Lighting Systems
(a) 277 V : 12 Lighting Circuits Requiring 1700 Feet of Conduit
and 3550 Feet of No. 12 AWG Wire
(b) 120 V : 24 Lighting Circuits Requiring 5000 Feet of No. 12 AWG Wire

10.12 Power Factor. Incandescent filament lamps operate at 100% power factor. Fluorescent and HID lamps operate with ballast circuits that generally regulate the current by reactive circuit elements. Because of this, the basic lamp ballast circuit operates at a power factor of generally less than 50%. However, practically all ballasts intended for general lighting applications have provisions to improve this power factor to 90% or better.

The most widely used type of fluorescent ballast (the two-lamp series type) has a slightly leading power factor. This is a desirable attribute since most other building loads tend to have a lagging power factor. The lighting designer should specify high power factor ballasts for the lighting applications.

Certain ballast circuits that are used in desk lamps, office copy machines, home appliances, etc., operate one or two low-wattage fluorescent lamps at a low power factor. The total reactive power recorded by these devices depends on the number of luminaires installed.

10.13 Temperature. The performance of incandescent lamps is relatively unaffected by ambient temperature. Light output and life remain normal in cold or warm weather. Performance is usually satisfactory even in the case of some of the confining luminaires. When extremely high ambient temperatures are encountered, as in ovens, special lamps should be used that have been manufactured with exhausting and sealing temperatures that are adequate for the intended service.

The starting characteristics of HID lamps and the starting and operating characteristics of fluorescent lamps are significantly affected by low temperatures. For satisfactory outdoor operation of these lamp types in cold weather, ballasts should be used that supply sufficient voltage to ensure reliable lamp starting. In addition, the ballast should be of a design that will withstand low temperatures if it is mounted outside. In the case of some building-mounted signs or security lighting equipment, it may be practical to remotely locate the ballast in a heated environment.

Rated light output from most fluorescent lamps is achieved with ambient temperatures of 21.1°–26.7 °C (70°–80 °F). Above this temperature range, light output is reduced about 1% for every two degrees Fahrenheit. In high ambient temperatures, circulating air improves the light output. This will be a problem in the design and application of most enclosed or recessed luminaires.

If the ambient temperature around the lamp is reduced below the 21.1°–26.7 °C (70°–80 °F) range, the loss of light at a rate of 2% per degree Fahrenheit occurs. This will be a problem in outdoor applications of fluorescent lighting.

In very cold weather, regular fluorescent lamps may not reach their full rated light output, particularly when subjected to air currents. However, when lamps are used in a closed fixture or are shielded from drafts, the ambient temperature around in a closed fixture or are shielded from drafts, the ambient temperature around the lamp may build up during operation, and light output will increase nearer to rated values. When sufficient voltage is available to start HID lamps in cold weather, they will gradually build up in light output to near normal values, even in open luminaires.

Fluorescent lamps that operate at higher currents (800 mA and 1500 mA) will, in properly designed multiple-lamp enclosed luminaires, maintain light output more easily than lamps of lower current in cold ambient temperatures. In addition, certain fluorescent lamps that operate at 1500 mA have been specially designed for outdoor application. One type is intended for use in open fixtures. In typical outdoor environments, the light output of the jacketed lamp is at its maximum in an ambient temperature of –23.3 °C (–10 °F). Due to the variation in outdoor conditions under which fluorescent lamps may be expected to operate, such as temperature, wind, and equipment, it may be desirable to seek the advice of luminaire manufacturers regarding the best choice of lamps and equipment for a specific climate.

Most ballast designs will start and operate fluorescent lamps satisfactorily down to a temperature of 10 °C (50 °F). Many will continue to provide reliable lamp starting below 10 °C (50 °F); but it is recommended that, when operating below 10 °C (50 °F), a ballast that is designed specifically for cold weather operation be specified. Such ballasts are rated to start lamps reliably down to –17.8 °C (0 °F) or –28.9 °C (–20 °F), depending upon design. They are available for slimline (430 mA), high-output (800 mA), and extra-high-output (1500 mA) fluorescent lamps.

High temperatures may shorten ballast life or, as with Class P ballasts, the thermal protector will open the circuit and turn off the lights. A provision for adequate heat dissipation should be provided for ballasts, both in the design and installation of luminaires. This is especially important for the higher VA rated ballasts.

10.14 Ballast Sound. Ballasts for fluorescent and HID lamps produce a very low level of sound when operated in the open on a heavy vibration-resistant base. However, when mounted in a luminaire, they induce vibrations into the luminaire. The large radiating surface acts as a sounding board and may radiate audible levels of sound. The sound is a distinctive tonal hum that may be distinguished from other background sounds in a given interior. When it is loud enough to become distinctly audible, some occupants may find this noise objectionable.

Whether or not annoyance is likely to be ascribed to a lighting system depends upon three factors

(1) The sound level radiated by the lighting equipment
(2) The tonal quality of the particular luminaires in question; that is, the distribution of sound power among the harmonics of 120 Hz that are being radiated
(3) The ambient sound level in the area that arises from other sources

In order of importance, the factors determining the level and tonal quality of the sound are

(1) The design and construction of the luminaire
(2) The design and construction of the ballast
(3) The VA rating of the luminaire and the illuminance level

(For a given lighting level, a large number of small luminaires generally yields a lower sound level than a small number of large luminaires.)

The ambient sound level in an area is determined by the activities in it. For re-lighting an existing area, sound level readings could be made. For new construction, the ambient noise levels listed in Table 77 may be used as representing typical experience. It should be pointed out that acoustical treatment has little bearing on ballast hum. This is true because the acoustical treatment will reduce the ambient noise level and the ballast hum by equal amounts. The whole question is whether the ballast hum becomes noticeably audible above the ambient sound level. Absolute sound level has no practical importance since it is practically never high enough to interfere with speech audibility or create any other objective problems.

10.15 Lighting Economics. The realm of lighting economics is multifaceted. It can be divided into the following categories:

(1) First costs
(2) Type and quality of lighting desired
(3) Energy costs
(4) Maintenance costs
(5) Effect on personnel

During the past 10 years, the order in which these aspects were considered has been altered. Also, from day to day, there is no fixed method to decide which of the above factors are the most important or should be considered first. The aspect of primary importance should be decided on a job-to-job basis, depending on the user's end needs, the type and amount of energy available, energy costs, maintenance, availability, and a number of intangibles, such as employee morale, health, comfort, and safety.

Table 77
Fluorescent Ballast Sound Ratings

Application	Ambient Noise Level (Measured with Standard 40 dB, Weighting Network) (dB)
Broadcast studio, church, country residence	20—24
Evening school, city residence, quiet office	25—30
Average residence, public library, study hall	31—36
Classroom, professional office	37—42
Noisy residence, business office	43—48
Store, noisy office, factories	49 and up

10.15.1 First Costs. In today's lighting market, first costs can be very misleading. Not only do first costs vary from fixture type to fixture type; but they also vary among fixtures of a given type. The variance in the first cost of fixtures within a category is due to a number of factors, that is, quality of workmanship, durability, attention to detail, reflector type, etc. A fixture may be selected on a first-cost basis to serve the client's needs and finances. However, care should be taken by the lighting designer when selecting a fixture in this manner. The most expensive fixture is not always the best for the task, and the least expensive fixture is not always the wrong choice. First costs should be weighed in the light of all other economic aspects.

With respect to first costs, incandescent lighting may appear to be less expensive to install than other more efficient sources when the desired illumination levels are relatively low. However, when the cost of the power distribution system is also considered, the more efficient lighting systems are often lower in first cost. When operating expenses are considered, the incandescent system is far more costly than fluorescent or HID lighting systems.

10.15.2 Type and Quality of Lighting Desired. In this aspect of lighting economics, the lighting designer should closely communicate with the client or the end user of the lighting system. A variety of questions should be answered at this design stage: "What are the characteristics of the visual tasks to be performed?" "What type of luminaire and light source will provide light of the right quality for good task visibility?" "How important is visual comfort?" "How important is lighting control?"

"How much light is required for the tasks at hand?" "What is the client's budget for lighting installation?" Once these questions have been answered, one can decide on the type (incandescent, fluorescent, HID) and the quality of lighting that should be installed.

10.15.3 Energy Costs. Energy costs vary from state to state and region to region. Energy efficiency is the dominant concern for most designers. However, to most clients, energy economics is the item of most concern. The lighting designer should weigh the cost of the lighting system versus its energy economics and then make an intelligent choice, with financial guidance from the client.

10.15.4 Maintenance. Lighting designers should obtain maintenance information from the client. How large is the maintenance department, if any, and how skilled are its members? Some end users do not have a maintenance department but call an electrical contractor to service lighting equipment. For these users, a low-maintenance lighting system is a wise choice. However, low-maintenance systems are sometimes more expensive than a system that requires considerable maintenance. In summary, maintenance costs should be weighed against lighting system costs.

10.15.5 Effect on Personnel. Various types and quantities of illumination will have different psychological effects on workers. These effects should be discussed with the end user so that the lighting designer can make a responsible decision as to the correct lighting system to use.

"Productivity" is usually the key word in most occupational environments. The lighting designer should be aware of the effect illumination has on productivity and call these effects to the attention of the client when the illumination budget is being prepared. The trade-off between production levels and illumination quality should also be evaluated.

In summary, it is suggested that the lighting designer choose the lighting system to be used on a life-cycle cost basis. This method of system economics addresses all lighting design aspects so that each aspect can make its proper contribution to the ultimate choice of the appropriate lighting system.

10.16 Illuminance Calculations. There are two principal approaches to the calculation of illuminance: one involves situations in which uniform distribution of illuminance is desirable, as in densely occupied offices or classrooms; the other is when non-uniform illuminance is desirable as a more energy-efficient way of providing for task performance.

The uniform illuminance method involves the use of utilization coefficients that are supplied by luminaire manufacturers, together with maintenance factors applied in a formula to determine the number of luminaires that are necessary to maintain the desired illuminance in a space. Then it is necessary to arrange the luminaires appropriately so as to provide the desired distribution of illuminance or to minimize veiling reflections in tasks, or both. As covered earlier, illuminance values will be substantially higher in the central portions of small and medium-sized rooms and lower near the walls on average. This knowledge can save energy, when task areas are confined to central areas of a room, by designing an average room for a lower illuminance.

The computation of non-uniform illuminance is more complex since it involves a calculation for the direct contribution of each luminaire at a particular point in a room plus the contribution of interreflected light from room surfaces to the point of interest. This process is repeated for each point in the space where illuminance information is desired. This is called the point-by-point method of calculating illuminance. Sometimes, the computation for interreflected light is omitted to simplify the computations. This may add 5%–10% to the task light that is available in small rooms, and somewhat more to the task light in larger rooms.

Because of the great many individual calculations required, computer programs have been developed to perform them. These programs are available through various computer software services and from manufacturers. Programs are also available for use with programmable hand-held calculators.

Refer to References [10] and [11], which contain considerable information on both uniform and non-uniform illuminance calculations.

10.17 Lighting and Thermal Considerations. Lighting energy in buildings can be used twice during the winter; one time for visual purposes (the only reason for its being in the building), and a second time to replace building heat losses when the outside temperature is below 18.3 °C (65 °F).

Electric lamps are 100% efficient as heat sources. Even the light from the lamps eventually becomes heat. When light or infrared rays from luminaires is intercepted by people or surfaces in a room, part of it is absorbed and raises the temperature of the surface. That which is reflected is bounced to another surface where another partial absorption takes place. In a brief instant, all the light and infrared rays entering a space from lamps or luminaires is absorbed and is useful in heating if the room needs heat at that particular time.

In the future, when buildings are designed, it will be necessary to evaluate the total impact of each subsystem on energy usage, as there may be a mandatory (or voluntary) energy (or power) budget with which to comply. Failure to do so could result in building designs that exceed their allotted budget, or, on the other hand, fail to function effectively and efficiently.

In order to compute the net effect of lighting on building energy usage, three variables should be checked

(1) The lighting energy used directly
(2) The heat gain that load lighting places on a cooling system
(3) The lighting heat gain contribution to the heating of a building

In winter, the lighting system maintains the temperature so that the room thermostat may be reduced from frequently cycling the permanently installed heating system and reducing the use of oil or natural gas, currently the most popular space heating fuels.

In single-story buildings, lighting units installed in a pattern across the ceiling automatically compensate for some of the heat lost through the roof, depending upon the insulation, lighting heat gain, and outside temperature. This compensation begins to take place when the outside temperature falls below 18.3 °C (65 °F), and, somewhere between 10 °C (50 °F) and 15.6 °C (60 °F), all of the lighting energy contributes to replacing heat losses in typical buildings. The luminaires within 10–

15 feet of the walls and windows at the building perimeter also compensate for some of the heat lost through these surfaces.

Low-rise buildings of one or two stories predominate in existing commercial and industrial building inventory. Over 90% of the existing area of commercial and industrial buildings is low rise.

Lighting is a low-temperature heat source. For example, 4 foot rapid-start fluorescent lamps operate with a bulb wall temperature of only about 40.6 °C (105 °F). The 1500 mA fluorescent lamps rise to a temperature of 60 °C (140 °F). For example, on a cold morning, the lighting system is not able to overcome a -12.2 °C (10 °F) overnight temperature setback within 30 minutes. A higher temperature conventional heating system (such as electric heating elements) should be used for this purpose.

Furthermore, after a fluorescent lighting system has been turned on for some time, most of the energy is stored in the luminaire. That is, the lighting fixture itself heats up and then the heat is transmitted to nearby surfaces, such as ceiling tiles, the air in ceiling cavities, and the building floor or roof structure. After several hours of operation, some of the building structure as well as the luminaires have temperatures well above that of the air in the occupied space, so they begin to radiate and convect heat into the occupied space. In mid or late afternoon, heat from the lighting system is entering the room at the same rate that it is generated, so the lighting is effectively contributing to the heating of the building.

At 5 p.m., when many building operations cease, people go home and lights are turned off after the evening cleaning operations. A building setback temperature may go into effect as well. However, the heat from the lighting system that is stored within the building structure continues to make itself felt, dissipating this energy within the building and delaying the thermostat's first turn-on of the furnace. When heating energy is not used overnight, the stored heat reduces the recovery energy required the next morning since the building will not have cooled down as much because of the storage effect.

In a multiple-story building, lighting on the top floor and around the perimeter of lower floors can replace some of the heat losses throughout the building shell. When the heat in the building interior is to be useful, it should be controlled and redistributed by a system designed for this purpose. Standard mechanical equipment is available to do this.

Sometimes after redistribution of interior zone heat has taken place, some energy is left over. This energy can be stored when insulated water tanks are available. There are a number of buildings employing this concept, which saves a considerable amount of energy and money. These storage systems can also be used in the summer to chill water during the electric utility's off-peak demand hours, which will only incur an operating expense without a demand charge. Then the chilled water can take the peaks off the cooling requirements during occupied hours.

The control and use of lighting heat in buildings has been treated in some detail because some of its aspects are not well known or understood. It is recognized that lighting creates a cooling load in buildings in warm weather and an allowance for refrigeration tonnage and some volume of air (or water) should be made in the mechanical system design of the building.

A discussion of the properties of air return luminaires that provide some advantages in controlling lighting heat in warm weather is included in 10.7.5.

In multiple-story buildings, the heat gain from lighting systems in the interior zones of a building sometimes requires conventionally designed cooling systems to run in very cold weather. In many buildings, this energy is wasted rather than recovered and redistributed to perimeter zones or stored for later use, as described earlier. However, in cold weather, economizer air systems that use outside air for cooling the interior zones of buildings can do away with the need for operating refrigeration compressors in winter. Such systems should be used in future buildings to reduce the requirement for refrigeration energy to handle lighting loads to the lowest possible level.

Now, and in the future, even though more efficient sources will be used for lighting and will be turned off promptly when not needed, and though less total energy will be used for lighting as a percentage of total building energy, the need to use lighting heat may become greater than at present. Designers and owners should take this into account in the thermal design of their buildings by redistributing, storing, and reusing lighting heat and other internal heat gains. This can prevent the wasting of building energy, so typical of past practice, and reduce the requirement for energy from new sources to satisfy space heating needs.

10.18 References. The following references shall be used in conjunction with this chapter:

[1] ANSI/NFPA 70-1990, National Electrical Code.

[2] ASHRAE/IES 90.1-1989, Energy Efficient Design of New Buildings Except New Low-Rise Residential Buildings.

[3] Chen K. and Castenschiold, R. "Selecting Lighting Controls for Optimum Energy Savings," *Conference Records*, 1985 Industry Applications Society Annual Meeting.

[4] CIE Report no. 29, "Guide on Interior Lighting."

[5] Flynn, J. E. "A Study of Subjective Responses to Low-Energy and Nonuniform Lighting Systems," *Lighting Design and Application*, vol. 7, no. 2, Feb. 1977, p. 6.

[6] Flynn, J. E. "Lighting Design Decisions as Intervention in Human Visual Space (the Role of CIE Study Group A)," Paper presented at Symposium—1974/CIE Study Group A, Montreal, Canada, 1974.

[7] Flynn, J. E.; Spencer, T. J.; Martyniuk, O.; and Hendrick, C. "Interim Study of Procedures for Investigating the Effect of Light on Impression and Behavior," *Journal of the IES*, Oct. 1973, p. 87.

[8] Flynn, J. E.; Spencer, T. J.; Martyniuk, O.; and Hendrick, C. "The Influence of Spatial Light on Human Judgment,"Compte Rendu, 18e Session, p. 75-03; CIE Congress, London, England, 1975, p. 39.

[9] *Federal Energy Conservation,* supplement no. 1, GSA Federal Management Circular FMC 74-1.

[10] *IES Lighting Handbook,* 1987 Edition (reference volume).

[11] *IES Lighting Handbook,* 1987 Edition (application volume).

[12] "Recommended Practice for the Specification of an ESI Rating in Interior Space When Specific Task Locations Are Unknown," Prepared by the Design Practice Committee of the IES, *Journal of the IES,* Jan. 1977, p. 111.

[13] "Selection of Illuminance Values for Interior Lighting Design (RQQ report no. 6), Prepared by the Committee on Recommendations for Quality and Quantity of Illumination of the IES (RQQ), *Journal of the IES,* Apr. 1980, p. 188.

[14] Thornton, W. A. "Difference in Color Vision," *Lighting Design and Application,* vol. 9, no. 2, 1980, p. 17.

Chapter 11
Electric Space Conditioning

11.1 General Discussion. This section deals with heating and cooling aspects of occupied spaces of commercial and industrial buildings. The concepts presented can also be applied to areas in which the control of ambient conditions is necessary for reasons other than human comfort.

Chapter 14 covers building management systems and Chapter 17 covers electrical energy management; both of these areas are vitally concerned with energy usage in commercial and industrial buildings. The largest source of energy consumption in commercial buildings is heating, ventilating, and air conditioning (together referred to as "space conditioning") and lighting. The heat produced by lighting has a material effect on the energy required for space conditioning.

The largest motors in commercial buildings are the compressors used for air conditioning. If central or distributed packaged air conditioning is utilized in a building, this load may be, in aggregate, the largest demand for the building (perhaps even exceeding the lighting load during periods of maximum usage) for buildings in temperate climates.

The costs of energy and related environmental considerations in generating electrical energy have reached the point where the decision to undertake the design of an all-electric building (which was becoming quite popular before the energy crises of the '70s) should be carefully considered. There are still a few areas and certain occupancies in which the all-electric building or premises is a viable choice.

(1) Buildings that have a large number of rooms with highly variable occupancy (e.g., hotels and motels) can usually be served most economically by an electric heater/air-conditioner unit in each room.

(2) In residential and smaller commercial buildings, the capital costs of construction can often be reduced considerably when an all-electric construction design can be undertaken. This is particularly true in remote areas where a fossil fuel source is expensive and/or not dependable.

(3) In an industrial plant in which there is co-generation or hydroelectric generation, outlying buildings may be heated most economically by electricity.

(4) In continuously warm areas where heat is seldom needed, energy cost becomes secondary to installation cost.

For the next decade, at a minimum, nuclear energy will not become available in sufficient additional quantities in the U.S. to materially affect decisions on all-electric buildings. Except for smaller, special designs where very high levels of continuous sunlight are available, the use of solar energy (or, for that matter, for generation of electricity by wind systems) will be limited by technology, high cost, poor records of maintainability, and storage techniques.

11.1.1 Space-Conditioning Control. The key to modern building energy systems is control. Electronic control systems make possible the optimization of energy usage in buildings or areas (see Reference [7].[74] For the smallest areas, thermostats with relatively elaborate features, such as time/temperature coordination, are available. Some also have provisions for incorporating the variable of outside temperature. A major advance has been made in the building management system, which today is practical for even medium-sized buildings. Building management, which is computer-based and involves extensive sensing and control systems for larger buildings, is covered in Chapter 14.

11.2 Primary Source of Heat. Space heating and air conditioning are the major components of an electric space-conditioning system. Engineers are using the planned environments approach more extensively today.

The primary source of heat for people is the heat generated within the body. The human body is a heat generating unit that is adjusted internally so as to maintain a temperature of 98.6 °F (37 °C) as long as the body is healthy. In a space-conditioning system, the object of it is to regulate the environment so that heat is not dissipated too rapidly for human vitality, well-being, efficiency, and comfort. All of the foregoing are closely interrelated. The building system does not warm the body because the body is much warmer than the surrounding air. The rate at which heat is lost from the body is dependent on the air temperature and rate of air movement around it. Comfort is a function, among others, of the humidity of the air. The surface temperature of the body is about 85 °F (29.4 °C), and the surrounding air is 10 °F (5.6 °C) to 15 °F (8.3 °C) lower. With this differential, the heat transfer occurs from the body outward. The same observations can be made in regard to air conditioning, in which the temperature is controlled so as to maintain the heat flow from within. In this case, the humidity assumes greater importance. It should be noted that the regulations regarding quantities of makeup air are based primarily on humidity and odor control, not on the buildup of respiration gases.

11.2.1 Electric Space Heating. Electric space heating is accomplished by resistance heaters or heat pumps. In resistance heaters, the heat is generated by the passage of electric current through the resistance offered by the conducting materials. Heat pumps accomplish the exchange of heat from one medium, such as air or water, to the space to be heated or cooled.

Heating equipment should be located so as to most effectively replace the heat lost to the outdoors and to counteract heat loss or to eliminate, as much as possible, any cold surfaces onto which body heat can radiate.

Wall panels and baseboard heaters are, therefore, placed on outside walls, preferably beneath window areas. In this position, they can effectively counteract and

---

[74]The numbers in brackets correspond to those in the references at the end of each chapter.

raise the temperature of the wall and inside glass surfaces. Convection currents that are set up within the room tend to move downward across the cooler window surfaces and then combine with infiltrating air to produce a constant downdraft across the glass. Heat rising from units beneath the window helps to neutralize this downdraft and prevents the cool air from circulating throughout the room. Locating the units at the baseboard or in the lower portion of the walls removes cool air from floors, while the natural rise of heat keeps ceilings warm. Most modern occupied spaces require cooling in interior spaces, even in the coldest weather, because lights, people, etc., provide more heat than the system loses. Thermostatically controlled perimeter heaters, which are off during much of the occupied period, do not offset window downdrafts, and a resultant pool of cold air may develop. Separate draft heaters are available to combat the downdraft, providing more heat than is absolutely necessary. Continuous perimeter heaters that have a capacity of 135–250 Btu/(hours × feet) [(40–73 W)/feet] should be placed beneath the entire length of the window. These are controlled by thermostats between adjacent heaters (mounted in line with the heaters), but that are isolated from radiation and conduction. The controls should operate independently of basic system controls to control downdrafts regardless of whether the basic system is heating, cooling, or off. Within the load carrying capacity of the control system, one thermostat may control all draft-barrier heaters on the same wall of a particular room. Often draft-barrier heaters on several adjacent floors may be controlled by a single thermostat between heaters that are beneath windows with the identical orientation.

Local control of temperature in each room is a great advantage of resistance heating equipment and room heat pumps; but particular attention should be paid to thermostat location, if optimum results are to be obtained. The thermostat should be mounted about 5 feet above the floor and on an inside wall in order to avoid the direct effect of the lower temperature on outside walls. It should not receive the direct output of a heater, and it should not be in a position to be affected by drafts when doors are opened. Direct heat from lamps or appliances will also cause erratic and inefficient operation.

Both line voltage and low-voltage thermostats are available, the latter operating in conjunction with a relay. The low-voltage thermostat enables tighter temperature control by building management.

A single heater installed as a supplementary source of heat or to serve a specific function, such as heating an entrance way or vestibule, may use a built-in thermostat to sense the temperature in that particular area. It is usually preferable to use a wall-mounted thermostat. A low-voltage thermostat is recommended when several heaters, with a combined electrical demand of 3 kW or more, are used to serve a room.

11.2.2 Heat Pumps. Heat pumps operate on the reversible cycle principle. They extract heat from a space during the warm weather. During cold weather, they are manually or automatically switched to introduce heat into a space, extracting heat from the air, water, or ground (see Reference [3]).

Units using water as their heat source are practical when water from a well, stream, lake, or river, which are substantially above freezing temperature, is available. Heat drawn from the ground by means of water pumped through embedded pipes is of relatively constant temperature; but the installation is expensive and sufficient ground area is not always available for laying pipes. Also, available heat varies with shifting soil conditions.

Units drawing heat from the air are most widely used because air is always available. The disadvantage here is that large equipment capacities are necessary in areas where the air temperature drops substantially below – 6.7 °C (20 °F), since a larger volume of air has to be handled in order to extract sufficient heat. Recent developments using compound compressors (two compressors in series) have made operation practical down to – 28.9 °C (– 20 °F); but the equipment is expensive and is not yet practical for small heating requirements.

Heat pump capacity is usually designed to handle the air-conditioning load. When outside temperatures become too low for the installed capacity to handle heating loads, supplementary resistance heaters are usually provided for additional heat. Heat pumps also have been developed for small rooms. These heat pumps permit room-by-room control, which is not practical with central heat pump systems.

Although higher in initial cost and maintenance costs than resistance heaters, heat pumps offer the advantage of lower overall operating costs due to the "free" heat extracted from external sources.

11.2.3 Resistance Heaters. Resistance heaters may be classified by type as follows:

    (1) Wall and ceiling units
    (2) Central furnace or boiler
    (3) Unit heaters
    (4) Infrared heaters
    (5) Heat storage equipment

11.2.3.1 Wall and Ceiling Units. Wall and ceiling units may be surface-mounted or recessed, incorporating heating coils, radiant glass, ceramic panels, or finned elements, with or without a built-in fan. Certain types of ceiling units may include built-in fluorescent or incandescent lamps and are used most often in bathrooms. Infrared lamps are frequently used as the heat source in these applications, and built-in fans are used for circulating the heat.

Baseboard units, as their name implies, are designed to be placed along the outside wall of each room at the location normally occupied by the baseboard. They are constructed in sections of about 2 – 12 feet in length and vary from about 3.5 – 10 inches in height. Sill-line heaters are similar to baseboard heaters; but they are intended to be mounted with the top of the heater enclosure at windowsill level.

The heating elements are rated at 180 W (600 Btu/hour)/feet to 600 W (2050 Btu/hour)/foot or more and can be selected to operate at nearly any common branch-circuit voltage. A variety of heating elements is installed in baseboard heaters, including glass panels, metal-alloy strips, ceramic, finned tubing, and metal-sheathed chrome wire types. One type uses a small electric boiler in which hot water circulates through the finned elements.

Provisions are made for wiring the heaters from the back, bottom, and ends, and for connecting two or more sections. The units may be equipped with receptacles for window-type or through-the-wall-type room air conditioners. Units may also contain convenience outlets. This permits wall outlets to be located below the heat discharge and avoids portable cord deterioration from exposure to the heat if the convenience outlets are located above the heater. Most baseboard heater installation instructions specifically prohibit placing wall outlets above heaters. Baseboard and sill-line heaters normally do not use fans and are of the convection type. Care should be exercised, particularly with the highest W/foot units, to assure that the high grill temperature (which can be over 93.3 °C [200 °F]) will not present a problem.

The heaters can be controlled by either a line voltage or a low-voltage wall-type thermostat, by built-in thermostats, or by thermostats installed in special baseboard sections designed for this purpose and matched in appearance with the baseboard.

Wall elements are similar to baseboard heaters, but are normally mounted higher than the baseboard type and on inside walls. Many types also include fans. They are generally available in larger sizes with higher ratings than baseboard heaters.

Radiant heating panels of glass, ceramic, and metal alloy are designed for recessed or surface wall mounting, similar to resistance-type wall heaters. The heating element may be made up of tempered glass into which is fused a continuous alloy grid. Some units utilize a metallic coating, which is fired to the back of the glass.

11.2.3.2 Central Furnace or Boiler. The electric furnace is a central heating system and is closest in operation to the fuel-fired furnace with resistance elements replacing the combustion chamber. The heat is distributed by means of blowers and ducts. This equipment has step controls to permit the energizing of successive sections of the heating element in accordance with the amount of heat needed. The same ducts may be used to distribute cooled air during the summer months from a central air conditioner, or heat pump coils can be employed in the furnace to provide normal heating and cooling. Resistance elements are then used to supplement the heat pump in very cold weather or when quick recovery is desired.

The central boiler operates similarly to the central furnace, only with the duct system replaced by a piped system. The resistance elements are used to generate hot water, which is pumped through the system to hot water coils, baseboard heaters, or radiators.

In either type of central heating system, the operating cost is usually higher than that of room resistance heaters. This is due to the loss of heat in the ducts or pipes through unused areas and to not normally supplying separate thermostats for each room.

A modification of the central furnace concept is to distribute unheated air through the ducts. Heating is accomplished by duct insert heaters located in the registers of each room or at some point in the ductwork.

11.2.3.3 Unit Heaters. Unit heaters, used chiefly for spot heating of industrial or commercial areas, employ a relatively strong fan or blower to force air through a heating element into the space to be heated. Louvers are usually provided for direction control. A distinctive form of heater is called the "schoolroom unit ventilator," although its use has not been restricted to schools. Its flexible operation permits cool outdoor air to be drawn in and directed through heating elements by a blower into the space to be heated. With the heating elements de-energized, the cool outdoor air may be required to ventilate when cooling rather than when heating. A return intake permits room air to be mixed with the outdoor air in various proportions before being passed through the heating coils.

11.2.3.4 Infrared Heaters. When very localized heating is needed, infrared radiant heating is the most commonly used type. It is typically installed in religious facilities, bowling alleys, warehouses, manufacturing areas, loading platforms, sidewalk areas adjacent to store windows, drive-in banks, and areas adjacent to concession stands where infrared heaters provide comfort for the participants, workers, and shoppers. Quartz lamps, which have a very high output in the infrared region, are popular as the heat source for such applications. NEMA HE1-1980 (Reaff. 1986), Manual for Calculating Heat Loss and Heat Gain for Electric Comfort Conditioning [2][75] provides useful design and application information.

11.2.3.5 Heat Storage Equipment. Heat storage equipment is available in two major designs

(1) Large single units serving a whole structure
(2) Small units approximately the size of regular water radiators that can be dispersed about the various rooms in a structure

Heat storage equipment may be used to take advantage of off-peak utility rates, since it can be charged during valley hours. A heater with a capability of storing 354 kWh is about 5 feet $\times$ 3.5 feet $\times$ 3 feet high. A heater with a 54 kWh storage capacity is about 3 feet $\times$ 2 feet $\times$ 1.5 feet deep. Heat is released by convection and radiation and, in some models, may be assisted by a small fan.

11.2.3.6 Heat of Light and Waste Heat. In the modern commercial building, from $2-4$ W/ft^2 are used for lighting offices, data processing centers, and public, sales, and other commercial areas. These areas usually represent from one-third to one-half of the annual building energy usage.

If this heat is permitted to enter the spaces completely, as it would be with pendant-type fixtures, in the winter, it could produce a major part of the heat required. In summer, it could represent a tremendous load on the air-conditioning system.

In very large areas, the heat generated by the lighting, along with other heat sources, such as occupants and equipment, will be so high that cooling (air conditioning or external air) will be required even in the winter. All but the smallest buildings use an air return system (usually a hung ceiling) that separates the return and supply air. By using ventilating control systems, which consist of fans and dampers, it is possible to mix the exhaust return air, the air through the air-conditioning or air-cooling systems, and the outside air. This mixing is necessary in order to recycle the excess heat in winter for heating and, in summer, for cooling. The effectiveness of the recycling can be enhanced by electronic control of the ventilation mixing and exhausting of the different air flows and by controlling the air temperatures involved.

It is important that the heat from all but the smallest motors be exhausted directly to the outside air except, possibly, when the heat can be utilized in winter, as indicated above. For this reason, machine rooms usually have separate ventilat-

---

[75]NEMA publications are available from the National Electrical Manufacturers Association, 2101 L Street, N.W., Washington, DC 20037.

ing systems in order to avoid the possibility of smoke seeping into occupied spaces in the event of equipment failure.

11.2.4 Types of Air-Conditioning Systems. The type of air-conditioning system chosen significantly affects the ability to control energy consumption in a building. For example, unit air-conditioning systems do not have the potential for control on a building basis; for power demand control; and for fine-tuning of building temperatures.

(1) Central-compressor-type air conditioners usually represent the largest motors in a commercial building. Table 17 in Chapter 3 lists the loads that are generally supported by various portions of the building power system. Control of the loads on the air-conditioning system is usually affected by controlling the temperature of the cooling water, which is circulated throughout the building to the heat exchange units in the individual areas for cooling. As a rough figure, one ton of air conditioning requires 1 kW of electrical capacity for the compressors alone. Typically, 40% more capacity is required for ventilation and auxiliary equipment for the air-conditioning system. Where higher pressure ventilating systems are used, as in high-rise buildings, an even higher percentage is required for these auxiliary loads. When low-cost excess or process steam is available, steam turbines may be used to drive the compressor.

(2) Absorption machines are also central units that provide chilled water for building cooling. However, no compressors are involved. The chilling is accomplished by changes in the physical state of certain compounds, with heat introduced by steam injection (usually waste steam from some other process or purchased excess steam from a utility) as the energy input, and cooling water as the output. In this type of system, the approximately 0.4 kW/ ton of air conditioning is required for auxiliary and ventilation systems in the building.

(3) Heat pumps were discussed in 11.2.2. Heat pumps are usually used in residential or smaller commercial building installations.

(4) Unit or packaged air-conditioning units are commonly used in private residences, small buildings, and spaces that were constructed without provisions made for air conditioning.

The advantage of using unit air conditioners is that the tenant has the responsibility for air conditioning the occupied space, including initial installation cost and operation. Unlike the systems in items (1) and (2) above, there is no need to apportion the coolant (chilled water) costs among the tenants. It does have drawbacks, which are associated with carrying electric power distribution facilities for these units throughout the building, of not permitting sophisticated central control (including temperature and power demand), and of lesser efficiencies associated with smaller units. It does have the advantage of avoiding coolant risers (chilled water loops and air ducts, in some cases) from a central system.

11.3 Energy Conservation. The term "energy conservation" conjures up many varied meanings. In its simplest terms, it means "minimizing the use of energy in a given application of energy." In the case of electricity, energy conservation often includes the minimizing of electrical demand (kW) as well as energy demand (kW $\times$ hour). In this context, the term "energy management" is often used. This should not be confused with the term "load management," which usually implies control by some party other than the end user (see References [5] and [6][76]).

As an engineering topic, energy conservation may be divided into two broad areas. First is the area of practical energy conservation, which has to do with the engineering decisions that influence energy consumption in a given building. The second area centers on legislative guidelines and limits for energy consumption based on the weighing of the relative importance of the social versus the economic factors involved in energy resource allocation.

11.3.1 Practical Energy Conservation. Clearly, this is a topic that could occupy several books without exhausting it. Many of the energy uses in buildings, while electrical in nature, are frequently the responsibility of the mechanical design team. Heating, cooling, ventilation, heat recovery, controls, etc., are in this group of energy uses. There is, however, considerable interaction between mechanical and electrical disciplines. For example, temperature setback during unoccupied hours could impose an abnormal recovery peak on the restoration of an electric heating system.

The following discussion begins with lighting, since lighting is a major use of electrical energy in many buildings. It is not intended to be a discourse on efficient lighting designs, but rather it is intended as an aid in the evaluation of alternate designs.

The cost of operating a lighting system involves much more than energy consumed by the light fixtures. The analysis of the energy consumption should include the lighting system's interaction with all the building systems. In many buildings, the lighting system may account for one-half the load on the building's air-conditioning system. Furthermore, lighting loads can vary from 5 Btu/ft^2 (1.5 W/ft^2) or less, to 15 Btu/ft^2 (4.5 W/ft^2) or more. Thus, the lighting designer has a major influence on the ultimate size of the air-conditioning system.

Energy consumption may be analyzed on the basis of total energy consumed or on the basis of operating costs. An evaluation based on consumed energy should include the cost of energy generation which, for fossil fuels, is approximately 3.3 times the Btu equivalent of the electrical energy consumed. The operating cost evaluation will, however, generally have the most immediate and direct application.

The interaction of the lighting system with the other building systems arises as a direct result of the heat produced by lighting. In the cooling mode, the building system should remove this heat. Due to the influence of the lighting system, this is equivalent to an increased air-conditioning cost of at least 33% for a large well-designed, central air-conditioning system. For smaller, less efficient, or badly designed systems, this overhead figure might be as high as 60%.

In the heating mode, the heat produced by the lighting system will represent a gain. In some buildings with an inefficient or lightly loaded furnace or boiler, the heating energy generated by the lighting system may be less expensive than that

---

[76]NIST publications are available from the National Institute for Science and Technology, Institute for Computer Sciences & Technology, Washington, DC 20234.

from the furnace or boiler and the illumination would essentially be free. For instance, this break-even point would occur with electricity at $0.06/(kW × hour) and fuel oil at $1.00/gallon and a heating plant efficiency of 40%. This assumption, however, may not hold true because the heat produced by the lighting system may not be in the area where it is required nor in the proper quantity at a given time.

In lighting system design, a number of factors bearing on energy conservation should be taken into consideration. These include the following:

(1) Use of the most efficient light source and luminaire for the particular application
(2) Provision of adequate local and zone switching
(3) Possible use of time clocks, photocells, and dimming systems
(4) Possible integration of lighting fixtures with air-conditioning systems
(5) Efficient lighting design relative to quality, quantity, and task.

11.3.2 Reduction in Demand for Space Conditioning. Electrical demand, or the rate of usage of electrical energy, is usually a major cost item for commercial buildings. New generation capacity (and, therefore, demand costs) is relatively expensive for utilities.

In an effort to reduce the need for additional capacity, some utilities offer significant incentives for users to reduce their demand through various techniques. These include the following:

(1) The use of time-of-day rates to encourage off-peak use of power.
(2) Reduce maximum demand by shifting time of operation of as many loads as possible to avoid concurrent operation. Maximum demand is the basis of most billing for capacity associated charges. Off-peak demand, which is less than maximum demand, is effectively charged in part for nondemand usage, energy, fuel, and fuel adjustment.
(3) An actual incentive may be paid by the utility to the customer to install demand saving systems. This incentive may be based on demand reduction or installed storage capacity.

The air-conditioning system is a significant load in most commercial buildings. It is a seasonal load in temperate climates, with peaks limited to a few hours per day. Two techniques that have been used effectively are to freeze ice in a central on-site plant that, in turn, is used to chill coolant water for use during high cooling demand periods or to chill cold water for storage in a large tank. Because of the heat created by fusion, the energy stored in a given volume will be about seven times greater for the ice than the water.

When economics justify ice or water storage, ice storage is normally preferred for smaller buildings ($<50\,000\,\text{ft}^2$), and water storage is used for larger buildings or for buildings that are being retrofitted. Larger water storage facilities can also be used as storage tanks for fighting fires (an advantage in terms of insurance rates).

The coolant water can be chilled either by refrigerant machines or by a separate ice plant or by both. A special analysis is required to evaluate the use of these systems, which some utilities are prepared to develop, or assist in developing, at a nominal fee.

The efficiency of a well-designed conventional cooling system is usually on the order of 0.7-0.9 kW/ton. This compares favorably with values of 0.9-1.2 kW/ton for a typical storage system. The lesser efficiency can be offset somewhat by taking advantage of the lower coolant temperatures.

The use of a storage system may be justified, even considering its usually lesser efficiency, by its greatly reduced electrical demand charges, by its operation of refrigeration machines during periods when the utility offers low energy (cost per kWh), and by incentives offered by the utility. It may be possible to reduce the size of a refrigeration plant if the refrigeration load is averaged over the entire day.

Some work has been completed on heat storage for electric heating systems using a ceramic storage element. In this type of system, the electrical energy is stored at a relatively high temperature and the heat is reclaimed by controlled air flow around the heated ceramic. This system has been used mostly in smaller buildings, and it is not popular in the U.S. at the present time.

Energy conservation measures have reduced overall utility demands and the rate of generation capacity growth, so that the justification for on-site energy storage, particularly for a retrofit of existing plant, has been limited. As demands increase, the advantages of such storage may be enhanced. In any new large building development, as a part of overall energy studies, in conjunction with utilities, the advantages of such storage should be studied. About 1000 ice/chilled water storage systems were installed through the year 1987 (see References [4][77] and [8]).

11.3.3 Standards and Codes. Most of the standards or guidelines in use today are prescriptive in nature; that is, they are specific design criteria that, when applied, will minimize the energy demand of a structure. ASHRAE/IES 90.1-1989, Energy Efficient Design of New Buildings Except New Low-Rise Residential Buildings [1][78] is a typical example of these standards. They do not usually deal with the performance of the various energy using components. There is a trend developing toward the use of performance standards for various classes of buildings. Such standards or guidelines will probably set limits on annual energy consumption based upon some common unit, such as per square foot of floor area. The use of performance standards should allow designers greater latitude in design and should, in the final analysis, result in buildings operated in a more energy-efficient manner.

The use of any of the computer programs that evaluate energy consumption is often essential in order for the designer to properly assess the various alternatives available.

11.4 Definitions

air conditioning. The process of treating air so as to simultaneously control temperature, humidity, and distribution to the conditioned space.

---

[77] EPRI publications are available from the Electric Power Research Institute, 3412 Hillview Avenue, P.O. Box 10412, Palo Alto, CA 94303.

[78] ASHRAE publications are available from the American Society of Heating, Refrigerating, and Air-Conditioning Engineers, 1791 Tullie Circle, N.E., Atlanta, GA 30329. IES publications are available from the Illuminating Engineering Society, 345 East 47th Street, New York, NY 10017.

air ventilation. The amount of supply air required to maintain the desired quality of air within a designated space.

ambient air. The air surrounding or occupying a space or object.

British thermal unit (Btu). The quantity of heat required to raise one pound of water 1 °F.

calorie. The quantity of heat required to raise one gram of water 1 °C.

coefficient of performance (heat pump). Ratio of heating effect produced to the energy supplied.

control. Any device used for regulation of a system or component.

degree day. A unit based upon temperature difference and time, which is used for estimating fuel consumption and for specifying nominal heating loads of buildings during the heating season. Degree days = number of degrees (°F) that the mean temperature is below 65 °F $\times$ days.

dehumidification. Condensation of water vapor from the air by cooling below the dew point, or removal of water vapor from air by physical or chemical means.

heat, specific. The ratio of the quantity of heat required to raise the temperature of a given mass of a substance 1° to the heat required to raise the temperature of an equal amount of water by one degree.

heat capacity. The amount of heat necessary to raise the temperature of a given mass of a substance 1°—the mass multiplied by the specific heat.

heat pump. A refrigerating system employed to transfer heat into a space or substance. The condenser provides the heat, while the evaporator is arranged to pick up heat from the air, water, etc. By shifting the flow of air or other fluid, a heat pump system may also be used to cool a space.

heating system, radiant. A heating system in which the heat radiated from panels is effective in providing heating requirements. The term "radiant heating" includes panel *and* radiant heating.

heating unit, electric. A structure containing one or more heating elements, electrical terminals or leads, electric insulation and a frame or casing, all of which are assembled into one unit.

humidity. Water vapor within a given space.

humidity, relative. The ratio of the mole fraction of water vapor that is present in the air to the mole fraction of water vapor that is present in saturated air.

infiltration. Leakage of outside air into a building.

isothermal. The process that occurs at a constant temperature.

lag. The delay in action of a sensing element of a control element.

load, estimated maximum. The calculated maximum heat transfer that a heating or cooling system will be called upon to provide.

radiator. A heating unit that provides heat transfer to objects within a visible range by radiation and by conduction to the surrounding air, which is circulated by natural convection.

recirculated air. Return air passed through the air conditioner before being supplied again to the conditioned space.

return air. Air returned from the conditioned space.

solar constant. The solar intensity incident on a surface that is oriented normal to the sun's rays and located outside the earth's atmosphere at a distance from the sun that is equal to the mean distance between the earth and the sun. The values for July and January are 415 and 445 Btu/hour $\times$ ft^2, respectively. The mean value is 430, and the sea level value in July, because of atmospheric attenuation, is close to (300 Btu/hour) $\times$ ft^2.

temperature, dew point. The temperature at which condensation of water vapor begins in a space. The dew point temperature is a function of pressure and humidity.

temperature, dry bulb. The temperature of a gas, or a mixture of gases, that is indicated by an accurate thermometer after correction for radiation.

temperature, effective. An arbitrary index that combines, into a single value, the effects of temperature, humidity, and air movement on the sensation of hot or cold felt by the human body. The numerical value is that of the temperature of still, saturated air that would induce an identical sensation.

temperature, wet bulb. The temperature at which liquid or solid water, by evaporating into the air, can bring the air into saturation adiabatically at the same temperature.

therm. A quantity of heat that is equal to 100 000 Btu.

thermal conductivity. The time rate of heat flow through a unit area of homogeneous substance under steady conditions when a unit temperature gradient is maintained in the direction that is normal to the area.

thermal diffusivity. Thermal conductivity divided by the product of density and specific heat.

thermal insulation. A material having a high resistance to heat flow and used to retard the flow of heat to the outside.

thermal resistivity. The reciprocal of thermal conductivity.

thermal transmittance (U factor). The time rate of heat flow per unit temperature difference.

thermostat. A device that responds to temperature and, directly or indirectly, controls temperature in a building.

ton of refrigeration. Is equal to 12 000 Btu/hour.

velocity, room air. The average sustained residual air velocity in the occupied area in the conditioned space.

11.5 References. The following references shall be used in conjunction with this chapter:

[1] ASHRAE/IES 90.1-1989, Energy Efficient Design of New Buildings Except New Low-Rise Residential Buildings.

[2] NEMA HE1-1980 (Reaff. 1986), Manual for Calculating Heat Loss and Heat Gain for Electric Comfort Conditioning.

[3] Ambrose, E. R. "Heat Pumps and Electric Heating," New York: John Wiley and Sons, 1969.

[4] "Commercial Cool Storage (Reduced Cooling Costs with Offpeak Electricity)," Electric Power Research Institute (EPRI) Management and Utilization Division, Palo Alto, CA.

[5] "Energy Conservation in Public Buildings," National Institute for Science and Technology (NIST) Committee Report, Jul. 1972.

[6] "Technical Options for Energy Conservation in Public Buildings," National Institute for Science and Technology (NIST) Committee Report.

[7] Roots, W. K. "Fundamentals of Temperature Control," New York: Academic Press, 1969.

[8] Tamblyn, R. T. "Thermal Storage: Will It Be Ice or Water?," *Heating/Piping/Air Conditioning*, Aug. 1988.

Chapter 12
Transportation

12.1 General Discussion

12.1.1 Use. Vertical transportation makes multiple-story structures feasible. Office buildings, multiple residences, hospitals, hotels, department stores, and many other types of buildings should be provided with elevators and escalators. Elevators are utilized in almost every building of two or more stories, other than private residences, to provide access for the handicapped. The diversity of transportation needs requires a variety of equipment to meet various types of service demands. The degree of refinement in control and operation may affect the amount of energy used. However, elevator equipment or systems should never be selected solely on the basis of reduced energy consumption simply in the interests of economy. Any such saving is insignificant compared with total elevator operating costs, while a possible deficiency in elevator service may impair the investment return on the entire building.

12.1.2 Design Factors. There is no simple formula for determining the characteristics of an elevator plant. The number of variables involved does not lend itself to some rule-of-thumb assessment that could be used as a measure by those unfamiliar with elevator application and practice. Proper elevator design requires a complete knowledge of not only elevator machinery, controls, and operations but of mathematical probabilities and human behavior as well. Only the experienced elevator engineer is prepared to give advice about the proper number, size, speed, control, and operation of the elevators that will be required to handle the traffic in a particular building adequately. The engineer is in a position to predict these requirements accurately, even if the building is not as yet erected, because of the wealth of available data from many different types of buildings in many different localities, and through the use of computer simulation techniques. Moreover, the engineer should be prepared to judge whether the elevators have the necessary characteristics for fast and efficient service, smoothness of acceleration and deceleration, accuracy of stopping, and overall ease of operation for building personnel and the visiting public. The reliability of the equipment and minimal maintenance are also important design factors. ASME A17.1-1990, Safety Codes for Elevators and Escalators [10],[79] ANSI/ASME A17.2-1988, Inspectors' Manual for Elevators

---

[79]The numbers in brackets correspond to those in the references at the end of each chapter. ASME publications are available from the American Society of Mechanical Engineers, 22 Law Drive, Fairfield, NJ 07007.

and Escalators [2],[80] and ASME A17.3-1990, Safety Code for Existing Elevators and Escalators [11] cover the design, construction, installation, operation, testing, maintenance, alteration, and repair of elevators, escalators, moving walks, dumbwaiters, material lifts, and dumbwaiters with automatic transfer devices, stairway chair lifts, and wheelchair lifts. An invaluable aid to the users of the ASME A17.1-1990 safety codes is ASME A17.1-1990 [10], which provides text, charts, diagrams, and photographs that explain and augment the code requirements. Because these standards have been prepared with due consideration paid to past accident experience, and are based on sound engineering principles, they are widely accepted by state and municipal jurisdictions. At a minimum, installations should conform to these requirements in order to ensure the safety of those who use, maintain, and inspect such equipment. The designer should be aware of local modifications to these standards, e.g., those that involve elevator control during a fire.

12.1.3 Consultations. The elevator manufacturer and the elevator consulting engineer are the logical consultants on matters pertaining to vertical transportation. Their expertise should be obtained in the preliminary design stage. When the building design has already progressed to the point where major structural or equipment changes can no longer be made, then elevator service may suffer and the building may have insufficient vertical transportation, which will result in detrimental building services and lower the building's marketability.

12.1.4 Efficiency and Economy. Transportation efficiency and economical operation are primarily a function of the equipment and systems. Automation has replaced the human element in elevator system control and operation. Acceleration, deceleration, floor leveling, and door operation are automatic. Today, through the development of automatic group supervisory control systems and adequate protection devices, large groups of high-speed elevators are operated entirely by passengers with complete confidence and safety.

12.1.5 Responsibilities. Regardless of the complexities of some elevator systems and the need for expert advice in applying elevator equipment, building owners are certainly entitled to know what they are buying and why they are buying it. Furthermore, the owner has the prerogative of selection, since it is the owner's money that is being spent. In order for the selection to be a wise one, this chapter will provide a better understanding of elevator systems, and will briefly review the components that make up a system, with particular emphasis on the advantages of each to be derived and power characteristics to be considered.

The economic success of a commercial building, particularly a busy office building, may very well be determined by the quality of the services rendered to building occupants. Elevators play a major role in these building services. It is the architect or elevator consulting engineer and, finally, the building owner and building manager who bear the responsibility of providing an adequate elevator plant and elevator service.

---

[80] ANSI publications are available from the Sales Department of the American National Standards Institute, 11 West 42nd Street, 13th Floor, New York, NY 10036.

12.2 Types of Transportation

12.2.1 Electric Elevators. All modern electric elevators are of the traction type, in which suspension ropes pass from the elevator car to a counterweight around a grooved driving sheave on a hoisting machine. The motion of the elevator and the counterweight in either direction depends on the friction created between the suspension ropes and the grooved sheave surfaces by the suspended weights. Elevators are inherently safe. The loss of rope tension results in the loss of traction. It is the loss of traction that prevents either the car or counterweight from being drawn into the overhead structure. Since winding drum machines do not prevent the loss of traction, ASME A17.1-1990 [10] prohibits the installation of drum-type machines except for limited travel, slow-speed freight elevators without counterweights. Traction machines consist of two types: geared machines with suitable gearing (worm gear or helical gear) between the hoisting motor and the driving sheave, and gearless machines in which the sheave is mounted directly on the motor shaft. Geared machine elevators are not widely used for car speeds that are above 450 feet/minute. It is quite common, however, to use gearless machines above 400 feet/minute. Gearless machine elevators have been installed for car speeds as high as 2000 feet/minute. Although design characteristics tend to classify the geared type as a lower car speed machine and the gearless type as a higher car speed machine, it is inevitable that there will be speed ranges in which either machine type is applicable. In making a selection, it should be kept in mind that, although the geared machine is somewhat less expensive, the gearless machine provides improved efficiency, smoothness and quietness of operation, and a considerably longer life.

Typically, traction equipment is located in a machine room directly above the elevator hoistway. Elevators are available with the traction equipment located below or at the side of the hoistway. These arrangements are typically called a "basement machine." Basement traction types of elevators may be used in buildings in which owners are considering the addition of future floors; in a retrofit of an existing building; or when it is desirable to eliminate a penthouse projection above the roof line. Bear in mind, however, that the overhead loadings on the building structure that are imposed by elevator installation are doubled with basement applications.

12.2.2 Hydraulic Elevators. Typically, hydraulic elevators have direct plunger driving machines, in which a plunger or piston, which is attached directly to the car, operates in a cylinder under hydraulic pressure.

The typical cylinder should project below ground as far as the elevator travels above. The costs entailed in providing this arrangement makes hydraulic elevators uneconomical for buildings above a certain height and may require supplementary plunger support.

In recent years, the holeless hydraulic elevator has been introduced for two- or three-stop applications. The cylinder is mounted above the pit floor, thus eliminating the need for a costly plumb hole below the pit level.

Hydraulic elevators are particularly suited to low-rise buildings in which overhead space and building load are limited, and in which passenger travel time is not a prime consideration. Hydraulic elevators are powered by a motor driven pump, which feeds the cylinder from a supply tank. The car speeds of hydraulic elevators are usually limited to 200 feet/minute.

ASME A17.1-1990 safety codes [10] recently approved the use of roped hydraulic elevators. This elevator would be provided with a vertical plunger or piston with a minimum of two ropes attached to same. The ropes would travel from the piston over a sheave at the top of the hoistway to the car. The roping ratio permitted by ASME A17.1-1990 [10] will not exceed 2:1.

The need for increased motor size and the resulting increased current demand should be taken into consideration when hydraulic elevators are to be located in a heavy usage environment.

12.2.3 Dumbwaiters. Electric dumbwaiters are used to transport material only. The dumbwaiter requirements of ASME A17.1-1990 [10] restrict car enclosure size.

The machine may be either the traction or drum type. The hoist ropes of the traction type extend from the car around the traction drive sheave on the machine to the counterweight. The hoist ropes of a drum-type machine extend from the car to the machine, where it is wound on a drum. A counterweight is not normally used for the drum type. Drum-type machine installations are limited to low-rise buildings because of the small amount of hoist rope that can be wound on the drum. Dumbwaiters can also be of the direct plunger or roped hydraulic type. See 12.2.2 for a description of these driving types.

Since dumbwaiters are strictly for material handling, they do not have buttons in the car enclosure and are always operated externally. Dumbwaiters that cover only two floors usually come with a "call and send" control system in which the car can be called to or sent from either landing. With multiple-floor dumbwaiters, a multiple-button control system is used in which floor control stations contain a call button and a send button for each landing served in order to send the car to the other landings. For intensive dumbwaiter service, more sophisticated control systems are employed, such as central station dispatching control systems.

12.2.4 Escalators and Moving Walks. Escalators and moving walks are applicable when it is desirable to move people continuously (see Figs 146 and 147). Department stores, shopping malls, office buildings, rail and air terminals, parking facilities, subways, and sports arenas are particularly suited for escalators or moving walks, or both. Escalators are generally operated at an angle of incline of 30° and are furnished in widths of 24, 32, and 40 inches and at speeds of 90–120 feet/minute. The two most common sizes are the 32 inch and 40 inch escalators. Although handling capacities for the 32 inch and 40 inch escalators are rated at 5000–6700 persons/hour and 8000–10 700 persons/hour, respectively; at the 90 feet/minute rate, actual observations indicate that the rated handling capacity is realistically 2500–3350 persons/hour and 4000–5350 persons/hour, respectively (see Reference [18]).

A departure from the conventional escalator is the modular escalator concept, in which the drive unit is contained in a truss, as opposed to in a drive machine, at the top end. One drive unit is required for approximately each 20 foot increment of vertical rise. This results in a reduction of the motor horsepower rating as well as in space.

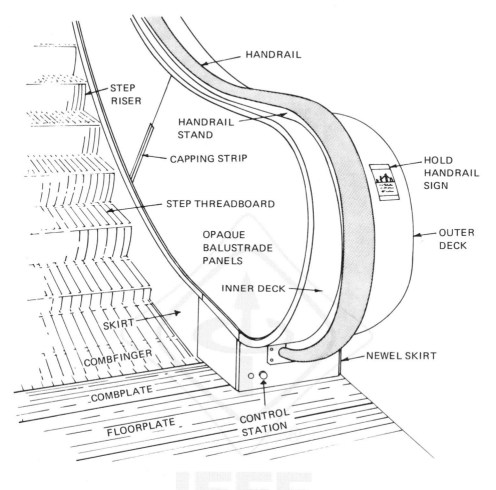

Fig 146
Opaque Balustrade Escalator

The minimum width of the exposed treadway of a moving walk is 22 inches. Maximum width is dependent on treadway slope and speed. Maximum width is also governed by the applicable code and is as shown in Table 78.

The maximum speed of the moving walk treadway is dependent on the maximum treadway slope at points on the treadway. This speed shall not exceed the lesser of the values determined in Table 79.

Moving walk capacities vary with the speed, width, and angle of the incline.

Access and egress areas that are near escalator and moving walk installations should be considered in order to avoid hazardous congestion. The entry and exit shall be provided with a safety zone kept clear of all obstacles. The width of the zone shall not be less than the width between the center lines of the handrails plus

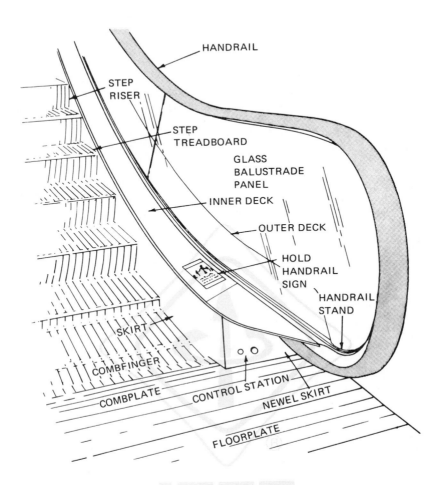

Fig 147
Transparent Balustrade Escalator

8 inches. The length of the zone, measured from the end of the newel, shall not be less than twice the distance between the center lines of the handrails. These dimensions are absolute minimums, and every consideration should be given to traffic patterns in order to provide adequate access and egress.

When the escalator or moving walk is installed outdoors, ASME A17.1-1990 safety codes [10] require that "it shall be constructed such that exposure to the weather will not interfere with normal operation."

12.2.5 Material Lifts. In addition to elevators and dumbwaiters, many vertical material handling systems are available for vertical transportation. Elevator related systems include elevators and dumbwaiters that automatically load and unload cars. When the code limitation of 9 ft^2 of platform area is exceeded, ASME A17.1-1990 [10] requires the material lift to be classified as an elevator rather than

Table 78
Moving Walk Capacities

Maximum Treadway Slope at Any Point	Maximum Moving Walk Treadway Width in Inches		
	90 ft./min. Maximum Treadway Speed	Above 90–140 ft./min. Treadway Speed	Above 140–180 ft./min. Treadway Speed
0°–4°	Unrestricted	60	40
Above 4°–8°	40	40	40
Above 8°–12°	40	40	Not permitted

GENERAL NOTE:
1° = 1.75 E –02 radian
1 in. = 25.4 mm

Table 79
Maximum Treadway Slope and Speed

Maximum Treadway Slope at Any Point on Treadway	Maximum Treadway Speed (ft./min.)
0°–8°	180
Above 8°–12°	140

GENERAL NOTE:
1° = 1.76 E –02 radian
1 ft./min. = 5.08 E –03 m/s

as a dumbwaiter. Elevator sized material lifts can also be equipped for automatic loading and unloading of carts and can be adapted to automatically handle carts or containers on overhead powered conveyor tracks, floor wire or optically guided self-propelled carts, and floor or chain vehicles. Elevator-type material lifts are often equipped with power operated hoistway doors that are automatically opened to receive or discharge the load and automatically closed when the car is ready to travel.

Conveyor related systems include selective vertical conveyors, which are continuous chain systems operating the full height of the building. These systems automatically load tote boxes on the up traveling chain and discharge them from the down traveling chain. Such systems can interface with horizontally traveling belts or roller conveyors and are completely automated.

Another conveyor-related system utilizes self-propelled electric carriers of tote box size. The box runs in a track on a powered chassis that receives electric current from a trolley in the track. Switching is provided to direct the carrier to any location. Horizontal travel is achieved by the powered wheels on the carrier, and vertical travel is achieved by the engagement of a pinion in the carrier into a rack gear in the track.

Elaborate automated storage and vertical conveying systems have been developed to handle anything from finished automobiles to aircraft freight containers. This is accomplished by using the combination of a vertical-elevator-type carriage with a material transfer device operating in a rolling tower, which indexes both horizontally and vertically to store or retrieve the article. Vertical travel energy consumption and power requirements are related to the system used, whether it is the counterweighted-elevator-type, or the hydraulic- or winding-drum-type hoisting machine. Horizontal travel requirements are similar to those of a railroad, rubber-tired vehicle, chain, or belt, depending upon the system used. It all adds up to the power needed to start, accelerate, run, decelerate, level or position, and stop, including the possible consequences of regeneration when the load is decelerated and stopped.

Conveyor related systems should conform to ANSI/ASME B20.1-1990, Safety Standards for Conveyors and Related Equipment [5]. It is important to maintain fire integrity when designing and installing conveyor equipment.

12.2.6 Manlifts. Manlifts are a vertical passenger conveyance consisting of an endless belt with attached passenger steps and grasps. These lifts are frequently used in commercial valet-type parking garages in large cities and in other buildings where it is desired to have high-speed employee travel to remote or high-vertical areas. Manlifts are restricted to use by qualified persons only, since there are certain hazards associated with their operation. Manlifts are not included in the scope of ASME A17.1-1990 [10]; but they are covered by ANSI/ASME A90.1-1985, Belt Manlifts [4].

12.2.7 Pneumatic Tubes. Early pneumatic message tube systems required a manual sorting area at which a person received the dispatched message carriers and then manually inserted them into their destination tubes. Each point of origin and dispatch was connected by two tubes, all of which collected at the sorting area. Next came the carrier readers that used single-tube loop systems with a limited number of stations. System loops were joined at automatic electromechanical sorting tables. These systems were cumbersome but did allow for automation.

Today, systems are computerized, allowing for a single-tube system throughout with pushbutton dispatching from any station to any other. The computer aligns the tube network to allow direct passage of the carrier from one station to another. In addition, today's systems can accommodate tube sizes up to 8 inches in diameter and 4 inches by 12 inches oval for special applications.

12.3 Elevator Control, Motors, and Motor Generators. In general, the principal functions of elevator control are to connect the source of power to the elevator hoisting motor; to determine the direction of motor rotation and, consequently, the direction of car movement; and to dictate the acceleration, deceleration, leveling, and stopping of the elevator at a floor. Such control also provides the necessary overcurrent protection for the motor, and includes provisions to handle negative or overhauling loads satisfactorily when the hoisting motor is actually being driven by the elevator. The term "control" should not be confused with the term "operation." "Control" is the system for regulating the elevator as indicated above. "Operation" is the method of actuating the control in some predetermined manner so that the elevator, or elevators, respond in accordance with the pattern most suitable to the particular traffic demand.

12.3.1 Control Systems. There are six basic electric elevator control systems
(1) DC rheostatic control
(2) AC resistance control
(3) AC servo control
(4) Generator field control
(5) DC direct drive control
(6) Variable frequency control

These systems are shown in simple diagrams in Fig 148.

12.3.2 DC Rheostatic Control. Today, dc rheostatic control systems have little application because of the cost associated with its control components and its poor performance level when compared to other less costly and better performing systems that are now available. This system was popular many years ago when the line power delivered to the elevator was direct-current.

12.3.3 AC Resistance Control. Resistance control is used with geared machines for low-speed elevators up to 150 feet/minute. AC controls can be one- or two-speed, with a fast-to-slow speed ratio of up to 1:6. AC resistance control traction elevators should not be used when maximum smoothness of performance and close accuracy of stopping at a landing are required.

The rheostatic control system incorporates a one- or two-speed ac squirrel-cage induction hoisting motor and a type of control that provides, in addition to the principal control functions, reduced voltage starting through the medium of starting resistances or reactances. Wound-rotor motors are rarely used because of their more elaborate construction and because they require more complicated control equipment than the squirrel-cage type.

12.3.4 AC Servo. Presently, SCR control of an ac motor is generally used with a geared machine and is usually restricted to 450 feet/minute.

The control system is a static control, closed-loop system that uses tachometer feedback to control an ac squirrel-cage induction hoisting motor. The ac motor

Fig 148
Basic Electric Elevator Speed Control Systems

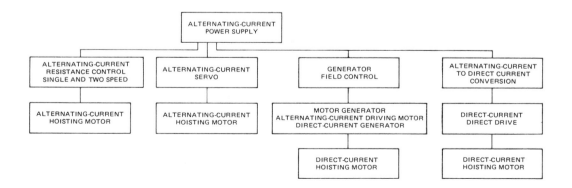

is controlled during acceleration, running, and deceleration by comparing the tachometer feedback signal to a speed reference signal that provides for smooth operation under all load conditions. This control system provides accurate stopping at landings. Also, this type of ac control is generally a nonregenerative type of control.

12.3.5 Generator Field Control. The generator field control system may be used for either geared or gearless machines without restriction as to car speed. The system incorporates a continuous running, constant speed motor-generator set for each elevator to supply variable voltage dc to the dc shunt-wound hoisting motor. Control of the hoisting motor is obtained largely through control of generator field excitation. This gives an inherently smooth performance since the inductance of the generator field reduces the motor armature current fluctuations and, consequently, eliminates noticeable shocks during the various steps of acceleration and deceleration. In addition to providing smoother and more refined control, it permits the use of smaller and more economical control components because small field currents, rather than large motor currents, are being handled. It also generates power back to the line during overhauling conditions.

12.3.6 DC Direct Drive. Solid-state technology is not only replacing the traditional relay controls; but SCR drives are also replacing the motor-generator sets on geared and gearless installations.

With SCR drives, energy generated by the motor during normal decelerating and regenerative braking conditions is fed back to the line. However, this is not necessarily an efficient use of power since the power factor may be poor. Static power conversion, which eliminates the motor-generator set and its inherent losses, is a more efficient drive system than the generator field control system. Motor armature control, using all static, solid-state modules, provides a positive, smooth response to all speeds and loads within the rating of the drive.

Solid-state innovation does not detract from safety; however, ASME A17.1-1990 [10] imposes different requirements on static systems.

12.3.7 Variable Frequency Control. The latest control is the variable voltage/variable frequency approach. It consists of a converter/inverter system that rectifies just the three-phase ac supply to dc power. A width modulation technique is used to chop the dc power into variable voltage and variable frequency. This method may produce harmonics that should be taken into consideration in the design of the supply transformer as well as in the design of other equipment (i.e., induction motors) on the secondary of the transformer.

Variable frequency power supply minimizes motor slip and flattens starting current peak. At full power, it places up to a 50% lower load on the power source in comparison to other drives. This relates to a reduction in supply wiring fusing, less need for machine room environmental control and cooling, scaled down emergency power systems, plus an improved power factor. The 0.95 power factor relates to a 35% improvement over other types of drives.

12.3.8 Hoisting Motors. Elevator hoisting motors, in general, and ac motors, in particular, are special-purpose motors and should be designed to provide high starting torque with comparatively low starting current. These motors should have the ability to withstand, and perform well during, repeated starting and stopping.

The motors are rated on the basis of the horsepower required when the car is carrying a full load in the up direction at full speed. As this occurs infrequently in actual operation, the time/temperature rating is intermittent (instead of continuous) on a 15 minute, 30 minute, or 60 minute basis for a given temperature rise in °C, which is based on the type of insulation used. The intermittent type of service permits the use of smaller motors that consume less energy. AC motors should function at a low noise level to avoid possible complaints by building occupants. DC motors should satisfactorily commutate the peak currents that are experienced during acceleration and deceleration under full-load conditions.

12.3.9 Rating of Motor-Generator Sets. Motor-generator sets should have an adequate capacity to handle the peak power demands of the hoisting motor with satisfactory regulation. AC motors should have low free-run losses, since they remain running while the elevator is stopped. They are rated on a continuous basis, usually lower than the hoisting motors with which they are associated, but never less than the rms hoisting motor horsepower for the elevator duty cycle. The time/temperature rating is the temperature rise, in °C, for a motor continuously running at nameplate rating, and is based on the type of insulation used. Motor-generator sets and elevator hoisting motors should be evenly matched to obtain the best results from each.

12.3.10 Starting Motor-Generator Sets. Motor-generator sets are usually started through resistance or wye-delta (open transition) switching. On three-phase ac power supplies, the wye-delta arrangement connects the driving motor in wye during no-load starting and in delta during running. When the traffic demand is intermittent, and especially with automatic elevator operation, the motor-generator set is shut down to reduce power consumption after the elevator has been idle for a predetermined time.

12.3.11 Control Design Fundamentals. There are certain fundamentals of control design that invariably produce more acceptable, more reliable, and more economical equipment. For example, dc contactors, switches, and relays are quieter and more reliable in operation than similar equipment of the ac type and may be timed by capacitors instead of dashpots. The necessary dc power for such equipment and elevator machine brakes can readily be obtained from solid-state rectifiers or supplementary excitation generator units. Electronic solid-state devices may be desirable, since they not only reduce the operating currents involved but can usually accomplish the same results more economically than contactors, switches, or relays. In addition, the introduction of static control has provided a way to accomplish more sophisticated operations because of its programming capability.

12.3.12 Rectifiers for Power Conversion. In areas where buildings were originally supplied with dc power, motor-generator sets were frequently installed to allow existing dc elevator motors to be retained when the utility converted to ac. In some cases, solid-state rectifiers were used; but their size usually discouraged their acceptance. In either case, unless there is a positive load imposed on the conversion units that is sufficient to absorb the energy created by negative or overhauling elevator loads each time, supplementary means should be provided to dissipate the regenerated power. Otherwise, the elevators may overspeed.

12.4 Elevator Horsepower and Efficiency

12.4.1 Horsepower Calculation. The required hoisting motor horsepower is a function of the rated load for a particular elevator car, the speed of the car in the up direction with this load, and the motor-to-load mechanical efficiency. Since traction elevators are provided with counterweights equal to the dead weight of the car, plus about 40%–50% of the load rating (roughly the average load), the horsepower formula may be expressed as follows:

$$\text{hp} = \frac{C \cdot V \cdot k}{33\,000 \cdot e} \qquad\qquad (\text{Eq } 30)$$

where

C = Rated load of car (including car weight) in pounds.
V = Full-load speed of ascending car in feet/minute.
k = Percentage of rated load that is unbalanced by counterweight, usually about 60%.
e = Motor-to-load efficiency, usually 50%–70% for geared elevators and 75%–85% for gearless elevators.

12.4.2 Ratings. ASME A17.1-1990 [10] requires that passenger cars be rated in accordance with the net area of the car platform in square feet. This is required to minimize overloading beyond the capacity of the elevator equipment. It likewise rates freight elevators on a square foot of net platform basis, but distinguishes between three separate types of loading: general freight, motor vehicle, and industrial truck. Detailed information on dimensional data in English and metric units for a wide range of passenger, hospital, and freight elevator requirements for electric and hydraulic system applications is covered in Reference [18].

12.4.3 Variations. For practical applications, there is a wide variety of loads, car speeds, and resulting horsepowers. In general, it might be said that geared machine motors vary from 7.5 hp at 1200 or 1800 rev./min. to 100 hp at 500 rev./min. and gearless motors from 20 hp at 60 rev./min. to 150 hp at 130 rev./min. Motor generators can attain speeds up to 3600 rev./min.

12.4.4 Efficiency. The line-to-load efficiency, that is, the ratio of the useful work done in moving the load to the input energy from the supply, varies within considerable limits, depending on the load in the elevator car and whether the car is ascending or descending. It is very low under a balanced load condition, at which time the weight of the counterweight exactly balances the weight of the elevator car with the load contained in it. The current at this point is only attributable to the losses in the motor-generator set, line, and friction in the hoistway.

12.5 Elevator Energy Consumption and Heat Release

12.5.1 Energy Consumption. Elevator energy consumption is measured in kWh/car mile and is a function of the type of equipment used, the load contained in the car, and the number of stops made by the car in 1 mile of travel. Since the latter is a measure of the intensity of elevator service when the elevator is in almost continuous use throughout the normal operating period, the average energy consumption of a particular car in a particular class of building can be estimated. Office building elevators, for example, usually average 150 stops/car mile for local service, and 75 or less stops/car mile when express runs are involved. By calculating the energy consumed in making the corresponding number of average runs in a mile, and adding the energy consumed (with generator field control) during the time that the car is stopped, an approximation of the number of kWh/car mile is obtained. Balanced load conditions are usually assumed for the calculation. Typical office building passenger elevators may consume from 3–8 kWh/car mile, depending on the type of equipment used, the character of the building, and the service demand.

12.5.2 Heat Release. About 80%–90% of the energy taken, over a period of time, from the input power lines for the operation of elevators is dissipated in the elevator machine room in the form of heat. With gearless machines and generator field control, approximately two-thirds of the heat released is from the motor-generator set; the remaining one-third is from the hoisting motor and control equipment. Since performance adjusted for normal temperatures becomes unsatisfactory at excessively high temperatures, it is the responsibility of the architect, consulting engineer, or owner to provide the means, when necessary, to maintain the ambient temperature of the machine room between 55 °F (12.8 °C) and 100 °F (37.8 °C). Spill air from the building air-conditioning system, separate air-conditioning units, or machine room ventilating fans are often used for this purpose. The heat released by the elevators can readily be calculated from the energy consumption in kWh/car mile and the estimated number of car miles/hour, which is averaged on the basis of the number of working hours that the elevators remain in service. The heat released from solid-state motor drives is approximately 40% less than the heat emission of a motor-generator set. With the introduction of more sophisticated microprocessor operational systems, machine room temperatures become more critical. Temperatures should not exceed 90 °F (32.2 °C) in order to assure operational reliability. Air conditioning of these areas is essential.

12.6 Elevator Conductors and Diversity Factor

12.6.1 Conductor Size. Elevator conductors should be capable of carrying hoisting motor, motor-generator set driving motor, or adjustable speed drive system currents without overheating and, in addition, be of sufficient size to limit the voltage drop to not more than 3%. Conductor sizes are determined by the requirements of ANSI/NFPA 70-1990, National Electrical Code (NEC), Article 620 [6],[81] which states that the "motor load ampacity should be based on not less that 140% of the motor-generator set nameplate current for conductors supplying a single continuously rated motor." Conductors supplying two or more motors shall have an ampacity of not less than 125% of the nameplate rating of the highest rated motor in the group plus the sum of the nameplate current ratings of the remainder of the motors in the group. For adjustable speed drive systems in which the power drive transformer is integral to the power conversion equipment, the ampacity is based

---

[81]NFPA publications are available from Publications Sales, National Fire Protection Association, 1 Batterymarch Park, P.O. Box 9101, Quincy, MA 02269-9101.

on the nameplate rating of the converter. When the transformer is not integral, the ampacity is based on the sum of the individual nameplate loads.

NOTE: It is critical to apply the ambient temperature correction factors and to consider voltage drop when selecting conductor insulations and sizes.

Controllers are required by the NEC, Section 430-8 [6] to have a current or horsepower rating. The ampacity of conductors feeding adjustable speed drives is determined in accordance with the NEC, Section 430-2 [6]. Elevator ac and dc adjustable speed drives may use a transformer to match the incoming ac line voltage with the required hoist motor voltage. This transformer may be located in the controller cabinet or outside of the cabinet in a separate enclosure.

To establish a separate location of the transformer, the NEC [6] requires that the ampacity of the conductor feeding the transformer be based on its nameplate rating. Because of the intermittent elevator duty cycle service, a smaller kVA transformer rating is often used, which indicates a current rating less than that actually required by the power conversion equipment. This may result in a lower conductor ampacity than needed, which will then result in undersized feeders.

This proposal would require that, for separately located power transformers, the ampacity of the conductors supplying the transformer be based on the rated input current of the drive reflected to the primary side of the transformer if this current is larger than the transformer rated primary current.

Examples

(1) Current rating of an elevator adjustable speed control system (ac-to-dc SCR four-quadrant, six-pulse, dual converter, three-phase)

$$I_{ac} = \sqrt{\frac{2}{3}} \frac{V_s}{V_p} I_{dc} + I_c$$

where

$$I_{dc} = \sqrt{I^2 + I_{ripple}^2}$$

(2) In the case of an ac power converter in which the ac input current is directly proportional to the current in the ac motor.

$$I_{ac} = \frac{V_s}{V_p} \cdot I_{motor}$$

Definitions for Examples

I = Elevator hoist motor dc full-load current.
I_{ripple}^2 = Ripple current, for a filtered SCR output; the ripple is generally neglected.
I_{dc} = Effective elevator hoist motor dc full-load current.
I_c = Other controller load currents.
V_s = Transformer secondary voltage rating.
V_p = Transformer primary voltage rating.

The elevator industry has been using various demand factors that are provided by each manufacturer; often the values differ widely for the same elevator system. The elevator industry standard for demand factors is as follows:

Assume

(1) All elevators have similar service.
(2) A 50% duty cycle

Method of Calculation

(1) $I_{max} = 1.25\, I_\ell + \sum\limits_{N=1}^{N-1} I_n$

(2) $I_{min} = \sqrt{(1.25\, I_\ell)^2 + \sum\limits_{N=1}^{N-1} (I_n)^2}$

(3) $DF = \dfrac{I_{min} + F(I_{max} - I_{min})}{I_{max}}$

(4) $F = \dfrac{1}{N+1} + 0.5$

where

N = Total number of elevators.
$N - 1$ = Number of remaining elevators (n).
I_ℓ = Largest elevator full-load running current.
I_n = Current for each elevator (n).
I_{min} = Theoretical minimum A is the square root of the sum of the squares of all the individual rated A.
I_{max} = Theoretical maximum A is the sum of all the individual rated A as stated in the NEC, Sections 620-13(b) and 430-24 [6].
DF = Demand factor.
F = Percentage of a number of elevators running at the same time out of "N" total elevators.

(a) F is defined as X out of Y chances that all cars will start simultaneously in the same direction.
(b) Assume that the minimum F is 0.5. That is, at any time
  (i) All cars are starting together in the same direction.
  (ii) No cars are starting together.
  (iii) Some cars are starting together.
  In (i), (ii), and (iii), the "average" condition would be 0.5.
(c) By definition, the probability of one car is 1.
(d) $F = \dfrac{(1)}{(N+1)} + 0.5$ will satisfy conditions (a), (b), and (c).

Example

Determine the demand factor for a six-car elevator bank where the full-load running current of the largest elevator is 100 A and, for the remaining elevators, 80 A each.

$$F = \frac{1}{6+1} + 0.5 = 0.643$$

$$I_{min} = \sqrt{(1.25 \times 100)^2 + 5(80)^2} = 218 \text{ A}$$

$$I_{max} = 1.25 \times 100 + 5(80) = 525 \text{ A}$$

$$DF = \frac{218 + 0.643\,(525 - 218)}{525}$$

$$= 0.79$$

12.6.2 Current Ratios. Typical ratios of starting or accelerating current to running current are 2–2.6 for elevators with generator field control, 1.6 or less for a low-speed, heavy-duty freight elevator with generator field control, and 2–3 for ac rheostatic control with reduced voltage starting. The power factor of the elevator starting peak is better than 90% with generator field control and usually more than 75% with rheostatic control. In busy buildings, passenger elevators may be in motion as much as 60% or more of the time they are in regular service. The time duty for other classes of elevator service ranges down to 25% or even less.

12.6.3 Diversity of Operation. When two or more elevators operate from the same set of feeders, it is not probable that they will have simultaneous current peaks very frequently. Therefore, advantage may be taken of the diversity of elevator operation in calculating the size of the feeder. For a group of similar cars, each with an rms current of I, the line rms current will fall somewhere between nI, when the peaks coincide, and $\sqrt{n}\,I$, when the peaks are in the most random pattern. The diversity factor is the ratio of the most probable current to nI and is a function not only of the number of cars on one set of feeders, but also of the type and intensity of the elevator service. The demand factor represents the percentage of the overall current that may be used in designing the feeder. Typical diversity factors might be 0.97 for a group of two elevators and 0.890 for a group of four elevators under specific operating conditions. The NEC, Article 620-15 [6] now lists demand factors for elevators.

12.6.4 Disconnecting Means. Each elevator is provided with a fused service switch or circuit breaker located in the elevator machine room, which constitutes the means for connecting or disconnecting the particular elevator equipment from the power lines. The fused disconnect or circuit breaker is used to protect the branch-circuit feeders that are attached to the elevator equipment.

The location of the disconnecting means for elevators shall conform to the requirements of the NEC, Article 620-51 [6], which states that "a disconnecting means shall be located within sight of the power converter or motor starter for elevators without generator field control. When the disconnecting means is not within sight of the controller or hoist machine, an additional manually operated switch shall be provided adjacent to the remote equipment and connected in the control circuit to prevent starting."

On elevators with generator field control, the disconnecting means shall be located within sight of the motor starter for the drive motor of the motor-generator set. When the disconnecting means is not within sight of the hoist machine, the control panel, or the motor-generator set, an additional manually operated switch shall be installed adjacent to the remote equipment and connected in the control circuit to prevent starting."

12.7 Elevator Operation

12.7.1 Manual and Automatic Operations. Operation is the method of actuating the control. It determines the manner in which the car is operated and its response to different traffic demand conditions. Operation may be broadly classified into two categories: (1) manual, where the elevator is under the control of an authorized attendant or authorized group of employees, and (2) automatic, where the elevator responds automatically and is, therefore, suitable for general self-service operation. The modern type of operation is automatic. The term "automatic operation" means a system wherein the elevator starts in response to momentary actuation of devices at the landings, or in the car identified with the landings, or in response to an automatic starting mechanism, wherein the elevator stops automatically at landings for which stops have been registered.

12.7.2 Single Automatic Pushbutton Operation. This method of operation incorporates one button in the elevator car for each floor served and one button at each landing. Pressing a button automatically initiates and completes car travel to the corresponding floor without interference from other buttons that might be subsequently pushed during the trip.

Single automatic pushbutton operation was the first type of fully automatic elevator operation. Today, it has little application in commercial buildings because it responds to one call at a time and ignores persons at floors passed during the trip who may wish to travel in the same direction.

12.7.3 Selective Collective Operation. This method of operation extends the usefulness of the automatic elevator by allowing the car to respond to several different calls on the same trip. It incorporates one button in the car for each landing served and up and down buttons at landings. Car or landing calls are registered by momentary pressure on these buttons and remain registered until answered by the arrival of the elevator. Calls may be registered in any sequence and at any time. The car stops automatically in response to up landing calls on the up trip, to down landing calls on the down trip, and to furthermost calls irrespective of the direction of car travel. Stops are made in the order that the particular floors are reached.

12.7.4 Duplex Collective Operation. This method of operation is the automatic operation of two selective collective elevators from common landing buttons. The buttons in the cars operate each elevator individually. Landing buttons register calls that will stop either elevator, if available, for service in the proper direction. The system moves the elevators only on demand, prevents both elevators from starting for the same landing calls, and ensures the satisfactory utilization of each car in overall service requirements.

12.7.5 Group Automatic Operation. This method is the automatic operation of several automatic elevators coordinated by a group supervisory control system and applied to modern passenger elevator systems. It provides fast and efficient service and adjusts to changing traffic needs as dictated by building occupancy. Cars start automatically after the elevator doors close and stop automatically in response to calls registered by call buttons in each car or by up and down landing buttons, which are common to the group. A landing call stops the assigned car that is approaching. Fully loaded cars are programmed automatically to bypass landing calls. Doors open and close automatically and are provided with devices both to detect a person in the doorway and discourage passenger interference with the door closing operation. Automatic dispatching is provided in which cars selected at designated dispatching points close their doors and depart in a regulated manner.

With the introduction of microprocessor-type controls, elevator operation has become more sophisticated. In the case of some manufacturers' group systems, these controls monitor the position of each car in the hoistway, all car and corridor calls, and car status. Then the controls compute the time each car can address each hall call. The car with the shortest arrival time is selected to handle the call. This is done approximately five times each second, and assignments are constantly upgraded to provide maximum performance. The next generation of controls is imminent, which includes artificial intelligence (i.e., prediction based on past performance, monitoring corridor calling to project system loading).

12.8 Quality of Elevator Performance. High-quality passenger/elevator performance is essential for most commercial buildings. Not only should the riding public be transported rapidly with a minimum of delay; but the entire performance should be smooth and acceptable to passengers. Landing stops should be accurate to provide for passenger transfer safety.

12.8.1 Passenger Comfort. Fast service, although partly a function of the top speed of the elevator, is limited by the sensitivities of passengers. The passenger usually has no different sensation at a steady-state speed of 1800 feet/minute than at lesser speeds. However, the passenger is quite sensitive to acceleration, deceleration, and rates of change of acceleration, which should be consistent with passenger comfort, independent of the top speed of the elevator.

12.8.2 Leveling. Accurate registration of the elevator platform with the floor landing is important for passenger elevators not only as a timesaver, but also to prevent the tripping of passengers upon entering or leaving the car. It is particularly important on hospital elevators in which wheeled stretchers or beds with patients are carried. It is also essential on freight elevators if small wheel trucks are used for handling freight. Consequently, automatic leveling is furnished on almost all elevators. Today, it is the rare elevator that does not have some means to automatically control floor positioning.

Leveling systems include one-way automatic leveling, which is corrected only when the car tends to fall short of the floor landing, and two-way automatic leveling, which provides correction for the tendency of the elevator to either fall short of or overshoot the floor landing. If the type of leveling includes a maintaining feature, it will not only level during the initial stop, but also while the car is at a floor and rope stretch causes the car position to change due to loading and unloading.

12.8.3 Handicap Requirements. In order to comply with ANSI A117.1-1986 [1], which covers the specifications for making buildings and facilities accessible to, and usable by, physically handicapped people, elevators should have accessible call buttons, special signals, tactile floor indications, minimum door operating times and platform sizes, and a myriad of other features. Comprehensive coverage of elevator requirements can be found in ANSI A117.1-1986 [1] and in Reference [13].

12.9 Elevator Doors and Automatic Door Operation

12.9.1 Interlocks. Corridors opening into an elevator hoistway are required to be protected by hoistway doors that are interlocked with the elevator control so that the car cannot leave a landing unless the doors are closed and mechanically locked. The car itself must have a door or gate that also requires closure before the elevator can move (see ASME A17.1-1990 [10]).

12.9.2 Power Operated Doors. Passenger elevator hoistway doors and car doors are almost always power operated. The opening operation is usually automatic and may be initiated in advance of car arrival at each floor to permit entrance or exit of passengers by the time the elevator has stopped level with the landing. The closing operation is automatic on self-service elevators and is usually controlled by a timing device. An arrangement of multiple light beams, electronic detectors, or safety edges provides door closing protection that is required by ASME A17.1-1990 [10], which specifies the requirements for a reopening device as well as the maximum torque and kinetic energy for door closing.

Power operation of freight elevator hoistway doors and car doors or gates is desirable, both from the standpoint of convenience and saving time.

12.9.3 Passenger Elevator Doors. Passenger elevator doors in commercial buildings are usually of the horizontal sliding or of the side opening or center opening type. The latter is preferred because passenger entrance or exit is accomplished more rapidly after the car comes to a stop.

12.9.4 Freight Elevator Doors. Freight elevator doors are usually vertical sliding and of the bi-parting variety. The top of the lower door panel is provided with a trucking sill, which is designed to carry the loads that will be encountered during the movement of freight onto or off of the car.

12.9.5 Fire Rating of Elevator Entrances. Elevator hoistway entrances usually have a 90 minute fire rating, which is tested by a recognized testing laboratory in accordance with ASTM E152-1981, Methods of Fire Test of Door Assemblies [12].[82] The actual fire rating is as required by the building code.

The interface of the hoistway door jamb and wall, whether it be masonry or gypsum board, is critical and should conform to the requirements outlined in ASME A17.1-1990 [10]. Model building codes (see References [13], [14], and [17]) specify where entrances are required to be labeled. The requirements for labeling

---

[82]ASTM publications are available from the Customer Service Department, American Society for Testing and Materials, 1916 Race Street, Philadelphia, PA 19103.

entrance components are specified in ASME A17.1-1990 [10]; usually, a label should be provided on the frame, panel(s), transom, and hardware components.

12.10 Group Supervisory Control. The typical commercial building, particularly the busy office building, has a "going to work" traffic peak in the morning, a "returning home" peak in the late afternoon, and also a peak at midday when people are going to or returning from lunch. These traffic peaks, and the in-between traffic as well, should be handled rapidly and with minimum waiting on the part of the public.

Group supervisory control provides the means for regulating each elevator bank and the equalization of service to the respective waiting passengers by automatically programming the response of the cars in order to match elevator service to the demands and direction of passenger traffic flow.

The supervisory control should be flexible enough to monitor the changing traffic patterns imposed on the elevator group and react to these changes with efficient service. Some traffic trends are predictable at certain times of the day; however, the system should also be able to monitor unpredicted changes. Each system should be tailored to the specific application. Very few buildings produce the same traffic flows, and efficient elevator operation should meet the needs of the building it serves.

12.11 Specification of Elevator Plant

12.11.1 Selection of an Elevator. The proper size, number, location, and configuration of a passenger elevator is determined by the vertical transportation demand of the building. Periods of peak traffic are the most critical. Elevatoring schemes are determined by reviewing the building's vital statistics. These include the amount and type of building population, the architectural arrangement (both horizontal and vertical), exterior transportation, geographic location, etc. After all these data are tabulated, the building's vertical transportation needs are formulated. Car number, sizes, speed, etc., are all critical factors that need to be determined. Sufficient elevators should be provided to move people swiftly to and from their destinations, even when some elevators are out of service. Size is also a critical factor when dealing with the movement of other than ambulatory people. It is extremely important to ensure that the elevator will transport hospital beds, hotel displays, museum paintings, etc., if this type of service is needed.

12.11.2 Car Capacity. Elevator cars are rated according to their maximum inside area as related to load and as established by ASME A17.1-1990 [10]. The capacity of an elevator is based on approximately 1.5 ft^2/person, whereas nominal passenger capacity is closer to 2.3–2.5 ft^2/person.

Passenger car ratings for commercial buildings range from 2000–5000 lbs., usually in 500 lb. increments. A rating of 3000 lbs. is the usual minimum recommended for office buildings, and 5000 lbs. is the usual minimum recommended for a vehicular service elevator in a hospital.

Cars for general passenger use need greater width than depth in order to facilitate the transfer of people at individual floors. Hospital cars require a greater depth than width because they are designed to carry beds, stretchers, etc.

Passenger elevators are often used for freight, especially in high-rise office buildings. They are often referred to as "service elevators" and must be designed in accordance with passenger loading requirements. They should also be reinforced to withstand the expected loading of hand or powered trucks. The size of freight elevators varies within wide limits due to the diversity of materials that may be carried. Standard ratings may be anywhere from 2500–20 000 lbs. Consideration should be given as to whether the freight to be carried is general in nature; loaded by industrial truck or whether the truck is just used for loading and unloading or carried on the elevator; or of a specialized nature, such as automobiles. Depending on the nature of the load and the loading method, consideration should be given to the building support structure in relation to the elevator guide rails.

12.11.3 Double-Deck Elevators and Sky Lobbies. In buildings with large floor areas and populations that require a large number of elevators or elevators that can carry capacities greater than 5000 lbs., several solutions to reduce the space required for elevators should be considered.

(1) Using double-deck elevators, that is, elevators having two decks or compartments mounted one above the other in a single hoistway arranged to serve two consecutive floors simultaneously.

To passengers, a double-deck elevator looks and operates just like a conventional single-deck elevator. The only obvious difference to prospective passengers is that, upon entering the building, they are directed to one deck for service to odd floors and the other deck for service to even floors. This arrangement permits simultaneous loading and unloading of both decks. Once the elevator leaves the main landing (upper and lower lobbies) and responds to a hall call, it is programmed to serve calls in the same way as a conventional elevator. This solution, when properly applied, provides equivalent elevator service in comparison to a conventional system, but requires less space and fewer elevators.

(2) Sky lobbies are best suited to structures that can be designed to have two or more separate zones or separate buildings, one above the other. The first floor of the lowest zone is designated as the main lobby. The first floor of the upper zone(s) is called a sky lobby. From a sky lobby, all floors in a zone are served by groups of local and express elevators in the same way as in a typical building. Elevator service from the main lobby to each sky lobby is provided by large express shuttle elevators, either single or double deck. Using this concept, passengers desiring service to the upper floors of an upper zone are required to take an express shuttle elevator to the sky lobby serving that zone and then to transfer to a local or express elevator to reach their ultimate destination. Using sky lobbies minimizes the space required for local and express elevators on the lower floors and should be studied for any building where three or more groups of elevators are required, or where multiple functions are expected to take place in a building, such as in a hotel above an office building. Conventional elevators may be used for both shuttle and local/express elevators. Double-deck elevators may be used as shuttle elevators, or as local or express elevators to obtain additional space savings.

12.11.4 Selection of Elevator Car Speed. Proper elevator speed for passenger cars is a function of the roundtrip distance to be traveled and the degree of service desired. In general, taller buildings require higher elevator speeds based upon having sufficient express runs to realize the full effect of the rated speed for a good portion of the distance.

The best speed for a passenger elevator is determined by considering the cost of a single elevator unit, the number of units required to satisfy traffic demand, and the service return expected from the investment.

Freight elevators usually range in speed from 50–200 feet/minute and are usually hydraulic in buildings that have up to five floors.

12.11.5 Quantity of Service. An elevator plant in a commercial building is expected to provide both quantity and quality of service. Quantity of service is measured by the passenger carrying capacity of the elevator group per unit of time and becomes important during peak traffic periods. Passengers measure quality of service in terms of how long is the wait for an elevator, and how long does it take to complete the ride?

The quantity demand on elevators is evaluated differently for different classes of buildings. In an office building, it is a function of the potential population who might use the elevators and the expected percentage of this population requiring service during the 5 minutes of heaviest passenger traffic. For a new building, the former is determined from the net rentable area and the anticipated population density in terms of square feet of area/person. The latter is an experience factor resulting from analysis of demand data from many existing buildings of the same type. When an existing building is modernized, in most instances, the demand data is known and does not have to be predicted.

12.11.6 Building Classes. Office buildings are classified as single-purpose (occupied by one or two large concerns) or diversified tenancy (occupied by a number of small and unrelated concerns). For either type, the normal population density may vary from 90–150 ft^2/person. Diversified tenancy buildings usually have a 5 minute elevator arrival rate of 11%–12% of the building's population. The 5 minute arrival rate for single-purpose buildings normally varies from 12%–20% of the building's population.

12.11.7 Quality of Service. Quality of service is most commonly evaluated by the time interval between succeeding cars, since this reflects passenger waiting time. Intervals of 20–30 seconds are considered excellent for office buildings in which time is usually at a premium during the arrival and departure peaks.

12.11.8 Number of Elevators Required. The number of elevators required in a building is basically determined by the relationship of the building quantity demand to the unit handling capacity, that is, the handling capacity of a single car of the type selected. The car's handling capacity is directly proportional to the number of people that can be accommodated in the car per trip and inversely proportional to the time required to make a roundtrip. The unit handling capacity of 5 minutes, the number of required cars, and the resulting time interval between succeeding cars may be expressed as follows:

$$HC = \frac{300 \cdot L}{RTT}$$

$$N = \frac{BD}{HC}$$

$$I = \frac{RTT}{N}$$

where

HC = Passengers carried by each in 5 minutes.
BD = Building's 5 minute critical passenger demand.
L = Passengers loaded per car each roundtrip.
RTT = Roundtrip time of one car in seconds.
N = Number of cars required.
I = Interval between cars in seconds.

12.11.9 Roundtrip Time. Roundtrip time is one of the most important factors in elevator plant calculations since it influences both unit handling capacity and interval. It represents the time between successive departures of elevators from a main lobby or the frequency with which elevators pass a particular point in the building. It is calculated on the basis of proven mathematical probabilities plus empirical data that are determined through experience.

12.11.10 Interval. Interval is the final check on the elevator calculation. If it is unsatisfactory, more elevators should be provided.

12.11.11 Grouping of Elevators in the Plant. In serving all or part of the floors in a building, one centralized group of elevators should be provided rather than a number of small and widely scattered groups that give parallel service. Elevators in small groups have inherently longer travel intervals and, consequently, longer passenger waiting times. Furthermore, it becomes difficult to equalize the demand for each elevator group because prospective passengers often prefer one or another group of elevators if they have a choice of groups serving their particular floor.

Elevators within a group should be located so as to minimize the walking distance required to register a call and board an elevator. An alcove arrangement is preferred for larger elevator groups with two cars opposite two, three opposite three, etc. No more than four elevators should ever be located in a single line unless such an arrangement is unavoidable.

For comprehensive information on elevator usage and traffic analysis for buildings, see Reference [16].

12.12 Regenerated Energy. When an electric traction elevator is operating with an overhauling load, such as a full-load down, energy is generated by the motor-generator set or the static SCR drive. AC drive motors, when operating under normal energy conditions, pump this regenerated energy back into the power company's feeders. If this regenerated energy is pumped back to a standby power generator and the generator does not have the capacity to absorb this regenerated energy, the generator or the elevator may overspeed.

This regenerated energy may be absorbed in the form of emergency lighting, appliances, or other loads that should always be connected when emergency power is being used, in the energy losses of the prime mover of the standby generator, or in dummy load resistances provided for this purpose. Typically, a motor-generator set can absorb approximately 20% of its rating in kW.

Regenerated power calculations for a variable voltage gearless machine are based on the following:

(1) Running full-load down = Approximately 40% of running full load up at a 40% negative power factor.
(2) Stopping full-load down = Approximately 50% of starting full load up at a 50% negative power factor.

A sample calculation follows:

(1) Assume that we have a 75 hp motor generator, 460 V, three-phase, 60 Hz, with the following characteristics:
  (i)   Nameplate marking   86 A
  (ii) Starting (full load)   186 A
  (iii) Accelerating (full load) 217 A
  (iv) Running (full load)   118 A
(2) Regenerated current stopping, full-load down is $0.5 \times 186 = 93$ A.
(3) Regenerated power stopping, full-load down is $93 \times 460 \times 1.73 \times 0.5$ (pF) = 37 kW.
(4) Therefore, a load equal to or greater than 37 kW should be provided to absorb the regenerated power.

12.13 Standby Power Operation of Elevators. Arrangements to permit one or more elevators to be operated when the main power fails are often provided and may be required by local codes. Standby power sources, such as gas or steam turbine alternators or diesel or gas engine alternators, are frequently used to supply power for elevators in high-rise buildings with offices, apartments, etc., and are used for practically all elevator installations in hospitals and health care facilities. A typical standby power control system will, upon the failure of the main power supply, permit one elevator at a time in each bank to descend to the main floor on power supplied by the standby system (see Fig 149).

12.13.1 Standby Power Requirements. The amount of power needed for these requirements and the cost of the necessary equipment vary widely, depending on the number of cars to be used, their rated load, intended running speed, and method of operation. Power demand should be calculated for each individual installation. In addition, provisions should be made to dissipate the power regenerated when the elevators are under negative or overhauling loads. This regenerated energy may cause overspeeding of the elevators unless a sufficient amount of power is dissipated by the emergency system, by the electric losses in the elevator equipment, or by some other positive load on the standby supply. There are situations in which increased standby power is required for all elevators. Equipment savings may be made by first supplying power during outages to one-half of all the elevators, provided that the traffic can be safely rerouted and the capacity of the operating elevators is adequate to handle the extra traffic. Power should then be

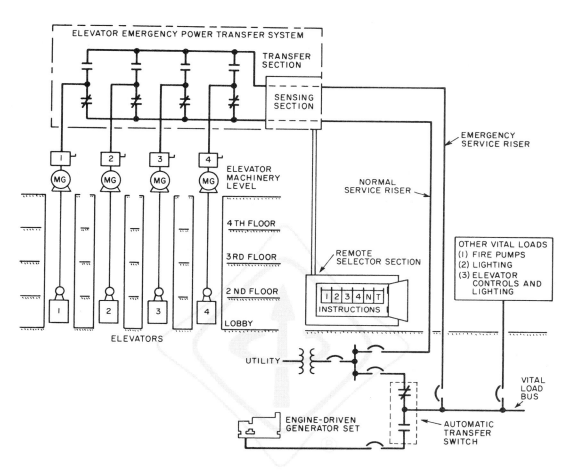

Fig 149
Typical Standby Power Transfer System

transferred to the second bank of elevators shortly thereafter to clear stalled elevators. Power may be left on in this bank until power returns to normal. When elevator service is critical, it is recommended that there be fully automatic generating plant starting and power transfer.

12.13.2 Elevator Standby Power System. Figure 149 depicts one scheme of an elevator standby power transfer system, which serves a bank of four elevators. This system consists of an automatic transfer switch for each elevator, a sensing and control panel, and a remote selector station. In this system, when normal power fails, a preselected elevator is powered from the standby power riser. An operator can select individual elevators from the remote selector station to permit complete evacuation of all elevators. Interlocks in the remote selector allow only one elevator to be connected to the standby power riser at a time. The generating set and the standby power riser need to be sized for only one elevator, which will minimize the installation cost.

A more commonly installed system uses one transfer switch to transfer the elevator feeder from normal to standby power. All of the elevators are energized, and the limit as to how many elevators operate at one time is established by the designer. The elevator controls are then provided to supervise the individual elevator operation in accordance with the system design. This method is more economical because all switching control is accomplished by pilot devices that are inside the elevator control. In addition, the elevator control can automatically supervise the return of each car and also switch cars to automatic service while providing interlocking protection and backup flexibility.

When SCR direct drive elevators are used, the designer should use caution. This type of elevator usually has isolation transformers, which, even if only one elevator is going to run, should all be energized. The standby power system should be sized to supply this inrush current. Close design coordination with the elevator supplier is critical to the proper selection and sizing of the standby power system.

12.14 Operation of Elevators During Fire Conditions. In the event of fire, elevators shall be made available for the use of emergency personnel. Elevator operation is handled in two phases.

(1) Phase I requires that all automatic operation elevators that have to travel 25 feet or more shall have a switch at the main floor and smoke sensors in each elevator lobby that, when activated, shall automatically return all elevators nonstop to the "main" or "designated" landing.

(2) Phase II requires that a switch be provided in each elevator that, when activated, allows emergency personnel to have control of the elevator. For a detailed description of elevator operation during a fire, refer to ASME A17.1-1990 [10].

12.14.1 Sprinklers in Elevator Machine Rooms and Hoistways. The ASME A17.1-1990, Code Rule 102.2(c) [10] requires the automatic disconnection of the main line power supply to elevator equipment prior to the application of sprinkler water in elevator machine rooms and hoistways. Other sprinklers in the building should not affect elevator power. There are three methods of complying with this requirement; the first method is probably the most economical.

(1) Rate-of-rise/fixed temperature heat detectors—Would be installed in the elevator machine room or hoistway or both, and arranged to initiate the automatic disconnection of the main line power supply once the Phase I recall was completed. These detectors would be placed near each sprinkler, and the sprinkler temperature rating would exceed the heat detector temperature ratings. The detectors would work independently of the sprinkler system. During a fire, the heat detector would operate first and initiate the shunt trip of the circuit breaker to disconnect the main line power to the affected elevators before water was released.

(2) Pre-action sprinkler system in the elevator machine room and the hoistway—Employs automatic sprinkler heads that are attached to a piping system containing air that may or may not be under pressure. A supplemental fire detection system is installed in the same area as the sprinklers. Actuation of a heat detector opens a valve, which permits water to flow into

the sprinkler piping system and then to discharge from any sprinkler head that is open. The heat detector controlling the flow valve in the sprinkler piping initiates the shunt trip of the circuit breaker to disconnect the main line power to the affected elevators.

The pre-action system would not be activated by a smoke detector, as this would shut down the elevator prematurely and would not allow the Phase I recall. A new requirement in ASME A17.1-1990, Rule 102.2(c)(5) [10] prohibits sprinkler activation by smoke detectors in elevator machine rooms and hoistways for this very reason.

(3) Dry pipe sprinkler system in the machine room or hoistway — A dry pipe valve would be installed in the sprinkler piping where it entered the elevator machine room or hoistway. The piping would contain air or nitrogen under pressure from the sprinkler head to the dry pipe valve, and water to the opposite side of the dry pipe valve. Heat from a fire would open the sprinkler head, allowing the air to escape and water to flow into the system. The dry pipe valve would open and initiate the shunt trip of the circuit breaker to disconnect the main line power to the affected elevators.

There is some question as to whether there is a discrepancy between the requirements in ASME A17.1-1990, Rule 102.2(c)(4) [10] and the NEC, Section 620-51(a) [6], which states, in part, "…nor shall circuit breakers be opened automatically by a fire alarm system." It is the opinion of the A17 Committee that there is no discrepancy because a sprinkler system is not a fire alarm system; it is a fire suppression system.

12.15 Emergency Signals and Communications. Elevators shall be provided with an audible signaling device and a means of two-way conversation between each elevator and a readily accessible point outside of the hoistway. This may be a telephone, intercom, etc. More stringent requirements are required in buildings in which an attendant or watchman is not continuously available to take remedial action. ASME A17.1-1990 [10] dictates these needs in complete detail.

12.16 Car Lighting. Elevator cars shall be provided with not less than two lamps that will shed a minimum illumination at the car threshold of not less than 5 fc for passenger elevators and 2.5 fc for freight elevators. The NEC, Article 53 [6] now requires that all elevators shall have a means for disconnecting all ungrounded car lighting conductors and this means shall be located in the respective machine room. In addition, passenger elevators shall be provided with an emergency lighting power source that shall be automatically switched on instantly after normal lighting fails in the car. The emergency lighting power source is required to be on the car; thus, the standby generator cannot be used. Refer to ASME A17.1-1990 [10] for complete details.

12.17 References. The following references shall be used in conjunction with this chapter:

[1] ANSI A117.1-1986, Buildings and Facilities — Providing Accessibility and Usability for Physically Handicapped People.

[2] ANSI/ASME A17.2-1988, Inspectors' Manual for Elevators and Escalators.

[3] ANSI/ASME A17.4-1986, Emergency Evacuation of Passengers from Elevators.

[4] ANSI/ASME A90.1-1985, Belt Manlifts.

[5] ANSI/ASME B20.1-1990, Safety Standards for Conveyors and Related Equipment.

[6] ANSI/NFPA 70-1990, National Electrical Code.

[7] ANSI/NFPA 101-1988, Life Safety Code.

[8] ASME A17 Interpretations, 1971–79.

[9] ASME A17 Interpretations, 1979–89.

[10] ASME A17.1-1990, Safety Codes for Elevators and Escalators.

[11] ASME A17.3-1990, Safety Code for Existing Elevators and Escalators.

[12] ASTM E152-1981, Methods of Fire Tests of Door Assemblies.

[13] *Minimum Passenger Elevator Requirements for the Handicapped*, National Elevator Industry, Inc., 185 Bridge Plaza North, Room 310, Fort Lee, NJ 07024.

[14] *National Building Code*, Building Officials and Code Administrators International, 4051 West Flossmore Road, Country Club Hills, IL 60477.

[15] *Standard Building Code*, Southern Building Code Congress International, Inc., 900 Montclair Road, Birmingham, AL 35213.

[16] Strakosch, George R. *Vertical Transportation, Elevators, and Escalators*, New York: John Wiley & Sons, 1983.

[17] *Uniform Building Code*, International Conference of Building Officials, 5360 South Workman Mill Road, Whittier, CA 90601.

[18] *Vertical Transportation Standard*, National Elevator Industry, Inc., 185 Bridge Plaza North, Room 310, Fort Lee, NJ 07024.

Chapter 13
Communication and Signal System Planning

13.1 General Discussion. This chapter outlines the engineering considerations that are involved in planning communication and signal systems in commercial buildings. Well-planned facilities, designed to accommodate both present and future needs, will pay dividends throughout the life of the building. All trends point to the continually increasing importance of communication, control, and signaling systems in commercial buildings.

Communication systems should be treated as building resources. Strong organization, especially in telephone and data cable plant systems, greatly increases the long-term value of these resources. The trend is to design cable support and systems to serve two or more generations of communication equipment and to design "backbone" systems to accommodate several years of growth and change at the branch distribution level.

It is important that the engineer investigate future communication needs while the building is in the planning stages. Planning the communication system for a commercial building should account for voice (telephone), data (computer applications), video, and other intelligent building systems. Connections occur over local (building), campus, and global networks. It takes expertise to analyze the needs of a particular user and to select an appropriate communications strategy. Communication system suppliers and independent communication consultants are prepared to offer assistance in planning the needed facilities; the importance of seeking consultations at an early stage is critical.

Communication system design criteria for each type of system should answer the following questions:

(1) Where are system components located?
(2) What is the topology of device interconnection?
(3) What are the wiring requirements of each type of device? What are the code requirements?
(4) What are the head-end equipment requirements? Size? Environment? Power? Wire management?
(5) How often will the system change? How fast will it grow? How will these both be accommodated?
(6) Who will install the system?
(7) When should the system be installed? Be commissioned?

13.1.1 Cable Plant. The cable plant provides the signal path between user terminals and the central control equipment of all communication and signaling systems. The term "cable plant" not only covers the insulated conductors themselves, but also the methods and materials used to support and protect the conductors. Data cable systems often parallel or share telephone cable systems. The space requirements for data facilities are less predictable and will often be greater than those needed for telephone requirements. When designing riser and distribution systems for integrated communication facilities, the overall space requirements may be more than double that for telephone facilities alone.

The choice of conductors used for the signal path is usually dictated by the signal characteristics of the communication system under discussion. Conductor size and number, insulation type and color, shielding, and jacketing are essentially system-dependent. The choice of methods for support and protection of these conductors is also dependent on system characteristics; but the engineer should give attention to selecting the physical spaces where the cable plant is installed before making this choice. The following discusses the basic methods for the distribution and protection of communication conductors in commercial buildings.

13.1.2 Riser Conduits and Sleeves. Riser conduits are not usually employed in buildings that exceed 12 stories or cover a large horizontal area because the advantages of having the cables protected by conduits is offset by the difficulties of installing and properly supporting the heavy cables in conduits, and making complicated splices in the splicing cabinets.

The number of riser conduits or sleeves is determined by the usable area in a building. Table 80 is a guide to determining the proper number of sleeves or conduits for telephone service. When designing riser and distribution systems for integrated communication facilities, the overall space requirements may be more than double that for only the telephone facilities.

The general design guidelines for riser sleeves and conduit are
(1) Minimum conduit size is 4 inches.
(2) Sleeves should be vertically aligned to permit the pulling of riser cables.
(3) Both ends should be reamed and threaded.
(4) Initial unused sleeves and conduits should be capped.
(5) Sleeves should be stubbed 2 inches above the finished floor.

Table 80
Riser Sleeves or Conduits for Telephone Service

Total Usable Floor Area to Be Served in Thousands of ft^2	Number of Sleeves or Conduits (Minimum Size 4 in.)
0–50	2
50–100	3
100–200	4
200–400	6

NOTE: For every additional 200 000 ft^2 or part thereof, add two additional sleeves or conduits.

(6) Sleeves should be located immediately adjacent to the wall and to the left of the cable terminating space. They should never be located in the center of a usable wall.

(7) In conduit runs, there should not be more than two 90° bends, and large radius bends should be considered.

13.1.3 Riser Shafts and Slots. Riser shafts are usually employed in larger buildings. The shafts should start from the main terminal room and extend to all floors of the building. For architectural reasons, if a shaft cannot be vertically aligned, conduits of adequate size should be provided to connect the bottoms of the higher floor shafts to the tops of the lower floor shafts.

Slots have become a problem for fire prevention, so they have generally become unacceptable. When used, they should be 4 inches clear minimum to 6 inches clear maximum in width with a minimum 2 inch high curb. All slots should be vertically aligned.

13.1.4 Underfloor Distribution Systems. There are important advantages to providing an underfloor telephone and data distribution system. The increased efficiency of raceways makes it possible for every desk or each piece of communication equipment to be served from adjacent floor outlets. Equipment changes that are made necessary by relocation of partitions or movement of personnel or furniture can be undertaken with minimum delay and inconvenience.

Using raceways, the need for surface wiring along baseboards or overhead is eliminated, thereby improving the appearance of the building. In addition, changes in telephone service that later require removal or relocation of wiring leave no unsightly areas. Telephone industry studies indicate that four out of ten business telephones are relocated every year.

13.1.5 Underfloor Raceways. There are two types of underfloor raceways in general use for communication cable distribution. They are known as "underfloor duct" and "cellular floor."

Underfloor duct is a system of parallel ducts running within the floor slab. The space between parallel runs may vary between 3 feet and 6 feet. Cross runs and junction boxes should be located every 40 feet or less, depending upon the layout of the area. (This conforms to most office requirements.)

The cellular floor consists of longitudinal cells, which actually support the floor slab and which are connected together by means of header ducts. The header ducts should be connected to the cell area at intervals that are necessary for maximum coverage, generally no greater than 50 feet. Alternating the power and telephone cells can provide a flexible layout. In many cases, it becomes desirable to use a third parallel cell for cables serving low-voltage signaling circuits. At least 1 in^2 of area should be provided for each 100 ft^2 of occupied space.

13.1.6 Ceiling Distribution Systems. Placing wires and cables in a false ceiling area close to the workstation is the most commonly used distribution system. Ceiling systems utilize one of two methods to reach the telephone and data terminal locations.

One method is to place the wiring in the ceiling space and poke up through the floor structure to the office above. This is commonly referred to as the poke-through method. The design of poke-through should be carefully considered by the engineer. Structural damage can be caused to the floors by drilling the holes. The holes, if not properly plugged, can allow the passage of liquids, fire, and dirt between floors. Disruption in occupied spaces will be caused due to the noise and debris of core drilling, and by cable installers working over desks and on ladders. This method is not always possible, but when combined with power poke-through outlets and effective management, it can be satisfactory. The availability of flush poke-through devices and flush permanently set floor boxes in various combinations of power and communication systems allows a "fixed" (but well-planned) grid of floor outlets to flexibly and economically serve many intermediate-sized spaces.

The second method is to place the wires and cables in the ceiling space and confine the service to the same floor. This method utilizes partition walls or utility columns to bring wiring down to user locations. The placement of utility columns is the owner's responsibility. Telephone wires, data cables, and electrical power wires can be placed in utility columns.

When the ceiling area is being used as an air plenum, a metallic raceway system should be designed for most wires and cables in the ceiling. This is a requirement of ANSI/NFPA 70-1990, National Electrical Code (NEC), Articles 300–318 [3].[83] Most states have adopted this NEC requirement. It is also a part of *OSHA Rules and Regulations*, Subpart 5, Electrical Section, 1910.308/309. Exceptions to the NEC [3] allow the use of cables specifically listed for air plenums or cable in those areas. The NEC, Article 800 [3] describes cable listing types for communication systems.

13.1.7 Undercarpet Wiring Systems. Flat conductor cables (FCC) are specially designed cables for power, telephone, and data that are installed directly on top of structural floors beneath removable carpet squares. The basic system consists of flat cables, junction boxes, terminal blocks, and pedestals with power receptacles, telephone, and data outlets. When FCC is used, careful installation is critical. Refer to the NEC, Article 328 [3] for special requirements.

13.2 Telephone Facilities. This section outlines space requirements for telephone facilities as opposed to communication facilities that include data, video, and other systems. A complete system for telephone facilities in a building includes provisions for the entrance cables, the main terminal room, the riser cable system, the distribution cable system for each floor, the cable distribution terminals, and the station cabling. Special consideration should be given to buildings that require private branch exchanges (PBXs), large public telephone installations, electronic switching systems, etc. If an additional private telephone system is planned, early consultation with the equipment supplier and communication consultants is essential. Early contact with the Building Industry Consulting Service (BICS) at the local telephone interconnect company is advantageous to the building designer regardless of the specific building tenants or type of telephone equipment being proposed.

---

[83]The numbers in brackets correspond to those in the references at the end of each chapter. ANSI publications are available from the Sales Department of the American National Standards Institute, 11 West 42nd Street, 13th Floor, New York, NY 10036. NFPA publications are available from Publications Sales, National Fire Protection Association, 1 Batterymarch Park, P.O. Box 9101, Quincy, MA 02269-9101.

13.2.1 Service Entrance Cables. This is the means by which the telephone company's cables cross the owner's property and enter a building. The type and location of the service depends upon telephone company distribution facilities, local practices, and codes. Service will normally be furnished from overhead lines or from underground cables. In most localities, service duct will be provided by the owner. Direct buried service may be allowed in some areas. The owner should provide the necessary trench work. Some areas require a utility easement for buried service. The local telephone company building industry consultant can be of great assistance in designing the service entrance.

The number of service entrance conduits required should be based on the ultimate number of telephone lines. For large commercial buildings, a 4 inch conduit for every 150 000 ft^2 of usable floor space with a minimum of two conduits is recommended.

In large or specialized installations, provisions for fiber-optic cables may be appropriate. These cables are smaller than large twisted-pair telephone cables. It is preferable to install fiber cables in smaller, dedicated conduits or in innerduct (protective tubes) within larger ducts. This may lead to requirements that are in addition to the conduits noted above. In some cases, the fiber cable will replace larger cables; but reducing conduit requirements is not recommended unless specific service plans are known.

A means of bonding, grounding, or insulating cable inside the building is necessary. The openings for cable entrance should always be kept sealed to avoid penetration of water and gas, and to avoid service interruption from flooding or equipment damage from dampness.

Whatever the type of entry, suitable provisions should be made for the appropriate openings in walls, roof, or below grade (with sleeves or conduit) well in advance of the required service date. General recommendations regarding conduit entrances are

(1) Use corrosion-resistant material.
(2) There should not be more than two 90° bends, and large radius bends are recommended.
(3) All ends are to be reamed, bushed, or capped, or all three.
(4) Sleeves through foundation walls should reach undisturbed earth to prevent shear of direct buried cables.
(5) Minimum depth should be 18 inches or deeper as required by local codes.
(6) Conduit placed on private property should not be terminated in joint use manholes with electrical cables or equipment.
(7) Provide a means for locating or identifying nonmetallic buried conduits from above the surface.

Consideration for a dual or alternate entrance provision should be given to the following types of buildings:

(1) Hospitals
(2) Airports
(3) Police stations
(4) Military installations
(5) Power generation or control installations

(6) Radio stations
(7) Television stations
(8) Transmitter sites
(9) Data centers

If the building support is served by overhead cable, the cable will normally be supported on poles and enter the building aerially. Proper safety clearance from other utility lines and above roadways should be provided. Refer to ANSI C2-1990, National Electrical Safety Code (NESC), Section 23 [1] and to the NEC, Article 800 [3].

13.2.2 Main Terminal Room. The heart of a building's communication system is the main terminal room. It is the main cross-connecting point between the telephone company's central office and the riser cable system for in-building telephones. The entrance cable usually runs directly to this main terminal room after it is brought into the building.

It is desirable to utilize conduit from the service entrance to the main terminal. Conduit simplifies future additions and protects the service cable. For some types of cable, metal conduit is required by code.

The terminal room should be accessible to communications personnel at all times. It should be located as closely as possible to the center of the riser cable distribution facilities. The room should be well lighted, cooled, ventilated, and equipped with electric outlets. Its floor support should be designed to support heavy terminal equipment. This room should be used only for communication equipment, and it should be capable of being locked for security reasons.

The main terminal room will contain the utility demarcation point. Telephone utility cables from the central office are routed through protective devices and terminated on special jacks called the "network interface." These jacks are the boundary between the utility and the premise's distribution system. Utility cable terminals and cable protectors (surge protection) are always utility property. Surge protection for data and communication cables may be required within the premises in addition to utility surge protection. The terminal room may also contain utility-owned switching and transmission equipment. The terminal room also contains terminals for the building's (or campus') distribution cables, usually labeled the main distribution frame (MDF). The MDF and distribution cables are usually owner property.

For large installations, floor-type main terminals are located in the main terminal room. The space allocated to this room should be adequate to meet the original telephone service needs of the occupants and to provide for later installation of additional terminals for growth. Depending upon communication requirements, batteries may have to be installed in or near the main terminal room.

In buildings that have rentable floor area of up to $70\,000$ ft^2, wall-type main terminals are commonly used. Each terminal provides connections for the desired number of pairs of cable for voice, data, and EDP equipment.

In some buildings (primarily single-occupant commercial and institutional types), the switching equipment that serves private branch exchange (PBX) or central exchange (Centrex) switchboards may be located in the main terminal room. In this event, the room's environment and electrical services should accommodate the PBX. See 13.2.6 for more information.

13.2.3 Riser Systems. The riser system is the backbone of a commercial building's telephone network. It provides facilities for bringing cables from the main terminal room to the various floors of the building.

The riser cables may be brought up either through riser conduits, sleeves, slots, or riser shafts, depending upon the type and size of the building. Local codes should be checked to determine who has the responsibility for fire prevention in both used and unused riser facilities.

13.2.4 Apparatus Closets. The equipment needed for electronic key telephone services (EKTS) is often housed in apparatus closets. EKTS systems are used in some facilities to extend the functionality of central office or PBX systems. Locating apparatus in closets eliminates unsightly wiring and contributes to improved office appearance. When maintenance or changes are required, communication workers may do their work without inconvenience or distraction to office personnel. Physical conditions within the building frequently dictate the number, location, and type of closets. In general, it is desirable to have one apparatus closet for each 10 000 ft^2 of usable floor area. This can vary for smaller areas and up to a maximum of 20 000 ft^2 of floor area, depending upon access and other layout considerations.

Apparatus closets should be lined with ¾ inch thick fire-resistant plywood to a height of 8 feet for mounting equipment. The floor, walls, and ceiling should be designed in accordance with fire codes, and proper lighting with switching should be installed. Ventilation is also required. One separately dedicated 20 A three-wire circuit run is necessary for at least two 120 V duplex receptacles or an electric plug-in strip. A size 6 AWG copper wire or 0.5 inch conduit connected to the building ground is required in all apparatus closets.

There are two types of apparatus closets, walk-in and shallow. The general specifications for these are given in Table 81. Table 82 gives an idea of the space required for typical centralized EKTS apparatus closets. Using Tables 81 and 82 together, one can determine the closet sizes needed.

When data network equipment will be accommodated in the apparatus closets, additional power and cooling may be required. The shallow telephone closet is inappropriate for most data network equipment. When designing integrated com-

Table 81
Apparatus Closet Specifications

	Walk-in Closet	Shallow Closet
Minimum depth	3 ft.	1 ft., 6 in.
Maximum depth	None	2 ft., 6 in.
Minimum width	5 ft.	3 ft.
Minimum height of doors	6 ft., 8 in.	6 ft., 8 in.*
Minimum width of doors	3 ft.	3 ft.†

*When shallow closets are used, the center post between double doors should be eliminated, if possible.
†Minimum for single door, 2.5 ft. for double doors.

munication facilities, allow for the depths shown for walk-in closets in Table 81. Note that spaces containing communication equipment that require power are often labeled communication equipment rooms (CERs) rather than apparatus closets.

13.2.5 Satellite Closets. Unlike apparatus closets, satellite closets do not contain electronic and power equipment. Their primary use is to provide cable terminating facilities, that is, connecting blocks for station wiring.

Satellite closets are often necessary because the inadequate space of a single apparatus closet fails to meet increasing needs, or the physical layout and shape of the building. Table 83 gives examples of the dimensions of satellite closets.

13.2.6 Telephone Equipment Rooms. In many buildings involving large, sophisticated installations, additional floor area and adequate floor support are necessary to accommodate the complex switching equipment associated with PBX or Centrex telephone services. This is in addition to the space required for apparatus and satellite closets. Since PBX and Centrex services will vary according to the needs of building tenants, each equipment room should be individually designed. Such rooms should be centrally located, well lighted, and well ventilated. They may require cooling, and they should be equipped with power and grounding facilities. PBX equipment can require as little as a 20 A single-phase dedicated circuit or as much as an 80 A three-phase connection. Specific requirements should be investigated in each case and for each supplier who may furnish a system. They should be as clean as possible, and located in areas that are not subject to dampness or flooding. The PBX supplier should be contacted for advice on designing individual equipment rooms.

Table 82
Centralized Electronic Key Telephone
Equipment Space Specifications in Linear Feet

Floor Area Served (ft^2)	Electronic and Power Equipment	Terminal Blocks	Total
Up to 5000	3 ft.	5 ft.	8 ft.*
5001–10 000	3 ft.	8 ft.	11 ft.*
10 001–20 000	6 ft.	9 ft.	15 ft.*

*If more than one wall is utilized, add 1 foot for each turn.

NOTE: This table is to be used in conjunction with Table 81.

Table 83
Satellite Closet Specifications

Floor Area Served (ft^2)	Linear Feet
Up to 2000	2 ft.
2001–4000	4 ft.
4001–6000	6 ft.

13.2.7 Public Telephones. Facilities for public telephones are generally needed in the lobbies, the arcades, and at truck loading docks for truckers and contractors of all commercial buildings. While telephone companies will install pre-manufactured telephone booths at suitable locations, built-in booths with a wide variety of designs, finishes, and colors greatly enhance the appearance of a building. Telephone companies are prepared to work with building designers toward providing facilities that will harmonize with a building's interior.

Public telephones should be located where they will be readily accessible to the general public and contractors. In many cases, the use of signs that harmonize with the architectural design of the building are desirable in order to call attention to telephone locations. Accessibility for the handicapped is now required when designing public telephone facilities.

There is no simple reliable rule for determining the most suitable number of public telephones that should be provided in the lobby, arcade, or street floor of large office buildings or other types of buildings frequented by the general public. For design assistance, the telephone company's public telephone business representatives should be consulted as early as possible in the design phase.

13.3 Data Facilities. The world of data systems is immense. There are many parallels between data systems and telephone systems; but data systems are associated with greater complexities in system design.

A data system may consist of terminal equipment, local distribution cable, transmission equipment, and data center equipment. A data system might also be a collection of computing machines linked to share resources without any central equipment. Or it might be a floppy storage disk carried between two personal computers. Although the expansion of telephone facilities to carry data is often very economical, it is absolutely necessary to review the owner's plans before making any assumptions about equipment locations or the wiring topology of a data network.

Data networks serve user terminals. The terminals may be workstations, personal computers, video display units (VDUs) or video display terminals (VDTs) (character or graphics terminals), printers, teleprinters, or facsimile machines.

Terminals may be linked to central systems via dedicated conductors or transmission equipment. They may reside on a wide area network (often connected with telephone lines) or on a local area network. Switched networks and various network bridges and mail systems support more complex interconnections.

Data transmission equipment is similar to computer equipment. It is electronic in nature and requires varying degrees of power service continuity, depending on the type of traffic (information) carried. Data switching equipment has similar parallels to PBX equipment.

Data centers (i.e., computer rooms) contain both the central communications equipment and data storage and processing equipment. They are usually associated with a concentration of data cable and exhibit several special design requirements.

13.3.1 Data Terminals. Most data terminals are based on cathode ray tube (CRT) technology. With a basic terminal, or VDU, a screen is usually associated with a keyboard and connected to a central processing unit (CPU). Personal computers and workstations tied to a network may be used as terminals to access remote computing power. Terminals, personal computers, and workstations vary over a wide range of sizes. Their power requirements range from 40 W to several thousand watts, most often in the range of 100–300 W. Lighting for comfort at a CRT terminal is difficult. A well-shielded or indirect system with a moderate illumination level (i.e., between 20–50 fc [200 and 500 lux]) is typically employed. Refer to ANSI/IES RP24-1989, Recommended Practice for Lighting Offices Containing Computer Visual Display Terminals (VDTs)[84] for additional design information.

Careful consideration should be given to the branch circuits feeding large numbers of data terminals. Harmonics generated by electronic power supplies can overheat neutral conductors and transformers on three-phase systems. As networks grow larger, the continuity of power systems becomes an issue. It is not uncommon to find a special power system (uninterruptible, regulated, or filtered) serving terminals at a work space that is purposely separated from the normal power system.

A number of specialty terminals including video projectors, monitoring system interface, network monitors, etc., may be part of the data system, which also includes teleprinter terminals.

Teleprinter terminals are connected to dedicated services through a public network. Teleprinter units come in various sizes and types. The unit is located so as to provide the operator with ample work and aisle space, with a lighting level of 70 fc (750 lux), 30 inches above the finished floor. The placement of the teleprinter should be considered in the context of its effect on adjacent offices. When one unshrouded unit is in operation, the noise level is 70 dB(A). The signal connections are accomplished by using two pairs of cables, each of which is about 3/16 inch diameter.

Power requirements are 120 V, 60 Hz, 2 A running; 8.2 A starting. Dedicated circuits are recommended to prevent loss of power to units that can be caused by accidents in other receptacles during normal office operation. Some installations require machines to be on line continuously to accept messages during unattended hours. Power shall be available for this type of operation. Standard building air conditioning or ventilation is sufficient for the teleprinter's operation. Care should be taken to ensure proper ventilation when a teleprinter is installed in small enclosed quarters or in a sound attenuation enclosure.

13.3.2 Cable Systems and Support Structures. Data terminals are connected via a communication cable. This cable may be a twisted-pair cable (sometimes identical to telephone cable), a shielded twisted-pair cable, a coaxial cable, or a fiber-optic (glass or plastic) cable. The cable may run to a computer room or to equipment in a nearby closet. Refer to Reference [8] for additional information.

Cable systems for local area networks may have a topology distinct from the telephone system. The classes of cable system topologies include bus, star, ring, and

---

[84] IES publications are available from the Illuminating Engineering Society, 345 East 47th Street, New York, NY 10017.

branching tree. These topologies resemble their names. A bus connects points along a line. A star connects points to a central location. A ring connects points on a loop. A branching tree is a complex (maybe hierarchical) system of interconnection. Logical transmission may be different from the physical topology, for example, a logical ring may be wired as a star (i.e., back to a central closet).

It is often advantageous to enforce a star topology for terminal wiring. It allows changes to the interconnections via physical cable changes at a central closet location. With appropriate data network equipment, a star topology can limit signal loss to a single terminal when an individual cable is damaged or disconnected.

Once the topology of data wiring is defined, its support structure may be designed using the same distribution systems used for telephone wiring. Some particular concerns, beyond those for telephone wiring, follow:

(1) Coaxial, fiber-optic, and, to a lesser extent, shielded cable require greater care than telephone cable. The minimum bending radius of these cables is greater than that for telephone cable.

(2) Space required for terminations, especially at outlet boxes, should be carefully considered for each type of cable.

(3) Data cables are generally larger in diameter than telephone cables.

(4) Data cables generally have more stringent distance limitations than telephone cables.

(5) Data cables are more susceptible to interference than analog voice (telephone) cables.

The backbone or riser cable for data systems may assume any of several forms. The simplest approach is to route individual terminal cables to one location in a building. On the other hand, it is not unusual to collect the signals from several data sources at a closet and route a single cable to the central building location. This is accomplished by various data transmission and network equipment. In this case, the backbone or riser cables may be lighter than the sum of the terminal cables they serve; the riser cables are usually coaxial or fiber-optic cables.

Data cable is often terminated with modular terminal systems, which allow quick cross-connects. Typically, data terminals require more physical space per cable than telephone cable terminals, which are quite compact. Data terminals are sensitive to moisture and dirt. Data connections should be clean and tight due to the nature of data signals. Typical terminal space dimensions may be misleading. Each project should be analyzed separately to define terminal space requirements.

13.3.3 Data Transmission and Network Equipment. Data equipment locations will depend on distance limitations and the economics of the particular data system utilized. In many cases, some of the data equipment will reside in apparatus closets or CERs and terminal rooms associated with the telephone system.

This equipment requires ventilation or cooling, power (with consideration for continuity), and illumination between 30–70 fc (300–750 lux). The illumination is needed for changes and maintenance; the equipment usually operates unattended.

Data equipment is often assembled on a rack with modular cable terminations; consideration should be given to access and floor space. It is usually appropriate to maintain data equipment adjacent to telephone facilities, but to designate separate telephone and data spaces within terminal rooms and CERs.

13.3.4 Data Center Installations. Data centers, otherwise known as computer centers or computer rooms, are the hub of various data storage and processing systems. It is unusual to find a self-contained computer room.

Three classes of outside connections occur. The first is to data terminals residing in the same building or on the same campus as the data center. The second is to widely dispersed terminals via the public telecommunications network. The third is to other users via point-to-point radio (microwave or satellite) links. These classes of connections are associated with different approaches to data cable access.

The first approach requires terminations for data terminal wiring in the building. This is usually a large bundle of small cables routed to racks in the computer room. The second approach requires a cable path from the computer room to the main distribution frame or utility network interface for access to the public network. The cable is usually a large multiple-pair cable routed to terminals allowing for a cross-connect to the data communication equipment. In a large facility, public network access may be via a fiber-optic cable. Typical data centers utilize a mix of connection classes.

Data equipment size and installation requirements will vary considerably, depending on the user's communication service needs. It is not unusual to have area requirements as small as 200 ft^2 or as large as tens of thousands of square feet. The specifics should be obtained from the user and the prospective equipment manufacturers, while taking into consideration potential growth. Without specific data, the following typical installation requirements should be used to design facilities of any size:

(1) Average live load of communication equipment is approximately 70 lb./ft^2.

(2) Ceiling height above the finished floor is 8 feet, 0 inch minimum.

(3) Lighting level of 70 fc (750 lux) at 30 inches above the finished floor. A more specific lighting approach is appropriate at lower levels with well-shielded fixtures at terminals, and at high levels at tape and disk drives. Dual switching with a fine grain zoning will allow for a flexible lighting response for computer equipment moves.

(4) Aisle space and equipment configuration is dependent on the physical restrictions of a building, as well as interconnect cable, maintenance, supervision, and manufacturing requirements.

Today, most systems have been designed so that power and signal cables can enter from below. Hence, raised flooring should be installed over the structural floor, creating a space to be used as a cable raceway. In addition, the raised floor can be used as an air plenum, which will provide the necessary conditioned air directly to the equipment's air intake louvers. The raised floor should have an adequate number of floor registers or perforated panels, or both, to provide the required air flow. Cutout trims should be provided to protect against the cutting of cables. Ramps should be provided to facilitate the movement of equipment and test gear. The height of a raised floor is dependent on available head room, number of planned cables, and pressure restrictions affecting air conditioning. For various requirements regarding the placement of wire/cable in underfloor plenum spaces, the NEC [3] and ANSI/NFPA 75-1989, Protection of Electronic Computer/Data Processing Equipment [4], and local codes should be consulted.

Consideration should also be given to acoustical treatment in equipment operating areas. Noncombustible acoustical ceiling tiles, and flame-retardant carpets, wall coverings, curtains, etc., provide for noise reduction and absorption. Proper design and equipment placement can reduce mechanical equipment noise. Duct lining, resilient equipment mounting, the location of the air handler unit, and the size of dampers are just some of the considerations.

Conditioned air should be supplied, which will limit the ambient temperature and humidity to within the limits specified by the equipment manufacturer. Most equipment today can operate without any degradation in performance at temperature levels in excess of 25.6 °C (78 °F). For energy conservation purposes, where supercooling is not required, the ambient should be maintained at human comfort levels. An adequately filtered backup ventilation system should be provided to limit the temperature rise to −12.2 °C (10 °F) above the outside ambient in case of main air-conditioning outage. Thermal shock, direct distribution of conditioned air to critical equipment, humidity control, and redundancy are all factors for the mechanical designer to consider.

Typical power requirements for a data facility are 25 W/ft^2 for communication and computer equipment, 2 W/ft^2 for lighting, and 15 W/ft^2 for the heating/ventilating/air-conditioning (HVAC) system. Service is normally 208Y/120 V_{ac}, three-phase, four-wire, 60 Hz. For some large computing equipment, 400 Hz is required and power is furnished via a frequency converter. Steady-state voltage requirements are typically +10%, −8% with a frequency variation of 0.5%. The duration and rate-of-change of any variation should be taken into account. Generally, a 20% transient that lasts longer than 30 ms, or a voltage outage that lasts longer than 15 ms, may cause errors, loss of information, shutdowns, or equipment damage. Equipment tolerances can be exceeded by numerous events, including utility or building power system faults, large switched loads, local utility network switching, lightning, reclosing operations, planned brownouts, or unplanned blackouts. These can occur several hundred times a year. Users have several systems available to protect their facilities against voltage variation or outages, or both. Constant voltage transformers, the mechanical inertia motor-generator set, power conditioning systems, and uninterruptible power supply (UPS) systems are just some types of equipment available to improve voltage regulation. Refer to IEEE Std 446-1987, IEEE Recommended Practice for Emergency and Standby Power Systems for Industrial and Commercial Applications (ANSI)[85] for additional information.

The system chosen shall be dependent on the type of operation, economics, and consequences that can develop if unregulated voltage is provided. The utilization of UPS systems for critical computer systems is increasing. Refer to 4.10 in Chapter 4 and Reference [9] for additional information.

Grounding for communication equipment should be provided. An independent ground riser from the building's earth electrode network shall terminate on an isolated ground bus and be sized on at least 2 kcmil/linear foot of cable run.

---

[85] IEEE publications are available from the Institute of Electrical and Electronics Engineers, IEEE Service Center, 445 Hoes Lane, P.O. Box 1331, Piscataway, NJ 08855-1331.

Single- or multiple-point grounding, or both, may be used depending on the communication equipment's frequency characteristics. Refer to FIPS Standard 94 [7] for additional information.

In some installations, the impedance of the ground system and its association with power system grounding may affect data system performance, especially data transmission equipment. System suppliers usually have documentation on specific system grounding requirements.

Satellite communication systems are used by certain large data processing users. Point-to-point microwave links between buildings are also used. These systems require special considerations that are not covered here. For further information, see Reference [8].

13.4 Fire Alarm Systems. Ideally, fire alarm systems for commercial buildings should provide early detection, accurate location of the alarm origin, fire department notification, and automatic control of the HVAC system, elevators, and other building systems that are necessary to make the building safer for its occupants. Fire alarm initiation devices include smoke and heat detectors, manual pull stations, water flow and tamper switches (associated with sprinkler systems), and other fire suppression systems. These devices are usually arranged by zone and monitored by a control panel. The control panel operates numerous devices including alarms, door holders, fan motor shutdown, elevator recall, and smoke control systems. Systems for large and high-rise buildings include voice communication in stairwells, and fan and elevator override controls for firefighters. Local signaling is provided by various types of annunciators. Automatic reporting to a listed monitoring agency or municipal fire department is recommended and often required. It should be noted that the fire alarm system may be a part of a facility-wide automation system.

13.4.1 Design Criteria. Fire alarm systems are required by life safety regulations. The engineer should meet with the local Fire Marshal to determine the legal fire alarm requirements. Although beyond the scope of this recommended practice, building codes require compartmentalization, fireproofing, sprinkler systems, and smoke control to provide adequate fire protection in commercial buildings. Electrical engineers are not charged with the responsibility for designing architectural, structural, and fire suppression systems; but they should integrate the fire alarm system design with the overall fire protection plan. The following three building codes are widely used:

(1) *Uniform Building Code*—Used by most cities west of the Mississippi.
(2) *Basic Building Code*—Used by many jurisdictions east of the Mississippi River.
(3) *Southern Building Code*—Used in most areas in the Southeast.

The basic standards of the fire alarm industry are developed by NFPA (National Fire Protection Association) and UL (Underwriters Laboratories). UL Standards are equipment testing standards that are used to determine if equipment meets the functional and operational requirements of the appropriate NFPA Standards, and to determine that the equipment does not present a safety hazard if properly installed and maintained. The following standards provide appropriate guidelines for the application, installation, maintenance, and use of fire protective signaling systems:

(1) ANSI A117.1-1986, Buildings and Facilities — Providing Accessibility and Usability for Physically Handicapped People.

(2) ANSI/ASME A17.2-1988, Inspectors' Manual for Elevators and Escalators.

(3) ANSI/NFPA 71-1989, Installation, Maintenance, and Use of Signaling Systems for Central Station Service.

(4) ANSI/NFPA 72E-1990, Automatic Fire Detectors.

(5) ANSI/NFPA 90A-1989, Installation of Air Conditioning and Ventilating Systems.

(6) ANSI/NFPA 101-1988, Life Safety Code — 1988 Edition.

(7) ASME A17.1-1990, Safety Codes for Elevators and Escalators.

(8) NFPA 72-1990, Installation, Maintenance, and Use of Protective Signaling Systems.

13.4.2 Power Supply. Fire codes have special requirements for fire alarm system power supplies. An emergency power supply is required and should be sized to operate under normal load for 24–60 hours plus 5 minutes in full alarm.

13.5 Security and Access Control Systems. Electrical and electronic devices that provide monitoring and security detection are used primarily to supplement passive physical and security staff monitoring provisions. Intrusion devices are used to monitor the security of property within the facility in areas where high risk of theft is probable, as well as to protect from employee theft through unmonitored exits. Current codes do not require security systems; but they may be justified for the personal safety of building users and the public (in parking garages) and to reduce property loss.

These systems utilize a host of detection devices to keep track of people moving in and out of secure building areas. Video cameras, motion detectors, magnetic switches, metal detectors, infrared sensors, acoustical sensors, pressure transducers, window film, etc., are generally wired back to a zoned central control system. The system may sound local alarms and transmit signals to listed monitoring agencies. Access control components, such as electric locks, cipher locks, door operators, card readers, parking gates, etc., are often interconnected to a central unit to record transactions at each device.

When security systems affect personnel safety or protect valuable items or information, the engineer should specify a supervised security system. The electronics of these systems indicate trouble when shorts, grounds, or open wires develop.

13.6 Audio Communication Systems. Audio communication systems are characterized by the presence of electrical analog signals, often imitating the human voice. Three energy levels are usually present. Low-level signals from microphones and crystal phonographs are signals in which the variations in signal amplitude are expressed in μV and are subject to low-level interference. The length of low-level conductors should be short, and shielded conductors are usually employed. Line-level signals have been preamplified. These signals are present between various audio processing equipment. Output from one device is often input to another, and the careful adjustment of equipment to match system line levels will greatly improve system performance. High-level signals are used to drive speakers. These lines have the potential for several hundred watts of power. Limitations on the installation of high-level lines are identified in the NEC, Article 725 [3].

Audio systems are comprised of input, signal processing, and output devices. Input may be a microphone, radio receiver, etc. Signal processing includes amplification, spectrum shaping, filtering, gain control, delay, and storage. Output is usually a speaker or several speakers.

In many cases, sound systems are combined with other systems. In movie theaters, the sound system gets its signal from the film track. In hospitals, tones generated by call systems may be broadcast over multiple sound systems to signal for staff assistance. In high-rise buildings and large assembly halls, fire alarm signals and recorded instructions may be broadcast over a sound system. When a sound system may also serve as an emergency system, especially where life safety is involved, the sound system will fall under stricter codes. Standby power may be required. Wiring may need to be installed in metal raceway. Speakers of special construction will be required to be listed for this purpose. In some cases, circuit supervision is required.

13.6.1 Public Address Systems. Public address (PA) systems are utilized to broadcast a message to a group of people in large buildings. The message may be a voice announcement, music, or both. PA systems often utilize ceiling- or wall-mounted speakers distributed evenly throughout the area of their operation. Input may be via a fixed microphone, a telephone handset, or an intercom system. Signal processing usually involves amplification and may also include spectrum shaping.

The PA system will involve speakers with conductors back to an amplifier location. Power and ventilation or cooling for the signal processing equipment is required as well as a cable to the source (i.e., the telephone switch, intercom equipment, or microphone).

In some facilities, the PA system has to overcome high noise levels, i.e., on factory floors or shops, in large warehouses, or in large stores. In this case, high-level horns are used or individual speakers are located very near machine operator locations.

PA speakers are usually daisy chained together and function as a single unit. In some facilities, the use of all-call (i.e., addressing everyone in a building, auditorium, etc., via a centralized speaker) would be obnoxious. These facilities might use a multiple-zone system. Additional wiring is required to home run each zone. Additional control and, possibly, additional amplifiers may be required to accommodate multiple zones.

Sound masking, which is the continuous broadcast of noise over speakers, is accomplished over a similar system except that speakers may be located above an acoustical tile ceiling and spaced at much greater distances. Although masking and PA announcements could be run over the same system, the technical requirements of each system suggest different solutions. It is recommended the systems be separately designed and installed.

13.6.2 Intercom Systems. Intercom systems allow a conversation between two or more people in different spaces. The typical intercom station has a handset and small loudspeaker. Sometimes the station is only a small loudspeaker and several buttons.

Intercom systems use two call systems. The first system initiates by depressing a button with a direct-connect system. The second system initiates a call by dialing a code (as on a telephone keypad) with a central switched system.

Direct-connect intercom systems are generally wired on a bus, i.e., one station is connected to the next in a line. This type of system is economical but usually is limited to one or two voice paths.

Central switched intercom systems are generally wired to a central unit, which is similar to a small telephone switch. Several manufacturers support a bus wiring option using digital signaling. This scheme reduces initial wiring costs; however, it increases the cost of additions and moves. In larger systems, it is appropriate to treat the station wiring as a resource and enforce a high degree of organization on the system. Installation concerns are similar to those of telephone systems.

Intercom systems are often connected to a public address system, allowing an all-call capability. Special amplifiers can be installed to allow answerback through some speaker systems.

Intercom systems can be connected to door access controls, i.e., in an apartment call system, tenants may have a door release button on their inside station.

13.6.3 Dictation Systems. Dictation systems may comprise a portable machine or a fixed, multiple-channel in/out system serving many people. Large systems are often used in large offices, hospitals, and medical clinics.

A portable system usually handles removable tape cartridges and requires a power outlet or batteries.

A fixed system usually collects its input via telephone, intercom, or fixed dictation stations. Dictation stations look like a telephone except with tape transport controls on the instrument. Wiring for stations is similar to telephone wiring. Connections to telephone or intercom systems are made with wire that is similar to telephone wire. A fixed system may serve several transcription stations. Transcription stations provide a headset and tape transport controls to a transcriptionist. They are connected to the dictation machine via dedicated conductors. The dictation equipment requires power and a clean, reasonably stable environment to protect the tape transport machinery.

13.6.4 Sound Reinforcement Systems. Sound reinforcement systems are utilized to improve the audibility of a speaker or performing artist in the performance space and in real time. This requires that the problem of feedback, usually a high-pitched oscillation of an unstable system, be addressed in system design.

Sound reinforcement systems may be classified into two major categories. The first employs a network of distributed loudspeakers placed throughout the area to be covered. The second employs a single central loudspeaker or loudspeaker array located as needed within the area to be covered.

The proper design of a sound system should take into consideration the following:
  (1) Acoustical environment of the area or space (volume, absorption in all of the varying uses anticipated, ambient noise levels, reverberations, etc.).
  (2) Type of program (speech or music) or information (advisories, warnings, life safety, or evacuation) to be reproduced.

(3) Transducer characteristics (loudspeaker and microphone directionality, coverage angles, sensitivity, and frequency response).

Audio compression, equalization, and delay are commonly used to tailor sound, systems to fit into tenant locations.

The requirements for sound reinforcement equipment are similar to those of public address equipment. The difference lies in the level of audio engineering that is required to satisfactorily serve a space. When a single speaker cluster is utilized, its mounting (structural support), exact location, and directional characteristics should be engineered.

A distinct variation on reinforcement systems is the meeting room or boardroom system. These systems are characterized by multiple-microphone locations and distributed speakers, which aggravate the tendency for the system to be unstable. Automatic microphone mixers are available to turn unused microphones off and to mute speakers that are nearest to live microphones. This requires additional wiring to speakers. Usually, every speaker is home run separately. In a meeting room, low-level microphone wiring becomes critical, due to accumulated noise and its affect on the automatic mixing equipment. Sometimes, provisions are required for microphone preamplifiers (and their power) to be situated in or below the meeting table adjacent to each microphone location.

13.6.5 Audio Production and Recording Equipment. The needs of audio recording equipment range from a single magnetic tape recorder or dictation machine to a complete professional recording studio where multiple-channel control consoles, multiple-track magnetic tape recorders, and disk cutting equipment are extensively used.

There are two major types of professional audio recording facilities. One is the production studio where the program is originated. These production facilities require detailed planning to achieve the desired acoustics, lighting, heating, ventilation, and air conditioning. The second type of recording facility produces duplications of recorded material. This may take the form of high-speed magnetic tape duplicating facilities and disk cutting equipment.

There are several special applications of audio recording equipment, for example, continuous recording of a voice communications channel.

System requirements include regulated power, environmental conditioning, and extreme care in space layout. The lighting should be appropriate for each task and be directly related to operator locations. Most often, consultants who work in the field are retained to design and specify these systems and coordination with them for power and lighting requirements is necessary in order to have a functioning installation.

13.7 Image Communication Systems. Image communication systems are characterized by a two- or three-dimensional pictorial image. These may be recorded and reproduced as video, cinema, or computer graphics. The most common systems employ baseband video and broadband television signals. These signals are typically carried on coaxial cable with special impedance characteristics. Splitters, terminators, impedance matching devices, and broadband amplifiers characterize this equipment. Image transmission systems are migrating to fiber-optic facilities, including fiber-optic cable and terminal equipment, especially for longer distance transmission.

Installation requirements are similar to other communication systems. Power, ventilation or cooling, physical space, and signal cable organization should be considered in designing these systems. Design issues that are unique to image systems, video in particular, include limitations on cable length, type, and terminations, which should be addressed to maintain good quality images. Image systems require particular attention to lighting both at input (camera) and viewing locations.

13.7.1 Video Systems. Video systems include closed-circuit television (CCTV), surveillance security, and information display. Security and information systems are generally black-and-white systems. Black-and-white video cameras with a broad range of illumination sensitivities are available. CCTV systems may be black-and-white or color. Color systems are typically used on college campuses and at corporate facilities for training and conferencing systems. Color cameras require more specific lighting systems with both level and lamp color restricted to within suitable ranges. A performing arts video studio will often have a stage lighting system. Video systems employ switchers, amplifiers, and various recording equipment. A baseband video signal requires between 4–25 MHz of bandwidth and is sensitive to cable lengths.

Video security systems require power and signal cables to camera locations and protection for the cameras from environmental conditions. They may require special raceway systems to ensure the security of video cables.

Information display systems, like the flight schedule display at airports, usually involve video signals derived from a computer-based data processing system. The distribution is treated just like one would be for a video system. Signals may be baseband video or modulated television channels.

Video conferencing facilities are designed to allow remote groups of people to converse and see pictures of each other and of other images. In some cases, full-motion video is available; in other cases, slow-motion video or still pictures are transmitted. These systems include microphone, video camera, speaker, and video display components and some control device that connects to a public or private network. Some systems also provide inputs for computer-generated text and graphic information that are similar to facilities for information display systems. Installation requirements for device interconnection are like those for similar audio and video devices. Of particular concern is the lighting, since both the imaging equipment (camera) and projection equipment are located in one area.

13.7.2 Television Signal Distribution. Distribution of television signals encompasses master antenna cable systems, cable utility distribution, CCTV, and, sometimes, security systems. In master antenna cable systems, one or several antennas are utilized to receive television signals. A broadband amplifier and channel selectors are used to level the relative signal strength of adjacent television channels. A cable system distributes the signals. Splitters and taps are used to connect branches and outlets of a tree-like distribution system. Additional broadband amplifiers may be distributed throughout larger systems.

Cable utility channels, modulated CCTV signals, and an FM radio band may be added to the signals available on a distribution system. Modulators and channel transducers are available to manage channel allocation so the full bandwidth capacity of a cable system may be utilized. Each of these systems are typically tied together with coaxial cables, although video transmission on fiber-optic cable is possible. Fiber-optic cable is primarily applicable to long runs (between buildings, etc.) because the use of splitters and taps for coaxial cable is more economical than that for equivalent fiber-optic cable.

13.7.3 Projection Systems. Projection systems may be of the front or rear screen variety. Common types of projectors include slide, film strip, motion picture, overhead transparency, overhead opaque, microfilm, microfiche, and television.

Lighting in viewing areas is critical to the success of projection systems, particularly video projectors, as are the maximum viewing distance and angle of viewing. Projection systems should also be designed for the required image brightness, evenness of screen illumination, and best edge-to-edge focus.

Projection systems are often associated with audio systems and usually require control system extensions. Film and slide projectors may need rear and front control jacks with wire to allow for remote control. Audio input to local sound reinforcement systems is desirable for film projectors.

Meeting rooms and classrooms often utilize motorized projection screens. Power for the screen motor and wiring to a control switch are required.

13.8 Nurse Call Systems. Usually found in hospitals and nursing care facilities, nurse call systems furnish patient-to-staff and staff-to-staff communication to support day-to-day operations. Devices include patient call, toilet emergency, nurse assist, emergency call, and staff and duty stations. These are wired individually back to a control panel. The nurse call control panel illuminates dome lights and sounds a tone when a patient calls. Many systems also furnish two-way voice communication between patients and staff. In most occupancies, nurse call systems should be connected to emergency system power. Refer to IEEE Std 602-1986, IEEE Recommended Practice for Electric Systems in Health Care Facilities (ANSI) [6] for additional information.

13.9 Pocket Paging Systems. Paging, other than voice paging over a public address system, is available via pocket pager devices in local and wide area systems. Wide area systems are available through paging services. The service is accessed by calling a telephone number. A digital (usually numeric) signal is received by the party wearing the pocket pager device identified with that phone number. Local paging systems operate in much the same way. Voice systems are available; but voice signals require more air time than digital signals, so most large systems support only digital signals.

Local paging systems consist of a local base station connected to a transmission facility. This facility may be a whip antenna or a low-level, long-wire antenna. Base stations require an FCC license.

The pager control is connected to a code control box or via a telephone system. The pager control modulates and controls a transmitter, which is connected to an antenna. A whip antenna should be located carefully to prevent interference with nearby occupancies and to cover the site that it serves. A long-wire antenna may be run in the ceiling space throughout the facility to furnish an evenly distributed low-level signal.

In addition to the concern for continuity of power, proper grounding for transmitter and antenna should also be addressed. Long-wire antennas should be routed and supported carefully, especially in crowded ceiling spaces.

13.10 References. The following references shall be used in conjunction with this chapter:

[1] ANSI C2-1990, National Electrical Safety Code.

[2] ANSI/IES RP24-1989, Recommended Practice for Lighting Offices Containing Computer Visual Display Terminals (VDTs).

[3] ANSI/NFPA 70-1990, National Electrical Code.

[4] ANSI/NFPA 75-1989, Protection of Electronic Computer/Data Processing Equipment.

[5] IEEE Std 446-1987, IEEE Recommended Practice for Emergency and Standby Power Systems for Industrial and Commercial Applications (ANSI).

[6] IEEE Std 602-1986, IEEE Recommended Practice for Electric Systems in Health Care Facilities (ANSI).

[7] Federal Information Processing Systems Standard 94 ("FIPS 94"), National Institute for Science and Technology, Washington, DC 20234.

[8] *The Intelligent Building Sourcebook*, Bernaden, John and Neubauer, Richard, editors; Johnson Controls, Inc.; Fairmont Press, Inc., publisher; 1988.

[9] Griffith, D. C. "Uninterruptible Power Supplies," New York: Marcel Decker, Inc., 1989.

Chapter 14
Facility Automation

14.1 Introduction. Facilities continue to become more complex, and the governing regulations continue to become more stringent, especially with respect to life safety. Because of this, as well as the increased cost of operating a facility, the use of automation systems to supervise, control, and monitor a variety of individual systems becomes economically advantageous.

For the purposes of this chapter, the automation system is termed the "facility automation system (FAS)." The FAS includes all of the hardware, software, wiring, and incidentals that are part of system supervision, control, and monitoring functions.

The scope of this chapter includes a wide range of commercial facilities, essentially everything except industrial, health care, and institutional facilities.

14.1.1 Adjuncts to the Basic System. The FAS includes such intangibles as
(1) Training of operating personnel
(2) Maintenance (equipment and software)
(3) System data (installation drawings, diagrams, test data; manufacturer's data; and instructions, parts lists, guarantees, service availabilities, and recommendations)
(4) Owner instructions (normal and catastrophic)

Caution should be exercised to ensure coordination with other systems found in a facility, whether or not it is covered in this recommended practice.

FAS design and coordination is an established, yet expanding and rapidly evolving, engineering discipline. Software, hardware, wiring, wiring space, and access can be expensive if system design economics are not understood.

14.2 Functions. Subsystems include
(1) Energy management
(2) Heating, ventilating, and air conditioning (HVAC)
(3) Fire (alarm systems and sprinklers)
(4) Security (intrusion detection, bomb detection, weapons detection, closed-circuit television [CCTV])
(5) Transportation/traffic (elevator, escalator, roadway, parking)
(6) Pollution (interior and exterior air quality, oil spills, toxic and hazardous waste)

(7) Electric
(8) Utilities (electric, fuel oil, natural gas, steam, hot water, chilled water, potable water, sewage)
(9) Communication (life safety, FAS maintenance and operation, coordination with general system) (see also Chapter 13)
(10) Miscellaneous (mail handling, pneumatic tube, conveyors, seismic, catastrophic)
(11) System performance surveillance (monitoring of system performance, component malfunction, diagnostic routines, operator nonresponse, improper operation)

The system level (degree of sophistication) should be established immediately after the need for a FAS has been recognized. It is suggested that the selection of a system level be made after consulting the following:

Facility Size	Major Components
(1) Small	Demand controller Programmable controller
(2) Medium	Central console with distributed processing (direct digital control)
(3) Large	Packaged or hybrid with distributed processing (direct digital control)

Function capabilities will be tailored to suit each individual requirement, and will generally be selected from the following:
(1) Supervisory
(2) Controlling
(3) Monitoring
(4) Data acquisition

To satisfy the overall needs of management on a daily, monthly, and annual basis, ancillary subsystems will be needed and will include some of the following:
(1) Status/control stations for individual systems, e.g., fire detection, alarm, elevators, mechanical, or HVAC subsystems
(2) Data input, e.g., selective manhours, graphics, alarms, logs

14.3 Establishment of System Requirements. In establishing the requirements for a FAS either (1) an audit of existing facilities, or (2) a thorough review of mechanical and electrical designs should be made to determine potential systems for interfacing. Life-cycle cost analyses of the overall FAS and of incremental control strategies may be required to obtain approval from management. Integration of functions other than those that facilitate energy demand and consumption savings, such as fire management and security, can also generate savings by eliminating duplicate central control components. Energy management and fire alarm functions can be integrated into a common "front end" if such a front end is approved for fire alarm functions.

The manufacturers of FAS equipment have developed systems and services that are compatible with practically every need, for today and also for the foreseeable future. The hardware is generally standardized; but these items and the subsystems are flexible, so that the production models can be coordinated with, or tailored to meet, any but the most unusual project requirements.

It is essential to note that manufacturers' production line models have advantages in cost, expediency, parts availability, and test by use over specially designed equipment. Use of the latter should be limited to those instances where a need has been proven or suitable equipment is nonexistent.

Most FAS equipment manufacturers (or their vendors) offer the software as well as the hardware. Software is available as production line items or customized to meet special needs. One of the FAS designer's tasks is the selection of the most suitable of these, alone or in combination. The FAS should be reliable and readily maintainable. The FAS that is not operable (or is only marginally functional) is far worse than none at all. All of the facilities, operating equipment, and systems depend upon the FAS for proper commands and functions. When the FAS or any of the facility's equipment, systems, or subsystems that are controlled or monitored by the FAS are in a failure mode, that condition should be made known by an appropriate alarm, readout, or signal. In the event of a failure, service by a qualified service organization should occur within a reasonable time, and the service organization should have an adequate supply of replacement parts. Furthermore, the response(s) of the facility's equipment or systems to FAS commands should be made known by an appropriate alarm, readout, or signal.

There is little margin for error in FAS design, installation, and maintenance. The safety, security, and energy management of the facility may be wholly dependent upon the control and monitoring provided by the FAS unless redundant manual equipment is available.

When the FAS is utilized for life safety functions, some form of standby power is essential to ensure continued operation of the FAS during any time that the normal source of electric power is unavailable. Consideration should be given to the extent of standby power. To ensure proper operation in the standby power mode, not only the FAS, but also the critical field devices that are monitored and controlled by the FAS should be powered by the standby source.

14.4 System Description and Equipment

14.1.1 Terminology. The following terms are used in this chapter and are frequently encountered in discussions of FAS equipment:

hardware — Equipment, such as processors, computers, peripherals, displays, and electronic circuits.

software — Systems programming necessary for operating the computer equipment.

central processing unit (CPU) — The hardware part of a computer that directs the sequence of operation, interprets the coded instructions, performs arithmetic and logical operations, and initiates the proper commands to the computer circuits for execution.

man/machine interface (MMI) or operator/machine interface (OMI) — The displays, keyboards, printers, etc., used to monitor or modify the operation.

field device — Equipment that receives or transmits a signal, such as controllers, sensors, and relays.

field interface device (FID) — A mid-level or lower level data processing unit that operates compatibly with the CPU but generally at a different location. These devices are sometimes referred to as satellite processing units (SPUs), remote terminal devices (RTDs), or slave stations.

intelligent multiplexer (IMUX) — A device that combines data from a number of points in the data environment and communicates on a single channel in the "report by exception" mode.

multiplexer (MUX) — A device that combines multiple signals on one transmission medium.

data transmission medium (DTM) — The pathway by which signals are transmitted within a FAS. Examples include wire, radio transmission, and fiber-optic links. A combination of several types of DTMs may be used, as appropriate, within a FAS.

14.4.2 General Description of a FAS and Its Components. A typical FAS consists of sensors for input data; field interface devices (FIDs), intelligent multiplexers (IMUXes), or multiplexers (MUXes) located in the data environment; data transmission medium (DTM) for transmitting signals between FAS components; the central processing unit (CPU); and the MMI components that make up the control console. The power supply to the CPU requires consideration of its reliability to ensure intended operation. It is possible that some form of power line conditioning, standby, or uninterruptible power supply (UPS) may be needed as a matter of choice or to conform to applicable codes, laws, or standards. Sensors may be analog or digital. An analog sensor provides an output signal whose amplitude is a function (usually proportional) of the sensed condition, such as temperature, pressure, humidity, voltage, or current. A digital sensor (in its simplest form) is a switch that operates a contact when a certain condition occurs, such as a door opening or closing, or a digitally coded output signal (usually some form of binary code) that indicates the numerical value of the sensed condition. A pulse transmitter accessory on a watthour meter that transmits a pulse when a certain value of energy has been consumed represents a type of digital sensor.

Output control functions can turn fans on and off, open or close dampers, etc., and are typically just a switching action of a relay or electronic logic element, in the case of digital output control, or a variable electronic signal, typically 0-10 V_{dc} or 4-20 mA current, for analog output control. The sensors and output devices to be controlled are connected to the FIDs, IMUXes, or MUXes, which act as collection points for sensor information and the output control commands.

The DTM connects the FID, IMUX or MUX, and the CPU. Traditionally, the DTM consisted of a twisted pair of wires carrying digital information between the FID, IMUX or MUX, and the CPU; but now, the selection of the transmission link requires more careful consideration in the design stage. That is, it should be coordinated with the FAS equipment and be capable of transmitting and receiving data and commands reliably without system performance degradation because of external influences, such as noise (e.g., electromagnetic interference [EMI]). The transmission link should also be able to handle the data at the required data transmission rate. And, lastly, the link needs to meet all of today's requirements and also those in the reasonably foreseeable future. Reliability and maintainability are inherent design considerations. In high-risk areas, some form of monitoring or supervision may be advisable to guard against tampering.

A method that essentially provides two paths is a looped link. In a more sophisticated system, two independent loops will be used (see ANSI/NFPA 72A-1987, Installation, Maintenance, and Use of Local Protective Signaling Systems [2],[86] and ANSI/NFPA 72D-1986, Proprietary Protective Signaling Systems, Style 7 Performance [4]). Another method that essentially provides two paths utilizes bidirectional communications.

NOTE: The ANSI/NFPA Standards referenced above limit the number of fire protection devices on multiplex lines; at least one FAS equipment manufacturer offers a coaxial or fiber-optic cable as the transmission link.

Other media used include fiber-optic cable, wideband coaxial cable, and radio-frequency transmission (including carrier signals on power lines). Each has its advantages and disadvantages that should be considered when selecting a transmission medium for a facility.

The CPU and the MMI devices will be discussed in more detail later in this chapter. Basically, the CPU is the *brain* of the system, taking information from the sensors and giving instructions to the MMI devices on the type and frequency of information to be displayed. It also sends control action commands to the output controls.

From an operating standpoint, a facility may require multiple operating centers. Security may be headquartered in a location other than that of the personnel who run the mechanical and electric systems. A fire command center should be readily accessible to the fire department and is usually on a lower floor; while the mechanical system control center may be located near the penthouse containing the mechanical systems. That is, the FAS design should often accommodate multiple operating centers for efficient operation of many facilities. Some FASs offer a telephone dial-out capability that alleviates the need for multiple operators during nonbusiness hours. The operating center may also be located at the police or security desk, or at the emergency operations center.

The size and complexity of the FAS may vary greatly. A system may contain only one of the mentioned functional subsystems, such as HVAC, fire management, security, access control, energy management, or maintenance management. For example, a simple fire management system may contain only detectors, manual stations, and alarms. This is usually accomplished by a standalone subsystem.

---

[86] The numbers in brackets correspond to those in the references at the end of each chapter. ANSI publications are available from the Sales Department of the American National Standards Institute, 11 West 42nd Street, 13th Floor, New York, NY 10036. NFPA publications are available from Publications Sales, National Fire Protection Association, 1 Batterymarch Park, P.O. Box 9101, Quincy, MA 02269-9101.

However, fire management systems, which may use the same FAS equipment as the mechanical systems for smoke control, are usually integrated into a common system for efficiency of operation, provided that Underwriters Laboratory (UL) listings and insurance underwriters' requirements are met.

Subsystems described in this chapter may be either used alone or combined into a large system. Small facilities can have the subsystems combined just as effectively as the large systems. The prime reasons for combining subsystems are efficiency of operation and sharing of equipment, to reduce the initial acquisition and installation costs, and to enhance operations and maintenance. In general, combined systems allow facility operators to perform more functions for less money. However, caution is required when considering the use of one CPU for all facility systems.

The advantages of distributed processing systems, which can isolate individual subsystems and thereby spread intelligence and decision-making capabilities to various parts of the systems, require more study from technical, reliability, and economic standpoints.

The distributed system has been made practical by the introduction of smaller, relatively low-cost microprocessors that provide the advantage of equipment diversity. Distributed systems may involve distribution by function; e.g., by fire, mechanical control, and operations. Distributed systems may also be distributed by location; e.g., by building areas or by building in a multiple-building complex. Distributed systems may provide for different degrees of reliability and sophistication for each of the distributed systems. The typical distributed system utilizes independent microprocessor subsystems to report to the console "main" CPU. In some instances, the distributed subsystem CPUs may be equipped with provisions for failure or test mode operation independently of the central or main console.

There are nationwide companies that offer remote computer and software services for facility automation, with data links to the user's facilities. These have various advantages and disadvantages that require assessment on a project-by-project basis.

14.5 Central Monitoring and Control Equipment. The heart of the FAS is the CPU (see Fig 150). The CPU may contain a microprocessor or minicomputer, either of which is controlled by software.

All system control is by way of the CPU, which continuously polls (or scans) all the connected equipment looking for changes. Upon detecting a change, the data are processed to determine if the change is a new alarm, status change, or an operator command requiring service and then to select an appropriate response. The CPU then transmits a signal to the appropriate location in the system to initiate the alarm and status changes in a useful form, or to execute other programmed output functions. The CPU also contains programs to detect analog alarm limits, time programs, special action programs triggered by a system event, and other application programs. There are numerous application programs already available, or a program may be customized to the facility.

All operator data are presented on the operator's console or printer, or both. All operator commands and control are executed by the operator's console. Commands should have a positive feedback in the display area of the operator's console to indicate that the command has been issued. Some functions may require a signal at the console that indicates that there has been a proper response to the command. Some systems respond to voice commands from the operator; however, these voice commands are usually restricted to noncritical/nonemergency functions.

Fig 150
Block Diagram of a Medium-Level FAS

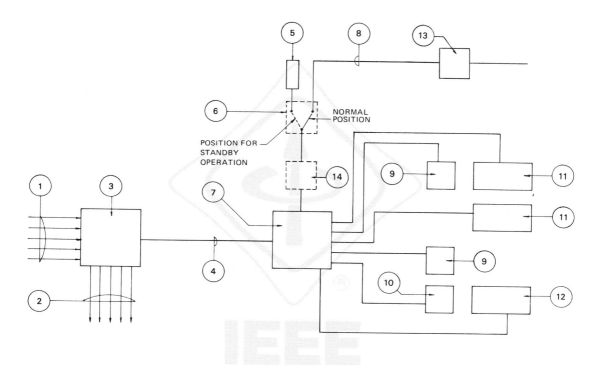

1. Sensor inputs
2. Output functions
3. Field interface device (FID) or multiplexer (MUX)
4. Data transmission media (DTM)
5. Standby power source
6. Power transfer device
7. Central processing unit (CPU)
8. Normal power source
9. Printer (prints complete data)
10. Printer (prints selected data only)
*11. Operator's consoles (all functions)
*12. Operator's console (selected functions)
13. Power line conditioning equipment
14. Power line conditioning equipment (alternate location)

*Man/machine interface (MMI) or operator/machine interface (OMI), full or selected functions (as needed)

Commands that a system will typically execute are as follows:
(1) Fire alarm subsystem
    (a) Start fire pumps.
    (b) Start, stop, or modulate ventilating fans in response to programmed fire management needs.
    (c) Notify police or firefighting services.
    (d) Test and reset remote fire alarm and security systems.
    (e) Test sprinkler systems.
(2) Security subsystem
    (a) Secure access security systems.
    (b) Lock and unlock remote gates and doors.
(3) Energy management subsystem
    (a) Turn lights on or off.
    (b) Start, stop, or modulate HVAC components.
    (c) Start, stop, or modulate remote motors and mechanical equipment.
    (d) Change status of remote control systems.
    (e) Change the setpoint of remote temperature, pressure, and humidity controllers.
    (f) Change the position of remote mixing and exhaust dampers.
    (g) Adjust load shedding setpoints.
    (h) Start, stop, load, unload, and test run standby engine or turbine generator(s).

From the console, the operator should be able to obtain a display of the status of any of the system inputs. Alarm conditions are displayed automatically and should initiate an audible signal to inform the operator that a condition requiring a response has occurred. The operator acknowledges the alarm condition on the console, silencing the audible signal and allowing the system to continue all functions. Fire and security functions also require operator acknowledgment whenever the alarmed point returns to the normal condition.

The Instrument Society of America (ISA) defines several arrangements of annunciator performance. The ISA specifications are recommended to be used when selecting such equipment.

"Human factors engineering" considerations should determine the arrangement of signals and manual control devices on the console. That is, the designer should separate (or cluster) HVAC, security, fire management, and energy management items into individual sections on the console, using distinctive colors and shapes for ready recognition of signals and operational parts. Different and readily recognizable sound signals should be used to indicate fire, smoke, security intrusion system, or subsystem, failures.

In addition to an audible sound, most of the ANSI/NFPA 72 Series standards require two forms of operator notification. One form of notification should be a printout. Printers are, therefore, an important part of the central monitoring section. Printers provide a hard-copy record of all system activity and should document (1) the point that alarms, (2) when it alarmed, and (3) when it was acknowledged by the operator. Printers may automatically provide the operator with action or advisory instructions so that proper responses are enhanced.

Several printers may be used in a system. A printer may be designated to handle only one kind of traffic, such as alarm only, security only, HVAC only, logs only, or various combinations of these based upon how the system will be operated.

Printers are electromechanical devices that require maintenance to ensure that they will remain operational. Many systems are designed with a backup printer that will automatically take over when one of the primary printers is not operating. This ensures that important information is not lost. Printers normally assigned to a given function, such as logging, may be programmed as a backup for another machine and will then print out the failed printer's information as well as its own. Backup magnetic tape systems may be used to record all activities entered and alarmed as well.

Many systems require control of all or selected functions from locations other than the primary central monitoring location. An example may be security monitoring, which is normally controlled from the security office and which may be remote from the main control room. To adequately fill this need, an operator's terminal and printer(s) can be located in the security office as well as in the control room.

As another example, the building superintendent may want to have access to the system. For the superintendent, access to the system should be restricted to only the HVAC or other mechanical systems and perhaps the fire detection and alarm equipment, not to access or control any of the security equipment.

The use of computers, whether microprocessors or minicomputers, allows flexibility in the programming of the system. The main program routine of scanning and basic control of the machine is contained in an operating system. Software programs that are developed by a manufacturer and used in many of the same manufacturer's systems are generally available.

The unique features of each system are accommodated by application programs written specifically for the facility involved. Application software defines the input points, output functions, and the logic of system operations. It is usually entered at the operator's console.

The system should be designed to allow for the modification of the application software from an operator's terminal. The cost of the terminal is usually justified by the flexibility and convenience of being able to update the software locally, rather than resorting to the manufacturer's facilities. For a fire system or other equipment used in an emergency situation, it is desirable to have a generic response to an incoming signal. For this reason, UL does not allow fire systems to have the programming flexibility afforded noncritical functions.

A large or important facility should have a color video display unit (VDU) with a keyboard as the operator's terminal. The keyboard has alphanumeric capabilities and dedicated function keys for easy operation. This type of terminal allows the system to communicate with the operator in the English language and the nomenclature of the facility itself. An alarm may be displayed as "Intrusion Alarm at the West Entrance" followed by the action that the operator should take, such as "Call Security on extension 537" (operator prompting).

The VDU can also generate graphics, such as a floor plan or a schematic of an air-handling system to show the operator exactly which equipment or sensor location is reporting a trouble indication or data readout. Color VDU displays further enhance the system and the functions it can perform.

It should be recognized that color graphic VDU displays and sophisticated features increase the hardware and software costs and the design and delivery time of a FAS. Complex systems may not be justified for many facilities; however, the continuing advances in electronic technologies will make state-of-the-art systems increasingly easy to justify economically in the near future.

Annunciators are often used in control rooms or other locations to provide a continuous visual status indication of selected system parameters. These annunciators can be as simple as a grouping of indicator lights with a printed legend for identification. An elaborate annunciator might be a backlit graphic layout of a complete building showing its floor plan details and lighting the various automatic detection devices on the floor plan when they are in the alarm mode.

CCTV allows the operator at the central monitoring location to select and view many remote areas for the purpose of security, access control, or equipment monitoring. Remote cameras can be selected and controlled from the console. CCTV monitors should display date and time data on the screen. Consideration should also be given to acquiring spare monitors. The console designer should limit the number of monitors that an operator should observe to six. This is the largest number that an individual can effectively watch without becoming confused or distracted. Combining other detection devices that will cause a camera to be switched to a given monitor and signal the operator is a better alternative.

Systems are available that permit one monitor (VDU) to be automatically sequenced so that it can display information from more than one CCTV camera. In the event of an outage of a CCTV camera, that part of the system can be switched to another VDU.

14.6 HVAC and Energy Management

14.6.1 HVAC Monitoring and Control. This function includes the automatic monitoring of HVAC equipment. It also provides operating personnel with information on the status of these systems and selected components. Temperature, dewpoint, humidity, pressure, flow rate, and other key operating parameters are continuously "monitored and displayed upon command or when any abnormal or alarm condition occurs.

An HVAC facility system also provides for the remote control of necessary functions for the operation of HVAC equipment. From the OMI/MMI, fans can be turned on or off, fan speed adjusted, dampers positioned, control valves positioned, pump speed controlled, equipment started and stopped, control points adjusted, and all other functions necessary to properly operate, monitor, and control the mechanical equipment of the facility.

HVAC facility systems are programmed for several operating modes. These programs are developed by the energy management and facility operating groups. Their data should be incorporated into the software to ensure HVAC operations meet code, occupancy, and energy conservation needs. These HVAC programs will account for seasonal needs as well as day, night, and holiday occupancy in each of the buildings, or building areas, that comprise the facility. In the case of large areas in which occupancy varies greatly over short time periods, the programs may have to include real-time or hourly control of HVAC components and possibly a major portion of the lighting.

The types of equipment typically supervised by an HVAC control system are

 (1) Air-handling equipment
 (2) Steam absorption chillers
 (3) Boilers
 (4) Electrically driven water chillers (compressors, reciprocating or helical screw)
 (5) Air compressors
 (6) Air-cooled condensers
 (7) Dampers
 (8) Evaporators
 (9) Fans
(10) Heat pumps
(11) Heat exchangers
(12) Liquid tanks
(13) Pumps
(14) Refrigerators
(15) Sump equipment
(16) Valves
(17) Control switches (electric/pneumatic; pneumatic/electric)
(18) Reheat devices
(19) Cooling towers
(20) Icemaking equipment

The HVAC conditions and quantities to be monitored or controlled may include

 (1) Optimized start
 (2) Supply air and water reset
 (3) Temperature dead-band operation
 (4) Enthalpy changeover
 (5) Demand limiting
 (6) Damper position
 (7) Flow rates
 (8) Fuel supply and consumption
 (9) Gas volume
(10) Humidity/dewpoint
(11) Real and reactive electric power demand and consumption
(12) Line current and voltage(s)
(13) Liquid level
(14) Equipment running time
(15) Equipment wear (revolutions or cycles)
(16) Leaks and oil spills
(17) Fan speed
(18) Degree day heating/cooling
(19) Power failures (main, auxiliary, control)
(20) Pressure
(21) Programmed start/stop operations

(22) Status of miscellaneous equipment and systems
(23) Temperature
(24) Toxic gases and fluids
(25) Combustible gases (e.g., methane in sumps and manholes, hydrogen in battery rooms)
(26) Valve position
(27) Wind direction
(28) Wind velocity
(29) Holiday scheduling
(30) Run-time reduction
(31) Night temperature set back
(32) Central monitoring
(33) Optimized fresh air usage
(34) Trend logging
(35) Pump speed
(36) Steam flow
(37) Amount of chilled water generated
(38) British thermal units (Btus) of heating or cooling consumed
(39) Amount of daylight
(40) Amount of solar energy
(41) Solar collector tilt angle
(42) Indoor air environmental quality

14.6.2 Direct Digital Control. Traditionally, mechanical systems for buildings have been designed with automatic temperature control (ATC) for HVAC systems. Building on this experience plus proven techniques from process control, microcomputer technology offers engineers a powerful tool for the control of HVAC systems.

14.6.2.1 Automatic Temperature Control. In closed-loop control, a *sensor* provides information about a variable (e.g., temperature to a *controller* that actuates a *controller device* (e.g., a valve) to obtain a desired *setpoint*. The output of the controller should operate the controlled device to maintain the setpoint (for example, by modulating a chilled water flow through a coil) even if air or water flow rates or temperatures change. This should happen on a continuous basis and should be fast enough to maintain the setpoint, in which case the controller is said to be operating "in real time."

The creation of a comfortable environment by heating, cooling, humidification, and other techniques is a real-time process that requires closed-loop control. HVAC systems require many control loops. A typical air-handling unit (AHU) needs at least three (one each for fresh air dampers, heating coil, and cooling coil) plus accessory control devices to make them all work in harmony. How these control loops operate has a major effect on the amount of energy used to condition the air.

14.6.2.2 Supplying Computers to Control. When the controller in a closed-loop system is a digital computer, then it is called "direct digital control."

This seems the obvious way to apply a computer to a control loop. However, most computers in control applications today are *not* applied in this way. Until recently, most computers were principally used as *supervisory systems* to supervise the operation of an independent control system.

A supervisory computer monitoring the ATC system and capable of resetting the controller setpoint has some very basic limitations.

(1) The most sophisticated supervisory computer cannot improve the operation of the control loop because the controller is really in command. Any deficiencies or inaccuracies in the controller will always remain in the system.

(2) Interfacing the computer to a controller that is frequently a mechanical or electromechanical device is expensive and inaccurate.

(3) The computer's sensor and the actual controller's sensor may not agree, leading to a good deal of confusion or a lack of confidence in one system or the other.

14.6.2.3 Direct Digital Control Software. A computer's power is in its software. When applied to automatic temperature control, properly designed software offers dramatic benefits.

(1) Control system design is not "frozen" when a facility is built. Alternative control techniques can be tried at any time, at little, if any, additional cost.

(2) With software configured control, all control panels can be identical, which facilitates installation, checkout, and maintenance. One standard DDC computer can control virtually any piece of HVAC equipment.

(3) The control system can be upgraded with improved software in the future. No additional equipment or installation will be required.

(4) Comfort and operating cost trade-offs are easily made by the flexibility to modify the operating parameters in the control system. Optimum energy savings can be realized without sacrificing occupant comfort.

Flexible software should allow changing not only setpoints, but control strategies as well. Control actions, gains, loop configurations, interlocks, limits, reset schedules, and other parameters are all in software and can be modified by the user at any time without interrupting normal system operations.

With DDC, an operator, via software control, may access all important setpoints and operating strategies. Accuracy is assured by the computer. Control loops can be reconfigured by revising the loop software, with no rewiring of control devices. Reset schedules can be changed just as easily. For example, heating setpoints and strategies can be set in the summer with complete assurance that the DDC system will perform as expected when winter arrives.

14.6.2.4 Direct Digital Control (DDC) Loops. Typically, DDC closed loops consist of sensors and actuators, in addition to digital computers, as the controllers. Certain design features should be used to obtain optimum performance from DDC loops. Sensors for DDC loops are very important, since the computer relies on their accuracy to provide the precise control that an HVAC system operator needs. A 1° change in some temperatures, such as chilled water, can affect energy consumption by a couple of percentage points, so that a control system with even 1° of error is not fully controllable in terms of energy use. So as not to waste the precision of the DDC, quality sensors should be used that do not require field calibration, and do not have to be adjusted at all to interface with the DDC computer. Control set-points are thereby achieved with absolute accuracy under all conditions, at all times.

With the computer performing DDC, the traditional problems of temperature fluctuations and inefficient operation can be eliminated. Proportional-integral-derivative (PID) control techniques provide for the fast, responsive operation of controlled devices by reacting to temperature changes in three ways: (1) the difference between setpoint and actual temperature (proportional), (2) how long the difference has persisted (integral), and (3) how fast the actual temperature is changing (derivative). PID saves energy and increases accuracy simultaneously by eliminating hunting and offset, and by decreasing overshoot and settling time.

All digital computers work with binary (either on or off) information. Since it is necessary to modulate controlled devices (e.g., motors that operate dampers or valves), a complicated interface device (transducer) is often employed. A better method to use, which has been perfected in much more demanding process applications, is pulse-width modulation (PWM). The computer's binary outputs are directly connected to a modulating device. PWM uses bidirectional (open/close) pulses of varying time duration to position controlled devices exactly as required to satisfy demand. Wide pulses are used for major corrections, such as a change in setpoint or start-up conditions. The pulse width becomes progressively shorter as less correction is required to obtain the desired control setpoint.

14.6.2.5 DDC Energy Management. Many strategies have been developed to effectively manage and save energy in HVAC system operation. DDC systems can be intelligently integrated with temperature control functions in the same computers, in such a way that energy reductions are achieved without compromising the basic temperature control functions. This will also eliminate the need to supplement a conventional ATC system with an add-on energy management system (EMS), which will save equipment, installation, and maintenance costs.

14.6.2.6 DDC Distributed Networks. Implementing DDC in an entire facility with numerous pieces of HVAC equipment can be accomplished with any number of computer and process control systems. Starting with a basic control loop, a system can expand to control an entire facility.

A DDC computer should be capable of handling a number of control loops (four to eight is typical). Accessory on/off control and monitoring functions should also be controlled by the same computer. Each computer should be capable of independent operation and be able to perform all essential control functions without being connected to any other computer. This suggests that each separate major piece of HVAC equipment (such as an air handler, boiler, or chiller) has its own DDC computer, in the same way that each would have independent conventional control panels. These are then tied together in what is called a local area network (LAN) for communications. This results in a truly distributed processing network in which each computer can perform all control functions independently.

Twisted-pair, low-voltage wiring (foil shielded) is an economical choice for the interconnections, although coaxial cable or fiber-optic systems can be used if they are installed in the facility to provide a variety of communication services.

Somewhere in this LAN, a "window" is required to allow for human interface with the DDC computers. This is accomplished with a different type of computer, connected to the network at any location, which provides access to the DDC computers. All control setpoints and strategies can be programmed from this access computer, and all sensor readings can be monitored.

14.6.2.7 System Integrity. A DDC system can be designed for high reliability and for much shorter mean time to repair (MTTR) than a conventional ATC system. The major design requirements are

(1) Independent control computers—In a distributed processing network, these computers ensure that the failure of one computer will not adversely affect the operation of other computer systems.

(2) Remote data link diagnosis—Allows the computer manufacturer's factory experts to dial into the DDC system and troubleshoot control problems.

(3) Universal computer replacement—Requires that all control computers be identical, regardless of the HVAC equipment being controlled.

Since the access computer in a distributed network is not capable of any real control, it does not need any special backup system. Remote data link diagnosis can quickly pinpoint an access computer problem, and repair does not have to be immediate to maintain environmental comfort. All HVAC systems are under the control of the independent DDC computers, which will continue to function normally.

System integrity considerations should also include what happens when a computer fails. A safe condition has to exist when this happens. Therefore, whenever a DDC computer is used, all standard safety devices (i.e., for overload, freeze protection, etc.) should remain in the system with the computer. These are usually very simple devices (certainly less complex than a computer) that have been proven in many years of HVAC system design, and are not rendered obsolete when a computer is used for direct digital control of the system.

14.6.3 Energy Management. The energy management function of the FAS, which is accomplished primarily through the control of HVAC equipment, is a major item in the reduction of operating costs and the use of energy. The issue of energy and the specifics of control are covered more extensively in Chapter 17. The FAS will provide the capability to implement the energy management plan. Here, especially, the FAS designer should have close coordination with the energy management engineers. During the preliminary design period, many things happen all at once. Fuel and utility costs change, codes change, techniques change, operating methods change, all in short, and sometimes overlapping, time spans.

The initial decision to implement a single or redundant CPU and to use distributed processing for HVAC should be made early, and backed up with cost and operational data. Thus, the FAS designer and the energy management group should develop a basic plan and get it approved quickly. Once that is done, the space requirements and other details can be distributed to all concerned. Selection of sensors and methods for responses can then be developed within their own timeframes. Similarly, shutdown alarms and other features can be developed as the design proceeds. To reiterate, the FAS designer and energy management team should develop the basic system early. Standards such as ANSI/UL 916-1987,

Energy Management Equipment [12][87] should be required reading before the equipment is selected.

14.7 Fire Management. Fire management, as used in this section, encompasses fire alarm systems, the functions of which are detection of fire (manual or automatic) and the sounding of a fire alarm signal for evacuation or other purposes. It may also include smoke control subsystems and more sophisticated occupant notification approaches, as well as other fire safety related control functions.

Fire alarm or management systems are required by code in most commercial facilities. The major objectives of these systems are to

(1) Detect a fire as early as possible.
(2) Notify the fire department.
(3) Notify the occupants.
(4) Notify in-house fire wardens (required if system encompasses fire protection signaling systems. See ANSI/NFPA 72D-1986 [5]).
(5) Appropriately control HVAC system to contain fire and smoke.
(6) Use HVAC system to create safe havens within the structure for occupants when evacuation is not practical.
(7) "Capture" the elevators, according to a preplanned scheme.
(8) Provide a fire command station to be used by fire fighters as a control center during the emergency.
(9) Provide an emergency two-way radio or telephone system, or both, for use by fire fighters and in rescue operations.
(10) Provide a voice communication system to direct occupants to safety.
(11) Start fire pumps.
(12) Supervise and back up all critical fire alarm components to assure their proper operation in an emergency.

The basic elements of a simple fire alarm system are initiating devices, control panel, and indicating devices. The other elements that make up a fire management system are "add-ons" to the basic system.

Initiating devices are the elements that sense the presence of smoke or fire and then inform the system. These devices are either manual or automatic, or both. Manual devices are typically manual pull stations that are strategically located throughout a facility and are intended to be operated by an occupant if he or she discovers a fire. Typically, codes require that a manual station be located at each legal means of exit (door or stair per floor), and at intervals not to exceed 400 feet along the path of egress.

Automatic devices, of which there are a number of types, sense a characteristic or a result of a fire. Detectors may be mounted in occupied areas or in restricted spaces (e.g., mechanical areas, electrical closets, plenums, hung ceilings, or ventilation ducts). The most common automatic detectors are

(1) Thermal detectors that sense heat.

---

[87]UL publications are available from Underwriters Laboratories, 333 Pfingsten Road, Northbrook, IL 60062.

(2) Smoke detectors that sense the visible (i.e., photoelectric) and invisible particles (i.e., ionization) generated by a fire.

(3) Flame detectors that sense the infrared or ultraviolet radiation from a fire.

(4) Rate-of-rise detectors that signal excess temperature rise in a given time period.

(5) Waterflow detectors that sense the flow of water in a sprinkler system.

(6) Tamper devices that signal an abnormal operation.

Each of the various detector types has subcategories. An engineer designing a fire management system should be fully familiar with the types and subcategories currently available and their correct application.

Traditionally, ten or twenty fire alarm detectors have been wired in parallel to constitute a fire alarm zone. When any one of these devices goes into alarm, the fire alarm control panel indicates that there is an alarm in the zone; but the exact device in alarm is not annunciated. The zone isolates a fire location to a particular part of the building, but not the exact location. Device malfunctions that can be caused by the dust contamination of particle ionization and photoelectric detectors can be difficult to locate with this configuration.

New fire alarm technologies allow the use of "intelligent" detectors. As many as 100–200 input/output devices can be wired in parallel on a single twisted pair of conductors. Power and data are transmitted simultaneously. Each fire alarm device has its own unique address in the system, which allows precise identification of a fire alarm point at the fire alarm control panel. Intelligent fire alarm devices may also have an analog output capability that allows the condition of the particle ionization or photoelectric head to be continuously monitored at the fire alarm control panel or central console. This also allows dirty chambers to be detected before a false alarm is generated and the building is inadvertently evacuated.

Indicating devices are used to notify the occupants that a fire condition exists. In the past, bells, gongs, or horns, or a combination of these three, have been the primary method used in fire alarm systems. Recent system designs use electronically generated fire signals that are transmitted by audio amplifiers and speakers. Systems of this design can also be used to broadcast voice messages to give occupants specific instructions, which cannot be given with a bell, gong, or horn. In many jurisdictions, codes require audio systems in high-rise construction. Visual signals are also required by many jurisdictions to signal individuals with impaired hearing. The use of prerecorded messages for directing evacuation or other instructions is controversial. There is the possibility of events that are not predictable and, therefore, not compatible with prerecordings. This is why such systems should be equipped with a microphone, and proper in-house procedures should be developed for its use.

The control panel is the computer of the system, taking the alarm information from the sensors, processing it, and activating the indicating and alarm devices. In addition, this control panel can also initiate other functions that are required in a fire management system, such as fire department notification, elevator capture, smoke control, etc. The control panel may include a device for recording the time and the location of any fire or smoke indication (a code requirement in some areas). Alternatively, this recorder may be separate and remote from the control panel.

Figure 150 shows a block diagram that describes a medium-level FAS in which the CPU performs the function of the control panel. Many systems use a microprocessor or a minicomputer, which are software-controlled to initiate the desired output functions. Newer systems use addressable/intelligent field devices that can pinpoint the exact location of an emergency situation. Additionally, many systems offer an alarm verification feature that greatly reduces the likelihood of false alarms from smoke detectors. These features should be evaluated on a project-by-project basis.

When an alarm occurs, these systems may be programmed to position dampers automatically and operate fans to create areas of positive and negative pressure to reduce the spread of smoke to other areas and to exhaust smoke from the building. The CPU can also initiate the "capture" of elevators, which sends them to a designated floor for use by fire fighters and authorized personnel.

Fire management systems should be as reliable as possible because their function is to protect life and property. To ensure that a particular system meets desired quality levels and the design has not been changed, approval for the equipment's intended purpose (indicated by a UL label) should be required when specifying it. Other laboratories or agencies also may be certified to do qualification testing; however, assure that the laboratory or agency is acceptable to the local code-enforcing authority. By specifying the UL listing, there is assurance that the equipment is capable of providing the system operation required by ANSI/NFPA 72 Series standards. However, the FAS designer is not relieved of responsibility simply by using laboratory listed products. Listed products are only useful when properly applied and thoroughly coordinated. Furthermore, the owner should be notified that scheduled maintenance and tests are required (see ANSI/NFPA 72H-1988, Testing Procedures for Signaling Systems [9]). Finally, the responses to signals, alarms, and communications require continual practice, upgrading, and monitoring by qualified personnel. The FAS designer may elect to participate in initiating these procedures.

The FAS designer should assure that electrical supervision, or monitoring integrity, of signal circuits is provided in accordance with the ANSI/NFPA 72 Series standards. In general, these standards require that a single open or a single ground trouble condition be signaled automatically to, and recorded at, the central supervisory station within 200 seconds of its occurrence.

14.7.1 Life Safety Communications. Some systems use prerecorded voice messages to automatically direct the occupants instead of sounding bells and horns (subject to the previously noted limitations). More elaborate versions of these systems use several messages that direct people to different areas, depending upon the location of the fire. For instance, occupants of the fire area may be directed to go up two floors and those below the fire area to go down two floors. Occupants of the floor that will be receiving people will be advised, and there may be a general building advisory. Special messages are broadcast to the occupants of captured elevators, advising them of the emergency and instructing them to leave the elevator car when it reaches the designated landing. The messages should be coordinated and sequenced properly, so that stairwells do not become overcrowded as people relocate. Usually, the messages for the elevators and fire floor are broadcast first.

In larger public buildings, staged or programmed notification (sometimes referred to as a pre-signal alarm) may be required to the building fire control staff, local fire department, central supervisory station, maintenance staff, and to the public as well as for evacuation.

These systems can also be used to transmit or broadcast special instructions by using a public address feature. In this case, a person in authority, usually from the fire department, will select the areas he or she wishes to address from the console, and his or her voice will be broadcast to those selected areas.

Emergency telephone systems are usually provided to give fire fighters a reliable and private two-way communication between the fire command center and each floor. The emergency phones may be permanently installed on the floor and elevators or phone jacks provided, in which case the fire fighter will carry a handset with a phone plug. If phone jacks and portable handsets are used, a highly visible storage area for a number of handsets should be provided at the fire command center.

For convenience, or for code compliance, buildings may include a fire standpipe telephone system. Essentially, this is a telephone at each standpipe hose or hose cabinet, at selected valves and at external high-pressure fire department hose connections. When feasible, and in accordance with codes, these telephones may be of the sound powered type.

14.8 Security

14.8.1 Security Systems. Security systems are an essential element in any facility. They share a common characteristic in that they continue to expand both in need and in technique. Although every security system should be specifically designed for each individual project, they are all based on the concept of providing safety for the facility occupants (employees, contractors, and visitors) and the protection of the contents of the facility. In areas where it is desirable to restrict access to authorized personnel, an access control system could be utilized (see 14.8.2). The type and depth of the security system varies with the number of functions performed at the facility. The security level in any one part of the facility may be entirely different from the security level provided in other areas. Banks with generally open or public access will need one type of system. Military installations or offices supporting military organizations have limited access, and, thus, other security needs. Research facilities, laboratories, and other places where commercial or industrial designs or developments are involved require in-depth security. Stores and normal commercial properties may require limited security precautions. However, if a facility is located in a known high crime area, the depth of the security system will tend to increase.

The concept of security is to make the system fit the need. Excess security is costly to install and maintain. Minimal protection can, of course, be the most expensive, simply because the owner believes that all necessary protection has been provided; but there will probably always be a weak spot in it. Security starts with protection at the perimeter of the property. A chain-link fence may be used to prevent a casual walk-in by neighborhood youngsters and will identify a specific property line. More protection at this fence can be provided by using outdoor perimeter detection devices that will detect intruders crossing the property. Infrared light beams, microwave, E-field, or covered and buried line detectors are just a few of the possibilities. At the perimeter of the building, magnetic switches can be used to monitor perimeter doors and other movable openings. Window foil, traps, pressure mats, and infrared devices may be used to protect other areas inside the building perimeter. Motion, audio, capacitance, infrared light, and vibration detection devices may be used to protect areas and objects within the facility.

Supervision of the data transmission media that extends to the sensors is important in a security system, just as it is in a fire alarm application. In the case of fire, a break in a wire is handled as a trouble condition. However, in a security application, the loss of supervisory current may be caused by an intruder attempting to compromise (defeat) the alarm system, and it should be treated as an alarm, not as a trouble indication. A $\pm50\%$ change in line current should indicate an alarm meeting UL Grade A requirements. This is adequate for most general security applications.

Communicating alarms in high-security areas should be by two channels. The telephone system may be used as the primary channel to the UL-listed central monitoring facility, and this should be backed up by a radio link.

When high security is required, random digital or high-speed digital interrogation and response are techniques designed to render the circuits that are most difficult to compromise. These techniques usually satisfy UL Grade AA requirements. CCTV can be used in manned operations to allow a single guard to observe many areas. Cameras can be controlled from the central monitoring location to view a larger area. Cameras are available that operate with very low ambient light levels (i.e., starlight) and still produce a satisfactory image. It should be remembered that the security office cannot watch a television continuously; thus the CCTV is an adjunct subsystem for specific and intermittent needs. Individual applications of CCTV cameras may require certain optional equipment or auxiliaries, such as

(1) Fixed focus camera
(2) Remotely controlled pan, tilt, and zoom lens functions
(3) Outdoor camera enclosures that may include heaters, windshield wipers, or sun visors. In some instances, pan and tilt limiters may be required to avoid directly aiming the lens into the sun.

Motion detection can be an integral part of a television system that is silently viewing an area. When motion is detected in an area viewed by a camera, that camera will be automatically switched on to a monitor for viewing, possibly using light or sound to attract the attention of the security officer. The trespass can be captured on videotape.

It is obvious that there are special operational considerations, as well as technical design problems, in the development of a CCTV installation plan. For the FAS designer, it is often advisable to obtain specialized support services to assist in the design of and user coordination with the CCTV system.

14.8.2 Access Control. Another security system allows access to a facility without the need for locks and keys. In many instances, keys and locks are too cumbersome and expensive a method to secure an area or building. Keys are lost or proliferate when people entrusted with them have copies made or fail to return them. Cylinder pins wear, and weather also affects conventional locks. A modern and very popular approach has been the card reader. A plastic card, similar to a bank or credit card, is encoded with a hard-to-reproduce cipher. Several different techniques for encoding are in use, such as magnetic stripe or bits, capacitance, photoelectric, and radio frequency. The encoded card is placed in (or near) a fixed card reader at a point of entry, which determines if the card is valid. If the card is valid for that door, time, and day, it will release the doorlock and allow entry. Systems that require entering a secret number, unique to each card, on a keyboard as well as inserting the card in the reader add another level of security. Other security systems are biological in nature, requiring a hand, eye, finger, or voice to be analyzed and matched to the database. Unauthorized entry attempts by individuals are processed as an alarm and indicated on the operator's terminal, other signal or display, and alarm printers. The system may also be instructed to record each entry transaction on a printer. The use of card access systems is increasing, and new innovative ideas, such as parking garage access; time, attendance, and resource control; and tracking badgeholders on a real-time basis throughout a facility, are being introduced. A properly designed card access system can be used to replace guards in low-traffic situations. The FAS designer is advised to research the available market to be assured that he or she is specifying and obtaining a state-of-the-art access control system.

14.9 Transportation and Traffic. Transportation and traffic control equipment that may require control and monitoring include the following:
  (1) Driveway and roadway traffic control
  (2) Loading dock traffic control
  (3) Elevator status
  (4) Escalator status
  (5) Moving walk status
  (6) Parking access control
  Whether these are to be programmed, manually controlled, sensor controlled, or simply monitored are matters for coordination with the discipline that has the design or operating responsibility.
  There are obvious precautions, such as
  (1) No remote control over escalators where starting or stopping remotely has the inherent possibility of danger to riders.
  (2) No possibility of a remote shutdown of an emergency system or safety device.
  (3) No countermanding of signals or directions by facility operators, unless a local authority requires emergency personnel to have that ability.
  (4) A procedure for advising police, fire fighters, or emergency medical personnel where to (or not to) enter the facility.

14.10 Pollution and Hazardous Waste. Monitoring and control of functions associated with pollution and hazardous wastes may be integrated into the FAS. Functions to consider including are monitoring of air quality (particulate level); monitoring and control of baghouse and scrubber equipment, such as pressure drop across air filters; monitoring of effluent water quality (e.g., percentage of dissolved oxygen); leak detection in hazardous liquid holding tanks; and monitoring of carbon monoxide in garages.

14.11 Electric Systems. Included in this category are the monitoring and control of lighting, normal and emergency power sources, power conditioning equipment, uninterruptible power supply (UPS) systems, and electric distribution systems.

14.11.1 Lighting Automation Systems. The ordinary control of lighting is performed by one or more of the following methods:

(1) Local manual switches
(2) Switches or circuit breakers in panelboards
(3) Time clock operation of contactors and relays
(4) Daylight sensors operating through relays and contactors
(5) Illuminance sensors controlling lamp light output through ballast dimming control circuitry

The FAS offers the opportunity to provide for the programmed control of lighting as a part of the energy management system. That is, programs are commercially available for the timed control of various lighting needs, such as

(1) Continuous light in stairwells and exit pathways
(2) Timed control of indoor, general light
(3) Sensor control of light adjacent to window areas
(4) Timed or light sensor control of light in vehicle parking areas and garage decks
(5) Continuous nighttime lighting for security
(6) Timed or sensor control of light used for advertising or display
(7) Notification of manual turning on or leaving on of lights during off-hours

14.11.2 Normal and Emergency Power Sources. Monitoring and control functions that are associated with power sources may include the following (refer to ANSI/NFPA 110-1988 Emergency and Standby Power Systems, Chapter 3 [10] and ANSI/NFPA 110A-1989, Stored Energy Systems [11]:

(1) Circuit breaker open/closed status
(2) Circuit breaker open/close command
(3) Static switch source 1-source 2 command/status
(4) Automatic transfer switch normal emergency command/status
(5) Remote emergency stop alarm
(6) Engine or turbine low-lube-oil pressure alarm
(7) Engine or turbine overcranking alarm
(8) Engine or turbine air shutdown damper alarm
(9) Generator on/off status
(10) Engine or turbine high temperature alarms
(11) Generator overtemperature alarms
(12) Frequency, and commercial supply and generator
(13) Kilowatts, kilovars, and voltage
(14) Power failure of supply or of power subsystems; operation of protective relaying

14.11.3 Uninterruptible Power Supply (UPS) Systems. Additional monitoring and control functions that are associated with UPS systems may include the following (refer to IEEE Std 944-1986, IEEE Application and Testing of Uninterruptible Power Supplies for Power Generating Stations (ANSI), Chapter 5 [13]):[88]

(1) Loss of synchronization (inverter not synchronized to alternate ac source, or alternate ac source unavailable)

(2) Low inverter voltage

(3) Protective device actuation

(4) DC bus undervoltage

(5) Overload

(6) Reverse transfer (load is supplied by the alternate ac source)

(7) Cooling trouble

(8) Alternate ac source trouble

(9) DC operation

(10) AC input disconnect device position

(11) DC input disconnect device position

(12) Static transfer switch position

(13) Output disconnect device position

(14) Maintenance bypass available

(15) Critical load on maintenance bypass

(16) Battery charging rate (float/equalize)

(17) Redundant power converter module unavailable

14.11.4 Electric Distribution Systems. The automation of electric distribution systems may be accomplished through the control and monitoring of power circuit breakers. Alternatives include electrically operated circuit breakers and fusible switches, shunt tripping of manually operated circuit breakers, and electrical contactor control.

14.12 Mechanical Utilities

14.12.1 Functions. The FAS may be called upon to monitor and control a number of mechanical systems, other than HVAC, including: water supply and distribution, sanitary and storm sewers, compressed air supply and distribution, de-ionized water, natural gas, and liquid fuel storage and supply. Functions and conditions may include

(1) Pressure (water, air)

(2) Level (water, sewage, fuel)

(3) Water acidity/alkaline concentration level

(4) Pump status

(5) Valve position

14.12.2 Integration. The monitoring and control of individual mechanical systems may require overall supervision by the CPU, rather than distributed control at each satellite FID location. An example where CPU supervision is required is in the implementation of electrical demand limiting.

---

[88] IEEE publications are available from the Institute of Electrical and Electronics Engineers, IEEE Service Center, 445 Hoes Lane, P.O. Box 1331, Piscataway, NJ 08855-1331.

14.13 Communications

14.13.1 Maintenance. A telephone jack or instrument should be installed at each remote FID and MUX location to facilitate both initial system checkout and point calibration as well as subsequent troubleshooting efforts.

14.13.2 Central Supervisory Station. A central supervisory station is an off-premises office, usually operated by an independent private company, for performing supervisory functions. Fire alarm and security operations are most often performed by these companies; however, many of the other alarm and control functions described in this chapter can also be performed by them. When security, life safety, and fire services are involved, the approval, authorization, and certification of these companies should be obtained from the authorities having jurisdiction over them (e.g., fire and police departments) and from insurance carriers.

The central supervisory station staff, as directed, will notify fire and police departments and other municipal agencies, building staff, listed company supervisory staff (paged, at home or at work), or perform such other activity as is prescribed whenever appropriate signals or levels are indicated. The central supervisory station may replace local building operations fully or partially for all or some of the FAS supervisory functions. During off-hours, the central supervisory station may be responsible for all supervisory functions.

Connections between the facility and the central supervisory station are achieved through a supervised telephone line. Provisions should be made for the connection of the central supervisory station to the facility's equipment, usually an auxiliary connection to the central console computer when one exists. For smaller facilities, it may be connected by an electronic interface.

14.14 Miscellaneous Systems. The functional categories of monitoring and control points that were already discussed are not intended to limit the range of options available to the FAS designer. Any number of facility process and support systems, from material handling and conveying to pneumatic tube, may be integrated into the FAS design.

14.14.1 Computerized Maintenance Management System (CMMS). The CMMS should include subsystems, as follows:

14.14.1.1 Equipment Subsystem. Equipment information includes specifications, ratings, spare parts, cross-referencing, manufacturer-of-record, installing contractor-of-record, maintenance contractor-of-record, model number, installation date, and other history.

Cross-referencing capabilities should provide reference to a piece of equipment by several common criteria.

14.14.1.2 Inventory and Purchasing Subsystems. An inventory subsystem should provide: (1) multiple warehousing or storage locations per item, (2) stock item usage information, (3) economic order quantity calculation methodology, (4) minimum and maximum order quantity calculation methodology, (5) reporting a notification to reorder specific items, and (6) cross-referencing.

The cross-referencing features shall enable the user to: (1) locate stock items by manufacturer, (2) locate stock items by manufacturer part number, and (3) locate stock items by generic name.

14.14.1.3 Work Order Planning Subsystem. A work order planning subsystem should provide the functions that help plan the work to be performed. The primary objectives of work order planning are
(1) To provide an efficient means of requesting and assigning work to be done by maintenance personnel.
(2) To provide an efficient means to requisition materials for the work order.
(3) To provide a method of transmitting written instructions on how work is to be done.
(4) To provide estimates and accumulate actual maintenance costs.
(5) To provide the information that is necessary for the preparation of management reports.

14.14.1.4 Work Order Subsystem. A work order subsystem should provide the mechanism for the entry, updating, and monitoring of work orders. The primary objectives of a work order subsystem are
(1) To provide for the entry and updating of work orders.
(2) To provide for various inquiries on work orders.
(3) To provide a method of reporting costs to the work order.
(4) To provide a means to capture historical data upon completion of the work order.

The work order subsystem should provide current work orders in the backlog according to specific criteria, including by a single cross-referencing item or by multiple items.

The cross-referencing procedure should allow
(1) Calling up the work order backlog of an individual craftsman or by a crew.
(2) Same as in item (1), except only high-priority work orders would be called up.
(3) Calling up the work order backlog on a specific piece of equipment or system.
(4) Calling up all work orders backlogged due to lack of parts or tools.
(5) Calling up all work orders issued to a specific subcontractor or maintenance contractor.

The work order subsystem should be capable of producing a maintenance schedule for each craftsman or crew.

14.14.1.5 Preventive Maintenance Subsystem. The preventive maintenance subsystem should identify all preventive maintenance activities within the facility for each major piece of equipment and system listed.

Preventive maintenance scheduling should utilize appropriate combinations of manually entered recommended maintenance intervals dependent upon calendar time (hour, day, month, quarter, year) and equipment or system operating parameters.

System operating parameters are defined as
(1) Running hour meters
(2) Clock hour meters
(3) Flow meters (ft^3/second, gallons/hour, etc.)
(4) Consumption meters (kWh, gallons, etc.)
(5) Load meters (percent of full load)
(6) Cycle meters (on/off cycles)

(7) Vibrations

(8) Noise

(9) Temperature

(10) Displacement

(11) Velocity

(12) Combinations of the above — "dynamic predicted maintenance," "signature analysis," etc.

(13) Other parameters

The preventive maintenance subsystem should be designed to minimize labor demands by scheduling maintenance only as required by the system operating parameter and in conformance with manufacturers' recommendations and generally accepted standards for maintenance.

The schedules should be available by crew or facility location or by major equipment systems. Summaries should display labor by craft, by facility locations, by crew, or in the entire facility.

14.14.1.6 Downtime Information Subsystem. The downtime information subsystem should provide downtime tracking for critical pieces of equipment within the facility. Summaries of outages and equipment availability should be instantly retrievable in order to isolate problem areas and evaluate the causes of downtime.

The downtime information subsystem should also provide for major overhaul or renovation work. The system should be capable of describing the step-by-step tasks, the scope of the work, and the craft requirements and weekly labor summaries by craft for labor forecasting.

14.15 FAS Design and Installation. The design of a FAS requires the coordinated efforts of a number of engineering disciplines including mechanical, electrical, and instrumentation and controls engineers. The following paragraphs address salient points about the design and installation of a FAS.

14.15.1 System Specification. A system specification is essential for all, or almost all, FAS systems. It spells out the requirements of the system, at an early stage and in sufficient detail for evaluation; in-house review; and preliminary discussions with approval authorities, consultants, and suppliers. Later, it will be developed to a point where it can be used for detailed design, incorporated into the technical specifications of a purchase (or lease) contract, or into furnish-and-install provisions of a contract. The system specification contains the following features:

(1) A list of all systems to be covered by the FAS, together with all functional requirements. Later in the design stage, these requirements will be detailed as to specific equipment, setpoint ranges, physical constraints, installation requirements, and the factors listed in this chapter regarding system features.

(2) Approval and review procedures — All of the operational, security, and general staff should approve the proposed installation at some point.

    (a) In-house and, when appropriate, consultant review by
        (i)   Owner's staff, if designed by others
        (ii)   Safety staff
        (iii)   Fire prevention staff
        (iv)   Security staff
        (v)   Law department (contract, claims)
        (vi)   Architects (location and finishes of alarm, display, and other devices)
        (vii) Operational staff
        (viii) Maintenance staff
        (ix)   Insurance staff
        (x)   Construction and inspection staff
    (b) Jurisdictional approval authorities, fire, police, code, and insurance underwriters

(3) Content and format of displays and hard-copy output to be developed in conjunction with all affected staff; general layout and characteristics of input and display devices; specification of the manner of operator input of variable system parameters, added devices, or system data as distinct from software modifications.

(4) Backup, emergency, and contingency modes of operation

14.15.2 Selection of Data Transmission Media (DTM). The FAS designer should select one of several DTMs in order to allow the central components to communicate with remotely located equipment. An economical solution often used to provide connectivity for a complex of buildings is existing telephone pairs. Other DTM to consider are

(1) Shielded or unshielded twisted pairs
(2) Fiber-optic cable
(3) Coaxial cable
(4) Power line carrier over existing wiring
(5) Two-way radio
(6) Point-to-point microwave
(7) Infrared light transmission
(8) Laser light transmission

Based on the owner's requirements, the designer may need to evaluate the reliability and speed versus first cost and life-cycle cost of the potential DTM.

14.15.3 Interfacing to Existing Equipment or Equipment Provided by Other Disciplines. One method that is available to provide a single identifiable interface point is the provision of a data terminal cabinet (DTC) in each mechanical equipment room or near each major system that is monitored or controlled. The DTC contains double-sided terminal strips, with one side connected to field wiring from instrumentation and controls, and the other side connected to a FID or MUX.

14.15.4 Procurement Documents. In preparing plans and specifications for a FAS, the designer should include a detailed point list of all functions monitored and controlled that includes the range of values expected for analog points. In general, performance-type specifications are preferred for hardware, in order to allow for differences in vendors' equipment. In specifying system software, a method for entering user data, such as adding and deleting points, changing temperature limits and start/stop times; the capability for user created programming; and the capability to design statements for DDC loops should be required.

14.15.5 Installation. Prior to equipment purchase, detailed construction drawings that show the connections to existing equipment, equipment locations, and all conduit runs should be provided. During the construction period, periodic tours of the job site should be made by the designer. Care should be exercised when making modifications to existing equipment so as not to defeat existing safety interlocks.

14.15.6 Testing. In general, two stages of testing are desired: factory tests and site acceptance tests. The factory test setup should include the CPU and all peripherals, representative FIDs and MUXes, and DTMs of each type to be provided in the system. Site acceptance testing should verify the operation of all points in the system and demonstrate that all analog inputs and limit alarms are within specified tolerances. Verification of system operation in failure modes (e.g., loss of power at a FID, loss of CPU operation) should be demonstrated. In addition, if the criticality of the FAS warrants it, an endurance test that continues for 2 weeks may be requested in order to demonstrate overall system reliability.

14.15.7 Warranty. A warranty period, which starts after system acceptance, should last at least 1 year. The warranty covers the repair of equipment and software that fails during the designated period, but does not include regular preventive maintenance.

14.15.8 Software. Commonly available software, some or all of which may be specified dependent on the size and complexity of the FAS, includes

(1) Alarm priorities
(2) Alarm inhibitions
(3) Analog alarms
(4) Integrations, e.g., energy consumption
(5) Totalization, e.g., summation of motor run times
(6) Time switching, including optimizing start and multiple-channel control
(7) Enthalphy control of HVAC systems
(8) Event initiated sequences, e.g., an alarm that initiates a specific sequence of operations
(9) Load shedding
(10) Load cycling
(11) Restart after power failure — prevents electrical overload on restart
(12) Process control, i.e., the use of system remote stations as the controllers for individual loops
(13) Optimum damper control (free cooling cycle)
(14) Security, e.g., patrol tours and card access
(15) Interlocking, i.e., the use of software instead of relays and timers, etc.
(16) Fire, i.e., alarms and specific event initiated sequences
(17) Maintenance management, i.e., the use of stored and real-time data to produce a work schedule for maintenance and servicing

14.15.9 System Documentation. The following documents should be submitted by the FAS contractor prior to the acceptance of the system:

(1) As-built drawings, including system block diagrams, central control equipment installation, schematic diagrams and physical layouts for FIDs and MUXes, and wiring diagrams of sensors and controls

(2) Test plan and step-by-step test procedures for factory, site verification, and endurance tests

(3) Operation and maintenance manuals covering
   (a) Functional design
   (b) Hardware
   (c) Software
   (d) Operator's instructions
   (e) Maintenance procedures

(4) Training program, including lesson plans and videotape training for both system operators and maintenance personnel

14.15.10 Programming. Most systems will be purchased with software already installed. This is the quickest and most reliable method of obtaining tested software that also meets the requirements of approval agencies. There are instances in which in-house programming will be undertaken, usually for specialized subsystems in a distributed-type system (e.g., parking control system). In other instances, there may be a requirement that the in-house staff have the capability of modifying the existing system. Under these circumstances, it is important that the following items be included in system specifications:

(1) The equipment should meet the programming needs of in-house staff. For example, an imbedded computer (special design) may not be practical, whereas a separate standard computer using a standard operating system (e.g., DOS, OS/2) and a specified higher level language (e.g., Basic, C, Pascal) might be suitable for the programming staff. When consultants may be called upon to perform initial and future programming, the same constraints hold true. The program should be furnished in a format that is usable by in-house staff or consultants; the object (machine-usable codes) assembly format is most often unusable by this staff.

(2) The software development should be checked frequently with the programming staff. Too often, the system specifications are written so loosely or so misunderstood that unsatisfactory programming is noticed only on final testing.

(3) The availability of appropriate staff should be assured before undertaking the programming. Often, the in-house programming staff will be unfamiliar with real-time programming; the safety redundancies, reliabilities, and legal requirements of life safety systems; and with the equipment being furnished. The costs associated with in-house programming often far exceed the cost of the computer and console equipment. If programming changes are required, the time for staff to familiarize themselves with the software may be extensive, especially if individuals not associated with the original installation are assigned to this task.

14.15.11 Physical Installation. There are installation procedures that enhance the reliability and security aspects of the FAS, including

(1) Equipment should be located in areas where it will be physically and environmentally protected. While some equipment is suitable for mounting in relatively unprotected non-air-conditioned areas (e.g., some programmable controllers), most are not. Most standard FAS equipment should be installed in air-conditioned rooms where the temperature, dewpoint, and air cleanliness are controlled. Conduit entrances into boxes should be sealed against the flow of air between areas (often an NEC [1] requirement). Terminal and equipment boxes should be properly gasketed and sealed. The enclosure protection should be the appropriate NEMA classification (e.g., water-resistant, dust-tight). The equipment should not be located in basements or mechanical areas if flooding is a possibility. In particular, where there is more than one control point, the failure of equipment at one location should not disable backup equipment.

(2) The equipment should not, as far as is practical, be located where the hazard that it is protecting against will disable the system. The DTM should be located in fire- and vandal-resistant areas. FAS wiring should normally be kept separate from other control wiring. Local power supplies should be separated from normal outlet wiring, as required, to meet reliability requirements. A completely separate conduit system provides a high degree of protection. In some instances, items such as terminal boxes might be lockable to prevent accidental or deliberate tampering. Except when security requirements militate against it, FAS wiring and enclosures should be clearly labeled. Codes and regulations usually require that exposed wiring be suitably protected and/or be of flame-resistant, high-temperature construction. (Refer to the NEC, Article 725 [1] for the appropriate cable types.)

(3) Chapter 1 contained a brief summary of electromagnetic interference (EMI) and the harmonics that can affect the performance of FAS systems. The power supplies to equipment should be of high quality, relatively free from noise, and, where required, use filtering, transformer isolation, and packaged power conditioners or converters to enhance power quality. UPS systems are effective in power improvement only to the extent that they contain these features. While twisted telephone pairs provide a high degree of immunity to certain types of interference, shielded or double-shielded cables provide much greater immunity. Perhaps the highest degree of immunity is provided by steel conduits and steel enclosures, which do have excellent magnetic shielding. Fiber-optic cable is immune to EMI, and, where available, photo-optical coupling for terminals provides excellent protection against common-mode noise.

14.15.12 Backup Modes. Fortunately, computer installations are becoming increasingly reliable; however, failures do occur, and routine maintenance is required. Software modifications or system reconfiguration may require outages or extended testing, particularly during the commissioning period (which may also be a concurrent operating period). The overall reliability of any system is affected by the number of fallback positions that are available in the event of the outage of any part of the system. Standby power and uninterruptible power supplies have been previously discussed. The following are possible modes of backup:

(1) Control Console — The control console may be designed so that the loss of one subsystem does not affect other subsystems. Redundant modules, such as multiplexers, may be automatically switched on or manually replaced. Software provisions may permit limited system control by using keyboard input in the event of control panel switch failure. When there is more than one console, the obvious choice is to use one of the alternate operating positions, and the system design should be developed to accommodate this. When one of the alternate positions does not normally have all of the system functions available, all should be made available when it is in backup mode. The transfer should be made so that all interconnections that might prevent operation in the backup mode are cleared. (Just switching off the power to a defective console may not be sufficient.)

(2) Distributed Computers — In the event of control console outage, the possibility of operating from distributed consoles has been discussed. Of course, control features should be provided if this is to be a possibility.

(3) Field Interface Devices (FIDs) — If control features (perhaps plug-in and portable) are available, another stage of backup can be located at the FID.

(4) The most fundamental system backup is manual operation or control of system devices at the device. This has the great disadvantage of loosing system control logic and might even involve some loss of safety features.

In any backup mode, the loss of telemetering functions is an important consideration.

14.16 Training. Unless maintenance and programming are to be performed exclusively under contract, training will be required. It is important that the training program for staff be spelled out completely in the original contract for equipment purchase; otherwise, separate charges will be encountered and the training will be at the convenience of the manufacturer. It is important that training for complex systems, particularly those that are nonstandard, be given at around the time of system commissioning because all the personnel responsible for system design and programming may not be available at a later date.

The contract should specify the number of people to be trained, the level of training for each group of people, and the length of time training is required. As part of the proposals submitted by the contractor, a complete outline of the training programs, the approximate times they are to be given, and the detailed material to be covered should be spelled out and approved by the user. There have been instances where training for computer systems was largely wasted because the training was very general in nature (emphasizing such general subjects as binary notation), given on different equipment from what was to be purchased, or provided on incomplete, nondebugged systems.

Except for simpler systems, training programs may be conducted at the factory or at the place of initial setup during the test period. Here, the staff will be away from their normal duties and can concentrate on studying the system. The operator's courses are usually fairly brief, but should also be attended by programming and maintenance personnel. For extensive systems, system maintenance personnel training may take many weeks. When budgeting the purchase of the system, the cost of living expenses and supplemental training expenses should be included, as well as travel expenses.

On-site training is required during the commissioning period. Here, the emphasis is primarily on hands-on training of the operators. The training should include simulated failures and the performance of all routines required (including any special operating system dumps); the opportunity should be taken to go through the complete range of operations with the operators. Maintenance personnel should be given the opportunity to work with the contractor's staff as the system is put into service, even though there is little opportunity to perform maintenance, since the contractor is primarily interested in getting the system on line and accepted.

Videotaped training programs are essential for recurrent training of facility personnel. These videotapes should be maintained within the facility for casual and formal reviews as needed. Training programs should be updated as the hardware and software are modified throughout the life of the facility.

When establishing the curriculum for on-site training, if it is not included in the contract, relatively high training personnel costs (including travel and living expenses) can be encountered.

14.17 Maintenance and Operation. Maintenance and operation may be performed by in-house personnel or by contractors. Because many new systems are microprocessor types, in contrast to minicomputer or mainframe types, there is a tendency toward in-house maintenance that is supplemented by contract or manufacturer maintenance for serious problems.

In any event, the effectiveness of in-house maintenance is a direct function of the training and skills of the staff. A technician or electrician will normally require special training to maintain these systems at any level. Even though a system may remain operational with relatively poor maintenance, it can be expected to degrade over a period of time to a level where it will be totally unreliable. Coordination of the maintenance programs covering the availability of competent personnel during off-hours and coordination with contract personnel is essential. A careful delegation of work responsibilities to the various classes of technicians will avoid the problem of a technician damaging equipment or aggravating an outage when attempting maintenance beyond his or her capability.

14.17.1 Maintenance Approaches. The peripheral devices that are used for sensing, controlling, and monitoring electromechanical systems may be electrical, electronic, or electronic/pneumatic. These are usually maintained on a first-level basis by the assigned electrical, mechanical, or specialized journeyman. The level of maintenance permitted should be commensurate with the journeyman's qualifications. For example, in the "off-shift," a general "mechanic" may be the only individual available and may be limited to functions as simple as manually operating equipment in the event of a failure.

Specialized sensing and electronic control equipment that may be located at the controlled or monitored devices may require more skilled maintenance. These devices generally incorporate a number of integrated circuits that should be field diagnosed, replaced on a modular basis, and returned to a central repair location or to a special service shop. The technicians performing this latter function, either electricians, mechanics, or instrument repairmen, will require specialized training.

The remote control station, also known as a field interface device (FID), slave station, data gathering panel (DGP), or remote terminal unit (RTU), which consists of relatively sophisticated electronic equipment, can be maintained by specialists or by specially trained building maintenance staff (often electricians). These terminals are usually arranged in a modular fashion so that cards, relays, and other devices can be replaced with identical elements following a prescribed maintenance procedure. Detailed work on the individual cards or other elements is usually avoided at the site. Rather, these elements are returned to a "central shop" or to a manufacturer's repair facility. Defective power supplies may be replaced as units by electrical staff.

The control distribution (data transmission) system that interconnects the main control rooms with remote terminals may consist of twisted pairs of telephone wires, shielded twisted pairs, or bundles of twisted pairs, coaxial cable, or fiber-optic cable. Typically, these could be maintained by qualified, trained electricians, unless delegated to specialized crews. The splicing and termination of fiber-optic cable usually requires specialized training, and the testing of coaxial cable will require special test equipment that may require the services of an electronic technician.

At the central control room, the distribution system will usually terminate in a special cabinet, which can also be maintained by qualified electricians. From here, interconnections are made to multiplex systems, computers, printers, mass storage units, interfacing, and other control devices. These units are often maintained by special maintenance personnel from the manufacturer or from an organization specializing in computer maintenance. The peripheral devices for output, display, and mass storage (e.g., floppy disks, hard disks, monitors, printers) are usually maintained by the computer system maintenance contractor. Some owners prefer to have only the most complicated equipment maintained by contractors and the less complicated modular equipment maintained by a specially trained in-house staff.

14.17.2 Operations. The system operator keeps the system running on a continuous basis by performing a prescribed set of routines, responding to alarms, keeping the system loaded with tapes or disks, keeping the printers in operation, and generally overseeing the inputs and outputs to and from the system. For larger systems, this may involve several levels of operation, including system supervision and equipment operations. In smaller facilities, the operator may have chores around the facility and will return to the central control room periodically or upon alarm. The operator may be a licensed watch-technician, who is completely familiar with the system to be controlled, or a trained console operator. If provided, the console may be continuously supervised by operators or for that part of the day when system operations are most numerous. When the system contains security devices, then remote printers or monitors, as well as alarms, may be located in the facility superintendent's area, at the security desk, at the watch-technician's office, at the fire control office, and at maintenance offices. The operator should be trained to notify special maintenance staff in the event of failures; security staff or municipal police, in the event of high security alarms; the fire department, in the event of smoke, sprinkler, or fire alarms; and the maintenance staff, in the event of building equipment failure. The operator should maintain, in a readily accessible computer or in written or printed logs, records that identify the locations where information, particularly mass storage information, is kept; should update records in mass storage media, such as floppy and hard disks and tapes; and should properly label and index this information. The operator is also responsible for assuring that adequate supplies of spare tapes and disks are available. The operator may be required to perform system dumps, restart (boot) the system, initialize the system, perform simple analysis of failure modes using prescribed routines, understand appropriate documentation, know how to switch to backup equipment when appropriate, and, in general, keep the system fully and continuously in operation.

In some systems, a number of the features described above are automatic: the switchover from one computer to another in the event of a failure or the display of diagnostic information when a failure develops. For the very small system that is well designed and where the capabilities of the operating staff are very limited, the system can be made almost completely automatic.

14.17.3 Maintenance Contracts. Many organizations require that the first year's maintenance be performed as part of the initial contract. This resolves the problem of extensive debugging. Even the best systems can develop extensive problems during the first year and, in particular, can be in conflict as to where the problems lie, whether in the central equipment, the distribution equipment, or the remote equipment. The existence of a built-in maintenance contract during the first year will, in effect, extend the debugging period and provide an opportunity to resolve many of the coordination problems. Thereafter, a separate maintenance contract is usually continued for a specified number of years, perhaps with an escalation clause for labor and material, and often with specified prices for special activities or part supplies. When the first year of maintenance is included in the installation contract, the contractor should be called upon to identify subsequent contract maintenance costs. The reason for segregating first year maintenance from subsequent years lies in the equipment warranties, i.e., no-cost replacement of parts, that are usually not in effect in the second, third, etc., years.

The maintenance contract should spell out the response time, the times when complete maintenance will be required, and the times when reduced maintenance will be available. In some cases, the user should be aware of nonstandard systems in which maintenance service may require several days notice. When special maintenance is required for unique systems, it is not unusual to have the field engineer travel thousands of miles or to wait weeks to obtain the services of an appropriate systems programmer. So the maintenance contract should spell out the maximum length of response time, if it is critical.

Complete maintenance service may be required 24 hours a day or may just be required during business hours, and the contract should spell out the type of maintenance service that shall be available for that period of the day. For example, all maintenance during the normal plant or building hours may be included as part of the regular contract; but call-in maintenance after hours may be on an hourly rate basis. There may be several periods during a day when different forms of maintenance are available and different response times are appropriate.

In some facilities, where very little maintenance is anticipated or where in-house maintenance can take care of virtually all problems, the contract may spell out only the cost of call-in maintenance. For large critical systems, this may be impractical unless a very capable in-house staff is on duty or available.

Notification of the contractor, access by the contractor, approval of contractor's personnel (particularly important when security considerations are involved), places for the contractor's personnel to report and sign in, and other considerations should be addressed as part of the maintenance contract.

Before undertaking a maintenance contract, the availability of contractor personnel, facilities, and the location of the maintenance office should be evaluated. This is particularly important when a system is built with nonstandard components, and in which the CPU is constructed with special design chips or other one-of-a-kind components that are supplied by a particular manufacturer.

14.17.4 Special Considerations. For most existing FAS systems operating in commercial buildings, the resources for making major system changes, major updates to the system, or software modifications are not available. The building owner may be a different individual from the building developer, owners may have changed, or maintenance contractors may have changed. In such cases, it is important that the system, once designed and configured, remain relatively unchanged. Local system modifications made by partially qualified in-house staff can result in minor disasters or in a system that is impossible to debug. Major changes to the system program, except from the manufacturer, can be extremely counterproductive. Usually, such local changes are improperly documented, and the resulting deviation from the standard may void the warranty, reduce obligations under maintenance contracts, and may make it difficult to obtain the services of a competent maintenance contractor.

Maintenance costs should be evaluated periodically, and engineering evaluations made for older systems. Economics may show that abandoning the central computer and multiplexing system is feasible, simply because the entire unit can be replaced by a physically smaller microprocessor with packaged programs that can be adapted to the existing system with a resulting lower maintenance cost. When such equipment is replaced, interfacing controllers are often available that enable the use of a new central computer with existing peripheral equipment.

Consideration should be given to communications links between the FAS CPU and the manufacturer's troubleshooting facility. Such a communications link enables the original system programmer to "take charge" of the local FAS from a remote location and perform debugging steps expeditiously.

14.18 References. The following references shall be used in conjunction with this chapter:

[1] ANSI/NFPA 70-1990, National Electrical Code.

[2] ANSI/NFPA 72A-1987, Installation, Maintenance, and Use of Local Protective Signaling Systems.

[3] ANSI/NFPA 72B-1986, Auxiliary Protective Signaling Systems for Fire Alarm Service.

[4] ANSI/NFPA 72C-1986, Remote Station Protective Signaling Systems.

[5] ANSI/NFPA 72D-1986, Proprietary Protective Signaling Systems.

[6] ANSI/NFPA 72E-1990, Automatic Fire Detectors.

[7] ANSI/NFPA 72F-1985, Installation, Maintenance, and Use of Emergency Voice/ Alarm Communication Systems.

[8] ANSI/NFPA 72G-1989, Installation, Maintenance, and Use of Notification Appliances for Protective Signaling Systems.

[9] ANSI/NFPA 72H-1988, Testing Procedures for Signaling Systems.

[10] ANSI/NFPA 110-1988, Emergency and Standby Power Systems.

[11] ANSI/NFPA 110A-1989, Stored Energy Systems.

[12] ANSI/UL 916-1987, Energy Management Equipment.

[13] IEEE Std 944-1986, IEEE Application and Testing of Uninterruptible Power Supplies for Power Generating Stations (ANSI).

14.19 Bibliography. The following documents can be useful in the planning, design, and operation stages of a facility automation system:

[B1] ANSI/ASHRAE 114-1986, Energy Management Control Systems Instrumentation.

[B2] ANSI/NFPA 75-1989, Protection of Electronic Computer/Data Processing Equipment.

[B3] ANSI/NFPA 90A-1989, Installation of Air Conditioning and Ventilating Systems.

[B4] ANSI/NFPA 92A-1988, Smoke Control Systems.

[B5] ANSI/NFPA 204M-1985, Smoke and Heat Venting.

[B6] ANSI/UL 38-1986, Manually Actuated Signaling Boxes for Use with Fire-Protective Signaling Systems.

[B7] ANSI/UL 268-1988, Smoke Detectors for Fire-Protective Signaling Systems.

[B8] ANSI/UL 268A-1985, Smoke Detectors for Duct Applications.

[B9] ANSI/UL 294-1986, Access Control System Units.

[B10] ANSI/UL 609-1990, Local Burglar Alarm Units and Systems.

[B11] ANSI/UL 611-1986, Central Station Burglar Alarm Systems.

[B12] ANSI/UL 864-1985, Control Units for Fire-Protective Signaling Systems.

[B13] ANSI/UL 1076-1988, Proprietary Burglar Alarm Units and Systems.

[B14] IEEE C37.1-1987, IEEE Standard Definition, Specification, and Analysis of Systems Used for Supervisory Control, Data Acquisition, and Automatic Control.

[B15] IEEE Std 960-1988, IEEE Standard FASTBUS Module High-Speed Data Acquisition and Control System.

[B16] ISA RP608-1978, Electrical Guide for Control Centers, Instrument Society of America, 67 Alexander Drive, P.O. Box 12277, Research Triangle Park, NC 27709.

[B17] NEMA SB4-1985, Training Manual on Fire Alarm Systems.

[B18] ACEC Publication 529, Guidelines for the Design and Purchase of Energy Management and Control Systems for New and Retrofit Applications, American Consulting Engineers Council, 1015 15th Street, N.W., Washington, DC 20005.

[B19] *ASHRAE Handbook* (1987 Edition), Chapter 51–87, "Automatic Control."

[B20] *The Intelligent Building Sourcebook*, Bernaden, John, and Neubauer, Richard, editors; Johnson Controls, Inc.; Fairmont Press, Inc., publisher; 1988.

Chapter 15
Expansion, Modernization, and Rehabilitation

15.1 General Discussion. In developing engineering criteria and plans for an expansion, modernization, or rehabilitation of an electric power system for an existing building, the electrical engineer or designer encounters problems over and above the electrical design considerations involved in new building construction. In commercial buildings, the electrical engineer has an opportunity to become involved in major electrical changes when the building or an area is rehabilitated, modernized, or expanded. Where several specialties are involved, the work is often spearheaded by architectural staff. For larger enterprises, the facility electrical engineering staff is in a position to recommend electrical changes for economic, environmental, conservation, maintenance, safety, and/or operational reasons. If the recommendations are based on essentially electrical requirements, in most cases, the need should be justifiable, compelling, and/or legally required; the engineer should make a good case based on a thorough investigation, effective report, and detailed presentation.

For a major building expansion, a designer is faced with the problem of whether to retain the original service equipment and expand it, replace the original service equipment with new equipment in the same location, install new service equipment in a new location, or install an additional new service point to cover the expansion. To a large extent, this chapter is concerned with the problems involved in making this decision and with the related problem of power system modernization (see Reference [3][89]).

The "grandfather clause" permits the retention of most equipment, even though it may not be allowed in new construction. Occupational Safety and Health Administration (OSHA) and other regulations do mandate that a few items of equipment should be replaced retroactively. If any changes are made in the installation, federal, state, other jurisdictional, and OSHA regulations may require that the equipment be replaced to meet the newer standards. OSHA regulations state that, in effect, every replacement, modification, repair, or rehabilitation of any part of any electrical installation or utilization equipment meet applicable requirements. Before undertaking a partial electrical rehabilitation, the full scope of required changes should be understood. The utility may require that changes, including

---

[89]The numbers in brackets correspond to those in the references at the end of this chapter.

increased load requirements, be made in accordance with their latest tariffs and regulations, and the extent of the changes may greatly exceed the original intent of the engineer. Construction, which may have been designed under a superseded code or that has been delayed, may result in construction that does not comply with the latest code. As a practical matter, most inspection agencies and OSHA offices will accept the installation according to the latest versions of codes; however, this should be verified with the appropriate agency.

15.2 Preliminary Study. The most desirable preliminary study is an engineering study that is undertaken prior to the development of firm system criteria. All too often a thorough evaluation of the existing system is neglected or is undertaken only when the project is well into the design stage. With a major building modernization or expansion, the preliminary study should be reported and issued before starting the actual design phase. A preliminary study final report should contain a full operational assessment of the existing system with emphasis on its reliability. Outlines of alternatives (i.e., retaining, replacing, rehabilitating, expanding, or supplementing the existing system) should be furnished together with preliminary or budget estimates (including annual owning and operating costs as well as capital costs). For commercial enterprises, the effect of changes on the financial value of the space will be important. When changes are mandated by law and recommended for the public good, such as environmental considerations, the basis for recommendations should be clearly spelled out for managerial evaluation. The initial studies should culminate in reports and presentations, as appropriate, to obtain approval, authorization to proceed, and the budget allocation for the project. The report should also state how dislocation of facility operations and tenant activity will be handled.

A preliminary report may well produce recommendations completely beyond the scope of the conditions that led to undertaking the study. Some items that warrant investigation in the modification of electric systems and that are not normally required in new building designs include the following:

(1) Age and condition of the existing equipment
(2) Maintenance costs and availability of parts for the existing equipment
(3) Automatic operation of the existing equipment as compared to manual control
(4) Voltage, frequency, and number of phases (grounding method) of the existing system as compared to a modern three-phase grounded neutral 60 Hz system
  (a) Availability of additional capacity (e.g., transformer and distribution limitations) from the utility
  (b) Economy of distributing at the existing system voltage
(5) Characteristics of the existing system
  (a) Interrupting, fault closing, and momentary ratings (including bus bracing)
  (b) Load current ratings
  (c) Coordination
  (d) Susceptibility to voltage disturbances

(e) Harmonic considerations (voltage and current)
(6) Compatibility of old and new requirements
(7) Safety of existing system (maintenance, operational, and public safety)
(8) Code and OSHA regulations regarding the existing system and proposed expansion thereof. Retroactive United States Department of Labor OSHA regulations that have not been met
(9) Availability of qualified personnel to maintain and operate the old system.
(10) Reliability of existing system and outage record including time required for repairs. It should be determined if the system or components are being run overloaded or operated at maximum capacity, e.g., are the fans on a fan-cooled transformer operating continuously?
(11) Energy conservation modifications, such as replacement, recircuiting, or provision of new controls for lighting systems; installation of more efficient power equipment; or provision of automated building control to replace relatively inflexible operational systems. Compliance with conservation regulations is referred to in Chapter 17.
(12) Availability of suitable quality power and avoidance of excessive electro-magnetic interference/radio-frequency interference (EMI/RFI) levels for sensitive electronic systems
(13) Obsolescence or unsuitability of equipment due to safety and environmental laws (i.e., asbestos jacketed cables and asbestos arc chutes and PCB [polychlorinated biphenyls] containing insulating fluids) are serious concerns. These items include: containment of insulating fluids from transformers and other oil containing equipment, requirements for fire containment around liquid filled transformers, and containment of electrolytes (and control of explosive gases) from large storage battery banks. These requirements are jurisdictional, environmental, by OSHA, and by insurance underwriters.
(14) Improvements relating to the economic viability of a commercial facility; making space more salable, enhancing tenant operations, improving building ambience; in terms of lighting, building services and provisions for tenant operations. Compliance with laws regulating accessibility and usability for the handicapped (e.g., elevator control).
(15) Changes in load characteristics, particularly the use of equipment generating high levels of harmonics, such as computer power supplies, electronic motor controllers, and other solid-state power equipment that may require resizing neutrals or transformer de-rating [3]. (See IEEE C57.110-1986, IEEE Recommended Practice for Establishing Transformer Capability When Supplying Nonsinusoidal Load Currents (ANSI) [2].[90])
(16) Proper grounding and segregation of neutral conductors from equipment grounds
(17) Standby and emergency power systems, especially those required for public safety. Changes in codes may require modernizing or replacing these sys-

---

[90]IEEE publications are available from the Institute of Electrical and Electronics Engineers, IEEE Service Center, 445 Hoes Lane, P.O. Box 1331, Piscataway, NJ 08855-1331.

tems. Control for these systems should be analyzed with regard to the adequacy of automatic features. Availability of power for data processing areas or other critical areas should be examined. The need for UPS systems should be considered.

The study will require a thorough field survey by qualified engineers familiar with field conditions and construction. Even a competent designer, who is thoroughly familiar with the fundamentals of modern electrical design for new buildings, may lack sufficient knowledge or experience in handling old or existing work. Data regarding the life expectancy of existing equipment including cables may be important in determining whether to retain such equipment. A history of operational problems and equipment failures over the years is invaluable in predicting the reliability of the existing system and the probable rate of future failures. Coordination of the study with other disciplines is essential. In the analysis of the retention and expansion of the existing system versus providing a new system, the operational and safety characteristics of the equipment should be compared. This includes comparison of certain items that are listed in Table 84. If a system is poorly documented, extensive load studies and measurements may be required. They may also require the use of recording instruments. Cable tests, on a sampling basis, may be required to assess the quality of the installed cable system. An inspection of the existing system is essential.

15.2.1 Examples of Poor Planning. An engineer might encounter previous rehabilitation work that was not properly planned. Typical examples follow:

(1) Because of a rash of failures, all wound-rotor motors were rewound or dipped and baked recently. A study now shows the replacement of the motors with new high-efficiency motors that have electronic controllers is the best solution.

(2) All the old knifeswitch panels were replaced with Q-type molded-case circuit breakers. However, a new service requires higher interrupting capacity than the replacement breakers are rated.

(3) Because of the substandard condition of a 30-year-old Type RW cable (particularly in the gutters of switchboards), building electricians recently completed replacement of all the cable runs. A study now shows that new higher capacity panelboards with larger feeders are required.

(4) All 120 V ballasts in the plant were changed several years ago to the rapid-start type, eliminating starters. A study now shows that a 480Y/277 V service is justified, and new ballasts will be needed or overall transformation required.

(5) De-lamping, removal of unused ballasts, and replacement of remaining ballasts with highly efficient lamp and ballast combinations was recently completed as an energy-saving measure. A full study shows that recirculating to provide more flexible control (50% lighting during off-hours, perimeter control, and control by area) will provide for greater savings; but most of the completed wiring will have to be redone.

(6) Expensive power conditioning equipment was installed when a modified grounding system for the electronic equipment would have solved the original EMI problem.

Table 84
Examples of Existing Conditions and
Replacement Strategies and the Justification of Remedial Action

Existing Condition	Replacement
	Justification for replacement
Open Knife Switches	Dead-Front Construction Personnel safety
System with Ungrounded Neutral	Systems Using Grounded Neutral Maintenance safety Standard transformer connections
Fuses with Unrated Interrupting Ratings	Class R, J, RK, or L Fuses Protection of equipment Safety Noninterchangeability
Fixed Switchgear	Drawout Switchgear Maintainability Safety
60 °C (140 °F) or 75 °C (167 °F) Rise Wire or Cable	Higher Temperature 75 °C (167 °F) or 90 °C (194 °F) Increased ampacity for given raceway, conductor (maximum standard lug rating is 75 °C [167 °F])
Open Buswork	Metal-Enclosed Switchgear Safety
Manual Close and Trip	Electrical Remote Close and Trip Permits centralized operations Operating efficiency Safety
Local Control	Supervisory Control Operating efficiency Building safety
Cascade Breaker Tripping	Fully Rated Circuit Breakers Safety and OSHA requirements
Medium-Voltage Air Circuit Breakers	Vacuum or SF_6 Circuit Breakers Maintainability Fault rating
Low-Voltage Power Breakers with Peak Sensing Solid-State Trips	Low-Voltage Power Breakers RMS Sensing Solid-State Trips Reduction of false trips
Low-Voltage Power Breakers Without Ground-Fault Protection (480Y/277 V)	Low-Voltage Power Breakers with Ground-Fault Sensing Safety Code Coordination
Low-Voltage Power Circuit Breakers Ground-Fault Protection on Main Circuit Breakers Only	Low-Voltage Power Circuit Breakers Selective Ground—Fault Coordination and Time- or Zone-Selective Unmaintainability Excessive trips

Table 84 *(Continued)*

Existing Condition	Replacement
	Justification for replacement
Medium-Voltage Air-Break Motor Contactors	Medium-Voltage Vacuum-Break Motor Contactors Maintainability

Medium-Voltage Load Interrupter Switches, No Single-Phase or Blown Fuse Protection	Modify or Replace to Provide Single-Phase or Blown Fuse Protection
	  Safety
	  Equipment protection

Low-Voltage Switches, Class H Fuses	Low-Voltage Switches, Replace or Modify Fuse Assembly to RK1, RK5, or J
	  Safety
	  Noninterchangeability

Induced Draft Cooling Towers with Two-Speed Motors	Induced Draft Cooling Towers, Solid-State Adjustable Frequency Drives with Reversing Feature
	  Energy savings
	  Ice buildup prevention

Medium-Voltage Cable, 100% Insulation Rating	Medium-Voltage Cables, 133% Insulation Rating Greater reliability

Service Entrance Equipment with No Transient Protection	Service Entrance Equipment Adds Transient Protection
	  Danger from lightning
	  Protect sensitive equipment

Motors Having Short Bearing Life	Motors Having Bearing Life of over 100 000 Hours Rated (25 000 Hours for Belted Loads)
	  Maintenance
	  Expense

Motor Controllers Insensitive to Low-Level Faults and to Single Phasing at Light Loads	Motor Controllers with Load Tracking Sensitivity, Ground-Fault Detection, and Single-Phase Protection
	  Maintenance
	  Coordination

Manual Control of Major Building Functions	Installation of a Building Management Control System
	  Fire codes
	  Insurance requirements
	  Manpower savings

Switch and Relay Building Control System	Installation of Modular, Computer-Based System for Control and Data Logging
	  Insurance
	  Flexibility
	  Expandability
	  Security
	  Life safety systems
	  Graphic capabilities
	  Data acquisition

Building Management System — Proprietary Older System	Direct Digital Control (DDC), Distributed Architecture
	  Maintainability of hardware
	  Programmability
	  Reliability
	  Operational ease and staffing

Table 84 *(Continued)*

Existing Condition	Replacement
	Justification for replacement

Large Telephone, Communication Cables	Fiber-Optic Systems
	    System expansion cost
	    Reliability
	    Smaller size
	    Fire considerations
	    Electrical noise reduction

Lighting — Four-Lamp Troffers with Two Ballasts	Lighting Fixture *Replacement* — Three-Lamp Solid-State Ballast
	    Maintenance cost
	    Light levels
	    Energy conservation
	    Operating cost

	Lighting Fixture *Modernization* — Installation of New Solid-State Ballasts, Separate Control to Each Ballast Group
	    Energy conservation
	    Greater light output
	    Maintenance

Poke-Through Floor Wiring	Raised Floor
	    Reduction of electrical noise
	    Improved ventilation of equipment
	    Wiring simplification

Computer Areas — Overloading of Neutrals Because of Harmonics	Reinforce Neutrals, De-Rate or Replace Transformers, Correct Harmonic Problem Prevent overheating equipment, cable, outages

While serious consideration should be given to retaining such previous improvements, the implementation of an overall plan of modernization may justify elimination of, or further modifications to, previous improvements.

15.2.2 Making Choices. If an area is rehabilitated or modernized primarily for architectural or operational reasons, then the engineer is faced with the constraints of the physical installation. Methods of dealing with these constraints are described in the following sections.

Major alterations, which are predominately electrical in nature, in a commercial building can seldom be funded unless some overwhelming need can be demonstrated. Such a need might be a legal requirement (e.g., retroactive code section), inferior working conditions (e.g., inadequate illumination), environmental regulations, obvious unsafe conditions, inability to maintain a reliable electric system, inadequate power resources, misoperation of electronic equipment due to supply or EMI, a need to reduce power or maintenance costs, an obvious public safety requirement, or a need to maintain a certain appearance standard.

In evaluating various approaches, economic considerations play a major role. An "own-and-operate" or life-cycle cost evaluation takes into account the initial costs and the operational and maintenance costs over the anticipated life of the system(s). The financial staff should be consulted in making this evaluation because factors, such as depreciation, financial costs, tax considerations (capital costs are treated differently from expense costs), and related items, may be involved.

Each approach should be evaluated, using the same criteria discussed in Chapter 1, regarding the involvement of others. If an expansion is to be undertaken, the character of the proposed new system should be considered and then compared to the existing system; as an extreme example, it could be incongruous (for many operational and safety reasons) to retain open medium-voltage equipment when new vacuum-type metal-clad circuit breakers are being installed in a large expansion.

When a project involves only the addition of new equipment and/or space, the existing service may have sufficient capacity to serve the additional load. ANSI/NFPA 70-1990, National Electrical Code (NEC), Article 220-35 [1][91] permits the additional loading provided that the metered maximum demand (e.g., 15 or 30 minute demands) for at least a 1 year period plus 25% plus the new load does not exceed the rating of the existing service.

15.3 Design Considerations. The cost of a rehabilitation project for both engineering and construction may run 50%–100% higher than a comparable project involving only new construction. When field conditions cannot be exactly determined, the estimate of extra work allowances will have to be liberal enough to provide for undeterminable construction difficulties.

15.3.1 Utility Considerations. Utility considerations are discussed in Chapter 7. The utility should be consulted prior to undertaking any changes that materially affect load requirements, major equipment installations, system redesigns, or power quality. Begin negotiations with the utility as soon as the need for major changes is determined. A review of their latest tariffs, rules, design details, available types of supply, and installation requirements may materially affect the design and the costs involved. It is appropriate to negotiate the most favorable terms associated with the changes with the utility. It may be appropriate to negotiate the selection of rate schedules (tariffs), allocation of costs, costs for items as extra service points, utility incentive plans, and the possibility of special contracts. Sometimes, it can be pointed out to a utility that it is to their advantage to provide an additional service without extra cost to the owner; the second service, if provided, would reduce costs for both the utility and the customer. Negotiations should be undertaken by individuals who are familiar with utility practices and tariffs.

Some utilities will provide special rates in depressed areas, where excess service already exists, or where the nature of the load is highly desirable. The question of availability of such rates should be raised. An example of a special rate is for service at existing low voltage from an existing underground system at a partially abandoned marine terminal area. One utility provides a rate incentive for the use of ice storage as an auxiliary conditioning coolant that significantly improves the load factor. Co-generation, as a means of improving power cost economies, is covered in Chapter 17. Tariffs are discussed in Chapter 7.

---

[91] ANSI publications are available from the Sales Department of the American National Standards Institute, 11 West 42nd Street, 13th Floor, New York, NY 10036. NFPA publications are available from Publications Sales, National Fire Protection Association, 1 Batterymarch Park, P.O. Box 9101, Quincy, MA 02269-9101.

The incentive for service rehabilitation may be increased reliability. Some utilities may provide a choice between first and second contingency reliability at a cost differential to the customer. In some cases, reduced reliability and lower power costs may be acceptable, at least for part of the load, by the purchase of interruptible or less reliable service.

15.3.2 Ground-Fault Protection. Installation of ground-fault protection is required for new or modified services for commercial buildings, when the service is rated over 1000 A in accordance with the NEC, Article 230-95 [1], and for each building or structure with a main disconnecting means rated 1000 A or more in accordance with the NEC, Article 240-13 [1]. To a large extent, this will alleviate the problem of sustained low-current arcing ground faults in which the ground fault is insufficient to trip the line overload elements. However, the installation of this ground-fault protection, which is required only at the service or building main disconnect, may bring about a rash of system failures (because of ground faults on subfeeders or branch feeders that do not have coordinated ground-fault protection) that the system is operationally unacceptable. Installing the required ground-fault protection may require major modifications to other system elements in order to obtain acceptable coordination. The details of this coordination are covered in Chapter 9. When each of the service switches (six are the maximum permitted at one service by the NEC [1]) is rated at less than 1000 A at 480 V, then the requirement for ground-fault protection may be eliminated; this may be desirable in systems where maintenance skills or manning levels will be low. Even in cases where there is no code requirement for ground-fault protection, good practice may justify it in some installations.

15.3.3 Sizing Feeders. Keeping connected load on individual feeds relatively as low as is practical may avoid the prospect of massive power outages on a single bus. The cost of large circuit breakers or fused switches rated at 4000–6000 A also may be disproportionately greater than, for example, the cost of several smaller devices rated at 1600 A. Even if the service connection is 5000 A, there is no reason why the individual switchboards cannot have much lower ratings. For example, if the service stab is rated at 5000 A, the local codes may require a single service switch: but this switch can serve perhaps three switchboards, each protected by a main device rated at 2000 A. If the local code does not restrict, the NEC [1] permits up to six service switches from one utility service.

Even though newer wire insulations permit higher cable ampacities, it is doubtful that, in a major rehabilitation or expansion, replacement of existing 60 °C (140 °F) or 75 °C (167 °F) cables with new 90 °C (194 °F) conductors can produce a substantial increase in load handling capacity. Note that cables that have the 90 °C (194 °F) rating cannot have an operating temperature at the point of connection higher than the rating of the equipment terminals. The UL listing for almost all low-voltage equipment is based on the use of 60 °C (140 °F) insulation for circuits rated 100 A or less, and 75 °C (167 °F) for circuits rated over 100 A, unless the equipment is marked otherwise. Some equipment rated 100 A or less is listed and marked for conductors having 75 °C (167 °F) insulation; but at this time, no equipment terminals are listed higher than 75 °C (167 °F). Conductors with insulation rated 90 °C (194 °F) or higher may be used; but when terminated in UL-listed equipment, conductors should be used at their 60 °C (140 °F) or 75 °C (167 °F) ampacity to avoid exceeding the UL listing and thereby violating the NEC, Section 110-3 (b) [1].

Since 120 V, 220 V, or 240 V power systems utilize 600 V cable, upgrading the system by voltage increase may include reuse of the original cables if the condition of the cables justifies it. In reusing main feeders, care should be taken that the neutral is not undersized. The neutral of a system feeding primarily incandescent lighting may have been designed to be smaller than the line conductors. Replacement with a fluorescent system calls for full capacity neutrals because of third harmonics (the NEC, Article 300-15, Ampacity Tables, Note (10) [1].) It has been found that certain circuits feeding a very high percentage of nonlinear loads that involve extensive phase firing, wave chopping, or other methods of wave shape control (particularly switching-type power supplies) generate such high levels of harmonics that an oversized (up to two times phase conductor size) neutral or the use of multiple neutrals from the same source may be required.

Newer cable insulations and/or jacketing are becoming available, which have the advantage of much lower flammability, lower toxicity under high-temperature conditions, and lower smoke emissions. When available and practical, it is desirable to use these cables in areas of public assembly and in high-rise buildings. Some state and local codes and rules may specify such materials be used when they are available.

15.3.4 Staging Operations. An important consideration is whether the building can be shut down for the rehabilitation project (perhaps after the old tenant moves out and before the new tenant moves in), or whether the building should be kept in operation during the project except for weekends or nights. The conflict between normal operations and rehabilitation work should be dealt with in the specifications. Items that have to be considered are

(1) Protection of building occupants against harm and equipment against damage.

(2) Ability to operate the building and of tenants to carry on their normal functions. A schedule, which may have to be modified as the work progresses, should be developed in conjunction with the contractor and in accordance with the terms of the contract in order to make work areas available.

(3) Because work may be going on while the building is occupied, special liability insurance provisions will be required on the part of the contractors.

(4) Designated staff, including individuals with safety and equipment integrity duties, should be assigned. For smaller jobs, this may be the normal operating staff and inspectors; for larger jobs, full-time individuals may be assigned.

(5) The responsibility and accountability for each assignment should be clearly defined. The design engineer cannot accept responsibility for those activities that he or she cannot control.

A well-developed program in which all aspects of the work are staged is essential; however, work in existing buildings is subject to delay (or to being speeded up) if unforeseen conditions develop. Some of these conditions may not be electrical in nature, for example, if disturbing asbestos is required before equipment can be installed. It is a good idea to have alternate work schedules ready in case the progress on any path of construction is delayed.

15.3.5 Scrap Materials. The salvage value of existing equipment should not be neglected. Old equipment, such as switchboards and conduits, is usually not worth much more than junk except for the copper that may be in it. An inventory of the value of old dc feeders may indicate a probable scrap value greater than the estimated cost of new feeders for a new ac system. Lead jacketed cables are valuable because of the easily recoverable lead in them. The contract should clearly state the disposition and estimated quantities of scrap material. In contracts where the disposition aspects are not covered, disputes between the contractor and client may arise concerning ownership of valuable materials.

Most scrap material will be of such little value that the cost of removing it is more than it is worth. In this case, the contract should require that the contractor store these materials in a specified manner and remove them from the site promptly.

15.3.6 Hazardous Scrap Materials. Hazardous materials should be handled in a manner acceptable to the EPA and other regulatory agencies, documented in accordance with regulations, and disposed of only by approved agencies. Asbestos, nonflammable liquids (askarels) commonly used in transformers, and capacitors that may contain PCBs, batteries, and other hazardous materials may require active inspection or supervision by the owner's representative. In some cases, the owner has the responsibility for this material beyond delivering it to a disposal agency. Older coil-and-core ballasts may contain PCBs in capacitors; if more than a specified number of ballasts are to be disposed of, then disposal should be made in the same way as with hazardous materials. In some jurisdictions, the disposal of large quantities of fluorescent lamps may require a special permit.

If existing equipment is to be expanded, it is essential that the necessary parts be available. For example, there may be four or five spare spaces in an old panelboard for additional circuit breakers; but such disconnects may not be available at acceptable prices or with timely deliveries. In attempting to expand an old system, the circuitry may become unduly complex. The resulting single-line and riser diagrams should clearly and simply illustrate the system operationally (this is important). Buildings have been disconnected because the operator, watch-engineer or technician could not understand the system and properly operate a critical switch. Key interlocks or electrical interlocks, while often necessary, should be carefully designed so as to not unduly inhibit or complicate system operations. They should also not be used in such a manner that an emergency operation will be inhibited.

15.4 Retaining Old Service Equipment. Although retention of old service equipment can minimize overall costs, there may be drawbacks to such retention. The original service entrance equipment may have an inadequate interrupting rating when compared to the available fault current capability of the utility service. It is not the general practice of local utilities to inform a customer when available short-circuit currents from their supply have increased. The customer usually discovers this predicament only when he or she investigates the need for additional building power. Older installations will seldom have the expansion capabilities that will be needed in the future, even if immediate power requirements can be satisfied.

In trying to expand an old service, problems will be encountered in matching old equipment with new equipment. It might cost almost as much to purchase a custom-made adapter or bus detail section to accommodate new busway to an existing service switch as to replace the switch. Installation of ground-fault protection on existing disconnects may be difficult and may even be impractical if the disconnect does not have an electrical trip; new disconnect devices can be purchased with low-cost ground-fault protection installed integrally. In the field, modifications to older equipment are generally much more expensive and may be less satisfactory than comparable factory-installed features. If the existing service equipment and the main switchboard disconnect are otherwise satisfactory and dependable, it may be possible to upgrade this equipment despite unmodified and insufficient fault handling capability. The installation of current-limiting reactors or busway between the utility service point and the service equipment may permit a reduction of the maximum short-circuit currents to values within the rating of the equipment. The reactor has not been popular because of its unfamiliarity in application, possible local code limitations, and because a new service protector or bolted pressure switch of adequate interrupting rating might be more simple to install and less expensive. It may be possible to substitute or install (ahead of the unit) current-limiting fuses to reduce the available fault current to the older switchgear. Reactors or fuses require careful engineering (including the possible use of special software programs) and usually consultation with the equipment manufacturer. The method of using equivalent rms let-through current to assure protection of lower rated (maximum interrupting rating) downstream devices as taken from standard curves should be confirmed by referring to actual tested values from appropriate standards and/or manufacturer's data. In order to determine the available fault currents at the downstream devices, a fault study may have to be conducted.

The operational characteristics of older circuit protective devices may present coordination problems. A service switch instantaneous device is most difficult to coordinate with downstream devices. Older dashpot time delay features are often erratic and cannot be depended upon for accurate timing. Historically, no interrupting rating was designated on older circuit breakers, and these circuit breakers may now be rated at no more than 5000 A by the manufacturer. If the existing service equipment embodies fuse protection, replacement of the older style of fuses may require certain bus adapter work and the installation of new fuseholders or fuse adapters to either increase the fuse size or interrupting rating or prevent insertion of improper fuses. If existing equipment is to be retained, the expense of a thorough overhaul should be provided in the overall cost estimate. Buswork bolts should be retightened, and, if signs of overheating (coloration) are present, refinishing or the equivalent may be required. Dashpots almost invariably need cleaning and refilling. Circuit breakers should be operated, observing possible points of malfunction, and, if possible, should have tripping characteristics checked by simulating overload and fault currents. Main contact springs on older circuit breakers or fuse clips may have lost temper due to overheating. Arc chutes may be cracked or chipped, requiring replacement.

At the work site, a factory representative may be required to perform major or complex field work or to provide technical assistance to the crew. The manufacturer will have to agree to schedule for this work. The manufacturer's quoted living and traveling expenses for time away from the home office by the service representative should be added to the cost for such work. Extended availability of such representatives may be limited, and it may be necessary to have the manufacturer train local electricians or technicians. If contract work is proceeding simultaneously with the overhaul, work rule regulations may require the involvement of local electricians in a manufacturer's overhaul. In one instance, the repair of an open switchboard circuit breaker by the manufacturer would have cost about one-third more than an installed new molded-case circuit breaker.

If the system is provided with standby or emergency power, care should be exercised in testing for ground faults as provided by the NEC, Article 230-95c [1] unless switching of the neutral conductor is provided. Feedback from an engine-generator can be hazardous.

15.5 Completely New Service Equipment. The installation of a new service point and new service equipment will avoid the many disadvantages of trying to retain and expand old service equipment. Outage time can be kept to a minimum if the original service point and service equipment can be kept in operation until the moment of reconnection to the new service equipment. If the new service point is adjacent to the old one, the old service equipment can be removed after the transfer of load and this space allocated for future electrical expansion. Installation of a new service point may permit increasing the service voltage to 480Y/277 V three-phase service (from a lower voltage system), which is today's standard. In this case, it would be necessary to install transformers to step down the voltage to the old parts of the building, if conversion of equipment in the old building to a higher voltage is uneconomical. Modern distribution methods may involve less in total cost for a completely new service than the cost of attempting to retain the old service and expanding it.

Utility service rules may require the customer to install a new service point with new characteristics when major changes for additions are to be made in the customer's electric power system. Utilities are often required by their state's Public Service Commission to continue supplying power to the premises of a customer at the originally installed voltage and frequency. If, however, the existing facilities are inadequate, utility rules may require filing a new application for power that presents the utility with the opportunity to require abandonment of obsolete power sources. Most utilities will actively resist an attempt to expand an existing 25 Hz, dc, two-phase, or other nonstandard service. To eliminate an obsolete supply characteristic, the utility has, in a few instances, been induced to bear part of the customer's conversion costs.

15.6 Additional New Service Point. An alternative to purchasing completely new equipment is to retain the existing service point and to install a new service point that is usually more centrally located (electrically) within the building. The additional service point will generally be located in the new area if a major area expansion of the building is involved. If the additional service point is to provide for an increase in capacity only, the new service point might very well serve only equipment such as motors, while lighting, for example, remains on the old service.

Although adding a new service point sounds like a simple solution, utilities may be loath to provide more than one service point without an excess service or other charge, or they may treat the second service as a separate service with a separate bill. The NEC [1] and local codes govern the use of more than one service, and approval of the local inspecting authorities, as well as that of the utility, is required to authorize exceptions. Chapter 4 describes the various types of multiple-metering installations and explains the advantages of each. In determining if it is better to undertake the cost of an additional service point, it is necessary to include the cost of the equipment, estimated maintenance costs for old and new equipment, power losses of cables and transformers (if significant), and the additional utility power cost for the additional service point (if applicable) (own-and-operate costs). The local fire department may require one device to initiate the opening of all services under emergency conditions.

15.7 Voltage Transformation. It may be advantageous to install a new service point and to reduce the new and usually higher voltage to a voltage that is usable by the "old" parts of the building. If the utility insists on retaining the customer's old lower voltage, it may be advantageous to install transformers to step up to 480Y/277 V for new equipment. A few hundred feet of busway in the higher current ratings can equal the cost of a transformer. In the typical medium-sized or large building, 480Y/277 V is usually adequate for distribution throughout the building. However, there are some cases where it may be desirable to transform to a higher voltage, such as 2400 V or 4160 V. These voltages are very practical for supplying medium-sized motors (in the sizes of 200 hp or above) and usually require much smaller raceways. When ordering transformers for step up, this fact should be specified to the manufacturer since a standard transformer is often not suitable in terms of connections and insulation ratings for such application. The NEC, Articles 450 and 710 (Over 600 V) [1] requirements for the protection of transformers should also be considered.

High voltage could be stepped down at the receiving end for loads other than motors. If medium voltage is transformed up and retransformed down at one sending end and one receiving end with no intermediate taps, and no danger of confusion of different feeders can exist, the engineer may elect not to install medium-voltage switches. Switches for overcurrent and short-circuit protection of feeders may be required by the NEC, Article 240-3 [1]. Low-voltage switches can serve as the disconnecting medium at both ends. However, when there are taps emanating from the medium-voltage line, fully rated load-break disconnecting switches are usually desirable for operational and safety reasons. The use of medium voltage is recommended only when adequate maintenance and operating personnel are available within the customer's own staff or from a contractor's electrical maintenance staff.

If multiple service entrances are used with different voltages or if transformation is involved with different voltages, it is undesirable to apply unlike voltages for identical types of usage in the same areas. For example, it is preferable to keep the fluorescent lighting at 208Y/120 V for an entire area rather than have a sporadic mixing of different voltages. There is no problem with providing different voltages for lighting in different areas, and there is also no problem in providing a different voltage for lighting than for power equipment in the same area or for standby and emergency systems. Conspicuous marking of voltage ratings is important. Identification of conductors, beyond the mandatory requirements of the NEC [1], may be desirable. If transformers with a delta primary to a grounded wye secondary transformation are used, ground-fault protection may be provided through the neutral of the wye connection of the secondary. Simple computations will show that primary protection only, on a delta-wye transformer, sometimes cannot be set to properly protect against a ground-fault condition on the secondary. A ground fault on the secondary load conductors will reflect only 57.7% of the current in the primary that an equal current three-phase secondary fault would develop.

15.8 Distribution of Power to Main Switchboards. A number of methods of distributing power are available, including cable in conduit, busway, cable bus, and cable with continuous rigid cable supports (cable trays). Perhaps the simplest of these methods to use for low voltage is the busway, particularly the types with single-bolt joint construction. In all cases, the service switch should protect the feeds to the main switchboards. When feeds are concealed, conduit and cable are commonly used.

With one cable per phase, the service entrance switch overcurrent device will serve as adequate overload and short-circuit protection. On the other hand, if several paralleled cables per phase are required, protection may not be provided to individual cables by the service entrance protective device. One way of eliminating this problem is through the use of fusible lugs known as "cable limiters" that are installed on each cable. To assure service continuity, it is necessary to place limiter lugs at both ends of an individual cable to prevent reverse current flow to the fault. It should be noted that a limiter lug should be specified to be the current-limiting type if it is to reduce available fault currents. The limiter lug is designed to protect an electric system against a catastrophic failure of individual paralleled cables, and are not to be considered as replacements for fuses. Should one of the limiter lugs open, the remaining cables may carry excessive current for extended periods. Therefore, it is desirable to periodically test them either with a clamp-on ammeter or a similar device.

Although busways and cableways are available for high ampacities (6000 A or more), the original source of power should be divided in such a way so that the use of busways over 2000 A or, in some cases, 3000 A (see the NEC, Article 300 [1]) in normal building installations is avoided. The more massive a bus structure is, the more difficult it is to protect. Cables may be installed in cable trays. In such a case, ground-fault protection (which is required in high-current services (see the NEC Article 240-13 [1])) is especially necessary for long cable tray runs. Ground-fault protection should be set low enough to positively detect arcing ground faults that could damage equipment, but high enough to avoid nuisance tripping. The engineer should consider the effect of a failure of a single feeder on other feeders before using any system involving a large number of conductors in a single enclosure or wireway (this also applies to controlling cable bundles).

Mixing different feeds from different services in the same raceways does not contribute to continuity of service. The local electrical code, for example, that of New York City, may be extremely specific as to the separation of 480Y/277 V multiple services and feeders. Where more than one set of service connections or takeoffs is provided by the utility for electric service, it is prudent to arrange the building distribution in such a way so that outage of a single service will create minimum disruption to building power. For example, if two panels are provided on a floor, one can be from service A, the other from service B. Half the power can be taken from one service and half from the other service. It would not be desirable, for example, to take all the air-conditioning compressor units from service A and all the air-handling units from service B. Some local building regulations do not permit the intermixing of feeds from separate services when such services are widely separated, as might be the case when an additional service point is provided. Some local codes require that distributed loads be divided or separately switchable, primarily for energy conservation implementation.

15.9 Existing Plans. Unless the plans of existing systems include recent as-built drawings, and reasonable assurance is available that later changes have not been made by building staff, the accuracy of such plans is suspect. It is, therefore, necessary to verify the accuracy of questionable plans on the job site and to record any additions or corrections. Close liaison with the building staff is required. It will be necessary to note building structural changes and to determine acceptable floor loadings for transport and placement conditions of switchgear, transformers, and other major equipment; and ceiling loadings for the suspension of heavy conduits and busways. The full size of conduits shown on old drawings may not be usable for expansion or modernization.

Floors with numerous core drillings may not be suitable for further drilling. Old conduits may be rusted to the extent that cable pulling problems are accentuated, EMT (thin-walled conduit) lengths may have pulled apart in the ceilings, sections of nonmetallic duct may have been crushed, and tile ducts may contain concrete obstructions. To check empty conduit runs, it is desirable that a clearance checking device be pulled through the conduit. Where there are operating cables in existing conduits, it is usually impractical to check such conduits prior to the proposed work, except for conditions that are determinable by visual inspection. In a large project, where data on the conduit condition are unavailable and there are no spare conduits, it may be worthwhile to remove (and replace, if necessary) cables in a few typical runs to permit determination of problems that may be encountered in rewiring.

The old system may very well contain code violations based on current standards as a result of code changes. If a rehabilitation is undertaken, the code violation may require correction. Field changes are always expensive when they are performed as contract extras. If a contractor estimates work in which the scope is indeterminate and he or she is required to give a fixed or lump sum price on such work, it can be assumed that his or her estimate will be predicated on the worst anticipated conditions.

In a rehabilitation project, knowledge of the exact location of embedded conduits may be essential because of the necessity to cut into concrete slabs for the extension or rerouting of conduits. Such conduits can be exactly located by x-ray if both sides of the floor or wall slab are accessible. However, this location work should be scheduled when personnel will be absent from the vicinity because of the radioactive hazards in the test area. When such work is required, it is essential to obtain the services of a firm specializing in radiographic location because of the laws, safety requirements, and special techniques involved. Metal detectors may also be used to locate conduits.

15.10 Scheduling and Service Continuity. Several well-known scheduling methods, such as the Gantt chart, PERT, and critical path (some of which may be computerized) are available to assist the engineer. However, the need for continuous electric service during construction accentuates problems when compared to new work. Consider some problems that have resulted from poor planning. In one case, workers were denied access to an office building at the beginning of the work day because electricians were completing electrical connections. In another case, power to a public terminal building failed during rush hour because temporary load reconnections that were coupled with system transients caused tripping of an old breaker with a defective overload unit. The work schedule should provide extra time for completion of scheduled power outages. For example, such work could be scheduled to start on a Friday night so that it can be completed over a weekend. An engineering, operations, inspection, and contractor chain of command communication is especially important during this work. Should critical decisions be required, the availability of engineering assistance or support is essential. The contractor's staff may not be in a position to make project engineering decisions. Consideration should be given to maintaining a continuous work flow, even though a portion of the work is delayed. For example, if new overload relays for the main circuit breakers are delivered and will not fit (a last-minute determination), the entire project should not be adversely affected.

The design of the job, if staged, should involve the development not only of old and new single-line diagrams together with the applicable plan, detailed drawings and specifications, but also detailed work phasing schedules. A design engineer should be aware of the steps involved in making complex power system additions and alterations. Drawings or sketches and work phasing schedules that show the step-by-step procedure should be developed, even though they are of only temporary value. The drawings or sketches should be modified at the construction site as the construction work is performed to show the true or as-built status. It is often advisable for the designer to specify a new or temporary replacement item rather than resorting to a field modification. Returning to the example of the main circuit breaker with new overload relays — if the overload relays fit perfectly, the job can be completed in one night; if not, a large section of the building could be out of service longer, additional overtime may be required, or the repair may be completed in an incorrect manner. An alternative plan of installing a temporary fuse cubicle or having one available for possible use could be prudent. In a project involving old construction, time contingencies are as important as cost contingencies. Scheduling a modernization project over several years (with perhaps contract escalation clauses for inflation) may be much more effective regarding cost and outage than an accelerated contract. Time permitting, problems can be minimized by keeping the day-to-day scope of work limited in an area. Without proper control, overtime costs can be disastrous and require supplementary budget authorization.

System rehabilitation, both in construction and in field investigations, often involves personnel and equipment safety procedures. These should be fully coordinated with facility staff, using existing safety rules, supplemented by rules and procedures that are necessary for the special nature of the rehabilitation. Reference may be made to these rules in drawings and specifications. New OSHA regulations require the detailed development of lockout/tagout plans for personnel safety when power circuit work may involve hazardous energy sources.

15.10.1 Example 1—Building Extension, Medium-Voltage Service. Consider an expansion that is almost double the area of a large terminal building including public, vehicular, office, and tenant areas. The question is whether to expand the medium-voltage power system as a primary-selective system or start from scratch and abandon the existing system. In this instance (see Fig 151), 13 200 V is provided to the building from four existing feeders. Two of these are standby feeders of the existing primary-selective system. If the primary-selective system is to be retained, two additional normal feeds are required to handle the building expansion. The two standby feeders are adequate for second contingency conditions (e.g., one feeder out of service and one subsequent failure). The installation of a four-feeder network permits the use of the four original feeders with all feeders carrying a

Fig 151
Example 1—Building Extension, Medium-Voltage Service
(After connection of the new substation, the old equipment [shaded] was removed.)

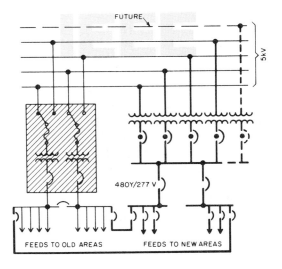

normal load, thus eliminating the need for two additional feeders. When computers are to be installed, the need for an uninterruptible power supply may be eliminated because the primary switching interruptions have been eliminated. Note that network primary feeders should, in general, have essentially the same receiving voltages and phase angles in order to provide proper load distribution between transformers. The old equipment remains in service until all loads have been transferred, one by one, to the new switchboards. Network transformers with high impedance (e.g., 7%) are specified to reduce available fault currents, provided increased voltage drop is no problem. (Current-limiting fuses may be considered as an alternative if standard impedance transformers are to be used.) Main circuit breakers (or fused switches) and switchboard buses are, in this instance, limited to 2000 A in order to limit equipment size, reduce downstream available fault currents, limit the extent of damage in the event of switchgear failure, and provide greater overall system reliability.

In Chapter 5, it was pointed out that the standard packaged network protector, when used in interior distribution, usually does not inherently meet NEC [1] requirements for system protection without supplementary protection. The use of standard drawout metal-enclosed switchgear that embodies the network-type relaying may be appropriate; the drawout features will provide for safer maintenance.

15.10.2 Example 2 — Building Extension, Low-Voltage Service. In Example 2, the idea is to increase the size of a large building by about a 50% in area. In the preliminary study, perhaps 20 schemes (seven of which are illustrated in Fig 152) are analyzed to obtain the most desirable solutions from both economic and operating standpoints. Discussions with the local utility assist in determining which schemes are acceptable. The problem in Fig 152 has been generalized and simplified for purposes of presentation. The existing building is fed from utility vaults at 208Y/120 V from a common network system. Air-conditioning compressor equipment in the older area is supplemented to provide chilled water for air conditioning that will serve the new area as well. A first analysis suggests increasing the capacity of the existing service entrances. This is accomplished by increasing the ratings of the existing service connections by adding new ones. Adding new service connections is preferable because it minimizes the fuse or circuit breaker sizes, which enhances fault protection.

The old equipment can remain intact, provided that fault handling ratings are adequate. If transformers and network protectors are owned and operated by the local utility, the utility has to concur in increasing vault capacity. This is probably the utility's first choice since adding network transformers to an existing network installation generally makes most effective use of the equipment. When using four transformers, two of them should be able to handle the full load under second contingency conditions. During a second contingency condition, the remaining two transformers operate at an emergency overload of perhaps 20%–30%, so that the maximum rating of the bank is between 60%–70% of the normal rating of four transformers. Therefore, adding one transformer to a bank of four increases the capacity of that bank by more than the normal rating of the transformer added.

Thought should be given to the physical problems involved, i.e., the advisability of carrying exposed heavy current busway adjacent to public areas. The availability of

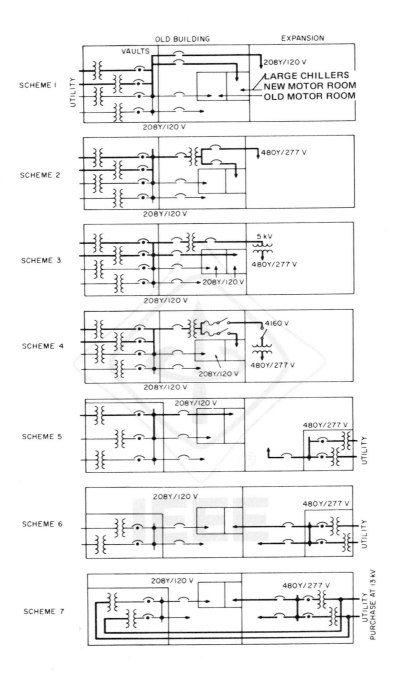

Fig 152
Example 2 — Building Extension, Low-Voltage Service, Simplified
Single-Line Diagrams of Seven Possible Ways of Handling a Major Expansion
(Heavy lines are new installations.)

space is important, and the value of the space for rental or other purposes should be a determinant when deciding to locate service entrance equipment in the old or new part of a building.

(1) Scheme 1 involves the expansion of the original networks by the utility and transmission of power at 208Y/120 V to the new area. With permissible voltage drops, the cost may be substantial (see Fig 152). The cost of 208 V feeds to the new large chillers would be excessive.

(2) Scheme 2 also involves expanding the 208Y/120 V network, stepping up to 480Y/277 V and supplying this voltage to the new part of the building and to the new compressor equipment in the old part of the building. Overall costs are substantially less than in Scheme 1.

(3) Scheme 3 involves stepping up to 5 kV at the 208Y/120 V expanded vaults and retransforming to 480Y/277 V in the expanded area. Since motor load in the expanded area consists mainly of smaller air handling units, 2400 V or 4160 V is not a practical utilization voltage. The motor room, when expanded in the existing building, continues to be fed at 208Y/120 V. As in Scheme 1, the cost of feeding the large air-conditioning chiller motors at 208 V is excessive. However, there are considerable savings in the feeds to the expansion and in the distribution at 480Y/277 V, which more than offsets the transformer cost.

(4) Scheme 4 involves 4160 V or 2400 V three-phase distribution in much the same way as Scheme 3. However, each individual load center is provided with its own transformation from 4160 V or 2400 V to 480Y/277 V, and the new air-conditioning compressors are fed at 4160 V or 2400 V. In Scheme 3, medium-voltage switches are not specified because there are no intermediate taps on the medium-voltage feeder cables.

Scheme 3 shows a 5 kV switch or circuit breaker, as required by the NEC [1]. However, in Scheme 4, medium-voltage switches are desirable for isolation purposes because each feeder serves several loads, and fused switches are used for transformer protection. When several feeder taps are involved, the provision of these medium-voltage switches enables operation of all but one substation under conditions of a transformer outage, simplifies fault finding, and enhances safety considerations. The first two considerations may not be given the same weight in making the decision if a highly qualified utility-type maintenance crew (not normally available to building maintenance staffs) with special fault finding equipment is to operate and maintain the system.

(5) Scheme 5 has the main disadvantage of requiring both the installation of a new service for the new part of the building and expanding the old vault. If no major construction is involved in expanding the old vault and the increased air-conditioning load is relatively small, this might be an acceptable scheme. There is a good possibility that the utility would balk at installing both the network expansion and new network installation at their own cost and may request additional capital and annual costs. Multiple services should meet code requirements.

(6) Scheme 6 is perhaps the best choice of all. Vault expansion is not required in the old building, 480Y/277 V is provided for the new motors in the old building without transforming. This scheme can be so desirable that even if conjunctional billing is not provided between the old and new service points, the total annual cost may be less than that for the other schemes. This does involve the question of multiple services to one building.

(7) Scheme 7 might be the best choice if sufficient differential exists between high- and low-voltage rates to justify owning the network systems.

In practice, two factors besides cost may work against the use of lower voltages for the expansion. There simply may not be enough space to carry a heavy conductor system (busways or multiple conduits) through existing areas, and, above certain sizes (approximately 1000 hp), chiller motor starters may not be available that can operate at 208 V. If utility considerations strongly favor expansion of the existing vaults and the expansion involves capacity on the order of 2500 kVA or more, Scheme 4 (medium-voltage distribution to the new equipment) would probably be the best choice.

15.10.3 Example 3 — Retention of Existing System and Expansion. In this instance, the availability of additional power is severely limited in a large 30-year-old building containing offices, commercial areas, freight handling facilities, and consumer facilities. A series of changes are made which, while not modernizing the existing electric system in its entirety, provide adequate power for continued expansion (see Fig 153). Power is available at 208Y/120 V, three-phase, from individual network transformers owned by the utility. These transformers are distributed throughout the building in vaults. The existing electric system is adequate beyond the service switches, except for new heavy load requirements. The anticipated load is expected to rise steadily over a period of years, perhaps 10% annually, and the distribution of loads on risers and the nature of the additional loads are such that it is entirely practical to retain the existing equipment and distribution system for the existing loads and to install new service equipment and switchboards for the new loads.

At some later date, older equipment could be modernized without affecting the new installation. A second consideration when retaining original equipment is the need to maintain full service during business hours throughout the power system expansion. By installing new equipment to pick up additional loads, the old equipment is maintained intact. After the new equipment is installed, loads are shifted from the original to the new circuit breakers (or fused switches) to relieve heavily loaded feeders and provide capacity in the exiting feeders for future normal expansion. For example, a feeder rising 15 floors is cut and fed as two feeders, one from the top floor and one from the basement. New service connections are provided by the utility, one for each service entrance equipment room.

15.10.4 Example 4 — Rehabilitation on a College Campus. This campus (see Fig 154) is comprised of three buildings — the main building, the library building, and the science center. The center section of the main building was constructed about 80 years ago, while the East and West wings were added about 10–20 years later. The library and science buildings are modern structures that were built within the past 10–15 years. The rehabilitation basically called for the enlargement

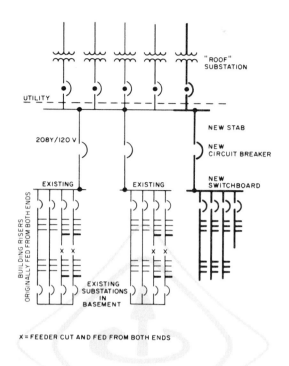

Fig 153
Example 3 — Retention of Existing System and Expansion
(Heavy lines are new installations.)

and renovation of the main building. In the course of investigating the electric service for this expansion and modernization, it was found that each of the buildings on the campus was separately metered and was also supplied at a different utilization voltage. An economic analysis indicated that it would be prudent to purchase primary power at a single service on the property and feed each of the existing buildings from that point. After discussing the plan with utility company representatives, it was decided to purchase power at 13.8 kV and to transform this voltage to 480Y/277 V at an outdoor substation on the property. This secondary voltage was suitable for the science building and for the planned rehabilitation of the main building. At the library building, a 480-280Y/120 V transformer was installed in the vault, which previously housed the utility company transformer. To simplify the switchover and to keep the outage time required for reconnecting the services for the science and library buildings to a minimum, the revenue meters and metering current transformers were purchased from the utility company, obviating any work in the metering cubicles of either service. The purchase price was considerably less than the cost of having the meters and transformers removed. In addition, it provided the owner with metering facilities for maintaining energy consumption records at each location for statistical and accounting pur-

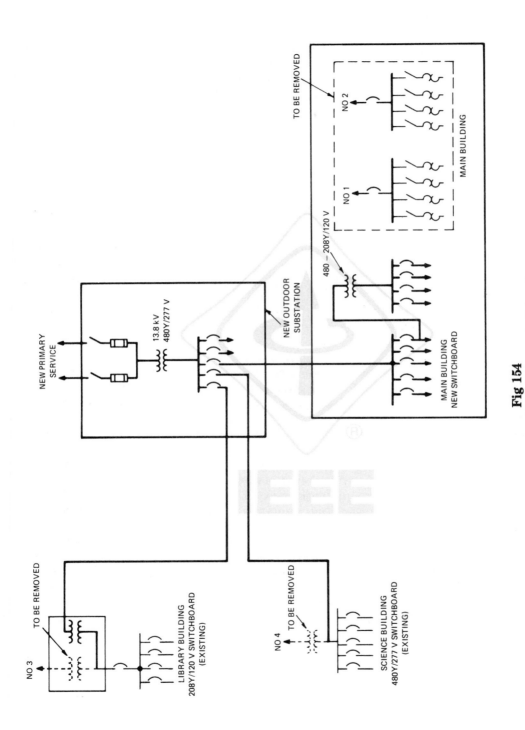

Fig 154
Example 4 — Rehabilitation on a College Campus
(Heavy lines are new installations.)

poses. A new switchboard and distribution system were installed in the main building. When new loads and all existing loads that were to remain were connected to this new service, the original two services were discontinued and the old switchboard was removed.

15.11 Wiring Methods. There are a number of wiring methods that lend themselves to building rehabilitation. While it is desirable to utilize existing conduits, as a practical matter, it may be necessary to abandon all or part of the existing conduit system. If a new architectural finish is being provided, an opportunity exists for running new concealed conduits. However, when physical or cost limitations preclude concealment, any one of the surface raceways that are code-approved may be used. When existing boxes are flush-mounted, it is possible to install commercially available box extensions that will permit connection of a conduit or a surface raceway. Almost all wiring equipment is available for surface or flush mounting, and where architectural considerations permit, the use of surface-mounted equipment simplifies rehabilitation.

When a new hung ceiling is being installed or an existing hung ceiling is being replaced, an excellent opportunity exists for installing a new lighting system. When the ceiling is to remain and is structurally adequate, either surface- or pendant-mounted fixtures may be installed without modifying the ceiling. It should be noted that the cost of removing, refurbishing, and relocating a lighting fixture in a hung ceiling might be more expensive than installing a new fixture.

When extensive underfloor wiring has to be provided, particularly for data processing, computer, and communication equipment, the use of a raised floor may be practical. Raised floors, which are floors mounted above the structural floor, permit wiring to be conveniently run to equipment from underneath the raised floor. The space created by the raised floor can be used to provide cooling air to equipment. The raised floor is usually made of manufactured modules of metal and simply assembled on raised pedestals anchored to the structural floor. It can provide a highly effective communication (signal reference) grounding system to supplement the normal "power" grounding system. For rehabilitation projects, there are many advantages to using a raised floor; it provides space for wiring to equipment, equipment ventilation, partial shielding from electrical noise, and improved appearance.

15.11.1 Smoke and Fire Considerations. It is essential to maintain the fire ratings of floors and walls in accordance with fire and building code regulations. The problem of controlling the use of unprotected utility openings during construction is more serious than in new construction because of existing occupancy. Part of the modernization investigation process is to determine if any such openings exist and to see that they are sealed, either by maintenance forces or as part of the construction contracts. In addition to the contractual requirements for control of fire hazards, it may be desirable to have a "fire warden" from the operating staff to check on such hazards. When existing enclosures (and some of them may be very old) are opened for temporary work, large amounts of flammable and potentially smoke producing material may be exposed; suitable fire precautions should be taken.

Floor outlets may also be fed from the ceiling below them by means of conduit and poke-through boxes; however, poke-through installations have the serious disadvantage of requiring provisions for retaining floor fire rating and requiring analysis to assure that the floor's structural integrity is not affected.

The engineer should be aware of changes that are taking place in the development of low-smoke, low-toxicity, and fire-resistant cables, particularly of the non-halogen types; various code changes may be made in coming years concerning or even mandating the use of these cables in certain applications. In a well-maintained modern system that meets all code requirements and that contains all the fire stops and seals required, the danger of fire or smoke spreading throughout the electric system is minimal. Correcting deficiencies in the old system is an important part of the modernization program. The engineer may consider it prudent to exceed minimum code requirements in protecting areas where a higher degree of protection seems warranted, such as in high-rise buildings or medical facilities.

15.12 References. The following references shall be used in conjunction with this chapter:

[1] ANSI/NFPA 70-1990, National Electrical Code.

[2] IEEE C57.110-1986, IEEE Recommended Practice for Establishing Transformer Capability When Supplying Nonsinusoidal Load Currents (ANSI).

[3] Goldberg, D. L. "Electrical Expansion, Modernization, and Rehabilitation for Buildings," *IEEE Transactions on Industry and General Applications*, vol. IGS-6, May/Jun. 1970, pp. 219–229.

Chapter 16
Special Requirements by Occupancy

16.1 General Discussion. This chapter covers specific considerations for the most common type of commercial facilities. The preceding chapters described in detail the basic electric systems and components that are necessary to meet the needs of commercial facilities insofar as modern power supply, electric distribution, transportation, lighting controls, and communication systems are concerned. However, in each classification of occupancy, there are certain items applicable to each occupancy that are not required in other occupancies.

In this chapter, the different classifications of commercial occupancies are listed and the special considerations for each are discussed. Most of the usual requirements, such as service entrance, substations, types of distribution, and communication systems, are described in detail in other chapters.

The designer of commercial facilities may group his or her design parameters into four basic categories:
(1) Minimum requirements of applicable codes
(2) Dependability of service and provision for contingency
(3) Flexibility for growth and changes
(4) System safety and efficiency, and user comfort

Good judgment and selling ability are required by the designer. The designer should demonstrate to the building owner and electric utility company the need to include, in the original installation, the service capacity for potential future loads, and to make provisions for facility additions at a later date. Once the construction of the facility is finished and all space is completed, the cost of installing additional equipment may be double or triple the cost that would have been incurred had provisions for these future loads and additions been included in the master plan.

Minimum requirements are generally covered by applicable regulatory codes (see Chapter 1). The design engineer should first determine the code enforcement agencies that have jurisdiction over the facility and communicate with them requesting specific information as to the codes, ordinances, and regulations adopted and enforced in that area. Copies of these documents should be reviewed for the specific interpretations that are applied by the code enforcement agency. Codes, ordinances, and regulations establish the legal minimum requirements for safety, and may not adequately address the needs of specific occupancies. The design engineer should meet the minimum requirements and expand from that point, as required, by the specific nature of the facility, occupant, and building owner.

The system should be designed to be maintainable at a high level of performance by assigned maintenance staff. Complex protective systems or voltage levels beyond the scope of the maintenance staff should be avoided.

16.2 Auditoriums. An auditorium may be described as "an indoor area for large gatherings of people for meetings, entertainment, expositions, or sporting events." Auditoriums may be a part of schools, office buildings, laboratories, churches, or any building where people worship, work, or play. Most schools have one large room for assemblies, plays, gymnasium, and various entertainment events. Some auditoriums, especially those used for sports events, may require special services. Many auditoriums have a stage, which requires stage lighting that can range from small systems with track lighting and a few spotlights to electronic dimmer systems with extensive use of spot, flood, and border lights. In all cases, stage lights should be provided in auditoriums to illuminate the faces of people on the front portions of the stage. The auditorium's general lighting should be controlled from a central location. Motion picture, slide, and overhead projection may be required, and service outlets should be provided as well as projection booth equipment, when used. Convenience outlets for cleaning and general maintenance should also be provided, as required. Building codes dictate the need for emergency aisle and exit lighting. A dimmer bypass should be provided for the house lights.

There may be requirements for greatly increased lighting intensities on the main floor for the illumination of a boxing ring, basketball court, or ice rink. Ice shows may require spotlights scattered throughout the main floor area. The lighting should be flexible so that it can accommodate the different uses of the main floor area.

The type of lighting control and the location of the control booth are important. Provision for communication systems between various lighting centers is necessary. Arrangements for television equipment may also be necessary. Lighting controls may be elaborate for color television, which requires higher intensities and good color rendition.

Attention should be given to the problems of glare. High lighting levels can become objectionable unless a great deal of care is used in the design of fixtures, shielding, and ventilation.

Auditoriums may be used for exhibitions, which may require a sufficient number of outlets for both three-phase power and 120 V lighting for the various exhibits. Power outlets may require several hundred amperes each. Central distribution centers should be furnished to supply portable distribution outlets. A grid of underfloor raceways should also be considered.

The area under the main floor is sometimes used for industrial exhibits, or for a menagerie, storage, or horse stalls (in the case of a rodeo or horse show). Good general lighting should be provided as well as ample convenience outlets for three-phase power and 120 V lighting. This area may have a low ceiling, which may require special attention be given to lighting fixture design to achieve the required lighting levels.

Air-conditioning loads, both for the main auditorium and exhibit halls as well as for smaller rooms, may be major power loads. Ice shows require large refrigerating plants. Continuous ventilation will be required to maintain air quality.

Large auditoriums may have several entrances. Each of these entrances should have an illuminated marquee and signs. The main entrance lobby and foyer may be very ornate and require careful study to provide a satisfactory plan of illumination and decorative lighting. Large signs should be provided so that notices of coming events may be read from a passing vehicle. Electric heating and cooling for the ticket booths may be required, depending upon the local area's climate.

Public address and music systems need special attention. Often, musical programs need good sound reproduction in an arena that may have poor acoustics. Amplifiers and special lighting for concerts often require relatively large amounts of power. Permanently installed radio and television jacks should be provided. Local radio and television stations should also be consulted as to their requirements. Microphone jacks should be positioned at likely points of activity. Telephone facilities and press communications may be required. Large numbers of public telephones and facsimile machines may be needed, especially in the lobbies and foyers. Electric timers and electronic scoreboards for indoor sports may be widely placed around the arena.

Power services should be of dual source supply, when practical. A power failure when large crowds of people are assembled in darkened areas could easily lead to panic. Emergency lighting is usually dictated by local codes; but these are generally minimum requirements. High-intensity discharge (HID) lighting is not considered acceptable for emergency lighting due to the excessive time it takes to produce an acceptable lighting level after a momentary power outage.

In auditoriums, the following checklist should be used as a guide:

(1) General lighting
(2) Stage lighting
(3) Exit and emergency lighting/emergency and standby power
(4) Public address system
(5) Signal and communication systems
(6) Radio and television facilities
(7) Projection and sound equipment
(8) Signs, scoreboards, and timers
(9) Telephone, telegraph, and facsimile facilities
(10) Copying facilities
(11) Air conditioning and refrigeration
(12) Ventilation and heating
(13) Special event lighting and power outlets
(14) Refreshment stands
(15) Elevators and escalators
(16) Parking
(17) Floodlights
(18) Snow and ice melting systems for walkways, ramps, and driveways
(19) Stage elevators

16.3 Automobile Areas. This classification covers a sales office, showroom, and service shop facilities on one floor, or a multiple-story building including storage and parking facilities. In the latter type, the automatic parking garages erected on small areas of high-priced real estate in large cities are becoming more numerous as the parking situation becomes more acute. Most parking garages used by tenants, employees, and the general public provide for self-parking with ramps between floors. Passenger elevators may be required for facilities with multiple floors.

16.3.1 Sales Offices and Showrooms. Normally, sales offices include many partitions for salesperson/client negotiations. Convenience and telephone outlets are required in each space. Showrooms require higher lighting intensities. Special lighting effects that are to be used for sales promotion material and to display special features of the new cars may be needed. Turntables showing the complete product in the entire 360° range will require motor driven platforms. A public address system to sales areas (including lots) is usually necessary. Special lighting may also be needed for video display areas.

16.3.2 Service Shops. In service shops, lighting intensities should be set in accordance with the *IES Lighting Handbook*, 1987 Edition [11][92] recommendations. Wattage may be limited by local energy codes. Surface-mounted raceways with convenience outlets located every 18 inches along the back of the benches can be used to distribute the necessary 120 V power for the numerous testing and analyzing equipment. Explosionproof systems are required in below-grade inspection pits and certain other areas.

Service shops have power requirements for hoists, machine shop tools, and air compressors. Welding outlets may be required if extensive body work is done. These welders are usually found in ranges up to 45 kW, single-phase and three-phase. Air compressors should be located near the electrical service entrance to reduce circuit costs. Compressed air lines are generally cheaper to run than conduit and cable. In the machine shop area, a single length of busway with plug-in overcurrent devices will enable the addition and movement of machine tools at will.

The power supply should be so designed that starting large motors (e.g., compressors or machinery) does not adversely affect sensitive electronic test equipment. This can be accomplished by the segregation of feeders, the use of isolation transformers, and the use of power conditioning for sensitive equipment.

When paint-spray booths are provided, lighting intensities require up to 500 fc with shadowless distribution. ANSI/NFPA 70-1990, National Electrical Code (NEC) [1][93] and local codes provide stringent rules covering such application. A 95% explosionproof installation is worthless. As a practical matter, it may be best to have all objects illuminated by light fixtures installed on the outside of wire glass and directed to the work plane. Consideration should also be given to prefabricated booths. Infrared drying, either portable or stationary, is used in many cases.

16.3.3 Parking Lots and Garages. Electronic systems may be provided for the collection of tolls, fees, and tickets. Such systems may involve automatic gates,

---

[92]The numbers in brackets correspond to those in the references at the end of each chapter. IES publications are available from the Illuminating Engineering Society, 345 East 47th Street, New York, NY 10017.

[93]ANSI publications are available from the Sales Department of the American National Standards Institute, 11 West 42nd Street, 13th Floor, New York, NY 10036. NFPA publications are available from Publications Sales, National Fire Protection Association, 1 Batterymarch Park, P.O. Box 9101, Quincy, MA 02269-9101.

ticket issuing machines, and electronic cash registers. Treadle-and-loop systems are used to count vehicles for auditing purposes and lot-full inventories. Large self-parking garages may require personal safety systems, including surveillance closed-circuit television and "panic stations," which would include pushbuttons that automatically alert building security, two-way communications, and an alarm siren. Larger parking lots or structures may require rooms for electronic equipment and perhaps computers. These systems may require isolated power supplies, regulated power, and, in some cases, standby power.

Telephone services should be provided in the waiting rooms for customers as well as for the personnel operating the garage. The following checklist should be used for garages in general:

(1) Illuminated directional signs
(2) Remote door controls
(3) Telephone systems
(4) Public address systems
(5) Signaling and communication systems (including personal safety systems)
(6) Fire alarm
(7) Lighting (general and localized)
(8) Power for tools and testing equipment
(9) Elevators and escalators
(10) Ventilation systems for enclosed garages
(11) Snow and ice melting systems for ramps and walkways
(12) Electric infrared heating for waiting customers and attendants
(13) Emergency power, especially if handicapped persons are involved
(14) Manlifts for use by parking attendants

16.3.4 Car Wash Facilities. Special electrical considerations include pumps, conveyors, fans, washing roll drives, water heaters, towel washers and dryers, soap- and wax-metered dispensers, and an extensive interlock system. Equipment located in the wash area should be watertight. GFCI protection should be installed in adjacent work areas when convenience outlets are provided and "wet" conditions could exist.

16.3.5 Automated Parking Facilities. Completely or partially automated parking facilities may be provided for commercial parking lots, building tenants, and car rental facilities. The designs for security and access control systems involve the garage designer, special computer and programming designs, and detailed equipment and communications design. A significant amount of design work may be required for the installation of automatic gates (for both people and vehicles), treadles, sensors, signals, and interconnections.

16.4 Banks. The design of bank buildings is apt to involve expressive architecture. Today, banks tend to use an open floor plan with relocatable partitions. The lighting, both outdoor and indoor, should complement the architectural design.

Outdoor lighting may include floodlighting the building for decorative purposes and may require provisions for supplying power away from the premises. The use of electric signs and directional instructions for night banking may require underground cable installations before walks and driveways are constructed. Remote teller drive-in islands for banking directly from automobiles may need lighting, communication, and closed-circuit television. Electric heating, ventilating, and air conditioning may also be needed. Teller exterior walkup windows may need lighting, communication, and electric infrared heating. Unattended automatic banking machines require power, lighting, communication, and electric heating and cooling depending upon whether they are located on the exterior wall of the bank or in a lobby. Although not part of the bank building, remotely located machines to permit customer transactions may require many of the same facilities as mentioned above.

Indoor lighting includes general lighting in the main area. High ceilings permit the use of spotlights and floodlights. Indirect lighting may also be used. The teller's counters should be lighted with care to avoid shadows and glare since visual tasks are demanding and computer monitor displays may be difficult to read. Provisions for closed-circuit television for banks with multiple branches should not be overlooked. Alarm and communication systems, both visual and audible, demand top priority in the modern bank. Security and alarm systems are normally not a part of the construction contract; but close coordination with suppliers selected by the bank is necessary to ensure that adequate and properly located raceways are included in the building. Extensive communications provisions should be provided for computer systems, facsimile machines, and modems. Special power requirements, including isolated grounds and dedicated grounds, may also be required. Provision will probably be required for computer terminal equipment at tellers' windows. Power and data circuits may also be needed. Closed-circuit television is used both for checking money and for guard surveillance and protection. Automatic still cameras are often used as well as videotaping.

In a safety deposit department that is to be used by the public, banks desire a pushbutton in each booth to call an attendant. Banks also need an intercom system and ventilation in the vault in the event that people are locked in the vault accidentally. A signal system for calling employees should also be provided.

The accounting department may have large numbers of business machines. An underfloor duct system or a raised floor (access floor) system that is at least 6 inches high provides an excellent means for installation and rearrangement of power, data, and communication cables.

A central air-conditioning plant may serve the whole building. In smaller buildings, individual units may be used. In central plants, control centers should be used to indicate the status of blowers, pumps, and compressor units, as well as temperatures, pressures, etc.

Some bank buildings have space available for rental. These spaces should be treated as office spaces just like in office buildings (see 16.18).

The checklist for banks should include the following items:

(1) General and decorative lighting
(2) Specialty lighting
(3) Exterior lighting
(4) Security and emergency lighting
(5) Fire alarm
(6) Burglar alarm and holdup systems
(7) Telephone and facsimile communications

(8) Intercommunication system

(9) Paging annunciators

(10) Electric door locks and controlled access systems

(11) Closed-circuit television

(12) Air conditioning, heating, and ventilating

(13) Business machines, data processing equipment, and cash machines

(14) Computer mainframe, peripheral stations, local area network (LAN), and tie to remote stations

(15) Power outlets

(16) Snow and ice melting systems for driveways, walkways, and ramps

(17) Emergency and standby power requirements

(18) Surge protection

NOTE: Remote transmission of alarm signals may involve radio links.

16.5 Brokerage Offices. Brokerage offices require information on display, teletype, and special telephone services. These services should be coordinated with the telephone companies in the local area. Brokerage offices, and stock and commodity exchanges have a heavy concentration of domestic and foreign communication facilities. Special wireways may be required throughout the building to house these wires and cables. Computer keyboards and displays are generally provided at each desk.

Special lighting is needed for the quotation boards. Automatic boards require specialized wiring techniques. Air conditioning is the same as in general offices. The checklist for brokerage offices is similar to the one shown in 16.4.

16.6 Places of Worship. The lighting of churches, synagogues, and temples requires a special study in keeping with the particular type of architectural design involved. Soft well-diffused lighting is recommended. Such lighting tends to underscore the architectural beauty of the edifice.

With a few exceptions, the design of the modern place of worship is moving away from the cathedral type of edifice to one of more simple design. The vast majority of places of worship are being built for service to the congregation and include recreational and educational areas as well as a main auditorium. In these types of places of worship, the general conditions for schools and auditoriums may be followed.

Gothic buildings with their high vaulted ceilings and great stained glass windows require direct lighting from well-designed pendant-lantern-type luminaires. Pendant-type lighting fixtures that are suspended from high ceilings should have an integral system for lowering the fixtures (or equally effective means) for relamping. Indirect lighting with a brightly lighted ceiling is not recommended for this type of edifice. Pinhole downlighting may be effectively used with high ceilings to provide sufficient illumination with spotlights. These spots should be carefully shaded to avoid glare.

The lighting of the altar can be accomplished by spotlights or floodlights concealed in the altar arch or behind ceiling beams. The lighting should be well spread out to avoid the theatrical spot effect. Dimmers may also be used to advantage.

Many places of worship have beautiful stained glass windows that can be illuminated so that they may be seen at night. High lighting intensities are needed to penetrate the density of the stained glass; therefore, sufficient lighting circuits should be provided for this task. Floodlighting the steeple, dome, or the front of the church may be accomplished with powerful floodlights. Photocells and automatic timers will probably be required to limit such lighting to a few hours each night.

Public address systems with good coverage throughout the church should be provided. The perimeter around the place of worship may be required for the crowds who remain outside for special services, funerals, or weddings. Portable public address systems and speakers may be needed for those occasions. Radio and television remote pickups may require special outlets and heavy feeders for the large lighting loads required by television cameras. Modern places of worship require extensive coverage of the altar area with microphones.

Some pipe organs are being replaced by electronic organs. Feeders will be required in the choir loft to handle this load. Tape recorders and record players may also be used. Provisions for connecting these to the central public address system should be made in the original plans.

In colder climates, the steps, sidewalks, and even the leaders and gutters may need to contain electric heat for melting snow.

Today, kitchen facilities for meeting rooms are usually provided. Electric ranges, hot water tanks, automatic dishwashers, and large appliances, such as toasters and coffee urns, have relatively high current ratings.

Certain religious groups require special circuits at certain locations. It is imperative that the owner be contacted relative to any such requirements.

The checklist for places of worship should, therefore, include the following items:

(1) Special interior lighting
(2) Special exterior lighting
(3) Floodlighting
(4) Emergency lighting
(5) Organ, piano, or music synthesizer
(6) Public address system
(7) Sound and video systems
(8) Snow and ice melting systems (where needed)
(9) Kitchen appliance service
(10) Air conditioning, heating, and ventilating

16.7 Athletic and Social Clubs. The general arrangement of the ground floor for larger clubs may be similar to that of a modern hotel. General lighting may be decorative. Outlets for reading and table lamps, and service outlets for maintenance equipment should also be provided. Some clubs have a special insignia that may require electric outlets. Swimming pools, gymnasiums, handball courts, and bowling alleys require special lighting. High bay lighting in a gymnasium may require sufficient lighting circuits to serve special events.

The upper floors of certain clubs have sleeping quarters for members. The power requirements for these floors are the same as for hotels. A central antenna, satellite, or cable for television as well as provisions for the air-conditioning needs of each room should be installed. Telephone, facsimile, or intercommunication systems, or all three, may also be installed in rooms. Saunas are standard equipment in most clubs. Sauna electric heaters may vary from 5–15 kW, depending on the size of the room.

16.8 Colleges and Universities. These facilities often have a wide divergence of uses. Occupancies, such as auditoriums, gymnasiums, hospitals, clinics, libraries, office buildings, etc., may need to reference those classifications elsewhere in this chapter.

16.8.1 Central Power Plants. A central power plant may be used to supply the heating, cooling, and hot water requirements for campus buildings. There is a trend to include at least some electric generating facilities on campus. It is generally a good location for the emergency or standby electric power plant, or both. Cogeneration is also becoming increasingly acceptable. This building may also serve as the distribution center for the electric system. Underground distribution is usually used between buildings. The designer should take into consideration probable ex- in all phases of power, light, and communication (including video) systems. Underground distribution with spare ducts is recommended for these systems.

16.8.2 Classrooms. Classrooms should be provided with an adequate level of illumination, properly diffused, to eliminate glare, eye strain, and objectionable shadows. Properly engineered fluorescent lighting is considered the best illumination system for general classroom use. Supplemental lighting on chalkboards and bulletin boards may also be required.

Public address or intercom systems and clock systems, as well as provisions for a closed-circuit television and master antenna system, may be needed in classrooms. The number of convenience outlets should be ample in order to be able to carry the largest of projection equipment loads for illustrated lectures.

16.8.3 Laboratories. The various sciences, as well as other curricula, may have laboratories for their particular needs. The flexibility and capacity of power circuits for special apparatus is important. Surge protection is generally warranted.

In electrical laboratories, the facilities may become more elaborate. A large range of voltages, frequencies, and dc power supplies may be required. Risers containing the various feeders may run the full height of the building. Power at frequencies over 60 Hz, such as 400 Hz power, requires short feeders and should be converted to the required frequency near the point of use. Special voltage requirements should also be converted locally. Heavy-duty portable cables terminating in switched receptacles or protected terminal posts may be used for feeding the heaviest loads. Isolation and power conditioners may also be required.

Room panelboards should include circuit breaker or switch- and fuse-protected circuits to large ampacity outlets for test or converter use. The tables and counters may warrant individual panelboards. Convenience outlets may be required across the back of each table. These outlets should be polarized for the various voltages and currents, and for grounding purposes.

Receptacle slot configurations for each type of service should be standardized throughout laboratory spaces so that portable equipment may be used in any location without reliance on adapters. NEMA standard configurations are recommended for this purpose.

Special panelboards and controls are frequently installed in laboratories. It is advisable to plan the locations of these with the department heads.

Power loads are heavy in modern laboratories. The use of wind tunnels for research may require drives of thousands of horsepower. These requirements need to be taken into account in the initial system design. In mechanical laboratories, large machines for testing and even manufacturing parts for research will produce load densities that will require an industrial type of power distribution system.

Special attention should be directed to any laboratory use that could cause a dangerous condition to occur or result in loss of many months of work if power is lost. Examples of this are chemical laboratory fume hoods or controlled environment studies where the emergency power supply should be considered. Ground-fault circuit interrupters are advisable when wet conditions exist, such as in chemical laboratories. Photobiological laboratories may require special lighting.

16.8.4 Dormitories. In a modern college, dormitories need many convenience outlets to discourage the dangerous use of "octopus" plugs that are fed from a single receptacle.

Today, study lamps have been supplemented by personal computers, stereos, and tape drives, small refrigerators and cooking equipment, clock radios, portable television receivers, and small space heaters for use in the fall and spring when central heating is off. Even though house rules may prohibit such devices, they still are used and these loads should be taken into consideration in the building design.

Laundry facilities should be provided on each floor, and outlets for steam irons and hair dryers should be installed. In the basement, drying equipment is often provided and the use of electric clothes dryers is common.

Public address, room annunciator, and fire detection alarm and suppression should be provided. Telephones (including private exchange, coin and calling card/credit card phones) should be included on each floor. Telephone outlets, cable television system outlets, and outlets for the college local area network (LAN) should also be provided in each dormitory room.

Lounges and cafeterias should be well lighted and may be provided with food warmers and steam tables to keep the food hot after it is received from a central kitchen. Lounges may have several television sets and will also need television and LAN outlets.

16.8.5 Miscellaneous Requirements. In addition to the principal types of power requirements, the following requirements should also be considered:

(1) Outdoor lighting should be provided for the many bikeways, streets, and sidewalks that may be a part of the campus. Because it is private property, this type of lighting is usually the responsibility of the college.

(2) Telephone and computer systems may be private systems connected to outside networks through a switching center.

(3) Public address systems permit rapid and efficient transmission of information between the administrative staff and the instructors.

(4) An electric clock system with automatic programming can sound the time signals for the various class periods.

(5) Central monitoring and control of the mechanical, electric, security, and fire alarm systems on the campus should be considered. An effective energy management system and computerized maintenance system can prove extremely cost effective for maintenance and energy conservation.

(6) Automatic ties to municipal fire alarm, sprinkler waterflow, and security systems are desirable and may be required.

(7) Snow and ice melting equipment

(8) Emergency telephones along campus pathways for student aid

16.9 Computer Centers. Raised floors are mandatory for most computer centers to allow for efficient and flexible installation and for the concealment of power and data cables. Specifics on computer equipment should be established in order to determine the following requirements:

(1) Acceptable temperature and humidity ranges

(2) Voltage and frequency constraints (including transients)

(3) Effect of power interruptions

Generally, computer spaces require air-conditioning and humidity systems that are separate from all other building spaces. Generally, normal voltage fluctuations can be tolerated; but transients (or voltage spikes) may upset computer operations. Surge protection is also desirable. Special isolating and voltage regulating transformers with electrostatic shielding and transient suppressors have proven effective in eliminating this problem. These transformers can also provide voltage regulation when unusual voltage variation can be expected or the equipment is voltage sensitive. Some computer systems are designed with the capability to pick up and continue after a power outage; others may suffer loss of valuable data. Standby power should be considered for any computer that is operated continuously or where utility power is undependable. When interruptions cannot be tolerated, an uninterruptible power supply (UPS) may be economical, as determined by a study that compares the costs of computer downtime and loss of data with the owning and operating costs of the UPS.

Grounding for equipment and computer power circuits requires detailed consideration. If isolating transformers are used, neutral grounding should be handled for a separately derived system according to NEC requirements [1]. Each computer should be grounded to a ground bus and then to an approved ground. Neutral conductors and buses should be insulated so that each system has only one point at which the system's neutral conductor is grounded.

Most microprocessor- and computer-associated electronic equipment in commercial buildings utilize switching mode power supplies, which generate a high percentage of harmonics (often on the order of 30%). Recent experience with installations where this type of power supply predominates (such as in computer rooms) has indicated the need to increase the size of the neutral above that of the lines. Some manufacturers of prefabricated partitions and furniture are using double-sized neutrals in electric wiring, or breaking the conventional multiple-wire systems from five-wire (including ground) into eight-wire systems: three lines, three neutrals (essentially three single-phase circuits), one equipment ground wire, and one electronic ground wire. The designer should be aware of the need to reinforce the neutral under such conditions and to de-rate the system transformer or specify transformers designed to operate under high-harmonic current conditions (standard transformers are not designed to operate with significant harmonic current).

16.10 Department Stores. In many respects, a department store is a combination of many small specialty merchandising establishments. Fundamentally, the difference lies in the fact that, since the whole building is usually occupied by the same store owner, expansion or reduction of the space occupied by one department can be effected by reducing or enlarging the space of others.

16.10.1 Distribution Systems. Distribution systems in department stores often resemble those for office buildings. Sufficient flexibility should be provided in the distribution system to accommodate increased needs for power and lighting caused by changes in the use of the space. No diversity factor should be used in calculating the lighting loads. Maximum electrical load occurs when the store is open in the evening, especially during the summer when the air-conditioning load is added to the power required for lighting. Panelboards should be sized to supply ambient lighting loads at up to 3 W/ft^2 of selling space and up to 1 W/ft^2 for spot, accent, and showcase lighting. It is a good idea to provide for an additional 0.5 W/ft^2 for additional nonlighting specialties, such as point-of-sale equipment, pharmacy refrigerators, etc. Most department stores require a building automation system that shuts down two-thirds of the lighting at closing. The remaining lighting (except pathway and security) as well as general-purpose outlets shut down at a later scheduled time. Panelboards should be located so that they are not accessible to the general public. Circuits supplying computers, essential refrigeration, and cleaning equipment outlets should remain energized.

Cash registers or point-of-sale systems that report back to central-verifying-type registers generally require clean power, and separate raceway systems should be considered. Surge protection should be included for power and data transmission circuits that are associated with data processing equipment. The refrigeration system circuits for perishable items should be fitted with lock-on clips.

Certain special areas in a department store require extra electric service, e.g., the electric appliance department. Provision should be made, with adequate circuiting, for the demonstration of electric ranges, microwave ovens, toasters, mixers, refrigerators, etc. In the case of table and floor lamp displays, a large number of receptacles is usually required. Television and sound system areas also require extensive outlets. Television and related equipment also require many program source (antenna, satellite dish, and commercial cable) outlets.

16.10.2 Lighting Systems. Lighting problems in a department store should be seriously considered because the public is involved and so much flexibility is required. The designer should consider the following points:

    (1) General overall lighting scheme
    (2) Accent lighting or spotlighting
    (3) Showcase lighting
    (4) Exterior show window lighting
    (5) Special areas
    (6) Exterior site lighting

(7) Security lighting

(8) Emergency exit and egress lighting code requirements

Intensities of 50–70 fc are typically provided in selling areas; however, energy conservation and codes may mandate lower levels. Efficient fixtures and light sources with showcase, accent, or display lighting to provide brightness and highlight detail may be used with high-reflectance surfaces to make more effective use of lighting at lower power consumption per square foot. This allows sufficient illumination for the appraisal of merchandise when it is taken out of showcases. Close cooperation should be maintained with the interior designer or architect for the store to achieve a pattern of lighting fixtures suspended from or built into the ceiling, which does not distract attention from merchandise on display, and which is in harmony with the architectural features of the building. The color of light sources is very important in selling and merchandising areas and requires special engineering attention.

The use of cove lighting to supplement general illumination is common practice. Illuminated coves serve very important functions. The cove-lighted curtain wall provides a terminal or stopping point for the eye when viewing the store generally; hence, it provides space definition. When direct lighting or downlighting is employed as the principal means of illumination, coves provide the indirect component of light that adds to the softness of the whole interior. Cove lighting usually provides only a small amount of the total illumination for selling areas.

Accent lighting or spotlighting is required by the display department of practically every department store. It is common practice to use reflector or projector lamps as the source of illumination for this purpose. Spotlights may be of several types, either concealed in the ceiling with adjustment for direction through the light aperture, or mounted on tracks or outlet boxes and exposed to view. The equipment should be placed in a modular pattern throughout the store to allow for the placement of sale merchandise and special mannequins or other displays wherever the store display personnel may desire.

Showcase lighting may be provided by using reflectors placed in the interior of the case to be illuminated. The intensity should be two or three times the intensity of the general lighting in the store. The fluorescent lamp is practically always used (except for jewelry) as the light source because its shape lends itself to this application. Recent improvements in the color rendition of showcase lamps helps reduce the need for incandescent lamps.

Show window lighting in a department store should be carefully considered, since it is at this point that the prospective purchaser obtains his or her first impression of the goods being offered by the merchant. The usual practice is to employ a spotlight every 12 or 18 inches along the length of the window front. Color filters may be used to show the merchandise to its best advantage. Recent installations also make use of reflector lamps or border light strips to illuminate the background of the window. These strips add interest by means of light or color contrast.

The heat generated in areas where lamps, particularly incandescent lamps, are near items being highlighted, or are in relatively small enclosed areas, such as display cases or show windows, can damage the displayed items. This is particularly true with plastic-type materials, such as appliance housings. Proper ventilation and area temperature control can contribute to alleviating the problem; however, radiated heat from light fixtures can be damaging. Display designers should be encouraged to creatively use bright surfaces and color techniques to reduce the energy requirements for display lighting and thereby reduce the possible damaging effects of this type of lighting.

Security lighting should include battery-operated lights for main pathways and spotlights for cash register areas.

16.10.3 Communication and Signaling Systems. Telephone and telegraph systems in department stores generally employ a relatively large private branch exchange with telephones in each department as well as in the administrative offices.

Order receiving equipment that involves a heavy concentration of telephone wires may also be required. Communication circuits for credit card verification systems and computerized cash registers are also commonly utilized.

Adequate provision should be made for public coin and calling card/credit card telephones at locations convenient to customers because stores do not generally handle customers' outgoing calls.

Television cameras are extensively used for surveillance in department stores; therefore, adequate power, video, and control circuits should be included in its electrical design.

Paging systems should be considered. In some large department stores, local paging systems are installed in various departments. A clock or dismissal system should also be considered. In public areas, all of these systems are sometimes combined into a decorative type of combination unit. These systems require signals or loudspeakers located on walls or columns 8–10 feet above the floor. Frequently, an electric service or outlet is required at these locations. Door protection systems are often installed. When there are many street display windows, portable telephones or two-way radios may be provided for intercommunication between the window decorator and an observer on the street. Merchandise theft control devices are often installed at floor exits.

16.11 Fire Stations. There are two classes of fire stations: those manned by a permanent force, and those manned by volunteers. In the first category, living quarters are generally provided for the firefighters who are stationed at the firehouse around the clock. Kitchen facilities should be provided, and radio and television outlets are needed. The alarm system should be tied into a central control headquarters, and special-purpose telephone services are usually connected. A cut-out contactor for ranges and other cooking equipment is frequently used to disconnect these items when the station personnel respond to a fire call.

The central control headquarters contains the central fire alarm system with provision for emergency power, fire siren, administrative offices, etc. A two-way radio for communication with the trucks and other mobile equipment is also located here. Fire alarm maintenance shops contain test and instrument repair equipment. Automatic door openers are installed in all doors for handling firefighting apparatus. A tie-in to the traffic signals is provided at many firehouses to flash the red stop signal and halt traffic. In rural fire stations, some automatic devices should be added. Battery charging on all the truck batteries should be operating continuously when the trucks are in the building. Quick-disconnect plug connectors should be used to disconnect the battery charging leads automatically when the trucks go out on a call. The lights should be automatically controlled.

16.12 Gymnasiums. Gymnasiums are primarily used for active sports, and, particularly in schools, may also serve as multiple-purpose areas. Particular attention should be paid to ensure that all electrical lighting and devices have adequate guards to protect against damage from thrown balls. Lighting should be designed to provide minimum shadow and lack of glare for participants. Lighting for different sports may require special aiming patterns or fixture construction. While metal-halide or high-pressure sodium lamps may provide an effective lighting system, the restrike time in case of a momentary power outage makes an alternate or emergency light source mandatory for evacuation purposes. Lamps and fixtures are available with standby tungsten halogen or quartz lamps and power failure relays that satisfy the requirements of the NEC [1] and ANSI/NFPA 101-1988, Life Safety Code [7]. Fluorescent luminaires may also be included in the system for this purpose. Lower wattage high-pressure sodium lamps with an instant restrike feature for emergency lighting are also available. Special power requirements include hoist drives for backboards and other gym equipment as well as drives for folding bleachers and dividing walls. If any of these controls use fixed switches, the switches should be of the key type. Scoreboard power with convenient control should be considered. Public address equipment requires special attention to ensure appropriate selected coverage for each sport and to prevent feedback in an acoustically live space. Floor outlets should be avoided, due to potential hazards when such an outlet could be in a sports playing area. Special fixtures with self-lowering devices are required for high-bay lighting or over swimming pools. Catwalks may also be used in high-bay areas for relamping.

16.13 Health Care Facilities
16.13.1 Hospitals
16.13.1.1 General Discussion. Hospital electric systems are complex in design, expensive to construct, and highly regulated by authorities that have jurisdiction over their installation. Because new medical technologies continue to arrive on the scene, hospital electric systems are constantly changing. Among the different building types, hospitals have unique electrical requirements because

(1) Patients are particularly vulnerable to electric shock.
(2) Hospitals have unique electric system reliability requirements.
(3) The hospital strategy for dealing with fire and other life-threatening emergencies is different from that of other occupancies. Hospitals use what is sometimes called a "defend in place" life safety strategy.
(4) Hospitals have many types of sensitive computer-based medical equipment and instrumentation. Sensitive electrical equipment should have disturbance-free electric service.

Hospitals also have many different types of information, communication, and signaling systems, some of which are specified by the design engineer to be provided by the contractor, and some of which are owner-furnished. Moreover, hospitals require special lighting techniques in laboratory, diagnostic, treatment, and patient care areas. Besides these special requirements, hospitals include virtually all of the other elements of a commercial building—including kitchens, dining areas, business offices, computer rooms, and housekeeping services.

Because of the very specialized requirements of hospitals, the reader should refer to IEEE Std 602-1986, IEEE Recommended Practice for Electric Systems in Health Care Facilities (ANSI) [10][94] for more detailed information. The following material is only a brief summary of health care recommendations and requirements based on the codes in effect at the time of the publication of this recommended practice.

16.13.1.2 Codes and Standards. Like most commercial buildings, hospitals come under the jurisdiction of local building officials, local and state electrical inspectors, local and state fire marshals; local and/or state elevator inspectors, and other authorities having jurisdiction over specific parts of the project. In addition, health care facilities generally come under the jurisdiction of a state health department or health care licensure authority. Hospitals that are accredited will be subject to review by the Joint Commission on Accreditation of Health Care Organizations (JCAHO).

The following codes and standards will be of interest to anyone involved in designing or operating health care electric systems:

    (1) ANSI/NFPA 70-1990, National Electrical Code (Article 517, "Healthcare Facilities") [1].

    (2) ANSI/NFPA 99-1990, Health Care Facilities (Chapter 3, "Electrical Systems" and Chapter 7, "Electrical Equipment") [6].

    (3) IEEE Std 602-1986, IEEE Recommended Practice for Electric Systems in Health Care Facilities [10].

    (4) *The Accreditation Manual for Hospitals (1990)*, published by the Joint Commission for the Accreditation of Health Care Organizations (JCAHO).

    (5) *The Plant, Technology and Safety Management Series—The KIPS Survey Guide (1989)*, vol. 4, 1988 series, published by the Joint Commission for the Accreditation of Health Care Organizations (JCAHO).

    (6) NFPA 110-1988, Emergency and Standby Power Systems [B6].[95]

    (7) NFPA 110A-1989, Stored Energy Systems [B7].

While there are a number of codes and standards that deal with hospital design and construction, the designer should be aware of the fact that many different locales have special requirements for the installation of health care electric systems. The prudent designer will spend some time with the authorities having jurisdiction over the project to determine what special rules and interpretations of the national standards, if any, that might apply. More importantly, local authorities should be included in the design process and made to feel a part of the project.

---

[94] IEEE publications are available from the Institute of Electrical and Electronics Engineers, IEEE Service Center, 445 Hoes Lane, P.O. Box 1331, Piscataway, NJ 08855-1331.

[95] The numbers in brackets preceded by a B refer to the bibliographic references that are at the end of this chapter.

16.13.1.3 Hospital Staff. Virtually every hospital will have a Vice-President of Support Services and/or a hospital engineer responsible for operating and maintaining mechanical and electric systems. Most hospitals will have a biomedical or clinical engineer responsible for monitoring, testing, and maintaining medical equipment and instrumentation. Larger, more sophisticated facilities may have a Director of Telecommunications. Hospitals engaged in extensive construction programs will have a Director of Construction on staff, and hospitals involved in the planning, design, or construction process may have a planner or architect who specializes in health care planning. Finally, each hospital will have a governing board and a safety committee that will make certain rulings that will affect the installation of electric systems in patient care areas.

Because so many people are involved, hospital electrical design is an interactive process involving many different members of the hospital staff, including physicians and nurses. For this reason, it is important to conduct interviews with staff members to ascertain needs and preferences with regard to design and specifications. Equipment should be specified based on staff preferences and the availability of service.

Special consideration should be given to the knowledge and capability of the hospital maintenance staff. Often, small facilities will not have trained full-time electricians on staff.

16.13.1.4 Patient Safety. The design of electric systems includes certain features for the protection of occupants, the public, and maintenance personnel. In general, the design strategy includes protecting people from energized or live parts and grounding metallic surfaces that are likely to become energized. In addition to the usual concerns, hospitals should be concerned with the electrical safety of patients. Hospital patients are particularly vulnerable to electric shock because

(1) Patients relinquish much of the control they normally have over their lives and safety to hospital personnel.

(2) Patients are exposed to, and even connected to, a variety of electrical equipment.

(3) Patients are catheterized with conductive electrical and nonelectrical catheters that penetrate the skin, which is the body's natural high-resistance barrier to electric shock. Cardiac catheters, in particular, can lead to cardiac fibrillation as the result of high-current densities that are transmitted through the heart muscle.

(4) Patients are often anesthetized or sedated. The loss of normal sensation renders patients wholly or partially defenseless.

Hospitals use three basic strategies to reduce the risk of electric shock. First, hospital codes require two ground paths for power circuits in patient care areas: a green ground conductor and a metallic conduit.

Second, hospitals use what used to be referred to as "equipotential grounding." While the term "equipotential grounding" has fallen out of favor in the health care industry, the principle still applies. By bonding exposed conductive surfaces (i.e., likely to become energized) together or to a common bus by means of ground conductors, the possibility of potential differences is minimized. Current codes call for a maximum potential difference of 500 mV in general patient care areas and

40 mV in critical care areas. Measurements are for frequencies of 1000 Hz and less and are taken under normal conditions across a 1000 Ω resistor.

A third method of reducing the risk of electric shock in health care facilities utilizes ungrounded isolation transformers. The transformer is actually grounded, to a minimal extent, through stray capacitance and insulation resistance. This "ground" is monitored using a line isolation monitor (LIM) that alarms when the leakage current exceeds 5 mA. The LIM is dynamic in that it alternately checks each line of the single-phase, two-wire ungrounded system to determine the potential leakage current in that line in the event that the other line becomes solidly grounded. Until recently, isolated power systems were required for all anesthetizing areas. They are now required for operating rooms that use flammable anesthetics and those considered to be wet locations as defined by the NEC [1] and ANSI/ NFPA 99-1990 [6]. Although rarely used in intensive and coronary care units, they can provide an additional measure of safety in areas that have critically ill and catheterized patients.

It is important to note that isolated power systems are recommended as an effective tool for improving the level of safety and the continuity of power in certain critical care areas of a hospital.

In the past, critical care areas were required to include ground jacks (referred to in the code as the "patient equipment grounding point") for the purpose of providing a redundant ground path for electrical (or conductive) equipment connected to (or near) the patient. Ground jacks are presently recognized as radio-frequency grounds for sensitive electronic equipment. They are permitted by codes, but not required, and they can be used for safety grounding, if desired.

16.13.1.5 Standby Generators, Essential Electric Systems, and Reliability Considerations. Hospitals are required to have very reliable electric systems. Reliability is achieved in three ways

(1) Through multiple utility electrical services, when available.
(2) Through standby generators, automatic transfer switches, and essential electric systems.
(3) Through prudent electric system design, selectively coordinated overcurrent and ground-fault protection, and reliable end-use devices, such as receptacles and fixtures.

In general, hospitals are required to have on-site standby electric systems with on-site fuel sources. Standby generators should be located above flood levels or otherwise situated so that localized catastrophes are unlikely to affect both the normal and standby systems. The essential electric system should consist of

(1) An emergency system, which consists of a critical branch and life safety branch — each with its own automatic transfer switch programmed to restore power within 10 seconds.
(2) An equipment system with delayed automatic or manual transfer switches.

In general, the life safety branch serves exit lighting, egress lighting, the fire alarm system, and other loads related to life safety. The critical branch provides critically important circuits in patient care areas, life support equipment, and other loads that are necessary for patient care. The equipment system serves

(1) Heating for critical care areas (always) and patient rooms (when specifically required)

(2) Clinical air compressors and vacuum pumps

(3) Air systems with special air change, filtration, or pressurization requirements

(4) Equipment, such as sterilizers, elevators, and sump pumps and other types of mechanical equipment, as required for hospital operation

Radiology and medical imaging equipment can be connected to the critical branch or the equipment system.

Hospital ground-fault and overcurrent protection systems should be selectively coordinated to prevent unnecessary power outages in critical areas or to critical systems due to a fault on an adjacent branch of the power system. The coordination study should consider the normal and standby fault current sources. Hospitals with ground-fault protection on their mains (1000 A or greater on 480Y/277 V systems) are required to have a second level of ground-fault protection downstream to ensure selective tripping. Because nuisance or nonselective tripping of ground-fault protection units can threaten patient safety, the coordination of ground-fault protection units with each other and the downstream overcurrent protection is very important.

Receptacles in hospitals are required to be tested periodically for contact pressure and ground conductivity. Depending on the codes in effect in a particular locale, hospitals are required to have listed hospital-grade receptacles in patient care areas. Recent codes have required normal system circuits to be installed in critical care areas. This is to prevent a transfer switch or circuit breaker failure on the emergency system from interrupting total power to an operating room, intensive care unit, or some other critical area.

Automatic transfer switches should be listed for emergency use, and consideration should be given to providing bypass/isolation switches that permit removing the automatic transfer switch for service.

16.13.1.6 Life Safety. Fire alarm systems for health care facilities are designed to alert the public and the hospital staff of a fire emergency and to annunciate the location. Hospital fire alarm systems can be coded systems that annunciate the fire location by using a series of coded sounds, or noncoded systems that annunciate the fire location at a central location (generally the telephone operator's location). When noncoded systems are used, the hospital relies upon an overhead paging system to announce the location of the emergency. In general, hospitals use a "defend in place" strategy in dealing with fire emergencies. Most patients are unable to evacuate, so the strategy is to contain and suppress the fire while leaving the patient in place or relocating him or her to a place of refuge within the hospital. Certainly, some medical procedures within the facility should continue in the event of a fire. Until recently, the JCAHO prohibited the use of pre-alarm fire alarm systems or systems that alarm only part of the hospital. JCAHO now recognizes systems that selectively alarm portions of the hospital.

In general, voice alarm systems are not recommended for hospitals because patients are unable to respond to the recorded message. Multiplex systems are becoming increasingly popular in large and medium-sized hospitals.

16.13.1.7 Medical Equipment and Power Quality. It is most important for the electrical design engineer to obtain data sheets, rough-in information, and vendor installation drawings before attempting to design electric systems for diagnostic and treatment areas. Sometimes, elaborate powering, grounding, and shielding will be required. Virtually every part of a modern medical center will have computers and computer-based diagnostic and treatment equipment. For this reason, care should be taken to provide proper grounding and to isolate critical loads from elevators, variable frequency drives, and other equipment that can generate voltage drop, noise, and transient overvoltages. Often, uninterruptible power supplies are required for computers, and transient voltage surge suppressors are needed to protect hardware. Indeed, there are many technologies available to solve power problems in hospitals; but it is very important for the designer to seek the advice of the medical equipment manufacturer before applying power conditioners or other correcting devices.

16.13.1.8 Communication, Information, and Signaling Systems. Hospitals have a proliferation of communication, information, and signaling systems. The major systems include

(1) Patient nurse call systems
(2) Emergency call systems (i.e., "code blue" systems)
(3) Telephone systems
(4) Paging and public address systems
(5) Radio paging (pocket pagers)
(6) Television (commercial and closed-circuit), including antenna systems and cable systems
(7) Security (intrusion, door alarm, pharmacy, and television surveillance)
(8) Departmental intercoms
(9) Pneumatic tube systems
(10) Dictation systems (cassette and tank systems)
(11) Emergency services radio
(12) Physiological monitoring (hardwired and telemetry)
(13) Hospital information systems, data processing systems, and data networks
(14) Interval timers or elapsed time indicators
(15) Central clock systems (hardwired and electronic carrier systems)
(16) Building or energy management systems
(17) Medical gas alarm systems
(18) Isolated power line isolation monitors
(19) Standby generator annunciators, trouble alarms, paralleling switchgear, and transfer switch status indicators
(20) Blood bank alarms
(21) Fire pump and fire alarm annunciators
(22) Satellite uplink and other systems for staff training
(23) Narcotics alarms
(24) Mortuary alarms

16.13.1.9 Hospital Lighting. Hospital patient care areas require special lighting. Operating rooms require operating room lights that are dimmable and provide thousands of footcandles of light in the surgical area. Ambient lighting in operating rooms should be sufficient to prevent visual contrast problems for those working under high-intensity light. Consideration should be given to special color rendering lamps (i.e., 5000 K fluorescent lamps) in surgical and laboratory areas. Consideration should also be given to providing the same special lighting for prep and scrub areas to give the physician time to adjust his or her eyes to the new light source.

Other areas of the hospital require special lighting as well. For example, infant nurseries (especially neonatal intensive and coronary care units) need lights with intensity controls to prevent damaging infant eyes. Post-op surgical areas require special exam lights for the physician to examine surgical wounds. Sometimes, germicidal lamps are required for infection control. Special consideration should also be given to corridors and holding areas where patients may be forced to stare at the ceiling for long periods of time.

16.13.2 Outpatient Facilities. Outpatient facilities include clinics, medical and dental offices, and ambulatory surgery centers. In general, those areas of outpatient facilities that are used to examine or treat patients should have power circuits and grounding that is designed in accordance with general patient care areas as defined by the NEC, Article 517 [1]. For design purposes, ambulatory facilities are divided into two types. The first type, which administer inhalation anesthetics for the purpose of performing surgery and/or have electrical life support equipment, are required to have standby generators and essential electric systems much like those required for hospitals. The second type, ambulatory facilities that do not use anesthetics or life support equipment, are permitted to have standby systems with a single transfer switch or battery powered equipment (i.e., egress lighting) in accordance with ANSI/NFPA 101-1988 [7]. Isolated power systems are not required for anesthetizing areas, and, indeed, provide little benefit for many outpatient procedures. However, it should be noted that isolated power systems can provide certain benefits and are required where the anesthetizing area is a wet location as defined by ANSI/NFPA 99-1990 [6] and the NEC [1].

16.13.3 Nursing Homes and Limited Care Facilities. All limited care, custodial care, and extended care facilities are basically divided into two types. The first type utilizes anesthesia and life support equipment. These facilities are required to have standby generators with essential electric systems consisting of an emergency branch and a critical branch. In general, the emergency branch serves loads, such as egress and exit lighting, which are required for life safety. The critical branch serves mechanical equipment, patient room heat (when required), and patient care areas. The second type of limited care and extended care facilities, such as nursing homes and custodial care facilities, that do not administer anesthesia or provide life support are permitted to have simple essential electric systems consisting of a standby generator and a single transfer switch to another power source or 90 minutes of battery power to provide egress lighting, exit lighting, fire alarm systems, etc., and other systems as required by ANSI/NFPA 101-1988 [7]. In general, those areas of nursing homes and limited care facilities that are used to examine or treat patients should have power circuits and grounding that is designed in accordance with general patient care areas as defined by the NEC, Article 517 [1].

16.13.4 Psychiatric Facilities. The term "psychiatric facility" is a broad term that can encompass mental hospitals, substance abuse treatment centers, custodial care facilities for the mentally handicapped, and other facility types. In general, psychiatric facilities that provide in-patient care for four or more patients are considered hospitals and have essential electric systems just like hospitals. On the other hand, local authorities might consider a particular facility, such as a custodial care facility for the mentally handicapped or a substance abuse clinic, as a limited care facility. For this reason, the designer should seek a ruling on his or her building's status from the local and state authorities. In general, those areas of psychiatric facilities used to examine or treat patients should have power circuits and grounding that is designed in accordance with general patient care areas as defined by the NEC, Article 517 [1].

Fire alarm systems in psychiatric hospitals will often have special provisions to keep patients from initiating false alarms. On the other hand, provisions should be made to open doors that are normally secured when a fire alarm is initiated. Likewise, the nurse call and communication systems will have special features to monitor patients and provide protection for nursing personnel. In general, depending on the acuity of care provided, psychiatric hospitals should have tamper-resistant fixtures, tamperproof receptacles, and other features to protect the patient from harm and the electric system from vandalism.

16.14 Hotels. The two general classifications are transient and residential. In the transient field, there are also two types: the commercial or business hotel and the resort hotel. The commercial hotel covered here will be the typical hotel used by the business traveler, rather than the vacationer.

Electric service and distribution systems in hotels will vary from 120/240 V or 208Y/120 V systems in small facilities to 480Y/277 V systems in medium-sized and large facilities. Some very large facilities may have medium voltage (4.16–34.5 kV) for supplying the principal substations in the building. Medium voltage should be considered when economics dictate substations at intermediate floors and/or the roof. All transformers should be sound-isolated from guest rooms.

Restaurants and lobbies will require illumination levels of at least 25–35 fc. Large meeting facilities, such as banquet rooms and ballrooms, generally have movable partitions for dividing a large room into smaller rooms. The lighting control system, usually a programmable dimming system, should have the capacity for dividing control capabilities accordingly for ultimate flexibility. Likewise, the sound system should have multiple amplifiers and speaker zones with the same flexibility. The meeting facilities may have small stage lighting systems, including controls, for highlighting head tables or certain events. The general lighting levels of meeting facilities will range from 60–70 fc for technical meetings and seminars to lower levels for dinner and entertainment functions. Emergency lighting may be controlled with the other lighting, but should have an emergency override in case of power failure. General-use receptacles should be plentiful, with some high amperage power outlets for use during trade shows and other exhibits. Television and telephone outlets should also be provided.

Many hotels have several stores with street frontages or in arcades. These stores may have separate meters depending upon utility submetering rules and policy. Hotels generally contain a hair styling salon, newspaper stand/gift shop, restrooms, and, possibly, a swimming pool and health club. All of these facilities require large numbers of outlets as well as specialty lighting.

In the residential type of hotel, the public rooms are usually not as numerous or as decorative as in a transient hotel. Power for individual air conditioners should be made available when a common system is not provided. A master antenna, radio, and television system may also be required. These hotels approach the apartment-house-type of building.

A checklist for hotels includes the following items:

(1) Medium- or low-voltage service entrance
(2) Primary or secondary unit substations, or both
(3) General lighting
(4) Special lighting in public rooms
(5) Night lighting
(6) Emergency and standby generators
(7) Building management system
(8) Fire detection, sprinkler waterflow, and alarm systems
(9) Electronic security and surveillance system
(10) Private telephone system
(11) Public telephone line system (phones, facsimile machine, computer, etc.)
(12) Swimming pool, spa, sauna, whirlpool, and health club
(13) Intercommunication system
(14) Public address system
(15) Central television and radio system
(16) Television program source provisions
(17) Bellhop annunciator
(18) Separate services for stores and shops
(19) Kitchen appliances
(20) Elevators, escalators, and dumbwaiters
(21) Message annunciators
(22) Lightning protection
(23) Safe and vault security systems
(24) Sports lighting systems
(25) Sign lighting
(26) Snow and ice melting systems
(27) Landscape lighting

16.15 Libraries. Libraries need well-distributed general illumination because reading and visual work is done throughout the entire area. Lighting should be engineered to provide the required horizontal illumination levels for reading and writing and the required vertical lighting levels in the stacks for identifying book titles. In the book stacks, which may be several tiers high, it may be necessary to provide outlets for motors on book lifts or dumbwaiters for transporting books and manuscripts to the top shelves. Special outlets are needed at checkout desks and for copying and facsimile machines.

In large libraries, and communication or telephone systems, or both, are needed between library stations. A closing time signal system may also be needed. Security systems for checkout control are becoming more important and raceway for such a system should be included in the building structure. This system normally involves screening equipment at exit points and door control for exterior doors. Provisions should be made for electronic catalog files, monitors, and microfiche readers.

16.16 Museums. The adequacy of lighting circuits and the convenient location of outlets are important in museums. The lighting of exhibits often requires outlets in unusual locations. Multiple-circuit track lighting on walls or hung from ceilings has proven quite effective. A high degree of flexibility for supplementary lighting should be provided. Special requirements include close temperature and humidity control. As these buildings are usually made of stone and decorative masonry, these outlets cannot be installed economically after the building is finished (however, some use of Type MI cable has been successful).

Some museums exhibit apparatus that may require large power outlets. Special types of power requirements may be served best by using a design for flexible power distribution that is similar to that suggested for laboratories (see 16.8.3).

An elaborate burglar alarm and fire alarm system as well as a temperature and humidity variation alarm may be needed, especially in large city museums where valuable paintings and other art collections are shown. These systems may tie in with proprietary, central station, or municipal police and fire departments. Transmission is usually accomplished by wire or radio link, which calls for a close liaison with these agencies.

16.17 Newspaper Buildings. Newspaper buildings are, in a sense, multiple-storied manufacturing plants, with large motor loads for the presses, conveyors, elevators, and moving stairways. The power distribution systems are large, with a power utilization voltage of 480Y/277 V fed from unit substations that may be supplied by an internal medium-voltage distribution system. Power service should be dual source, if possible.

Large systems of telephone, telegraph, teletype, computer facilities, and radio services are required. Television may be added with facilities available for both reception and transmission. Wire services may be connected with not only the entire United States but also foreign services. Each wire service usually has distinct requirements that may call for careful study of specifications as received from the various companies. Radio and television systems will require special shielding to prevent electromagnetic interference. Microwave facilities may also be included.

Local telephone requirements are greater in newspaper buildings than in ordinary buildings. Telephone outlets may be provided in floor duct systems on most of the floors or by the use of newer flat cable systems that are laid under carpet tiles. Ducts leading to the switchboard should be oversized to accommodate changes in office arrangements.

A checklist for newspaper buildings includes the following items:
(1) Large power distribution system at medium voltage or 480Y/277 V
(2) Special lighting in many departments
(3) Emergency lighting

(4) Emergency and standby power systems
(5) Clock systems
(6) Special requirements for telephone, telegraph, radio, television, computers, facsimile machines, and photography
(7) Intercommunication system
(8) Private telephones
(9) Helicopter landing field on the roof or in the yard or short takeoff and landing (STOL) airplane facility
(10) Fire alarm systems
(11) Microwave and satellite links

16.18 Office Buildings. Office buildings range in size from the smallest one-story building to the tallest high-rise office building. The interior space of the typical office, for similar tasks regardless of building size, has essentially the same requirements for lighting, ventilation, space conditioning, receptacle power, and appliances. Only smaller office buildings are generally provided with a 240/120 V or 208Y/120 V service. Except for large area and high-rise buildings, the voltage service is normally 480Y/277 V, many with the utility vault in or adjoining the basement. The large area and high-rise buildings are frequently served at 4.16–34.5 kV. The utilization voltages are predominately 120 V for receptacles and task lighting, 277 V for general-purpose lighting, and 480 V for power.

When these buildings are located on power networks where only 208Y/120 V power is available, the design engineer should analyze, along with the utility company, the economics of supplying such buildings with a medium-voltage service. Such analysis should include the following items:

(1) Power service
  (a) Size, location, and composition of expected loads
  (b) Possible transformer location(s)
  (c) Primary voltage and feeder arrangement (network or radial)
  (d) Primary wiring by utility or building owner
  (e) Transformers supplied by utility or building owner
  (f) Any utility costs to be assessed to building owner
  (g) Transformer vault and access requirements
  (h) Special metering requirements
(2) Load considerations
  (a) Interior—Heating, ventilation, air conditioning, lighting, other loads
  (b) Exterior—Sign and parking lot lighting, snow melting, electrically operated gates
(3) Special systems
  (a) Fire alarm system, including communication and other systems control that is required for high-rise buildings
  (b) Telephone systems
  (c) Remote equipment status and control systems (central monitoring)
  (d) Security systems
  (e) Lighting and power control systems for energy conservation

(4) Emergency and standby power systems
  (a) Load requirements
      (i) Pathway and exit lighting
      (ii) Elevator
      (iii) Fire pump and booster pump
      (iv) Stairway exhaust and supply fans
      (v) Data processing
      (vi) Other desired loads, such as sump pumps, house pumps, sewage ejector pumps
  (b) Location
      (i) Type (internal combustion engine or combustion turbine) — Combustion turbine units generally do not qualify for emergency power due to their long start-up time.
      (ii) Intake, exhaust, and unit silencing
      (iii) Fuel supply and storage
      (iv) Ambient temperatures and heat rejection
      (v) Maintenance
  (c) Potential for peak shaving and load curtailment
  (d) Closed transition capability for parallel operation

Office space arrangements are continually changing. Partitions are moved, task areas are shifted, and office equipment is changed or relocated. It is important that the lighting system layout be flexible and that power and convenience outlets can be added or subtracted quickly and easily. Frequently, these changes should be made with a minimal interruption to full daily occupancy of the space. Low-voltage switching is often used to change switching patterns. Prefabricated flexible wiring systems should also be considered.

Underfloor raceway systems provide one practical way of providing for system flexibility. The two most common types are the prefabricated raceways that are cast in the floors, and a cellular system that utilizes the steel form under the concrete slab as part of the raceway construction. In either system, outlets can be provided at the time of construction, or, with special location and drilling rigs, outlets can be installed at any time after construction. These raceways may have dividers to separate power and communication or telephone circuits. The outlets are usually a combination of telephone and power outlets. It should be noted that some codes require fireproofing around any poke-through assembly. When heavy load densities are to be expected in both communication and power circuits, the use of conduit on the bottom sides of slabs to serve the floor above usually is a poor choice for new buildings. Utilities that are run as described above are usually spaced in convenient modular groupings, for example, a 6–8 foot spacing. The raceways are usually fed from header ducts that are fed from conduits or prefabricated assemblies connecting to the electrical, telephone, and communication closets. Flat cables for telephone, data, and convenience power outlet branch circuits to be laid under carpet tiles are now being used (where permitted) instead of underfloor systems or ceiling modular pole systems. When the density of power, communication, and data cables is heavy, consideration should also be given to raised access floor systems.

Office buildings may be occupied by one tenant; however, in most cases, multiple-tenant occupancies are to be expected. Provisions should be made for metering or including the cost of the electricity in tenants' bills. Tenants may take raw space with only base utilities provided and design their own interiors, in which case, building standards should govern the tenant space design, and construction should be monitored to ensure compliance.

16.18.1 Large and High-Rise Office Buildings

16.18.1.1 Layout of Electric Systems. The use of multiple systems that are electrically and physically separated is desirable and should be considered if economics permit. For example, from the service entrance(s), separate electric systems A and B that run to each floor can enhance reliability. Rather than using direct electrical ties, transfer switches (with switched neutrals) taken from protected circuits can provide power for critical needs. This approach can provide for half (or part) service in any area, even upon loss of one of the systems. Some local codes require separation of lighting into groupings; although, in this case, the intent of the code is energy conservation.

The risers and electrical closets for these independent systems should, when the size of the project and the degree of protection to the public warrants, be physically separated; perhaps on separate sides of the building and protected against damage by suitable fire ratings of enclosures, or by fireproofing. This also applies to critical control wiring including fire alarm systems, emergency communication systems, and smoke and evacuation control systems.

It is common practice to install major substations and other major power installations on mechanical (machinery) floors or in walled areas, thus protecting building occupants from the effects of catastrophic failures of these systems. In this case, the installation should be such that the failure of the mechanical systems will not damage electrical equipment (e.g., flooding) or prevent access (e.g., smoke) to the electrical areas. Ventilation of electrical areas should be such that smoke is not carried into areas of occupancy.

In large or tall buildings, the mechanical systems are usually zoned into individual system-designated areas. It is important that the zoning of the electric systems be compatible with that of the mechanical systems. If mechanical or electric systems should be out of service to any given zone, then such service should not be deprived to any other zone because of improper segregation of services.

New low-smoke, low-flammability, and low-toxicity cables (products of heat or combustion) have been developed. These cables are often referred to generically as nonhalogen types or polyolefin types. The designer should consider the use of these cables because they are available and approved in installations where safety conditions are critical. Even small amounts of smoke emitted from conventional cables of the halogen (conventional) type can damage computer-type equipment because of its high acid content (corrosivity). While high-temperature cables, such as the silicon-or teflon-type, have been available for many years for use as exposed wiring to fire alarm devices; however, this type of cable is very expensive and is not generally used for normal power wiring.

In Chapter 5, Tables 37 and 38 provide the acceptable sound levels for transformers and suggest methods for minimizing them. In a commercial building, this may be most important; lower noise transformers (including special mountings) are available and the manner of their installation can reduce the effects of noise. Flexible connections from transformers to heavy-current busways is also important in reducing noise.

16.18.1.2 Maintenance of Electric Systems. The sealing of openings between floors and between adjacent areas is essential. Major fires have been transmitted between adjacent areas because of improper sealing. Fire ratings should be maintained between these areas. The building staff should assure that contractors and others doing work in the building maintain such seals. It is easier to maintain seals and fire ratings in conduit sleeves than large slots. When outlets are installed in a poke-through-type system (such as in cellular-type underfloor systems), the installation of seals (particularly the intumescent type, which swells to provide a fire and smoke seal) may be required. Seal integrity in telephone and other communication installations, even though installed and maintained by nonbuilding staff, should be assured by building maintenance staff.

16.18.1.3 Special Systems. Smoke purge or smoke control systems are automated systems that divert ventilation air into areas that are to be occupied during a fire condition. These areas, often called "refuge areas," which include specified stairwells, corridors, and other safe locations, are maintained at a positive air pressure, while other areas, in which smoke may be generated, are maintained at lower pressure. This is usually accomplished automatically by the control system, with the initiation of this action either manually from a control center (usually the console) or automatically as a result of the operation of smoke or fire detectors in the ventilation fan and duct system. See Chapter 14 for more information.

16.18.1.4 High-Rise Buildings. Perhaps two of the most distinguishing characteristics of high-rise office buildings are limited egress (evacuation capability) and limited access by fire department equipment to upper floors. Special precautions for the safety of building occupants, vertical transportation systems (see Chapter 12), complex building management and security systems (see Chapter 14), medium-voltage distribution (see Chapter 4), and emergency and legally required standby systems are often required.

It is physically impractical to quickly evacuate most occupants from very tall buildings that are on fire or in other emergency conditions because of the danger of panic and confusion, physical infirmities of occupants, the sheer distance of the stairwells, and limited accessways. The elevators are normally not used for such evacuations, unless under fire department supervision. In a well-designed, properly operated, and adequately maintained high-rise building, the protection of building occupants, who may number in the thousands and might have difficulty in leaving the building because of fire or other emergency, is assured by the structural barriers provided by floors, walls, and partitions; ventilation, communication, and smoke control systems; fire alarm systems; and other electric systems described below. Life safety codes, OSHA, the NEC [1], local codes, fire and police department regulations, governmental regulations, and insurance underwriter requirements prevail for such structures.

Typically, a 30-story building may have substations in the basement and on the top floor or roof. Higher buildings might have substations every 20–30 floors, with intermediate floor substations supplying floors above and below. Panelboards are usually installed at each floor. When practical, it is advisable, at 480Y/277 V, to limit the ampacity of the busway risers to limit areas of outage in case of busway failure.

Two choices are available for the installation of general-purpose transformers that will provide transformation of 480 V to 208Y/120 V. One choice, and the most common, is to install the general-purpose transformers on each floor or every third floor that is fed from the 480 V risers. The other choice is to install a 208Y/120 V riser, running physically parallel with the 480 V riser and serving each floor.

The NEC, Article 300 [1] indicates the vertical lengths of cable that can be run without intermediate supports. The spacing of intermediate supports is also given in a table in the NEC, Article 300 [1]. Aluminum cables can usually be run longer distances than copper (but for the same ampacity, the aluminum cables will be larger). Cable supports, for more than several floors, may be of the manufactured-wedge-type. When very long runs of medium-voltage cable (up to hundreds of feet) without intermediate support are required, a steel armored cable often referred to as "bore-hole" cable may be used. This cable requires the use of special fittings at the top and, sometimes, at the bottom.

Provisions should be made for cable pulling and other operations that require lots of space; very heavy vertical pulls may also be required. Consideration should be given to the sizing of equipment (i.e., transformers and switchgear) so that it will fit into the larger passenger or freight elevators. This will avoid having to use riggers to lower equipment either outside the building or in an elevator shaft (not in the car), should replacement be required.

The elevator recall system automatically brings elevators down from their landings at the time of a fire or other emergency (such as a shift to emergency power or loss of the utility power system). The elevators are usually returned to the lobby (unless the fire is there) and held there for possible operation by fire department staff. When elevators are operated from emergency generators, controlled switching is provided so that only one or a limited number of elevators can be operated at one time to prevent overloading the generators. See Chapter 12 for more information.

Transformers in high-rise buildings are usually the ventilated-dry-type (or possibly encapsulated for the very smallest ones). Safety and maintenance considerations usually indicate this choice. The only use for liquid filled transformers would be in vaults, perhaps in a below-grade or outdoor main substation. The fire hazard and maintenance considerations of even the "less flammable" and "nonflammable" (see the NEC, Article 450 [1]) liquids or fluids work against their selection for use within the structure. Even dry-type transformers can fail in a mode that generates smoke; therefore, whenever possible, all but the smallest transformers should be located or ventilated in such a manner that the effects of such a failure will be limited.

When the building temperature is markedly different from the outside temperature (i.e., during heating or cooling periods), air pressures between lower levels and upper levels can be relatively high, resulting in airflow between areas; this can result in moist air, dust, and smoke being carried between areas. The NEC [1] generally requires the sealing of raceways where such conditions exist.

16.18.1.5 Heavy Current Considerations. In 16.18.1.4, it was indicated that smaller feeders are to be desired; this reduces the effects of outages related to a single failure, simplifies coordination, and avoids massive busways and cable assemblies. The disadvantage of the reduction in diversity of smaller feeders is subordinated, in this case, to system safety and reliability.

The network systems that are described in Chapter 4 are usually selected for very large commercial buildings. A network system has the advantage of avoiding an interruption of service upon the loss of a medium-voltage feeder. The network system may be installed by the building owner, in which case, it should meet the requirements of all the applicable codes and regulations, including those of the NEC [1]. If the utility installs the network according to utility practices, while still meeting the requirements of applicable authorities; then it should be in isolated vaults that completely protect the occupants. Chapter 4 contains a discussion of the special considerations of nonutility-owned network systems. The network bus is usually a heavy current bus and should be adequately protected with relaying and may be of high-integrity construction (i.e., segregated bus construction). It should be noted that the NEC [1] does not deal specifically with network design; but system requirements should be interpreted from appropriate articles of the NEC or other codes.

16.18.1.6 Emergency Systems. Emergency systems are described in IEEE Std 446-1987, IEEE Recommended Practice for Emergency and Standby Power Systems for Industrial and Commercial Applications (ANSI) [9]. It is important to arrange these systems so that, if one or more generators fail to support the emergency loads, automatic load shedding will reduce the load to the capability of the remaining generator(s).

Most high-rise buildings have at least one computer center and some have complex operations, such as commodity exchanges. Standby systems may be required for data processing (computer and communications) handling systems; these may be installed by building tenants. A clear understanding of utility power system reliability and quality and of the availability of emergency power to tenants should be understood and incorporated into tenant leases. Tenants should understand the extent to which they are entitled to emergency generator power, whether they have to pay separately for this feature, and whether they have to install additional generation capacity (including space for such an installation). Tenants should determine the system reliability required in order to make a decision on the need for tenant-owned UPS systems.

16.18.1.7 Building Management Systems. Building management systems are covered in detail in Chapter 14. While the building management system may incorporate all or most of the building control functions, several design objectives should be achieved

(1) Have backup locations for operations (e.g., auxiliary consoles, operation at the remote stations, or at the equipment to be controlled).

(2) Provide diversity in equipment through the use of distributed systems (distributed by function or by operating areas).

(3) Assure that the communication path has redundancy or is highly reliable (e.g., loop-type distribution, multiple feeds, high-reliability coaxial systems).

(4) Use of redundancy and modularity to assure reliability.

A modern building control system may control or telemeter hundreds or thousands of points. Data acquisition and logging are important aspects of the building management system. One of the most important considerations in the design of such systems, comparable to reliability, is to provide a system that is conducive to operator performance, especially under emergency conditions.

16.19 Parks and Playgrounds. Unless parks and playgrounds are supervised, particular attention should be paid to making the electrical installation as vandalproof as possible. Pathways open to pedestrian, equestrian, or bicyclist use at night should be lighted in such a way that the pathway ahead is always clearly defined. Sports area playgrounds should have lighting as recommended in various publications, such as the *IES Lighting Handbook*, 1987 Edition [11], and should have key, coin box, or remote control. Swimming pools and fountains are covered elsewhere in this chapter.

16.20 Piers, Docks, and Boat Marinas. The electrical installations for these facilities should take into account that moisture, particularly salt water, will cause rapid deterioration of metal parts. Rigid hot-dipped galvanized conduit has proven satisfactory, provided all surface damage, such as form cutting and threading, is galvanized coated. Plastic coated conduit (with joints that are plastic covered) or rigid nonmetallic conduit (where permitted) can also be recommended. Mineral insulated, copper sheathed (Type MI) cable is also available for this environment, as well as metal-clad (interlocked armored) cable with inner- and outer-extruded protective jackets. The use of heavy galvanized metal boxes, cast bronze boxes, or nonmetallic boxes (such as molded fiberglass) is advisable. A grounding conductor should be included in all raceways located over water. The design should account for possible pier or dock movement and the motion of any floats that are fitted with power. Feeders to floats or to flexible joints in docks can best be handled by the use of Type SO cable with proper grounding. Transformers exposed to the weather should be of the encapsulated type in corrosion-resistant enclosures. Service to piers, docks, and boat moorings should have lockable power disconnects at the shore end for use by pier, boat security, and fire department staff. Design coordination should be made with the fire department for disconnecting service in the event of an emergency.

Lighting should be designed for pedestrian safety. U.S. Coast Guard obstruction lights may also be required. Exterior lighting is required for night loading and unloading as well as for protection against pilferage. Interior lighting should be designed to accommodate the high stacking of commodities.

Power outlets at docks and piers for ship use should be designed for the power requirements of the ships that are expected to visit the dock or pier. Ships requiring large amounts of power generally have cables with lug connections. Smaller ships (i.e., tugs, etc.) usually have cables with plugs. Tug companies often standardize voltage, current, phasing, and wires for their plug configurations. The engineer should investigate the shore-side power requirements for ships that are expected to dock. Telephone and cable television facilities should also be considered. Piers and docks may also have power requirements for cranes, conveyors, hoists, winches, battery charging for electric truck batteries, pumps, electric doors, escalators, and elevator service. Cable tracing (heating) should be provided for piping that would otherwise freeze in winter. Two-way radio and public address systems may also be required.

Baggage and materials handling facilities will require power outlets, often at 480 V (see 16.34).

16.21 Police Stations. Police stations are general-purpose buildings; but they require some special services not usually found in other commercial buildings.

Communication facilities are a vital part of the services needed. Two-way radio and transmitting equipment is required to keep the station in touch with radio equipped patrol cars and foot patrols. This system requires standby power equipment in case of power failure. Fire alarm connections with local and adjacent county departments should be included, as well as facsimile tie-ins with other state and federal agencies. In large cities, this system is often interconnected with other station houses. In some areas that have community cable television systems, a means is provided for citizens to summon police, fire, and ambulance service. These types of facility connections may even be required.

In stations that have confinement cells, a multiple-circuit radio wired to each cell, as well as a public address system, should be provided. A closed-circuit television for cell observation may be required. All items located in cells should be totally vandalproof. A projection booth may be used for showing slides and motion pictures. Electric door locks on individual cells and cell blocks may also be provided. Electric door lock controls should be in a continually manned and secure location. Adding these locks to thick walls will be costly if they are overlooked in the original design. Standby power and manual backup for electrically operated cells should be provided.

A garage may be required for the storage and servicing of patrol cars, patrol wagons, emergency vehicles, and repair units. Electrical service that is usually found in garages are required.

Traffic control may be installed in the police station. The latest electronic type of control may be complex, and the cable to the various systems may require fairly large wiring installations. Space should be provided for these systems.

16.22 Prisons. Prisons and correctional institutions may include several types of commercial buildings

    (1) Main administration
    (2) Cell blocks
    (3) Manufacturing buildings
    (4) Recreational buildings
    (5) Dining halls
    (6) Hospitals
    (7) Power plant

(8) Warehouse

(9) Garage

(10) Barn and poultry sheds

All parts of the electrical installation that are accessible to prisoners should use materials that are designed to prevent damage. Boxes, fixtures, and fittings should also be designed to prevent opening by other than authorized personnel.

16.22.1 Administration Buildings. The administration building is usually outside of the prison's walls or is part of the outer wall surrounding it. The administration building contains general offices and the central systems for communication and security. Direct connection to outside law enforcement agencies is maintained by radio and telephone systems.

Controls for security facilities, such as outer wall lighting, cell block supervision, and main gate controls, are contained in the main building. Particular attention should be paid to the location of these security controls to prevent them from becoming accessible to unauthorized people. Consideration should be given to the fact that the control room may be seized by prisoners, and a means of backup control should be provided in a second secure area. It may well be that recognized modern methods of electric power distribution in a building will be sacrificed for security reasons.

16.22.2 Cell Blocks. Cell blocks have radio or intercom systems for each cell that are controlled from the administration building. Electric door locks and an elaborate interlocking system ensures the security and safety of guards. Lighting in the main corridors should be dual service to preclude the possibility of any outage. All controls should be well away from any possible access by inmates.

In addition to life safety lighting, smoke evacuation systems that are connected to the emergency power system may be required.

Closed-circuit television may be required to scan the various cell block areas, enabling the guards to maintain surveillance over the entire area. Regulations should be checked to ensure that such surveillance is acceptable and does not abridge the inmates' rights. Lighting levels in these areas should be adequate to ensure good television images. Camera requirements should also be checked.

16.22.3 Manufacturing Buildings. Manufacturing buildings are a part of many correctional institutions. Power distribution to them is similar to that used in industrial plants (see IEEE Std 141-1986 (ANSI) [8]).

16.22.4 Dining Halls. When centralized, dining halls may be used as auditoriums. Stage lighting, such as that required for a medium-sized high school, may be needed. A public address system may be required for entertainment purposes and announcements. Large kitchens will need the usual electrical accessories. A projection booth for movies, videos, and 35 mm slides may be required when the dining hall is used as an auditorium.

16.22.5 Hospitals. Fully equipped hospitals to take care of an emergency and any type of operation are generally required. The power and lighting system is similar to that of a general hospital (see 16.13). Panelboards are located according to security regulations.

16.22.6 Power Plants. Power plants may be included to generate electric power and to produce the steam, hot water, chilled water, and heating for the entire institution. Emergency and standby power will also be required. Normal electric power may be generated at the power plant; but a connection to an outside utility in case of local failure or sabotage is mandatory.

The power plant should be in a secure location. Power should be distributed underground. The size of the prison determines the distribution voltages required.

Site lighting inside the fence or outer walls, and floodlighting of the outer walls with searchlights in the guard towers will constitute a significant load. UPS equipment should be considered to maintain a minimum level of site lighting during transition to an emergency or alternate power source. The guard towers may have electric heaters and air conditioners as well as telephones and other intercommunication systems.

16.22.7 Warehouse Buildings. Storage of supplies is a requirement in correctional institutions. This function is often combined with power plant functions. Normal warehouse features may be required, such as loading docks, exterior lighting, high-bay shelving and lighting, and battery chargers. In addition, coolers may be required for the storage of refrigerated foods.

16.23 Radio Studios. Radio studios should be soundproofed; therefore, all ventilation, both heating and cooling, should be of extremely low-noise level design. Air-conditioning loads are generally high in all parts of the building.

The following points should be considered:
 (1) Air-conditioning loads (centralized or individual units)
 (2) Shielded wiring for EMI/RFI/EMP protection
 (3) Low-noise levels of transformers
 (4) Quiet fluorescent lighting ballasts (or remote-mounted ballasts)
 (5) Clock systems
 (6) Paging annunciator
 (7) Private intercommunication systems
 (8) Special connections to networks
 (9) Stage lighting for audience shows
(10) Services for telephone news coverage
(11) Special sign lighting

Some stations have the transmitter in or adjacent to the studio. A provision for removing heat from the large transmitters is required. Power requirements are large, and dual reliable services may be required. Standby power equipment with automatic transfer switchgear is also needed in cases of emergency.

Transmitter towers generally need obstruction lights for warning aircraft in accordance with FAA 70/7460-1, Obstruction Marking and Lighting Advisory Circular [B11].[96]

16.24 Recreation Centers. Recreation centers are generally mutiple-purpose facilities. Outdoor requirements can involve parks and playgrounds (see 16.19) with various athletic fields, swimming pools (see 16.30), skating rinks, and parking lots.

---

[96]FAA publications are available from the Federal Aviation Administration, Department of Transportation, 800 Independence Avenue, S.W., Washington, DC 20591.

Indoors, recreation centers often include auditoriums (see 16.2) and gymnasiums (see 16.12). In smaller recreation centers, a single multiple-purpose room will serve as both an auditorium and a gymnasium. Other spaces designed for particular sports, such as handball and racquetball courts, squash courts, weight training rooms, etc., may also be included. Some recreation centers also include craft shops; when craft shops are included, attention should be given to outlets for power tools, kilns, etc.

16.25 Residential Occupancies (Commercial). Apartment buildings and condominium projects can vary greatly in size, number of floors, and number of dwelling units. The trend in some cities is toward larger housing developments with buildings up to 40 or more stories. The power distribution risers in such buildings approach the size of those in large office buildings. Pumps, fans, elevators, central space heating, water heating, and air conditioning can be large loads. Air-conditioning loads of the window or through-the-wall types should be anticipated when a central system is not provided. Clothes washers, dishwashers, and other residential-type appliances are significant loads. People doing housework, such as washing and ironing, at night, add to peak lighting loads and, thus, increase maximum power demand. The following checklist is for apartment buildings and condominiums:

(1) Metering (master or individual) — utility, submetering, rent inclusion
(2) Exit and emergency lighting
(3) Telephone system
(4) Security systems, which may include closed-circuit television, vehicle entry gate control, apartment intercom and entry door(s) release, restrictive elevator controls, and identification/coded card doorlock systems.
(5) Television signal distribution, either master antenna or from the cable utility company
(6) Individual or central laundry facilities
(7) Exterior lighting
(8) Air conditioning heating, ventilating, and water heating (central or individual)
(9) Special appliances, such as garbage grinders, compactors, and dishwashers
(10) Fire alarm systems and, for high-rise, special life safety systems, such as smoke detectors may be required for each occupancy.
(11) Signal systems for handicapped and senior citizens
(12) Parking and garage facilities
(13) Recreational facilities, such as swimming pools and spas, fitness facilities, and recreation rooms
(14) Elevators (passenger and service)
(15) Restaurants and lounges
(16) Retail spaces
(17) Lightning protection
(18) Snow and ice melting systems for walkways, ramps, and driveways
(19) Cooking

16.26 Restaurants. There is substantial electrical load concentration in restaurant kitchens. When gas is used for cooking, a $50\ W/ft^2$ connected load is not uncommon. Connected loads of over $100\ W/ft^2$ have been used in all-electric fast food takeout kitchens (including ventilation and air conditioning).

Kitchen layouts are frequently based on a certain brand of equipment. It is important that the engineer check the submittal data of the equipment actually furnished to ensure that power load provisions and outlet locations are acceptable.

Electrical power in restaurants should fulfill the primary needs of the particular type of building it is in. It is used to preserve and prepare food and provide appropriate lighting.

The following points are important:
(1) Adjustable lighting from high to low levels
(2) Air conditioning
(3) Electric (or gas) cooking and baking
(4) Food warmers
(5) Serving tables
(6) Conveyors and dumbwaiters
(7) Public address systems
(8) Provisions for background music
(9) Plug-in telephone outlets
(10) Decorative lighting
(11) Parking lot lighting and lighting for drive-through facilities
(12) Television outlets
(13) Kitchen exhaust equipment including fire detection equipment, special ventilation, such as restroom and smoking area exhaust fans
(14) Electric hand dryers
(15) Outlets for janitorial equipment
(16) Business machines in main offices
(17) Exit lighting
(18) Emergency lighting and power
(19) Snow and ice melting systems for ramps, sidewalks, and driveways

16.27 Schools (Kindergarten Through 12th Grade). The general requirements for schools (Kindergarten through 12th grade) are similar to those for colleges and universities (see 16.8). In the general school building, all activities are normally under one roof. The use of individual or connected buildings is common for large schools.

Today, greater effort should be directed toward the more efficient use of electrical energy. Except for selected education programs, power requirements have been fairly stable or decreasing in recent years. In lighting, more efficient sources and better lenses have enabled designers to achieve reductions in W/ft^2 that is required for good vision. Special outlets or extra floor outlets are required in home economics, vocational training, business training, computer skills, and foreign language spaces. Requirements for auditoriums, gymnasiums, and libraries are covered elsewhere in this chapter.

16.27.1 General Requirements. In most areas, lighting should involve at least two levels of control. Special systems should include fire detection and alarm, security control, and program (announcing) systems. Both commercial and closed-circuit television, telephone, and central monitoring/control for mechanical systems may be included.

16.27.2 Mechanical Systems. Fans, pumps, chillers, cooling towers, and air conditioners should be considered. Sometimes, electricity is also used for space and water heating. Heat pumps, heat recovery systems, and energy-efficient motors reduce the required power use.

16.27.3 Laboratories. Science laboratories require facilities with semi-flexible power distribution features. Convenience outlets and plug-in power outlet raceways may be needed on benches. The electric requirements for laboratories should be checked in the early stages of design because, frequently, special equipment may have to be manufactured. This type of equipment may require a long delivery time. See 16.8.3 for more information on large laboratories.

16.27.4 Manual Training Departments. A manual training department may contain large motor powered machine tools and welding equipment. Courses may be revised and new tools and skills taught; flexible power distribution systems should be provided. Appropriate shop and tool lighting should also be provided. A master cutoff of power for each shop is recommended for instructor convenience and safety. Emergency stop buttons for operating undervoltage release devices that disconnect all shop tool power should also be provided.

16.27.5 Kitchen Facilities. School lunch programs have expanded the power requirements of kitchens and cafeterias. Ovens, ranges, mixers, freezers, and exhaust systems require large blocks of power. Heavy-duty power outlets for hot food carts may also be required.

16.28 Shopping Centers. A shopping center is a group of stores concentrated in a compact area surrounded by vast parking lots. The majority of these are located on the outskirts of cities and towns, and power is generally supplied at medium voltages. The parking areas should be well lighted by lighting standards that are generally controlled by photoelectric cells and time clocks, which turn the lights on automatically at night and off after closing hours. The various stores are each wired to the owner's requirements and normally metered separately. See the tables in Chapter 2 for applicable load figures.

New shopping centers are being constructed as malls, and older ones are being enclosed to provide indoor malls. Because of the height of the malls, high-intensity discharge (HID) lighting is often economical and appropriate for this type of installation (see Chapter 10). The design is usually such that the lighting from the storefronts, especially display lighting, should dominate the visual scene.

16.29 Supermarkets. The supermarket is a fast growing and changing institution. The small margin of profit and the great volume of business causes the supermarket to attempt new merchandising ideas. These require a flexible electric system. See the tables in Chapter 2 for applicable load figures.

The lighting may be 50–90 fc for general illumination. Auxiliary lighting is also generally used. High-intensity incandescent fixtures over the produce and meat sections have been used to give better color to the products when color is important. Valance lighting around the perimeter of the store has also been used and is also frequently used for advertising products as well as for illumination.

The front areas of the supermarket may be highly illuminated to attract attention, and the rear of the store may be brilliantly lit to bring the back "closer" to the front for advertising value and to attract the customer to the rear of the store.

The open freezer and refrigerated cases cause the refrigeration load to be very high. These loads usually have several banks of compressors at several locations in the store, and refrigerant is run to the cases on the sales floor. Three-phase power is usually run to the compressor location, and an additional 120 V circuit is run to the case location for the lighting and small ventilation fans within the case.

The freezer cases usually contain electric defrost elements, and the wattage for these elements exceeds that of the refrigeration unit. These defrosting elements may operate on one phase of the motor circuit feeder when the compressor motor is not in operation. The defroster loads are generally considered to have 40%–60% diversity when the service entrance is being calculated.

The checkout systems are designed to move the customers out quickly. Usually, one circuit is provided for each checkout. A computer with laser bar code readers is frequently used. These counter readers display the price and then enter the price and article name on the tally slip. The entire group of purchases is totaled, including tax, and then customer change is displayed and the tally slip is printed. Raceway for data cables from the central computer should be provided. The central computer keeps the inventory record, product flow, and suggested reordering schedule. When credit cards are accepted, card readers and verifier circuits should be considered.

Coffee grinder loads have increased. These may range from 1–5 hp, three-phase or single-phase. Choppers and grinders may have two motors running simultaneously, which may be 10–15 hp three-phase motors. Hot iron sealing machines usually require one circuit per machine. Meat saw motors range as high as 10 hp three-phase. Electric door openers, meat tracks, rotisseries, and moving displays should be considered in the final survey of loads.

Communication systems range from the simple single pushbutton system and single bell and chime system to systems that have speaker boxes located at convenient points in the store where the customer may request the location of items. This information is transmitted to the office, and the locations of such items are transmitted back to the customer. This system is usually interlinked with cash register locations, enabling the cashiers to communicate with the office when change, stock clerks, baggers, security personnel, etc., are needed.

Refrigeration failure alarms, burglar alarms, closed-circuit television surveillance, water leak alarms, and other security systems should also be considered.

Signs are an important part of the load and are sometimes elaborate, with running and neon lights. Some markets prefer to use 35–50 foot pylons with large lettering and moving parts.

Special consideration should be given to future electrical loads: conveyors to deliver the groceries from the cashier location to the customers' automobiles, conveyors to deliver the product from the shelf to cashiers, pushbutton shopping, automatic meat processing, automatic stocking, etc., are all likely additions to the future supermarket, and all depend, in one way or another, on electricity.

Provisions should be made for a sound system to provide music and announcements to public areas. Empty conduits should be installed for private and public coin and calling card/credit card telephones. Power for vending machines, ice sales storage, snow melting systems, and other miscellaneous loads should also be checked.

16.30 Swimming Pools and Fountains. When swimming pools or fountains use electric power for motor driven water pumps, underwater lighting, surface lighting, overhead lighting or outlets for pool cleaning, etc., a serious life safety hazard can develop. The designer should avoid the use of metallic pipes for water intake and for water outlets of pools and fountains, that is, between the pools or fountains and the motor driven pumps, regardless of the voltage applied to the motors and electrically operated valves.

The quality of wire, fixtures, etc., should be the best obtainable for such wet conditions. All electric pumps and valves should be isolated from the public.

The grounding system of all electrical devices, and pump and valve equipment should be of minimum impedance.

Electrical potential should be prevented from developing in any portion of the water. If and when the grounding system or the electric insulation system fails or deteriorates, it should be removed, replaced, and tested. The entire system should be periodically tested to ensure that acceptable insulation and ground resistance values are maintained. As a minimum, the installation shall comply with the NEC, Article 680 [1].

The pertinent rules and approved equipment lists apply to deck boxes, lighting luminaires and circuiting, bonding of metal structures, grounding connections, ground-fault interrupters, and low-voltage lighting.

The trend toward low-voltage lighting, generally 12 V, is gaining attention due to the availability of lighting equipment, two-winding transformers, and ground-fault circuit interrupters. When low-voltage lighting is applied, design details on sizing equipment and wiring are of prime importance. Placing transformers and interrupters in the vicinity of pool lighting luminaires has proven to be feasible and economical. Lighting fixtures are installed 18 inches below normal water level and mounted in pool walls. Wet niche types of lighting fixtures are preferred.

If above-water lighting is used, it is imperative that it be designed to reduce surface reflection to ensure that a lifeguard can easily see the entire bottom of the pool.

Effective bonding of all metallic parts of a pool's structure, including lifeguard stands, ladders, diving board stand, reinforcing steel, skimmers, and other metal parts, can be accomplished with a solid copper conductor that is not smaller than AWG No. 8.

16.31 Telephone Buildings. In addition to conventional offices, portions of commercial buildings are dedicated to communications switching equipment. Newer facilities use electronic, rather than electromechanical, switching; however, power requirements remain high (per unit area) due to the miniaturization of equipment and its resulting higher density. Telephone switching equipment is typically powered from dc power supplies feeding the equipment and a battery bank that floats on the line. When a utility service interruption occurs, the batteries assume the load until an on-site ac generator can be brought on line. The chargers are typically sized so that the battery bank can be recharged at the same time the communications load is served. Battery rooms require special treatment, such as spill containment and venting. Batteries, chargers, and the load should be located in close proximity because of the high currents involved.

Air conditioning is critical and will probably need to be served from the standby power bus. Air from the battery room should be exhausted directly to the outside.

Lighting in equipment rooms will need to be arranged to suit the narrow maintenance aisles. Equipment aisle lights are typically a part of the communications equipment racks, which makes designing equipment rooms less difficult. Since only small areas need to be lighted at any particular time, each aisle section should have its own switch.

When underground cables enter a building cable vault, special precautions need to be taken against natural gas leakage and sewer gas leakage into the phone raceway system and then flowing into the vault. Explosionproof lighting, sniffers (gas detectors), fire detectors, etc., may be required. It is essential that all cable penetrations into the vault be sealed to prevent the spread of fire.

Maintenance outlets may be required throughout equipment rooms for portable equipment, such as soldering irons and instrumentation.

Considering the critical nature of the equipment room during floods, earthquakes, wind storms, and other natural disasters, one design goal should be to maximize the equipment room's potential for survival in the event of a catastrophe.

16.32 Television Studios. The electric system for this type of building is similar to that for radio studios (see 16.23) except for the lighting loads. General lighting for office buildings should suffice for the offices, corridors, foyers, etc.

In order to produce a quality picture, a television studio needs large amounts of concentrated lighting. The lighting designer should also take into consideration the requirements of color television. Every show has its own lighting engineer. This engineer's preferences for lighting should be considered.

Studio buildings should be designed for color television productions. Transformers, ducts, and cable raceways should be planned and installed during original construction to meet these requirements.

Television studio spotlighting and floodlighting fixtures are mounted on a grid of pipe supports for maximum flexibility. They are wired by means of pigtails, connector strips, and patch panels to a central control console in each studio. Special attention should be given to ventilation and air-conditioning loads that are produced by the power density of heavy lighting. Special wiring and shielding may be required to avoid interference with television signals. Telephone and teletype communications may be needed as well as provisions for network pickup and service. Other considerations include provisions for microwave pickup from off-site trucks, special security systems, videotape editing and reviewing rooms, emergency power systems, isolation of technical and broadcast power from motor loads, and extensive grounding grids.

16.33 Theaters. Theaters are of two general classifications, the motion picture theater and the legitimate or stage production theater. Because certain areas in both types of theaters are similar, they will be covered as a single unit in this section to avoid duplication. Generally, theaters have a marquee with illuminated signs and decorations, in addition to a concentration of lighting outlets on the soffit to illuminate the entrance to the theater and to attract the eyes of prospective patrons. Electric signs that show the name of the theater are also frequently placed on the front of the building.

The entrance lobby and foyer are generally decorative in nature, and the engineer should study the architectural details so that the lighting will blend properly with the architectural treatment and color scheme. The minimum wattage per square foot used in the foyer may be about 2.5 W/ft^2 and, in the lobby, about 3 W/ft^2. Ticket booths may require a telephone, a signal system, and, possibly, a special outlet for an electric heater, air conditioner, and fan.

The interiors of motion picture theaters are of a rather plain design, although some of them, if the pictures are shown along with a stage show, may more nearly approach the legitimate theater for interior treatment and illumination. The designs are of such a varied character that it is impractical to establish any clear formula for an illumination plan. However, provision should be made for aisle lights, exit lights, and orchestra and emergency lights.

The projection booth should have provisions for a minimum of two projection machines, each supplied by a separate circuit, floodlights, spotlights, a rewinder, exhaust fans, an intercommunication signal system, and a dc power supply for projection arc lamps. Depending on the size of the theater, provisions may be made at several locations for floodlights or spotlights.

Theaters may have passenger elevators and escalators to serve the patrons who are seated in the mezzanines and balconies, and, in a number of installations, orchestra lifts, stage lifts, and turntables have been provided.

In legitimate theaters and large motion picture theaters, auditorium lighting is controlled by dimmers on the stage switchboard. Border lights, floodlights, and spotlights are controlled individually.

The wiring system may be complicated, depending on the lighting effects required to suit the type of entertainment provided. The stage requires stage pockets, several rows of border lights, outlets for special electric effects, and motors for operating the fire curtain, contour curtain, roll curtain, heavy drops, ventilators at gridiron, and numerous other items.

16.33.1 General Lighting Systems. General lighting is required for the gridiron, fly galleries, dressing rooms, etc. The power load for air conditioning is a major item that deserves careful consideration in the preparation of the electric system. Convenience outlets should be provided for cleaning appliances and general maintenance work.

16.33.2 Stage Lighting Systems. All lighting fixtures (instruments) involved in a stage presentation are controlled through patch panels and dimmers to achieve the desired lighting effects. Extensive wiring is required because each fixture or group of similar fixtures is carried back to the patch panel as an individual two-wire 120 V circuit. Common neutrals are not used because phase connections vary. If possible, the patch panel should be located so that it has a view of the stage. Dimmer switchboards (generally SCR control units) produce heat and require ventilation, and should be located in a sound-isolated area.

A lighting control booth with a good view of the stage is necessary and should contain a stage lighting control panel from which all of the production lighting can be controlled. The control panel controls the many dimmers and contactors to provide the desired stage lighting for each scene. Each scene will require a different illumination configuration, and the sequence of illumination settings is typically repeated for each production. All of this may be preset and sequenced by a microcomputer in the control panel. An emergency panic switch is usually incorporated and is used to immediately restore the theater egress lighting upon an indication of a threatening condition. The booth also will normally contain follow spots, and slide and motion picture projectors. A sound control booth is frequently located near the lighting control booth.

16.34 Transportation Terminals. Transportation terminals consist of passenger terminals for railroads, buses, and aircraft that interface with platforms, boarding ramps, ticket and reservation areas, as well as concessions or stores, theaters, and entertainment areas for the travelers' convenience. Concession areas require general provisions as described in 16.10.

Provisions should be made for supplying metered or rent-included power to each tenant. In some cases, such as in airports, the terminal may very well be a shell in which most of the areas are built by tenants, which conform to tenant-established building standards and approved construction methods.

16.34.1 Requirements for All Terminals. The following general requirements apply to all terminals:

(1) Fire alarm systems, which may include smoke detection systems, are usually interconnected to a central station for transmission of alarms to the fire department. In very large complexes, the functions of the central station may be handled in-house. The usual fire alarm pull-boxes may be supplemented by a number of telephones that may be used for information, porter call, and police and fire emergencies.

(2) Sprinkler supervisory and alarm systems are tied into the fire alarm system for transmission of alarms to the fire department.

(3) Public telephones and other telephones served by the local telephone company usually require only conduits. If the system is owned by the phone company, it will install equipment and cables. If the building owner purchases the system, the equipment and cables will be installed by the supplier or contractor. A room for telephone services, which meets the standards of

the telephone company, should be provided. The telephone company will install coin and calling card/credit card public telephones to meet various architectural requirements; however, close consultation with the telephone company is required to establish design standards. Provisions should be made for local private systems, which may be needed by individual transportation companies.

(4) Terminals require public address systems that may be controlled from more than one location and that may require zoning so that announcements can be made in different loading, lounge, and baggage areas.

    (a) One of the major problems in designing public address systems is the use of high-powered speakers that are too widely spaced, which results in distorted announcements. The use of lower powered speakers closely spaced, particularly where noise levels are high, with automatic variable volume control based on ambient sound levels is desirable.

    (b) Provision should be made for background music, which is often transmitted from commercial specialists.

    (c) Provisions may be required for automatic departure announcements when established schedules require very frequent departures, such as bus terminals.

    (d) Consideration should be given to extending the public address system into the restrooms, outside loading areas, embarking areas, and, possibly, parking areas.

(5) Master clock systems may be used to ensure that all clocks are maintained at the same time, and that manual resetting is not required after temporary power outages, or for changes related to daylight saving time. Carrier frequencies (generally 3000 Hz) for master clock circuits are often superimposed on power distribution systems for clock correction and program signals.

(6) Arrival and departure signs may be displayed on closed-circuit television or variable message signs. Both systems are usually driven by microprocessors in newer installations and may contain elaborate storage information systems that involve schedules. Both systems can include an automatic roll-down feature in which messages are kept in sequence regardless of changes and, as items are removed or added, the proper sequence (usually established on a time basis) is retained.

(7) In a modern terminal, security systems will usually include closed-circuit television surveillance of public areas, taxi loading areas, and other locations where the public should be protected. Closed-circuit television usually includes a provision for zooming and scanning (pan and tilt). The special-purpose telephones that were described in 16.11 may also form part of the terminal security system.

(8) Directional signing may consist of backlit or frontlit signs with off/on control. Sign colors and designs, usually designed in conjunction with graphic specialists, are essential to provide adequate terminal control. In some cases, advertisement signing is combined with directional signing; the former often being an important source of revenue for the facility. Outlets should be

provided for other signs (for example, Christmas displays), which may be temporarily required.

(9) Outlet systems should be adequate to serve vending machines, water coolers, temporary displays, and for cleaning and maintenance equipment. The latter two items may require 480 V outlets. Codes and good practice may require the use of ground-fault circuit interrupters for 120 V outlets to ensure the protection of personnel.

(10) A modular system of local wiring, outlets, and metering to accommodate future change in rentable space is desirable.

(11) Electronic dispatching control and surveillance systems may be used to indicate the departure times of vehicles, trains, or aircraft, to observe such departures, and to direct traffic through the use of special signing systems. The location of vehicles in docks may be detected through loop presence detectors, treadles, or similar sensors. When complex systems are involved, dispatching boards may be used. The dispatching boards may be manually or automatically operated to indicate the status of loading platforms and traffic. Closed-circuit television surveillance of areas where traffic congestion may develop should be provided. Ramps into or out of terminals usually also require close surveillance.

(12) Computer systems are frequently used to control terminal and tenant systems. When separate rooms are required for computer installation (as distinct from certain microprocessor intelligent terminals that may be installed in open areas) raised flooring, separate air-conditioning systems, special fire and smoke detection systems, flame suppression agent systems, and separate emergency circuits may be required.

(13) Emergency power systems require the use of engine- or turbine-driven alternators. Uninterruptible power supplies are usually only required for special systems, such as computers and alarm systems; however, UPS systems are often furnished as part of the special equipment installation.

(14) The building supervisory system may be as simple as an annunciator or as complex as a completely computerized building management system. The latter system includes alarm detection and indication points for items such as escalators, elevators, fans, pumps, chillers, cooling towers, temperature detectors, and other similar devices. The status of operation, such as up or down for escalators, emergency trip-out of equipment, overtemperatures, area lighting status, and similar alarms, will be audibly and visually indicated, usually with a reset provision for audible alarms and always with a provision for indicating multiple alarms. The building supervisory system may also be used to automatically control heating, ventilating, and air-conditioning equipment and to interface with building electrical owner demand equipment to optimize overall power utilization.

(15) Other communication systems, some of which may be owned by tenants, can include portable radio systems and pagers for larger areas, special building antennas and loops, and automatic printer systems, such as teletype machines, facsimile machines, and annunciators. Provisions may also be made to communicate with special staff by the use of coded announcements

on the public address system. It is usually not wise to announce uncoded emergency situations on the public address system.

(16) Baggage handling systems are fairly simple, involving the use of manual labor for the most part except at airports where systems become quite complex (described below). A major consideration is to provide telephones at convenient locations for obtaining porter service and to provide some form of alert signal so that porters will know that they are being called.

(17) It may be desirable to use special spaces, such as shafts, utility tunnels, or trenches, to provide a practical means of handling cabling at minimum cost. It is often possible to utilize cable trays, particularly for communication wiring. Communication wiring may consist of coaxial, fiber-optic, multiconductor control, signal, telephone, and data cables.

(18) Because new terminals are usually individualized, highly creative architectural designs, the lighting design should be closely coordinated between the engineer and the architect or lighting consultant. When high ceilings are involved, HID sources are usually selected as the primary source of lighting. Energy conservation and maintenance considerations have created a tendency to move away from incandescent lighting to more efficient sources of lighting regardless of ceiling height.

(19) The following types of systems are often found in terminal buildings:
  (a) Snow and ice melting systems for ramps, sidewalks, and driveways
  (b) Kitchen equipment
  (c) Radiant heating
  (d) Supplementary air conditioning
  (e) Fare collection or change issuing machines
  (f) Smoke purge control systems
  (g) Emergency escalator and moving walkway shutdown systems
  (h) Lightning protection systems
  (i) Dispatch booth power, lighting, and control
  (j) Infrared heating

16.34.2 Railroad Terminals. Some railroad systems may use an electronic reservation system for trains that is very similar to the system used in airline terminals. Train departures are often displayed on large boards with flip-type disks, which make up letter matrixes and use the rolldown feature. Newer terminals often contain extremely large illuminated wall-to-wall advertising displays. Lighting levels are usually kept fairly low, perhaps 8–15 fc, except in ticketing areas, areas where pedestrian safety and security is involved, and in reading areas of lounges.

16.34.3 Bus Terminals. Extensive ventilation systems are required to exhaust fumes from bus engines when buses load and unload in closed terminals. Complex systems of traffic controls are usually required in larger terminals. Information telephones are often distributed throughout the terminal, and a large information booth is usually incorporated into the terminal-type design. Lighting levels in bus terminals may typically range from 10–30 fc, with levels as high as 50 fc in limited areas where highlighting is desired. Security provisions for bus terminals are usually fairly severe, requiring the use of closed-circuit television and emergency telephones. Lighting, when used properly, can be conducive to the feeling of security amongst patrons. Today, bus terminals are often located in urban areas as part of a general area renewal strategy. In such cases, lighting the facade of the bus terminal becomes an important part of improving the overall ambiance of an entire area.

16.34.4 Airports. Airports may range from simple terminal buildings to huge complexes that are associated with major urban areas. A large airport will have most of the facilities of a small city, such as hotels, shopping areas, bus handling systems, and extensive traffic and roadway control systems, many of which have already been discussed in this chapter.

Automated baggage handling systems of the carousel or moving belt type are supplied from extensive conveyor systems that lead to the airport apron areas. For handling aircraft cargo, very extensive electronically controlled systems, some using linear motors, are utilized. Such cargo handling systems use industrial material handling technology.

Most ticketing and reservations operations are handled through computerized systems; however, the main or central computers are often remotely located in an airline building off the airport grounds. Extensive communication wiring systems are required for ticketing and communications between baggage, apron, ticketing, and gate areas. Extensive signing is required at each gate position as well as signing to indicate the location of gates from which flights will be departing. Provisions have to be made for security control, particularly for the installation of detection equipment and control stations, which are usually at the entrance to each gate or group of gates in order to detect the presence of weapons and explosives.

While terminal area lighting may be held to relatively low levels, perhaps 10–25 fc, much higher levels are utilized in the ticketing areas.

The control tower design, which will include provisions for air traffic controllers, ground traffic controllers, and possible facility surveillance, should be designed in conjunction with the appropriate air traffic control authorities. Special expertise is required in the design of these airport systems and of the related runway and taxiway lighting systems. Airport control systems are specified in detail in FAA 150/5360-4B, Utility Airports Advisory Circular [B12], FAA 150/5360-9, Planning and Design of Airport Terminal Facilities at Non-Hub Locations Advisory Circular [B13], FAA 150/5360-11, Energy Conservation for Airport Buildings Advisory Circular [B14], FAA 150/5360-12, Airport Design Standards Advisory Circular [B15], and FAA 150/5360-13, Planning and Design Guidelines for Airport Terminal Facilities Advisory Circular [B16].

At airline gates, which may be some distance from the ticketing booths, 500 kW or more may be required for 400 Hz alternators that provide power to the planes while they are parked. Emergency fuel shutoff and alarm systems are provided when fuel piping systems are used. Requirements for plane control at gate positions are usually specified by the airline. Provisions should be made for apron lighting when work will be going on while the aircraft is being loaded, serviced, and refueled.

Hangar lighting is almost always of the HID type supplemented with emergency lighting for power outages. A typical large hangar may require in excess of 1000 kW of light. Grounding systems have to be provided for the aircraft both in the hangers and on the aprons, and at loading and fueling positions.

The power distribution system for larger airports is almost always at medium voltage with energy supplied at individual buildings either from spot-network systems or from medium-voltage primary transfer switches. For reliability considerations at larger terminals, multiple-feed selective systems are almost always used.

Extensive vehicle parking areas are required for short- and long-term parking. Where parking charges apply, toll plazas should be established. Newer designs utilize unmanned gate entrances with ticket issuing machines and manned exits that may contain provisions for automatic fee calculations. The audit system that is utilized should provide verification of the presence of vehicles, time of fee collection, and vehicle counters that are tied into the auditing device in each toll booth. This may consist of a special cash register or an automatic fee calculation and audit device. Today, the tendency is to move toward the installation of "intelligent" terminals that perform all fee calculations and report to a central computer. Parking lot signing that utilizes changeable message signs and lighting fixed signs is essential when traffic is directed to different lots.

16.34.5 Rapid Transit Stations. A rapid transit, subway, trolley, or elevated station is usually an extended platform that provides access to trains. Such stations have public address systems that announce the arrival of trains, safety precautions, and train delays. Lighting levels are relatively low except in the toll collection areas and at points of embarkation. Higher levels of lighting may be used in limited areas where, at night, a feeling of security on the part of the patron is desired. Simple signing systems will indicate which trains should depart first, if several trains are parked in a multiple-platform station. Where trains are normally parked, warning lights may indicate that trains will depart at a preset time interval. Elevators and escalators are frequently used at larger underground stations.

16.35 References. The following references shall be used in conjunction with this chapter:

[1] ANSI/NFPA 70-1990, National Electrical Code.

[2] ANSI/NFPA 72A-1987, Installation, Maintenance, and Use of Local Protective Signaling Systems.

[3] ANSI/NFPA 72B-1986, Auxiliary Protective Signaling Systems for Fire Alarm Service.

[4] ANSI/NFPA 72C-1986, Remote Station Protective Signaling Systems.

[5] ANSI/NFPA 72D-1986, Proprietary Protective Signaling Systems.

[6] ANSI/NFPA 99-1990, Health Care Facilities.

[7] ANSI/NFPA 101-1988, Life Safety Code.

[8] IEEE Std 141-1986, IEEE Recommended Practice for Electric Power Distribution for Industrial Plants (ANSI).

[9] IEEE Std 446-1987, IEEE Recommended Practice for Emergency and Standby Power Systems for Industrial and Commercial Applications (ANSI).

[10] IEEE Std 602-1986, IEEE Recommended Practice for Electric Systems in Health Care Facilities (ANSI).

[11] *IES Lighting Handbook*, 1987 Edition.

16.36 Bibliography. The references in this bibliography are listed for informational purposes only.

[B1] ANSI/NFPA 20-1990, Centrifugal Fire Pumps.

[B2] ANSI/NFPA 70B-1987, Electrical Equipment Maintenance.

[B3] ANSI/NFPA 70E-1988, Electrical Safety Requirements for Employee Work Places.

[B4] ANSI/NFPA 88A-1985, Parking Structures.

[B5] ANSI/NFPA 88B-1985, Repair Garages.

[B6] ANSI/NFPA 110-1988, Emergency and Standby Power Systems.

[B7] ANSI/NFPA 110A-1989, Stored Energy Systems.

[B8] ANSI/NFPA 910-1985, Protection of Library Collections.

[B9] ANSI/NFPA 911-1985, Protection of Museums and Museum Collections.

[B10] ANSI/NFPA 912-1987, Fire Protection in Places of Worship.

[B11] FAA 70/746-1, Obstruction Marking and Lighting Advisory Circular.

[B12] FAA 150/5360-4B, Utility Airports Advisory Circular.

[B13] FAA 150/5360-9, Planning and Design of Airport Terminal Facilities at Non-Hub Locations Advisory Circular.

[B14] FAA 150/5360-11, Energy Conservation for Airport Buildings Advisory Circular.

[B15] FAA 150/5360-12, Airport Design Standards Advisory Circular.

[B16] FAA 150/5360-13, Planning and Design Guidelines for Airport Terminal Facilities Advisory Circular.

Chapter 17
Electrical Energy Management

17.1 Energy Management Requirements. The establishment of a successful electrical energy management program is dependent upon the interest and encouragement of top management, and a formulated company policy that is committed to saving both energy and the financial resources that are associated with energy and demand savings.

The supervision of an energy management program should be delegated to that staff within the organization that will commit the time and resources necessary to it. The program will not be successful if it is assigned as a part-time duty to staff members whose primary responsibilities lie in other areas. In some instances, sufficient in-house staff may be available to develop a program; however, the management of commercial buildings will not have access to a staff that is sufficiently conversant with such a program. As additional services are needed, they may be obtained through the use of consultants and engineering firms that specialize in energy conservation technology. In order to ensure the success of the program, periodic reports should be provided to management. Projects showing the energy and cost savings, and payback that have resulted should be included. The report should be reviewed and commented upon by top management. Without the cooperation of the entire staff, and especially the building maintenance staff, the program will not be successful.

The topics discussed in this chapter represent three basic facets of energy planning:

(1) Administrative aspects, such as organizing for conservation, embarking on a conservation program, understanding electric rates, and managing loads and evaluating losses for purchasing decisions.

(2) Descriptions of electrical devices that are used in metering and lighting, and a discussion of equipment efficiencies in relation to the electrical and mechanical environment.

(3) On-site generation, which encompasses co-generation, peak shaving, and load curtailment.

17.2 Energy Conservation Opportunities. There are two recognized methods for determining energy conservation opportunities

(1) Energy audit

(2) Project lists

17.2.1 Energy Audit. The first method is to make a complete energy audit that lists major energy-using equipment with nameplate data relating to energy, any efficiency tests, and estimates of or measured hours per month of operation (which will include monthly utility data, amounts, and total costs). National Weather Service monthly degree days for heating and cooling should also be included. Generally, a 1 or 2 year compilation of data is used. The crucial part of the audit is an intelligent appraisal of energy usage, what is done with energy as it flows through the processes and facility, and how this compares with accepted known standards. This study can lead to the development of a prioritized list of projects with an acceptable rate-of-return.

17.2.2 Project Lists. The second method is to look for specific projects to reduce energy and costs. Shopping lists of projects are obtained from associates, newspapers, U.S. Department of Energy (DoE) publications, the local utility, or trade groups. These lists include the readjustment of thermostats for heating, cooling and hot water; removing lamps in lighting fixtures; installing storm windows and doors; caulking; additional insulation; etc. An excellent reference is a do-it-yourself guide call "Identifying Retrofit Projects for Federal Buildings by the Federal Energy Management Program (FEMP), Report 116."

National standards have been issued by the American Society of Heating, Refrigerating, and Air-Conditioning Engineers, Inc. (ASHRAE) as part of ASHRAE/IES 90.1-1989, Energy Efficient Design of New Buildings Except New Low-Rise Residential Buildings [1],[97] which was issued jointly by ASHRAE and IES and covers national recommendations of building energy performance standards (BEPS) for power and energy budgets for new construction. Another helpful document is "Total Energy Management," a practical handbook on energy conservation and management developed jointly by the National Electrical Contractors Association (NECA) and the National Electrical Manufacturers Association (NEMA) in cooperation with the U.S. DoE. Energy codes have been published by a number of states, some of which specify materials similar to those specified in the ASHRAE 90 Series, and others that specify permissible usage on various bases, such as W/ft^2, or allowable fc levels. This information is likely to be available from a state energy office.

17.3 The Energy Management Process

17.3.1 Obtain Management Approval and Commitment. One key to success in an engineering effort to manage energy effectively is the advance approval of upper management and supervision. This approval is even more important in a conservation plan because expenditures may not have an immediate effect on the profits for the facility.

Goals and guidelines should be established so that management knows what to expect. Furthermore, mutually agreed upon criteria will help an engineer properly

---

[97] The numbers in brackets correspond to those in the references at the end of this chapter. ASHRAE publications are available from the American Society of Heating, Refrigerating, and Air-Conditioning Engineers, 1791 Tullie Circle, N.E., Atlanta, GA 30329. IES publications are available from the Illuminating Engineering Society, 345 East 47th Street, New York, NY 10017.

direct these efforts. There have been cases in which corporate savings were unnoticed and unappreciated because staff failed to keep management properly informed.

The conservation program should also include appropriate organizational changes including the establishment of an energy committee composed of engineering, purchasing, accounting, tenant, and labor representatives. This committee should have the responsibility and commensurate authority to perform its job of conserving energy effectively. Clearly defined and accurately written goals (in energy units) and guidelines can properly guide the committee's efforts. Regardless of the breadth of experience of the energy team (which could consist of only one person), all affected staff should be consulted before preparing any plans. Results and proposals should then be completely communicated to all interested and involved staff.

17.3.2 Embarking on an Energy Conservation Program. It is important, from several standpoints, to establish the existing pattern of electrical usage and to identify those areas where energy consumption could be reduced. A month-by-month record of electricity usage is available from electric bills, and this usage should be carefully recorded in a format (possibly graphic) that will facilitate future reference, evaluation, and analysis. The following list of items (where appropriate) should be recorded in the electric usage history:

(1) Billing month
(2) Reading data
(3) Days in billing cycle
(4) Kilowatthours (or kilovoltamperehours, if billed on this basis)
(5) Billing kW demand (or kVA demand, if billed on this basis)
(6) Actual kW demand (or kVA demand, if billed on this basis)
(7) Kilovars (actual and billed)
(8) Kilovar hours (actual and billed)
(9) Power factor (average or peak, as billed)
(10) Load factor (average use compared to peak use)
(11) Power bill (broken down into the above categories along with fuel cost)
(12) Occupancy level
(13) Heating or cooling degree days
(14) A electricity usage history, including appropriate remarks (such as vacation periods)

A listing of building operations, equipment, and energy conservation opportunities (ECOs) will also provide both a usage history and a basis for evaluating future improvement. The listing of this information, along with electricity usage, is part of the energy audit. In general, there are four categories of ECOs. These four categories are as follows:

(1) Housekeeping Measures — Easily performed (and usually low-cost) actions (for example, turning lights off when not required; cleaning or changing air filters; cleaning heat exchangers; keeping doors shut; and turning off redundant motors, pumps, compressors, and fans)
(2) Equipment Modification — This is usually more difficult and more expensive because it involves physical changes to the electric system (for example, the

addition of solid-state, adjustable speed drives; reducing motor sizes on existing equipment; removing light fixtures; adding automatic controls to reduce lighting in unoccupied areas; and modifying heating and cooling systems).

(3) Better Equipment Utilization — The use of natural lighting as much as possible, the redirection of warmer air to cooler parts of the building during the heating season, and staggering starting times for tenants to reduce energy demand or consumption, or both.

(4) Changes to the Building Shell — Improving the insulation quality of the building to reduce energy losses to the outside environment (for example, adding insulation, reducing infiltration, controlling exhaust/intake, etc.) and reducing heat gains in the inside environment by using reflective materials, shading, and insulation.

Housekeeping and low-cost measures should be undertaken without delay. The larger and more expensive ECOs generally take longer to initiate and should often be performed after low-cost measures are completed. However, there may be cases when obvious equipment modification improvements can be made concurrently with low-cost improvements. Chapter 15 deals with electrical rehabilitation and includes discussions about the implementation of building changes including those for energy conservation purposes.

17.3.3 Equipment Audit. The analysis of equipment and its efficiency can be a very simple task that requires only a careful tour of a facility. The analysis can also involve the documentation of all equipment and associated efficiencies. In any case, the results usually have a direct relationship with the degree of effort expended. Regardless of the intensity of the energy audit, the above four conservation categories should be considered. In all cases, there may be other nonenergy related criteria that should be included in an evaluation of each ECO.

Studies show that more effective use of energy can be achieved by using waste heat or redirecting waste heat. Other applications of energy reuse include recapturing heat for space conditioning from lighting systems. Available waste steam (even if purchased) can also be used for heating, air conditioning (using a turbine refrigeration drive or absorption chillers), or used for building or area heating, thus avoiding the use of electricity or another energy source.

In some instances, constraints (particularly regarding energy availability) may require the expenditure of capital to increase the existing energy efficiency in order to have sufficient electrical capacity for future commercial building expansion. This may occur when either utility capacity is severely limited or the costs associated with increasing service size would be so high that expansions would essentially be made from current energy conservation load reductions.

Commercial buildings can be divided into five equipment/process categories by function. Each function can then be examined in the aforementioned four conservation categories. The five functions are lighting, HVAC (heating, ventilating, and air-conditioning) systems, motors and drives, electrical distribution equipment, and the building's environmental shell. A list of specific actions (or inquiries) is presented in IEEE Std 739-1984, IEEE Recommended Practice for Energy Conservation and Cost Effective Planning in Industrial Facilities (ANSI) [B1].[98]

17.3.4 Tracking Progress. An energy audit reveals information about a building's energy usage pattern and ECOs. While this information is extremely important, the building's current status is also important. Therefore, the establishment of a recording and monitoring system is needed to maintain control over electricity usage and to gauge the results of implemented ECOs.

There are five important aspects to tracking progress.

(1) Meters should be installed at various important load centers.
(2) These meters should be read and the data recorded on a regular basis (preferably weekly or by shift). Recording meters will enhance the analysis.
(3) The data should be analyzed to determine the need for action.
(4) In terms of energy units, the information should be kept on a common base, such as W/ft^2.
(5) When the power consumption of the load is constant, such as for a given lighting installation, once the power has been measured or calculated, the use of elapsed time meters and recorders that show on/off times may be sufficient.

17.3.5 Overall Considerations. There are several considerations that set the stage for action on ECOs, and all of them are important.

(1) The responsible staff should recognize the in-house limitations for designing and implementing an ECO. It may be necessary to call in an energy expert or a consultant to prepare the specific design under staff's guidance. The use of equipment that cannot be properly maintained or is overly complex should be avoided. For example, the use of very sophisticated demand controllers requires relatively complex settings. In some cases, the general maintenance personnel in a small commercial building may not be able to make complex settings. Hence, the use of general maintenance services could lead to the abandonment of sophisticated controllers.
(2) Careful consideration shall be given to appropriate codes and standards. Some prominent codes and standards are developed and published by the National Fire Protection Association (NFPA), the Environmental Protection Agency (EPA), state energy codes, and the Occupational Safety and Health Administration (OSHA).
(3) The building's energy balance is an important consideration. There may be cases where an ECO action would merely shift the energy source from one point (or fuel) to another. (For example, 15 kW of lighting was turned off during the winter months in an electrically heated building. This meant that the heating system had to make up the 15 kW of heat that was supplied by the lights. In this case, it may have been better to leave the lights on.)

Lighting adds to the air-conditioning load in the summer so that 1 kW less of light may also reduce air-conditioning power requirements in excess of 1 kW. Hence,

---

[98]The numbers in brackets preceded by a B refer to the bibliographic references that are at the end of this chapter. IEEE publications are available from the Institute of Electrical and Electronics Engineers, IEEE Service Center, 445 Hoes Lane, P.O. Box 1331, Piscataway, NJ 08855-1331.

energy balance requires scrutinizing the entire system or building shell. Redirecting excess heat from lighting or any other source from the air-conditioning, cooling or ventilating system can significantly reduce the cooling load. Typically, one ton of air-conditioning equipment (rule-of-thumb) requires one kW of input power to the refrigerant compressors and perhaps another 0.4 kW in ventilation equipment. Because the efficiency of the air-conditioning system is significantly less than unity, it would take more than 1.4 kW to remove 1 kW of heat that is inserted into the air-conditioning load.

17.4 Lighting

17.4.1 Introduction. Fluorescent lighting has been the most widely used lighting system in the last several decades, especially in commercial establishments. Some 900 million fluorescent sockets are in place in the U.S. alone. Any short-range reduction in lighting energy depends on the modification of fluorescent lighting systems that can be readily made without major conversion costs. With this in mind, in the '80s, lighting manufacturers developed a line of energy-saving lamps and various types of energy-efficient ballasts and luminaires. Today, numerous possible combinations of these can be selected to suit any particular application need.

17.4.2 Energy-Efficient Light Sources

17.4.2.1 Incandescent. The addition of krypton gas to the bulb fill improved efficiency, which reduced wattage at equal light output by about 10%. But krypton gas is a rare, expensive gas, which created a negative impact on the cost of the lamps.

Reflector-type lamps are the most popular after "A"-type lamps. The first lamp to effectively reduce wattage was the elliptical reflector lamp, which was a replacement for standard reflector lamps used in baffled downlights. Depending on the area being lit, as much as 50% in energy savings can be realized as shown in Table 85.

Table 85
Approximate Illumination Footcandles at 10 Foot Mounting Height
(Elliptical versus Standard Reflector)

Distance from Center of Beam	60ER30	75R30/FL	75ER30	150R/FL	120ER40	300R/FL
0 ft.	12	4	16	12	24	21
1 ft.	11	4	14	10	23	20
2 ft.	9	4	11	10	18	18
4 ft.	3	3	4	6	6	12
6 ft.	1	1	2	4	3	7

The current trend in energy savings combines incandescent and tungsten halogen technologies. A series of lamps was introduced for accent, display, and downlighting for new construction and renovation. The tungsten halogen lamps provided excellent lumen maintenance, and twice the life ratings at the same wattage and lumen output of their equivalent incandescents.

Lower wattage halogen sources were developed in special bulbs, similar to incandescent-type bulbs, for improved results at reduced energy usage. This is an excellent example of the improvements in incandescent technology over the past 15 years. The 150 PAR was initially reduced to 120 W for a 20% energy reduction. Further development reduced the wattage to 90 W for a 40% energy reduction.

17.4.2.2 Compact Fluorescent Lamps. Fluorescent technology was developing at the time of the energy crisis in the '70s. Fluorescents were the most popular source of light in commercial and industrial applications.

The industry was put under great pressure when the federal government took the radical step of requiring the removal of half of its fluorescent lamps. The industry reacted quickly to develop a reduced energy lamp for most fluorescent circuits.

The first energy-saving lamps resulted in a 5 W savings in the most popular F40T12 model and 15 W savings in the second most popular F96T12 slimline model. Further development resulted in an additional 1 W improvement in the energy savings of the 40 W lamps so that it is now rated at 34 W.

In a relatively short period of time, a complete line of energy-saving lamps was on the market. Today, over 30% of the fluorescents used are of this type. Table 86 shows the energy-saving fluorescent lamps developed and watts saved per lamp with respect to the standard lamp.

Today, there are many types of compact fluorescent lamps. They use bent tube geometrics that can be adapted for incandescent socket use. Developments continue on these fluorescent lamps. The use of compact lamps, which utilize electronic circuitry, provides the light output of a 75 W incandescent lamp using only 18 W. Its life expectancy has increased to 10 000 hrs.

Virtually all compact fluorescent lamps use the "rare earth" phosphors for good color rendition and lumen maintenance characteristics. The wattage range for these compact lamps is 5–40 W. Generically, one classification of these lamps is

Table 86
Energy-Saving versus Standard Fluorescent Lamps

Energy-Saving	Standard	W Saved/Lamp
F30T12/RS/ES	F30T12/RS	5
F40/RS/ES	F40/RS	6
F40/PH/ES	F40/PH	6
F90T12/60/ES	F90T17	6
F48T12/ES	F48T12	9
F96T12/ES	F96T12	15
F96T12/HO/ES	F96T12/HO	15
F96T12/VHO/ES	F96T12/VHO	30

called "twin tubes," while another classification is called "quad tubes." Various fixtures are being developed for these compact lamps (see Fig 155).

While new products were being developed, considerable improvements were also being made in the standard straight tube lamps. It was discovered that the human visual system responded best to primary colors. By changing the proportions of the primary colors, white light can be made to match the color appearance of any fluorescent lamp.

The addition of rare earth phosphor to the standard fluorescent line increased luminous efficacy by 5% and improved the color rendition index (CRI) from the 50–70 range to the mid-80s.

A major breakthrough occurred with the introduction of the standard 40 W fluorescent bulb. Utilizing advanced phosphor technology with the optimization of bulb diameter, 40 W lamps are now available that can be retrofitted in a F40 preheat or rapid-start circuit that provides

(1) 3700 lm—a 17% increase over standard cool white.
(2) CRI 80—a improvement over the 50–70 CRI for standard lamps.
(3) 24 000 hours average life—a 20% increase over standard lamps.
(4) T10 bulb size—more compact.

**Fig 155
Fluorescent PL Lamp with Socket Adapter**

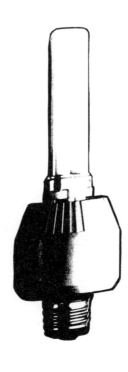

This new lamp could save energy and improve color rendition. It could also provide for useful application wherever a W/ft^2 requirement has been legislated.

17.4.2.3 High-Intensity Discharge (HID) Lamps. Today, HID lamps include mercury vapor, metal-halide, high-pressure sodium, and low-pressure sodium lamps. Lamp development has occurred mainly in high-pressure sodium and low-pressure sodium products during the last decade. Metal-halide lamps offered the best opportunity from a color acceptability point of view. The industry has introduced a series of lamps down to 32 W. Low wattage metal-halide systems have been positioned successfully to replace incandescent applications in the commercial marketplace wherever color is important. High-pressure sodium lamps offer the highest luminous efficacy in environments where colors need to be distinguished. New developments have led to the a lower wattage range that is down to 35 W. The improved color rendition versions of high-pressure sodium lamps were also developed with a CRI of over 80 in the lower wattage range.

Since HID lamps have had very few application problems, they are likely to experience further development in the years ahead.

17.4.3 Energy-Efficient Ballasts

17.4.3.1 Fluorescent. It is well known that fluorescent lamps that are driven at high frequency are more efficient. Electronic ballasts are now available for the F40T12, the slimline, the new T8 lamps, and other energy-saving fluorescent lamps on both 120 V and 277 V circuits. Operation of an electronic ballast involves the use of transistor circuitry to rectify the 60 Hz ac to a dc component, and then invert it back to an ac sine wave component having the frequency range of 10–30 kHz. When the frequency of the on/off operation of the mercury arc within the lamp is increased from 60 Hz to many times that value, the lamp efficacy can be raised by nearly 12%. At the same time, with the absence of the magnetizing losses within a core-coil ballast, the relative efficiency of the ballast is increased. Although the electronic ballast costs more than the standard core-coil ballast, operating factors should reflect an appreciable reduction in life-cycle costs for a lighting system.

There are two types of dimming ballasts: core and electronic. Core ballasts require auxiliary switching equipment to reduce the duty cycle or limit the current. These systems are energy efficient when lamps can be operated well below 100% for a substantial portion of the time. High-frequency electronic ballasts can readily be used to dim fluorescent lamps over a wide range of light levels. No major auxiliary equipment is required. All external control wiring is low voltage or fiber optic.

Some manufacturers have introduced ballasts that are optimized for energy-saving 35 W lamps, and these systems are slightly more efficient than when operated with 40 W ballasts. However, in applications where ballasts need not be replaced, the slight increase in efficiency does not justify the cost of refitting these ballasts. Table 87 summarizes the input watts for typical fluorescent lamp ballasts. The reduction in input watts for new energy-saving ballasts versus standard ballasts are also shown.

For incandescent light sources, major trends have developed since the energy crisis of the '70s. The first is toward a decrease in wattage size for many common applications. The second is to lower wattage and utilize reflector-type lamps that direct light toward a specific task.

Table 87
Typical Input Watts for Fluorescent Lamp Ballasts

Lamp Type	Nominal Lamp Current	Nominal Lamp (W)	System Input (W)				Circuit Type
			Standard Ballasts		Energy-Saving Ballasts		
			One-Lamp	Two-Lamp	One-Lamp	Two-Lamp	
F20T12	0.380	20	32	53	—	—	Rapid start, preheat lamp
F30T12	0.430	30	46	81	—	—	Rapid start
F30T12, ES	0.460	25	42	73	—	—	Rapid start
F32T8	0.265	32	—	—	37	71	Rapid start
F40T12	0.430	40	57	96	50	86	Rapid start
F40T12, ES	0.460	34/35	50	82	43	72	Rapid start
F48T12	0.425	40	61	102	—	—	Instant start
F96T12	0.425	75	100	173	—	158	Instant start
F96T12, ES	0.455	60	83	138	—	123	Instant start
F48T12, –800 mA	0.800	60	85	145	—	—	Rapid start
F96T12, –800 mA	0.800	110	140	257	—	237	Rapid start
F96T12, —ES, 800 mA	0.840	95	125	227	—	207	Rapid start
F48 –1500 mA	1.500	115	134	242	—	—	Rapid start
F96 –1500 mA	1.500	215	230	450	—	—	Rapid start

17.4.3.2 High-Intensity Discharge (HID). The choice of a ballast depends on economic considerations versus performance. A mercury lamp will operate from metal-halide ballasts; but the converse is not always true.

There are several different types of ballasts for high-pressure sodium lamps.

(1) Reactor or lag ballast — Inexpensive, with low power losses and is small in size.

(2) Lead ballast — Fairly good regulation for both line and lamp voltage variation.

(3) Magnetic regulated ballast — Provides the best wattage regulation with the change of either input voltage or lamp voltage. It is the most costly and has the greatest wattage loss.

(4) Electronic ballast — Has a steady, constant wattage output with changes in the source impedance as well as excellent regulation. During the life of a high-pressure sodium lamp, it can save 20% more energy by maintaining a constant wattage output in addition to the 15% intrinsic energy saving when compared to an equivalent core-coil ballast. Table 88 summarizes the input watts for typical HID lamp ballasts.

17.4.4 New Luminaires for Energy-Efficient Light Sources

17.4.4.1 High-Pressure Sodium Lamps. Proper luminaire design is the key to lighting efficiency. Newly developed luminaires use prismatic glass reflectors that are especially made for high-pressure sodium lamps. In addition to achieving maximum light utilization, they redirect the intense light source with excellent light cutoff and high-angle brightness control.

Luminaire manufacturers recommend aluminum reflectors for all general-purpose industrial applications, and glass coated reflectors where maintenance practice is compatible with servicing glass.

Table 88
Typical Input Watts for HID Lamp Ballasts

Lamp Type	ANSI Desig-nation	Watts	Reactor	Ballast Type			
				High-Reactance Auto-transformer (LAG)	Constant Wattage Auto-transformer (CWA)	Constant Wattage Regulated (CW)	High-Reactance Regulated (Regulated Lag)
Mercury	H46	50	68	74	74	—	—
	H43	75	94	91–94	93–99	—	—
	H38/44	100	115–125	117–127	118–125	127	—
	H39	175	192–200	200–208	200–210	210	—
	H37	250	272–285	277–286	285–300	292–295	—
	H33	400	430–439	430–484	450–454	460–465	—
	H36	1000	1050–1070	—	1050–1082	1085–1102	—
Metal-halide	M57	175	—	—	210	—	—
	M58	250	—	—	292–300	—	—
	M59	400	—	—	455–465	—	—
	M47	1000	1050	—	1070–1100	—	—
	M48	1500	—	—	1610–1630	—	—
High-pressure sodium	S76	35	43	—	—	—	—
	S68	50	60–64	68	—	—	—
	S62	70	82	88–95	95	—	105
	S54	100	115–117	127–135	138	—	144
	S55	150 (55 V)	170	188–200	190	—	190–204
	S56	150 (100 V)	170	188	188	—	—
	S66	200	220–230	—	245–248	—	254
	S50	250	275–283	296–305	300–307	—	310–315
	S67	310	335–345	—	365	—	378–380
	S51	400	463–440	464–470	465–480	—	480–485
	S52	1000	1060–1065	—	1090–1106	—	—

17.4.4.2 Fluorescent Lamps. A trend for lighting new buildings is the increased use of reflectorized fixtures. This trend may be traced to an increase in the number of state and federal lighting efficiency standards in recent years. However, these fixtures can create a "teardrop-like" distribution that may eliminate glare on a computer screen, but that also reduces light to other areas.

Optical reflectors are currently made from silver laminates, anodized aluminums, and enhanced aluminums. Each reflector exhibits different specular and diffuse reflectivity characteristics. Silver laminate is usually selected as the preferred optical reflector material due to its superior reflectivity characteristics. Most film manufacturers now use a thermoset adhesive process, which heats and chemically bonds the reflective material, aluminum backing, and the adhesive. This improved bonding method helps to promote the use of reflectorized fixtures.

The design of the optical reflector is another important performance consideration. At the present time, complex multiple-plane designs that use up to 30 reflective planes can be created using CAD/CAM technology, which permits excellent lighting distribution as well as improved fixture efficiencies.

A study of input energy requirements indicates that the total energy consumption required to operate the lamp and ballasts with silver reflector is 84.5 W and that, with a white paint interior, the total energy consumption is 126.6 W. Higher fixture input wattages produce higher operating temperatures, which effect lamp and ballast performance. The use of reflectorized luminaires has been economically justified on the basis of life-cycle cost for new installations.

17.4.4.3 Specular Retrofit Reflectors for Fluorescent Troffers. Reflectors are available in two basic types: semi-rigid reflectors, which are secured in the fixture by mechanical means, and adhesive films, which are applied directly to the interior surfaces of the fixture. Either silver or aluminum may be used as the reflecting media.

In general, reflectors increase the percentage of lamp lumens that reach the workplane. When the visual appearance of a delamped fixture is unacceptable or when the original fixture has an unusually low efficiency, reflectors may be used as an effective remedy.

17.4.5 Energy-Saving Lighting Techniques

17.4.5.1 Fluorescent System Considerations. Fluorescent lamps are sensitive to ambient temperatures. By using reduced wattage lamps or low-loss ballasts, less heat will be generated and the operating temperature point of the lamp will probably change. The critical area is the coldest spot on the bulb surface. Most fluorescent lamps will peak in light output at around a 100 °F (37.7 °C) cold-spot temperature. For enclosed luminaire types that ordinarily operate the lamps at higher temperature, replacing standard lamps with high-efficiency reduced wattage lamps may result in a net increase in luminaire output even though the reduced wattage lamps are rated for less output than are standard lamps.

17.4.5.2 Economics of Energy-Saving Fluorescent Lamps. Within the limits of conventional design practices, low-watt lamps can provide illuminating engineers with a new flexibility so that they can more closely tailor illumination levels to specific standards or requirements while simultaneously reducing owner's life-cycle costs.

In order to make a fair economic evaluation of more efficient fluorescent lighting systems, it is logical to group the present available products into four different categories in order to consider alternatives to a four 40 W lamp system.

(1) Energy-efficient fluorescent lamps
(2) Electronic ballasts
(3) Power reducing devices
(4) Improved fluorescent reflectors

For categories (1) and (2), detailed discussions have already been presented in 17.4.2 and 17.4.3. Category (3) represents a power reducing device that regulates the flow of current to the bulb once the lamp is lit. It can reduce the lumen depreciation rate and prolong lamp life. There are two variations, namely, respective savings of 30% and 50% of energy consumption. When excessive illumination exists in a building, power reducing devices may be considered as an alternative. Category (4) has been already been discussed in 17.4.4.

With the above four categories in mind, the relative savings and merits of the following options are compared:

(1) Using four energy-efficient fluorescent lamps and keeping the standard ballasts.
(2) Using four energy-efficient fluorescent lamps and new electronic ballasts.
(3) Installing improved reflectors, removing two lamps, relocating existing lamps in the fixture, and keeping one standard ballast.
(4) Installing a power reducing device and keeping the standard lamps and ballasts.

The study results are shown in Table 89.

The annual worth or equivalent net dollar savings per year can be calculated to measure the relative merit for each alternative. This means that a higher number is preferred over a lower number if everything else (other factors as lm/W, color rendition, building occupants' opinion, etc.) is equal. However, if everything is not equal, then specific considerations should enter into the final decision.

17.4.5.3 Using Daylight. Daylight entering a space may be analyzed in terms of the quantity and quality of the light. Daylight may be adequate in quantity to reduce the electrical lighting level; however, its quality should also be analyzed. Poor quality daylight may lead to discomfort and a loss in visibility, which may result in a decrease in human performance and productivity.

Assuming suitable daylight control, a southern exposure is preferred to optimize the daylight contribution into a space. The longhand design procedure involves two steps.

(1) Determination of the quantity of illumination coming to the window surface.
(2) Using that quantity to determine the daylight contribution to the interior space.

Once the contribution of illumination to the window surface has been calculated, two methods are available for determining the illumination contribution to the space, i.e., the point-by-point method, and the lumen method.

Table 89
Energy Savings and Cost Comparisons
for Four Alternative Fluorescent Systems

Alternative System	Total Wattage	Energy Savings (%)	Approximate Cost in 1990 Dollars			
			Lamp	Ballast	Other	Total
(0) Four standard lamps and two ballasts	174	0	$2.27	$15.00	—	$39.00
(1) Four EE lamps and two ballasts	155	10.9	$3.15	$15.00	—	$42.60
(2) Four EE lamps and two electronic ballasts	119	31.6	$3.15	$23.00	—	$58.60
(3) Reflector with two standard lamps and one ballast	87	50.0	$2.27	$15.00	$40.00	$59.54
(4) Power reducing device with four standard lamps and two ballasts	122	30.0	$2.27	$15.00	$30.00	$69.00

17.4.5.4 Lighting Controls. In order to reduce energy consumption, it is essential that minimum acceptable lighting levels be used during off-hours, cleaning periods, periods when only security lighting is required, and for other nonpeak periods as is practical. Energy codes, as contrasted with power oriented codes, may require finer control of lighting; and certain state codes mandate that large area lighting be divided into separate control zones.

The ultimate system of control would be to remotely control every fixture and to program the mode of operation; but, at the present time, this is impractical. Solid-state dimmers are available, or ballasts can be circuited in separate groupings. Three-ballast, three-lamp fixtures are available that will allow for three lighting levels per fixture; but the overall efficiency suffers slightly because of the three ballasts. Solid-state controls are available for dimming entire areas of ballasted lights; but special ballasts are required and the controls are expensive.

Manual control of a lighting system is often the least expensive, but also may be the least effective, alternative.

Automatic controls vary from a simple time clock to a sophisticated computer controlled lighting system. A price versus benefit cost analysis will be required for each installation. However, the system should be programmed for normal operation and have a local manual override. Safety considerations should be made for the possibility of someone being caught in an area when lights are extinguished.

Lighting should be switched in distributed groups (as contrasted to strings) so that areas can be lighted or darkened as conditions change. For further discussion on lighting controls, see Chapter 10.

17.4.5.5 Task Lighting — Other Considerations. One of the most important considerations in building energy conservation is the extent to which task lighting, i.e., lighting from luminaires designed to illuminate only the specific work in progress, can reduce the required levels of general area lighting. Task levels are available in the *IES Lighting Handbook*, 1987 Edition [2].

Task lighting, in addition to significantly reducing the levels of general lighting, can be turned off when not needed; whereas, general area lighting is often continuously required.

The walls and ceilings should be light in color to reflect, rather than to absorb, light rays. A periodic cleaning and lamp replacement program should be considered in the design stage. Many buildings have reduced lamp wattage by as much as 50% with more intensive cleaning programs. Maximum use should be made of natural light, but with the careful consideration of heat loss. The heat from ballasts and lamp losses should be recirculated or removed.

17.4.6 Lighting and Energy Standards. In 1976, the Energy Research and Development Association (ERDA) contracted with the National Conference of States on Building Codes and Standards (NCSBCS) to codify the ASHRAE 90-75 Series. The resulting document was called "The Model Code for Energy Conservation in New Buildings." The model code has been adopted by a number of states to satisfy the requirements of Public Laws 94-163 and 94-385.

There have been several revisions to the ANSI/ASHRAE/IES 90-75 Series since 1976. All were included in the lighting portion of ANSI/ASHRAE/IES 90A-1980,

Energy Conservation in New Building Design, and in EMS-1981, IES Recommended Lighting Power Budget Determination Procedure.

ASHRAE/IES 90.1-1989 [1] is a useful and practical standard for energy-conserving building design and operations. The U.S. Department of Energy published in the *Federal Register* of May 6, 1987, a proposed interim rule entitled "Energy Conservation Voluntary Performance Standards for New Commercial and Multi-Family High Rise Residential Buildings." When issued, this rule will be mandatory for all federal buildings, and a voluntary recommendation for nonfederal buildings.

17.5 Calculating Energy Savings

17.5.1 Introduction. An understanding of utility rates (or tariffs) is important because the cost of electric service is considered when evaluating an ECO. Once the energy saving is determined, it should be assigned a dollar value. In most cases, the average cost of electricity does not accurately reflect the savings per kW or kWh. Therefore, this section describes the important terms, concepts, and application of rates.

17.5.2 Tariffs. Virtually all tariffs have two sections: general rules and regulations, and rate schedules. Written documents are available through the public utility commission or the electric utility. An energy analyst should study the rates of the utility serving the building. However, it is only necessary to thoroughly understand the particular schedules applicable to the buildings involved in the investigation. A careful analysis of the alternate rate schedules and the rules will often, in itself, result in financial return.

The rules and regulations include information regarding billing practices, available voltages, customer/utility responsibilities, voltage regulation, balance and reliability, line extension limits, and temporary service requirements and availability. The rate schedules give the minimum (and maximum) values of usage to qualify for each rate along with the procedure for calculating the cost of electricity. Often, there are charges that apply to all rates, which are listed separately in another section and are called "riders." Some common rates are residential, commercial, industrial, low-load factor, time-of-day, and area lighting. Some common riders are fuel cost, special voltages, and charges for extra facilities, such as redundant services or transformers. Reduced rates may be available for areas in which the utility may have excess facilities (such as in depressed areas) or if the customer undertakes serious demand control implementations, such as ice storage for air-conditioning loads.

17.5.3 Billing Calculations. Each specific rate will contain a means to determine the cost for any or all of the following: kW, kWh, kVA, kvar, load factor, and power factor level. The charge may vary with time-of-day or time-of-year, and minimum service or customer charges may also be included.

The textbook definition of a kilowatt is "a measure of the instantaneous power requirement" (that is, the instantaneous rate of energy consumption). The comparable unit of energy is the kilowatthour (kWh). The billing kW or demand is the highest average rate of energy usage over the billing cycle. Usage is generally averaged for a 15 or 30 minute demand period. The demand for a 15 minute period is determined by multiplying the kWh used in that period by 4.

Some electric rates contain a ratchet clause (which establishes the minimum billing level for a metered quantity following the month of measurement, which typically continues for 3–11 months). The ratchet is generally applied only to the demand portion of the bill.

17.5.4 Declining Block Rate and Example. A declining block rate is one in which energy usage and/or demand is divided into discrete blocks in which the unit costs in each block are constant. Unit costs decrease in blocks representing larger units or usage.

For example, assume that a building used 150 000 kWh and had a peak demand of 250 kW. In a declining block rate, the first 50 kW costs $5/kW or $250, leaving c00 kW for the remaining blocks. The next block takes 150 kW at $4.50/kW or $675, leaving only 50 kW for the next block, which can accommodate 200. The last 50 kW is then billed at $4/kW, which costs $200. The total demand charge is the sum of the charges from each block, or $1125, calculated as follows:

Block 1	50 kW × 5.0 $/kW	= $ 250
Block 2	150 kW × 4.5 $/kW	= $ 675
Block 3	50 kW × 4.0 $/kW	= $ 200
Total	250 kW	$1125 or $4.50 kW average

The charges for kWh are calculated by using the following formula:

	kWh × Dollars/kWh = Charge	
Block 1	50 000 kWh × 0.07 $/kWh =	$3500
Remaining	= 150 000 – 50 000 = 100 000	
Block 2	50 000 × 0.065	$3250
Remaining	= 100 000 × 50 000 = 50 000	
Block 3	50 000 × 0.060	$3000
Total	150 000 kWh for $9750 or 6.5¢ per kWh average	

The total bill is then $1125 + $9750, or $10 875. The average cost of electricity can be expressed in terms of kW or kWh. The electric cost can be expressed as $43.50 dollars/kW or 7.25 cents/kWh. It is important to note that averages cannot be used to determine energy savings. If the demand of this building is reduced by 25 kW without an accompanying kWh reduction, the effect on the bill will be seen only in Block 3 of the demand charges (this rate of charge is referred to as the "tail rate"). The correct energy savings would then be 25 × $4 or $100. By using the average total cost per kW of $43.50, an erroneous savings of $1088 would be shown.

For the economic evaluation of energy conservation purposes only, the additional (rider) costs per kWh, particularly fuel costs, (which may at any time represent a significant percentage of the basic energy cost) may be estimated and added to the energy block rate. During the energy crisis of the '70s, fuel costs became a major issue, although ultimately, on a long-term basis, the utilities usually try to incorporate the major portion of this cost into the basic electric rates.

Most rates include a charge for power factor either

(1) By assuming a power factor in the kW charge
(2) By charging for kVA
(3) By charging for power factors below a given value
(4) By charging for kvars (reactive demand)

In any case, the utility is capturing its cost for supplying vars to its customers. The subject of var flow is beyond the scope of this recommended practice; those interested in more information should refer to Reference [B20].

17.5.5 Demand Rate. A more complex rate is the demand usage block rate in which the size of certain kWh blocks is determined by the peak demand. This rate allows smaller energy consumers to take advantage of the lower kWh charge when their energy usage is high. (The load factor is a measure of energy usage relative to demand. This factor is the ratio of kWh to kW demand times hours per billing cycle. Its value varies between the limits of 0 and 1.0.) The demand usage rate allows a utility to reduce the number of rates and encourages a more consistent level of electric usage. The concept is best described by the following example:

Suppose that a building uses 500 000 kWh and has a peak demand of 1000 kW. The kWh section of their building's rate schedule is shown in the following data:

Blocks	Charge
(1) First 50 000 kWh	7 cents/kWh
(2) Next 200 kWh/kW	6 cents/kWh
(3) Next 300 kWh/kW	5 cents/kWh
(4) All excess	4 cents/kWh

Since the demand is 1000 kW, the amount of kWh in Block 2 is 200×1000 or 200 000; similarly, the amount in Block 3 is 300 000. Once these block sizes are determined, the following calculation (identical to the previous example) can be used:

Block 1 $50\,000 \times \$0.07 = \$\ 3500$
$(500\,000 - 50\,000 = 450\,000 \text{ remaining})$
Block 2 $200\,000 \times \$0.06 = \$12\,000$
$(450\,000 - 200\,000 = 250\,000 \text{ remaining})$
Block 3 $250\,000 \times \$0.05 = \$12\,500$

Total   500 000 kWh $\$28\,000$

Average cost is 5.6 cents per kWh.

17.5.6 Time-of-Use Rate. The cost of new generating capacity for a utility is generally several times that of the cost of existing generating capacity. The difficulty in siting and the long lead time for construction (typically 10–15 years) required for new power plants considered together with the cost may make it desirable for the utility to offer inducements for customer demand reduction. Even for utilities with sufficient existing capacity, economic plant operating practice favors the highest load factors. The time-of-day rate utilizes different rate schedules for different times of day. "Peak load" and "shoulder" (high, but less than peak) loads would have higher cost rate schedules than those of night, weekend, or holiday loads. Economic incentives may also be offered for allowing the utility to automatically shed loads, such as water heating and air conditioning.

Electric meters are available with three sets of registers that record electric usage during various periods. Electronic-type meters permit even finer graduations. Meters equipped for remote (central reading) or local reading (electronically), and meters that generate a recorded tape of usage permit complex rate computations to be produced directly from a computer-based billing system. The utility may offer lower rates for loads that are automatically controlled by the utility on a time-of-day basis, such as for water heating.

17.5.7 Time Value of Money. Most engineers do not have an unlimited budget and, therefore, often need to make evaluations of various options. The energy savings of one or more projects have to be weighed against the "own-and-operate" costs. Since the equipment will usually function for many years and a future savings (or cost) is also involved, the "time" value of the money should also be considered. In order to properly evaluate an ECO, the installed cost of the equipment as well as its operating, maintenance, and energy expenses should be determined on an annual basis over the life-cycle of the equipment. Each annual expenditure (or savings) should be inflated and then discounted to the same base by the appropriate multiplier. If the engineer is unfamiliar with the process; then the anticipated costs by year should be developed by working with a corporate accountant to determine the value of each option to the company. For major projects, the internal rate-of-return, the financial position of the company, and operating considerations may affect the feasibility and scope of any energy-related project.

17.5.8 Evaluating Motor Losses. The energy-saving techniques employed in motor design add to its price. The value of this added price cost is offset by energy savings. The cost of losses evaluation is the process of determining how much additional investment is justified for each kilowatt of losses saved. This section provides a general, simplified approach to this evaluation.

Most commercial building motors have a life expectancy of approximately 10 years if they are constantly operated at or near their rated horsepower. Therefore, the evaluation of losses should cover a comparable period of time. Low-loss motors should have a longer life due to their construction (and lower operating temperatures); but insufficient data are available at this time to include the effect of additional motor life in the loss analysis.

The present value of future loss costs is determined by discounting the cost of 1 kW of losses in each of the years to year 0. The present value of an annuity factor is used for this discounting. The factor is as follows:

$$\text{present worth of an annuity} = \frac{(1 + i)^n - 1}{i(1 + i)^n}$$

where

i = Discount rate in per unit (a 10% rate = 0.1 per unit).
n = The number of years (5 years for motors).

The tail rate (energy or demand rate for the last applicable block from the rate tables) should be used in the evaluation of load and no-load losses. The annual demand cost is easily calculated by using the following formula:

Yearly cost = (kW loss) $\times$ ($/kW) $\times$ 12 months $\times$ (diversity effect)   (Eq 31)

NOTE: This formula assumes a constant demand charge throughout the year, which is rare.

Motor losses are composed of two types of losses: no-load losses, which stay fairly constant, and load (or copper) losses, which vary as the square of the load. Detailed loss information can usually be obtained for large motors (at least several hundred horsepower in size); but only generic information will normally be available for smaller motors. Loss information may consist of total losses at various loads rather than be separated into load and no load. The cost of no-load losses is simple to determine: it is the product of the energy cost times the number of hours that the motor is operated per year. Load losses are more difficult to determine for two reasons.
(1) The load losses vary as the square of the motor load.
(2) The motor load needs to be determined. The motor's load cycle can either be measured directly or be calculated. When total losses are given as a function of load, there is no need to do a no-load loss calculation.

Example— Suppose that a test shows that a motor is used for three 8 hour shifts/day for 50 weeks/year. During the day, the motor is shut off completely for only 2 hours (between shifts). The remainder of the day, it spends 6 hours with no-load connected, 12 hours at 50% loading, 6 hours at 70% loading, and the remaining 4 hours at 100% loading. The tail rate is $0.05/kWh, and the discount rate is 20%.

The no-load loss energy cost is determined by the "on" time.

annual no-load loss energy cost = (22 hours/day) $\times$ (7 days/week)
$\times$ (50 weeks/year) = $\times$ 0.05 $/kWh

annual no-load loss energy cost = $385/kW

The load loss cost is determined by the load cycle which, in this case, is repeated daily.

(1) Duration	(2) Load	(3) Load^2	(1) $\times$ (3) Per Unit Loss
$^6/_{24}$ (0.25)	0	0	0
$^{12}/_{24}$ (0.50)	0.50	0.25	0.1250
$^6/_{24}$ (0.25)	0.70	0.49	0.1225
$^4/_{24}$ (0.1667)	1.00	1.00	0.1667
$^{24}/_{24}$ (1.0)		Total	0.4142

The per unit losses are 0.4142/kW/day. Therefore, the annual load loss can be calculated as follows:

annual load loss energy cost = (24 hours) $\times$ (7 days/week) $\times$ (50 weeks/year)
$\times$ 0.4142 $\times$ 0.05 $/kWh

annual load loss energy cost = $174/kW

The worth of losses for this particular application (ignoring demand cost effects) is found by multiplying the annual costs by the present worth of the annuity factor as follows:

Value (over 5 years) of 1 kW reduction of

(1) No-load loss cost $= \dfrac{(1 + 0.2)^5 - 1}{0.2\,(1 + 0.2)^5} \times \$385 = 2.99 \times 385 = \$1151/\text{kW}$

(2) Load loss cost $\quad= \dfrac{(1 + 0.2)^5 - 1}{0.2\,(1 + 0.2)^5} \times \$174 = 2.99 \times 174 = \$\ 520/\text{kW}$

Assuming equal load and no-load reductions, it would be worth $1671 to increase the efficiency of a 125 hp motor from 92% to 94% (assuming 2 kW loss reduction) in this example.

17.5.9 Transformer Losses. Transformers can be manufactured with efficiencies as high as 98%–99%. Most transformer manufacturers offer a variety of loss designs with associated differences in cost. The manufacturer can determine the optimum design for a given value of losses, which makes it beneficial to include the cost of losses in the bid package. Both load (coil) and no-load (core) loss costs should be included since they each affect design parameters differently.

Transformer losses are determined at 100% load and at a winding temperature of 85 °C (185 °F) for 65 °C (149 °F) rise transformers (75 °C [167 °F] for 55 °C/ 65 °C [131 °F/149 °F] rise) for oil filled units. Corresponding average winding temperatures for 80 °C (176 °F), 115 °C (239 °F), and 150 °C (302 °F) temperature rise transformers are 120 °C (248 °F), 135 °C (275 °F), and 170 °C (338 °F), respectively. The winding loss varies approximately as the square of the load (and varies slightly with operating temperature). The transformer efficiencies at various levels are normally available from the manufacturer.

Had the previous example been a transformer loading situation, the annual load loss value would have been the same (at $174); but the no-load loss value would increase because transformers are energized 365 days/year for 24 hours/day. The new annual no-load losses would be $438/kW ($24 \times 365 \times \$0.05$). The kVA load, and not the kW load, should be used in determining the load losses. Since transformers normally last decades, at least a 10 year evaluation period should be used. Manufacturers have realized the value of reducing no-load losses, and recent developments in core steels have reduced core losses significantly. Even if a complete cost analysis is not performed, it is important to specify the low core losses for the typical commercial transformer that is continuously energized.

17.5.10 Evaluating Losses in Other Equipment. In general, losses associated with currents are a function of the load squared. Magnetic losses in iron core reactors, large magnets, or solenoids are a function of (approximately) voltage squared. It is possible to increase wire sizes strictly for the purpose of reducing loss costs. For example, a 100 foot run of No. 1/0 AWG aluminum has a resistance of 0.021 Ω and a rating of 120 A. The annual cost of losses using the motor example load pattern is as follows (using kW $= I^2R/1000$ and the $174/kW load less cost):

$$\$174/kW \times \frac{150^2}{1000} \times 0.021 = 174 \times \frac{22\,500}{1000} \times 0.021 = \$82/\text{year loss cost}$$

For a three-phase, three-wire circuit, the cost of losses would be three times this value or $246. Use of No. 3/0 AWG aluminum wire at 0.0133 Ω/100 resistance would reduce the three-phase losses to 0.0133/0.210 $\times$ $246 = $156 for a savings in excess of $90/100 feet. It is seldom practical to increase wire size solely for the purpose of energy savings; however, when heavy loads and long runs are involved, it is often necessary to increase the size of conductors for the purpose of reducing voltage drop. In systems with a neutral, if the loads on a feeder were high in harmonics (as with discharge lighting or with the use of silicon controlled rectifiers), then the power losses in the neutral should be considered (as an approximation, the losses would be multiplied by 4, where the neutral is effectively loaded).

17.6 Load Management

17.6.1 Introduction. Any energy-conscious engineer should attempt to exert control over a building's energy usage (kWh) and the rate of its usage (kW). The simple fact is that no energy is used when equipment is shut off. Therefore, one of the first jobs of an energy engineer is to make sure that unused, redundant, and idling equipment is shut off.

Early demand controllers were tied into a meter-pulse system and began shutting down equipment when it appeared that a preset demand would be exceeded. While this procedure can significantly reduce electric costs, there is some question as to the amount of energy saved (or added). A second generation of controllers has increased the effectiveness of this system by shedding or delaying operation of all nonessential load in addition to keeping the demand under a preset level.

The following sections briefly explain various types of controllers and how to design a proper system. While this discussion does not include any details about the hardware needed to implement an energy conservation program, this chapter does provide sufficient information to properly direct the efforts of engineers through the use of publications listed in other chapters of this book, and the appropriate codes and standards.

17.6.2 Controllers. The control function can be performed by many types of systems, which range from a simple manual system to a sophisticated computer system. The energy engineer should match system needs with equipment capabilities to determine the optimum choice. In many cases, significant energy, demand, and cost savings can be achieved by prudent operation of equipment or mechanical interlocking. The demand controllers may operate to automatically reduce load, for example, to directly control the air-conditioning compressors.

The simplest form of controls includes time clocks that switch loads based on a predetermined time schedule. Photocells are light activated controllers that can be used in conjunction with a time clock or other devices. Environmentally controlled switches and sensors that include infrared or ultrasonic presence detectors are also effective. Devices can either work independently or jointly to control energy usage. However, the following considerations should be made when applying a controller:

(1) The operation of the equipment in an automatic mode should not endanger anyone near the equipment or inadvertently interrupt any process.

(2) The controller should be periodically checked to see that it is operating as planned and has not been defeated.

(3) In the case of time clocks, the time should be checked and the time control adjusted to compensate for changing seasons and conditions.

(4) The controlled equipment should be capable of withstanding the planned number of starts and stops.

There are five types of demand controllers, four of which require a utility demand meter pulse. In any case, one should first recognize that any controller or meter calculates the maximum demand by averaging the kWh over a set interval (a 15 minute demand interval would indicate the kWh for 15 minutes multiplied by 4 since there are four 15 minute periods in an hour). The five controller schemes are noted below (see IEEE Std 739-1984, IEEE Recommended Practice for Energy Conservation and Cost Effective Planning in Industrial Facilities (ANSI) [B1] for a complete description).

The five types of demand controllers are

(1) Instantaneous — Controls loads at any time during an interval if the rate of usage exceeds a preset value.

(2) Ideal Rate — Controls loads when they exceed the set rate but allows a higher usage at the beginning of the interval.

(3) Converging Rate — Has a broader control bandwidth in the beginning of the interval but tightens control at the end of the interval.

(4) Predictive Rate — The controller is programmed to predict the usage at the end of the interval by the usage pattern along the interval and switches load to achieve the preset demand level.

(5) Continuous Interval — The controller looks at the past usage over a period equal to (or less than) the demand interval. Loads are switched so that no time period (of an interval's duration) will see an accumulation of kWh that exceeds the preset value. This controller needs no utility meter pulse.

Before any of the above controllers can be installed, a load survey should be made. This survey is, in essence, an equipment/process audit and can be done in conjunction with the audit described in the next section.

Demand controllers may be incorporated into the facility master control system. In this case, signals have to be sent or telemetered to the control system. Various software packages are available to perform the energy control functions, data logging, and visual alarm and display. Chapter 14 covers the building control system.

17.6.3 Equipment Audit and Load Profile. Each system and piece of equipment shall be surveyed to find out which loads can be switched off and to what extent they can be switched. The engineer shall evaluate any loss of equipment life or mechanical problems associated with switching each load. The survey consists of, but is not limited to, the listing of equipment into the following four categories:

(1) Critical Equipment — This equipment is required at all times or it needs to be controlled in the present manner for occupancy, safety, or other reasons. In commercial buildings, where public safety is involved and where life safety

considerations are required, the integrity of the power and control systems overshadows energy-saving considerations.

(2) Necessary — While this equipment is required for occupancy (or other reasons), it can be shut down at some measurable financial loss during extreme conditions.

(3) Deferrable — This equipment is important but can be turned off for varying periods of time. Some load may even be switched, virtually at will, provided that some minimum on time is allowed.

(4) Unnecessary — This equipment has usually been used or left on even though it is not needed. This equipment should be shut off and periodically checked. Sometimes, equipment is used only occasionally and the user fails to de-energize after use, so an indicator or semiautomatic controller may be applied to control the equipment.

Once the loads are recorded and analyzed, a proper control method can be established with the help of a load profile. The load profile is developed with a graphic meter for at least a week (continuous). The profile is then analyzed along with the equipment audit to determine the target demand. Kilowatt recorders are available for rental or purchase (since the use of continuous monitoring is highly desirable). A recording ammeter can also be a good indicator if the power factor is known. The off-peak load pattern can be as revealing as the peak load pattern. Recording may have to be performed several times each year to account for seasonal weather or operation patterns.

In the absence of a load profile, some judgment needs to be made as to the amount of equipment that should be kept running at the estimated time(s) of peak load. This estimate is then made a target or target demand for the controller. However, the designer should also plan to control the use of equipment so that it is de-energized when it is not needed.

17.7 Efficiencies of Electrical Equipment

17.7.1 Losses. All electrical equipment has some type of loss associated with its use. There are five different types of losses that should be considered in determining the optimum operating point for a piece of equipment.

(1) Resistive Losses — Are associated with the flow of current through a copper or aluminum conductor. These losses are generally a function of the square of the current, which can be calculated by using the equation $P = I^2R$. However, the energy engineer also has to recognize the temperature component of the conductor because increased current will invariably increase the operating temperature of a device. Solid-state devices have a constant voltage drop when they are conducting; so the power current function is essentially linear. It should be noted that resistive losses are increased because of the skin effect of conductors.

(2) Magnetic Losses — Are associated with motors, transformers, reactors, regulators, and solenoids. These losses are usually a function of the voltage squared (approximately) and consist of hysteresis, eddy current, and mutual induction losses.

(3) Motion Losses — Are produced as the equipment operates. These losses include friction loss from bearings, windage, and system restrictions.

(4) Mechanical Losses — Are reflected in the electric circuit's power requirements. These losses include inefficiencies associated with transmissions, eddy current clutches, and speed control devices (which can even be in the electric circuit).

(5) A combination of factors will cause additional or unnecessary losses if a piece of equipment is operated above its design limits. Operating above rated capability can cause overheating (and associated loss costs) as well as result in the destruction of equipment. Operating equipment too far below rated capacity wastes capital dollars, causes an increase in the no-load portion of the losses, and lowers the power factor. The key to energy engineering is to match the device to the load and the power supplied to the device.

17.7.2 Efficiency. Textbooks define efficiency as "the power (kW) output divided by the power (kW) input at rated output." The percent efficiency is 100 times this value. This calculation method is not appropriate for most energy evaluations. The efficiency of a device for any energy engineering effort should be considered over its entire cycle of operation. In an energy evaluation, the following expression applies (with the assumption that the output is converted to kWh):

$$\% \text{ energy efficiency} = \frac{\text{kWh out (over operating cycle)}}{\text{kWh in (over operating cycle)}} \times 100 \qquad \text{(Eq 32)}$$

Most equipment is given an efficiency rating at the nameplate or full-load condition. The device may then be used under different conditions, which makes the nameplate value of efficiency incorrect for the applied device.

17.7.3 Oversizing Electrical Equipment. The cost of losses may be high enough to justify the installation of wiring or equipment that exceeds the ampacity requirement of a particular circuit in oversize motors or transformers. In many cases, there will be virtually no change in the cost of the feeder over current device, the conduit, the pull boxes, and the receiving panel. However, in other cases, the added cost of equipment upsizing will be significant.

17.7.4 Motors. Motors represent the most significant portion of the energy consumed in commercial buildings. Motors are fairly efficient devices at rated load. In general, three-phase motors are more efficient than single-phase motors, and larger motors are more efficient than smaller ones. There is only minor improvement in efficiency above 200 hp, and the knee of the efficiency versus size curve occurs at about 10–15 hp. The peak efficiency of a motor occurs at full load with about 105% (of nameplate) balanced voltage at its terminals. However, as load is reduced from nameplate, optimum efficiency occurs at a lower voltage (see IEEE Std 739-1984 (ANSI) [B12] for further information).

Motor voltage unbalance will increase motor losses, due to the negative-sequence voltage that causes a rotating magnetic field in the opposite direction of motor rotation. A 2% voltage unbalance will increase losses by 8%, a 3.5% unbalance will increase losses by 25%, and a 5% unbalance will increase losses by 50%. The power factor of most three-phase motors is between 80%–90% at full load and decreases as load is reduced. The installation of power factor correction (to 95% or so) at motor terminals will accomplish two tasks.

(1) An improved power factor will decrease current requirements, thereby reducing I^2R losses in the supply line.

(2) More importantly, the use of capacitors at the motor will improve voltage regulation by increasing the voltage level when the motor is in use.

Large banks of unswitched capacitors can cause problems from several different aspects (see IEEE Std 739-1984 (ANSI) [B1] for further information) and are, therefore, not recommended as a first choice. If large banks of capacitors exist (or are planned), they should be switched as a function of building load. Switching of capacitance causes transient overvoltage that can adversely affect sensitive electronic equipment.

New high-efficiency motors are available, and their cost (20%–50% higher than standard models) is usually justified. However, each application should be evaluated. These motors achieve a higher efficiency by using better grade steel, special low-friction bearings, larger copper windings, closer tolerances, and smaller air gaps. These motors have the added benefit of a longer life because they run cooler than low-efficiency models.

Motor speed control takes several forms. The earliest methods of motor speed control involved the use of resistors to reduce the amount of voltage in motors. The losses of this type of system are easily determined and usually quite high. Modern techniques include voltage control with thyristors, such as semiconductor controlled rectifiers. These devices are more efficient; but they supply a somewhat distorted voltage supply, and the devices themselves have losses. The most sophisticated speed control is an adjustable speed drive that varies the frequency and voltage levels of a synthesized ac voltage wave. This drive does not synthesize a pure voltage wave; so losses do occur as a result of the harmonic content. The peak efficiency (typically, 86.5% for motor plus drive) of the new systems generally occurs at full speed and full load on the motor, and efficiency declines to 25% at 25% of rated speed. Electronic variable speed devices now range in cost from $180–$500/hp at 100 hp; but costs are steadily decreasing. Mechanical variable speed techniques include eddy current couplings and variable ratio belt drives that now cost from $50–$70/hp at the 125–150 hp level. Efficiencies of eddy current drives approach 90% at rated speed and 25% at 30% of rated speed. Variable ratio belt drives are limited to a 3:1 speed ratio range and maintain an 85%–90% efficiency over that range.

In any case, a controlled rectifier system causes voltage harmonics (including those from notching) on any power system. The use of an oscilloscope or a power quality instrument is recommended in any thyristor or switching inverter power supply application to evaluate the need for filtering at the source of the notching. Harmonics can cause capacitor and other equipment failure when large amounts of capacitance are on the system or the system is at a low-load condition, or both.

All devices that generate harmonics, such as discharge lighting or controlled rectifiers, on a system containing motors will cause increased motor losses because

(1) The higher frequency components are generally associated with higher losses; the effective resistance of conductors is even increased.

(2) The apparent power factor of the circuit is reduced when any form of "wave chopping" occurs.

(3) The harmonics perform no "work" in machinery, but incur losses and consequent heating and, for certain harmonics, may develop negative-sequence action (placing a heavier "load" on the motor).

17.7.5 Transformers. Transformers are very efficient devices. Their loss evaluation was covered in 17.5.9. The load losses are a function of the square of the load kVA to nameplate kVA ratio (modified for the change in resistance as the load increases). The no-load losses are a function of voltage squared.

The losses on double-ended substations, which are designed for normal feed from both sources, is reduced by using only slightly larger than normal transformers. In most installations, both transformers are evenly loaded, and the time that one transformer has to handle the full substation load is only in the event of contingencies. Transformers will handle overloads for hours with some loss of expected life. (See the IEEE C57 (ANSI) collection for information on load versus loss of life calculations.)

During the periods when one transformer is handling the entire substation load, the increased losses over the course of a year are relatively small and the effects of the overload can be reduced by as much as 50% by the use of supplementary fan cooling.

Double-ended substations have been used for many years to increase reliability. The decreased load loss from shared load is illustrated as follows:

1 unit of full load $\quad$ = 1 per unit loss

2 units at half load $\quad = \left(\dfrac{1}{2}\right)^2 + \left(\dfrac{1}{2}\right)^2 = \dfrac{1}{4} + \dfrac{1}{4} = \dfrac{1}{2}$ per unit losses

17.7.6 Capacitors. The use of capacitors can be a significant energy saver if they are properly applied. A capacitor bank is also a load (a 750 kvar bank draws almost 1000 A at 480 V), so it should be disconnected when var support is not required. If a fuse blows on a large capacitor, an unbalanced voltage will occur along with resultant increases in motor losses. Therefore, the fuse integrity of capacitor banks should be monitored. High harmonic content in the power supply has been known to cause either capacitor failure or unplanned operation of protective devices.

17.7.7 Equipment Overview. Equipment should be operated as nearly as possible to its nameplate rating, in terms of both load and voltage. The voltage should be well balanced and well regulated in an efficient manner whenever possible. Capacitors should be carefully applied and switched. Unnecessary and underloaded equipment should be removed or replaced. The system should be checked for voltage level, balance, and harmonic contact. Energy-saving devices should be considered if their worth has been scientifically established. Their function should also match the concepts discussed in this section.

17.8 Metering. Metering provides the opportunity to monitor and control the rate of energy consumption in an electric system. Accurate and complete kW, kWh, and power factor information is needed for a good energy audit as well as for continued measurement and control for an energy conservation program. It is worth noting that good metering is virtually worthless without a program of meter reading, data recording, data analysis, and a guide for action. (High-technology, solid-state devices are available for reading and summarizing a broad range of electrical parameters, such as power, voltage, current, and harmonics. Use of these devices makes periodic readings unnecessary.)

Meters themselves save no energy; but thoughtful use of metering information does save energy and money. Ideally, metering should be installed as the electric system is being built. However, metering is often an afterthought. There are several cautions that should be observed from an engineering standpoint:

(1) Relay-rated potential transformers and current transformers shall not be used for metering because their accuracy is not sufficiently high at normal loads. When metering is utilized for energy conservation purposes as distinct from revenue functions, then the required accuracy is usually on the order of several percentage points lower than that needed for billing purposes.

(2) The meters should be installed at a point where they can be easily and safely read.

(3) The meter should be out of the way of normal traffic (and even of doors used only on occasion) so that their glass covers, which normally protrude 18 inches from the surface, are not broken.

(4) Clamp-on current transformers may be required in areas where 24 hour production is required because the system cannot be shut down long enough to install donut-type current transformers.

(5) Current transformers and potential transformers have insulation voltage ratings that exceed the anticipated voltage level of the system on which they are being installed.

All appropriate safety requirements shall be considered before purchasing or installing the equipment. Portable instruments can be used to provide the input versus output energy for various machines and processes. Since the slip is essentially a linear function of the load on an induction motor from no load to full load, a digital readout tachometer can be used to determine the output horsepower of a motor. The electrical input effect should be translated to nameplate conditions (see IEEE Std 739-1984 (ANSI) [B1]) to acquire more accurate information. Recorded metering information will reveal differences due to the varying energy demands of the electric system. The metering results should be plotted (as noted in 17.3.4), and the reasons for variation should be sought. Valuable information can be gleaned by meter reading as frequently as every day. However, usage trends tend to be more indicative of improvement or lack of improvement in energy conservation.

Single reading variations may indicate one of the following:

(1) Abnormal conditions
(2) Faulty or failing equipment
(3) Defeat of energy-conserving controls
(4) Meter error
(5) Meter reading error

(6) New occupancy requirements

(7) Installation and operation of new equipment

Lack of any changes in the kW or kWh rate may indicate that the demand indicator was not reset or that the meter is not being read or is defective. In any case, the meter should be checked and readings verified.

Changes in the trend of usage or usage that is different from planned targets can signify any of the following:

(1) The targets are too high or too low.

(2) Something other than activity level affects energy usage.

(3) The energy conservation methods are (or are not) working.

Regardless of the metering plans used, all metering information should be recorded under the following (or similar) headings:

(1) Date

(2) Previous meter reading

(3) Present (new) meter reading

(4) Difference

(5) Meter multiplier

(6) Kilowatthour consumption

(7) Demand indicator reading

(8) Indicator constant

(9) Peak kW for current (just read) period

Extensive metering data are best handled as part of a computerized data base. Any of the modern commercial data base systems can be used to accumulate and analyze the data, as well as to prepare presentation materials (including graphs).

17.9 Operations. An energy conservation program should be developed in commercial buildings. Building designs should incorporate provisions for energy savings. Some of the items that should be considered in developing designs for commercial buildings follow.

17.9.1 Metering of Tenant Areas and Building Operations. When tenants are to occupy commercial facilities, it is highly desirable to meter tenant areas directly. In some states, submetering or the resale of electricity is permitted; while in other states, only check metering is permitted. Where submetering cannot legally be used to charge tenants, periodic load surveys may provide for the adjustment of utility billing based on power consumption, provided that the lease contains such provisions. Tenants who pay directly for energy will be less likely to waste it and will be more likely to monitor the use of appliances by their employees. In many cases, it is desirable to meter sections of buildings or operations so that charges can be allocated to the various areas in order to emphasize the need for responsible energy usage of all groups.

17.9.2 Interlocking or Key Switches. To prevent unauthorized use or tampering with security lighting and power circuits, the use of interlocking or key switches should be considered. On the other hand, in individual office areas, use of key switches may prevent turning off lights by the people that use the offices. Placing switches in locked areas or electrical closets is equivalent to using key switches and can negate the desire of an energy-conscious employee to help save energy.

17.9.3 Load Control. Programs of minimum, maximum, and off-peak operation of major items of equipment should be developed. Plans for the most energy-conserving usage of air-conditioning machines and other electrically driven equipment, including air-handling units, should be developed. In some cases, systems are essentially divided into multiples that permit the loss of a single unit without the overall loss of the system. In these cases, it might very well be practical, as part of a major energy-conserving effort, to shut down one of these multiple units for extended periods as an enforced energy-saving measure. Many areas of buildings are very lightly used or unused during portions of the day. Provisions should be made for automatically (or in some case, manually) shutting off power to nonessential areas as part of the normal duties of a watch engineer.

Often the control of such systems can be incorporated into the building management system control computer, which is generally accepted today as a standard in large buildings. These computers are equipped with pre-programmed packages that can be adapted to almost any use in a modern commercial building. For smaller installations and where technical assistance is available, the programmable controller, in conjunction with suitable interfacing and remote supervisory equipment, forms an excellent method for providing very flexible building control. The programmable controller has the advantage of utilizing the relatively simple development of software by untrained personnel for systems up to medium complexity. To ensure continued proper operation, the computer and its subsystems should have provisions for minimizing the effects of power loss or other power supply problems.

17.9.4 Load Shedding. The building staff should prepare for power reductions, as required by either the utility or governmental energy agencies. In the event that brown-outs become part of our way of life in an effort on the part of utilities to reduce electrical consumption, arrangements can be made for shedding noncritical loads on a normal operational basis. Load shedding involves dropping a number of loads, perhaps in stages, from the system on a planned basis. A well-developed plan of operation for load shedding will make it possible for operating personnel to routinely reduce load requirements without a panic situation developing, without serious degradation of service to the public, and without the loss of critical equipment.

17.9.5 Load or Peak Shaving with On-Site Generation. Until recently, much of the responsibility for energy management has been directed toward load shedding. However, the economic advantages to the consumer that results from the loss of these loads are sometimes limited. An alternate or additional approach is to utilize on-site power. This approach should be undertaken only if utility rules and other regulations permit, and limitations (operating and equipment) are followed. The demand reducing loads are almost always transferred to on-site generators. No attempt should be made to parallel the on-site generators and the utility in this type of operation.

The practice of utilizing on-site generator sets to reduce electric utility demand charges is becoming an economically attractive proposition in some regions of the United States. Previously, motor-generator sets were purchased primarily for their value in an emergency situation, with little consideration or expectation of achieving operational savings. Now, installation of motor-generator sets is being considered on strictly an investment basis. In addition, the existence of emergency systems has increased considerably during the past decade.

Peak load transfer is one method of improving the load factor by selecting loads to be automatically transferred from the utility source to an on-site generator. The amount of power shaved from the utility peak is dependent on the load or loads connected to the selected transfer switch(es). Care should be exercised in evaluating the optimum demand setting to ensure adequate peak shaving with minimum loading of the on-site generator. This transfer method, by its nature, requires no interface or coordination with the utility power source, since there is never a possibility of interconnection.

Figure 156 shows a peak load transfer system that provides power to two mechanical loads, each fed by its own automatic transfer switch. The emergency system is comprised of two motor-generator sets paralleled onto a common emergency bus. The system also feeds two other transfer switches that are designated as emergency loads. The system may be designed so that the peak shaving loads are not emergency loads. If this is the case, and a normal failure occurs during a peak demand mode, the peak shaving loads are immediately disconnected from the emergency bus, which allows the emergency loads to be transferred.

17.10 Energy Conservation Equipment. There are a variety of energy conservation devices currently on the market. The designer should be aware that many of

Fig 156
Peak Demand Reduction Using
Multiple On-Site Generators

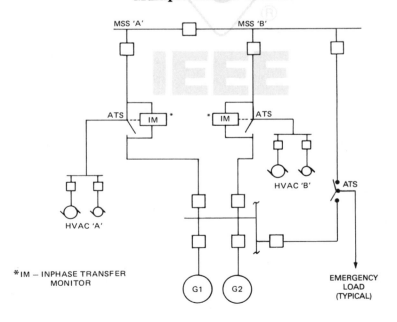

these devices are poorly designed, unrealistic in application, or even fraudulent. A designer should not attempt to assemble complex components, under any circumstances, into a functioning system unless he or she has specific knowledge of the application of each device. Makeshift application of computers frequently turns out to be disastrous.

One of the more difficult areas of design is to obtain suitable interfaces between the application devices, such as motors, heating units, and other utilization devices, and the relatively low-energy control systems. The interfacing of remote control systems and application devices requires careful coordination between the mechanical equipment or motor control center, manufacturer, and the manufacturer of the supervisory control equipment. One major problem is the compatibility of control devices, the controlled equipment, and the signal system.

When information is transmitted to remote power equipment, it may be wise to check to find out if the device has functioned as required. It is good to know if someone turned the lights on in a local area, or if a piece of equipment was started locally without computer or other automatic control intervention. Therefore, feedback from the controlled areas or equipment to the motor control center is desirable.

Listed below are some energy conservation devices and concepts that may be utilized. Some are basically energy demand reduction devices, while others are more concerned with overall energy conservation.

(1) Load limiters or demand limiters are devices that are programmed to operate building loads in sequence or in a manner such that the billing demand remains at an optimized value. Such devices can be used to provide alarms when the rate of energy usage exceeds established levels.

(2) Use of automated devices for shutting down or reducing the level of operation of nonessential equipment. Multiple speed or variable speed equipment with regulator or feedback control can materially reduce energy requirements. This can be combined with other simple devices, such as photocell, infrared, or ultrasonic control of lighting, and then integrated with the computerized controls of the heating and ventilating system.

(3) Use of waste heat, including that from lighting fixtures, as part of the space-conditioning system.

(4) Energy can be recovered in vertical transportation equipment by utilizing regenerative systems. A descending elevator, for example, can feed back energy into the power system.

(5) Use of high-efficiency motors, drives, belts, and power factor correcting ballasts will minimize line and equipment losses. Power factor correcting equipment (i.e., capacitors, synchronous motors) and the proper sizing of induction motors will serve to maintain the facility power factor at high values with minimum losses.

17.11 References. The following references shall be used in conjunction with this chapter:

[1] ASHRAE/IES 90.1-1989, Energy Efficient Design of New Buildings Except New Low-Rise Residential Buildings.

[2] *IES Lighting Handbook*, 1987 Edition.

17.12 Bibliography. The references in this bibliography are listed for informational purposes only.

[B1] IEEE Std 739-1984, IEEE Recommended Practice for Energy Conservation and Cost Effective Planning in Industrial Facilities (ANSI).

[B2] Bonnett, A. H. "Understanding Efficiency in Squirrel-Cage Induction Motors," *IEEE Transactions of Industry Applications*, vol. 1A-16, no. 4, Jul./Aug. 1980, pp. 476–483.

[B3] Brown, R. J. and Yanuck, R. R. "Life Cycle Costing," Commonwealth of Pennsylvania, 1979.

[B4] Chen, K. "The Energy Oriented Economics of Lighting Systems," *IEEE Transactions on Industry Applications*, Jan./Feb. 1977.

[B5] Chen, K. and Lally, W. "Update: Fluorescent Lighting Economics," *IEEE Transactions on Industry Applications*, May/Jun. 1983.

[B6] Cunningham, R. L. "Harmonic Pollution on an Industrial Distribution System," *Conference Record*, IEEE Industry Applications Society Annual Meeting 1980, no. 80CH1575-0, pp. 405–408.

[B7] Gibbons, W. P. "Analysis of Steady-State Transients in Distribution Low Voltage Power Systems with Rectified Loads," *IEEE Transactions on Industry Applications*, vol. IA-16, no. 1, Jan./Feb. 1980, pp. 51–59.

[B8] Grant, E. L. and Ireson, W. G. "Principles of Engineering Economy," Fourth Edition, New York: The Ronald Press Company, 1960.

[B9] Harkins, H. L. "Cogeneration for the 1980s," *Conference Record*, IEEE Industry Applications Society Annual Meeting 1978, 78CH1346-61A, pp. 1161–1168.

[B10] Helms, R. N. "Illuminating Engineering for Energy Efficiency," New York: Prentice-Hall.

[B11] Hickock, H. N. "Electrical Energy Losses in Power Systems," *IEEE Transactions on Industry Applications*, vol. 1A-14, no. 5, Sep./Oct. 1978, pp. 373–386.

[B12] IES LEM-1-1982, Recommended Procedure for Lighting Power Limit Determination.

[B13] Kovacs, J. P. "Economic Considerations of Power Transformer Selection and Operation," *IEEE Transactions on Industry Applications*, vol. 1A-16, no. 5, Sep./Oct. 1980, pp. 595–599.

[B14] Linders, J. R. "Electric Wave Distortions — Their Hidden Costs and Containment," *IEEE Transactions on Industry Applications*, vol. 1A-IS, no. 5, Sep./Oct. 1979, pp. 458–471.

[B15] McCully, R. A. "Development History of More Efficient Lamp Design," *Energy Engineering*, vol. 87, no. 1, 1990.

[B16] NEMA/NECA Publication "Total Energy Management—A Practical Handbook on Energy Conservation and Management," second edition, National Electrical Manufacturers Association and National Contractors Association, 1979.

[B17] Sieck, R. E. and Schweizer, P. F. "An Assessment of Energy Cost Trends on Electrical Equipment Selection for Petrochemical Industry," *IEEE Transactions on Industry Applications*, vol. 1A-16, no. 5, Sep./Oct. 1980, pp. 624–632.

[B18] Smith, C., editor. "Efficient Electricity Use—A Reference Book on Energy Management for Engineers, Architects, Planners, and Managers," New York: Pergamon Press, 1978.

[B19] Stebbins, W. L. "Implementing an Energy Management Program," *Fiber Producer*, Atlanta, GA: W. R. Smith Publishing Company, vol. 8, no. 5, Oct. 1980, pp. 44–47.

[B20] Stevenson, W. D. "Elements of Power System Analysis," Fourth Edition, New York: McGraw-Hill, 1982.

17.13 Projects. The following documents will be helpful to the reader in locating specific projects to help reduce energy and costs (see 17.2.2):

[P1] *ASHRAE Journal*, American Society of Heating, Refrigerating, and Air-Conditioning Engineers, Atlanta, GA (monthly).

[P2] *Consulting-Specifying Engineering*, Cahners Publishing Company, Des Plaines, IL (monthly).

[P3] *Electrical Construction and Maintenance*, Intertec Publishing, 888 7th Avenue, 38th floor, New York, NY 10106 (monthly).

[P4] *Energy Engineering*, Association of Energy Engineers, Atlanta, GA (quarterly).

[P5] *Energy Insider*, Department of Energy (DoE), Office of Public Affairs, Washington, DC (weekly).

[P6] *Energy User News*, Fairchild Publications, 7 East 12th Street, New York, NY 10003 (weekly).

[P7] *Lighting Design and Application*, Illuminating Engineering Society, 345 East 47th Street, New York, NY 10017 (monthly).

[P8] *Plant Engineering and Power Engineering*, Technical Publishing, 1350 East Tougy Avenue, P.O. Box 5080, Des Plaines, IL 60017-5080 (bi-monthly and monthly).

[P9] *Power Engineering*, McGraw-Hill, 1220 Avenue of the Americas, New York, NY 10020 (monthly).

Index

A

AC motors, speed control of, 291-292
AC/DC resistance ratio that exists in a 400 Hz
    system, 340
adding to existing systems, 91-92
air-conditioning systems
  general discussion of, 83-85
  heat generated from, 553
  types of, 553
aluminum
  conductor connection, procedures for, 351
  properties of, 318
apparatus closet
  general discussion of, 595-596
  specifications, 595
application techniques, lighting
  non-uniform, 507-513
  task-ambient, 513-519
  uniform, 506-507
approved appliances and equipment, use of, 53
arcing fault currents in per unit of bolted values,
    approximate minimum values of, 409
audio communication systems
  dictation systems, 605
  general discussion of, 603-604
  intercom systems, 604-605
  public address systems, 604
  sound reinforcement systems, 605-606
automatic load transfer devices, 255-256
automatic transfer
  circuit breakers, 255
  switches, 254-255

B

ballasts
  energy-efficient
    fluorescent, 498-499, 731-732
    high-intensity discharge (HID), 501, 732
  grounding of, 499
  low-loss, 499-500
  sound of, 539
  temperature of, 500
  voltage of, 500
braids, properties of, 332
building
  access, 60
  lighting subsystems, 503

load, estimation of, 94
loading, 60
building extension with
  low-voltage service (example), 667-670
  medium-voltage service (example), 666-667
burnouts, lamp, 500
bus bracing requirements, 471-472
busway
  construction, 377-378
  feeder, 378-380
  field testing of, 390
  general information about, 377
  installation, 389-390
  layout, 387-388
  lighting, 380-382
  over 600 V (metal-enclosed bus), 391-392
  parameters (typical), 389
  plug-in, 379-381
  ratings as a function of short-circuit power
    factor, 383
  selection and application of, 383-386
  standards, 382-383
  thermal expansion in, 386
  trolley, 382
  versatility, illustration of, 379
  voltage drop in, 384-386
  welding loads, 386
  when crossing building expansion joints, 386

C

cable
  bus, 343-344
  design, 322-328
  elevated ambient temperature, 339
  general construction, 316-328
  hot-spot temperature criteria, 339
  installation, 339-347
  low-voltage, 322-324
  medium-voltage, 324
  outer finishes, 328-332
  ratings, 332-339
  single-conductor and multiconductor
    construction, 331
  specification, 376-377
  systems, 315-316
  systems and support structures, 598-599
  termination criteria, 339
  terminations, 354-361

tray, 342-343
uprating for short-time overloads, 336
cable connectors
  jacketed and armored, 361-362
  performance requirements of, 362
  separable insulated, 362
cable dc test voltages (kV) for installation and
  maintenance, 370
cable faults
  equipment and methods, 374-376
  influence of ground-fault resistance, 373-374
  locating, 373-376
cable installation
  aerial, 341-342
  cable bus, 343-344
  cable tray, 342-343
  conduit, 344
  direct burial, 344-345
  hazardous (classified) locations, 345-346
  layout, 341
  open runs, 342
  open wire, 341
  procedures, 346-347
cable limiters (protectors), 245
cable ratings
  conductor selection, 332-333
  emergency overload criteria, 335
  fault current criteria, 337
  frequency criteria, 337-339
  load current criteria, 333-334
  voltage, 332
  voltage-drop criteria, 335
cable systems
  grounding of, 365-366
  protection from transient overvoltage, 367
  sheath losses in grounded, 366
cables, service entrance, 593-594
calculation of
  voltage dips, 138-139
  voltage drops, 124-133
Canadian voltage standard, 106
ceiling distribution systems, 591-592
ceiling system providing illumination, air supply,
  air return, space partitioning, and acoustical
  treatment, 504
central monitoring and control equipment,
  616-620
central supervisory station in a FAS, 634
circuit breaker protection, 472
circuit breakers
  current-limiting, 476
  low-voltage power air-type, 234
  molded-case, 239-241
  secondary transfer using, 191
  service transfer using, 190
  time current curves for 125-600 A molded-
    case, 449
  time current curves for 600-4000 A power,
    450
  tripping characteristics, selection of, 234-235
circuit cables, power limited, 324
circuits, low- and medium-voltage, 300-303
closets
  apparatus, 595-596
  satellite, 596

CMMS, see computerized maintenance
  management system (CMMS)
commercial buildings
  general description of, 43
  grouping of, 45
communication and signal system planning
  cable plant, 590
  general discussion of, 589
  riser conduits and sleeves, 590-591
communication systems
  audio, 603-606
  image, 606-608
communications, life safety, 628-629
computerized maintenance management system
  (CMMS), 634-636
conductor sizes, minimum for indicated fault
  current and clearing times (in AWG or
  kcmil), 338
conductor temperature ratings, rated, 321
conductors, elevator
  current ratios of, 576
  disconnecting means of, 576-577
  diversity factor for, 573-577
  operation diversity of, 576
  size of, 573-574
conductors
  comparison between copper and aluminum,
    317-318
  general, 316-317
  insulation, 318-322
connected load and maximum demand by
  tenant classification, 96
connectors
  for aluminum, 348-352
  for cables of various voltages, 352
  operating requirements, 352-353
  performance requirements, 352
  types available, 347-348
contactors, remote control, 257-260
contracts, 60
control and monitoring of pollution and
  hazardous waste, 631-632
control equipment
  central, 616-620
  transportation and traffic, 631
control systems
  general discussion of, 276-281
  security and access, 603
controllers
  fire pump, 284
  low-voltage, 281-282
  medium-voltage, 284-285
  multiple-speed, 282-284
  programmable, 289-290
  protection, 271-276
  special features of, 276
  starting, 266-271
coordination
  examples, 448, 451-460
  mechanics of achieving, 447-451
  of protective devices, 447
  study, preliminary steps in, 447
coordinating fusible unit substation,
  464-466
copper, properties of, 318

correction of voltage regulation and power
    factor, 207-208
current charts, peak let-through, 469-471
current-limiting
    action of a current-limiting fuse, 442
    busways, 442-444
    circuit breakers, 476
    fault, 305-307
    fuse peak let-through curves, typical, 443
    fuses, 441, 465-469
    reactors, 441-442

D

data
    center installations, 600-602
    facilities, 597
    terminals, 597-598
    transmission and network equipment, 599
daylighting and electric lighting, 525
DC motors, speed control of, 290-291
DDC, *see* direct digital control (DDC)
definitions used throughout IEEE Std 241-1990,
    63-71
design influences, interrelated utility and
    project factors, 156-157
devices, load transfer, 252-256
dictation systems, 605
direct digital control (DCC)
    distributed networks, 624-625
    energy management, 624
    loops, 623-624
    software, 623
distribution
    required for 277 V and 120 V lighting systems,
        comparison of, 537
    single-line diagram, typical, 448
    switchboards, metal-enclosed, 235-236
    transformer taps to shift the utilization
        voltage spread band, using, 114-116
distribution circuit arrangements for
    primary-selective feeders, 173-174
    radial feeders, 170-173
    secondary networks, 175-183
    secondary-selective feeders, 174-177
distribution systems
    ceiling, 591-592
    underfloor, 591
double-deck elevators, 581
drawings
    by designer, 57
    by manufacturer or shop, 58
drawout switchgear, 233-234
dumbwaiters, 564
duty calculation methods, comparison of, 413

E

effects of transformer failures, 218, 220
efficacies for the lm/W range of commonly used
    lamps, initial, 492
electric
    distribution systems, automation of, 633

lighting and daylighting, 525
    rates, 154-156
    space conditioning, general discussion of,
        547-548
    space heating, 548-549
electric heaters
    central furnace or boiler, 551
    heat pumps, 549-550
    heat storage equipment, 552
    infrared, 552
    resistance, 550
    unit, 551-552
    wall and ceiling unit, 550-551
electrical closets, 313-314
electrical design
    codes and regulations
        local, state, and federal, 47-48
        National Electrical Code (NEC), 46
        National Fire Protection Association
          (NFPA), 47
    considerations, 54-55
    coordination, 55-56
    drawings, 57-58
    elements, 44
    flexibility of, 56
    handbooks, 50-51
    manufacturers' data, 52
    periodicals, 51-52
    specifications, 56-57
    standards and recommended practices,
        48-49
electrical equipment efficiencies
    capacitors, 748
    equipment overview, 748
    losses, 745-746
    motors, 746-748
    oversizing of the equipment, 746
    transformers, 748
electrical equipment, reliability data for, 208
electrical fields
    conductor on ground plane in uniform
        dielectric, 327
    shielded cable, 326
electrical load equipment, types of, 92
electromagnetic hazards, 91
elevator car speed selection, 582
elevator doors and automatic door operations
    freight, 579
    interlocks, 579
    passenger, 579
    power operated, 579
elevator entrances, fire rating of, 579-580
elevator operation
    duplex collective, 577
    group automatic, 578
    manual and automatic, 577
    operation during a fire, 586-587
    selective collective, 577
    single automatic pushbutton, 577
    standby operation of, 584-585
elevator performance, quality of
    general discussion of, 578-579
    handicap requirements, 579
    leveling, 578-579
    passenger comfort, 578

elevator plant specification
  building classes, 582
  car capacity, 580-581
  car speed selection, 582
  double-deck elevators and sky lobbies, 581
  elevator selection, 580
  grouping of elevators in the plant, 583
  interval, 583
  number of elevators required, 582-583
  quality of service, 582
  quantity of service, 582
  roundtrip time, 583
elevator speed control systems, basic electric,
  569
elevators
  car lighting in, 587
  conductors and diversity factor for, 573-577
  control, motors, and motor generators in
    ac resistance control, 569
    ac servo, 569-570
    control design fundamentals, 571
    dc direct drive, 570
    dc rheostatic control, 569
    general field control, 570
    hoisting motors, 570-571
    motor-generator sets, rating and starting of,
      571
    rectifiers for power conversion, 571
    variable frequency control, 570
  doors and automatic door operations in,
    579-580
  electric, 563
  emergency signals and communications, 587
  energy consumption and heat release of,
    572-573
  fire rating of entrances to, 579-580
  group supervisory control of, 580
  horsepower and efficiency of, 572
  hydraulic, 563-564
  operation of, 577-578
  performance quality of, 578-579
  plant specifications of, 580-583
emergency and standby power systems, 185-192
enclosure monitor, 182
energy and lighting standards, 736-737
energy audit, 724
energy conservation
  embarking on a program of, 725-726
  equipment, 752-753
  general discussion of, 553-554
  opportunities
    energy audit, 724
    project lists, 724
  practical, 554-555
energy consumption
  and heat release of elevators, 572-573
  metering, 748-750
energy management
  in a direct digital control (DDC) system, 624
  in a facility automation system (FAS), 625-626
  process
    embarking on an energy conservation
      program, 725-726
    obtaining management approval and
      commitment, 724-725

overall considerations in, 727-728
  performing an equipment audit, 726-727
  tracking progress, 727
  requirements, 723
energy savings
  billing calculations, 737-738
  calculating, 737
  declining block rate and example, 738-739
  demand rate, 739
  evaluating losses in other equipment, 742-743
  evaluating motor losses, 740-742
  introduction to, 737
  tariffs involved in, 737
  time-of-use rate, 739-740
  time value of money, 740
  transformer losses, 742
energy, regenerated, 583-584
energy-saving lighting techniques
  economics of energy-saving fluorescent lamps,
    734-735
  fluorescent system considerations, 734
  lighting controls, 736
  task lighting, 736
  using daylight, 735
energy-saving operations
  interlocking or key switches, 750
  load
    control, 751
    or peak shaving with on-site generation,
      751-752
    shedding, 751
  metering of tenant areas and building
    operations, 750
environmental
  considerations, 61
  quality, 91
equipment
  central monitoring and control, 616-620
  ratings, 260-261
escalators
  general discussion of, 564-566
  opaque balustrade, 565
  transparent balustrade, 566
estimate, preliminary, 58
estimation of building load, 94
existing system plans during expansion,
  modernization, or rehabilitation, using,
  664-665
expansion
  adding new service point during, 661-662
  design considerations, 656-659
  examples of poor planning concerning,
    652-655
  feeder sizing during, 657-658
  general discussion of, 649-650
  ground-fault protection during, 657
  hazardous scrap materials from, 657
  installing completely new equipment during,
    661
  making choices during, 655-656
  power distribution to main switchboard
    during, 663-664
  preliminary study of, 650-652
  retaining old service equipment after, 659-661
  scrap materials from, 659

service continuity scheduling during, 665-666
staging operations during, 658
using existing system plans during, 664-665
utility considerations, 656-657
wiring methods used during
  general discussion of, 673
  smoke and fire considerations, 673-674

F

facilities, data, 597
facility automation system (FAS)
  adjunct to the basic system, 611
  and its components, general description
    of a, 614-616
  backup modes, 640-641
  block diagram of a medium-level, 617
  communications in, 634
  design and installation, 636-641
    data transmission media (DTM) selection,
      637
    installation, 638
    interfacing to existing equipment or
      equipment provided by other
      disciplines, 637
    procurement documents, 637-638
    system specification, 636-637
  documentation, 638-639
  energy management in a, 625-626
  establishing system requirements for, 612-613
  fire management, 626-628
  functions, 611-612
  introduction to, 611
  maintenance and operation
    approaches, 642-643
    contracts, 644-645
  mechanical utilities of, 633
  physical installation, 639-640
  programming, 639
  software, 638
  special considerations, 645
  system description and equipment
    terminology, 613-614
  testing, 638
  training, 641-642
  warranty, 638
FAS, see facility automation system (FAS)
fault currents
  limiting, 303-305
  line-to-ground, 427-439
fire alarm systems
  design criteria, 602-603
  general discussion of, 602
  power supply, 603
fire protection loads, 87-88
floodlighting
  guidelines for good results from, 525-527
  light sources for, 527-529
fluorescent lamps
  and temperature, 495
  ballast sound ratings of, 540
  colors of, 494-495
  dimming of, 495
  energy-saving versus standard, 729

radio interference from, 501
reduced wattage, 493-494
types of, 493
fluorescent PL lamp with socket adapter, 730
fluorescent systems, energy savings and cost
  comparisons for four alternative, 735
fuse
  applications, 227-228
  coordination, 460
  current-limiting characteristics, 465-469
  device maintenance, 475
  ratings, 225-227
  selection to provide short-circuit protection
    and backup protection for motor starters,
    473
  selectivity ratio tables, 460
  time current characteristic curves, 463
fuses
  application of, 471
  class letters of, 243-245
  classification of, 243
  coordination of, 460
  electronic, 228
  I^2t values for coordination, 463-464
  low-voltage, 241-245
  maximum continuous current and
    interrupting ratings for current-limiting,
    227
  medium- and high-voltage, 223-225
  NEC categories of, 242-243
  selectivity of, 463
  solid material, boric acid
    fuse units, 226
    refill units, 226
  Type RK5, typical total clearing time versus
    current curves for, 462
fusible unit substation, coordinating, 464-466

G

generators, peak demand reduction using
  multiple on-site, 752
ground-fault
  circuit interrupter (GFCI), 169
  current detection, 477
  protection, 306, 477-479, 657
  relaying, 478
grounding
  cable systems, 365-366
  low-voltage system (600 V and below), 168
  medium-voltage system (over 600 V), 168-169
group supervisory control of elevators, 580

H

hardness versus temperature, typical values of,
  320
harmonic
  interference, reduction of, 146
  producing equipment, 145-146
  voltages, 143-146
harmonics
  effect of, 144-145

from adjustable speed motor controls, 292
nature of, 143-144
hazards, electromagnetic, 91
heat
generated from air-conditioning systems, 553
of light and waste heat, 552-553
pumps, 549-550
heaters, electric
central furnace or boiler, 551
heat pumps, 549-550
heat storage equipment, 552
infrared, 552
resistance, 550
unit, 551-552
wall and ceiling unit, 550-551
heating systems, 86
heating, ventilating, and air-conditioning
(HVAC) systems
and energy management, 620
direct digital control of, 622-625
monitoring and control of, 620-622
HVAC systems, *see* heating, ventilating, and air-
conditioning (HVAC) systems
HID lamps, *see* high-intensity discharge (HID)
lamps
HID lamp ballasts, *see* high-intensity discharge
(HID) lamp ballasts
high-intensity discharge (HID) lamp ballasts
fusing, 501
grounding of, 501
radio interference from, 501
voltage of, 501
high-intensity discharge (HID) lamps
dimming of, 533-534
high-pressure sodium, 497-498
low-pressure sodium, 498
mercury and metal-halide self-extinguishing
types of, 496-497
mercury-type, 495-496
metal-halide-type, 496
high-pressure sodium lamps
general discussion of, 497-498
used for floodlighting, 529
horsepower and efficiency of elevators
calculation for, 572
ratings, 572
variations, 572
hospitals, special requirements for occupancy in
codes and standards, 690
communication, information, and signal
systems, 694
general discussion of, 689-690
hospital lighting, 694-695
hospital staff, 691
life safety, 693
medical equipment and power quality, 694
patient safety, 691-692
standby generators, essential electric systems,
and reliability considerations, 692-693

ICEA dc test voltages (kV) after installation
1968 and later cable, 370
ICEA dc test voltages (kV) after installation pre-
1968 cable, 369
illuminance
calculations, 541-542
recommended for use in selection of values for
interior lighting design, 490-491
specific, 491
illumination
quality, 485-488
quantity, 488-491
illumination footcandles at 10 foot mounting
height (elliptical versus standard reflector),
approximate, 728
image communication systems, 606-608
impedance
component values of, 415
data, approximate, 420
diagram, 429
equivalent diagram, 415
preparing diagrams of, 415
use of per unit, percent, or ohms in, 417-419
impedances
adding series, 416
busway, 421-422
combining in parallel, 417
combining of, 415-417
other circuit, 424-425
shunt connected, 425
system, 430-432
incandescent lamp flicker caused by recurrent
voltage dips, 137
incandescent lamps
color of, 492-493
dimming of, 493
effect of voltage variation on, 122
general discussion of, 491
life and efficacy of, 492
tungsten halogen lamps, 493
incoming lines, 293-297
Industry Applications Society (IAS), 45
initial efficacies for the lm/W range of commonly
used lamps, 492
input watts
for fluorescent lamp ballasts, typical, 732
for high-intensity discharge (HID) lamp
ballasts, typical, 733
installations, data center, 600-602
insulating materials, commonly used, 319
intercom systems, 604-605
interlock systems, 256-257
interrupters, single- and multiple-pole, 398-399

J

jackets, properties of, 332

L

lamp
burnouts, 500
fusing, 500

I

I, I^2, and I^2t, effects of, 468

removal, 500
switching, 500
lamps
  dimming and flashing of, 533
  fluorescent, 493-495
  high-intensity discharge (HID), 495-498
  incandescent, 491-493
life safety
  communications, 628-629
  general discussion of, 52-53
light sources
  incandescent lamps
    color of, 492-493
    dimming of, 493
    effect of voltage variation on, 122
    general discussion of, 491
    life and efficacy of, 492
    tungsten halogen lamps, 493
lighting
  and energy standards, 736-737
  application techniques, 505-529
  automation systems
    electric distribution systems for, 633
    general discussion of, 632
    normal and emergency power sources for,
        632
    UPS systems for, 633
  control of, 529-534
  economics, 539
    energy costs, 541
    first costs, 540
    maintenance costs, 541
    type and quality of lighting desired,
        540-541
  electric and day, 525
  energy-efficient compact fluorescent, 729-731
  heat and waste heat from, 552-553
  high-intensity discharge (HID) lamps, 731
  incandescent, 122, 491-493, 728-729
  introduction to, 728
  maintenance, 534-535
  objectives, 481-482
  outdoor and sports area, 525
  power factors, 537
  psychological effects of, 541
  regulations, 482-483
  temperature of, 537-538
  terminology, 485
  thermal considerations of, 542-544
  voltage of, 535-536
lightning and system transient protection,
    250-252
limited care facilities, special requirements for
    occupancy in, 695
line-to-ground fault currents
  calculation of, 440
  determination of, 427-439
lines
  incoming, 293-297
  over and on buildings, 294
load
  characteristics
    lighting, 81-82
    general-purpose receptacles for appliance
        loads, 82

space-conditioning and associated auxiliary
    equipment, 82-86
considerations, total, 92-96
current and short-circuit capacity, 301-302
estimates, 75-76
management
  controllers, 743-744
  equipment audit and load profile, 744-745
  introduction to, 743
tabulation, 76-78
loads
  for data processing systems, 88-89
  for food preparation, 89-91
  for miscellaneous or special, 90-91
  for transportation systems, 88
  profiles, 95
  transfer devices, 252-256
  typical, 84, 85-88, 90
looped primary system, 183-185
low availability utility kVA, effect of, 439
low-pressure sodium lamps, 498
low-voltage motors, 118
luminaires
  and air conditioning, 503-505
  for energy-efficient light sources
    fluorescent lamps, 733-735
    high-pressure sodium lamps, 732
    specular retrofit reflectors for fluorescent
        troffers, 734
  general considerations about, 501-502
  special lighting distributions for, 502-503

M

main switchboards during expansion,
    modernization, or rehabilitation, power
    distribution to, 663-664
main terminal room, 594-595
maintenance, 54
manlifts, 568
maximum demand, comparison of, 94
medium-voltage cable, shielding of, 324-328
mercury and metal-halide lamps
  self-extinguishing types of, 496-497
  starting characteristics of, 497
mercury high-intensity discharge (HID) lamps,
    495-496
metal-enclosed bus (over 600 V)
  construction, 391-392
  field testing of, 392
  ratings, 391
  standards, 391
  voltage, insulation, continuous current, and
      momentary current ratings of
      nonsegregated phase, 391
metal-enclosed, low-voltage 600 V power
    switchgear and circuit breakers, 233-235
metal-halide high-intensity discharge (HID)
    lamps, 496
meter
  location, 158
  mounting, control, and associated equipment,
      158-160

metering
  energy consumption, 748-750
  for energy conservation, 162
  types, 160-163
miscellaneous subsystems in a FAS
  downtime information, 636
  equipment, 634
  inventory and purchasing, 634
  preventive maintenance, 635-636
  work order, 635
  work order planning, 635
modernization
  adding new service point during, 661-662
  design considerations, 656-659
  examples of poor planning concerning, 652-655
  feeder sizing during, 657-658
  general discussion of, 649-650
  ground-fault protection during, 657
  hazardous scrap materials from, 657
  installing completely new equipment during, 661
  making choices during, 655-656
  power distribution to main switchboard during, 663-664
  preliminary study of, 650-652
  retaining old service equipment after, 659-661
  scrap materials from, 659
  service continuity scheduling during, 665-666
  staging operations during, 658
  using existing system plans during, 664-665
  utility considerations, 656-657
  wiring methods used during
    general discussion of, 673
    smoke and fire considerations, 673-674
molded-case circuit breakers, 239-241
momentary voltage variations (voltage dips), 135-138
motor circuit, protection for typical, 476
motor controls
  AC, 291
  DC, 286, 290
motor running overcurrent protection, 474-475
motor starters
  selection of fuses to provide short-circuit protection and backup protection for, 473
  short-circuit protection, 472-473
  synchronous, 285-286
motor starting, effect of
  cables and busways, 139
  on distribution systems, 138
  on standby or emergency generators, 138
  on transformers, 138-139
motor-generator sets, rating and starting of, 571
moving walks
  capacities of, 567
  general discussion of, 564-566
multiconductor and single-conductor construction, cable, 331

N

network
  principles, 308-309

protectors, 248-250
vaults for high-rise buildings, 282, 308-311
nominal system voltages
  nonstandard, 112-114
  standard, 113
nonstandard nominal system voltages, 112-114
non-uniform lighting, 507-513
nurse call systems, 608
nursing homes, special requirements for occupancy in, 695

O

occupancy, special requirements by
  athletic clubs, 682-683
  auditoriums, 676-677
  automobile areas, 677-678
    automated parking facilities, 679
    car wash facilities, 679
    parking lots, 678-679
    sales offices, 678
    self-parking garages, 678-679
    service shops, 678
    showrooms, 678
  banks, 679-681
  brokerage offices, 681
  colleges and universities
    central power plants in, 683
    classrooms in, 683
    dormitories in, 684
    laboratories in, 683-684
    miscellaneous requirements in, 684-685
  computer centers, 685-686
  department stores
    communication and signal systems in, 688
    distribution systems in, 686
    lighting systems in, 686-688
  fire stations, 689-690
  general discussion of, 675-676
  gymnasiums, 689
  health care facilities, 689-696
  high-rise office buildings
    building management systems in, 704-705
    emergency systems in, 704
    heavy current considerations in, 704
    layout of electric systems in, 701-702
    maintenance of electric systems in, 702
    special safety precautions in, 702-704
    special systems in, 702
  hotels, 696-697
  large office buildings
    building management systems in, 704-705
    emergency systems in, 704
    heavy current considerations in, 704
    layout of electric systems in, 701-702
    maintenance of electric systems in, 702
    special systems in, 702
  libraries, 697-698
  museums, 698
  newspaper buildings, 698
  office buildings, 699-701
  parks and playgrounds, 705
  piers, docks, and boat marinas, 705-706
  places of worship, 681-682

police stations, 706
prisons
    administration buildings in, 707
    cell blocks in, 707
    dining halls in, 707
    hospitals in, 707
    manufacturing buildings in, 707
    power plants in, 707-708
    warehouse buildings in, 708
radio studios, 708
recreation centers, 708
residential occupancies (commercial), 709
restaurants, 710
schools (Kindergarten through 12th grade)
    general discussion of, 710-711
    general requirements in, 711
    kitchen facilities in, 711
    laboratories in, 711
    manual training departments in, 711
    mechanical systems in, 711
shopping centers, 711
social clubs, 682-683
supermarkets, 711-713
swimming pools, 713
telephone buildings, 714
television studios, 714-715
theaters
    general discussion of, 715
    general lighting systems, 715-716
    stage lighting systems, 716
transportation terminals
    airport, 720-721
    bus, 719-720
    general discussion of, 716
    railroad, 719
    rapid transit, 721
    requirements for, 716-718
water fountains, 713
operational considerations for electric systems, 54
outdoor and sports area lighting, 525
outer finishes, cable
    metallic, 328-331
    nonmetallic, 328
outpatient facilities, special requirements for occupancy in, 695
overcurrent protection, motor running, 474-475
overhead service, 293-294
overvoltages, transient, 146

P

panelboards, 237-239
partial load calculation for an office building, sample of, 96-98
peak demand reduction using multiple on-site generators, 752
peak let-through current charts, 469-471
per unit reactance in an electric utility system, 422-423
phase relations of voltage and short-circuit currents (medium-voltage generator feeding a distribution line), 403

phase voltage unbalance
    causes of, 139-141
    effect of, 143
    in three-phase systems, 139-143
    measurement of, 141-142
pilot devices
    automatically operated, 288-289
    manually operated, 286-287
pneumatic message tube systems, 568
pocket paging systems, 608-609
pollution and hazardous waste, control and monitoring of, 631-632
power company, relation to, 78-81
power conversion, rectifiers for, 571
power distribution to main switchboards during expansion, modernization, or rehabilitation, 663-664
power distribution transformers, 212-223
power source, selection of, 149-152
power supply disturbances
    electromagnetic interference (EMI), 63, 251-252
    harmonics, 62-63
power supply, fire alarm systems, 603
power system configuration
    for 60 Hz distribution, 197-204
    for 400 Hz distribution, 206-207
    with 60 Hz UPS, 204
power transmission and distribution on utility systems, principles of, 106-108
primary distribution system, utility service supplied from, 117-118
primary system, looped, 183-185
primary-unit substations, 236
principal transformer connections, 105
professional registration, 46
projection systems, 608
properties of copper and aluminum, 318
protection
    circuit breaker, 472
    controllers, 271-276
    ground-fault, 306, 477-479
    lightning and system transient, 250-252
    motor starter short-circuit, 472-473
    transformer fuse, 473-474
    wire and cable, 472
protective devices, coordination of, 447
protectors
    cable limiters, 245
    network, 248-250
    service, 246
psychiatric facilities, special requirements for occupancy in, 695-696
public address systems, 604
public telephones, 597
pushbutton control circuits, typical, 287

R

ratings, equipment, 260-261
reactance
    of busways, cables, and conductors, 423
    of motors rated 600 V or less, 424
    of rotating machines, 424

of transformers, 423
per unit, 422-423
rectifiers for power conversion, 571
reduced voltage starters, cost comparison of, 274
reflectance, surface
color, 487-488
psychological factors, 488
recommended in offices, 487
room finishes, 486-487
reflections, veiling, 485-486
regenerated energy, 583-584
regulator compensation on primary distribution
system voltage, effect of, 107
rehabilitation
adding new service point during, 661-662
design considerations, 656-659
examples of poor planning concerning,
652-655
feeder sizing during, 657-658
general discussion of, 649-650
ground-fault protection during, 657
hazardous scrap materials from, 657
installing completely new equipment during,
661
making choices during, 655-656
on a college campus, an example of, 670-673
power distribution to main switchboard
during, 663-664
preliminary study of, 650-652
retaining old service equipment after, 659-661
scrap materials from, 659
service continuity scheduling during, 665-666
staging operations during, 658
using existing system plans during, 664-665
utility considerations, 656-657
wiring methods used during
general discussion of, 673
smoke and fire considerations, 673-674
reliability data for electrical equipment, 208
remote control lighting, typical components of,
531-532
remote control switching relays and lighting
contactors, 530-533
removal, lamp, 500
resistance heaters, 550
retention of existing system and expansion, an
example of, 670-671
riser
conduits and sleeves, 590-591
shafts and slots, 591
systems, 595
rotating machines reactance, 399-401, 424

S

sample partial load calculation for an office
building, 96-98
secondary transformer connections, 166-167
secondary-unit substations, 236-237
security and access control systems, 603, 629-631
security systems
access control in, 630-631
general discussion of, 629-630
selective coordination, 446

selectivity schedule, typical, 458
service
laterals, 295-299
protectors, 246
rooms, 311-312
transfer using circuit breakers, 190
underground, 295-296
service continuity scheduling during expansion,
modernization, or rehabilitation, 665-666
service entrance
cables, 593-594
conductors within a building, 296-297
installations, 297
service entrances
number of, 297-298
physical arrangement of, 298
service equipment, vaults and pads for, 306-308
sheath losses in grounded cable systems, 366
shielded and nonshielded cable constructions,
commonly used, 329
shielded cable
electrical field of, 326
typical taped splice in, 364
short-circuit
calculations
general discussion of, 408-409, 433, 439-440
on load side of main switch, 399
typical, 402
comparison of duty calculation methods, 413
computer-generated calculation results,
434-437
current calculations, basis of, 410-413
currents
determination of, 425
longhand solution of, 425-427
detection equipment, 409-410
per unit representation of, 419-422
symmetrical current basis of rating, 412-413
total current basis of rating, 411-412
short-circuit current
asymmetrical, 401-402
and voltage in a zero power factor circuit,
405
DC component of, 404-406
decay of dc component and effect of
asymmetry of current, 406
effect of distribution circuit lengths on, 441
examples of reducing available, 444-446
reducing available, 441
sources of, 399
symmetrical, 401
and voltage in a zero power factor circuit,
404
from four sources combined into total, 407
system conditions for most severe duty,
414-415
total, 406-407
utility source of, 401
signal distribution, television, 607-608
single-conductor and multiconductor
construction, cable, 331
sky lobbies, 581
sodium lamps
high-pressure, 497-498
low-pressure, 498

sound reinforcement systems, 605-606
space heating, electric, 548-549
space-conditioning
  control, 548
  definitions, 556-559
  primary source of heat from, 548-553
  reduction in demand for, 555-556
  standards and codes for, 556
specific 480Y/277 V network served building,
    single-line diagram of, 428
specification, cable, 376-377
speed control of
  AC motors, 291-292
  DC motors, 290-291
splices
  preassembled, 365
  taped, 363-364
splicing devices and techniques, 362-365
spot network
  basic, 178
  four-unit, 179
sprinklers in elevator machine rooms and
    hoistways during fire conditions, 586-587
standard nominal system voltages and voltage
    ranges, 103-104, 113
standby power operation of elevators
  power requirements, 584-585
  standby power system, 585-586
starters
  autotransformer, 269-270
  low-voltage, 281-282
  medium-voltage, 284-285
  part-winding, 266-267
  reduced voltage, 274
  resistor or reactor, 267-269
  series-parallel, 271
  solid-state, 271
  wye-delta, 270-271
storage of technical files, 62
stores, special lighting considerations for,
    519-524
stress-relief cone, 356
substations
  primary-unit, 236
  secondary-unit, 236-237
surface reflectances in offices, recommended,
    487
switchboards, metal-enclosed distribution,
    235-236
switches
  automatic transfer, 254-255
  bolted pressure, 247-248
  enclosed
    application, 247
    National Electrical Manufacturers
      Association (NEMA) requirements for,
      246-247
  high-pressure contact, 247-248
switchgear
  automatic control devices, 229
  auxiliary equipment and features, 229-230
  capability required of, 230
  drawout, 233-234
  metal-enclosed 5-34.5 kV circuit breaker,
    230-232

metal-enclosed 5-34.5 kV interrupter, 228-230
system
  grounding, 167-169
  protection and coordination, general
    discussion of, 397-398
system reliability analysis
  and total owning cost, 209
  general, 208-209
system requirements for
  commercial buildings, 43
  institutional buildings, 43
  residential buildings, 43
system voltage
  classes, 102-105
  nomenclature, 114
  terms, 101-102
  tolerance limits, development of, 108-111
systems
  HVAC, 620-626
  lighting automation, 632
  nurse call, 608
  pocket paging, 608-609
  projection, 608
  riser, 595
  security, 629-631
  video, 607

T

taped splice in shielded cable, typical, 364
task-ambient lighting
  advantages of, 513
  disadvantages of, 513-519
  general discussion of, 513
technical file storage, 62
telephone
  equipment rooms, 596
  facilities, 592
  service, riser sleeves or conduits for, 590
telephones, public, 597
television signal distribution, 607-608
tenant classification of connected load and
    maximum demand, 96
terminal room, main, 594-595
terminations
  cable, 354-361
  classes of, 355-360
  definitions of, 353-354
  devices and methods of, 360-361
  nonshielded cable, 354
  purpose of, 353
  shielded cable, 354-355
terminator
  typical Class 1 molded-rubber terminator (for
    solid dielectric cables), 358
  typical Class 1 porcelain (for solid dielectric
    cables), 357
  typical Class 3 molded-rubber terminator (for
    solid dielectric cables), 359
testing
  in the field, 369-372
    line voltage fluctuations, 372
    megohmmeter test, 373
    procedure, 372

resistance evaluation, 373
wiring systems
  AC versus DC, 368
  application and utility, 367-368
  DC corona and its suppression, 372
  in the factory, 368-369
three-phase systems, phase voltage unbalance
  in, 139-143
through-fault protection curves for liquid
  immersed or dry-type Category II
  transformers, 453
time current curves
  for 125-600 A molded-case circuit breakers,
    449
  for 600-4000 A power circuit breakers, 450
  fuse, 463
  showing complete ground coordination, 461
  showing complete phase coordination, 459
  showing primary fuse and transformer
    protection criteria for delta-wye, solidly
    grounded on both primary and
    secondary, 456
tolerance limits
  in volts, 115
  standard fluorescent lamp ballasts in volts,
    116
  standard three-phase induction motors in
    volts, 116
total clearing time versus current curves for
  Type RK5 fuses, typical, 462
total load considerations, 92-96
total short-circuit current equals sum of
  sources, 400
transfer
  devices
    automatic, 255-256
    load, 252-256
  switching methods, typical, 189
transformer
  connections
    principal, 105
    secondary, 165-166
  construction, 220-223
  failures, effects of, 218, 220
  fuse protection, 473-474
transformers
  approximate impedance values for, 217
  for dry-type (in dB), 219
  for single-phase and three-phase oil cooled (in
    dB), 219
  general discussion of, 454
  insulation temperature ratings in °C for, 217
  preferred kVA ratings for, 215
  showing primary fuse and transformer
    protection criteria for delta-delta, 455
  showing primary fuse and transformer
    protection criteria for wye-wye
    transformer solidly grounded on both
    primary and secondary, 456
  sound levels, 218-219
  specifications for, 215-220
  types of, 212-215, 220-223
  typical per unit R and X values, 418-419
transient overvoltages
  general discussion of, 146

wiring system protected from, 367
transient protection, lightning and system,
  250-252
transmission and network equipment, 599
transportation
  and traffic control equipment, 631
  consultations, 562
  design factors, 561-562
  efficiency and economy in, 562
  general discussion of, 561-562
  responsibilities, 562
  types of, 563-568
  use, 561
transportation types
  dumbwaiters, 564
  electric elevators, 563
  escalators, 564-566
  hydraulic elevators, 563-564
  manlifts, 568
  material lifts, 566-568
  moving walks, 564-566
treadway slope and speed, maximum, 567
tungsten halogen lamps, 493
types of metering, 160-163

U

undercarpet wiring systems, 592
underfloor
  distribution systems, 591
  raceways, 591
uniform lighting, 506-507
uninterruptible power supply (UPS) system
  configurations
  "cold" standby redundant, 195
  dual redundant, 200
  hot tied bus, 203-204
  isolated redundant, 197, 201
  nonredundant, 194
  parallel capacity, 199
  parallel redundant, 196, 199
  parallel tandem, 201-202
  single-module, 198
  super-redundant parallel, 204
uninterruptible power supply (UPS) systems
  application of, 197
  distribution systems for, 204-206
  monitoring and control functions for, 633
unshielded cable on ground plane, 327
UPS systems, *see* uninterruptible power supply
  (UPS) systems
use of approved appliances and equipment,
  53
utility
  grounding on ac services, 157
  metering and billing, 157-158, 163-165
    by service voltage characteristics, 158
    by type of premises, 158
  policy for supplying tenants in commercial
    buildings, 118-119
  power generation, transmission, and
    distribution system, typical, 106
utility service
  planning for, 152-154

supplied from primary distribution system, 117-118
supplied from transmission lines, 118
utilization equipment, effect of voltage variation on, 120-121
utilization voltage of 600 V and below, selection of, 116-117

V

vaults
  detailed design of, 309-311
  preliminary design of, 309
  network, 308-311
VCP, *see* visual comfort probability (VCP)
veiling reflections, 485-486
video systems, 607
visual comfort probability (VCP), 503
voltage
  classes, 42, 103-104
  control in electric power systems, 106-116
  harmonic, 143-146
  insulation classes and dielectric tests, 218
  levels, 41-42
  profile
    limits for a regulated distribution system, 111-112
    standard for a regulated power distribution system (120 V base), 110
  ratings
    for utilization equipment, 119-120
    of standard motors, 119
  selection, 116-119
  system driving, 425
  systems outside of the United States, 106
  transformation during expansion, modernization, or rehabilitation, 662-663
voltage dips, incandescent lamp flicker caused by recurrent, 137

voltage-drop
  calculation, 124-133
  considerations in locating the low-voltage power source, 134-135
  tables
    busway, 127-132
    cable, 125-127
    transformer, 130-133
  values
    for three-phase busways with aluminum bus bars, 388
    for three-phase busways with copper bus bars, 387
voltage regulation and power factor correction, 207-208
voltage variation
  effect on
    fluorescent lamps, 121
    high-intensity discharge lamps, 121
    incandescent lamps, 122
    induction motor characteristics, 121
    synchronous motors, 121
    utilization equipment, 120-121
  momentary, 135-138
volts, tolerance limits in, 115-116

W

wall and ceiling unit heaters, 550-551
weighting factors to be considered in selecting a specific illuminance, 491
wiring
  methods during expansion, modernization, or rehabilitation
    general discussion of, 673
    smoke and fire considerations, 673-674
  methods for hazardous locations, 345
  systems
    testing, 367-373
    undercarpet, 592